Epigenetics

Epigenetics

A Treatise on Theoretical Biology

SØREN LØVTRUP

Department of Zoophysiology,
University of Umeå,
Sweden

A Wiley–Interscience Publication

JOHN WILEY & SONS

London · New York · Sydney · Toronto

Published by John Wiley & Sons Ltd.

Library of Congress Catalog card No. 72-5719

ISBN 0 471 54900 2

Printed in Great Britain by
Garden City Press Ltd.
Letchworth, Herts.

Til Minde
Om Min Mor

ACKNOWLEDGEMENTS

After my graduation, when all roads were closed for the pursuance of an academic career in my native country, I had the good luck to get employment in Sweden. I am indebted to the University of Göteborg, its Medical Faculty and the Department of Histology and its head, Professor Holger Hydén, for the hospitality with which I was received in Göteborg, and for the appreciation subsequently accorded to my work. Without this support the present book would never have been written.

I also acknowledge my gratitude to the Hargitt Fellowship Committee, and its chairman Professor R. Bruce Nicklas, at the Department of Zoology, Duke University, for awarding me a fellowship that enabled me to write the last chapter of the book, while being relieved of the duties which normally reduce the time available for studies and creative work.

Several friends and colleagues have participated in discussions concerning the subject of the present book, and some have been kind enough to read parts of the manuscript. In this way I have obtained valuable information and, still more important, received criticism which unquestionably contributes significantly to my work. Whatever shortcomings remain are, of course, all my responsibility. I am indebted to the following persons, listed in alphabetical order, for assistance in this respect:

Professor John R. Gregg, Dr. Nils Olof Lindh, Professor A. Lindroth, Dr. Clarence A. Loeffler, Dr. Huguette Løvtrup, Dr. Kjell Hansson-Mild, Professor Andrzej Pigon and Professor Marianne Rasmuson. Special thanks are due to Professor John G. Lundberg for compiling the data in Table 12.

The fact that the book has been written in Umea, where the University, and hence the University Library, are very young, has meant that most of the literature had to be borrowed from other libraries. The efficient and kind service given me by the staffs of the University Library in Umea and of the other scientific libraries in Sweden is gratefully acknowledged.

My thanks are also due to the staff of John Wiley & Sons for the valuable help I have received during the editing and printing of my book.

PROLOGUE

'In the beginning was the Word'
(St. John, 1.1*)*

An essential aspect of scientific endeavour is to discriminate events in the 'real' world, and to invent abstracts for their classification. For this purpose it is necessary to name both kinds of phenomena. Since naming is a much easier task than discrimination or invention, it very often happens that names float around which do not correspond to any significant concrete or abstract concepts, or that several names apply to one of these. It may also happen, however, that one name covers several concepts, and some phenomena may remain unnamed.

I have no ambition whatever to contribute to the terminological confusion but, contending that many important distinctions are not made in our present embryological vocabulary, I have been obliged to introduce a number of new concepts. As far as possible, I have used existing names, properly redefined, but in some cases it has been necessary to introduce new names. In dealing with the problem of terminology I shall take the opportunity to pay homage to C. H. Waddington who has coined the felicitous word 'epigenetics' which I have used as the title of the present book.

As the text is not addressed particularly to systematists, I feel that the systematic nomenclature ought to adhere to a standard reference. Since the sole modern work which can assume this function is Rothschild's *A Classification of Living Animals* (1965), I have, whenever possible, followed this text, changing the names in the original publications when required for conformity. Except where specific distinction was of significance, I have given only the name of the genus.

Finally, although my own contribution to the present subject matter is modest, I have still found many occasions to quote my own work. In order to minimize the appearance of my name in the text, my own papers are cited simply by the year of publication in all those cases where this expedient does not lead to ambiguities.

CONTENTS

'There are biological theories in abundance but no theoretical biology. There are books in plenty on experimental embryology but none on theoretical embryology. Why is this?'
(J. H. Woodger, 1930, p. 4)

INTRODUCTION

There are probably many relevant answers to the question posed in the above quotation, but obviously the only satisfactory one consists in writing books on theoretical biology. Being concerned with the theory of epigenesis, the present book claims to meet this demand.

Current literature dealing with 'theoretical biology' shows that no unanimity exists regarding the scope of this subject, and it may therefore be expedient to begin the introduction with an outline of my own view in a brief discussion of the nature and purpose of scientific theories. Subsequently some attention will be devoted to existing biological theories, and finally an analysis of the phenomenon of ontogenesis will be presented, which forms the basis of the approach followed in the present book.

SCIENTIFIC THEORIES

Within any discipline of science, the field dealing with the theoretical aspects forms part of a hierarchy, the nature of which is stated in the following citation:

'Every branch of natural science has, it seems, three aspects: first and foremost is the investigatory aspect which is concerned with discovery; secondly is what we may call the speculative aspect, whose business it is to discover means for systematizing the knowledge furnished by investigation, issuing in constructive theories. The third aspect is the critical, which has to do with investigating methods of interpretation from the standpoint of logic and epistemology' (Woodger, 1929, p. 477).

The domain of theoretical science corresponds to the second aspect, the main task thus being the formulation of 'constructive theories'.

A scientific theory consists of an assemblage of statements, arranged hierarchically in such a way that a statement of any order is a logically necessary consequence of a statement in the adjacent higher order. When a scientific discipline is organized in this way it forms a deductive system. The deduction can proceed only in one direction, viz., from the statements of a higher towards those of a lower order, never in the opposite direction.

Formerly, it was usual to distinguish between the 'deductive' theoretical sciences and the 'inductive' empirical sciences. In the former one may start with certain axioms, corresponding to statements of highest order, and derive various conclusions which thus represent statements of lower order. What are given at the outset in the latter case are certain empirical observations which, in the theoretical hierarchy, stand for the conclusions, whereas the statements of higher order, called generalizations and hypotheses, are supposed to be derived from the observations through inductive inference. Owing mainly to the work of Popper (cf. 1968) it is now largely, if not unanimously, accepted that this distinction cannot be upheld, primarily because inductive reasoning cannot be logically justified. It follows that empirical sciences must be organized in deductive systems just as are the theoretical ones. The important difference between theoretical and empirical sciences is that whereas specific rules are laid down for the deductive procedure, there are no guidelines to aid the process leading from observations towards explanatory hypotheses of higher order. The actual situation is that 'the process of discovering new hypotheses is one of constructive guessing or inventive imagination. It always involves a leap in the dark. It is for that reason that the process is so difficult and that important successful hypotheses are so rare' (Woodger, 1948, p. 354).

A number of demands must be met by a scientific hypothesis, among which we shall mention only one, stated by Woodger in the following terms: 'For in all cases it is required not only that the statements to be explained should be among the consequences of the hypothesis, but that it should have other consequences which can be tested experimentally [or otherwise]' (1948, p. 355). Through this testing procedure the prediction may be either verified or falsified. It is important to note, as stressed by Popper (1968), that the latter case leads to falsification of the whole theory, which hence must be rejected, whereas verification of a prediction only ensures temporary survival of the theory, since it does not preclude that subsequent testing of other predictions may entail their falsification and thus that of the theory.

It is of importance in the present context to specify some demands with which hypotheses need not comply. At their origin, most scientific theories have been based on quantitative relations, which often were relatively simple from a mathematical point of view. This holds generally for physical and chemical theories, and in biology for Mendel's theory of heredity. There is no doubt that this circumstance has facilitated the invention of the hypotheses, but it has furthermore entailed the advantage that the language of mathematics could be used in their formulation, which certainly helped in deriving predictions that could be tested experimentally. However, as stressed on many occasions by Woodger (cf. 1930; 1931; 1939), there are many natural phenomena which do not lend themselves to quantitative treatment, either because they are too complex or because their nature does not allow for it. For the exact formulation of a theory it will in such cases be necessary to construct an abstract logical language which can account for the qualitative relations revealed by observations. Admitting that the ultimate test of the soundness of any theory is that it can be expressed in a language of this kind, I venture to state that unless a ready-made deductive system like mathematics can be applied directly, it is necessary to develop quite clear notions about the nature of the processes covered by the theory before the latter can be formulated in an abstract language. In other words, it may be necessary to divide Woodger's second aspect in two steps, the first one trying to incorporate a set of observations into a deductive—or at least quasi-deductive—system by using an ordinary language, and the second one consisting in the construction of a formal logical system to embody the theory. Actually, each of these tasks seems to be sufficiently specialized to make it unlikely that one person can accomplish both.

In the formulation of generalizations and hypotheses it may be necessary to introduce concepts and hypothetical entities of various kinds. It cannot be emphasized enough that the value of these does not depend upon their being experimentally demonstrable. On the contrary, as long as they are unambiguously defined, and their effects are predictable and confirmable by observation, they are doing full service to the theory. To bear out this point the concepts of 'gravity', 'atom' and 'gene' may be mentioned. Essential information concerning the properties of the two latter has been gained a long time after formulation of the theories in which they occur, while the first has so far proven refractory to any further resolution.

After this discussion of the general nature of scientific theories, we may briefly consider their usefulness, since it is the latter which justifies the endeavours spent on their invention. The value of a scientific theory is manifested in different ways. Thus, when a body of knowledge is arranged in a deductive system, a number of previously unsuspected relations between observations and statements of higher orders may be revealed. It also becomes easier to unravel contradictions in prevailing thought and to distinguish between relevant and irrelevant, or at best unimportant, observations. Altogether these various expedients permit a notable integration of the accumulated knowledge in the field concerned.

However important this achievement may be, the most valuable contribution of theories is still the fact that they allow for planned research. This point was stated in the following way by Pirie: 'Some people think that the philosophy a scientist accepts is not of very much importance; his job is to observe the phenomena. This is a gross oversimplification and it involves the subsidiary hypothesis that all scientists are fully equipped with serendipity. A sensible philosophy controlled by a relevant set of concepts saves so much research time that it can nearly act as a substitute for genius . . . A scientist can have no more valuable skill than the ability to see whether the problem he is investigating exists and whether the concepts he is using are applicable' (1952, p. 280).

Thus, in the absence of a comprehensive theoretical superstructure each research project will be 'a leap in the dark', and the collected observations must of necessity be of varying significance to progress in the particular field. The situation changes completely if good hypotheses are available, because they will always give plenty of ideas for continued research. The history of science contains numerous illustrations of the contention that rational research is possible only on a theoretical basis. I believe the truth of this statement is best illustrated by referring to the spectacular achievements of physics and chemistry, little of which would have been realized if these sciences were not based on well-founded theories.

BIOLOGICAL THEORIES

The most important and most general biological theories are the Mendelian theory of heredity and the Darwinian theory of evolution, whose founders 'introduced key hypotheses of a relatively high level which produced order where formerly there was chaos, and set in motion new lines of investigation which created new branches of science' (Woodger, 1948, pp. 352–353). One further theory deserves mention in the present context, viz., the cell theory of Schleiden and Schwann, which is an integral part of all other biological theories, both the ones mentioned above and the theory of ontogenesis which is dealt with in the present book. Darwinism and its successor, neo-Darwinism, are not of direct import for the problem of ontogenesis, but the converse relation holds since the theory of ontogenesis is one of the premises of the theory of evolution. It is therefore expedient to postpone a discussion of the Darwinian and the neo-Darwinian theories to the last chapter where the bearing between epigenesis and phylogenesis is scrutinized.

Contrariwise, the theories of Mendel and of Schleiden and Schwann have direct implications for the theory of ontogenesis, and it is therefore necessary here to present a survey of these relations.

Mendelism

The ideas advanced by Mendel have developed into a theory of heredity, resting on a solid basis of observations, crowned recently with the discovery of the genetic code. As stated above, the initial success of the theory may be traced back to the fact that it involved quantitative relations, which could be formulated mathematically. Much genetic research has been concerned with problems that may be expressed in biochemical terms, a circumstance which had the consequence that it could be pursued on the basis of chemical concepts. This has been of enormous importance for the progress of the work, since chemistry is a science in which it is possible to carry out research based on deductions made from prevailing hypotheses. However, the answers obtained within this most successful union between genetics and biochemistry, called molecular biology, can only be couched in chemical terms. If the information stored in the genome had not been concerned with control of a chemical event, protein synthesis, this approach would never have led to meaningful results.

The modern theory of genetics is concerned with the inheritance of specific patterns of protein synthesis, and the mechanism through which they are effectuated. These phenomena are aspects of ontogenesis, but not the only ones, and this theory is therefore inadequate as far as epigenesis is concerned. This is no shortcoming by itself, as long as claims to the opposite are not made. However, on this background the neo-Darwinian dogma that phylogenesis is the outcome of cumulative gene mutations may be regarded with some scepticism, and by the same token it can be questioned whether the ardent attention paid by some developmental biologists to molecular-biological problems may not be somewhat misdirected.

At this point it must be emphasized that it is not my intention to advocate the view that protein synthesis, and thus the genes, are of no significance for evolution and development: the opposite is quite obviously true. The crucial point is not *whether*, but *how*, the genes are involved in these phenomena. It is here the importance of 'epigenesis' becomes evident, for it is through the process connoted by this word that the hereditary factors exert those actions which are of consequence for both ontogenesis and phylogenesis. There seems to be no possibility of enhancing our apprehension of the influence of genes and gene mutations on these events before we understand the mechanism of ontogenesis.

The cell theory

The cell theory, as advanced by Schleiden and Schwann, states that all plants and animals either are cells or are composed of cells. This theory, which has been corroborated by all subsequent observations, is one of the cornerstones of all biological research. Among the many inferences which it allowed for is 'that the key to every biological problem must finally be sought in the cell; for every living organism is, or has at some time been, a cell' (Wilson, 1928, p. 1).

This implies that any kind of organic form must be related to the form of cells or cell products and that any kind of biological activity in the last analysis must be referred to activities exerted by cells. It will be shown in the following pages that these statements are of great importance for a rational analysis of the phenomenon of ontogenesis.

ONTOGENESIS

It is a trivial truth, unfortunately not always adhered to, that no problem, whether theoretical or practical, can be solved before it is distinctly formulated. In the present context this implies that we must begin by analysing the implications of 'epigenetics'. This word traditionally stands for one of two alternative ways to account for the mechanism of embryogenesis. If we temporarily accept this view, it follows that, since embryogenesis is nothing but the earliest part of ontogenesis, the epigenetic mechanisms, which constitute the subject of the present book, may be involved in the latter event.

Ontogenesis is a very complex process, and almost certainly cannot be mastered except on the basis of a theory. It is therefore remarkable that no comprehensive theory of ontogenesis, or even of embryogenesis, exists. To be sure, many embryological hypotheses have been advanced, but they generally cover only selected aspects and in some cases they are mutually exclusive. The primary object of this work is to outline the foundations of a theory of ontogenesis. For practical reasons the attention will be confined to the earliest stages, and so far it may be considered to deal with the theory of embryogenesis. However, even though the mechanisms involved in the later phases of ontogenesis may be both quantitatively and qualitatively different from those active in embryogenesis, I think it is important to stress that development from fertilization to death is one continuous chain of causally related events.

Ontogenesis is a process in which the cooperation of various factors leads to the transformation of the fertilized egg into an adult organism. Our first goal must be to establish the nature of these factors. Subsequently we shall consider the antithesis between preformation and epigenesis. The reason for this approach is that the resolution of this classical problem furnishes a clue to the proper understanding of the contribution of epigenesis to the process of embryogenesis. After the discussion of these points it will be possible to formulate a rational program for the theoretical analysis of the contribution of epigenesis to the event of ontogenesis.

Embryogenetic factors

The following quotation from Woodger (1929) may serve as a first approximation to the establishment of the nature of the phenomenon of embryogenesis: 'The individual living organism is an event—a spatio-temporal happening. For human experience it is *known* as a certain perceptual object, namely a given animal or plant of everyday life. And this perceptual object is a character of an event. We *know* such an event *as* a persisting thing, that is to say we know the event which constitutes the organism as having a certain permanence of characterization of a certain type. It is of the type which requires time in which to display itself, since successive temporal parts of the event are differently characterized . . .' (p. 339). Consequently we realize 'that what develops—what changes—is the *mode of organization* of the event which is the organism' (p. 341).

Thus, according to this view, the fertilized egg and the adult are different temporal parts of the same 'spatio-temporal happening', the organism, embryogenesis being the earliest stages of this event. The circumstance that the developing organism is subject to changes implies that ontogenesis is distinguished by the occurrence of certain processes, the outcome of which is the observed changes. We shall, without specification at the moment, designate these as *ontogenetic* or *embryogenetic processes*. The material substrate on which these processes exert their action will be called the *ontogenetic* or the *embryogenetic substrate* (cf. Needham, 1942). The latter is identical with the embryo or the organism at any time. The embryogenetic substrate and the embryogenetic processes are the embryogenetic factors which, through various kinds of interaction, accomplish the process of embryogenesis. In the following sections a preliminary analysis of these factors will be attempted.

EMBRYOGENETIC PROCESSES

The cell theory leads to the inference that events which occur during embryogenesis must be the outcome of cellular activities. As a first approach the analysis of the embryogenetic processes might thus be confined to the establishment of the nature of cellular functions. However, among these only the following are of direct importance for embryogenesis: propagation by division, engagement in various kinds of synthetic activities and execution of movements and participation in aggregate formation. Consequently we may infer that corresponding to each of these types of cellular faculty a kind of embryogenetic process exists.

The first of these to be observed in the developing embryo is *cell division*, in some cases replaced by nuclear multiplication. In spite of occasional claims to the contrary, this event is an unconditional requirement for the occurrence of the other two kinds of embryogenetic processes. Synthesis of proteins and other substances normally occurs in association with cell division, as required to preserve the extensive properties of the dividing cells. Being an inherent part of the process of cell division, it is trivial from the present point of view. It may be mentioned that the earliest cell divisions in developing embryos generally are accompanied with little or no synthesis, thus leading to a reduction of the volume of the ensuing cells.

During the initial stages the cells exhibit many similarities, even if differences in size and cytoplasmic composition often are observable. However, as development proceeds, new and more spectacular differences appear between cells in various parts of the embryo. These modifications may concern the shape and behaviour of the cells, as well as the products of their synthetic activity. The creation of this spectrum of cell-types may be referred to the pattern of protein synthesis assumed by the cells; it thus owes its existence to the second cell function listed above. This kind of change in cell properties is called *cell diversification*.

As mentioned, the differences may in some cases be related to the shape and the behaviour of the cells. These properties may be influenced by the environment, but from isolation and cultivation experiments it is known that various types or *orders* of cells exist, identifiable by characteristics of this kind. It appears that, apart from those that remain in the original cell order, the first few steps of diversification to which the cells are subjected concern the formation of cells belonging to new cell orders. For various reasons these processes assume a unique position in the hierarchy of cell diversification, which may therefore warrant giving them a special name; they will henceforth be called *cell transformations*. The remaining diversification processes are distinguished by the fact that the embryonic cells cease dividing and begin to synthesize products of various kinds which are either stored in the cells or secreted; some of the latter substances may be deposited as extracellular elements. This activity is clearly different from that associated with the cell transformation, and we may therefore distinguish these processes by the name of *cell differentiations*. A number of cell differentiation patterns exist, each of which is characterized by the chemical nature of the substances produced by the cells.

When cells engage in movements and aggregate formation, modifications in the shape of the embryo may result, a phenomenon to which the name 'morphogenesis' is applied. These changes thus result from the third kind of cell function, but it must be pointed out that some of these activities are made possible only through interaction between cells and extracellular elements, and the latter may even in some cases by themselves contribute to morphogenesis. These various interactions between substrate elements which give rise to changes in their spatial arrangement, and thus in the shape of the embryo, may be called *morphogenetic processes*. The cellular contribution to morphogenesis comprises two aspects, partly the activities they display as substrate elements, and partly the elaboration of extracellular elements. The former depend upon the cell-order affiliation, the latter on the differentiation pattern. It is thus evident that both cell division and cell diversification are prerequisites for morphogenesis.

EMBRYOGENETIC SUBSTRATE

The activities discussed in the preceding section have been called embryogenetic processes. It is evident, however, that they are common to all living organisms, and they become embryogenetic processes only when they occur in an embryo. Hence it is a further premise of embryogenesis that the material substrate in which the processes mentioned occur must be of a special nature, that it must be an 'embryogenetic substrate'. The latter concept may be defined as the sum total of all the material components which make up the organism at any time during embryogenesis. The fertilized egg represents the first stage of the embryogenetic substrate. Some people consider that this characteristic may be attributed more properly to the unfertilized egg. It is difficult to meet this argument unequivocally, since parthenogenesis occurs as a natural phenomenon. The view adopted here may be supported by reference to the discussion of biological parts and wholes by Woodger (1929), or to the characteristic difference in nuclear ploidy obtaining between the gametes and the egg, but it may suffice to direct attention to the undeniable fact that with few exceptions embryogenesis is initiated by the process of fertilization.

The egg and the embryo are often surrounded by one or more envelopes or membranes. Most often their main function is protective, but some, especially those in direct contact with the egg, may also influence the morphogenetic process, and must therefore be considered as a part of the embryogenetic substrate. During embryonic development the egg cell is subdivided into a gradually increasing number of cells. At some stage these cells begin to produce substances of various kinds, which in many cases may be secreted by the cells to the surroundings, where they in various ways may influence the shape of the embryo.

It is thus seen that the embryogenetic substrate comprises three types of material components or *elements*, viz., *envelopes* and *membranes*, *cells* and *extracellular substances* produced by the cells.

Preformation and epigenesis

The history of preformation and epigenesis as descriptions of the mechanism of embryogenesis dates back more than two thousand years, the former view being advanced in a Hippocratic treatise, the latter by Aristotle (Needham, 1959). Considered as antitheses, they have been acclaimed or rejected in the following centuries in accordance with prevailing thinking. The ideas embodied in these two interpretations of development have been aptly formulated by P. Weiss (1939, p. 110) in the following way: '[Preformation implies] essentially that there is preformation of parts in the organism from its very first beginning . . . and that development consists fundamentally of nothing but a gradual conversion of latent into manifest differences without attending increase of complexity or emergence of new properties.

'*Epigenesis* . . . presumes that developing systems start from a rather primitive, homogeneous, chaotic or, at least, lowly organized condition into which increasing complication and real . . . diversification come in progressively as development proceeds; . . . the later parts of the organism are not as such pre-existing in the germ but are gradually established by a process of *individualization*'.

In the crudest sense of the word 'parts', meaning morphological parts similar to those present in the adult organism, the two hypotheses are antithetical, and the notion of preformation refutable, as shown by Aristotle. However, as time has passed, more subtle interpretations have been introduced, with the consequence that the two notions are no longer antitheses. We shall briefly discuss this reconciliation in the following section, and subsequently proceed to outline a more exact analysis of the component phenomena involved in embryogenesis.

RECONCILIATION BETWEEN THE HYPOTHESES

When the word 'preformation' is used in contemporary discussion it rather stands for predetermination in general. That no strict preformation obtains in this sense of the word is obvious since experimental embryology is based on the fact that interference with embryogenesis is possible. Yet, that something is 'preformed' at the outset of each individual case of ontogenesis is so evident that it seems incredible that it has sometimes been thought necessary to supply experimental evidence in support of this point. Thus, although the preformation hypothesis clearly is not sufficient to explain embryogenesis, it nonetheless appears to be necessary.

As regards epigenesis, it is quite clear that it cannot be accepted without modification. The shortcoming of this concept, which has been realized by adherents as well as by opponents, is that the passage from the 'lowly organized condition' supposedly represented by the egg to the high level of organization exhibited by adult animals, even the most primitive ones, corresponds to the creation of order. According to the second law of energetics this phenomenon cannot occur spontaneously. The advocates of epigenesis have tried to solve this dilemma in two different ways. The first approach, now largely abandoned, was to invoke the participation of some supernatural force. Secondly, it has been suggested that external factors may supply the negative 'entropy', in the widest sense of this concept, required to bring the epigenetic hypothesis in agreement with the second law of energetics. Without taking stand on this issue it may be presumed that in either case the agents exert their effect on some preformed matter. It must therefore be acknowledged that epigenesis is not sufficient to account for the mechanism of embryogenesis, but a wealth of observations demonstrate that it is necessary.

Thus, in its modern interpretation, the classical antithesis has been partially resolved, insofar as both hypotheses are necessary for the explanation of the mechanism of embryogenesis. As appears from the lecture by Whitman (1895), many outstanding biologists at that time shared this view, and although it may not always be stated explicitly—there is a tendency to neglect the aspect of preformation—it seems to be the prevalent notion among biologists today.

If the two hypotheses are the only logical alternatives, they must also be sufficient. Before we can establish this point it is necessary to investigate the nature of the external or environmental forces which, according to the adherents of the hypothesis of epigenesis, are instrumental in embryogenesis.

Environmental factors

We shall in the present discussion neglect the conditions obtaining in viviparous animals in which the embryonic environment is highly specific. This circumstance rather impedes a general discussion, but there is no reason to believe that any principal differences obtain between the mechanism of embryogenesis in this case and when the embryos develop in an inorganic milieu.

Two kinds of influences are often classified as environmental, those exerted by the fertilizing sperm and those imparted by the external medium proper. From the standpoint adopted here, that embryogenesis is initiated by the fusion between the egg and the species-specific spermatozoon, the act of fertilization is thus not 'epigenetic', but rather 'progenetic'. To the unfertilized egg, the sperm is an external factor, but those portions of it which enter the former are integral parts of the fertilized egg, and the possible changes it may have imposed on the organization of the latter are inherent properties of the initial embryogenetic substrate. This does not imply, of course, that they are of no significance.

As regards the influence of the external medium, it is true that exchanges and interactions of various kinds occur between embryo and environment. It is also true that certain external factors (tonicity, oxygen tension, temperature, etc.) must be kept within certain limits to allow normal development. It may be mentioned that the discovery by Herbst (1895) that Li^+ causes vegetalization of the sea urchin embryo is among the first demonstrations of the fact that the external milieu may alter the course of embryogenesis. Observations of this kind do not support the contention that environmental factors specifically participate in embryos. As regards Li^+, this ion is known to influence the development of many different embryos. To be sure, the morphological effects may differ from case to case, but this does not warrant the supposition that they are specific. As long as this view has not been experimentally corroborated, it seems simpler to presume that Li^+ in all cases affects the same embryogenetic mechanism, and that the observed variations in the effect may be referred to differences in the other embryogenetic factors involved, which exert their effect through interaction with the one upset by Li^+.

There are many observations, notably the circumstance that embryos of many different species may undergo normal development in the same external milieu, which suggest that the effects of external factors are rather unspecific. Thus, although the importance of the environment for embryogenesis is indisputable, no evidence has ever been forwarded which even remotely suggests that it participates in the way presumed by the hypothesis of epigenesis (cf. Whitman, 1895; Woodger, 1929; 1930; 1931; P. Weiss, 1939). If the latter is dependent upon this contribution it thus obviously must be rejected.

I believe that a proper notion of the possible action of the milieu on embryogenesis may be gained by a comparison with external influences upon the genome. At a certain stage it was a generally accepted tenet in genetics that the heritable factors could not be affected by the environment. The discovery by Muller (1927) that mutations may be induced by X-rays led to a revision of this view. The fact that most experimentally provoked mutations are deleterious has not shaken and need not shake the belief that the establishment of advantageous mutations is the principal aspect of the mechanism of phylogenesis. Similarly, although experimental changes in the milieu most often have adverse effects, it cannot be excluded that penetration into new environments during the course of evolution may have entailed changes in the embryonic development, which have led to beneficial modifications of the phenotype. This rather ignored possibility appears to be of great interest from a phylogenetic point of view; unfortunately, it is as difficult to verify as the involvement of gene mutations in phylogenesis.

PREFORMATION, PREDETERMINATION AND EPIGENESIS

We shall begin the present discussion by considering briefly the consequences of the fact that scientific explanations must be in conformance with the causal postulate. Simply formulated, the latter states that if differences obtain between two entities, e.g. organisms or parts thereof, it follows that the entities were either different from the outset, or else that they have been exposed to different environmental conditions (cf. Woodger, 1929; 1930; 1931). As far as the latter are concerned, I have above expressed the opinion that the external environment cannot be specifically involved in embryogenesis to any really significant extent. It therefore follows that any difference prevailing within the embryo at any stage of development must be causally related to differences present in the fertilized egg. So far the principle of preformation obviously retains its validity. However, it is not necessary to presume a one-to-one relationship between the differences obtaining at successive stages; slight variations, possibly only quantitative, may during the process of embryogenesis be transformed into qualitative differences. It will be realized that under such conditions the environmental influences alluded to above may refer to the internal environment residing in different parts of the embryo.

As regards the question of the relative contributions of preformation and epigenesis to the mechanism of embryogenesis, it appears that the first serious attempt to solve this problem was made by Woodger, who arrived at the following proposition: 'Thus all that is preformed . . . is the number and nature of possible modes of intracellular elaboration which a cell may undergo in a given individual of a given species, and also certain developmental processes which depend upon the intrinsic properties of the ovum, e.g. cleavage pattern. Within those limits the rest is "epigenetic"' (1929, p. 383).

In this statement two important points are implied, discussed in more detail elsewhere in Woodger's book. The first one pertains to the fact that, although each cell contains a complete set of genes, which control the 'modes of intracellular elaboration', i.e. patterns of histological differentiation, it will, during the course of normal development, assume only one of these patterns. This process of selection is, according to Woodger, an epigenetic process. The other point is that the level of organization of the developing organism is raised by the various activities going on. Woodger mentions as an example the progress of cell division, which entails that 'at the end of, say, gastrulation, we have a type of organization which was in no sense "preformed" at the period with which we started, since this type of organization involves cellular parts, and such parts did not exist during that period. On the other hand, what did exist then, and in that sense was "preformed", was the cell-type of organization, which still exists now, but as characterizing not the whole but certain repeated parts' (1929, p. 341). In this quotation the contention is advanced that certain processes by their mere occurrence contribute to increasing the complexity of organization.

Woodger thus evidently considers that preformation and epigenesis in concert are necessary and sufficient to account for embryogenesis, provided that the classical concepts are redefined. This view contains important clues to the solution of the problem even if, as indicated by the heading of the present section, I intend to advocate the necessity of distinguishing between three different phenomena, which may be called *preformation*, *predetermination* and *epigenesis*, respectively. The contribution of each of these will be considered in the following pages.

Preformed substrate

It may appear a cheap, but certainly a logically unassailable, solution of a complicated problem to maintain that what is preformed in embryogenesis is the initial embryogenetic substrate, the fertilized egg, nothing more and nothing less, but maybe this notion will become acceptable once its implications are brought out.

The importance of the *preformed substrate* may perhaps be epitomized by the statement that it contains information relating to structural order on various levels of organization. Thus, beginning with the highest one, the cellular level, the egg, being an entity possessing the 'cell-type of organization' (Woodger, 1929), embodies the structural information concerning this kind of organization, which certainly must be required for further development if Virchow's dictum '*omnis cellula e cellula*' is to be taken seriously. The fertilized egg typically exhibits inhomogeneities, pertaining to the distribution of cytoplasmic components like yolk and hyaloplasm or to cortical properties. These inhomogeneities confer a kind of polarization on the egg structure, and are therefore usually called *polarities*. Regional differences of this kind represent a sort of information which is of great consequence for the course of embryogenesis.

The next highest level may be represented by particulate bodies. Among these I shall mention only the mitochondrion and the centriole. The fact that they are self-replicating structures indicates that they contain the information required for their own reproduction. It seems to hold for both of these particles that their particular organization incorporates information of great importance for embryogenesis.

On the high-molecular level the nuclear genome carries the most important structural information, in the form of DNA molecules specifying the composition of the RNA and the protein involved in the development as well as the function of the organism. Many other high-molecular substances in the egg may hold structural information, of importance at least during the early embryonic stages, but they need not concern us here.

On the low-molecular level many different compounds are undoubtedly involved. We shall mention only the reserve material, mainly present in the form of yolk granules, which contain the chemical information required for a substance to be used either for synthetic purposes or for the supply of energy.

Predetermined mechanisms

Embryogenesis is initiated as soon as the formation of the substrate is completed by the process of fertilization. This circumstance demonstrates that the embryogenetic substrate, besides the structural aspects just described, also embraces information concerning the execution of certain dynamic activities, namely those called above embryogenetic processes. The faculty to carry out these operations is thus somehow programmed in the fertilized egg. Since it seems inadequate to use the epithet 'preformed' in association with dynamic events, we shall instead talk of *predetermined mechanisms*, but *programmed operations* would, of course, suit as well (cf. Waddington, 1968).

For all we know, the most important aspect of the information needed for cell division to occur is embodied in the structure of the centriole. If one is present in the fertilized egg, the process of cleavage will start, and thereby prepare the ground for the other two kinds of processes. But a centriole will not be able to instigate a cell division unless it is located in a cell, so the organization of the cell is part of the overall predetermination of cell division, and a number of other conditions must also be fulfilled. It may be mentioned that the information embodied in the centriole entails the predetermination of a number of other operations besides cell division. This point will be dealt with later.

Available evidence suggests that all cells, and certainly all blastomeres in the early embryo, contain a full complement of genes. If this endowment were the only prerequisite needed for the exploitation of the information concerning protein synthesis stored in the DNA molecules, then it might be envisaged that synthesis of a complete set of proteins would start immediately after fertilization. Under such conditions the process of cell diversification could not be realized. Nothing of the kind actually happens; as we shall discuss later, protein synthesis generally is very modest during early stages of embryogenesis, and when it gathers speed, it turns out that different cells assume different patterns of protein synthesis. This state of affairs has been explained by invoking the principle of *nucleo-cytoplasmic interaction*. This predetermined mechanism comprises a nuclear and a cytoplasmic aspect. As far as the former is concerned, it seems that the genome contains a number of genes, whose function it is to control the transcription of the structural genes. These master genes appear to be in turn steered by cytoplasmic factors. The cytoplasmic composition allowing for this activity is only gradually established. The circumstance that the cells vary with respect to the pattern of diversification, or protein synthesis, which they assume must ultimately be referred to differences in the cytoplasmic composition of the cells in question.

Morphogenetic processes are the outcome of cellular activities and of interactions between cells and extracellular elements. As far as the contribution by the cells is concerned, it must rely upon their possession of the cell-type of organization, and this is predetermined by the fact that they are descendants of the egg which is itself a cell. As mentioned above, the various cell-orders arise through transformation processes, and since the extracellular elements are differentiation products, it follows that the nature of the participants in the morphogenetic processes, and thus also of the processes proper, is predetermined by the genome.

Epigenetic events

When, after fertilization, the formation of the embryogenetic substrate is completed, the predetermined mechanism embodied in the centriole enters into action, answering for the embryogenetic process of cell division. Gradually also the other predetermined mechanisms begin to function, with the result that cell diversification and morphogenetic processes take place. As development progresses, various kinds of interaction occur between the embryogenetic processes and between the latter and the substrate. All these changes are obviously not preformed. If they were predetermined, it would be impossible to change their course, and the well-established existence of this possibility excludes the application of this epithet. Under these circumstances it appears unavoidable that *everything, without exception*, which happens in the embryo after fertilization must be classified as *epigenetic events*. This definition is in agreement with the sense of the word 'epigenesis', even though it does not coincide with the classical one quoted above as far as the implications are concerned.

The embryogenetic processes are continuously affected by the substrate and by the other processes, and may be changed by external influences. Thus, it is known from the rules of cell division that generally the two daughter cells arising from this process will be of equal size and composition. In the cleaving egg this rule is often violated from the first division, and cases where inequalities do not obtain at the third cleavage, if existing at all, are found only in the most primitive animal phyla. The cleavage pattern is the outcome of interactions between the substrate and the mechanisms of cell division, and since the former usually embodies inhomogeneities or polarities of various kinds, the cells located at different regions in the embryo will vary with respect to size, cytoplasmic composition and rate of division. The assertion that the cleavage pattern is epigenetic is at variance with the view expressed by Woodger as quoted above. The epigenetic nature of this event is borne out, however, not only by its conformance with the definition advanced here, but also by the fact that it can be modified experimentally by interference with the properties of the substrate.

As a consequence of the interaction between the substrate and the process of cell division, a number of cells arise which differ with respect to their cytoplasmic contents. Through the mechanisms of nucleo-cytoplasmic interaction these differences may lead to the creation of an assortment of cells possessing different diversification patterns. The epigenetic modification of the cleavage pattern outlined here may thus lead to an ordered spatio-temporal distribution of diversification patterns. The nature of these is determined by the substrate, i.e. the genome and the cytoplasmic composition of the fertilized egg. The spatial distribution is partly determined by the polarities, and partly by the process of cell division, while the latter in cooperation with the cytoplasmic composition is responsible for the temporal pattern.

The principal difference between preformed mechanism and epigenetic event is easily grasped also in this case. Many means obtain by which the nature as well as the spatio-temporal distribution of differentiation patterns may be changed. Since alterations of this kind cannot be accomplished by modifying the preformed genome or the predetermined mechanisms of nucleo-cytoplasmic interaction, it follows that interference with the process of cell diversification can occur only through modification of the substrate. This expedient must ultimately involve changes in the cytoplasmic composition of the cells which affect the diversification processes.

With the appearance of the products of cell division and cell diversification, i.e. cells and extracellular substances, morphogenetic processes are initiated. The mechanisms responsible for these processes are predetermined, depending upon properties inherent in the cells and the extracellular substances, but morphogenesis itself is an epigenetic event which, as experience shows, may be modified by various means. However, as far as experimental interference is concerned, the situation is slightly different from that obtaining with respect to cell division and cell diversification. We have just shown that because the mechanisms of these processes are predetermined, their course can be altered only by changing the embryogenetic substrate. As concerns the morphogenetic processes, the executive agents are themselves elements of the substrate, and as such they may be affected by external factors, with a consequent change in the progression of morphogenesis. The morphogenetic mechanisms are therefore not predetermined in the absolute sense as are the other embryogenetic mechanisms.

Summarizing, we may state that epigenesis is a series of causally related events, the nature of which is determined by information prevailing in the embryogenetic substrate, partly in the form of structural information and partly in the form of rules of instructions pertaining to the embryogenetic processes (cf. Waddington, 1968). The main task of the embryologist, or epigeneticist, is to make out the nature or content of this information, and how it contrives to direct the transformation of the egg into an embryo, larva, etc., in accordance with the causal postulate. The fact that epigenesis almost always follows the course thus specified may easily convey the impression that it is 'predetermined'. For reasons stated above this adjective should be avoided in this context because it conveys a wrong notion about the nature of epigenesis. As it is, the information contained in the substrate may be changed, with consequent modifications of the embryogenetic process. Needless to say, the causal postulate is obeyed even in this case, and the modifications will be predictable if the impact of the external influences on the information is known.

In spite of honest efforts I may not have succeeded in bringing out clearly enough the implications of epigenesis as defined here, and therefore the following section will be devoted to the description of an analogy which may clarify the issue, also with respect to the consequences of epigenesis for the process of phylogenesis.

An analogy of epigenesis

As I see it, there are many processes which may serve to illustrate the phenomenon of epigenesis, but one example which combines actuality and lucidity is the working of a computer. Neglecting the hardware, two sets of factors are required, a 'predetermined' program outlining a number of operations, and a 'preformed' set of data on which the operations are to be performed. When both are presented to the computer, a process will occur which possesses the characteristics of an epigenetic process. Thus, it is a series of causally related events, which are the outcome of interaction between the program and the data. Computer operations, like epigenetic events, are predictable in every detail when the program and the data are known, yet the particular process has never been programmed, and it is for this reason that it must be wrong to consider the epigenetic events as predetermined or programmed. The course of the process may be changed by modifications of either the data or the program. This is a point which is of great importance for our understanding of the process of embryogenesis, particularly of the means available for its modification as required for phylogenetic evolution.

Thus the course of embryonic development may be altered by changing the data, i.e. the substrate. These variations may be concerned with mutations of the genome, as envisaged by neo-Darwinism, through which means the number and nature of the diversification patterns are changed. However, they may equally well affect cytoplasmic and cortical properties of the fertilized egg, which directly or indirectly may influence the spatio-temporal distribution of the diversification patterns. This latter possibility, the consequences of which may entail important evolutionary innovations, has been largely neglected in the past.

Ontogenesis may also be modified by changes in the predetermined mechanisms. As far as cell division is concerned there seems to be little chance of variation. Alterations of cell diversification patterns must be preceded by gene mutations; in order to impress the effect of the latter upon the phenotype, and thus be amenable to the mechanism of selection, a predetermined mechanism of nucleo-cytoplasmic interaction is indispensable. Presumably, mutations generally lead to modification of existing diversification patterns, and it may thus usually not be necessary to provide new mechanisms of nucleo-cytoplasmic interaction.

New predetermined morphogenetic mechanisms are established with the appearance of new cell-types and new extracellular substances. With respect to the former it appears that all cell-orders are present in every phylum except the lowest ones; little innovation has therefore been possible on this point. Somewhat more variation obtains with respect to the extracellular substances, but even here a remarkable conservation appears to be the rule in the animal kingdom. Variations in the spatio-temporal distribution of the morphogenetic elements, as dictated by the nature of the substrate, seem to be of at least as great, if not greater, importance for phylogenetic progression than changes in the predetermined mechanisms.

The computer analogy may also bear on the question of the information content of the adult organism as compared to that of the fertilized egg (cf. Waddington, 1968). As far as I gather, the final result delivered by the computer is considered to contain the same amount of information as the data and the operation instructions combined. If this is accepted to hold for the process of embryogenesis, the dilemma of the classical hypothesis of epigenesis is solved; the amount of information, or order, is the same at the outset as at the termination of ontogenesis, neither increase nor decrease takes place. It must be emphasized, however, that the comparison cannot be made between the genotype and the phenotype, as suggested by Waddington, since the fertilized egg contains much more information than is embodied in the genotype, when the latter word is used in its ordinary sense as identical with the genome.

We may conclude this section by pointing out that the computer analogy also fits to the extent that a very complicated computation may be based on a relatively simple set of data and a limited number of operation instructions.

Theoretical analysis of epigenesis

From the above discussion it appears that all the processes involved in embryogenesis and in ontogenesis may be classified as epigenetic events. As a matter of fact, it becomes difficult to distinguish between the concepts of 'epigenesis' and 'ontogenesis' unless the latter is considered to be a purely descriptive account of the course of development and the former a causal explanation of the phenomenon. Without attempting to justify logically this terminology, it will be adopted in the present book. This may serve to explain the wording in the headings of this section and of the following chapters.

On the preceding pages I have quoted almost exclusively publications by Woodger. This may be unfair insofar as other outstanding contributions have been made to the problem of theoretical embryology. It cannot be helped, however, that the work of Woodger has been a particularly important source of inspiration and encouragement for the present book. Also, when it comes to outlining the approach to be pursued here, Woodger has advanced some significant propositions which deserve mention. Woodger has pointed out that 'in biology our fundamental notions are for the most part so vague and lacking in precise definition that it is impossible to work out their logical consequence' (1930, p. 5). This point is amply brought out by the embryological literature, which abounds with concepts like 'canalization', 'competence', 'determination', 'differentiation', 'evocation', 'field', 'germ layer', 'gradient', 'individualization', 'individuation', 'induction', 'morphogenesis', 'morphogenetic substance', etc., the exact meaning of which is anything but clear. If scientific progress depended only upon the naming of observed phenomena, embryology would be an advanced science.

Woodger has also sketched the outlines of a theory of embryology, as shown from the following quotations: 'Thus the consideration of the sort of further hypotheses required leads us to an analysis of what might be called *elementary* processes of development, and takes us into the field of general physiology' (1948, p. 359), and 'In order to take certain further steps in embryology we are *compelled* to invent hypotheses concerning submicroscopical cell structure which will be explanatory of the behaviour of cells during development' (1948, p. 360). The program proposed here may be called the *analytic approach*.

As far as the first point is concerned, we have already commenced the analysis by establishing the existence of what has been called the 'elementary embryogenetic factors' rather than the 'elementary processes of development'. In the detailed analysis which follows the second recipe of Woodger will be adapted. Thus, an attempt will be made to establish the nature and the properties of the preformed substrate and the predetermined mechanisms, down to the subcellular level. The hypothetico-deductive approach will be applied as far as possible, i.e. generalizations and hypotheses will be advanced, and their consequences tested by reference to observation records.

Particular attention will be paid to the mechanisms of interaction between the embryogenetic factors. It will be realized that although remarkably few factors are involved during a considerable part of the developmental period, the number of possible modes of interaction very soon becomes so large that the process of embryogenesis becomes extremely complex. For this reason alone it will be necessary to confine the present work to the earliest developmental stages.

Once the analytic approach has been completed, it should be possible to adopt the *synthetic approach* in an attempt to establish the importance of epigenesis for the processes of ontogenesis and phylogenesis. At our present state of knowledge the analytic approach cannot be brought to an end. Nevertheless, for the initial phases of embryogenesis the elements involved are known in sufficient detail to warrant an attempt towards synthesis. Therefore, with the purpose of demonstrating the validity of the analysis presented here, the various ideas will be put to test through application to certain selected cases of embryogenesis.

The relation between phylogenesis and ontogenesis was formulated in the following way by Garstang (1922, p. 98): 'The phyletic succession of adults is the product of successive ontogenies'. Since it is postulated that the process of ontogenesis is the outcome of epigenetic events, it is immediately obvious that the latter must be intimately engaged also in the realization of phylogenetic evolution. It is, of course, impossible to discuss phylogenesis on the basis of epigenetic mechanism purportedly concerned only with early embryogenesis. Yet, as stated above, the views advocated here appear to be valid beyond this range, and in the last part of this book an attempt will be made to establish their bearing on the problem of phylogenesis.

'If, furthermore, we define *the task of developmental mechanics* so that it shall include the fewest mysterious concepts . . . *we must reduce the processes of organic formation to the fewest and simplest modi operandi. . . . Accordingly, in any given case, we must trace back each* individual formative process to the *special combination of energies by which it is conditioned*, or, in other words, to its *modi operandi*.'
(W. Roux, 1895, pp. 151 and 152; translated by W. M. Wheeler)

EPIGENESIS: THE ANALYTIC APPROACH

As outlined in the introduction, the epigenetic events, the sum total of which constitute ontogenesis, are the outcome of the interplay between the embryogenetic substrate and the embryogenetic processes as directed by the predetermined mechanisms. The actual cases of embryogenesis observable in nature, even in the lower animals, are usually so complex that some kind of resolution—experimentally as well as intellectually—is necessary before it is possible to disentangle the workings of the forces involved. As appears from the quotation above, this point was clearly realized by Roux, and the same holds for many of his contemporaries. Progress towards this end has been made since then, but the outcome is rather meagre, considering the lapse of time. I believe this will be borne out by the attempt in the following chapters to analyse, on all relevant levels of organization, the properties of the four embryogenetic factors, and the rules governing their interactions.

EMBRYOGENETIC SUBSTRATE

The embryogenetic substrate is at any developmental stage identical with the organism proper. It is thus no constant entity; starting with the fertilized egg it alters with time according to a regular pattern, known in more or less detail for many animal species. These changes are accomplished by the embryogenetic processes of cell division and cell diversification; while morphogenetic processes change the form of the embryo, they do not contribute to the modification of the nature of the substrate. As a result of the changes mentioned the organism comes to consist of a number of morphologically discernible entities, the morphogenetic or the substrate elements. Although it would appear more logical to start with the initial substrate, the fertilized egg, and pass on to the elements, it turns out to be expedient to deal with the latter first; this approach is consequently followed here.

Elements

The morphogenetic elements comprise living and non-living components. In the former case an element may be defined as the smallest unit which can persist as a discrete living entity; this definition obviously applies only to cells. As regards the non-living elements, they are identical with the extracellular substances discussed in the introduction. They are differentiation products, secreted by cells or otherwise derived from the activities of differentiated cells which, being chemical substances, are best characterized by their chemical composition. The hyaline membrane surrounding certain eggs occupies a position between the extracellular elements and the egg envelopes. The reason why it cannot be included among the former is that it is secreted by the egg which, according to the notions advanced, cannot be classified as a differentiated cell. In spite of this it will be dealt with in the present section.

In the following discussion of the substrate elements particular attention will be paid to those properties which are known or suspected to be of importance for embryogenesis. As will be seen, these include morphological and chemical, but also, and to a remarkable extent, physical features.

CELLS

Cells may be defined as entities possessing the 'cell-type of organization'. The implication of this expression, borrowed from Woodger (1929), appears from the following quotations: 'The real significance of the cell concept lies in the recognition of the fact that . . . this is the simplest type of organization which is capable of characterizing a whole (and hence independent) organism, but in virtue of its spatial repeatability it also constitutes the type of organization which characterizes certain of the *parts* of those organisms whose organization is above this level' (p. 310), and 'Thus the cell-type of organization appears as the irreducible minimum which forms the starting point [of embryogenesis]' (p. 340).

It was shown in the introduction that all embryogenetic processes ultimately may be referred to activities carried out by cells. It is therefore of great importance to analyse the cell-type of organization, with special reference to those features which enable cells to exert functions that are instrumental in the embryogenetic processes. It was also pointed out that animal cells may be grouped into a number of cell orders, characterized by differences in shape and behaviour. As the cellular contribution to embryogenesis depends upon their cell-order affiliation, it will be necessary to devote a section to the specification of the features distinguishing each of the cell orders. Finally, we shall discuss in more detail those cellular properties which are of particular importance for morphogenesis.

The cell-type of organization

The expediency of the concept of the 'cell-type of organization' lies in the fact that it can be employed in cases where the application of the word 'cell' is disputed, e.g. in the case of the Protozoa (often considered as acellular organisms) and of the gametes (considered here as potential parts of a developing organism). The property of spatial repeatability, i.e. reproduction by division, is not a feature unique to the cell-type of organization; certain cell parts like chromosomes, centrioles and, presumably, mitochondria also possess this faculty, and conversely cells may lose their reproductive capacity under certain conditions. The characteristics of the cell-type of organization mentioned in the statements by Woodger quoted above are of great importance for embryogenesis. However, a number of other cellular properties are involved, the nature of which it is our objective to establish. I propose to approach this pursuit by a closer scrutiny of two cell constituents which, as it turns out, may account for many of the cellular contributions to the embryogenetic processes. The subcellular morphology is also of some importance for embryogenesis, not least as regards the distribution of the constituents mentioned, and the present section will therefore be concluded with a discussion of some morphological subunits of the cell-type of organization.

Two important cell constituents

The constituents to be discussed here represent classes of substances, rather than single chemical compounds. However, in each class all the members are distinguished by the possession of certain physico-chemical properties, and since only these are of importance from the present point of view, we may neglect the question of differences in chemical composition. The first constituent is a protein, distinguished by the ability to polymerize into structures of varying shape, in which a tension may prevail. By this characterization it is realized that we are dealing with a class or genus of structural proteins which, owing to the property just specified, may be called *tensible proteins* (TP), i.e. proteins capable of becoming tense. As will be discussed later, these proteins appear to be distinguished by their similarity to actin. For this class of proteins, Mazia and Ruby (1958) have suggested the name 'tectins', but it seems to me that 'tensilins' might be preferable, because it more distinctly emphasizes the fact that the proteins in question may be involved in various dynamic activities. I shall nonetheless refrain from adopting it in the present context.

As concerns the second constituent, it is above all characterized by its ability to form hyaline matrices. This feature suffices to refer it to the classes of mucopolysaccharides and glycoproteins. The former are probably of more importance than the latter in the phenomena dealt with here, and for that reason, as well as for convenience, we shall talk of *mucopolysaccharides* from now on, without thereby excluding that glycoproteins may be of decisive influence in some cases.

Tensible proteins

Among the various 'theories' advanced in the last century concerning the structure of cytoplasm, certain considered the fundamental structure of the cytoplasm to be 'delicate' fibrils, either separate or forming a connected network, traversing a homogeneous ground substance (cf. Wilson, 1928). It is particularly noteworthy that Flemming, one of the exponents of the first alternative, called the filar theory, advanced the hypothesis 'that the essential energies on which life depends have their seat in the fibrillae' (Wilson, 1928, p. 68). This point of view has been advocated again and again, but it has been rejected just as often. Also Wilson admits that the notion is correct to some extent, but in his opinion many observations cannot be accounted for in this way. With the advent of the electron microscope it has been possible to develop much clearer ideas about the organization on the subcellular level. In combination with numerous biochemical and biophysical observations, the knowledge thus gained has fully borne out Flemming's hypothesis, insofar as it seems generally agreed that all kinds of movements, on the supracellular, the cellular and intracellular level, may be referred to the action of proteins similar to actin, organized in polymers of various kinds (cf. Gibbons, 1968).

In the present context I shall adopt this contention, postulating that in an undifferentiated cell, as encountered for instance in the early embryo, the only kind of structural protein prevailing is of the type which is here called 'tensible proteins'. It must thus be responsible for the formation of all the structures like microfilaments, microtubules and membranes, the existence of which has been demonstrated by the electron microscope. In differentiated cells other structural proteins may exist side by side with this type of protein. According to the properties ascribed to the tensible proteins, a tension may under certain conditions reside in these structures, and the force thereby generated is proposed to be the only means by which cells can carry out mechanical work.

We shall in the following sections discuss first some biophysical aspects of the formation and function of various subcellular structures made up by proteins of this kind and subsequently their contribution to the morphology of the cell.

Biophysical aspects. It is a general biological notion that, when mechanical work is exerted, a contraction is involved, and it is commonplace to talk of 'contractile' forces, proteins and structures. It is also usually assumed that whenever a tension prevails in a biological object, the tension is elastic. Before we can embark upon the actual problem, we shall discuss the validity of the concept of 'contraction'. Later in the section we shall also raise the question of whether the tension prevailing in animal cells rightfully can be considered an elastic tension.

It seems a good working hypothesis to presume that the physical forces displayed by living organisms are 'physical' in the sense that they belong to the class of phenomena dealt with by the science of physics. As it turns out, 'contraction' is one of the many poorly defined biological concepts which is unknown to classical physics. Contraction appears to presuppose the existence of tension, elastic or otherwise, in a structure suspended between two frames. If either of the latter cannot develop enough force to counteract the tension, they will yield, with the consequence that the distance between the frames, the length of the tensile structure and the tension proper are reduced. Although a shortening thus takes place, the phenomenon is not, and cannot be, described physically in terms of a contraction.

This situation may, of course, represent a deficient development of physics, but it is nevertheless advisable for biologists to avoid describing physical phenomena with concepts unknown to this branch of science. For this reason the word 'contraction' will be replaced by 'tension' in the present text; as appears, this expedient may occasionally require rather awkward wording, but this is mainly a reflection of the fact that our thinking and phrasing traditionally are based upon the concept of 'contraction'. If this substitution is made, several advantages are gained: not only are we dealing with a defined phenomenon, but furthermore this approach will in certain cases lead to a much more consistent appreciation of some of the most usual processes of 'contraction'.

From the preceding discussion it will be grasped that the most well-known representative of tensible proteins is actin, first isolated from muscles. It is quite common that proteins from other sources exhibit properties, morphological, physical and chemical, closely similar to those of actin, and they have consequently been called 'actin-like'; there even seems to be a tendency to take one more step and call them 'actins'. Other names have been suggested, for instance 'tubulin' for the protein forming microtubules (cf. Adelman, Borisy, Shelansky, Weisenberg and Taylor, 1968). As we shall presently see, it is questionable whether the formation of disparate kinds of polymers necessarily implies the existence of different kinds of proteins. It may rather be imagined that in a given cell only one kind of protein exists which, according to the prevailing conditions, may be used to build up several different structures. If this is true, then obviously no consistency is achieved if the proteins are named after the structures in which they are found. Amino acid determinations on tensible proteins of widely different origin have shown the composition to be very similar to that of actin in many cases (cf. Hatano and Oosawa, 1966; Renaud, Rowe and Gibbons, 1968; Adelman *et al.*, 1968). However, manifest disparities have also been established: thus it was shown by Stephens and Linck (1969) and Stephens (1970) that actin and microtubule protein from the same animal (*Pecten*) differ in several respects (molecular weight, amino acid composition, etc.), and actin has been found to be immunologically different from the protein in other TP structures (cf. Ishikawa, Bischoff and Holtzer, 1969).

As far as such differences are concerned, they seem to be the natural outcome of mutations which have occurred in the course of evolution. It is thus to be expected that the TP species found in different animal taxa vary with respect to their chemical composition, although it may be envisaged that certain amino acids in crucial sites are identical. To the extent that the syntheses of the various proteins are controlled by different structural genes, differences may arise even within the same animal. This situation is likely to prevail with respect to actin, which may be characterized as a differentiation product, whereas the microtubule-forming protein, using a term that will be defined later, presumably must be classed as a 'non-repressible protein'.

The physical properties of TP have been studied on isolated cytoplasmic structures (cf. Hoffmann-Berling, 1960; Mazia, 1961) and on extracted and reprecipitated preparations (cf. Sakai, 1962). Among the results obtained it will be mentioned only that the development of tension depends strongly upon environmental factors like ATP, ions, particularly divalent cations, substances which interfere with disulphide bonds, etc. Although some cases of ATP-independent tension phenomena have been described, the laws of energetics clearly demand that supply of energy occurs at some stage in each tension–relaxation cycle, and ATP appears to be the normal source of energy (cf. Hoffman-Berling, 1960). As concerns the molecular mechanism involved in the development of tension, this question is beyond the scope of the present treatise, and we shall therefore leave this interesting problem without notice (cf. the suggestions advanced by Astbury, Beighton and Weibull, 1955, and by Jarosch, 1964).

The TP structures observable by means of the electron microscope are polymerization products of subunits, dimers and monomers, the latter of which have a molecular weight of 45–60,000. As far as the process of polymerization is concerned, there 'is a growing belief, which may soon attain the status of dogma, that structures built from a single species of protein will aggregate spontaneously under the proper conditions and that the geometry of the structure is specified by the spatial distribution of interacting sites on the protein subunit' (Adelman *et al.*, 1968, p. 1192). In the regular structures formed, all subunits are in an equivalent or quasi-equivalent position relative to the surrounding ones; this arrangement is supposed to represent a state of minimum free energy.

Oosawa and Higashi (1967) have published a very interesting article on the statistical energetics of the polymerization of tensible proteins. They consider this phenomenon to be similar to crystallization, and that equilibrium may obtain between monomers and polymers. The authors suggest that small molecules of various kinds may influence the polymerization. One kind, called *regulators*, is supposed to influence the state of equilibrium, another type, called *selectors*, may determine the geometry of the polymerization product.

As regards the mechanism of action of the regulators, I shall not quote any details, yet it seems necessary to emphasize that the laws of energetics do not allow for deviation from an established state of equilibrium in the absence of energy supply. The notion of regulators is important because it emphasizes the reversibility of the polymerization process, a characteristic which is definitely suggested by the remarkable lability displayed by TP polymers, as will be discussed later.

The existence of some kind of selectors is a prerequisite for the contention advanced above that one kind of monomer may be made responsible for the formation the various polymers present in a cell. Oosawa and Higashi suggest that the selectors become bound to the monomers. It is in this connection worthwhile to notice the apparently specific affinity exhibited by linear polymers towards the purine riboside triphosphates, ATP and GTP. There is no evidence that these substances act as selectors, but their importance for the formation of TP structures is borne out by the fact that polymerization in salt solutions is almost impossible in their absence.

Oosawa and Higashi (1967) also contend that subjection to tension may lead to the transition from one to another kind of polymer. This phenomenon need not affect the geometrical structure, but it may influence other properties, for instance the equilibrium constant. This possibility is of great interest in view of the fact that many observations, some of which will be discussed later, indicate that linear polymers become stabilized when they are suspended between two frames. It will be realized that only under these conditions may a tension prevail in the polymer.

An exhaustive discussion of the nature of the tension developed by TP polymers involves knowledge of some properties which will be dealt with later. However, being a biophysical problem, it is most naturally taken up in the present section. Our main concern is to compare the tension in structures composed of TP with elastic tension, in order to decide whether the two phenomena are identical. In a body subjected to the deforming influence of external forces, elastic forces are developed to counteract the latter. For obvious reasons the body in question must be suspended between two frames or otherwise acted upon by two oppositely directed forces in order to produce an elastic force or tension. It may also prevail when the material has the form of a continuous sheet, as for instance in a rubber balloon. These stipulations hold also for the tension exerted by TP polymers, being the basic requirements that must be fulfilled in order that a force may be deployed.

There are, however, some manifest differences between the two kinds of phenomena, one observable in microtubules, the other in membranes. Thus, if a microtubule is suspended between two frames which are able to resist the tensile force, the latter may stabilize the polymer, as envisaged above. If, on the other hand, the frames yield to the force, the result is not a reduction of tension, but collapse of the polymers. This behaviour is mainly inferred from observations on filipodia (Gustafson, 1964) to be discussed later. Yoneda (1964) has studied the cortical tension in the sea urchin egg. Without at the moment discussing the location of this parameter, it will be contended that formally it may be considered a membrane tension. Yoneda made the remarkable discovery that compression of the egg does not change the tension, although this operation evidently causes an increase in surface area. This observation is not compatible with the notion that the tension is elastic.

I believe that both the reported observations may be accounted for by the lability of TP polymers. The first example shows that shortening of the linear polymers, microtubules, which constitute the core of filopodia, with ensuing lowering of the tension, cannot take place. We may presume that the collapsed polymers are replaced by new ones, shortened to fit the altered environmental conditions. The tension exerted by the microtubules in the new filopodium is probably the same as in the old one even though, as we shall presently see, the force exerted may be weaker. The second example suggests that stretching of a membrane does not occur, for in that case elevation of the tension should necessarily follow. Rather, we may envisage that new monomers are incorporated proportionally to the enlargement of the surface.

It thus seems that TP polymers cannot be deformed to any significant degree. This statement may, in principle, lead to the conclusion that the tension residing in them cannot be elastic, but for the sake of clarity we shall illustrate the case by means of two models, one representing a perfectly elastic material as approximately exemplified by a steel wire, and the other an imperfectly elastic material, typified by a rubber band. In the former case the tension may be substantially elevated by an infinitesimal increase in the distance between the frames of suspension. In the TP polymers it is possible that a certain increase may occur through this expedient, but as soon as the threshold value is reached, representing the force with which the monomers are kept together, the structure will yield with the consequence that new monomers are added and the original tension restored. The apparent plasticity displayed by TP polymers recalls the properties possessed by the latter model, but as we have seen, it does not depend upon elastic deformation, but upon the structural lability which allows for supply and withdrawal of material. Since neither of these models fits the actual case, it seems questionable whether it is warranted to maintain that the tension developed by TP polymers is elastic.

In order to predict the contributions of TP polymers to vital phenomena it is necessary to formulate generalizations concerning their properties. Some can be derived only from observations on living cells, and must therefore be postponed to a later occasion, but it is possible to propose the following four ones on the basis of this brief survey of the physico-chemical properties of tensible proteins.

(1) The tension residing in TP polymers enables living cells to deploy mechanical force within their own confines and in interaction with the environment.

(2) Owing to the reversible equilibrium between polymerization and depolymerization, structures consisting of TP polymers display a high degree of lability. It is postulated that many kinds of activities displayed by animal cells depend on this lability. Under certain conditions, for instance when they are subjected to tension, stabilization of the structures may take place and, as we shall see later, association with lipids may have a similar effect.

(3) Assuming that the development of tension is due to properties inherent in the monomers, and that the contribution is the same for each of them, it follows that the prevailing force must be proportional to the number of monomers. As an example illustrating this tenet it may be mentioned that many short microtubules should be able to balance a smaller number of long ones. This proposition has been advanced by Boveri (1888) as concerns the action of spindle fibres.

(4) From the postulate that the same kind of monomer participates in the formation of different TP polymers present in a cell, it follows that convertibility obtains between the various structures.

Since the tensible proteins are very similar to actin, it is to be presumed that the principles outlined here are valid also for the activity displayed by muscles. If this is granted, it may be necessary to revise some of the current notions concerning muscular performance. Thus, the development of tension as such does not involve work being executed, and hence energy expenditure may not *a priori* be anticipated in association with isometric 'contractions'. This circumstance can possibly explain that smooth muscles in general and the adductor muscles in Bivalvia in particular may be 'active' for long periods of time without any appreciable consumption of energy (cf. Hanson and Lowy, 1960). The shortening and lengthening of the muscles should thus be referable to the lability of the TP polymers.

The situation is different in striated muscles. These can perform both isometric and isotonic 'contractions'; in the latter case shortening of the muscle occurs and mechanical work is performed. The change in length results from the actin filaments penetrating deeper into the interstices between the myosin filaments (Hanson and Huxley, 1955; Hanson and Lowy, 1965). It seems that this mechanism of 'sliding filaments' is an expedient allowing for changes in the length of the striated muscle. In this way the collapse of the polymers is avoided which otherwise may be needed to accomplish a shortening of tensile structures. It is seen that it is possible to retain the word 'contraction' in this connection as synonymous with 'shortening', but that according to the view presented here, no implications as concerns the generation of tension or force are acceptable. It appears to me, however, that it would be better to discontinue the use of this ambiguous word.

It follows from these these ideas that the activity of striated muscles cannot be associated with changes in the length of the actin filaments, and that the maximum tension generated by each of the polymers is constant. The former point is brought out by electron microscopical studies (Hanson and Huxley, 1955; Hanson and Lowy, 1965), but the latter seems to be contradicted by many observations. However, a closer scrutiny of Figure 1 shows that the tension developed is proportional to the number of attachment points between each end of an actin filament and the corresponding myosin filament. When two opposing myosin filaments are attached to the same level of an actin filament, the tension is reduced. If this interpretation is correct, it must be concluded that the tension generated does not depend upon the mechanical properties of the actin filaments, but rather upon the tenacity with which the latter are attached to the frames represented by the myosin filaments. Hence, the actin filaments may never exert their maximum tension.

Isometric 'contractions' in striated muscles are associated with a notable consumption of energy, presumably owing to the incessant junction and disjunction of the attachments between actin and myosin filaments. It seems that this waste of energy is the price that has to be paid for the speed with which the striated muscle can perform mechanical work.

Structures formed by tensible proteins. It has been proposed above that in an undifferentiated cell all structural elements are built up by tensible proteins. Two classes of TP structures may be distinguished, linear or one-dimensional polymers, represented by microfilaments and microtubules, and two-dimensional polymers, present in the form of membranes.

Electron microscopic investigations have shown that considerable variation obtains with respect to the diameter of microfilaments. It appears that there are at least two classes of these polymers, thin microfilaments, the width of which has been reported to lie in the range 4–8 nm, and another species with a diameter of 10 nm. The two types may be distinguished by the fact that only the thin filaments bind heavy meromyosin

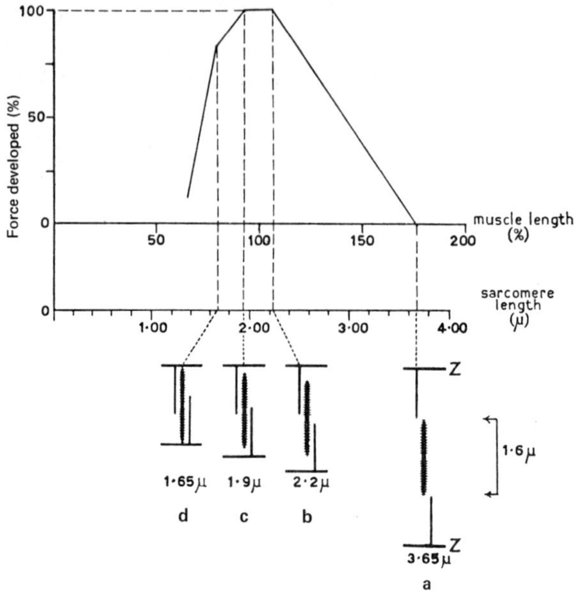

Figure 1. Relation between the force developed by a frog muscle fibre and the extent of overlap between actin and myosin filaments. The upper part of the figure shows the force plotted against the fibre length and against the sarcomere length, as determined by means of an interference microscope. The lower part of the figure shows the relative positions of the myosin filaments (thick lines) and the actin filaments (thin lines) at four selected sarcomere lengths. It is seen that, going from a to b, the force is increased in proportion to the number of attachment points on the myosin filaments to which the actin filaments are opposed. The shortening associated with the change from b to c does not change either the number of attachment points or the force. Going from c to d the force is reduced as the actin filaments become attached to the same rather than to opposite ends of the myosin filaments. (Reproduced from J. Hanson and J. Lowy, *Brit. Med. Bull.*, **21,** No. 3, 264 (1965), by permission of the British Council, London)

(Ishikawa *et al.*, 1969). The diameter of microtubules is about 24 nm. Thin microfilaments are supposed to be helical polymers, consisting of two strands (Figure 2); whether this holds also for the 10 nm filaments does not seem to be established. Microtubules are cylindrical structures, whose walls consist of parallel strands of monomers (Figure 2), the number of which may be 13 (cf. Porter, 1966). These polymers are characterized by their rigidity, a property which is of great biological importance.

The protein molecules isolated from the two kinds of polymers are distinguished by their ability to bind purine trinucleotides; generally one molecule is bound by each monomer. A characteristic difference obtains in this respect between microfilaments and microtubules, insofar as adenosine triphosphate (ATP) is involved in the former, guanosine triphosphate (GTP) in the latter case (cf. Stephens, Renaud and Gibbons, 1967).

As discussed in the preceding section, the formation of the polymers is supposed to

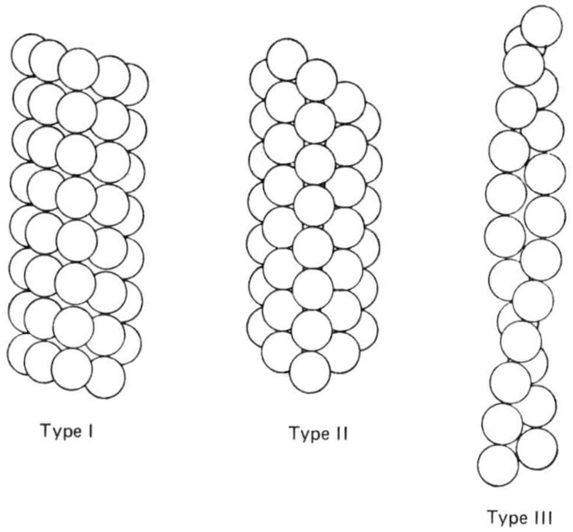

Figure 2. Linear polymers. Types I and II, models of a microtubule; type III, model of a microfilament. (Reproduced from F. Oosawa and S. Higashi, *Progr. Theor. Biol.*, **1**, 486 (1967), by permission of Academic Press, New York)

occur spontaneously, and it therefore follows that the presence of the various kinds of polymers in no way can be referred to the existence of preformed templates directing the pattern of polymerization. If, as proposed by Oosawa and Higashi (1967), the latter is determined by low-molecular selectors, then it would be natural to presume that in a given cell only one kind of selector, and thus one kind of polymer, may be found. It will be shown later that it is possible to classify cells on the basis of the presence or absence of microtubules in their cytoplasm, but in spite of this the above expectation is not generally fulfilled, for in certain cases all three kinds of linear polymers discussed above are found together in the same cell. It must therefore be admitted that the mechanisms regulating the formation of these structures remain obscure. It may be mentioned that certain factors residing in the nucleus must influence the polymerization of microtubules, as evidenced when cells which normally possess only microfilaments in preparation for cell division begin to elaborate the microtubules required for the formation of the mitotic apparatus. This event presupposes that a different kind of selector pervades the cytoplasm, and it may be inferred from the convertibility postulated above that microtubules gradually come to replace the microfilaments. It appears, however, that the conversion is not complete, and that consequently the two kinds of polymers may exist side by side, as is clearly observed in many eggs.

It has been suggested above that the lability, i.e. the incessant formation and collapse, of the linear TP polymers may be an important aspect of their ability to exert dynamic activities. It follows that impediment in either formation or breakdown of the polymers must check this faculty, and hence that means permitting interferences of this kind may be valuable tools in the elucidation of the mechanisms involved. Colchicine has for a long time been known to suppress cell division through dissolution of the mitotic spindle. As will be described later, the latter is composed of microtubules, and it thus appears that colchicine acts through inhibition of the polymerization of microtubules. Deuterium oxide conversely prevents the collapse of the polymers, but both substances repress processes involving microtubules in the gastrulating sea urchin embryo (cf. Tilney and Gibbins, 1969). This finding strongly supports the suggested relation between polymer lability and dynamics. It may be noted that neither of the agents mentioned affects either the microtubules present in the flagellated ectodermal cells of the sea urchin embryo or the cytoplasmic microfilaments. A group of substances, the cytochalasins, have recently been found to interfere with a number of dynamic cell activities (S. B. Carter, 1967). Thus, in cultured cells, cytokinesis, but not karyokinesis, is suppressed, a circumstance which leads to the formation of multinucleate cells. The same author also found that the adhesive properties of the cell surface are changed. Wessells, Spooner, Ash, Bradley, Luduena, Taylor, Wrenn and Yamada (1971) have studied the effect of cytochalasin on various cellular processes, and recorded inhibition in a number of cases, e.g. cell movements, axonal outgrowth in embryonic nerve cells, cytoplasmic streaming, contraction in smooth and cardiac muscle and in the cortex of *Xenopus* eggs and various morphogenetic phenomena. Specifically, they observed disintegration of the network of thin microfilaments which normally is present in undulating membranes, a finding which may explain several of the recorded effects. The mechanism of action may, in analogy to that of colchicine, be interpreted as involving interference with the formation of thin microfilaments. In this context it must be emphasized, however, that the microfilaments present as a sheath inside the plasma membrane and extending into the filopodia of various fibroblasts (cf. Buckley and Porter, 1967) are not affected by cytochalasin according to Wessells *et al.* (1971). This circumstance suggests that the cytochalasins interfere with the dynamic interaction between plasma membrane and microfilaments, rather than with the formation of the latter.

I believe that the propositions made here concerning linear polymers are generally acceptable. The same does not hold as regards the contention that membranes are two-dimensional TP polymers. I might make the case easy for myself and, using the logical approach, point out that this follows from the postulates made above and the facts that membranes are cellular structures and that phenomena like membrane tension and membrane movements actually have been observed. However, there are some few observations that more directly support this notion. Thus, it has been found that mitochondrial membranes are made up by repeating units, the structural proteins of which are estimated to have a compound molecular weight of 150,000, thus close to that of the dimers found in the linear polymers (cf. Green and Goldberger, 1967). The fact that mitochondrial membranes can develop a tension which, through the mechanism of reversible swelling, enables them to perform mechanical work should not be forgotten in this context (Lehninger, 1965). However, the most convincing evidence is probably the observations that a main constituent in the erythrocyte membrane is a protein with properties similar to those of actin (Mazia and Ruby, 1968), which furthermore can be polymerized into filaments (Marchesi and Steers, 1968).

Disappearance of cytoplasmic membranous structures in association with a copious formation of linear polymers has been recorded by many electron microscopists (cf. Porter, 1956a; Rebhun, 1961). The convertibility suggested by these findings also favours the view advocated here. I realize that the evidence mobilized in its support is sparse, but I hope it may gain in credulity when its consequences are worked out and compared with actual observations in the following chapters. I cannot help expressing the opinion that some supporting evidence may have been overlooked, owing to the fact that the currently accepted membrane concept, the unit membrane of Robertson (1959; 1967),

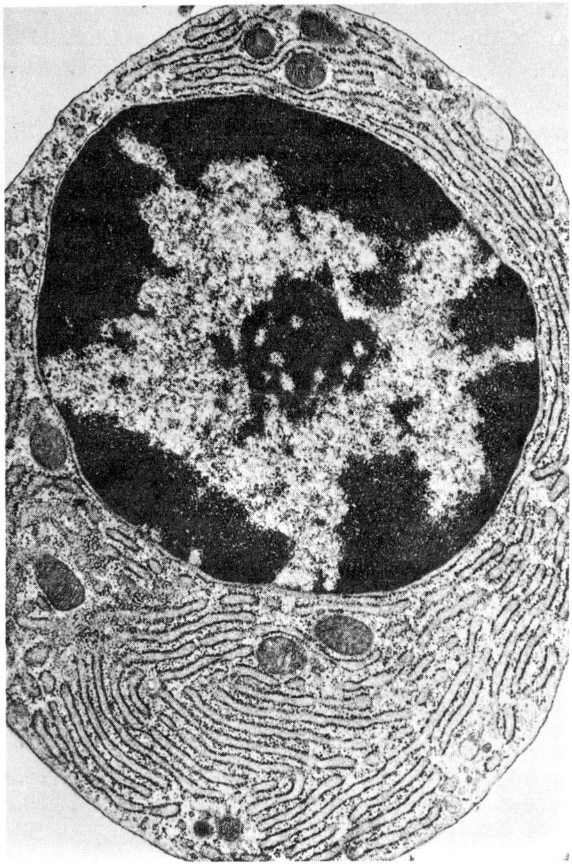

Figure 3. Plasma cell from guinea pig bone marrow, as visualized with the electron microscope. The influence of the nucleus, the cortex and the mitochondria on the otherwise parallel alignment of the cytoplasmic reticulum is clearly indicated. (Reproduced from D. W. Fawcett (1966), *An Atlas of Fine Structure. The Cell Its Organelles and Inclusions*, by permission of W. B. Saunders Company, Philadelphia)

a modification of the model of Danielli and Davson (1935), is incompatible with a membrane the main structural component of which is TP monomers.*

Compared to the linear polymers, membranes appear to be rather stable formations. Since membranes may persist even when apparently no tension prevails, it seems that other factors must be responsible for the permanence in this case. Nothing is known about this question, but it may be significant that spontaneous formation of membranous vesicles may occur from the subunits of membrane proteins in the presence of lipids (cf. Green and Goldberger, 1967). These authors suggest that the membrane proteins and the lipids form a complex, in which the latter serve to impregnate the membrane, and it may be imagined that a stabilization is accomplished at the same time.

*Note added to the proof, May 1972. The situation has changed since this statement was written in 1968, for recently a number of reports have been published which propose a new membrane model that bears some similarity to the one advocated here.

It is possible that intracellular membrane complexes like the endoplasmic reticulum arise through fusion of small vesicles. Evidently the arrangement of the complexes is subjected to influences from the environment, as exerted by nucleus, mitochondria and plasma membrane (Figure 3). The alignment of the membranes may possibly be referred to the affinity between mucopolysaccharides and TP polymers, which is discussed in the following section. The formation of extensive membranes like the plasma and nuclear membranes probably may require the guiding influence of a phase boundary.

The generalizations advanced in the preceding section were mostly based on observations of linear polymers, but it is contended that they are valid also for membranes, with the modification that membranes are relatively stable. Their contribution to cellular dynamics therefore cannot depend upon a structural lability, as observed with the linear polymers. It rather seems that the feature distinguishing membranes is that they are able to conduct a tension–relaxation wave. Closer consideration will reveal that this phenomenon bears some relation to the tension–stabilization–relaxation–collapse cycle in linear polymers. Thus, if tension obtains at one point of a membrane, this will entail that surrounding monomers become 'suspended', acquiring a tension with which they can act on monomers situated further away. It is also envisaged that this tension must leave a relaxation in its wake. As will be shown in the following pages, this membrane property is borne out by many observations.

Mucopolysaccharides

One of the most typical properties of mucopolysaccharides is their ability to form hyaline matrices by imbibition of water. This phenomenon is of biological importance in different ways, some of which we shall have the occasion to discuss later. Here the attention will be focused on the interplay between mucopolysaccharides and tensible proteins.

Two kinds of interaction may be distinguished:

(1) The hyaline cytoplasmic matrix, the *hyaloplasm*, formed by mucopolysaccharides, seems necessary, or at least favourable, for the polymerization of tensible proteins. Several examples are known from the embryological literature where part of an egg or an embryo is more or less deprived of hyaloplasm, with the consequence that formation of microfilaments or microtubules is suppressed.

(2) A remarkable affinity obtains between mucopolysaccharides and tensible proteins, entailing that the former attach themselves tenaciously to the surfaces of TP structures. This phenomenon is borne out by many observations, for instance by the presence of a 10 nm hyaline layer around microtubules (Porter, 1966). The hyaline core of the latter also suggests the presence of mucopolysaccharides.

This relation between the two kinds of substances has some interesting and important consequences. First, since linear polymers, as previously discussed, can exist in a dynamic equilibrium between formation and collapse, which may involve extension and withdrawal of the structures, it follows that an accumulation of mucopolysaccharide occurs around the region where the polymers originate. We shall presently discuss some cases illustrating this effect. It may be presumed that mucopolysaccharide molecules may attach themselves between two different polymers. As a consequence alignment may result, especially of the rigid microtubules, but also, as intimated above, of membranous structures.

As regards the interaction with tensible proteins, two kinds of mucopolysaccharides may be distinguished. It is presumed that both types adhere to TP polymers, but one of them is characterized by acting as an adhesive, cementing the structures together. From now on the adjective *adhesive* will be used to designate this kind of substance, as contrasted with the *non-adhesive* mucopolysaccharides which attach to TP polymers without assuming the function of an adhesive. The chemical basis for this difference is not definitely established, but there is some reason to presume that, in the higher animal taxa at least, the adhesive substances are sulphated mucopolysaccharides. It may be mentioned that the latter probably reduce the lability of the linear TP polymers.

Morphological subunits

No attempt will be made to propose a definition of the cell-type of organization; it may suffice to mention that, apart from some rare exceptions, cells are characterized by possessing a nucleus, embedded in a cytoplasmic matrix, which in turn is bounded by a cortex. Each of the three subunits can be resolved into a number of elements, of more or less importance for the process of embryogenesis. We shall here outline the most influential elements, with particular reference to the distribution of mucopolysaccharides and TP polymers.

Nucleus

Most of the information required for the protein synthesis taking place during ontogenesis is stored in the nucleus in the form of DNA molecules, the consequence being that cell division associated with cytoplasmic growth, and cell differentiation, depend on the presence of the nucleus. Apart from the involvement in protein synthesis, the nucleus also participates in the process of cell division. This contribution does not depend on its structural integrity, since the nuclear membrane is dissolved prior to each cell division, liberating the chromosomes to the cytoplasm. As observed by Boveri (1896), the presence of the latter seems to be indispensable for normal cell division. This rule has been borne out, I believe, in spite of the fact that exceptions occasionally may be observed (cf. Wilson, 1928).

Cytoplasm

The cytoplasm contains a variety of structures and particulate bodies, as well as many high- and low-molecular substances. Some of these participate in those chemical processes, synthesis of proteins and other components, energy supply, etc., which form an essential part of the mechanism of embryogenesis. Among the formed bodies the centriole is of central importance for embryogenesis, being a prerequisite for the process of cell division, the mechanism of which will be discussed later.

Numerous observations have demonstrated the presence of a hyaline, relatively viscous ground substance in the cytoplasm. This very fact suggests that the cytoplasmic matrix contains, or consists of, mucopolysaccharides, but although this point has been borne out repeatedly by histochemical methods (Holmgren, 1940; cf. Monné and Slautterback, 1950), it seems that only in the very recent past have the implications of this circumstance been subjected to serious discussion. It has already been mentioned that the hyaline matrix, the hyaloplasm, constitutes the medium in which cytoplasmic TP polymers are formed. It appears that certain kinds of cells are distinguished by possessing a hyaloplasm containing the adhesive type of mucopolysaccharides. The consequence of this feature for the shape and behaviour of the cells will be discussed later.

It has been suggested above that the incessant formation and collapse distinguishing unstabilized linear TP polymers, in combination with the adhesion between mucopolysaccharides and TP structures reported above, may give rise to accumulation of hyaloplasm around the location from which the polymers originate. The mitotic apparatus, the dynamic nature of which has been demonstrated by Inoué (1964), affords an excellent example of this mechanism, since electron micrographs have shown that the hyaline inner regions of this structure are crowded with microtubules. A similar phenomenon, accomplished by microfilaments, will be discussed in the following section.

Hyaline formations of this kind are distinguished by their high viscosity, hence the epithet *plasmagel* usually applied to them. It may be inferred that this property is a consequence of the presence of TP polymers. The hyaline nature of plasmagels depends

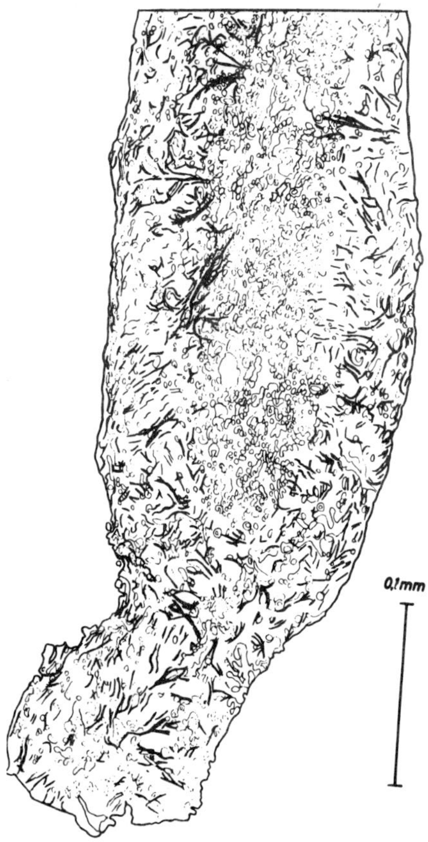

Figure 4. Median section through a protoplasmic strand of *Physarum* showing fibrils (microfilaments) in the ectoplasm. (Reproduced from K. E. Wohlfarth-Bottermann (1964) in *Primitive Motile Systems in Cell Biology*, (Eds R. D. Allen and N. Kamiya), by permission of Academic Press, New York)

on the fact that most of the particulate matter of the cytoplasm is excluded. This cannot be due only to a physical occlusion by the meshwork of the TP polymers of particles above a certain size, for some large bodies, for instance pigment granules, may be present in the plasmagels. It seems most likely that particles which adhere to the surface of the polymers gradually become accumulated in the plasmagel, whereas all others remain in the cytoplasm.

Besides the linear polymers, the cytoplasm also contains a number of membranous structures, the endoplasmic reticulum, the Golgi complex, etc. These formations presumably are involved in embryogenesis, but their contribution appears to be rather indirect, and we may therefore leave them aside in the present context.

Cortex

The last cellular subunit to be dealt with is the cortex, in which we may distinguish the *ectoplasm*, the cell or *plasma membrane* and the *surface coat*.

The ectoplasm is a hyaline layer which, in some cell-types, is present just beneath the cell membrane. The formation and the nature of the ectoplasm have been a riddle until it was shown recently to be massed with microfilaments (Figure 4, cf. Wohlfarth-Bottermann, 1964a; 1964b). The ectoplasm is a typical plasmagel, which obviously is present only in cells provided with microfilaments. Granting the adhesion between the mucopolysaccharides of the hyaloplasm and the TP polymers, then only one more assumption is required to account for the observed accumulation of hyaline material, viz., that the microfilaments may become attached to the cell membrane. Generalizing this presumption, it may be stated that linear TP polymers can become attached to the plasma membrane. It is a remarkable fact, which will be corroborated by many different observations discussed later, that this phenomenon usually occurs only at free surfaces, i.e. surfaces exposed to an external medium. The presence of linear TP polymers allows for the deployment of tensile forces by the ectoplasm.

The proposition that the plasma membrane is a two-dimensional TP polymer suggests that a tension may reside also in this cortical component. As a TP polymer, the plasma membrane may be expected to exhibit the dynamic properties characterizing this kind of structure. It is likely, particularly in view of the affinity between mucopolysaccharides and tensible proteins, that the phase boundary between the hyaloplasm and the external medium acts as a kind of template for membrane formation. This implies that, whenever a naked matrix of mucoprotein is exposed to a salt solution of proper composition, a membrane will be formed at the interphase through the surface precipitation reaction (cf. Heilbrunn, 1956). The inelasticity ascribed to TP polymers entails that the changes in surface area associated with various cellular activities, for instance formation of pseudopodia, must involve formation or resorption of plasma membrane. The chemical nature of the plasma membrane suggested here makes it easy to explain the availability of material for membrane formation or repair, for TP monomers are ubiquitous in the cytoplasm.

In passing, it may be mentioned that if the views advanced here are correct, it is possible to visualize the origin of the cell-type of organization. If, namely, once a 'nuclear' organism existed, bounded by a layer of mucopolysaccharide, which could also produce and secrete tensible proteins, then a membrane might be formed spontaneously at the surface of the hyaline matrix.

It remains to deal with the surface coat, the presence of which has been propounded by many authors, first by Rhumbler (1898). The initial failure to demonstrate it with the electron microscope raised doubts as to its existence, but of late it has become evident that most or all cells are covered by a thin layer of mucopolysaccharide. Owing to the postulated adhesion between TP polymers and mucopolysaccharide, the existence of this layer is readily accounted for, except as concerns its presence on the outside of the cell. In many eggs, a special kind of surface coat is formed when the contents of the cortical granules are emptied to the perivitelline space during fertilization. There are probably several other ways in which mucopolysaccharide may arrive at the outside of the cell membrane, but we shall leave this question open. Fibrous proteins (microfilaments) have occasionally been observed imbedded in the surface coat. It may reasonably be presumed that hyaloplasm and surface coat are made up of the same material; this implies that both will contain either the adhesive or the non-adhesive type of mucopolysaccharide.

We are going to deal with the importance of the surface coat later, but it may be mentioned here that it appears to be instrumental in preventing the fusion which may take place between 'naked' cells when their plasma membranes are brought in contact.

Cell orders

A considerable number of different cell-types prevail in the adult organism and since all of them derive ultimately from the egg it follows that at various stages of development certain cells must be subjected to changes which, it may be presumed, are passed on to their descendants. It was proposed in the introduction to adopt the name 'cell diversifications' for the processes responsible for such changes. It appears that they occur in a stepwise fashion, and the characters displayed by a given cell at a certain ontogenetic stage may thus be the outcome of a series of cell diversifications which have taken place in earlier cell generations. I have on several occasions (1965a; 1966a; 1967a; 1967b) claimed that it is possible to distinguish between the initial and the subsequent steps of cell diversification because the former are concerned with the formation of cells belonging to certain basic orders of cells, whereas the latter impose various kinds of synthetic activities on the cells without thereby changing their cell-order affiliation. For that reason it was suggested that the former should be called 'cell transformations' and the latter 'cell differentiations', without thereby implying that the two kinds of processes are fundamentally different.

The contention that cells may be grouped into a limited number of orders rests upon two facts, first that cells representing disparate differentiation patterns often exhibit striking similarities, suggesting some kind of kinship, and second that, on the basis of their appearance, cultured cells from many different sources have been found to belong to a few distinct groups. The traits distinguishing the cells are concerned with their shape and behaviour. In the following pages it will first be attempted to establish in a preliminary way the number and the nature of cell orders, and the cell transformations required to account for their genesis. This operation will show that the existing nomenclature is rather unsatisfactory, and a proposal for a new terminology will therefore be advanced. Subsequently, the most important properties of the cell orders will be outlined. This discussion will to some extent anticipate that of the following section, dealing with various important cell attributes, an expedient which seems inevitable, since familiarity with the notion of cell orders is needed in that context.

Cell orders and cell transformations

In his stimulating work on the evolutionary importance of cells, Willmer (1960) has adopted a cell classification founded primarily on studies of cultured cells. According to Willmer there are three basic cell-types, viz., amoebocytes, epitheliocytes and mechanocytes. The latter group comprises two separate kinds, the ordinary fibroblast and a spindle-shaped cell. These two cell-types are by Willmer called myxoblasts and myoblasts, respectively. In culture cells from almost any source assume the appearance corresponding to one of these cell-types. Cells isolated from the nervous system, neurons and glia cells, are exceptions to this generalization, and Willmer therefore considers them to represent special cell classes. One further cell-type is acknowledged by Willmer, viz., the flagellated cell, regarded as a special form of an epitheliocyte.

I have on previous occasions adopted Willmer's scheme with slight modifications (1965a; 1966a; 1967a; 1967b), but it seems necessary to introduce further changes in order to bring the classification into agreement with various observations, previously overlooked. For this reason it may be justified to reiterate the arguments which led to my cell classification. First of all, I do not believe that the neurons and glia cells represent special cell orders; rather, they are differentiated forms of cells belonging to some of the fundamental cell orders. The unique feature of these cells is that they cannot revert to the more primitive form. This point will be taken up later. Leaving aside flagellated cells for the moment, it will be seen that if we follow Willmer there are four different basic cell-types, viz., amoebocytes, epitheliocytes, myxoblasts and myoblasts. According to my view these represent four, and not three, cell orders. Granting this, as well as the possibility that a cell may pass through more than one step of cell transformation, it follows that two processes of this kind are necessary and sufficient to account for four different cell orders. These will then comprise one order of untransformed cells, two in which the cells have passed through one, and one in which they have passed through two, steps of transformation. Adopting Willmer's nomenclature for the time being, it is suggested that the untransformed cell-type is represented by the amoebocyte, the two one-step transformed cells are the epitheliocyte and the myxoblast, respectively, while the myoblast belongs to the order of cells that have undergone two stages of transformation.

In order to justify the classification without going into detail, it may be mentioned that one of the transformations, that leading from the amoebocyte to the epitheliocyte, and from the myxoblast to the myoblast, is concerned with the adhesivity of the cells. The other one, leading from the amoebocyte to the myxoblast and from the epitheliocyte to the myoblast, interferes markedly with the shape of the cells. This feature is not always a very satisfactory criterion, since the cell shape may be subjected to changes under the influence of environmental factors. However, as we shall discuss later, the shapes assumed by myxoblasts and myoblasts require a particular kind of cytoskeleton, the possession of which thus distinguishes these two cell-types.

Returning to the flagellated cells, it has been mentioned that Willmer (1960) considers these to be modified epitheliocytes. I have previously accepted this view, but maintained that two kinds of flagellated cells exist, one corresponding to the epitheliocytes, and one to the myoblasts. I no longer hold this opinion, having arrived at the view that it is not possible to treat the flagellum in isolation from the other type of vibratile organelle, the cilium. Thus I propose that there exists a third kind of cell transformation, through which members of each of the four cell orders presented above may acquire one of the two types of vibratile organelles. Without entering into a discussion of the distribution of flagella and cilia between the cell orders, it may be mentioned that, according to the ideas presented here, the existence of eight orders may be envisaged.

The reader has presumably realized by now that, to the extent they exist, the names used for the various cell orders are rather inadequate and inconvenient, so before we proceed further it may be expedient to introduce a new nomenclature.

Nomenclature

It is well known that epitheliocytes may form other things than epithelia (e.g. glands of various kinds) and that the differentiation products of myxoblasts may comprise other things than the 'jelly' or 'slime' suggested by their name (e.g. skeleton, melanin pigment, etc.). Likewise, the cell belonging to the order of myoblasts may, in my opinion, give rise to other cell-types than muscle cells, a point which will be borne out by examples in a later context. It is obvious that ambiguities arise if names are adopted for the cell orders which refer to particular differentiation patterns assumed only by certain members of the order. The only logical course is to introduce names which refer to order-specific properties, and preferably also indicate the generic relation between the cell-types. It is not possible to comply fully with these demands, but the proposal advanced here may be an approximation.

The first transformation confers upon solitary cells (amoebocytes and myxoblasts) the property to form aggregates (epitheliocytes and myoblasts); it thus converts *solocytes* into *colligocytes* (from Latin *colligere*, to collect or gather). In the solocytes the second transformation is concerned with the transition from loboform to filiform pseudopodia, and even if the corresponding colligocytes do not form pseudopodia, there is good reason to believe that the difference in shape between the two cell classes may be referred to the elements responsible for each of the two kinds of pseudopodia. It may therefore be justified to suggest that the second transformation is concerned with a conversion from *lobocytes* to *filocytes*. The third transformation leads to the formation of either *flagellocytes* or *ciliocytes*. Since the common feature in these cell-types is the possession of one or more vibratile organelles, they may be classed together as *vibriocytes*. I have been unable to find an expression which characterizes cells without vibratile organelles, and it is therefore necessary to adopt the name *non-vibriocytes* for this case.

Neglecting for the time being the vibriocytes, it is seen that amoebocytes are *solo-lobocytes*, myxoblasts *solo-filocytes*, epitheliocytes *colligo-lobocytes* and myoblasts *colligo-filocytes*. These names are not simpler than those used by Willmer, but they have several advantages: thus they indicate the generic relation between the cell classes, they refer to transformation and not to differentiation patterns and they may be used in a convenient abbreviated form, viz., *sl-cells*, *sf-cells*, *cl-cells* and *cf-cells*, respectively.

Actually, each of these cell-types can exist in two forms, so that for instance, the sl-cell must be either an amoebocyte, a *non-vibrio-solo-lobocyte*, or a flagellocyte, a *vibrio-solo-lobocyte*. These names may be abbreviated to *nsl-cells* and *vsl-cells*, respectively. For most purposes it is of little importance in the present context whether or not the cells are vibriocytes, and for the sake of convenience reference to the third transformation will therefore generally be omitted in the following, except when the discussion is directly concerned with vibriocytes.

The eight cell orders can be arranged in six classes, each containing four orders. The names of these classes are solocytes, colligocytes, lobocytes, filocytes, vibriocytes and non-vibriocytes, respectively. As will be seen, some of these expressions are of practical use. By adopting this nomenclature names for the transformations may also be coined. Thus we may call the process leading from lobocytes to filocytes the *lf-transformation*, the one giving rise to colligocytes from solocytes the *sc-transformation* and the one converting non-vibriocytes to vibriocytes the *nv-transformation*.

Cell orders and cell properties

Cells are classified on the basis of their shape and their behaviour. We shall briefly consider some of the factors influencing these properties, because this expedient will, among other things, contribute to justify the genetic relation between the cell orders suggested above.

Regarding cell shape it may be said that when the cortical tension is preponderant the cell will be spherical if the tension is uniform. For purely logical reasons it appears that deviations from the spherical form must involve the deployment of further forces, exerted either by a cytoskeleton or by environmental factors. As far as the first possibility is concerned, it has been shown by Byers and Porter (1964) that oriented microtubules are responsible for the elongation of the cells in the lens primordium of the developing chick embryo. As we shall presently see, elongation may occur also in cells devoid of microtubules, suggesting that even microfilaments may participate in the formation of a cytoskeleton. When deviation from spherical symmetry obtains, a polarization generally prevails in the cell. For reasons which will be discussed later, it appears that the polarization determined by a cytoskeleton of microfilaments is labile, and may be changed by external influences, whereas a cytoskeleton composed of microtubules confers a more stable polarization on a cell. It may therefore be proposed that cells containing microfilaments are intrinsically apolar, i.e. spherically symmetrical, whereas cells containing microtubules are intrinsically polar.

As far as cell behaviour is concerned, the distinguishing features in connection with cell classification are motility and aggregate formation. It cannot beforehand be excluded that these two traits are causally related, for aggregation will certainly contribute to immobilization of the cells, but as we shall see later, the properties responsible for aggregate formation may also influence the power of locomotion more directly. The motility exhibited by the solocytes depends mainly on the formation of pseudopodia, sl-cells moving by means of *lobopodia*, sf-cells by *filopodia*. Electron microscopical observations have established that the two cell orders are distinguished by the kind of linear TP polymers present in their cytoplasm, sl-cells containing only microfilaments, sf-cells both these and microtubules. It seems that microfilaments are actively engaged in the formation of lobopodia, and although filopodia contain both kinds of linear TP polymers, there can be no doubt that their peculiar form may be referred to the microtubular rigidity. Since each type of polymer is responsible for the characteristics of the pseudopodia distinguishing a cell order, and since the shape of the solocytes to a large extent depends upon their pseudopodia, it appears to follow that the differences between the cell orders as regards shape and motility ultimately depend upon the nature of the linear TP polymers present in the cytoplasm. It is proposed that the difference thus established between the two kinds of solocytes is a distinguishing feature between all lobocytes and filocytes, implying that the sl- and cl-cells contain only microfilaments, whereas sf- and cf-cells contain microtubules besides.

In the colligocytes the properties of the respective solocytes are combined with the presence of the adhesive type of mucopolysaccharide in the cytoplasm and in the surface coat. That residing in the latter location is responsible for the most characteristic trait of the colligocytes, their participation in the formation of cell aggregates.

When present in the cytoplasm, the adhesive mucopolysaccharide will act as a cement between TP polymers. It thus follows that the microtubules which, according to our generalization, are present in cf-cells must be aligned and kept firmly together. By this expedient the formation of filopodia is suppressed, and this means of locomotion is consequently lost. The rigidity of the microtubules ensures that a sturdy cytoskeleton is formed, which evidently must lead to an elongation of the cell. The spindle shape observed in single cells and in many aggregates depends in the first place upon the presence of the nucleus in the middle of the cell; some tapering may also be due to a gradual reduction of the number of microtubules. It is possible that the latter occasionally may continue into a flagellum-like projection at one or both ends of the cell. Presumably a structure of this kind will be a *stereoflagellum*, lacking the typical beat of normal flagella. In the presence of an adhesive surface, cf-cells may participate in the formation of two-dimensional cell aggregates. As a result the cell shape will be somewhat deformed, but the presence of the cytoskeleton is revealed by the fact that columnar epithelia are formed.

Microfilaments are not particularly rigid, so although they may become aligned under proper conditions, they cannot be expected to give rise to particularly efficient cytoskeletons, particularly since, as we have seen, to a major extent they are located in the ectoplasm. They may, however, in the presence of the adhesive mucopolysaccharide presumed to prevail in cl-cells, participate in the formation of the cytoskeletal core of the microvilli, which seem to be typical of this cell order. The microvillus appears to be the microfilament homologue to the stereoflagellum. When the intrinsically spherical cl-cell forms two-dimensional aggregates, a cubic epithelium will ensue. The question of aggregate formation will be dealt with in greater detail in the following section.

It was mentioned above that the adhesive type of mucopolysaccharides probably are sulphated, but this proposition cannot be of general validity, for in Actiniaria, known to possess well-developed epithelia, it is not possible to demonstrate any sulphated mucopolysaccharides (Katzman and Jeanloz, 1970). It would be tempting, from a phylogenetical point of view, to submit that one attempt to produce sulphated mucopolysaccharides would involve the sulphatation of hyaluronic acid, thus giving rise to mucoitin sulphate. I am well aware that the existence of this substance is disputable (cf. Brimacombe and Webber, 1964), although its presence is occasionally being reported, thus for instance in Holothuroidea (Maki, 1956; Maki and Hiyama, 1956; quoted by Hunt, 1970). At any rate, there is some reason to presume that the substance, or group of substances, the possession of which distinguishes the class of colligocytes, and which must thus be transformation products, in some way differs from the other sulphated mucopolysaccharides (chondroitin sulphates, etc.) which are differentiation products.

Vibratile organelles contain microtubules, a circumstance which suggests that they may be anticipated only in filocytes. This prediction is not borne out, for vibriocytes may be formed from all four orders of non-vibriocytes. It must be pointed out that this fact does not necessarily violate the generalization concerning cytoplasmic linear TP polymers, since vibratile organelles cannot truly be considered to be cytoplasmic structures. Hardly anything is known about the transformation leading to the formation of vibriocytes, but since the activity of the organelles appears to require the presence of acetylcholine or 5-hydroxytryptamine (cf. Willmer, 1960; Manukhin and Buznikov, 1965), it may be presumed that the enzymes required for the synthesis and breakdown of either of these substances form part of the transformation pattern. As pointed out, vibriocytes are distinguished by the possession of either cilia or flagella, flagella being characteristic of the vibrio-solocytes, cilia of the vibrio-colligocytes.

The flagellocytes are characterized by the possession of one, or in some cases a few, flagella. In contrast to the cilia, which originate in special kinetosomes, probably derived from centrioles, the basal body of a flagellum is presumed to be the centriole proper. Two kinds of flagellocytes were anticipated above, one corresponding to each of the non-vibrio-solocytes. From the arguments advanced above, it may be envisaged that one possesses an approximately spherical, the other an elongated, body. The two forms are easily recognized in various Protozoa, the vsl-cell thus in *Naegleria*, the vsf-cell in *Trypanosomas*, *Herpetomonas*, etc. (Figure 5). It is an interesting fact that the flagellum

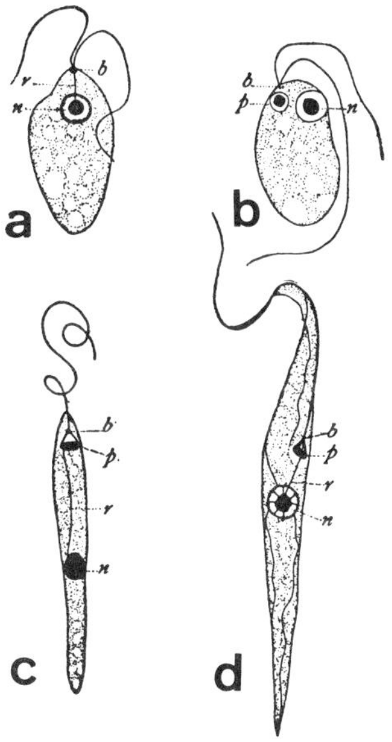

Figure 5. Flagellocytes. (a), *Naegleria*; (b), *Prowazekia*. These two Protozoa are thought to be lobocytes, belonging to the order of vsl-cells. (c), *Herpetomonas*; (d), *Crithidia*. These two Protozoa are thought to be filocytes, belonging to the order of vsf-cells. (Reproduced from E. B. Wilson (1928), *The Cell in Development and Heredity*, after C. W. Wilson, Swezy and McCulloch, The Macmillan Company, New York)

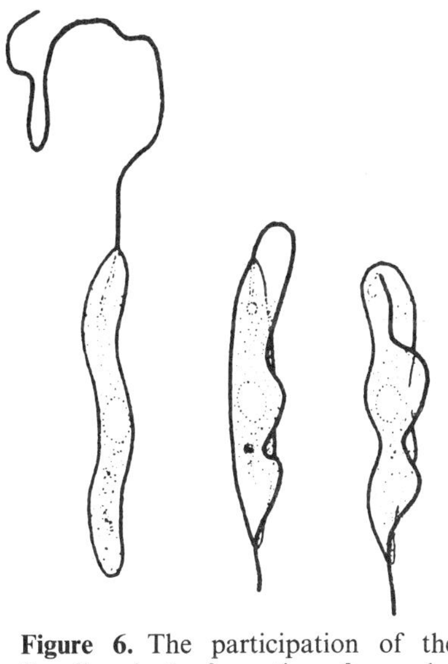

Figure 6. The participation of the flagellum in the formation of an undulating membrane in *Herpetomonas*. (From D'Arcy W. Thompson: *On Growth and Form*, 1942 after Mackinnon © Cambridge University Press)

in the latter case sometimes may participate in the formation of an undulating membrane (Figure 6). In Metazoa the spermatozoon very often is a vsl-cell, and this cell-type also seems to be present in the blastula of many animal species. In some embryos the apical plate is made up of elongated, flagellated cells, but as we shall discuss later, these cells exhibit features suggesting that they are colligocytes, and since the outgrowing process apparently is a stereoflagellum, it is most likely that they are ncf-cells. The nv-transformation generally occurs only in colligocytes participating in the formation of two-dimensional aggregates, in which the cilia will be present at the free surfaces. It should be noted that ciliated surfaces have lost their adhesivity towards the surfaces of other cells.

All cells in Metazoa are ultimately derived from an egg which, possessing the cell-type of organization, presumably must belong to a cell order. It might be envisaged that it represents an order of its own, and also that eggs from different animals belong to separate orders. As there is no evidence whatsoever supporting these possibilities, they may be rejected without more ado. This leaves us with the task of deciding to which of the eight cell orders the egg may be assigned.

Except when preparing for a cell division, the egg contains no microtubules, but rather microfilaments, usually located in the ectoplasm. The egg is thus no filocyte, and since it is not flagellated either, only two cell orders remain, the nsl-cell (amoebocyte) and the ncl-cell ('epitheliocyte'). Willmer (1960), who has taken an interest in this problem, is of the opinion that the egg is an 'epitheliocyte'. This notion is based on various kinds of evidence, among others the fact that flagellated cells are present in many early embryos, and according to Willmer's classification these are 'epitheliocytes'. As

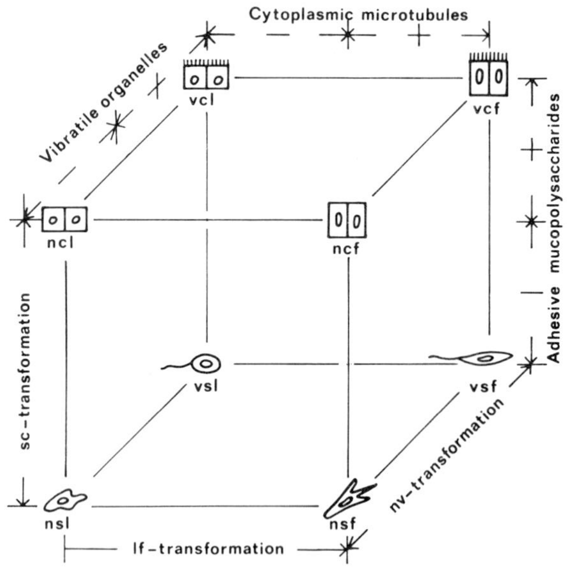

Figure 7. The cell cube. Typical representatives of the eight cell orders are located in the corners of the cube. The colligocytes are shown as forming two-dimensional aggregates. As indicated, each median plane divides the cell orders into two cell classes, distinguished by the presence or the absence of a particular feature representing one of the three cell transformations

appears, I consider flagellated cells to represent special cell orders which already have been excluded. On this basis, Willmer's classification thus cannot be accepted. It is true that oocytes very often possess microvilli, a feature otherwise typical of cl-cells. They disappear at fertilization, and subsequently most eggs and blastomeres do not exhibit any features which are characteristic of 'epitheliocytes'. The only remaining possibility is that the egg is an amoebocyte, an nsl-cell. This contention is, as a matter of fact, borne out by numerous observations of 'amoeboid' behaviour of eggs recorded in the literature, some examples of which will be discussed later. As it turns out, this classification is the only one acceptable, if the nsl-cell, as suggested above, is considered to be the untransformed cell-type.

The visualization of the genetic relations between the cell classes is facilitated by means of an illustration in which each cell order is placed in one corner of a cube. It is seen that this *cell cube* has the property that each of the three median planes will divide the cells in two classes, in which the members of one have, and those of the other have not, been subjected to a certain cell transformation (Figure 7).

As far as the reversibility of the transformations is concerned, it suffices to mention in the present context that various kinds of evidence suggest that the sc- and the nv-transformations are reversible, in contrast to the lf-transformation which appears to be irreversible.

Cell properties involved in morphogenesis

Cells participate in morphogenesis as substrate elements, building up the embryonic body, which implies that their contribution depends in part upon their bulk or shape. This property may be subjected to changes as a consequence of cell aggregate formation, but the ensuing shape is generally quite stable. Cells may also contribute to the dynamic events involved in morphogenesis, owing to their motility. Associated with cell movements changes occur in the shape of the cells, but these are transitory and do not bear upon the basic shape of the cell. It appears that the cell properties most important for morphogenesis are those mentioned here, cell shape and cell movements. Before we embark upon a discussion of their contributions, it may be pointed out that the latter obviously to a large degree depend upon the cell-order affiliation, the aggregate-forming colligocytes being responsible for the more static, the motile solocytes for the more dynamic, aspects of the various morphogenetic activities.

Cell shape

The shape of a cell depends ultimately upon its own intrinsic properties, but among these some may be distinguished which are more or less autonomous, and others which can exert their effect only through interaction with the environment. In the former group may be reckoned the cortical tension and cytoskeletons. Ignoring temporary changes associated with cell movements, environmental interference with the cell shape may be referred to the nature of the mucopolysaccharide present in the surface coat; if it is the adhesive type, strong interactions may be anticipated, whereas the force developed by the less adhesive type in general is too weak to make any significant impression upon the cell shape. We shall here discuss in turn the effects of these three factors, cortical tension, cytoskeletons and surface coat, on the shape of cells.

Cortical tension

The notion that a tension obtains in the cell surface dates far back. Until rather recently, this 'surface tension' was considered to be a true surface tension, similar to that responsible for soap bubble formation. Support for this idea was found in the fact that soap bubbles form aggregates which exhibit certain similarities with some types of cell aggregate. The circumstance that many observations could be accounted for only through the deployment of considerable ingenuity did not cause any discouragement (cf. Thompson, 1942). We shall in a following section discuss the differences between aggregation of soap bubbles and cells; suffice it to mention here that the main flaws of the analogy lie in the fact that in the latter case the tension involved is no surface tension, and that two separate forces, tensile and adhesive, respectively, are involved. In order to avoid confusion, it is now customary to talk about *cortical tension* in the case of cells. It has often been assumed that it is an elastic tension (cf. Mitchison and Swann, 1954; Selman and Waddington, 1955), but as related above, the observations of Yoneda (1964) have cast doubt on this contention.

The cortical tension must reside in the ectoplasm or in the plasma membrane, or in both. As an envelope enclosing the contents of the cell, the plasma membrane of necessity outlines the shape of the latter. In this process the surface may passively trace the influence of other factors, or it may be actively involved. The latter case requires that a tension prevails, and that it is strong enough to hold its own against other possible forces. If this occurs and the tension is the same in every part of the membrane, the law of minimum surface area will be obeyed and the cell become spherical. If the tension is different in one region from that obtaining elsewhere, or if other forces, for instance gravity, are able to exert their influence, the law of minimum free energy imposes upon the cell a shape deviating from that of a sphere, with a concomitant increase in the surface area. The possibility of non-uniformity distinguishes the cortical tension from the surface tension. Cell shapes more complex than those mentioned here cannot be accomplished by the cortex alone, but require the cooperation of an internal cytoskeleton.

Cytoskeletons

As mentioned above, both kinds of linear TP polymers may participate in the formation of cytoskeletons. The influence depends upon the mechanical properties of the polymers, and as these are quite disparate, the resulting shapes are very dissimilar.

Four different cases may be distinguished where microtubules influence the shape of cells. The first one is represented by the transitory formation, the mitotic apparatus, present in all dividing cells. Here the microtubules, originating in the two centres and anchored at the cortex, may temporarily cause the cells to round up, irrespective of their initial form. Since this change of shape is dictated by the tensile force of the microtubules rather than by the cortical tension, it does not represent a true case of the law of minimum surface, as is shown also by the fact that an elongation occurs at a later stage. This mechanism will be treated in detail on a later occasion. The second case is represented by the vibratile organelles, cilia and flagella. These originate in basal bodies situated in the cortex and, being covered by plasma membrane, they contribute to an extensive enlargement of the surface area and, of course, to a modification of the shape which, however, is of little influence on the main body of the cell. These structures are of importance for locomotion, and to some extent also for morphogenesis. The third case is represented by the filopodia found in the sf-cell. These labile formations, being continuously extended and withdrawn, are of great importance for various dynamic events, as will be discussed presently, and they profoundly affect the cell shape. In the fourth case, represented by the cf-cell, a great number of cytoplasmic microtubules are present which, owing to the presence of the adhesive type of adhesive mucopolysaccharide, are aligned parallel to each other, thus giving rise to an elongation of the cell body.

It is doubtful whether an ectoplasmic plasmagel alone may cause a deviation from the spherical shape, but when microfilaments are present in the cytoplasm they can form a plasmagel which, in cooperation with the ectoplasm, may influence the cell shape. As we shall discuss later, it seems that this phenomenon is responsible for the sausage shape assumed by migrating amoebae. Microfilaments may also participate in the formation of undulating membranes and microvilli, thereby substantially increasing the surface area, even though the cell shape is little affected.

Surface coat

Interaction with the environment resulting in change of the cell shape presumes the exertion of force. It is evident that in order to be effective, the force responsible for effects of this kind must reside in the outermost layer of the cortex, the surface coat. It has already been suggested that the latter consists of mucopolysaccharide, and that it may be either of the strongly adhesive sulphated or of the weakly adhesive non-sulphated type. The cells covered with the first kind, the colligocytes, are characterized by adhering to various substrates and to other cells. Those covered with the other type, the solocytes, do adhere to surrounding surfaces, but the forces are not strong enough to interfere significantly with the shape and the movements of the cells. In the present section we shall deal only with the strongly adhesive forces, concentrating particularly on certain physico-chemical aspects of the phenomenon of adhesion, and the influence it may have on the aggregation of cells.

Biophysical aspects of adhesion. I have stated unreservedly that cellular adhesion may be referred to properties of the surface coat. Much attention has been devoted to this question in recent years, but no unanimous agreement obtains as to the mechanism involved. Pethica (1961) and L. Weiss (1967) have enumerated a number of physico-chemical forces that perceivably may be responsible for adhesion, without arriving at a clear-cut decision between the various possibilities. Two groups of authors have advanced more specific opinions concerning this question. The first of these, to which belong Steinberg (1963b) and Curtis (1966), considers that an intercellular cement is not required to explain adhesion between cells, interaction of physical forces sufficing to keep two apposed plasma membranes together. The other group claims that cells are kept together by a cement in which the active component is a sulphated mucopolysaccharide (cf. Moscona, 1962; Mercer, 1964). To this group also belongs K. Meyer (1957), who expressed the view that sulphated mucopolysaccharides like chondroitin sulphate may function as an intercellular cement. As far as I can see, there are many observations which strongly support the contention that a specific substance confers adhesivity on the cell surface. In the present context it may suffice to mention the fact that only certain kinds of cells participate in aggregate formation. It seems very difficult to account for this difference in cell behaviour without assuming that the cell surfaces vary with respect to certain chemical properties.

Adhesion phenomena obey the general energetic law which states that equilibrium obtains when the free energy is minimum. This condition is reached when the total work done by adhesion is maximum, i.e. when the cells are arranged so that the mean force with which each cell adheres to the surrounding surfaces is the largest possible (cf. Steinberg, 1963a). This implies that the adhesive forces will strive to enlarge the areas of contact between the adhesive surfaces. It will be realized that this endeavour may result in an enlargement of the total surface area, and it is thus possible that the adhesive forces in cell aggregates counteract the tensile forces, which tend to ensure minimum surface area. The resulting equilibrium may under certain conditions bear some similarity to that accomplished by surface tension in aggregates of soap bubbles. However, the possible configurations are in the latter case limited by the fact that the tension is the same in each of the three surfaces meeting in an edge in a soap bubble complex; the angles must therefore be 120°. In cell aggregates much more variation is possible, because the two responsible forces may vary independently.

An adhesive cell may be subjected to three kinds of forces which all must influence its shape: firstly the cortical tension which tends to reduce the surface area, secondly a cytoskeleton which, when present, to varying degrees will counteract the effect of the cortical tension, and thirdly the adhesive forces striving to establish maximum surface contact with surrounding surfaces. It seems impossible to account quantitatively for the influence of cytoskeleton on the cell shape, but even the effects of the other two factors have largely been ignored so far. The situation is, of course, not facilitated by the fact that the cortical tension and the adhesive forces oppose each other.

We shall try here to deal in a rather qualitative way with the simple example represented by a spherical cell, i.e. a cell in which the cortical tension surpasses gravity. In this case a flattening of the lower surface will occur if adhesive forces obtain between the cell and the substrate. The area of contact, and thus the shape of the cell, will be determined by the relative strength of the tensile and the adhesive forces. This example has not yet been worked out theoretically, but there is some reason to believe that the free surface will be part of a sphere. When the tension exceeds that of adhesion, the shape of the cell will be in the range between a sphere and a semisphere; when they are equal, the latter form will be assumed (Figure 8). When the adhesion is stronger than the

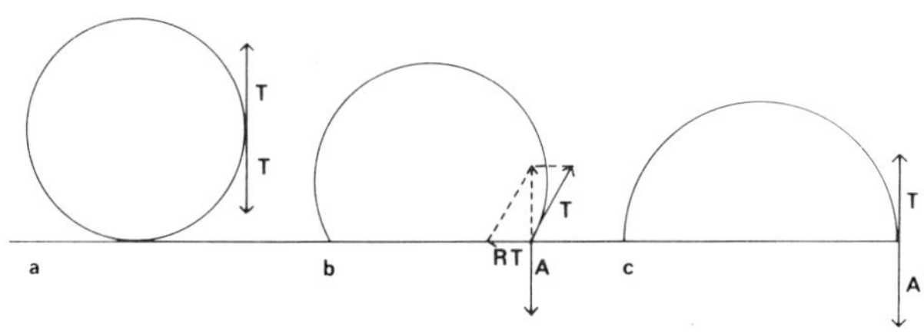

Figure 8. Interaction between tension T and adhesion A in a spherical cell. (a), $A=0$; (b), $0<A<T$. RT is the reduced tension in the cell surface in the plane of the substrate required to balance the tension and the adhesion in the circumference of the attachment area. (c), $A=T$

tension no equilibrium can be established, and it may therefore be presumed that the cell will flatten completely on the substrate. It is realized that if, as proposed, the tension is inelastic, it will not be changed as a consequence of the increase in surface area associated with the form change.

From the equation $P=2T/R$, where P is the internal pressure, T the cortical tension and R the radius, it follows that for unchanged tension the internal pressure increases as the radius decreases. Since this pressure opposes deformation, large cells should be more pliable than small ones. This fact may have some importance for the shape of cleaving cells, thus it is possible that the morula shape assumed by many embryos illustrate this case. When further subdivision occurs, the internal pressure may rise further, but the cells are then too small to influence the shape of the embryo.

Cell aggregation. The most important outcome of the adhesivity residing in the cell surface is the formation of cell aggregates. Of these two kinds may be distinguished, two-dimensional ones, in which the cells adhere partly to each other and partly to a common substrate, and three-dimensional ones, in which only intercellular adhesion obtains. It can be presumed that cells participating in aggregate formation must assume shapes conforming to an equilibrium between the tensile and the adhesive forces. Very little is known about this question, but we shall here try to outline some general rules which can be derived on this presumption.

As far as two-dimensional aggregates are concerned, we may infer that, according to the law of maximum surface contact, a single cell placed on the surface of such an aggregate must penetrate into it, if possible, and establish contact with other cells and with the substrate. The relative strength of the adhesive forces between the cells and between the latter and the substrate may be expected to influence the shape of the aggregate. Thus, it may be anticipated that if the former is stronger than the latter, the equilibrium distinguished by maximum work of adhesion is reached only when all the cells have parted their association with the substrate and formed a three-dimensional aggregate. Contrariwise, if the intercellular adhesion is the weakest, the cells must flatten out on the substrate. If the available space does not permit that all the cells establish contact with the substrate, a multilayered aggregate may possibly be formed. These considerations lead to the conclusion that cells with low intercellular adhesion may form flattened aggregates (meso- and endothelia) on an adhesive substrate, provided their cortical tension does not prevent flattening from occurring. Conversely, two-

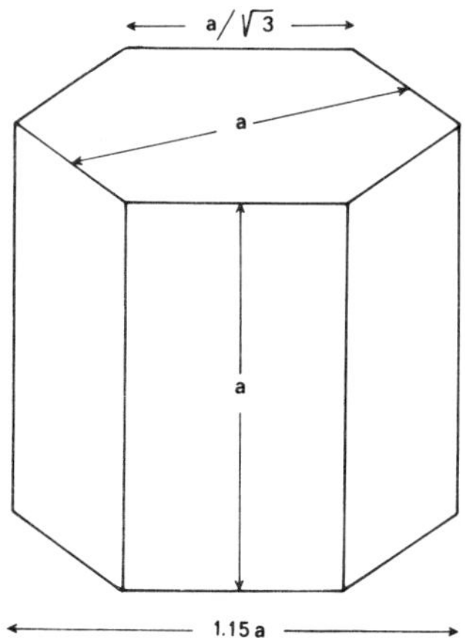

Figure 9. The equilateral hexagonal prism with minimum surface:volume ratio

dimensional aggregates, in which adhesion obtains between the cells (epithelia), are stable only if the strengths of the adhesive forces between the cells and between the latter and the substrate are identical. This notion agrees well with the fact that the substrate *in vivo* most often is a basement membrane, the matrix of which is generally considered to be mucopolysaccharide secreted by the epithelial cells.

It is possible to estimate the shape which must be assumed by an intrinsically spherical cell in a two-dimensional aggregate in order to establish balance between the forces exerted by the cortical tension and by the surface adhesion. To achieve the maximum surface contact required by the adhesive forces, the cells must assume a form allowing for packing. Among the triangular, rectangular and hexagonal prisms which meet this demand, the last gives the smallest surface:volume ratio under otherwise equal conditions. The fact that this form is found in epithelia may indicate that the tensile forces are decisive on this point. However, equilateral hexagonal prisms may be tall and slender or flat and large, and in both cases the contact with neighbouring adhesive surfaces is large. It may be shown that between these extremes a prism is found in which the total surface is minimum. If we put the height of this prism to a, the sides of the hexagon are $a/\sqrt{3}$, and as seen from Figure 9, the width in cross section lies between a and $1 \cdot 15\,a$. This result fits so well with common observations on 'cubic' epithelia that it may be presumed that adhesive and tensile forces are involved in the way suggested here. It is interesting to note that the law of minimum surface area definitely has the upper hand in deciding the shape of the cells.

It follows from the ideas outlined above that, on the assumptions made so far, only one kind of two-dimensional aggregate is possible, as exemplified by the usual cubic epithelium. The existence of columnar epithelia cannot be accounted for by variations

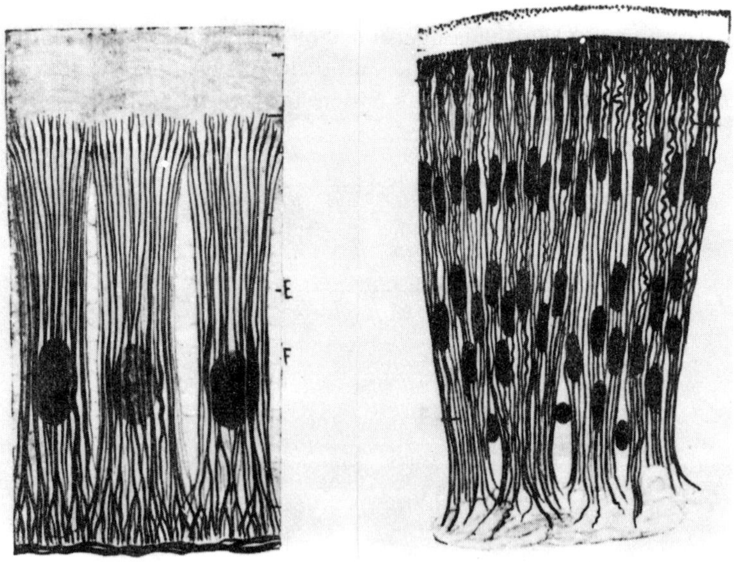

Figure 10. Columnar cells in epithelia from the digestive tract of various invertebrates, showing protoplasmic fibrils aligned parallel to the longest axis of the cell. (From P. del Rio-Hortega, *Trab. Lab. Invest. Biol. Madrid*, **15** (1917))

in the tensile or the adhesive forces, and they therefore must be formed by cells which are inherently elongated, owing to the presence of a cytoplasmic cytoskeleton. This contention is actually borne out by many classical observations (cf. Figure 10), and electron microscopical studies suggest that cells in columnar epithelia typically contain microtubules (Byers and Porter, 1964).

We may now turn to a brief consideration of three-dimensional aggregates. It follows from the rules that spherical cells, in order to unite maximum surface contact with minimum surface area, must assume the shape of a rhombic dodecahedron (Figure 11),

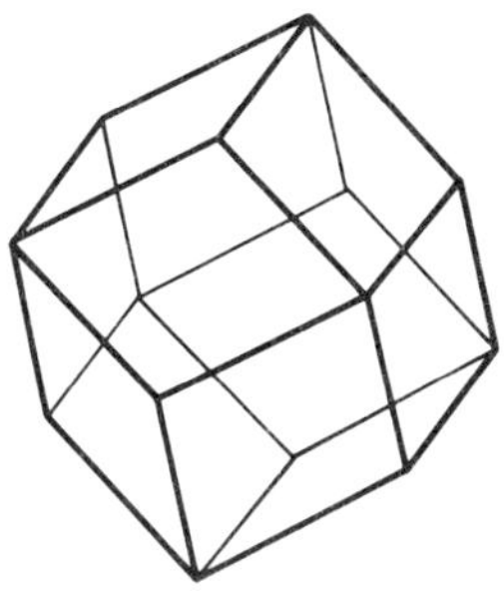

Figure 11. A rhombic dodecahedron. (From D'Arcy W. Thompson: *On Growth and Form*, 1942 © Cambridge University Press)

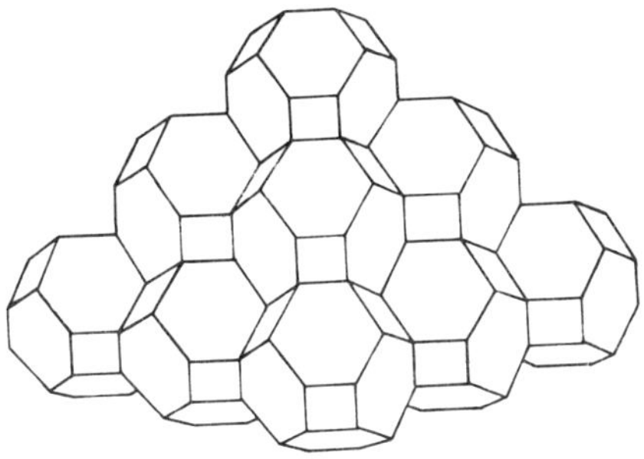

Figure 12. Packed orthotetrakaidekahedra. (From D'Arcy W. Thompson: *On Growth and Form*, 1942 © Cambridge University Press)

or better still the orthotetrakaidekahedron of Thompson (1887), shown in Figure 12. It is possible that this arrangement obtains in small regions of various parenchyma, but generally it is upset by the presence of non-adhesive surfaces (cf. Holtfreter's atypical epithelium, Figure 13), other kinds of cells or extracellular substances. Elongated, spindle-shaped cells may theoretically be packed in a three-dimensional aggregate

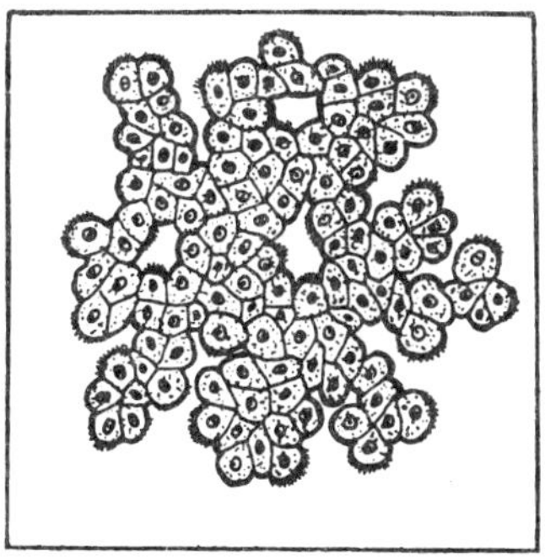

Figure 13. Three-dimensional aggregate formed by isolated amphibian epidermis cells. (Reproduced from J. Holtfreter, *Arch. Exptl. Zellforsch,* **15** (1934), by permission of Gustav Fischer, Jena)

without any extensive change of shape, as shown in Figure 14. An approximation to this type of aggregate may be found in some smooth muscles.

Squamous epithelia containing several layers of cells represent a transition between two- and three-dimensional aggregates. It may be mentioned that the flattening of the

Figure 14. Theoretical packing of spindle-formed cells into a three-dimensional aggregate

external cell layers commonly observed represents a deviation from the law of minimum surface area. It appears that a cytoskeleton, in the form of the so-called 'tonofibrils', is responsible for the cell shape in this case.

Cells with high cortical tension are often found in embryos. As an example may be mentioned the blastomeres in certain echinoderm embryos, which are known to be almost spherical and with a limited contact surface with the other blastomeres and the hyaline membrane. During the preparation for division, when the cortical tension increases, the area of contact is further decreased (Figure 15). At the other extreme may

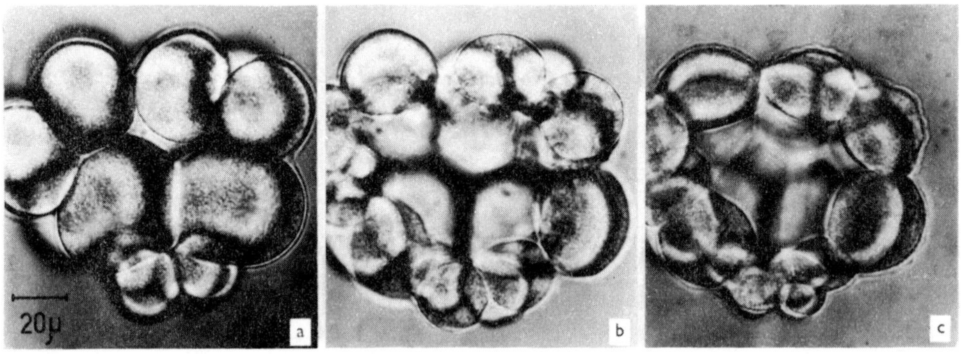

Figure 15. Changes in cell shape during the fifth cleavage division in the sea urchin embryo. (a), 16 cell stage; (b), the blastula near the end of the division, micromeres not yet divided. Note the rounding up of the cells with the resulting small contact area between them; (c), interphase. (Reproduced from L. Wolpert and T. Gustafson, *Exptl. Cell. Res.*, **25** (1961), by permission of Academic Press, New York, by courtesy of Professor Tryggve Gustafson)

be mentioned the pyramidal blastomeres in certain crustacean embryos (Figure 16), which hardly comply with the law of minimum surface area, suggesting that the cortical tension is extremely low. The blastulae of many spirally cleaving embryos represent

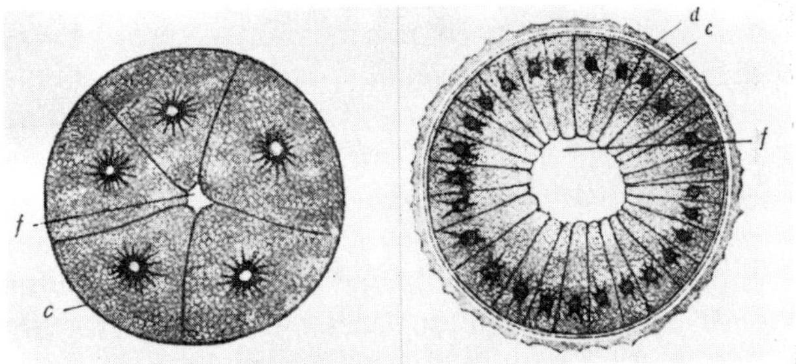

Figure 16. Pyramidal cells in an egg of *Branchipus* during cleavage. c and d, egg envelopes; f, blastocoele. (Reproduced from E. Korschelt and K. Heider (1936), *Vergleichende Entwicklungsgeschichte der Tiere*, after Brauer, by permission of Gustav Fischer, Jena)

intermediate cases. In permanent aggregates various structures like desmosomes, terminal bars, etc., are developed, which further the coherence between the cells. Formations of this kind cannot be made responsible for the primary aggregation of the cells, since they are formed only after the cells have established intimate contact. The circumstance that treatment of columnar epithelia with colchicine does not entail drastic morphological changes has led Tilney and Gibbins (1969) and Pearce and Zwaan (1970) to the conclusion that microtubules are not responsible for the maintenance, but rather for the development, of the elongated cell shape. I interpret this result to demonstrate that factors like basement membrane and desmosomes prevent the adjustment of the cell shape, a contention which is borne out by the observation by Tilney and Gibbins (1969) that when these stabilizing agents are suppressed, the individual cells assume a spherical shape.

Cell movements

Biological literature abounds with expressions like 'migration of nucleus', 'streaming of cytoplasm', etc. Phrases of this kind are unfortunate, for either they are meaningless, or else they imply the existence of self-propelling nuclei, pressure heads prevailing inside the cell and similar improbable events. The advantage of the generalization introduced above, provided of course that it is valid, is that whenever dynamic events occur, on whatever level of organization, the primary *agens movens* must be sought in some TP polymer.

We have previously outlined the most important properties of the TP polymers, but we shall here discuss in more detail the biophysical aspects of their participation in dynamic activities. Just as cell shape may be influenced by intrinsic factors and by interaction with the environment, two kinds of dynamic events may be distinguished, intrinsic or intracellular movements, and processes in which interaction with the environment takes place. The latter instance entails that the cells may change their location or, expressed in another way, that they possess motility. As will appear from what follows, both kinds of movements are of importance for embryogenesis.

Biophysical aspects of dynamic activities

Movements and other dynamic activities correspond to the performance of mechanical work, involving the deployment of a force for a certain distance. Through the tension developed in TP polymers, the cell may exert a force, but this is in itself a static phenomenon, through which no work is performed. It is therefore necessary to link the generation of tension with other processes; traditionally, it is presumed that a shortening, a contraction, of the tensible structure occurs, as a consequence of which work may be performed. We have above rejected contraction as a likely event, and it is therefore necessary to look for alternative explanations.

Apart from the development of tension, the properties which account for the dynamic activities of linear polymers are their lability and the fact that the ends are adhesive and may become attached to appropriate surfaces. As already discussed, tension will prevail when a linear polymer becomes suspended between two frames. If the latter are able to resist the tensile force, a stable situation may arise, in which no work is carried out. For this to occur, it is an indispensable condition that the frames yield to the tension, by which expedient an infinitesimal amount of work is done. As postulated above, the active polymer will collapse as a consequence, but may be replaced by others which are shorter and therefore able to develop tension under the new conditions.

It has been proposed that the presence of the adhesive type of mucopolysaccharide possibly may stabilize linear TP polymers, and if this suggestion is correct it follows that the polymers cannot carry out mechanical work in colligocytes without some kind of labilization.

Membranes are generally much more stable than the linear polymers, even if it has been necessary to presume that new plasma membrane is formed or resorbed when changes in surface area occur. However, certain observations suggest that under special conditions the plasma membrane may collapse in a more or less extended region of the surface of a cell. Normally, membranes appear to carry out mechanical work by the spreading of a tension–relaxation wave as discussed above. The conduction of such a wave involves the performance of mechanical work; if allowed to interact with the environment it may give rise to locomotion.

Intracellular movements

Extensive translocation of cellular material, cytoplasm as well as ectoplasm, may occur without being accompanied by any locomotion. The most familiar example of this kind is shown by the changes taking place during cell division, but many similar processes are known. No doubt they occur in all kinds of cells, but those observable in animal eggs are particularly conspicuous. For this reason, and because at least some of them are of great importance for embryogenesis, the discussion will be confined to intracellular movements in eggs. In the cytoplasm microfilaments, microtubules and membranes may be present; all these elements as well as the plasma membrane may be expected to participate in these activities. We shall discuss the subject under two headings, cytoplasmic translocations and cortical movements.

Cytoplasmic translocations. Cytoplasmic membranes may presumably be responsible for translocations of various kinds, but little is known about this. We shall therefore restrict the discussion to effects from the activities of microfilaments and microtubules.

In a general way it may be stated that translocation of some element requires that it adheres to the surface of tensible structures. We have already had the opportunity to discuss the fact that mucopolysaccharides adhere to microfilaments and to microtubules. The ectoplasmic layer was shown to owe its existence to this mutual relation, together with the fact that the microfilaments may fasten to the inside of the plasma membrane. It is possible to explain the accumulation of hyaloplasm in the central region of the mitotic apparatus by the same mechanism, with the difference that microtubules have taken the place of the microfilaments.

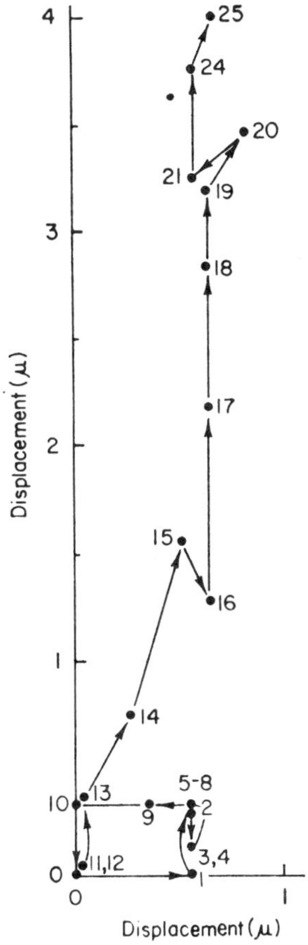

Figure 17. Motion of single echinochrome granule in an unfertilized egg of *Arbacia*. The numbers at each point represent successive times in seconds. (Reproduced from A. K. Parpart (1964) in *Primitive Motile Systems in Cell Biology* (Eds R. D. Allen and N. Kamiya), by permission of Academic Press, New York)

Observations of the movements displayed by various cytoplasmic inclusions have shown that they do not follow the statistics predicted for Brownian motion. Since the typical feature of these non-Brownian movements is that the particles in question after periods of rest may suddenly be subjected to extensive translocations, the phenomenon has been designated *saltatory movements*. Experimental evidence shows that cytoplasmic microfilaments and microtubules are responsible for these translocations (cf. Rebhun, 1964).

Of particular interest in the present context are studies by Parpart (1964) on the saltatory movements of echinochrome granules in *Arbacia*. In the unfertilized egg pigment granules are seen to move about in a 5–10 μm ectoplasmic layer near the surface (Figure 17). One to two minutes after fertilization the majority of the granules rapidly accumulate in the outermost region of the cytoplasm, and the movements cease. If it is granted that the particles are moved by cortical microfilaments, these observations lend themselves to an interpretation which comprises some interesting features. First of all, it follows that the echinochrome granules must adhere to the filaments. The accumulation of pigment granules in the cortex of the eggs of many different species shows that this is no unique feature. Secondly, it is apparent that cortical microfilaments are already present before fertilization, but clearly some radical change occurs in association with this event which profoundly affects their activity. Two possibilities present themselves: either the filaments are changed with respect to number or to physical properties, or else the surface available for attachment of filaments is increased.

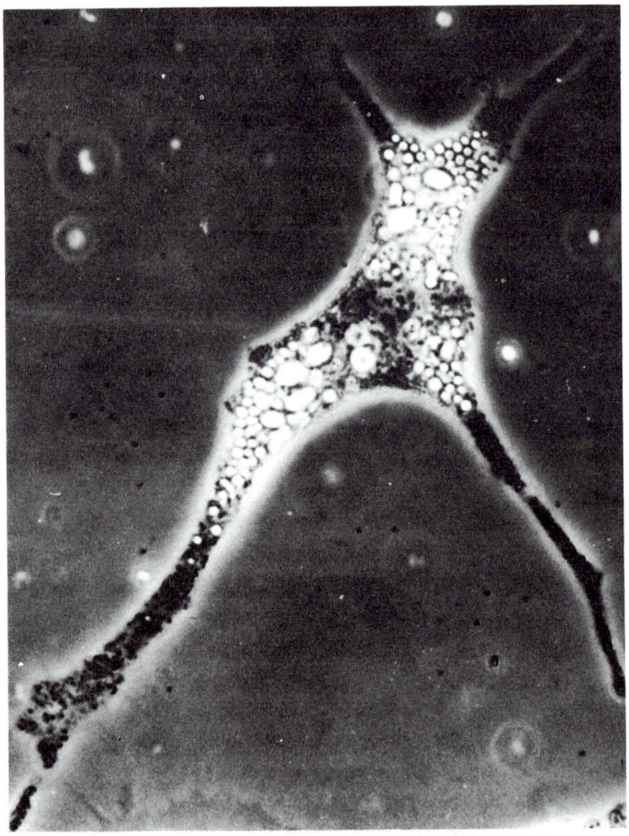

Figure 18. Pigmented sf-cell, isolated from an amphibian gastrula. Most of the pigment granules are located in the filopodia

Considering the extensive modifications to which the egg membrane is subjected during fertilization, the latter suggestion is highly probable, but the former cannot be excluded. The immobilization of the granules suggests that an increased density of microfilaments in the outermost layer leads to gelation of the ectoplasm.

It may be mentioned that even before fertilization, the association between microfilaments and pigment granules prevents the dislocation of the latter by centrifugal forces (Marsland and Landau, 1954). In contrast, cytoplasmic particles like yolk granules and mitochondria which are excluded from the ectoplasmic layer in *Strongylocentrotus* may be spun down by centrifugation (Mazia, quoted by Mercer and Wolpert, 1962).

A phenomenon similar to that described here, namely cytoplasmic streaming in *Nitella* and *Avena*, has been observed to be repressed by cytochalasin (Wessells *et al.*, 1971). Since presumably in either case an interaction between plasma membrane and microfilaments is required, it may be envisaged that the chemical acts by interference with this cooperation.

Cellular inclusions may also be moved around by microtubules, as shown in Figure 18 which depicts an sf-cell isolated from an amphibian gastrula. It is seen that the pigment granules to a large extent are accumulated in the filopodia. From the pigment distribution observed in the figure it may be inferred that little ectoplasm is present in sf-cells and that most of the cytoplasmic linear polymers are microtubules, to which the pigment granules adhere. In the egg the pigment is located in the ectoplasm in the animal hemisphere. The fact that it does not move to the newly formed cell walls in the blastula follows from the generalization that linear TP polymers, *in casu* microfilaments, become attached only to free cell surfaces. When the blastomeres are isolated, the pigment becomes evenly distributed in the cortex, as might be anticipated. The deviations from this rule observed in migrating cells will be discussed later.

Accumulation of particles around the mitotic apparatus has often been observed, but the central parts always remain hyaline. It is conceivable that the density of microtubules here is too great to allow for any inclusions. Those particles located near the nucleus when cell division begins will sooner or later be pushed out by the microtubules, whether or not they adhere to the latter. In the former case they may be expected to be accumulated outside the hyaline region, together with particles originally located more peripherally. Among the particles which are known to behave in this way are mitochondria (chondrokinesis, cf. Wilson, 1928) and certain granules which stain metachromatically with toluidine blue (cf. Pasteels, 1958; Rebhun, 1964).

It must be emphasized that adhesion between microfilaments and mitochondria is no general phenomenon, although it has been reported in many cases. An example of special embryological interest is encountered in Ascidiacea. In certain cases, e.g. *Styela*, the ectoplasm of the unfertilized eggs is tinted yellow, and it has been shown that this colour is due to mitochondria, which must owe their ectoplasmic localization to an affinity with the microfilaments (cf. Reverberi, 1961). At fertilization, the ectoplasm is accumulated at the vegetal pole, but as the spermaster begins to migrate along the inner surface towards the animal pole, it is dragged along, subsequently giving rise to the yellow crescent (Conklin, 1905). This phenomenon is predictable if the mucopolysaccharide of the ectoplasm and the mitochondria adhere to the astral mictrotubules (cf. Figure 46).

These results suggest that the adhering particles show no preference with regard to the two kinds of linear polymers. This is also borne out by the observation of Lillie (1901) on the eggs of *Unio*, in which yolk granules are found in the ectoplasm and around the mitotic apparatus. Echinochrome granules in *Arbacia* eggs may also be found in the latter location. The presence of pigment granules in the filopodia mentioned above is another illustration of this case; incidentally, in the latter instance, we are evidently witnessing a demonstration of the basic mechanism of melanophore action.

These few examples suffice to show the importance of microfilaments and microtubules for the organization of the egg, and for other cells as well, and also for the redistribution of cytoplasm which in some cases forms part of the mechanism of embryogenesis. In either case the dynamic activity must be referred to a continuous formation and collapse of the tensible structures.

Cortical movements. As discussed above, TP polymers are present in two cortical elements, namely, the ectoplasm and the plasm membrane, whereas evidence is wanting that this kind of material is present in the surface coat of normal cells. Consequently, it may be inferred that only the former can be involved in dynamic activities. It has been suggested that membranes are able to conduct a tension–relaxation wave. This phenomenon may in various ways interact with the environment, thus conferring motility upon the cells, but since the changes occurring in the membrane are transitory and reversible, it cannot be presumed that any intracellular movements, entailing modifications in the cytoplasmic organization, can come out of this activity.

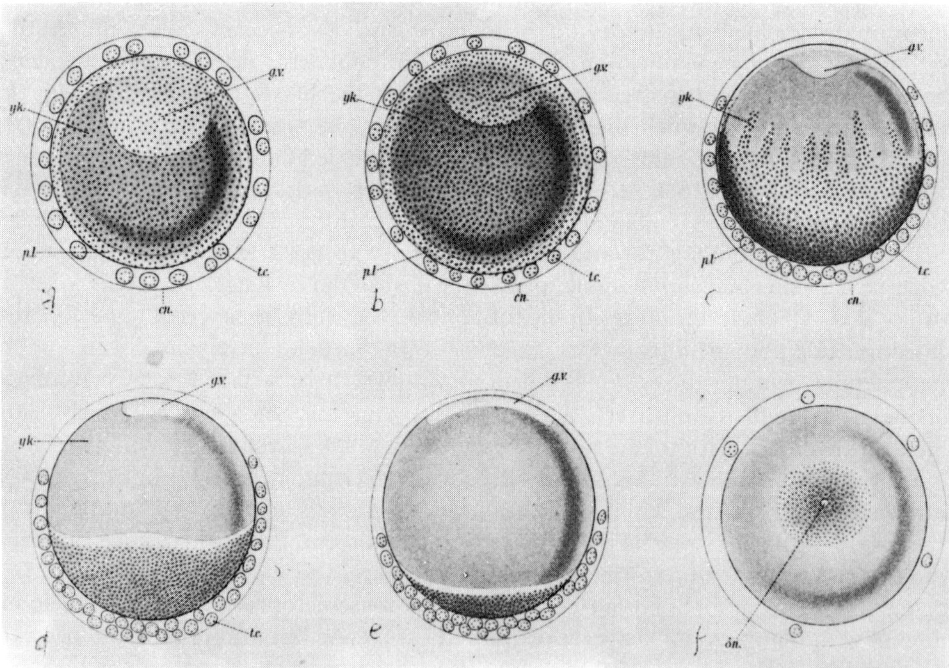

Figure 19. Fertilization reaction in *Styela*. (a), unfertilized egg before breakdown of the germinal vesicle; (b), unfertilized egg after disappearance of the nuclear membrane; (c), egg five minutes after fertilization, showing movement of yellow ectoplasm (dotted) towards the vegetal pole; (d) and (e), successive stages showing that the test cells participate in the movement; (f), egg viewed from the vegetal pole. cn., egg membrane; g.v., germinal vesicle; n, male pronucleus; p.l., ectoplasm; t.c., test cells; yk., yolk. (From E. G. Conklin, *J. Acad. Nat. Sci. Philadelphia,* [2], **13** (1905))

It is known, however, that in certain eggs ectoplasm may become accumulated in some part of the egg, at the animal or the vegetal pole, or at the point of sperm entrance. It might be supposed that only the ectoplasm is involved in this activity, but this is contradicted by the fact that the whole cortex is known to participate in the movements, as demonstrated for instance by the translocation of test cells occurring in association with the formation of the yellow polar plasm in *Styela* (Figure 19) and of electrolytic marks during the formation of the grey crescent in *Rana* (Ancel and Vintemberger, 1948). Owing to the inelastic properties ascribed to the plasma membrane, it follows that such extensive movements are entirely incompatible with the presence of an intact plasma membrane. The circumstance that a drastic increase in the permeability has been observed repeatedly in association with events of this kind intimates that a temporary breakdown of the plasma membrane occurs. We shall return to this question in a later section.

It has been observed that the movements always are elicited by some specific stimulus, sperm penetration, polar body formation, etc., and that they are directed towards the point of stimulation. If, as proposed, the plasma membrane is incapacitated it must be presumed that the ectoplasm is the active agent, the remaining parts of the cortex trailing passively behind. This suggestion is supported by studies of a similar phenomenon, the Ca^{2+}-induced cortical 'contraction' in *Xenopus* eggs (Gingell, 1970). It has been observed that in association with this phenomenon ectoplasmic microfilaments are condensed around the point of stimulation, and this reaction is suppressed by cytochalasine (Wessells *et al.*, 1971). As will be shown presently, the mechanism of cortical movements outlined here bears some resemblance to that responsible for amoeboid movements. It may be mentioned that ectoplasmic translocations are very important for the establishment of polarities in fertilized eggs.

Another kind of cortical reaction has been described in *Rana* by Ancel and Vintemberger (1948). It is associated with extensive cortical movements, as revealed by displacement of the pigment border, but the process is more or less reversible (cf. Figure 35). This phenomenon occurs immediately after fertilization. Elastic or quasi-elastic forces appear to be involved, and it therefore cannot be accounted for on the basis of generalizations advanced above. However, considering that the egg surface is at this time still intimately associated with the egg membrane, it may be suggested that the latter structure is wholly or partly responsible for the phenomenon. This contention rests upon the fact that the egg membrane apparently contains linear TP polymers (microfilaments) which may exert the dynamic activity observed, yet is not itself a TP polymer. It therefore does not have to conform to the generalizations applying to the latter.

Locomotion

Motility is acquired when the dynamic activities in the cells are able to interfere with the environment. From what has been discussed above, it follows that cells may utilize either one- or two-dimensional TP polymers for this purpose. It appears that the lability of the former can be exploited only by the solocytes, microfilaments being involved in amoeboid movements and microtubules in the formation of filopodia. All cells have plasma membranes, and it may therefore be anticipated that tension–relaxation waves, the dynamic activity distinguishing membranes, may be involved in locomotory activities. This prediction seems to be true, but two different cases must be distinguished.

In non-vibriocytes the cell surface, in which the waves prevail, is applied directly towards a substrate in the environment. In vibriocytes an intimate cooperation between microtubules, present as the core of vibratile organelles, and the plasma membrane results in conveying motility to cells suspended in a fluid medium.

Amoeboid movements. The adjective 'amoeboid' appears to be used indiscriminately by some authors to characterize all movements exerted by solitary cells. However, according to the commonly adopted terminology, amoeboid movements always occur on a solid substrate with the participation of lobopodia. This kind of locomotion is thus characteristic of amoebocytes, or sl-cells. From the particular properties attributed to this cell order, and from the general notion concerning dynamic activities propounded here, it follows that only plasma membrane and microfilaments may be participants in amoeboid movements. Several theories have been advanced to account for this phenomenon (cf. reviews by Wohlfarth-Bottermann, 1964a; Wolpert, 1965). We shall not discuss these in detail, but mention only that the so-called 'plasmasol–plasmagel transformation', so easily observed in amoebae, has generally been invoked as part of the mechanism of amoeboid movement.

As a result of the recent electron microscopical investigations discussed above we know that the hyaline plasmagel consists of a matrix of hyaloplasm, in which microfilaments are imbedded. It is important in this context to emphasize that although a certain stabilization of the polymers possibly obtains in the plasmagel, they are potentially labile, just like the microtubules in the sf-cells. The rigidity of the plasmagel enables it to serve as a cytoskeleton. The usually elongated shape of migrating amoebocytes, as well as the branching resulting from the extension of accessory lobopodia, may thus be referred to the plasmagel.

In view of this circumstance it is interesting to note that Goldacre (1964) has been able to procure a much more detailed picture of the plasmasol–plasmagel distribution than has hitherto been possible. As seen from Figure 20, there is a central tube of

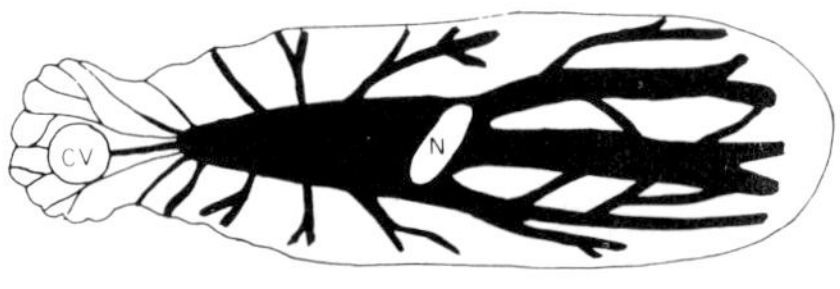

Figure 20. Distribution of plasmagel (white) and plasmasol (black) in *Amoeba proteus*. CV, contractile vacuole; N, nucleus. (Reproduced from R. J. Goldacre (1964), in *Primitive Motile Systems in Cell Biology* (Eds R. D. Allen and N. Kamiya), by permission of Academic Press, New York)

plasmagel from which branches stretching forward establish contact with the ectoplasm and, presumably, with the plasma membrane. If tension prevails in these branches, they may pull the cortex backwards, and the cytoplasmic plasmagel and plasmasol forwards. When this occurs, a dynamic state may be envisaged, in which plasma membrane is continuously formed at the anterior, and decomposed at the posterior, end. The accumulation of hyaline ectoplasm in the lobopodium is accounted for in this way, as well as the formation of the uroid. Provided this mechanism is correct, it affords a strong support of the idea that membranes and intracellular tensible structures are made up of the same monomers, for in this way the supply of material for their continuous formation, membrane at the anterior, and plasmagel at the posterior, end, is easily explained. The forward movement of the amoeba is a consequence of the retrograde transport of the cortex. This interpretation coincides in all essentials with the view advanced by Goldacre, who has been able to corroborate it experimentally in a number of different ways. Plasmagel formation being a completely reversible phenomenon, it follows that the polarization it imposes upon the cell should be labile. This prediction has been confirmed by Goldacre (1964) and Kamiya (1964), who showed that the direction of migration may be changed by experimental means.

The preceding discussion has neglected several questions which may be important for the definitive solution of the question of amoeboid movement, and Wolpert (1965) has indeed brought out various observations which are difficult to reconcile with the mechanism advocated by Goldacre. In the present context my aim has been to show that the latter fits well with the ideas adopted here.

Holtfreter (1948) has shown that cells isolated from embryos of Amphibia may assume an elongated shape and migrate by amoeboid movements. Under these circumstances the pigment granules accumulate at the posterior end. Assuming these granules to represent the remnants of a pigmented surface coat, Holtfreter suggested that the position of the cells in the embryo imposes a polarity on their organization. It will be realized that, in the light of Goldacre's mechanism, the accumulation of pigment granules may be a consequence of the transport of cortical material towards the rear end.

It is evident that this kind of amoeboid movement is intimately associated with an elongation of the cell body. The question arises whether a similar kind of activity may prevail in spherical cells. It will be realized that the ectoplasm is the only plasmagel which under these circumstances may be involved. Cortical movements, involving the rotation of lobopodia along the circumference of a spherical cell, have been reported by Holtfreter (1948) in cells from amphibian embryos (Figure 21) and by Rebhun (1963) in *Spisula* eggs. It is obvious that the ectoplasm takes part in these movements, and it is likely that it is the active, and the plasma membrane the passive, partner, as should be the case in order to correspond to the mechanism of amoeboid movements outlined

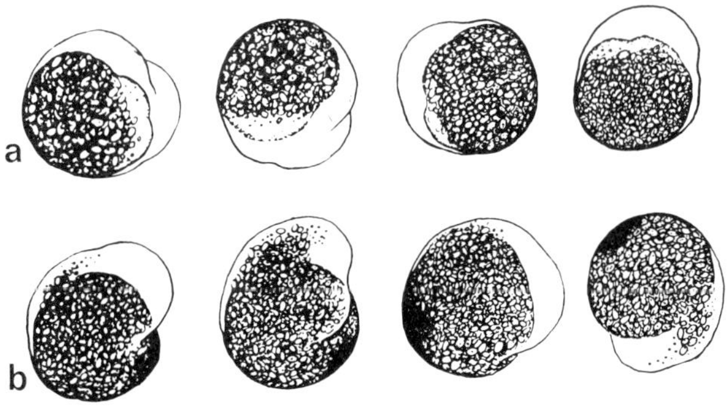

Figure 21. Rotating ectoplasmic movements in cells isolated from an amphibian embryo, in the absence (a), and in the presence (b), of endoplasmic sol–gel formation. (Reproduced from J. Holtfreter, *J. Morphol.*, **79** (1946), by permission of The Wistar Press, Philadelphia)

above. It is doubtful whether this activity is of great importance for locomotion in single cells, but in aggregated cells it may be of great embryogenetic importance, as will be shown in a later chapter.

Filopodia. Possession of filopodia is characteristic of the cell order for which the name solo-filocytes, or sf-cells, was introduced above. The formation of filopodia in sf-cells has been observed and described for at least a century. As will be shown later, their participation in embryogenesis has even been depicted on many occasions, yet their contribution was not clearly understood before its dynamic aspects were elucidated by the time-lapse cinematographic studies on gastrulation in sea urchin embryos (Gustafson and Kinnander, 1956; Dan and Okazaki, 1956).

The core of a filopodium consists of microtubules and microfilaments embedded in a hyaline matrix. It was suggested above that the properties of filopodia are determined by those of the microtubules, and this assumption was made use of to derive some of the properties of linear TP polymers from the observations of Gustafson (1961) on the behaviour of the filopodia formed by the secondary mesenchyme cells in the sea urchin gastrula.

The filopodia are rigid, as shown by the fact that they may be extended through the blastocoele for considerable distances, and they are very labile formations, which are continuously pushed out and withdrawn. The tips of the filopodia are adhesive, and if they happen to come in contact with a proper surface, they may become attached. This occurrence has two consequences: the filopodia develop a tension and they become stabilized. These four properties, rigidity, lability, generation of tension and stabilization through the latter event, were all ascribed to individual microtubules. There is no reason to doubt that the microtubular tips may become attached to the plasma membrane, but from the observations of Wessells *et al.* (1971) on the outgrowth of nerve axons, it appears that the activity in the growth cone may be referred to an undulating membrane arising from cooperation between the plasma membrane and a subjacent network of microfilaments. That the initial attachment of the mesenchyme cell pseudopodia to the blastocoele wall is accomplished by a mechanism of this kind is suggested by the fact that the authors mentioned have found suppression of gastrulation by cytochalasin in various invertebrate eggs.

By attaching themselves between separate microtubules the hyaloplasmic mucopolysaccharide molecules may ensure that several polymers are assembled in a filopodium. The mucopolysaccharide cannot be of the adhesive type, since that presumably would lead to immobilization of the microtubular complexes. It is for this reason that filopodia are formed only in the sf-cells.

If it is accepted that bundles of microtubules may be formed in the cytoplasm we may next consider what will happen if such a structure arises spontaneously in the cell. It may supposedly be directed at random and will grow in length in the cytoplasm until it reaches the surface. At this stage it will have to overcome the cortical resistance, and this is impossible unless it is supported at the other end. Thus, if no support in the cytoplasm was available, filopodia must always form at two opposite sides of the cell simultaneously and they would of necessity collapse at the same time. Such double filopodia would possess little flexibility, and no observations indicate that they exist. Considering that the filopodia of a fibroblast taken together form a stellate complex, it is reasonable to accept the old suggestion by Watasé (1895) that all the filopodia originate in or around the centriole. This idea has been confirmed by electron microscopical observations. Under these circumstances the microtubular bundles, being stretched out in different directions, may cooperate in establishing the thrust required for each of them to push out the cortex. Thus, the centriole functions as a kind of pivot around which the filopodia may swing, ensuring among other things that they may sweep over an extensive range. The function of the centriole evidently is to provide the information required for producing a stellate complex of microtubules. It is significant that in higher plants, where centrioles are absent, sf-cells are unknown. The polarization of the cell is determined by the axis passing through the centriole and the nucleus, the former being located at the anterior end. The direction of cell migration is accounted for in this way: since the nucleus will shield off a large part of the space from extension of filopodia, a greater number will be present anterior to the centriole, and this may determine the orientation of the movement.

As a general rule formation of filopodia does not occur in salt solutions, apparently because this process requires a space containing a medium which possesses some particular physical properties. Formation of filopodia must entail a considerable increase in surface area which, referring to the laws of adhesion discussed above, suggests that the 'affinity' between the medium and the filopodia involves adhesive forces. This point may be corroborated by the fact that the presence of certain kinds of mucopolysaccharides will condition aqueous media in such a way that filopodia can be formed. It is easily envisaged that if the cortical tension is elastic, the formation of filopodia must be opposed by an elastic force that increases with the length of the pseudopodium. This circumstance alone is a strong argument against the supposition that an elastic tension prevails.

Formation of filopodia is not suppressed on a two-dimensional substrate. This is particularly well demonstrated by the outgrowth of axons from neuroblasts in cell culture observed by Harrison (1910). In this case the directions of the axons are influenced by tensions and other factors in the substrate, a phenomenon called *contact guidance* (P. Weiss, 1934).

Two aspects are of decisive importance for the execution of movements by filopodia.

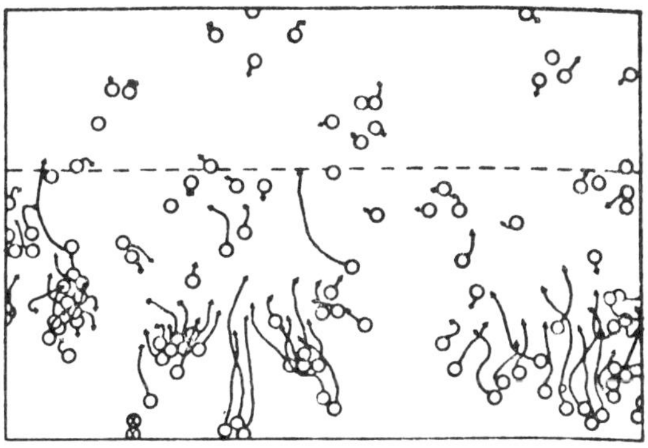

Figure 22. Movements of individual sf-cells on an adhesivity gradient. The circles represent the origin and the lines trace the paths followed by the nuclei of the cells. The dashed line represents the transition between high uniform (above) and graded adhesivity (below). (Reproduced from S. B. Carter, *Nature*, **208** (1965), by permission of Macmillan, London)

The first is the presence of surfaces to which the tips may adhere. This point has been stressed by Gustafson (cf. 1963), and S. B. Carter (1965) observed that when sf-cells are exposed to a substrate varying gradientwise with respect to adhesivity, the cells will move in the direction of increased adhesion (Figure 22). This figure clearly suggests that when the cells are exposed to a surface of uniform adhesivity, the movement will be random, resulting in a negligible displacement. If, on the other hand, the filopodia extended in one direction are able to secure a firmer hold on the substrate than their counterparts, the former will be able to ensure migration of the cells. The other factor is the strength of the tensile force deployed by the filopodia. As we have seen, the latter is proportional to the number of TP monomers involved. This generally implies that the cell will be shifted in the direction where the longest filopodia can be formed. We shall later discuss examples illustrating the importance of this mechanism for embryogenesis.

It appears that the two principles outlined here are involved in the phenomenon of *contact inhibition* discovered by Abercrombie and Heaysman (1953). In cultures of sf-cells it may happen that filopodia from neighbouring cells become attached to each other's surfaces. Apparently the force of this attachment is stronger than that with which the filopodia adhere to the substrate, and as a consequence the cells are pulled together. When undulating membranes are involved in the cell movements, this activity may temporarily be inhibited. As the two cells are apposed, the filopodia will be very short and consequently also weak, and it thus follows that the force developed by undulating membranes or by filopodia extended into a free space may suffice to break the contact between the cells. The phenomenon of *contact paralysis* (Gustafson and Wolpert, 1967) appears to be identical with the stabilization of linear TP polymers through development of tension described above.

It has been observed by Buznikov, Chudakova and Zvezdina (1964) that during the early developmental phases of embryos from various animal classes distinct oscillations occur in the content of 5-hydroxytryptamine (5-HT). Initially the variations are perfectly synchronous with the cell divisions; later on this correlation is lost. Variations still occur, and since in the sea urchin embryo these correspond very well with the phases of filopodial activity described by Gustafson, this author tested the proposition that 5-HT stimulates filopodial activity, an idea which was convincingly confirmed (1969). As the activity of the filiform pseudopodia is essentially a question of formation and collapse of microtubules, these observations strongly suggest that 5-HT is somehow involved in the formation of microtubules. I should mention, in passing, the fact that this substance, as well as another neural mediator, acetylcholine, are of importance for another kind of microtubule, viz., those present in the vibratile organelles.

The ability of filopodia to traverse a space is a unique feature among morphogenetic mechanisms; normally cell to cell contact is required for interactions to occur. For this reason the filopodia may, and must, be made responsible for various complex morphogenetic events. They are of great importance for the enormous variation characteristic of animal evolution.

Undulating membranes. Movements by means of undulating membranes were first described by Abercrombie and Heaysman (1953), and later studied in detail by Ambrose (1961). Cultured fibroblasts (sf-cells) exposed to an adhesive substratum will flatten out and assume a fan-like shape. The leading edge is hyaline, and is subdivided into a number of short pseudopodia. Both of these phenomena suggest the presence of tensible structures, and the observations of Wessells *et al.* (1971) have indeed demonstrated that the dynamically active agent in undulating membranes is a cytochalasin-sensitive complex between the plasma membrane and a network of microfilaments. At the rear end of the cell a pseudopodium is formed, bearing some resemblance to a filopodium; this structure may function to balance the thrust at the leading edge.

The absence of filopodia can be explained by lack of 'affinity' to the culture medium, whereas the flattening of the cell indicates some adhesion between cell surface and substrate. Intimate contact is not established, however, for undulating waves are constantly initiated at the leading edge, whence they travel backwards towards the pseudopodium (Figure 23). These undulations, which enable the cell to migrate on the

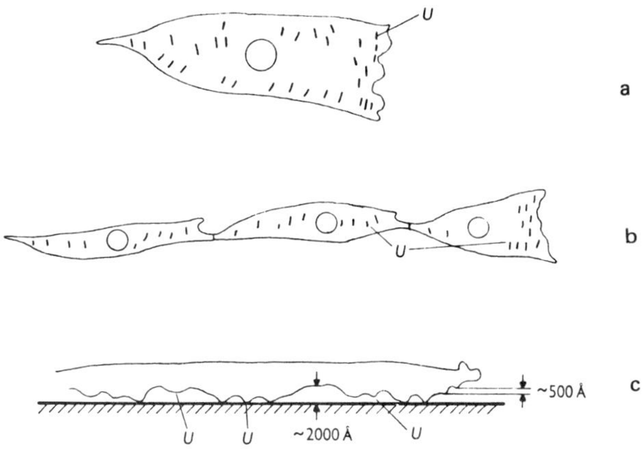

Figure 23. Undulating membranes in sf-cells. The lines represent areas which appear illuminated in the dark field surface contrast microscope, supposed to be in continuous undulating movement. (a), single cell; (b), streaming group; (c), interpretation of image. U, undulating contacts. (Reproduced from E. J. Ambrose, *Exptl. Cell Res., Suppl.* **8** (1961), by permission of Academic Press, New York)

substrate, are presumed to reside in the plasma membrane, an assertion which evidently implies that this structure is able to undergo rhythmical changes in tension. This activity appears to comprise not only the conduction of a tension–relaxation wave, a property which above has been contended to be typical for all biological membranes, but also the continuous generation of new waves.

If the latter was a general membrane characteristic, undulating membranes might be expected in all cells. As far as the sl-cells are concerned, it may be mentioned that Rebhun (1963) has seen peristaltic waves pass over the surface of eggs of *Spisula*, deprived of their vitelline membrane. This observation shows that conduction of tension–relaxation waves may take place, but undulating waves were not observed, and various other observations on sl-cells rather support this point. On the contrary, Ambrose (1961) contends that undulating membranes may be observed in sheets of epithelial cells, in which case they may contribute to an expansion of the area covered by the cells. Spreading of epithelia is a phenomenon encountered in many embryos, and it is possible that undulating membranes may be instrumental in this process. It may indeed be the only means of locomotion available to unciliated colligocytes, in which the nature of their cytoplasmic polysaccharide prevents the formation of pseudopodia.

Cilia and flagella. Various features, for instance length, number and pattern of beat, are often invoked to characterize the two kinds of vibratile organelles, cilia and flagella. It does not seem possible to attain unanimity by these criteria, and in the present context the distinction will be based only on the number present. Thus, if a cell possesses one, or exceptionally a very limited number of organelles of this kind, we shall call it a flagellocyte; if it contains many, it is a ciliocyte. As discussed above, flagella are formed in solocytes, whereas the possession of cilia is a distinctive colligocyte characteristic.

The proposed definition corresponds well to commonly held opinions, but it has one consequence which is not generally accepted. In many embryos vibratile organelles develop at a certain stage; often two kinds may be distinguished, very long ones at the animal pole, and shorter ones over the remaining surface. It is customary to call the former flagella, the latter cilia; presumably it is the difference in length which is decisive for this terminology. It is, however, a well-known fact that in either case each cell possesses only one organelle (Figure 24). It thus appears that, according to the terminology proposed here, only flagella are present, and the embryo must be character-

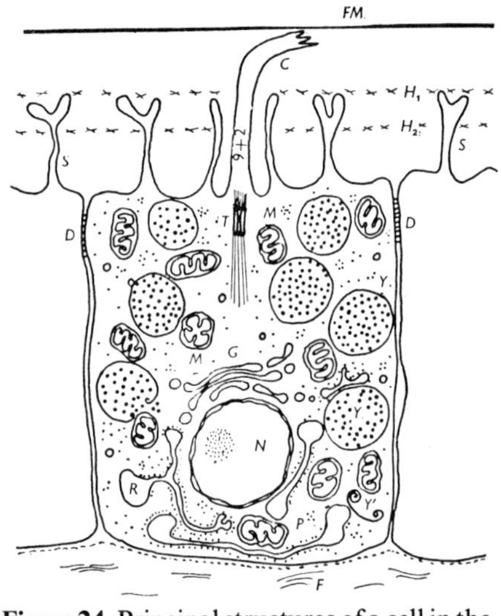

Figure 24. Principal structures of a cell in the wall of a sea urchin blastula, just prior to hatching. D, sections of septate 'desmosomes' which stretch around the circumference of the cell; F, fibrils of the inner surface layer bordering the blastocoele; FM, egg membrane; G, Golgi apparatus; H_1 and H_2, two condensations of the hyaline membrane; S, cell processes extending into the latter; M, mitochondria; N, nucleus; P, ribosomes; R, endoplasmic reticulum; T, basal body of the flagellum (C); Y, yolk granules. (Reproduced from L. Wolpert and E. H. Mercer, *Exptl. Cell Res.*, **30** (1963), by permission of Academic Press, New York)

ized as flagellated, not ciliated. As a matter of fact, it has been mentioned above that the flagellated cells at the animal pole may be ncf-cells, and the flagella stereoflagella, which are not vibratile organelles at all.

Cilia and flagella originate in a basal body or kinetosome. It was suggested by Henneguy and Lenhossek at the beginning of this century that the kinetosomes are centrioles (cf. Wilson, 1928), and in spite of certain differences in fine structure, the close relationship has been corroborated by electron microscopical studies. A flagellum may arise from the centriole normally present in a cell, whereas formation of cilia cannot occur without centriole replication. The vibratile organelles and centrioles are very similar in their general organization, cross sections showing them to be cylindrical structures containing nine groups of microtubules (cf. Fawcett, 1966). This circumstance had led to the notion that the centrioles act as templates for the formation of cilia and flagella. If this view is correct, it follows that whereas microtubule formation may be a spontaneous formation, their organization in a complex of a higher order such as that found in the centrioles requires structural information. Being information-carrying entities, two demands must be made on the centrioles, viz., that their information may be transcribed, and that they possess the power of replication, both of which are known to be fulfilled. As we have seen above, the information contained in the centrioles may be used for purposes other than copying their own structure, namely, for the formation of stellate complexes of microtubules.

Single cells and small organisms possessing either type of vibratile organelle can use these for three-dimensional propulsion in a liquid medium. In tissues consisting of layers of ciliated cells the concerted action of the latter may serve to transport objects. Both of these mechanisms may be involved in the translocations needed to unite egg and sperm, and certain embryos and larvae are motile owing to the possession of flagella or cilia. Yet, although vibratile organelles thus play a definite role in the overall mechanism of reproduction, it seems that as far as the aspect of locomotion is concerned, they are not of particular importance for embryogenesis. However, the possession of either cilia or flagella imposes a certain polarization on the cell body and, as we shall discuss later, this circumstance may be of importance for certain morphogenetic events.

Several kinds of beat are known in vibratile organelles, but here we shall consider only two, the pendular movement in which the organelle bends at the base during the effective stroke, and is subsequently straightened from the base to the tip, and undulating movements, in which waves of bending, originating at the base, pass along the organelle (Figure 25). The first pattern, most frequent in cilia, may correspond to the formation–collapse cycle found in cytoplasmic microtubules. The stroke would then correspond to the tension developed upon stretching between two frames. It is difficult to envisage a mechanism to account for this phenomenon, but it seems likely that a cooperation occurs between the plasma membrane and the microtubular structure. It will be noted that the 'collapse' does not involve degradation of the latter. Undulating movements are more usual in flagella; in accordance with the tenet advanced above, the wave-like spreading of a tension is typical of membranes, and in this case it therefore seems necessary to invoke the participation of the plasma membrane. These considerations suggest that the plasma membrane may be a very important participant in the activities displayed by vibratile organelles. Colchicine does not affect the microtubules in flagella, as might be anticipated from its purported mechanism of action and the circumstance that no depolymerization is involved in the flagellar movements. I have not come upon

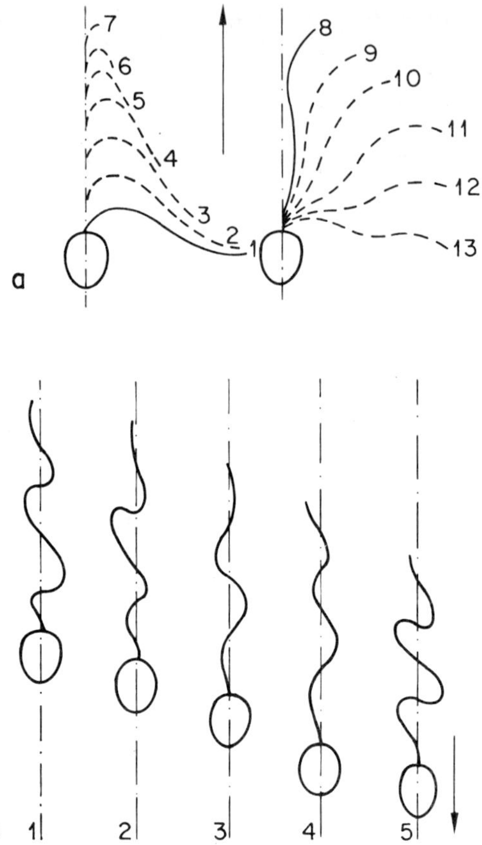

Figure 25. Flagellar movements. (a), pendular beat. 1–7, straightening of the flagellum; 8–13, the stroke. (b), undulating beat. The direction of displacement is indicated by arrows. (Reproduced from B. J. Krijgsman, *Arch. Protistenk.,* **52** (1925), by permission of Gustav Fischer, Jena)

information dealing with the effect of colchicine upon cilia, but since even in this case no structural disintegration occurs it is probable that it is without influence. According to Wessells *et al.* (1971), cytochalasin does not interfere with either ciliary or flagellar function, indicating that the particular kind of cooperation between microtubules and plasma membrane involved in this activity is insensitive to this substance.

Another kind of cooperation between a flagellum and a membrane is represented by the 'undulating membranes' observed in certain Protozoa, classified above as vsf-cells (Figure 6). It seems most likely that in this case the flagellum acts as a supporting skeleton, whereas the membrane is involved in the generation of the undulating movement. The collar in choanocytes may be formed in a somewhat similar fashion. To account for the particular shape of the membrane it must be presumed that, beneath the basal body, there is another 'centriole', lying at right angles to the former. This does not seem to be a very implausible proposition.

EXTRACELLULAR ELEMENTS

In adult organisms a great variety of extracellular elements, like cuticles, shells, cartilage, bone, etc., participate in the shaping of the body. During the early embryonic stages with which we deal here, a very limited number of extracellular components are involved in the morphogenetic processes. Among these are the hyaline membrane, which is part of the preformed substrate, and three elements formed epigenetically through the activity of differentiated cells, namely, oncotically active mucopolysaccharides, collagen fibrils and mineralized structures. One more preformed extracellular element may be distinguished, viz., the egg envelopes, which in some cases are of importance for morphogenesis; these will be considered in association with the discussion of the egg.

It may to some extent be misleading to consider these components as elements, since they always occur in intimate association with cells. However, the actions exerted ultimately depend upon properties possessed by these components, so the classification seems justified. In this section we are going to characterize the extracellular elements with particular reference to those features which are of importance for morphogenesis.

Hyaline membrane

The nomenclature relating to the surface layers of eggs and cells is very confusing, and it may therefore be justified to introduce this section with a terminological digression. As outlined above, it seems that all or almost all cells are covered by a hyaline layer of mucopolysaccharide, called the 'surface coat'. In eggs it is possible to distinguish two hyaline structures closely applied to the egg surface. One of these, present in the unfertilized egg, often called the 'vitelline membrane', is usually separated from the egg as a consequence of fertilization. After the occurrence of this event another surface layer may be formed at the egg surface, and this structure is the *hyaline membrane* which is the subject of the present section.

The basis for the distinction between 'coat' and 'membrane' made here is that the latter, in contrast to the former, may become separated from the cell (egg) surface. It seems that this difference can be explained by the presence of fibrous proteins in the membrane matrices. For the hyaline membrane this phenomenon has been demonstrated in *Psammechinus* (Wolpert and Mercer, 1963), *Spisula* (Rebhun, 1962) and Amphibia (Wartenberg and Schmidt, 1961). There is no direct evidence supporting the claim that these fibres are TP polymers, but if, as contended above, no other type of structural protein is present in undifferentiated cells, it seems necessary to infer that they are microfilaments. It has already been pointed out that the membranes proper under no circumstances can be considered as TP polymers.

The existence of a hyaline membrane has long been recognized in the sea urchin egg, due to the fact that the blastomeres during the first divisions may separate from the former. This separation is obviously possible because the tension in the cells may overcome their adhesion to the membrane. Since this situation may not always prevail, it is likely that a similar structure exists in many other eggs without being observable as a separate entity. The fact that it is hyaline makes it difficult to observe, and this may explain the discussion which has been carried on (cf. Dollander, 1951; Wartenberg and Schmidt, 1961) about the existence of the hyaline membrane (surface coat) in the amphibian egg first postulated by Holtfreter (1943a). Its presence has now been demonstrated electron microscopically (Figure 26).

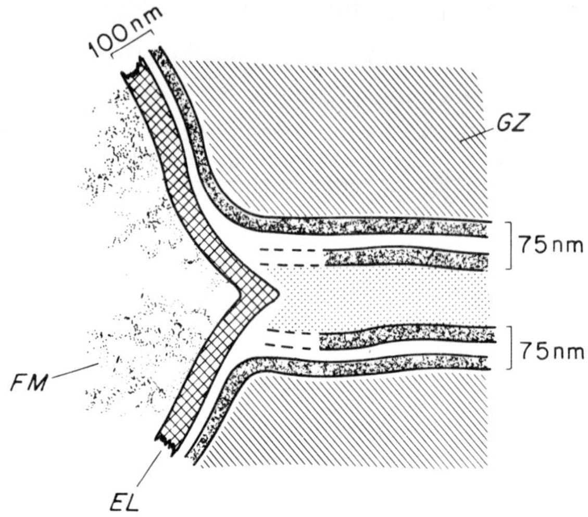

Figure 26. Diagram of a cell junction from the blastopore of *Triturus*, showing the supracellular hyaline membrane (EL). FM, fibrillar material. (Reproduced from M. M. Perry and C. H. Waddington, *J. Embryol. Exptl. Morphol.*, **15** (1966), by permission of The Company of Biologists, Cambridge)

The hyaline membrane is formed at fertilization by part of the contents of the cortical vesicles, when the latter are emptied into the perivitelline space. This is not a general phenomenon: in certain cases the cortical vesicles remain intact, in other cases they are ruptured, but their contents do not participate in membrane formation. The main component of the hyaline membrane in sea urchins appears to be mucopolysaccharide (Monné and Slautterback, 1950; Vasseur, 1952; Runnström, 1966), but, as mentioned, fibrous elements may also be present. It is likely that the matrix of the vitelline membrane is the original surface coat of the oocyte, which is replaced by a hyaline membrane when the former is separated from the egg at fertilization.

As far as the morphogenetic importance of the hyaline membrane is concerned, it suffices to mention here that it depends upon the fact that it is a supracellular structure to which the embryonic cell may adhere.

Oncotically active mucopolysaccharides

Various cavities in embryos are filled with a hyaline, jelly-like material, the main component of which is mucopolysaccharide. The latter term cannot be used to characterize this particular kind of substance, for although most mucopolysaccharides are able to bind water, it appears that only certain classes are able to bind the copious amounts of water required for the swelling observed. It is not easy to devise an appropriate name for these particular compounds, and the one employed here, *oncotically active mucopolysaccharides*, is somewhat ambiguous and rather inconvenient. In many instances the substance responsible is unquestionably hyaluronic acid, secreted by a special kind of sf-cells (mesenchyme cells), but the chemical composition may well vary from one animal group to another.

It is important to note that the presence of this substance in various cavities does not imply that it tends to occupy 'empty spaces'; the causal connection is rather the opposite, viz., that wherever the mesenchyme cells have a chance to secrete their product, its water-binding power will be reflected by the formation of cavities. In contrast to osmotic swelling, this oncotic swelling may be spatially confined; for this reason it is of great consequence for the shaping of various structures in the embryo. Besides the morphogenetic effect of this spatially confined swelling, oncotically active mucopolysaccharides are of significance also because the filopodia of sf-cells may span a cavity filled with this material. Under these circumstances the space is made available for the morphogenetic activities of these versatile cells. In certain cases the cavities may also contain sulphated mucopolysaccharides with the consequence that cartilage formation can occur.

Collagen fibrils

Collagen fibrils are a differentiation product, secreted by a certain type of sf-cells. In the adult body they are found in many locations, but in the embryo they are mainly confined to sheaths and membranes surrounding aggregates of colligocytes. Collagen is a protein, characterized by a high content of glycine, alanine, serine and proline-hydroxyproline, and generally also by the presence of hydroxylysine. The function of the collagen fibrils depends to a large extent upon the nature of the morphological structures to the formation of which they contribute. In general it may be stated that the mentioned membranes tend to stabilize the various aggregates. Because of their great mechanical strength and slight extensibility collagen membranes are important for the shaping of the embryonic body (stretching, etc.), as well as for its maintenance. Collagen fibrils also are an important constituent of cartilage.

Mineralized structures

Incorporation of mineralized elements in the body, most often in the external coverings, but also in internal skeletons, is a phenomenon encountered throughout the animal kingdom. The structures thus formed possess considerable rigidity, and consequently they may serve partly for protection, and partly for the shaping of the body. Skeletons are mainly involved in the latter function, whereas shells and other external coverings may have both functions. It appears that the crystallization involved in the formation of mineralized structures requires a matrix in which collagen or related types of protein form an essential component. A number of different compounds may be deposited, the most frequent being calcium carbonate and calcium phosphate.

The fertilized egg

We are now ready to deal with the preformed embryogenetic substrate, i.e. the fertilized egg. As intimated above, this entity possesses a number of characteristics which, through interaction with the processes executed by the predetermined mechanisms, may influence the course of development. Since some of these substrate features arise during maturation and fertilization, it is most expedient to discuss first the general organization of the unfertilized egg, and then the changes occurring as a result of these processes. In the fertilized egg the fixation of the various polarities has occurred, and we shall therefore conclude this section with a discussion of the material basis of the polarities.

THE UNFERTILIZED EGG

The unfertilized egg is most often covered by egg envelopes. Being located outside the egg, these structures are truly extracellular elements. They may, however, influence the course of embryogenesis by imposing polarities on the egg, which implies that they may be part of the material substrate of the polarities. This circumstance justifies their inclusion in this section. After the discussion of the envelopes we shall try to outline some of those features of the egg which are known to be of importance for embryogenesis.

Egg envelopes

With the exception of Porifera, where membranes are absent, and Cnidaria, where they are rare, eggs are surrounded by at least one and often several envelopes, deposited by the egg proper, by cells in the follicular epithelium or by cells in the walls of the oviduct. The envelopes vary widely with respect to their physical and chemical properties, as well as to their functions, although it probably is correct to assert that they serve mainly for protection of the egg and the embryo. The innermost envelope, the only one with which we shall be concerned here, may be protective, but it can also be of importance for osmoregulation. As will be shown later, it may even be of morphogenetic significance.

As to the name of this structure, various authors have shown profuse generosity, thus 'it is known as the *vitelline membrane* in insects, molluscs, amphibians and birds; in tunicates and fishes it is usually called the *chorion*. In mammals the membrane of an exactly similar nature is called the *zona pellucida* . . . The zona pellucida thus takes the place of the zona radiata . . .' (Balinsky, 1965, p. 90; cf. also Wischnitzer, 1966). The situation is not improved by the fact that 'chorion' is used in other senses as well. Sometimes, but not consistently, the name changes to 'fertilization membrane' in the fertilized egg. It does not seem necessary to employ different names for this structure, and believing that clarity may be gained by using the same name for a homologous structure in different eggs, I shall, as already stated, adopt the old name *egg membrane* (cf. Conklin, 1897) for this innermost egg envelope. By some authors this expression is employed to signify the plasma membrane of the egg, but I do not think that this improper usage should prevent its acceptance in the sense proposed here.

One circumstance which has contributed to the terminological confusion is that some egg membranes are hard and rigid (chorions), others soft and plastic (vitelline membranes). This difference may undoubtedly be referred to differences in chemical composition, but it seems that in spite of this variability, mucopolysaccharides and fibrous proteins are the main constituents (cf. Rosenbaum, 1958; Balinsky, 1965).

Egg membranes are found in cases where no follicular envelope is present, which shows that they must be formed by the egg. As a matter of fact, as suggested above, the mucopolysaccharide matrix of the membrane may be nothing but the surface coat of the oocyte, reinforced with fibrous proteins.

Due to the presence of these proteins, the plastic membranes possess a certain tension, which ensures that they become closely applied to the egg surface. Most often the membranes are reasonably uniform, consequently they must impose a spherical shape upon the egg, provided the tension suffices to overcome gravity. Rigid membranes are very often plastic initially, hardening only upon contact with the external medium. Under such circumstances it may be anticipated that the membrane is spherical, although occasionally a slight elongation may prevail. Rigid egg membranes are known which are shaped like straight or curved cylinders, or otherwise deviate markedly from spherical symmetry. In such cases hardening occurs before the eggs are laid, the particular shape being impressed upon the egg by the maternal body. Membranes of this kind furnish a means to establish embryogenetic polarities on an egg, provided, of course, that the egg is apposed to the membrane. This situation usually prevails in eggs deposited on land, but in aquatic eggs it often happens that material derived from the cortical vesicles, present in the perivitelline space, swells by binding of water. In this case an oncotic pressure develops, counteracted by the mechanical strength of the membrane. As the pressure is uniform in the perivitelline space, the egg will be spherical. It may be mentioned that a similar situation obtains when many eggs are imbedded in a water-binding matrix in a cocoon.

In some eggs a small opening, a micropyle, is found in the egg membrane. The location of the micropyle always bears a definite relation to the animal–vegetal polarity of the egg, being situated either at the presumptive animal (e.g. Echinoidea, Pisces) or at the presumptive vegetal pole (e.g. certain Mollusca). Very often, but not always, the egg membrane is impermeable to spermatozoa, which therefore can establish contact with the egg only by passing through the micropyle. By this expedient the spatial orientation of the fertilization reaction is predetermined. Micropyles arise through inhibition of the growth of the egg membrane, either because of the attachment to the ovarian wall (vegetal pole) or because a follicle cell establishes contact with the egg surface (animal pole).

The egg

Much speculation has been devoted to the properties which confer upon the unfertilized egg the developmental potential displayed in the course of ontogenesis. As maintained above, I believe this is a wrong formulation of the problem: in actual fact the egg only acquires its potential through the act of fertilization. This does not exclude, of course, that certain properties inherent in the unfertilized egg are important for embryogenesis. In the following pages we are going to consider some of the most pertinent features, dealing with the egg only as a substrate, thus without reference to those components like centriole and nucleus which are essential parts of the predetermined mechanisms. Cytoplasm and ectoplasm will be discussed under separate headings.

Cytoplasm

Generally, the egg is considered to be a cell, even though it occasionally has been claimed that the organization of the egg is too complex to warrant this view. It seems that the former view must be accepted (Gregg, 1959), but under any circumstances we may, with Woodger (1929), claim that the egg has the 'cell-type of organization'. From this point of view, the size is probably the most prominent characteristic of the egg. With respect to this parameter, eggs cover a wide range, even if allowance is made for the yolk, which in many cases constitutes the bulk. However, with few exceptions it is true that even the smallest egg is very large compared with the cells in the adult. This unusual size bestows upon the egg the faculty to undergo repeated cell divisions, uninterrupted by intermittent periods of 'growth', a phenomenon which is a *sine qua non* as regards subsequent development. It may be inferred that ways and means to elaborate a very large cell must represent the first step in the evolution of Metazoa. How this occurred originally is entirely unknown, but observations have revealed that even today different methods are employed by Nature to achieve this end, the more primitive ones involving 'cannibalistic' engulfment by the egg of kindred cells or of specialized nurse cells of various kinds, whereas in higher animals the egg is nourished by follicle cells forming an epithelium around the egg.

Another unusual feature of eggs is the accumulation of nutritive reserves. Great variation obtains in this respect, and it is interesting to note that phylogenetic evolution in the Vertebrata has been accompanied by a gradual increase in reserves, reaching a maximum in Reptilia and Aves, followed by a drastic reduction in Mammalia. Generally, some of the reserves are present as yolk platelets, containing lipo-phospho-proteins, but lipids may be present as droplets and glycogen in granules. The reserve material is used for two purposes, to wit, supply of energy and synthesis of the products required for elaboration of the embryonic body.

It appears from studies on amphibian embryos (1953a; 1959b) that during normal prefeeding development only carbohydrates and lipids are used for energy supply to any significant extent, and in this order, whereas the nitrogen-containing compounds are preserved for synthetic purposes. Only when the embryos are starved do the latter become mobilized for combustion. The yolk proteins thus contain a reservoir of amino acids which are used mainly for protein synthesis. It further seems that deoxyriboside-containing substances are present which can be mobilized for DNA synthesis (Hoff-Jørgensen and Zeuthen, 1952; Hoff-Jørgensen, 1954). It may be mentioned that in certain eggs the cytoplasm is unevenly distributed along the animal–vegetal axis, thus constituting a cytoplasmic polarity.

Ectoplasm

On the basis of the yolk distribution it is usual to distinguish between *isolecithal* (*oligolecithal*), *centrolecithal* and *telolecithal* eggs. The first contain so little yolk that it does not significantly affect the cytoplasmic organization. The centrolecithal type apparently owes its existence to the presence of a copious amount of ectoplasm, from which the yolk granules are excluded. Actually, the difference between this and the isolecithal type may depend upon the paucity of yolk in the latter, a circumstance which tends to obscure the transition between ectoplasm and cytoplasm. Telolecithal eggs also contain ectoplasm, but generally a distinct polarity obtains in the animal–vegetal direction, as visibly displayed by the eggs of most Amphibia, where the pigmented ectoplasm is absent from the vegetal cortex.

As a feature which is specific for the ectoplasm of most eggs, we may mention the so-called *cortical granules*, located in vesicles beneath the cell membrane. By cytochemical methods they have been shown to contain mucopolysaccharides (Monné and Hårde, 1950b). In many cases the vesicles have been observed to collapse during fertilization, their contents being emptied into the perivitelline space, its place being occupied by yolk.

As suggested above, attachment of microfilaments to the plasma membrane is instrumental in the formation of the ectoplasm. It was also postulated that linear TP polymers can adhere only to free surfaces. In this connection some observations made on growing oocytes of *Lymnaea* are of great interest (cf. Raven, 1966). It was found that initially the follicle cells are in intimate contact with the surface of the oocyte, but later a 'cavity' appears between the cell surfaces at the apical and lateral sides. At ovulation this cavity, presumably containing the mucopolysaccharide matrix of the egg membrane, extends over the whole surface of the egg except for a small area, the micropyle, through which passes the stalk of attachment. It was found that ectoplasm is formed beneath the plasma membrane only where the latter faces the cavity, i.e. where the surface is free.

It may be anticipated that pigment granules and cortical granules become accumulated in the cortex with the advent of ectoplasm formation. This seems to be borne out by the observations by Motomura (1941) and by Runnström and Monné (1945), showing that in the sea urchin the presumptive cortical granules, scattered at random in the cytoplasm of the oocyte, accumulate in the cortex of the mature egg (Figure 27). Ectoplasm clearly is formed at a later stage of oogenesis in this case than in *Lymnaea*, as discussed above.

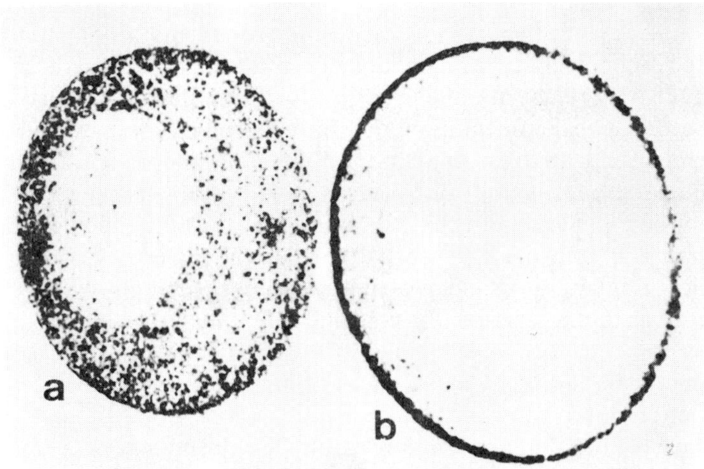

Figure 27. Cortical granules in *Arbacia*. (a), in the oocyte the prospective cortical granules are scattered in the cytoplasm; (b), in the mature egg the granules are located in the cortex. (Reproduced from L. Monné and S. Hårde, *Ark. Zool.* [2], **1** (1950), by permission of Kungl. Vetenskaps-akademin, Stockholm)

In many instances the matrix of the egg membrane is occupied by microvilli, extended from the surfaces of the egg and the follicle cells, respectively (cf. Raven (1966) for literature on Mollusca; Kemp (1956) and Wartenberg and Schmidt (1961) on Amphibia). It has been suggested previously that microvilli may be formed from ectoplasmic micro-filaments, provided adhesive mucopolysaccharides are present on both sides of the plasma membrane. The follicle cells may reasonably be classified as cl-cells, which are characterized by the presence of microfilaments and the adhesive type of mucopoly-saccharide, and they therefore ought to be able to form microvilli. Being an sl-cell, the egg should not contain the mucopolysaccharide required for formation of microvilli.

We are here facing a problem which cannot be properly resolved for the time being. There are various histochemical observations suggesting that both egg membrane and cortical material in Echinoidea and Amphibia contain sulphated mucopolysaccharides (cf. Monné and Slautterback, 1950; Immers, 1962; Rosenbaum, 1958). Taken at their face value, these results invalidate the cell classification of the egg cell proposed above, but more must be known about the origin of the mucopolysaccharides before a proper decision can be made on this point. It is important in this context to point out that the egg mucopolysaccharides appear to be different from those present in adult tissues (cf. Immers, 1962), and hence it may not be warranted to relate these findings to the scheme of cell classification.

MATURATION AND FERTILIZATION

The phenomena of maturation and fertilization are often regarded as embryogenetic processes, and in the widest sense of the adjective, this stand cannot be disputed. However, if it is accepted that embryogenesis is a series of epigenetic events initiated only after fertilization, then the two processes are progenetic events, serving to transform the oocyte into the original embryogenetic substrate which is the fertilized egg.

Because of the importance of these processes for development, but also because they are easily accessible to experimentation, they have been the subject of extensive studies. It must be emphasized that the experimental material comprises a relatively limited number of animal groups, and since Nature is more resourceful than human imagination, it cannot be excluded that current notions, and thus the following presentation, lacks in generality. In spite of all the work devoted to the subject, and this includes not a little theoretical thinking, it would probably be an exaggeration to suggest that clear concepts prevail as to the nature of the contribution of the processes to the mechanism of embryogenesis.

This statement does not apply to the nuclear phenomena associated with maturation and fertilization; it is well known that the meiotic divisions serve to generate the haploid female pronucleus, and that fertilization ensures the formation of a diploid nucleus through fusion of the two gamete pronuclei. These nuclear events are of particular interest from the point of view of genetics and cytogenetics. Without in any way contending that they are of no importance for ontogenesis, it must nonetheless be acknowledged that they may be dispensable, as amply demonstrated by various observations of parthenogenesis. As will appear from the discussion in a subsequent chapter, the contribution of the nucleus to the early development is modest insofar as only a limited part of the information stored in the genome is required to carry the embryo through those stages which are the subject of the present work.

The other partial events, those associated with the cytoplasm and the cortex, turn out to be of much more direct importance, both for the initiation of development and for the execution of various embryogenetic mechanisms. For this reason, attention will be concentrated on the latter in the present context. Maturation and fertilization are generally closely associated, but they are all the same sufficiently different with respect to their nature and their significance to warrant that they are treated separately. The present section will be concluded with a comparison of the phenomena of fertilization and activation.

Maturation

The phenomenon of maturation begins when the oocyte receives the stimulus which causes rupture of the germinal vesicle and ends with the formation of the second polar body; it comprises nuclear, cytoplasmic and cortical events. As stated in this definition, some kind of stimulation is required to initiate the process of maturation. In Vertebrata this is accomplished by pituitary hormones, in some eggs the fertilization affords the releasing stimulus, and in many other cases contact with the external medium upon shedding suffices to set off the process. It is possible that still other mechanisms may serve to initiate maturation, but this question is beyond the scope of the present work.

Polar body formation, the most conspicuous feature of maturation, results from a process of cell division, which generally may be characterized as being extremely unequal. This event comprises nuclear and cytokinetic aspects; the former are not dealt with here, the treatment of the latter must be postponed to the discussion of the mechanism of cell division.

The great variation obtaining with respect to the properties of eggs entails that a wide spectrum of reactions may be associated with the process of maturation. It is probably impossible to propose any completely general pattern of response to the initiating stimulation, except as regards the nuclear phenomena. The following presentation may represent a majority of the known cases, but various exceptions are known. Among the reactions observed the most usual are an increase in the cortical tension, an enhanced activity displayed by the ectoplasmic microfilaments and formation of the maturation spindles. Superficially, these cortical and cytoplasmic phenomena do not appear related, but in view of the recent discoveries discussed above, they do have one thing in common: they are dynamic events which, according to our premises, must be carried out by TP polymers. It may thus be envisaged that stimulation of the activity of these structures is an integral part of the mechanism of maturation.

In the following sections we shall discuss the cytoplasmic and cortical reactions associated with maturation. For obvious reasons the phenomenon of maturation has been studied most extensively in those cases where it is initiated after the egg has left the maternal body. Under these conditions contact with the medium, or some other stimulus, starts off a chain of events, which presumably are causally connected. In many cases it is impossible to establish their mutual relation, even if the temporal sequence occasionally may shed some light on this question. As regards the cytoplasmic reactions, they may contribute to the polarity of the egg. The cortical phenomena will in many cases serve to prepare the egg for the act of fertilization, but this is not generally valid.

Cytoplasmic events

During maturation the egg cytoplasm is distinguished by the presence of microtubules in the meiotic apparatus (maturation spindle). Recalling that members of the order of sl-cells, to which the oocyte is proposed to belong, contain no microtubules, it appears that one cytoplasmic change which results from the initiation of maturation is the acquisition of the capacity to form microtubules.

Dating back to the classic discovery by O. and R. Hertwig that no sperm aster is formed if the spermatozoon enters a sea urchin egg before the rupture of the germinal vesicle, and supported by many later observations (cf. Wilson, 1928), it has been envisaged that some effect emanating from the nuclear sap is necessary to establish a condition of 'cytoplasmic maturity' required before fertilization can occur. Although some experimental evidence suggests that astral activity may prevail before the rupture of the germinal vesicle, and that it even may be involved in this process, there nevertheless appears to be some justification for the contention that the nucleoplasm contains a factor necessary for microtubule formation. It must be pointed out that a causal relation of this kind may be a general phenomenon in all lobocytes, where breakdown of the nucleus and formation of the microtubule-containing mitotic apparatus are intimately associated. If this is the case, then the specific feature of the phenomenon of maturation lies in the initiating stimulus, which certainly in general is quite different from that which normally sets off a cell division.

In spite of some scattered claims to the effect that polar bodies are formed elsewhere, it is usually contended that they are expelled at the animal pole. It will be realized that this proposition presumes that the animal pole is always predetermined. Although this no doubt most often is true, the meiotic apparatus is known in some cases to remain quiescent in the cytoplasm until fertilization has occurred, showing that the determination, or at least the realization, of the animal pole is an outcome of this process. Hence it may be more correct to invert the relation and define the animal pole of the egg as the point where the polar bodies are formed.

It is a typical feature of the meiotic divisions that the microtubules of the meiotic apparatus become attached to a narrow region around the animal pole. If we make the assumption that whenever and wherever microtubules are able to adhere to the inside of the cortex they will become attached, it follows that in immature eggs microtubules do not easily adhere to the cortex outside the animal pole, or else that the meiotic apparatus is unable to form microtubules which span the volume of the egg. It is well known that after fertilization astral microtubules generally become attached to all parts of the cortex, indicating a change in cortical properties or in the, by now, mitotic apparatus. As far as the former are concerned, fertilization is often associated with an accumulation of ectoplasm at the animal and vegetal poles. It therefore seems possible to advance the assumption that the ectoplasm impedes the establishment of contact between microtubules and plasma membrane. We shall later discuss this mechanism in more detail.

Cytoplasmic 'streamings' of various kinds have repeatedly been described to occur in association with maturation and with fertilization. As pointed out already, autonomous cytoplasmic movements cannot be accounted for; events of this kind must be the consequence of either microtubular or cortical activities.

Cortical events

In many oocytes the surface is rather irregular, but with the onset of maturation it becomes smooth. This change seems to be a prerequisite for fertilization, as suggested by the generalization of Just (1939) that fertilizability is associated with surface smoothness. This phenomenon indicates an increase in tension, which may reside in egg membrane or plasma membrane, or both. This change can be initiated without any other stimulation than that to which the egg is submitted when it leaves the maternal body. It may thus be envisaged that contact with the external medium can induce the observed increase in tension. In this connection reference may be made to the measurements of 'stiffness' or tension on amphibian eggs (Berntsson, Haglund and Løvtrup, 1964), showing that exposure to a hypotonic medium leads to a considerable increase in this parameter. Eggs of marine animals are not shed into hypotonic media, but it may be presumed that in this case the ionic composition as such influences the tension of cortex and egg membrane. In those instances where other agents are excluded, it is possible that this cortical reaction affords the stimulation required for the initiation of maturation.

The accumulation of cortical granules associated with the maturation in the sea urchin egg, which was mentioned above, suggests that in this case a stimulation of the microfilamentous activity takes place.

Various cortical movements are associated with maturation, as revealed by ectoplasmic translocations. Two kinds have been distinguished, reversible reactions, as found in eggs of Amphibia, which must be ascribed to the egg membrane, and irreversible ones in which the plasma membrane may be involved. We have already previously pointed out that cortical movements of this kind require a stimulation, and that they lead to accumulation of ectoplasm towards the point where the stimulus is applied. During maturation the stimulus arises from the expulsion of the polar bodies, thus by definition at the animal pole, and accumulation of ectoplasm accordingly ought to occur at this location.

The formation of the so-called *polar plasm*, which results from this kind of cortical movement, was first observed in *Clepsine* by Whitman (1878). This phenomenon has been observed in many spirally cleaving eggs, notably in Annelida (e.g. *Tubifex*, cf. Schleip, 1929; Lehmann, 1956). It is a very typical feature, observed already by Whitman, that animal polar plasm takes the shape of a ring around a core of hyaloplasm, the latter being accumulated by the meiotic apparatus (Figure 43). In principle one might anticipate cortical movements associated with each of the maturation divisions. I have found

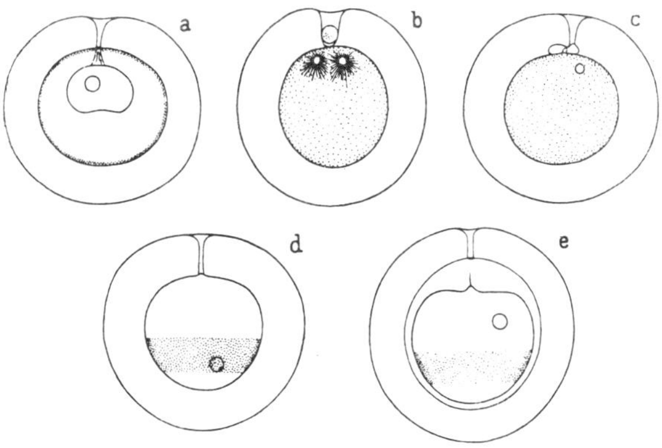

Figure 28. Maturation and fertilization in *Paracentrotus*. (a), optical section of oocyte; (b), surface view of the egg after the formation of the first polar body; (c), mature egg; (d), same after the cortical reaction has occurred with formation of the pigmented ring; (e), fertilized egg after formation of perivitelline space. (Reproduced from W. Schleip (1929), *Die Determination der Primitiventwicklung*, after Boveri, by permission of Akademische Verlag, Leipzig)

no information on this point, which possibly may be related to the fact that only the first one is associated with a visible accumulation of ectoplasm.

One of the most famous examples of cortical movements during maturation concerns the pigment ring formation in *Paracentrotus* observed originally by Selenka and by Boveri (cf. von Ubisch, 1950). In this case the oocyte contains ectoplasmic pigment granules evenly distributed over the surface. Immediately after the formation of the polar bodies the ectoplasmic layer is retracted from both poles, most extensively at the animal pole, thus giving rise to a pigmented ring stretching from slightly above the equator to some distance from the vegetal pole (Figure 28). Apparently this phenomenon involves retraction from, rather than movements towards, the poles. That no violation of the generalization is involved is evident from the fact that withdrawal from one pole automatically implies approach to the opposite one. This particular cortical reaction obviously presumes 'stimulation' at both poles; that at the animal pole is already accounted for, that at the vegetal pole may somehow be related to the presence at this location of the stalk through which the egg is attached to the ovarian wall (Jenkinson, cf. von Ubisch, 1950).

The results discussed here indicate that the process of maturation is associated with a lively activity displayed by the cortical TP polymers. In some cases the ectoplasmic layer is formed only at this stage, and may even disappear again (*Sabellaria*, Fauré-Fremiet, 1924). As was discussed above, this event occurs earlier in *Lymnaea* (Raven, 1966).

Fertilization

Temporally, fertilization is generally considered to last from the moment the spermatozoon establishes contact with the egg proper until the two pronuclei have copulated. As a definition of fertilization this specification is unsatisfactory, for fertilization may have begun already when the sperm penetrates the egg membrane, and some of the extranuclear phenomena may go on beyond the stipulated time interval. For the purpose of definition it may rather be stated that fertilization is a series of events which, triggered by the fertilizing sperm, initiates the process of embryogenesis. In some cases maturation is also released by the spermatozoon, and it is therefore necessary to specify that the events mentioned do not include the meiotic divisions. It must be emphasized that penetration of a spermatozoon in no way ensures that fertilization occurs, for if the egg has not reached the state of fertilizability, numerous sperms may enter the egg without releasing the fertilization reaction.

Applying the suggested definitions of maturation and fertilization, it appears that various degrees of overlapping may occur between these two phenomena. On the basis of observed differences in temporal relation, a number of fertilization types may be distinguished. In the present section we shall first outline the characteristics of these types. Among the complex of events initiated by the spermatozoon, the most conspicuous is the union of the male and the female pronuclei. However, besides this nuclear phenomenon, a number of cytoplasmic and cortical changes occur in association with fertilization which are of utmost importance for embryogenesis. As for maturation, it holds that to the extent these processes are dynamic, they may be referred to the activity of TP polymers present in cytoplasm, cortex and egg membrane. We shall here discuss separately the cytoplasmic and cortical elements of the fertilization reaction.

Fertilization types

It appears that Wilson (cf. 1928) was the first to point out that different fertilization types may be distinguished on the basis of the temporal relation between maturation and fertilization. The extreme cases, according to Wilson, were represented by the egg which is fertilizable only after the expulsion of both polar bodies, and the one in which the maturation does not begin, except possibly for the rupture of the germinal vesicle, before the sperm has fertilized the egg. Accounting for the various intermediate cases, Dalcq (1957) has established the following four fertilization types:

I. Fertilization before breakdown of the germinal vesicle.
II. Fertilization when the first meiotic apparatus is attached to the animal pole cortex.
III. Fertilization when the first polar body has been formed, and the second meiotic apparatus is attached to the animal pole cortex.
IV. Fertilization after the second polar body has been formed.

In the discussion of maturation it was pointed out that occasionally the animal pole may not be determined in the egg before fertilization. If rupture of the germinal vesicle and formation of the meiotic apparatus occurs under these circumstances, the latter must remain in the cytoplasm until the animal pole is established. This situation represents a special case, situated between the types I and II listed above; it is therefore possible to distinguish five fertilization types, which from now on will be designated by the arabic numerals 1 to 5 (Figure 29).

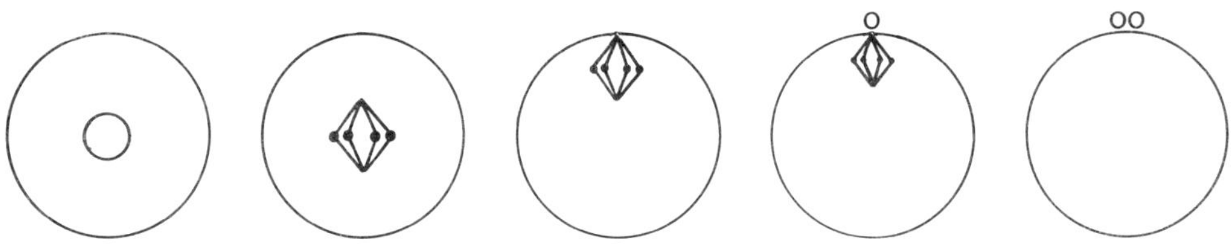

1 2 3 4 5

Figure 29. Fertilization types. The figures represent the stage at which the egg is fertilizable. The sperm has not been indicated because no general rules can be established with respect to the point of the egg surface where the spermatozoon enters in the various cases

Before we consider the five types in more detail, it must be pointed out that the nuclear aspect of the fertilization reaction, the fusion of the pronuclei, obviously cannot occur before the maturation is completed. Thus, in the types 1–4 the sperm nucleus either lies quiescent in the cytoplasm or migrates towards the animal pole while the maturation is brought to an end.

It may often be difficult to distinguish between type 1 and types 2 and 3. Thus, in certain Mollusca, fertilization normally occurs before the rupture of the germinal vesicle, but the latter phenomenon may occur in the unfertilized egg (cf. Raven, 1966). If delayed fertilization suffices to ensure the formation of a normal meiotic apparatus, the egg obviously belongs to type 2 or 3, even if fertilization usually takes place before the breakdown of the germinal vesicle. It may be mentioned that in *Ostrea* exposure to sea water may initiate maturation, but no meiotic apparatus is formed (Inaba, 1936); this egg thus belongs to fertilization type 1. Granting that exposure to the surrounding medium may cause the breakdown of the germinal vesicle and other maturation reactions in types 2 and 3, it is evident that some eggs may belong to type 1 because fertilization is internal.

Fertilization type 1

In this case the spermatozoon stimulates the inception of both maturation and fertilization. The elevation of the cortical tension occurring in many eggs during maturation thus cannot be a prerequisite for fertilization. Since this phenomenon may correspond to a charging of the plasma membrane, necessary for the execution of the various cortical movements associated with fertilization, the latter may not be anticipated, unless the required tension obtains at the outset. As a matter of fact it appears that often ectoplasmic movements are inconspicuous in this type of fertilization.

Ectoplasmic translocations elicited by the fertilizing sperm may be instrumental in the establishment of the animal pole. In the absence of this kind of process, it may be presumed that at least the animal pole is determined in the egg. The presence of the animal pole before fertilization may be demonstrated by the attachment of one pole of the first meiotic apparatus to the cortex. In fertilization type 1 this possibility is excluded, and since excentric location of the nucleus may be incidental, it is difficult to demonstrate a polarity unless it is revealed by visible cortical or cytoplasmic features. In the classical example of this fertilization type, *Parascaris*, predetermination of the animal–vegetal polarity appears to obtain (cf. Schleip, 1924), involving among other features a micropyle at the animal pole.

Fertilization type 2

When no distinct cytoplasmic polarity obtains, the first meiotic apparatus will arise near the centre of the egg. If an animal pole is present, one of its ends must become attached here, according to the contention advanced above. The present fertilization type is distinguished by the fact that the meiotic apparatus does not move from its original location, indicating that no animal pole prevails. On the basis of the preceding discussion we may conclude that in this case the cortex contains a more or less uniform layer of ectoplasm, impeding the attachment of microtubules to the plasma membrane.

This state of affairs obviously must be changed by the cortical reactions set off by the fertilizing spermatozoon. Formation of an animal pole requires that the ectoplasmic layer is removed from one part of the cortex, and since the cortical movements generally are directed towards a point of stimulation, it follows that formation of an animal pole implies that fertilization must occur at the diametrically opposite point of the surface, i.e. at the vegetal pole. In spherically symmetrical eggs the animal–vegetal polarity may be determined in this way, the sperm entering either at a random point of the surface, or if the egg membrane is impermeable to the spermatozoon, at a place predetermined by the location of a micropyle.

The former case may be anticipated to prevail in various Annelida, where the eggs develop in the body cavity in clusters, the central cell of which becomes the ovum at the expense of the surrounding cells. It appears that under these circumstances the external influences which can determine an animal–vegetal polarity are missing until the sperm establishes contact with the surface. The second possibility is realized in *Unio*, as appears from the following quotation: 'The egg of *Unio* is most definitely oriented. The orientation is not due to the arrangements of the deutoplasmic materials, for the yolk-granules have no definite abiding-place in this egg. Yet, . . . not one was found in which the polar globules [bodies] were not formed opposite to the micropyle, which thus occupies the center of the vegetative pole. It is here, of course, that the spermatozoön enters' (Lillie, 1901, p. 228). This observation certainly suggests that the orientation, i.e. the predetermination of the animal–vegetal polarity, involves only one point, the location of the sperm entrance, all the other aspects of the polarity being consequences of this circumstance. It may be mentioned that Lillie's illustrations clearly show that following fertilization a retraction of ectoplasm occurs at the presumptive animal pole (Figure 30).

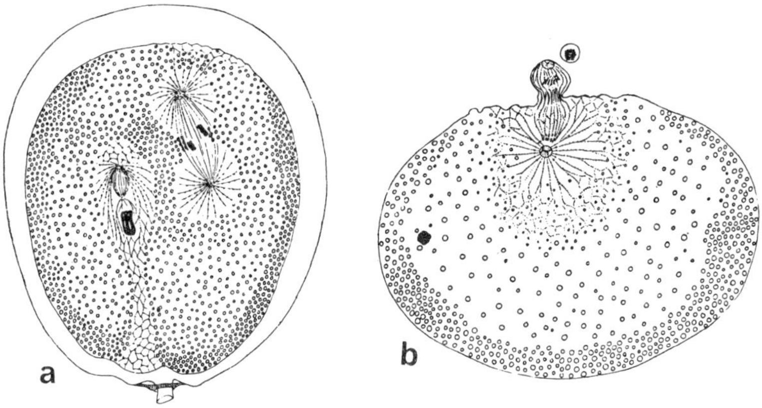

Figure 30. Fertilization and maturation in *Unio*. (a), early fertilization, showing the egg membrane with micropyle, the first meiotic apparatus and the sperm pronucleus, preceded by an amphiaster; (b), formation of the second polar body. An accumulation of yolk granules towards the vegetal pole has occurred between the two stages. Unfortunately, no illustration exists of the egg before fertilization is initiated. (Reproduced from F. R. Lillie, *J. Morphol.*, **17** (1901), by permission of The Wistar Press, Philadelphia)

The cases discussed are probably rather exceptional; most often a polarity is predetermined, the sperm acting only as a triggering mechanism. The location of the fertilization may consequently occur at any place at the surface except, for some unknown reason, not too close to the animal pole. This is the situation obtaining for instance in *Crepidula* (Conklin, 1902).

When fertilization occurs at the vegetal pole, formation of vegetal polar plasm may be anticipated, followed possibly by accumulation of animal polar plasm in association with the expulsion of the polar bodies. If the sperm penetrates elsewhere, the possible formation of vegetal polar plasm must be referred to a predetermined polarity in the egg.

This fertilization type has not previously been distinguished from the following one; I believe that the present analysis will reveal that they are fundamentally different. It is possible that type 2 is just as frequently encountered as type 3.

Fertilization type 3

If an animal pole prevails in the unfertilized egg, the first meiotic apparatus becomes attached to this location. If formation of polar bodies does not occur, it must be presumed that the plasma membrane at the animal pole is somehow immobilized, otherwise there is no obvious reason why the meiotic division should not begin. It is reasonable to assume that microvilli, extending from the egg surface into the egg membrane, are responsible for the cortical immobilization. The latter is abolished when the association between egg and membrane is severed at fertilization (cf. Wartenberg and Schmidt, 1961; Rebhun, 1962). Whether it is possible to substantiate the generalization that all eggs in this group possess microvilli cannot be decided for the time being.

It was suggested that in the preceding fertilization type spherical symmetry may obtain before fertilization. This possibility is very unlikely in type 3, since this would imply that microtubules might adhere to any point of the cell cortex. As we shall discuss in the following chapter, attachment of microtubules to a narrow cortical region is an unconditional requirement for the extremely unequal cell division leading to the formation of the polar bodies; if the microtubules can adhere everywhere, the cell division will be more or less equal.

As the animal pole, and probably most often the animal–vegetal polarity, is predetermined in this fertilization type, the location of the sperm entrance is not of importance for the determination of this parameter, and it may therefore occur at or outside the vegetal pole, as the case may be.

A large number of invertebrate eggs, notably those of Annelida and Mollusca, belong to either type 2 or 3. It is interesting to note that in some cases cortical movements, leading to formation of an animal pole, occur after deposition of the egg, but before fertilization; the egg thus passes from type 2 to type 3. This is illustrated in *Chaetopterus* (Mead, 1898; Lillie, 1906), where ectoplasmic translocation towards the vegetal pole takes place (Figure 31). The initial animal–vegetal polarity which must be responsible for the ectoplasmic movements is clearly visible in the unfertilized egg.

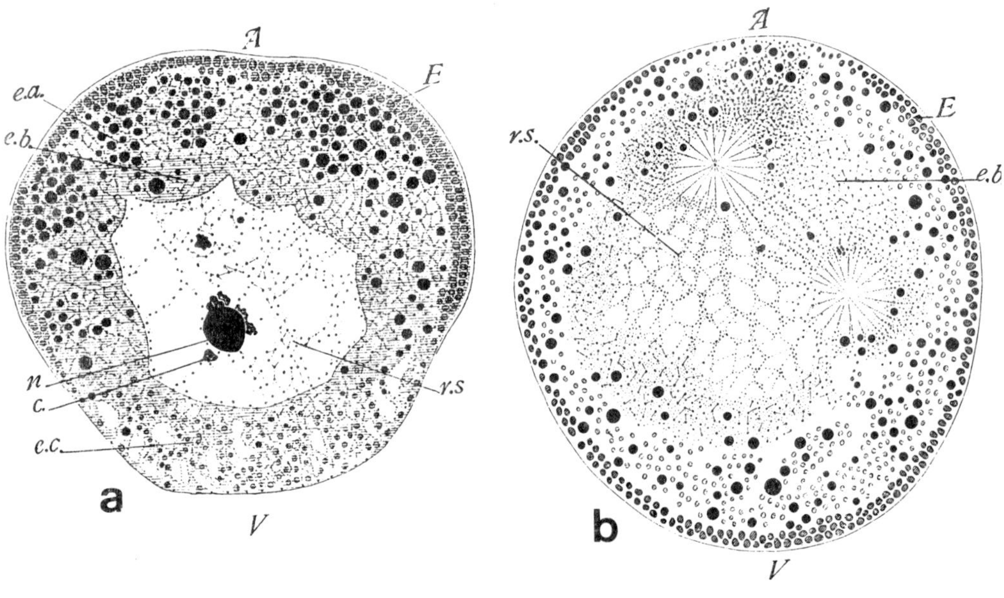

Figure 31. Fertilization in *Chaetopterus*. (a), oocyte located in the ovarian epithelium; (b), oocyte three minutes after fertilization. The illustrations show clearly that the particles of the ectoplasm (E), initially located in the animal part of the egg, after fertilization are moved towards the vegetal pole (V), leaving a region free beneath the animal pole (A). c., chromatin; e.a., e.b., and e.c., various cytoplasmic constituents; n, nucleolus; r.s., residual substance of germinal vesicle. (Reproduced from F. R. Lillie, *J. Exptl. Zool.*, **3** (1906), by permission of The Wistar Press, Philadelphia)

Fertilization type 4

In this type the animal pole must be determined before fertilization. It may arise as a result of cortical movements provoked during the incipient phases of maturation, but the fact that this process becomes arrested after the formation of the first polar body rather suggests that the cortex is relatively immobile. This fertilization type is especially characteristic of the Vertebrata, in the eggs of which the cortical immobility may safely be referred to the presence of microvilli.

The main difference between this and the preceding fertilization type appears to be that in the latter the cortical movements are completely suppressed, so that not even formation of the first polar body is possible. The comments on polarity and location of sperm entrance are valid even in this case.

Fertilization type 5

The completion of the meiotic divisions shows that the animal pole is predetermined, and that no inhibition of cortical movements prevails. In the sea urchin, where this fertilization type is prevalent, the egg membrane is extremely thin, about 10 nm. This circumstance may possibly explain the lack of cortical immobilization, in spite of the presence of microvilli (cf. Afzelius, 1956). Although a micropyle is present, the sperm may enter at any point of the surface, except at the poles. A survey of the distribution of the fertilization types among various taxa is presented in Table 1.

Table 1
Distribution of the fertilization types among the animal phyla*

Type 1	Types 2 and 3	Type 4	Type 5
Porifera	*Platyhelminthes*	*Chordata*	*Porifera*
Platyhelminthes	*Nemertina*	(Cephalochordata,	*Cnidaria*
Aschelminthes	*Mollusca*	Vertebrata)	*Echinodermata*‡
Mollusca	*Sipuncula*	Certain Invertebrata	
Echiura	*Annelida*		
Annelida	*Arthropoda*		
Arthropoda	(Insecta)		
(Crustacea)	*Chordata*		
Chaetognatha	(Tunicata)		
Certain Mammalia†			

*Based on data in Wilson (1928), Just (1939), Rothschild (1956), Dalcq (1957) and Runnström, Hagström and Perlmann (1959).

†cf. Rothschild (1956).

‡In the references quoted, Asteroidea are classified as belonging to type 1. However, although in this case maturation is not initiated in the ovary, as in Echinoidea, it is carried to completion upon shedding into sea water (Monroy and Tyler, 1967). It therefore belongs to type 5 like the other Echinodermata.

Cytoplasmic events

At this stage of the exposition the reader will be familiar with the notion that under the present heading only one kind of activity can be expected, viz., that executed by microtubules. We also know that in eggs usually the microtubules are present in the form of asters, and that these owe their formation to centrioles. Under normal conditions the latter are duplicated once in every cell division cycle, thus ensuring that each daughter cell contains a centriole. At fertilization a number of complications upset this simple pattern. We shall devote the first section to a discussion of this intriguing question. In association with fertilization the asters, or more generally the microtubules, are concerned mainly with moving around the pronuclei; these events also are of great interest and will be treated in the second section.

Centrioles

Provided both egg and sperm have a centriole, the fertilized egg will possess two, and if they duplicate in the normal course of events, it follows that each cell of the adult organism, including presumptive eggs and spermatozoa, must have two centrioles. The consequence of this possibility for subsequent generations permits the conclusion that at some stage before the first cleavage one centriole is lost or otherwise inactivated, or else that one step of duplication is suppressed.

It must be emphasized that, in agreement with the contention advanced previously, it is presumed that no real aster is ever formed in the absence of a centriole. When a meiotic apparatus is formed in the absence of centrioles, it must be anastral. According to Wilson (1928), an anastral meiotic apparatus is found in Nematoda and Copepoda (fertilization type 1), in Insecta and Urochordata (types 2–3) and in Vertebrata (type 4), whereas the meiotic apparatus is astral in Porifera (types 1 and 5), Cnidaria (type 5), Platyhelminthes, Mollusca and Annelida (types 1, 2 and 3) and Echinodermata (types 1 and 5). Obviously no clear-cut correlation obtains between the nature of the meiotic apparatus and the fertilization type.

One way to account for the loss of the egg centriole is to assume that it becomes attached to the cortex at an early stage of oogenesis and that, failing to divide before the first maturation division, it is lost with the first polar body. At this location the centriole may be instrumental in creating a cytoplasmic polarity by accumulating hyaloplasm near the animal pole. It may also be responsible for transporting the germinal vesicle towards the animal pole; an indication of this kind of activity in *Sabellaria* has been reported by Waterman (1934). If, in contrast to the normally prevailing situation, the microtubules can adhere to the cortex outside the animal pole, a tension may develop

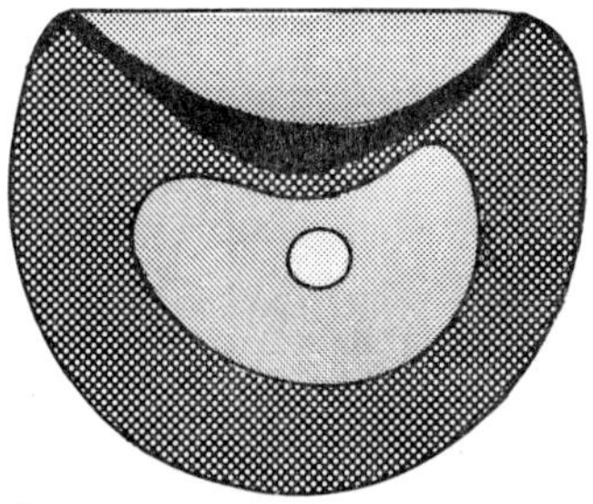

Figure 32. Egg of *Urechis*, showing the peculiar concavity at the animal pole, which disappears when the maturation division begins. (Reproduced from A. Tyler, *Biol. Bull.*, **60** (1931), by permission of Marine Biological Laboratory, Massachusetts)

tending to pull the pole inwards, since all the microtubules originate in this point. The pecular shape of the egg of *Urechis* may be accounted for by this possibility (Figure 32).

A number of expedients may be envisaged to ensure that the normal state prevails as regards the centrioles:

(1) The egg contains no centriole, an alternative which, if the meiotic apparatuses are astral, is compatible only with fertilization type 1. After appropriate replication the sperm centriole may be responsible for the formation of all the asters present in the maturation and cleavage spindles. This mechanism has never been observed. Another possibility is that the two meiotic apparatuses are anastral, while the sperm aster answers for the formation of the mitotic apparatus of the first and subsequent cleavage divisions.

(2) The egg contains one centriole which, as mentioned, is located at the animal pole, whence it ends up in the first polar body. It is immediately evident that the inner branch of the first, and both branches of the second, meiotic apparatus must be anastral. This mechanism cannot prevail in fertilization types 2–3, and in some instances of type 1, where an astral meiotic apparatus is formed in the cytoplasm. Probably this mechanism is common among Vertebrata.

(3) The third possibility is that the centriole divides normally preceding the first maturation division, whereas the second time it fails to divide. Under these conditions it may be anticipated that the inner branch of the second meiotic apparatus is anastral. I have found no records of this mechanism, but the other possible alternative, viz., that the role is taken over by the sperm centriole, has been observed in *Lymnaea* (cf. Raven, 1966).

(4) Common to all the cases described so far is the fact that the sperm centriole must be made responsible for the formation of the mitotic apparatus. If, however, the egg centriole divides prior to the second meiotic division, one will be present in the mature egg. Occasionally this centriole may even divide during the maturation division. The presence of a centriole in the mature egg has been reported for eggs of several phyla, belonging to all fertilization types except, apparently, type 4.

Under these circumstances three alternatives exist. The most usual one, reported for many different eggs, is that the egg centriole, or aster, disappears without a trace, the sperm aster forming the amphiaster of the first cleavage division. The second possibility is that the sperm centriole is eliminated. In *Unio* a sperm amphiaster is formed, but it disappears before cleavage begins, supported apparently by a mitotic apparatus formed by the egg centriole (Lillie, 1901). Finally, if the centriole of the proximal pole of the second meiotic apparatus and the sperm centriole both fail to divide, they may join forces to form the amphiaster of the first cleavage division. This situation has been reported for *Crepidula* by Conklin (1902).

(5) A final possibility is that no sperm aster is ever formed, as reported for *Cumingia* and *Limax* (cf. Raven, 1966). This situation may also prevail in the mouse (Rothschild, 1956), and is here supported by the fact that no rotation of the sperm occurs. It is evident that this observation implies that the mouse egg cannot conform to the generalization according to which the meiotic apparatus in Vertebrata is anastral. The circumstance that the polar bodies in the mouse are rather large is interesting in this context because, as will be discussed in the next chapter, this phenomenon generally presumes the presence of an astral meiotic apparatus.

The proposition that the egg centriole may end up in the first polar body has some consequences which have been borne out by experimental observations. Tyler (1932) found that upon exposure to dilute media the eggs of *Urechis* become activated, as an outcome of which two, one or no polar bodies are formed. The two latter cases are rare, but the remarkable fact is that when no polar bodies are formed, the eggs may develop parthenogenetically into completely normal trochophores. It will be realized that, according to the suggested mechanism, the latter are the only eggs in which a centriole is present that may take over the function normally exerted by the sperm centriole. Similar observations have been made on eggs of *Cumingia* (Morris, 1917) in which, as mentioned above, it has been reported that no sperm aster is formed. It seems difficult to reconcile these findings. In cases where replication of the egg centriole occurs, i.e. where the meiotic apparatuses are astral, polar body formation in association with activation should have no influence on the developmental potential. This expectation has been corroborated in many cases, e.g. *Mactra*, *Thalassema*, *Asterias* (cf. Tyler, 1931).

It must be emphasized that attachment of the centriole to the animal pole does not necessarily entail that it ends up unreplicated in the first polar body. The observations on micropyle formation in Echinodermata, involving the formation of a minute protuberance with intracellular fibres radiating from its base, strongly suggest the presence of a centriole (Figure 33; for further references, cf. Lindahl, 1932a). If, as stated by Wilson

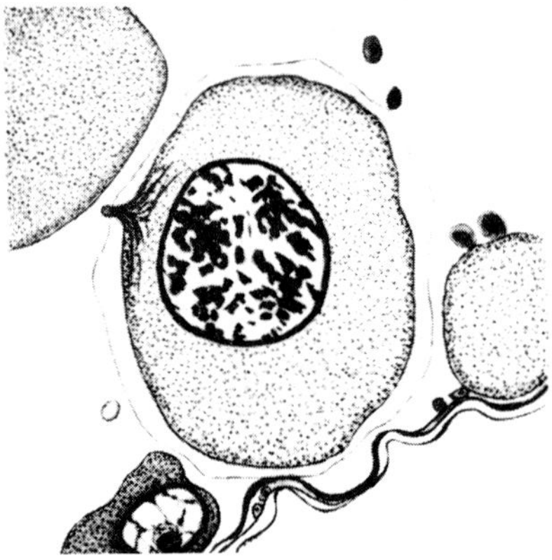

Figure 33. Micropyle formation in an oocyte of *Paracentrotus*. The structure located at the animal pole is strongly reminiscent of a centriole. (Reproduced from P. E. Lindahl, *Arch. Entw.mech.*, **126** (1932), by permission of Springer, Berlin)

(1928), astral meiotic apparatuses are formed, it must be presumed that the cortical attachment in this case does not interfere with the replication of the centriole.

Pronuclear movements

Under the conditions reigning in the oocyte in which maturation has been initiated, centrioles will give rise to the formation of asters. The forces generated by the astral microtubules jointly act upon the centre, and whenever the resultant force is different from zero, the position of the latter will be changed. If, as contended above, the force developed is proportional to the sum of monomers involved, i.e. to the number and length of the microtubules, the equilibrium position will in a spherically symmetrical egg lie in the geometrical centre, provided the cortex is uniform with respect to attachment of microtubules. In radially symmetrical eggs the centre must end up in the line connecting the poles; if the polarity involved properties interfering with microtubular activity, the centre must be displaced towards the end where the density is highest. It is suggested that this approach to the equilibrium position of the aster is the only mechanism—apart from the formation of filopodia and microfilamentous plasmagels—through which cytoplasmic movements may be achieved. It is easily realized that this activity comes to an end when the equilibrium position has been reached. It follows that all other dynamic events, movements of hyaloplasm, mitochondria, pigment granules, nuclei, etc., result from the fact that they become attached to the microtubules. Nuclei in general, and the pronuclei of the recently fertilized egg in particular, always come to lie near the astral centre, and the translocation of the former therefore rather closely corresponds to that of the latter.

When the sperm enters the egg, its centriole will be located behind the presumptive pronucleus. It may be envisaged that when microtubules are first formed, they will push the sperm head inwards perpendicular to the egg surface along the so-called *penetration path.* As the microtubules come to span a larger and larger surface, it may happen, either fortuitously or because the cortex is not uniform, that the pull in one direction is stronger than in another, giving rise to the often noticed *sperm rotation.* Thus, according to this notion, the sperm pronucleus may be moved about in the absence of a sperm aster, for instance by the egg aster, but rotation will never occur under these conditions.

From the preceding discussion it follows that the sperm aster, depending upon the circumstances, may migrate either towards the centre of the egg, or towards the region of the egg, generally at the animal pole, which is privileged with respect to formation and adhesion of microtubules. In the former case, the microtubular contact with and the ensuing translocation of the egg pronucleus will vary, depending upon its initial location relative to that of the sperm entrance. The diagrams published by E. L. Chambers (1939) correspond to the movements predicted by the mechanism outlined here (Figure 34). The displacements of the pronuclei occurring from the end of the sperm rotation until their union represent the *copulation path.* If they meet outside the equilibrium point of the aster, they may be translocated together to this position along the *cleavage path* (cf. Wilson, 1928).

The definition of 'copulation path' suggested here is more general than the usual one, according to which it signifies the path 'along which the sperm-nucleus moves towards the point of union with the egg nucleus' (Wilson, 1928, p. 424). Apart from the word 'moves', which ought to be replaced by 'is moved', the criticism which must be raised against this proposition is that it gives the impression that the course of the translocation of the sperm pronucleus is directed towards the egg pronucleus. This is correct only in the absence of a sperm aster; if the latter is responsible for the movements, they, and thus the translocation of the male pronuclei, will be directed towards its own point of equilibrium. When this happens to coincide with the location of the egg pronucleus, the impression may be gained that the male pronucleus is attracted to its partner, as a male to a female moth. The truth of the matter is revealed, however, when the position of equilibrium differs from that of the egg pronucleus; under these circumstances it may be seen how the female partner is callously dragged to the place of meeting.

In some cases no sperm aster is formed, its function being taken over completely by the egg aster. As an animal–vegetal polarity usually obtains, the movements of the

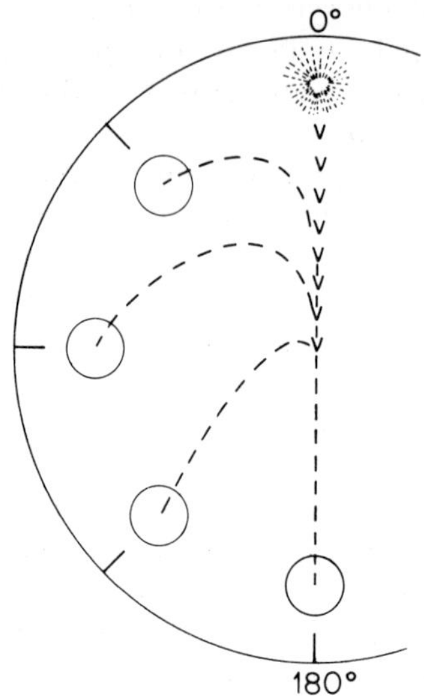

Figure 34. Diagram showing how the copulation path of the egg pronucleus (broken line) varies according to its position relative to the point of sperm entrance. The arrow heads signify the copulation path of the sperm pronucleus. (Reproduced from E. L. Chambers, *J. Exptl. Biol.*, **16** (1939), by permission of The Company of Biologists, Cambridge)

sperm pronucleus will be directed straight towards the animal pole where the egg pronucleus and the centre of the egg aster are situated. In some cases both asters may be present, and interaction may consequently occur. An example of this kind, described by Lillie (1901) as prevailing in *Unio*, is discussed in the following chapter.

The question of the forces responsible for the movements of the pronuclei in the fertilized egg has been the subject of heated discussions, which date back to the last century. The notion that the asters are responsible for this activity has been proposed many times, but just as often rejected. I shall not enter into details of the arguments advanced; references may be found in Wilson (1928) and E. L. Chambers (1939). It must be emphasized once more, however, that the standpoint taken here is the logical consequence of the postulates advanced above. According to these no other means than microtubules can be invoked as the active agent; as we have seen, the actually observed movements can be adequately accounted for on this assumption.

Cortical events

The fertilizing spermatozoon elicits a number of reactions in which the cortex is involved. Among these, the mechanism of sperm entrance is an interesting cell physiological problem which has recently experienced a renaissance. From an embryogenetic point of view it appears to be of less importance, hence it is not covered in the following discussion. Before we deal with the nature of the cortical reaction, it is necessary to call attention to the fact that of the three cortical elements, ectoplasm, plasma membrane and surface coat, the latter is in most eggs replaced by the egg membrane. In general, the egg membrane does not seem to be considered part of the cortex, presumably because it may become separated from the egg surface. It is, of course, solely a matter of terminology whether or not the egg membrane is regarded as part of the cortex, but it may be of some importance because in the latter case it cannot, by definition, be involved in the cortical reactions. I believe that this usage may have had the consequence that the possible contribution of the egg membrane to the cortical reactions has been largely, if not completely, neglected. Contending that this is a regrettable mistake, I propose in the following pages to consider the contribution of all three elements.

Egg membrane

It has been observed in many eggs that part of the fertilization reaction is a cortical process involving 'a slight deformation of the surface layers of the egg, which gives the impression that some sort of tension is being progressively released, or that a local band of contraction and expansion is passing around the egg' (J. C. Dan, 1950, p. 402). If in this statement the words 'contraction' and 'expansion' are replaced by 'tension' and 'relaxation', respectively, it is seen that we may be dealing with a phenomenon similar to that postulated to be characteristic of biological membranes. It may be presumed that this reaction presupposes the cortex to prevail in a state of tension prior to the application of the releasing stimulus. As discussed above, it appears that a cortical 'charging', revealed by the rounding up of the egg and the smoothening of the surface, occurs during maturation, conferring fertilizability upon the egg.

The only structural elements which can be invoked to account for this phenomenon are the egg membrane and the plasma membrane. If the fibrous proteins present in the egg membrane are microfilaments, both structures contain TP polymers and it may therefore be presumed that both can be responsible for the generation of the tension. The decision as to the relative contribution of each is hampered by the fact that the two membranes are intimately associated. When it is possible to accomplish a separation, as for instance in the amphibian egg, it turns out that the tension of the egg membrane constitutes an essential fraction of the total (Berntsson *et al.*, 1964).

It appears that the cortical reactions occurring in association with maturation and fertilization may be visualized microscopically by means of dark ground illumination. Thus, it was observed by Runnström (1928a) that the cortex of the echinoid oocyte is white, but that a colour change, passing through yellow to orange, and originating in the vegetal pole, takes place during the phase of maturation. There is good reason to presume that this reaction corresponds to the elevation of the cortical tension discussed above. At fertilization another colour change occurs, from orange to silver-white, spreading over the surface from the point where the fertilizing spermatozoon establishes contact with the surface (Runnström, 1928b). Shortly after this second cortical reaction the egg membrane begins to separate from the egg surface, and this process also starts at the point of sperm attachment, passing progressively over the egg surface (cf. Kacser, 1955).

Before fertilization the egg membrane is tightly joined to the egg surface by means of microvilli. This union must be severed, and it appears that the cortical reaction just described is responsible for this accomplishment. It is easily envisaged that disruption between the two structures may ensue if a 'contraction' wave is conducted in one of them, but nothing would presumably happen if both of them react simultaneously. It is therefore reasonable to presume that only one of them is involved. Rebhun (1962) found that the separated egg membrane is crowded with microvilli, torn off from the egg surface which is perforated correspondingly. To me it seems most likely that under these circumstances the egg membrane is the active, the plasma membrane the passive, component. Incidentally, if the fibrous proteins in the egg membrane are microfilaments, it is easily envisaged that the membranes of the microvilli may be transformed into this other type of TP polymer.

Ancel and Vintemberger (1948) have observed that in *Rana* a radially symmetrical 'contraction' occurs immediately after fertilization, visualized by the fact that the pigmented ectoplasm is moved towards the animal pole. This process is reversible (Figure 35), and according to the views advanced above it therefore cannot be executed

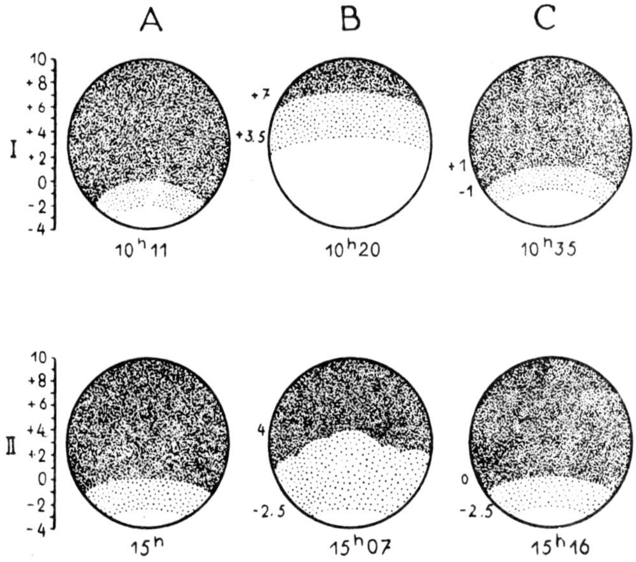

Figure 35. The first radially symmetrical cortical reaction in *Rana*, occurring shortly after fertilization. I and II represent different eggs, A, B and C different stages, the time of which is indicated. (Reproduced from P. Ancel and P. Vintemberger, *Bull. Biol., Suppl.* **31** (1948), by permission of Bulletin Biologique, Paris)

by a TP polymer. It may be presumed that even in this case the egg membrane is the executive agent. It is interesting to note that in contrast to the cortical reaction in the sea urchin egg, it is here oriented in accordance with the polarity of the egg, rather than with the point of sperm entrance. The mechanism proposed here presupposes that contact obtains between plasma membrane and egg membrane. As a matter of fact, if the rotation of the egg within the egg membrane may be taken as an indication of the severance between the membranes, then this point is confirmed, since the egg is free to move only after the occurrence of the second radially symmetrical cortical reaction which leads to the formation of the perivitelline space (Ancel and Vintemberger, 1948; cf. 1965c).

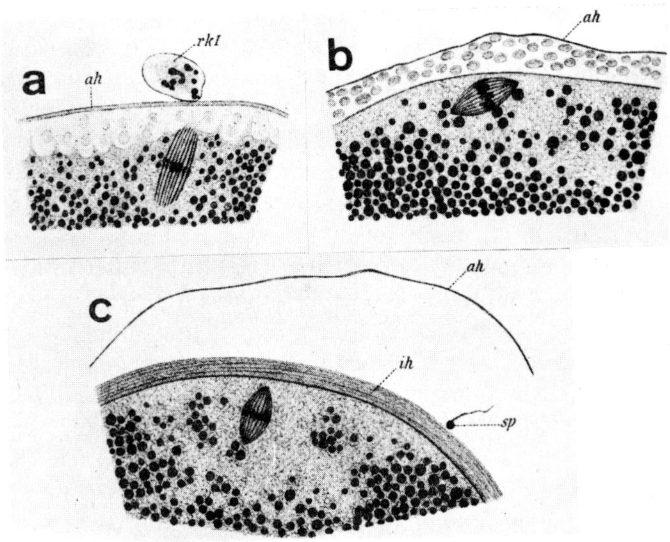

Figure 36. Expulsion of the cortical granules and their transformation into a hyaline membrane in *Branchiostoma*. (a), unfertilized egg with the cortical granules in the ectoplasm; (b), recently fertilized egg with the cortical granules located in the perivitelline space; (c), later stage in which the cortical granules have disappeared, being replaced by a hyaline membrane. ah, egg membrane; ih, hyaline membrane; rkI, first polar body; sp, spermatozoon. (Reproduced from J. Sobotta, *Arch. Mikrosk. Anat. Entw.gesch.*, **50** (1897))

In the discussion of the fertilization types it was observed that maturation may be arrested before the first or the second meiotic division (types 3 and 4). This circumstance shows that the egg surface remains in a state of inhibition, the physical cause of which was suggested to be the association with the egg membrane. It will be realized that in the absence of an egg membrane, no arrestment may obtain, and consequently it might be anticipated that after initiation of the maturation, development may continue without fertilization, provided, of course, that the sperm aster is not responsible for the cleavage divisions. It may therefore be natural to expect that in eggs without membranes, and which are not discharged to the environment, the maturation is set off by the spermatozoon. It is in agreement with this reasoning that the eggs of Porifera belong to fertilization type 1, and it might be expected that some groups of Cnidaria would have eggs belonging to this type also. The arrestment observed in fertilization type 5 involves only the cleavage division, but obviously even in this case an egg membrane must be anticipated, whose association with the egg is severed at fertilization. The idea that the egg membrane may immobilize the egg surface was first expressed by Heilbrunn (1920).

Plasma membrane

The separation between the two membranes may, in some cases at least, be associated with accumulation of microvilli in the egg membrane and perforation of the plasma membrane. Observations reporting the anticipated change in permeability may be found in the classical embryological literature (cf. Dalcq, 1928; Rothschild, 1956). Objections may be raised against some of the methods used earlier, but the question has been attacked with modern methods, which agree in showing that in Echinodermata the permeability to phosphate (Abelson, 1947) and to K^+ (Tyler and Monroy, 1959) is temporarily raised, and corresponding changes were recorded for membrane potential and resistance (Tyler, Monroy, Kao and Grundfest, 1956); in Amphibia changes in the permeability to isotopic water have been recorded (1962).

In many eggs the cortical granules disappear at fertilization; this observation was reported first in *Branchiostoma* by Sobotta (1897, cf. Figure 36). It is generally assumed that the collapse of the plasma membrane discussed above has the consequence that the vesicles containing the granules are ruptured, bringing the latter in contact with the external medium. In this connection it must be pointed out that Hagström (1955) has shown that the reaction observable with dark field illumination can be separated in time from the disappearance of the cortical vesicles. The contents of the granules may contribute to an enlargement of the perivitelline space, by oncotic uptake of water, thus elevating the egg (fertilization) membrane. It should be mentioned that, depending upon the relative strength of the tension or rigidity of the membrane, and the oncotic and osmotic pressures involved, the increase of the perivitelline space may involve either uptake of water from the medium or from the egg. In addition, it has been found that in the sea urchin the material from granules contributes to reinforcement of the egg membrane, and to the formation of a hyaline membrane on the surface of the egg (Figure 37, cf. Motomura, 1941; Endo, 1952; Runnström, 1966). In certain eggs the

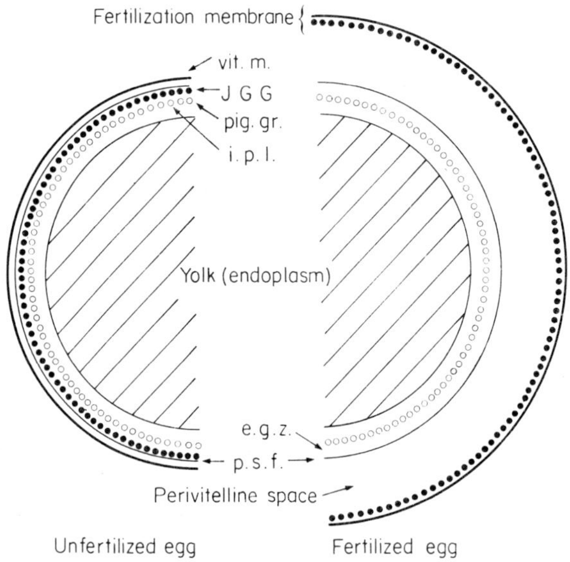

Figure 37. Cortical changes associated with fertilization in the sea urchin egg. It will be observed that all the cortical granules are indicated to participate in the reinforcement of the fertilization membrane; in fact some of their contents form the hyaline layer. e.g.z., i.p.l., ectoplasm; Fertilization membrane and vit.m., egg membrane; JGG, cortical granules; pig.gr., pigment granules; p.s.f., plasma membrane. (Reproduced from J. Motomura, *Sci. Repts. Tôhoku Imp. Univ.* [4], **16** (1941), by permission of Tôhoku University, Japan)

cortical granules do not disappear at fertilization, rather they may become scattered in the cytoplasm. This pattern seems to be typical of Mollusca (cf. Raven, 1966) and may be usual in all spirally cleaving eggs.

Normally, the attachment of the fertilizing sperm initiates a change in the surface which ensures that further spermatozoa cannot enter the egg. Allen (1958) has observed that the spermatozoon does not penetrate the egg until 25 seconds after attachment, at a time when the egg membrane elevation has passed an essential part of the way around the egg (Figure 38). This observation suggests that the block to polyspermy is somehow

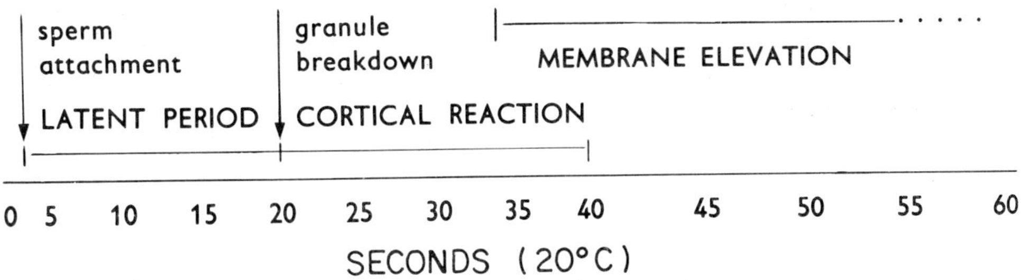

Figure 38. Schematic representation of the time sequence of events associated with fertilization in *Psammechinus*. (Reproduced from R. D. Allen and J. L. Griffin, *Exptl. Cell. Res.*, **15** (1958), by permission of Academic Press, New York)

related to the formation of the perivitelline space, even if it appears to exist before this process is completed (cf. Rothschild, 1956). This interpretation is confirmed by the fact that refertilization may occur if the egg membrane and the incipient hyaline membrane are removed (Sugiyama, 1951; cf. also Monroy and Tyler, 1967).

Polyspermy is no uncommon phenomenon; it is encountered most frequently in

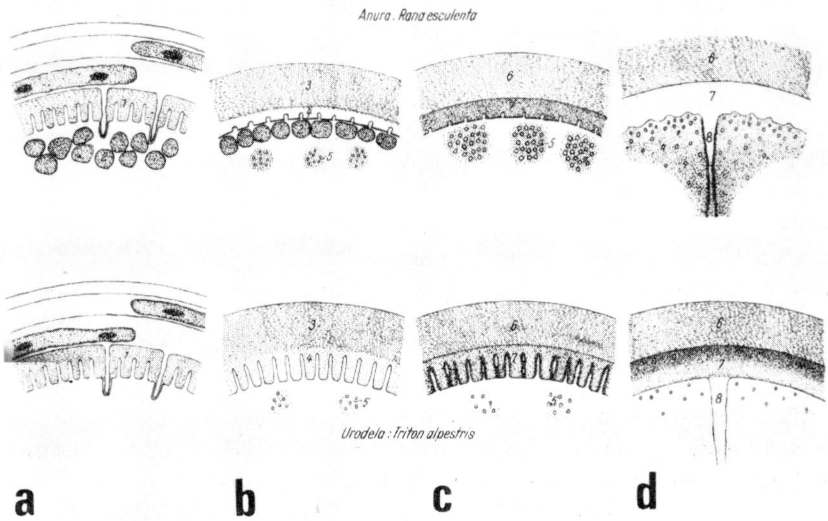

Figure 39. Cortical changes associated with fertilization and cleavage in Amphibia. (a), oocyte; (b), unfertilized egg; (c), fertilized egg; (d), cleavage egg. (Reproduced from H. Wartenberg and W. Schmidt, *Z. Zellforsch.*, **54** (1961), by permission of Springer, Berlin)

animals with internal fertilization, such as Caudata and certain Mammalia. It is possible that this correlation may be referred to the fact that the charging of the cortex ensuing from contact with the medium is wanting, which may have the consequence that the cortical reaction is absent, or less extensive and less rapid than elsewhere. Caudata have no cortical granules, but a hyaline perivitelline space is still formed (Wartenberg and Schmidt, 1961; cf. Figure 39).

No membrane elevation occurs after treatment with trypsin, but yet subsequent fertilization may lead to opening of the cortical vesicles (cf. Runnström, 1966). This observation might weigh heavily against the mechanism advocated to account for the breakdown of the vesicles, if it was not for the 'observation of Afzelius and Minganti . . . [showing] that the vitelline membrane is still present in electron micrographs after trypsin pretreatment before fixation. Apparently trypsin . . . weakens the structure of the membrane so that the cortical granules can be expelled through it . . . , causing "fenestrated membranes" ' (Allen, 1958, pp. 35–36).

Ectoplasm

On a previous occasion we have discussed the phenomenon of cortical movements which involve a simultaneous translation of plasma membrane and ectoplasm. It was observed that movements of this kind are initiated by some stimulation, as for instance penetration of the sperm, and that they are directed towards the point of stimulation. In association with fertilization, accumulation of ectoplasm at the animal pole is observed in Pisces (blastodisc formation), at the vegetal pole in *Styela*, Annelida, etc., and at a point outside the poles in Amphibia.

In some cases ectoplasmic accumulation occurs at locations other than the sperm entrance point, as is best evidenced by the cases where accumulation occurs at more than one point, e.g. at both the animal and vegetal poles. The most common case is probably represented by fertilization type 2, where fertilization usually occurs at the vegetal pole, first giving rise to the formation of vegetal polar plasm, and to an animal pole, at which the meiotic division may subsequently take place, entailing a stimulation which results in the accumulation of animal polar plasm.

The direction of the movements may be determined by an inherent polarity in the egg, independent of either the place of sperm entrance or of polar body formation. This situation possibly obtains in *Chaetopterus*, where accumulation, but not fertilization, occurs at the vegetal pole. Finally, we should mention the well-known case of *Paracentrotus*, where two movements occur simultaneously, with the result that the movements appear to be directed away from, rather than towards, the poles (Figure 28).

Fertilization and activation

Natural parthenogenesis is known in the animal kingdom, but it is a rare occurrence. It has been found, however, that development may be set off by a number of experimental means. Artificial parthenogenesis, or activation, seldom leads to normal embryogenesis, but the phenomenon has nevertheless attracted great attention, because it affords a means by which various aspects of the fertilization reaction may be studied in isolation. An enormous amount of work, particularly in the first quarter of the present century, has been devoted to the study of the phenomenon of activation, and although some of the experimental approaches are unsatisfactory from a modern point of view, there is nonetheless reason to believe that much useful information is hidden in the enormous literature on the subject. It remains to be seen whether improved insight into, among other things, the nature and behaviour of tensible proteins may lead to incorporation of some of these data into an up-to-date theory of fertilization. In the present context it is impossible to discuss the early observations; attention will here be devoted exclusively to a comparison between the phenomena of fertilization and activation, in order to pin down as closely as possible the reasons why activation may in some cases lead to successful development, in others not. A brief discussion of the mechanism of activation will end this section.

Differences between fertilization and activation

Apart from the nuclear aspects, the importance of maturation in many eggs lies in the charging of the egg cortex in preparation for fertilization, and the creation of such cytoplasmic conditions that microtubules may be formed, this latter phenomenon presumably being a result of the breakdown of the germinal vesicle. In all fertilization types except the first these events occur before fertilization. Successful activation of eggs belonging to fertilization type 1 evidently requires that the activating agent also can induce the maturation process.

No male pronucleus is present in the activated egg, and therefore no pronuclear fusion occurs. The cell nuclei consequently contain a haploid genome, a circumstance which in many cases, if not always, is responsible for the developmental block that usually occurs before gastrulation sets in. In certain cases regulation may take place, resulting in diploid and even triploid nuclei (Parmenter, 1940).

We have seen above that sometimes the egg centriole is lost during the maturation divisions, the sperm centriole being responsible for the formation of the mitotic apparatus of the first cleavage division. Under these circumstances development obviously cannot be anticipated to follow upon activation. The predicament may be theoretically resolved if the activating agent leads to the formation of cytasters or if the meiotic divisions are prevented. Both possibilities seem realizable, but the cell divisions effected by cytasters generally are abortive. On the contrary, as we have discussed earlier, suppression of polar body formation may ensue in more or less normal development in *Urechis* and *Cumingia*.

The sperm may elicit a number of reactions, particularly those involving ectoplasmic movements, which influence the organization of the egg. In some cases it may function only as a triggering factor, the orientation of the reaction being determined by inherent properties of the egg. When this situation obtains, there is no reason to believe that the sperm cannot be replaced by an activating agent, the effect of which is exerted on the whole surface of the egg. If, however, the direction of the cortical movements is oriented towards the point of sperm entrance, then it can hardly be expected that activation by dissolved substances can achieve anything but very abortive development. This holds for those eggs of fertilization type 2, in which the animal pole is determined by the fertilizing sperm, and equally well in all those cases where the bilateral symmetry stands in a fixed relation to the point of sperm entrance. Anticipating the discussion of the polarities, it may be stated that it is extremely unlikely that unfertilized eggs ever possess bilateral symmetry. It is rather more likely that the latter arises either by chance, or else under the influence of the fertilizing spermatozoon. I therefore wholeheartedly agree with Tyler (1941) when he expresses the opinion that in activated eggs bilateral symmetry arises by chance. This may be the normal course of events after fertilization, but if participation of the sperm is necessary for the establishment of bilateral symmetry, it may be anticipated that radially symmetrical embryos occur with exceptional frequency after activation, as actually observed in *Urechis* by Tyler (1932). This complication may, of course, be circumvented by the use of locally applied activation, for instance by pricking with a needle (cf. Bataillon, 1929). This statement does not intimate that the success of this expedient depends solely on the restricted localization of the stimulus.

Mechanism of activation

Apart from the aspects discussed in the preceding section, the main function of fertilization seems to be the severance of the association between egg membrane and plasma membrane, accomplished primarily by a contraction of the former, but aided by the oncotic swelling of the contents of the cortical vesicles, spilled into the perivitelline space. It may be expected that this outcome may be achieved by substances or conditions which interfere with the tension or other physical properties of either of the membranes or which influence the volumes osmotically. Evidently, the separation requires either reduction of the egg diameter, or increase of that of the egg membrane, thus apparently through opposite reactions in the two membranes. Under these circumstances it is no wonder that a bewildering range of means have been found to induce activation, including hyper- and hypotonic media, changes in ionic composition and temperature, electric shocks and narcotics as well as many other organic substances.

Loeb (cf. 1964) maintained that cytolysis is a normal aspect of fertilization, a standpoint which, in view of the perforation of the plasma membrane observed to occur, must be accepted as essentially correct. If, as suggested above, this phenomenon results in the opening of the cortical vesicles, and subsequent swelling of the perivitelline space, it is plausible that cytolytic agents may function as activators by causing a partial dissolution of the plasma membrane.

Many authors have been of the opinion that the unfertilized egg exists in a state of chemical inhibition, from which it is released at fertilization. Since the latter may be replaced by activation, it follows that if a mechanism of this kind prevails, the mobilizing effect of the sperm is completely unspecific.

POLARITIES

As discussed in the introduction, the causal postulate states that if differences obtain between two entities, then they either have been different from the outset, or else they have been subjected to different environmental conditions. If we trace backwards the fate of related organs and tissues in a vertebrate embryo, e.g. those of the digestive tract, we shall come to a stage where they are represented by a few endodermal cells, identical to all appearances. If the same were done for epidermis and nervous system, we should end up with another group of cells, constituting the ectoderm. These cells would be identical among themselves, but different from those of the endoderm. From this example it may be inferred that most differences existing in the larval or adult body have arisen in the course of development, but some differences prevail in the earliest stages, and it is easily realized that these are an absolute requirement for embryogenesis, for if the egg was a spherical body, homogeneous in all respects, the outcome of its subdivision would be an aggregate of identical cells which, at best, might form a hollow vesicle. The material basis of the so-called 'polarities' consists of such spatially confined differences in the egg which may influence the course of embryogenesis. In the following pages we are going to present a general discussion of the geometric relations of the polarities, to be followed by a survey of their material basis.

Geometric relations

The polarities are generally designated according to their orientation in the egg, e.g. the animal–vegetal polarity, or in the presumptive larva or adult, e.g. the dorso–ventral polarity. This nomenclature is not consistent, and although it works well in general, difficulties are encountered in some special cases, particularly in non-spherical eggs.

It may therefore be convenient to use names which bear more directly upon the geometric relations established by the polarities. If we start with a spherical, homogeneous body, then a localized difference at any point will define an axis originating on the surface at the specified point, which after passing through the centre will end on the surface at the opposite side. Since the difference lies at one side of the centre, it follows that a vector is determined, and this vector may be considered a formal representation of polarity. It may thus be stated that a polarity obtains whenever a vector is determined by conditions like those outlined above. For reasons which will become evident later it is practical to let the vectors end at the centre of the egg (Figure 40a).

In spherical eggs such a polarity may be determined by the animal pole; in general, the vegetal pole is located at the opposite point of intersection between the polarity axis and the egg surface. Polarities of this kind might be called polar or axial polarities. The former of these names may be rejected for phonetic reasons, but also because the poles in certain eggs, notably non-spherical ones, may remain undefined. The latter argument also applies to the second term, and it is therefore preferable to designate it by the expression *apical* polarity.

It will be noted that if, in a spherical egg, another difference is located on the axis at the opposite side of the centre, a second vector is defined, but no new axis (Figure 40b). According to the definition given above a second apical polarity obtains in this case. This possibility is realized in many eggs; in order to distinguish the two polarities they will be called the *animal–apical* and the *vegetal–apical* polarity, respectively. An apical polarity is seen to impose radial symmetry on a spherical egg.

When an apical polarity is established, any localized difference outside the vector will determine a polarity, represented by a vector passing from a point in the meridian involved to the centre of the egg (Figure 40c). This polarity, which defines a median plane, and thus a plane of bilateral symmetry, may be called the bilateral, or better still the *median*, polarity.

Occasionally a median plane may contain two identical and oppositely directed vectors (Figure 40d). Since another median plane obviously is defined perpendicular to that in which the polarities lie, biradial symmetry is conferred upon the egg.

In the absence of an apical polarity, two different vectors in the same plane will establish bilateral symmetry unless, as in Figure 40(d), they lie on the same axis. This situation might be represented by two vectors which do not end at the centre, but at the central axis (Figure 40e). The two polarities may be specified by using the rather unsatisfactory adjectives *dorsal–median* and *ventral–median*, respectively.

A vector perpendicular to the median plane will define a *dextro–sinistral* polarity which, as a matter of fact, obliterates the bilateral symmetry (Figure 40f). An asymmetry around the median plane may also be caused by circular and spiral 'vectors' in the surface of a sphere and, as we shall see, there is reason to believe that in some cases the latter representation may correspond more closely to actual conditions. However, since the effect always involves a right–left asymmetry, the name proposed above is expedient.

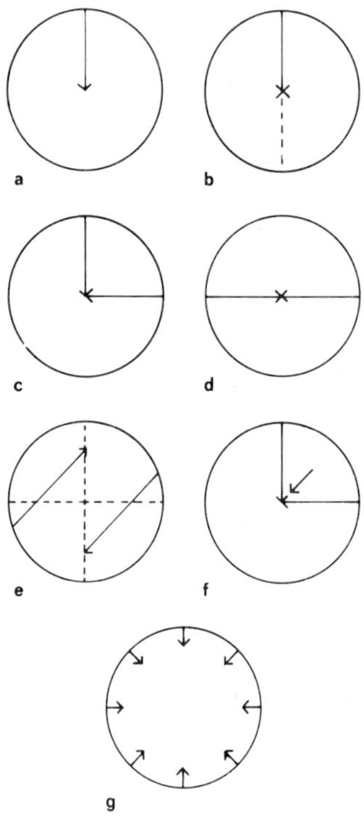

Figure 40. Relations between polarities and symmetry. (a), radial symmetry established by an apical polarity; (b), radial symmetry established by an animal–apical and a vegetal–apical polarity; (c), bilateral symmetry established by an apical and a median polarity; (d), biradial symmetry established by two identical and oppositely directed median polarities; (e), bilateral symmetry established by two median polarities (dorsal–median and ventral–median, respectively) lying on separate axes; (f), bilateral asymmetry established by an apical, a median and a dextro–sinistral polarity; (g), normal polarity

In centrolecithal eggs the yolk distribution defines a polarity which may be represented by vectors issuing from any point on the surface towards the centre. To avoid confusion with the other polarities the vectors must be shorter in this case (Figure 40g). This polarity may be called disto–proximal or radial, but the former term is inconvenient, and the latter is applicable only in spherical eggs. Since the polarity may be presumed perpendicular to any point on the surface, the expression *normal* polarity may be the most appropriate, even if the adjective has connotations which can give rise to misunderstandings. A normal polarity is found in certain eggs at fertilization. It will be realized, however, that this polarity also defines the inside–outside orientation with respect to the surface of the egg. In the uncleaved egg this orientation may not be of much importance, but once the blastula has been formed, the cells may be exposed to different influences at the two sides. This polarity thus does not obtain before the blastula has been formed; it is brought into existence epigenetically through the process of cell division. Here is one instance where it would be futile to look for the material basis of a polarity in the fertilized egg, unless it is considered completely formally to be represented by the surface of the egg.

To the reader acquainted with embryological literature it will be evident that the polarities, as defined here, correspond to what are very often called 'gradients'. This name implies the deployment of some kind of action in a quantitative way along an axis. As we shall presently see, it is possible that the effect of polarities sometimes may be of this type, but most generally they are qualitative, acting in an all-or-none fashion.

In contrast to customary usage, the polarities have been defined without reference to the morphological axes. The reason for this expedient is that it seems warranted to employ two sets of names, one referring to the polarities of the egg and the early embryo, the other to the morphological relations in the late embryo and in the adult. In the latter case it is practical to use the notion of *axis*, specified by the adjectives *antero–posterior*, *dorso–ventral* and, where it is necessary, also *proximo–distal* and *dextro–sinistral*, respectively. It appears that by adopting this nomenclature complications are avoided when, occasionally, the polarities and the axes do not coincide, as happens to be the case in *Branchiostoma* (cf. Conklin, 1932).

The polarities, barring the ubiquitous normal polarity, are of importance for the symmetry relations in the larval and the adult organism. Thus, if an apical polarity obtains in the egg, an antero–posterior axis may prevail in the ontogenetic end product. Similarly, a median polarity ensures the presence of the ultimate differential which, according to the causal postulate, may account for all subsequent differences between the dorsal and the ventral side. Hence, if bilateral symmetry obtains in an animal, it may be inferred that, at some early stage of development, a median polarity has prevailed in the embryo. It is interesting and important to note that this statement does not hold for the fertilized egg, the reason being that epigenetic conversion of polarities may occur. It is easily grasped that polarity conversion does not violate the causal postulate, whereas creation *de novo* certainly does. We may therefore state that the number of axes in the adult organism can never be larger than the number of polarities in the fertilized egg. The possibility exists that the potential effects of a polarity remain unexploited, so that the adult is less complex than allowed for by the organization of the egg; this is a rare occurrence but it seems to be realized in animals where metamorphosis is associated with extensive disorganization of the larval body, as for instance in Polyzoa. It deserves mentioning that Huxley and de Beer (1934) have published an excellent survey of the early work on embryonic polarities.

Material basis

The polarities may be impressed upon the egg through the external force represented by the egg envelopes, or they may reside in the egg proper, in which case they may be bound to either cortex or cytoplasm. Some polarities do not prevail in the unfertilized egg, but rather arise during the process of fertilization. It may be emphasized once more that this circumstance does not warrant the statement that the environmental factors are responsible for creating the polarities, since the starting point of embryogenesis is the fertilized egg, in relation to which the spermatozoon is an internal rather than an external factor. We shall in the present section discuss the polarities established by the egg envelopes, the cortex and the cytoplasm, respectively.

Egg envelopes

It has already been pointed out that egg envelopes, particularly the egg membrane, may influence the shape of the egg. If the geometry of the envelope deviates from that of a sphere, it will impose one or more polarities upon the egg. The best-known examples of this kind are found in eggs of Insecta and Cephalopoda.

Insect eggs may be elongate and almost 'cylindrical', in which case one or two apical polarities may be established. It is well known that the antero–posterior axis and the general form of the larva are determined by the shape of the egg envelope. In certain cases the egg is curved (Figure 41); apical polarities may here be represented by curved

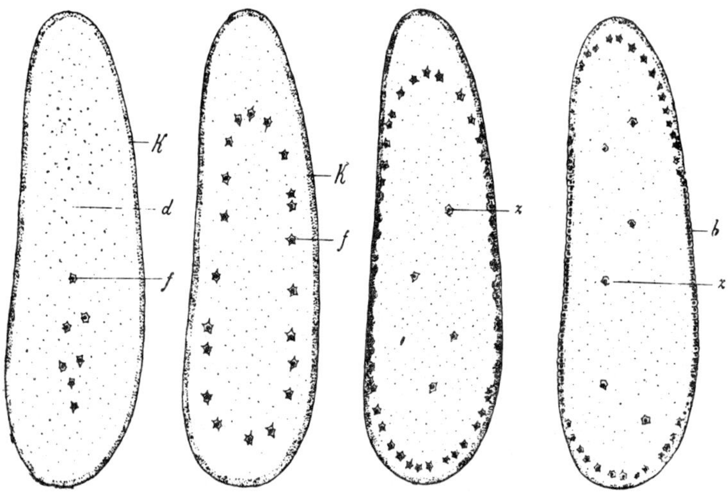

Figure 41. Four different stages of acellular cleavage in *Hydrophilus*. b, blastoderm; d, yolk; f, cleavage nuclei; K, ectoplasm; z, yolk nuclei. (Reproduced from E. Korschelt and K. Heider (1963), *Vergleichende Entwicklungsgeschichte der Tiere*, by permission of Gustav Fischer, Jena (after Heider))

vectors originating at the apices of the egg. Independently of the number of apical polarities, the median plane will be determined by the shape of the egg membrane, and thus also the median polarity and the presumptive dorso–ventral axis.

When, in Cephalopoda, the fertilizing sperm enters the egg through the micropyle at the animal pole, cortical movements are initiated, leading to accumulation of ectoplasm at this location. The germinal disc thus formed is bilaterally symmetrical, extending further down at the longest side, i.e. the side having the smallest radius of curvature (Figure 42). It is seen that through the cortical movements two polarity conversions are

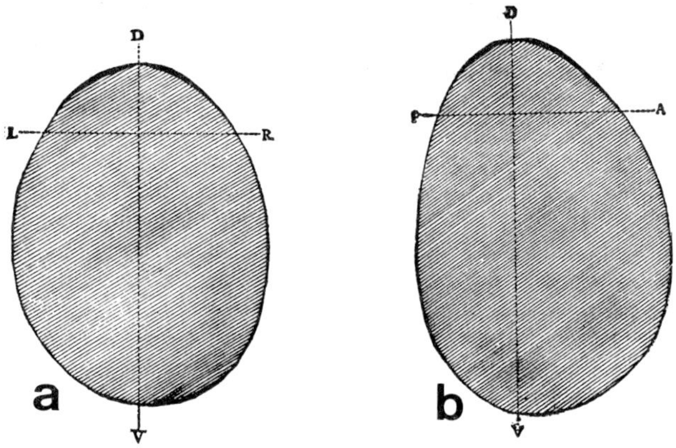

Figure 42. Fertilized egg of *Loligo*. (a), view along the median plane; (b), view perpendicular to the median plane. AP, presumptive antero–posterior axis; DV, presumptive dorso–ventral axis. (Reproduced from S. Watasé, *J. Morphol.*, **4** (1890), by permission of The Wistar Press, Philadelphia)

accomplished in one stroke: the normal polarity, dictated by the ectoplasmic distribution in the unfertilized egg, has become an apical polarity at the same time as the bilateral symmetry of the egg membrane has conferred a median polarity upon the germinal disc. In eggs with discoidal cleavage the apical polarity determines the dorso–ventral axis, the median polarity the antero–posterior axis, whereas usually the inverse relation holds.

It is interesting to note the difference between the two cases discussed here. In Cephalopoda, as in many other eggs, cortical movements are involved in the determination of the polarities. This is impossible in Insecta, where no cortical movements take place, the normal polarity of the ectoplasm being an essential aspect of embryogenesis. The peculiar shape of insect eggs may be a necessity dictated by the suppression of the cortical movements.

The dextro–sinistral polarity was defined as a vector perpendicular to the median plane. As appears from the present discussion, I believe that, with very rare exceptions, unfertilized eggs are never bilaterally symmetrical. Rather, it seems that the median polarity, as a general rule, becomes established at or after fertilization. From this it seems to follow that no dextro–sinistral polarity can exist before fertilization, for in that case it would be a median polarity. However, if the material basis of the polarity is some circular or spiral 'pattern' in the cortex, then it may exist before fertilization, even if it does not become a dextro–sinistral polarity before the median one is established.

Of the various cases of bilateral asymmetry known, only the one observed in Gastropoda has proved accessible to analysis. As observed by Crampton (1894) and Kofoid (1894), the symmetry of the adult depends upon the orientation of the mitotic apparatuses in the third cleavage division (Figure 43). It is seen that whichever force can influence the orientation of these spindles will also determine the symmetry of the adult snail. It

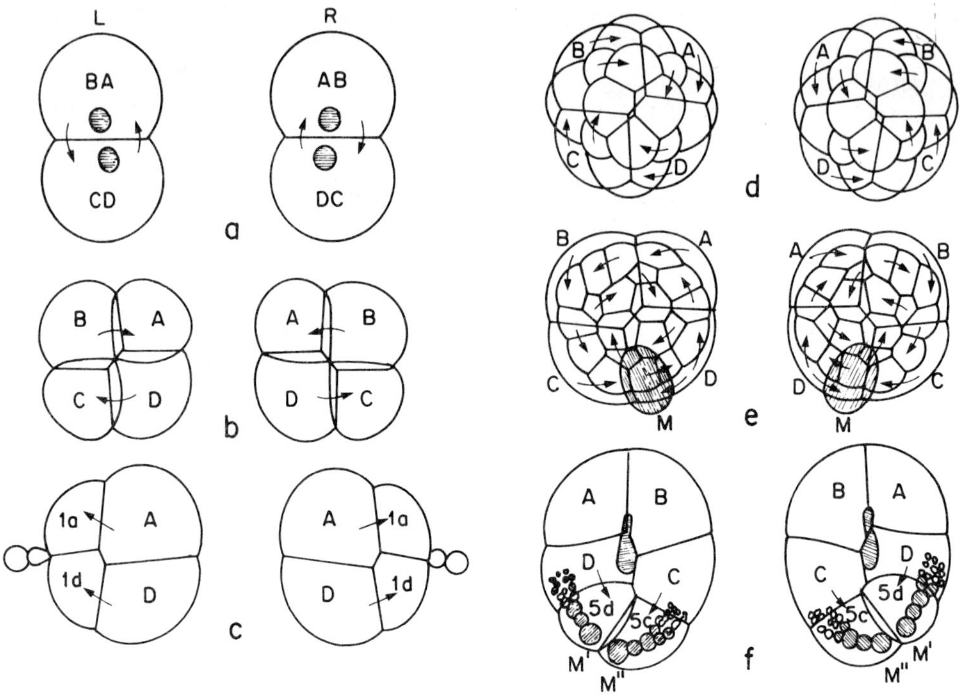

Figure 43. Levotropic (L) and dexiotropic (R) cleavage. (a) through (f), successive developmental stages, seen from the animal pole, except (c), which is seen from the side, and (f), which is seen from the vegetal pole. The arrows indicate the direction of cell division, i.e. the orientation of the mitotic apparatus. A, B, C and D, macromeres; 1a and 1d, the first micromeres; M, the blastomere 4d; M′ and M″, mesodermal cells. (Reproduced from E. G. Conklin, *Anat Anz.*, **23** (1903), by permission of Gustav Fischer, Jena)

may be imagined that a slight torsion in the egg membrane can exert an effect of this kind, provided that the association between egg and membrane is preserved. This situation seems to exist in spirally cleaving eggs, at least during the early stages (Pasteels and de Harven, 1962; Rebhun, 1962). Possibly such a torsion arises in the egg when it is propelled down the oviduct. This point has not been verified, but Conklin (1897) noted that a spiral pattern prevails in the outer shell of the egg of *Urosalpinx* and *Crepidula*, which at least lends some support to the suggested mechanism. Further corroboration may be found in the fact that dextro–sinistral polarity apparently is found only in animals possessing an oviduct. On the basis of hybridization experiments in *Lymnaea* (Boycott and Diver, 1923), Sturtevant (1923) has arrived at the conclusion that the symmetry is determined not by the genes of the egg, but rather by those of the mother. These findings, and particularly the fact that unexplainable exceptions to this rule were observed, may be accounted for if it is assumed that the maternal genes are responsible for the direction in which the ciliary bands in the oviduct are wound, and that chance may to some degree determine the orientation of the eggs in the oviduct.

In the present context may be mentioned some very interesting observations made by Kochav and Eyal-Giladi (1971) on the bilateral symmetry in the chick embryo. It has generally been assumed that the intrauterine rotation to which the egg is subjected is responsible for the determination of this symmetry. The said authors found that by placing the blastodisc in an oblique position normal bilateral symmetry is established. They therefore concluded that gravity is the primary factor in the process of determination, rotation acting indirectly by conferring upon the blastodisc an oblique orientation. It will be realized that this phenomenon may also be made responsible for the determination of the bilateral asymmetry of the embryo. It must be stressed, however, that in cases where gravity can exert effects of this kind there is reason to believe that the substrate of the polarity is to be found in the cytoplasm rather than in envelopes or cortex.

Cortex

The distribution of hyaloplasm plays an important role in the establishment of polarities. As has been discussed above, hyaloplasm may be accumulated in the ectoplasm of the cortex by means of microfilaments, but it may also be condensed in the cytoplasm by means of astral microtubules. Polarities in which hyaloplasm is involved may thus be either cortical or cytoplasmic, and as a general rule they will be classified according to the type of linear TP to which they owe their existence. Conversion from cortical to cytoplasmic polarities becomes possible under these circumstances.

In most unfertilized eggs the ectoplasm defines a normal polarity, and in those cases where its absence allows for the attachment of the first meiotic apparatus to the animal pole, it also determines an apical polarity. Except for the Arthropoda, in which the ectoplasmic layer is stationary, it is usually, as the result of the processes of maturation and fertilization, accumulated in the form of polar plasm at either or both of the poles (Figure 44), where it determines one or two apical polarities as the case may be.

It may be established as a general rule that the vegetal–apical polarity, when present, becomes converted into a median polarity. Usually this is achieved by influencing the course of the first cleavage division; anticipating the discussion of the mechanism of

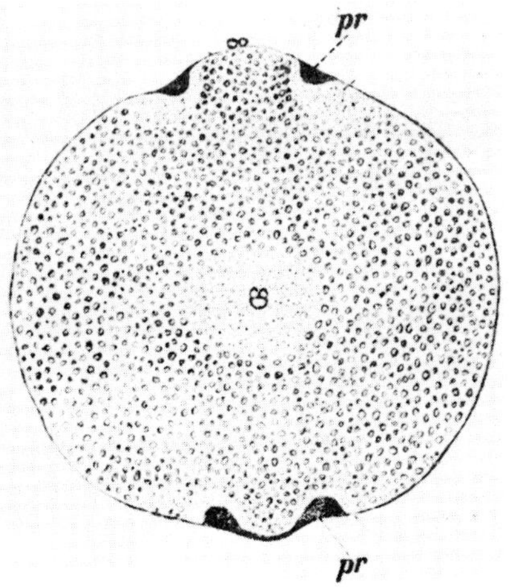

Figure 44. Polar plasms in the fertilized egg of *Clepsine*. pr, ring-shaped animal polar plasm. (Reproduced from C. O. Whitman, *Quart. J. Micros. Sci.*, **18** (1878))

cell division, it may be stated that if interference occurs with attachment of microtubules to the cortex the polarity is cortical, if it affects the formation of microtubules it is cytoplasmic.

Cortical polarities are always concerned with the distribution of ectoplasm, but the latter need not be accumulated at the poles; thus in the sea urchin egg the ectoplasm is withdrawn from the poles, as indicated from the pigmentation in *Paracentrotus* discussed above (Figure 28). This ectoplasmic distribution outlines three distinct cortical regions; we may therefore infer that two apical polarities are present. The influence of these 'gradients' has been borne out in many different ways, not least through the work of Runnström (1929), Hörstadius (1935) and Lindahl (1936).

Certain sea urchin eggs assume an ovoid shape after fertilization, suggesting that the cortical tension is higher in a region corresponding to the pigment ring in *Paracentrotus* (Figure 25). The fact that in eggs of many species neither pigment accumulation nor change of shape can be observed does not necessarily mean that the polarities are absent, for it is not contended that the pigment granules proper must be of embryogenetic importance or that the cortical properties responsible for the polarities must entail differences in cortical tension. It is, however, certain that these phenomena demonstrate the presence of an animal–apical and a vegetal–apical polarity, both bound to the cortex. This is not the place to discuss the mechanism by which these polarities exert their function, but it may be mentioned that they obviously are responsible for the typical cleavage pattern in the sea urchin egg.

The cortical polarities discussed so far have been apical, but median polarities also obtain, the most well known of which is found in the amphibian embryo. The material basis of this polarity has been subject to intensive studies. It has often been asserted that it derives from an unequal distribution of cytoplasm in the unfertilized egg (e.g. Lehmann, 1945), but the extremely careful experimental observations made by Ancel and Vintemberger (1948) have corroborated earlier evidence showing that this polarity is determined by the fertilizing sperm, the median plane passing through the point of sperm entrance and the polar axis. It is apparent that the sperm elicits a cortical reaction which results in the elevation of the pigment border and the consequent formation of the grey crescent at the side opposite to its entrance. Ancel and Vintemberger found that translocation of cytoplasm may be associated with this event, but many observations suggest that the cortical changes are more important than the cytoplasmic ones, even if the latter also may be of significance (cf. Pasteels, 1964).

Cytoplasm

The best-known example of an apical cytoplasmic polarity is found in telolecithal eggs, where the major part of the yolk is found at the vegetal end of the egg. This localization obtains already before maturation, being determined by the position of the egg in the ovary, the vegetal end being the one turned towards the ovarian wall. Whether one or two apical polarities are present can only be decided by observations of the subsequent epigenetic events.

In some cases the cytoplasm of the oocyte may be relatively homogeneous, but the microtubules of the asters formed during polar body formation, as well as those of the sperm aster, may collect hyaloplasm below the animal pole, thereby creating a differential in cytoplasmic composition corresponding to an apical polarity, which may be responsible for the establishment of the antero–posterior axis. A conversion into a median polarity appears to occur in *Unio*, when the egg aster, migrating from the animal pole towards the equator, drags along hyaloplasm to the latter location (Figure 45). For

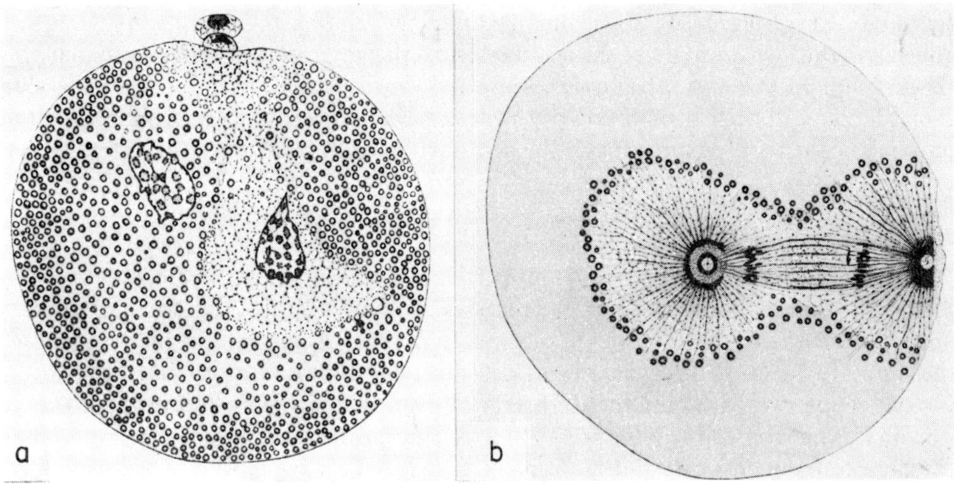

Figure 45. Movement of the egg pronucleus in *Unio*. (a), the pronucleus, surrounded by hyaloplasm, is displaced towards the cortex in a direction opposite to the one occupied by the male pronucleus; (b), the asters of the first mitotic apparatus are of unequal size, one centre being located close to the cortex. (Reproduced from F. R. Lillie, *J. Morphol.*, **17** (1901), by permission of The Wistar Press, Philadelphia)

reasons to be discussed later, the ensuing polarity may give rise to the formation of two cells of different size, which is an overt demonstration of the existence of a median polarity, passing through the centres of the two blastomeres. It will be realized that even if the cells were of equal size, the difference in cytoplasmic composition would suffice to warrant the existence of a polarity, provided it was of embryogenetic importance.

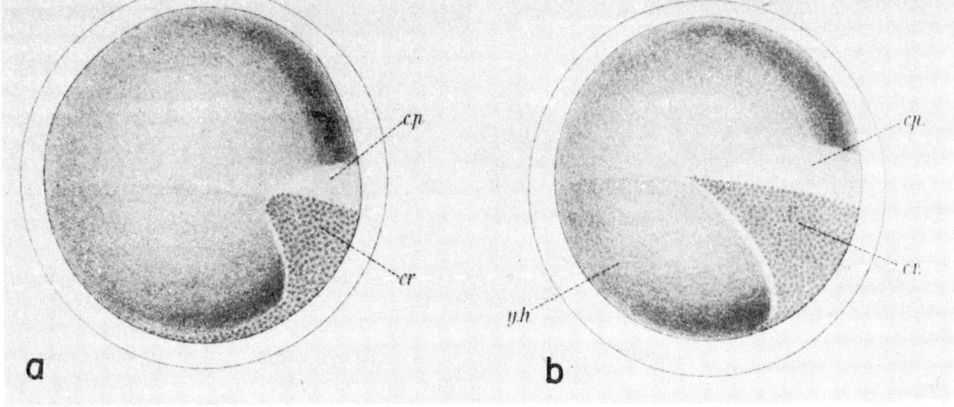

Figure 46. Formation of yellow crescent in *Styela*. (a) and (b), two successive stages, as seen from the side. c.p., clear ectoplasm; cr, yellow crescent; y.h., yolk hemisphere. (Reproduced from E. G. Conklin, *J. Acad. Nat. Sci. Philadelphia*, [2], **13** (1905))

A very instructive example of a conversion of a vegetal–apical cortical into a cytoplasmic median polarity is found in Conklin's description of the development of *Styela* (1905). In this case the yellow ectoplasm, distributed uniformly in the oocyte, is collected at the vegetal pole as a consequence of cortical movements initiated by the fertilizing spermatozoon (Figure 19). From this location the ectoplasm is displaced because it adheres to the microtubules of the sperm aster as this migrates towards the animal pole (Figure 46). In this way a yellow crescent is formed at the presumptive dorsal side, representing a dorsal–median polarity, whereas the grey crescent opposite the yellow one may be the visible expression of the rare phenomenon of a ventral–median polarity.

CELL DIVISION

After the preceding account of the properties of the embryogenetic substrate, we may direct attention to the embryogenetic processes. Among these, cell division is of primary importance because the subdivision of the fertilized egg is an absolute requirement for the normal course of the processes of cell diversification and morphogenesis. In the present chapter we are going to deal first with the mechanism of cell division, trying to explain this phenomenon on the basis of the cellular properties outlined in the preceding chapters. Subsequently, it will be attempted to account for the most important cleavage patterns as the outcome of interactions between the embryogenetic substrate and the process of cell division.

Mechanisms

Before we begin to tackle the problems proper, a terminological digression may be warranted. In all cell divisions a fibrillar body arises in close association with the chromosomes. When present during a normal mitotic cell division, it is called a *mitotic apparatus*. This entity consists of two *asters*, stellate fibrillar complexes, originating each in a *centre*, in the centre of which a centriole is located (Figure 47). Most of the fibres end

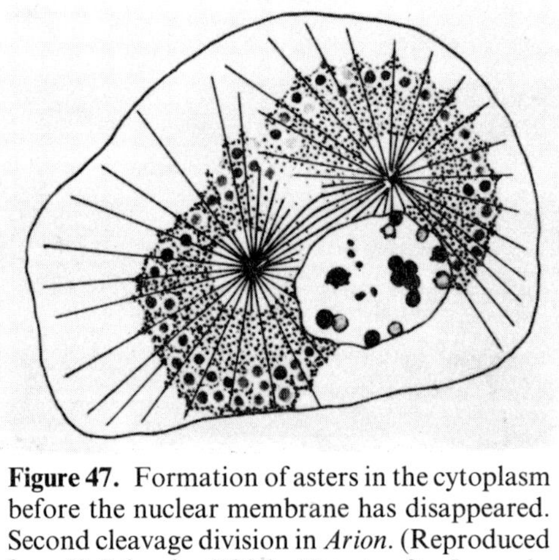

Figure 47. Formation of asters in the cytoplasm before the nuclear membrane has disappeared. Second cleavage division in *Arion*. (Reproduced from F. Schrader (1953), *Mitosis*, after Lams, by permission of Columbia University Press, New York)

in the cell cortex, but from a certain spatial section of the asters the fibres end in the opposite centre or in special structures in the chromosomes called *kinetochores* (Figure 48). The complex formed by the chromosomes and the two astral sections is called the

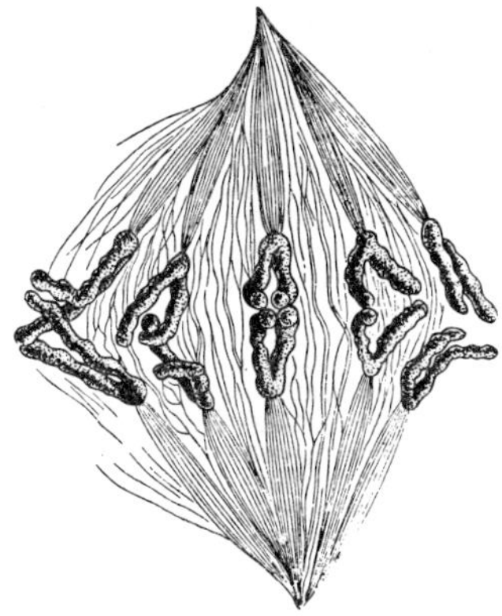

Figure 48. First maturation division in pollen mother cell of *Lilium*, showing that the fibrils, representatives of chromosomal microtubules, are attached only to a limited region of the chromosome, corresponding to the location of the kinetochore. (Reproduced from F. Schrader (1953), *Mitosis*, by permission of Columbia University Press, New York (after Mottier))

spindle. The cortical fibres are called 'astral rays', those uniting the centres 'spindle', 'pole-to-pole' or 'continuous' fibres, and the third type 'chromosomal' fibres. As appears, the centres are also called the *poles* of the spindle. Electron microscopical studies have revealed that the fibres seen in the light microscope are or represent microtubules (cf. de Harven, 1968). This circumstance warrants that the expressions 'fibre' and 'ray' are replaced by *microtubule*. When distinction with respect to the point of attachment is necessary, this may be achieved by means of the adjectives *cortical*, *central* and *chromosomal*, respectively. The structure described here may be called an *astral cell division apparatus*, and, when relevant, a *mitotic apparatus*.

In meiotic divisions this type is sometimes encountered, but as we have discussed above, centrioles are not always present in the unfertilized egg, with the result that no asters can be formed. The cell division apparatus is consequently anastral, the core consisting of chromosomal microtubules, which may unite to form bipolar, multipolar or barrel-shaped complexes (Figure 49). In either case, but presumably with particular reference to the anastral type, the cell division apparatus is called the maturation spindle. This terminology seems rather unfortunate, for when it is astral it comprises more than a spindle, and when it is anastral it does not conform to the general notion of spindle, as defined above, nor is it necessarily spindle-shaped. It therefore appears more consistent to use the expression *meiotic apparatus*, specified when necessary with the epithets 'astral' or 'anastral', as the case may be. As we have seen, the chromosomal microtubules in an

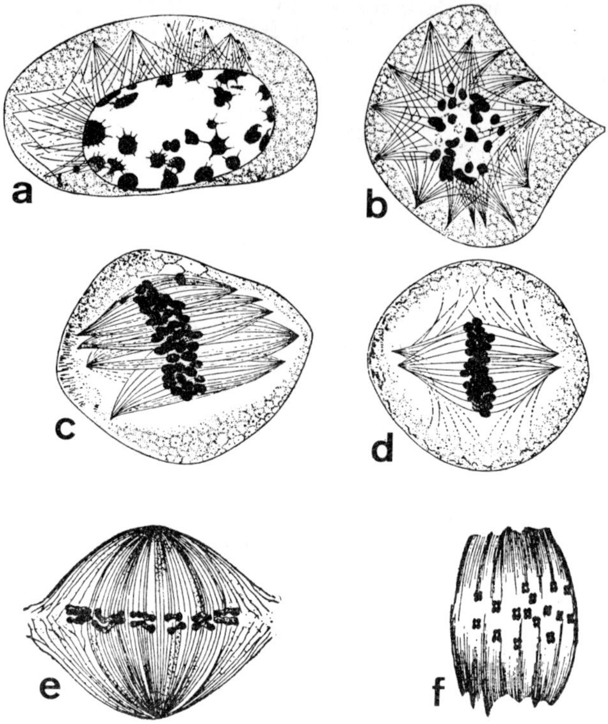

Figure 49. Anastral meiotic apparatuses. (a)–(d), division of spore mother cells in *Equisetum*, showing transition from a multipolar to a quadripolar state; (e), bipolar meiotic apparatus in *Triton*; (f), barrel-shaped meiotic apparatus in *Heterocope*. (From E. B. Wilson (1928), *The Cell in Development and Heredity*, after Osterhout, Häcker and Carnoy and Lebrun, by permission of The Macmillan Company, New York)

anastral meiotic apparatus may end in two or more poles. It seems likely that microtubules originating in the latter, particularly in proximal ones, may be attached to the cortex, so even in this case cortical microtubules may occur.

Mitotic and meiotic cell divisions are known to differ as concerns the nuclear aspects, but as appears from the preceding discussion these two phenomena exhibit fundamentally specific traits in other regards also. For this reason it is necessary, after a discussion of the mechanism of normal cell division, to devote a separate section to the maturation divisions. Certain cleavage laws, or rules of cell division, have been advanced to account for various cell division patterns, especially those observed in early embryos. We shall scrutinize the validity of these on the basis of the mechanisms of cell division advanced here.

MITOSIS

Cell divisions, mitotic and meiotic alike, may be subdivided into the processes of karyokinesis and cytokinesis. The former is concerned with the fate of the chromosomes, the latter with the remaining parts of the cell. These two processes are to some extent independent, since it is possible experimentally to separate one from the other. However, in normal cell division some overlapping obtains as regards the mechanisms involved, and we must therefore deal with both phenomena here, even though karyokinesis is of less direct importance for embryogenesis except, of course, insofar as it ensures that each cell is endowed with the same genome. After the discussion of these events, an attempt will be made to outline an overall mechanism for mitosis. However, before these problems are attacked, we shall first analyse in closer detail the nature of the mitotic apparatus.

The mitotic apparatus

The theoretical approach adopted here turns out to be particularly useful for a rational treatment of the problem of cell division. This process, being obviously a dynamic event, must, according to the postulate advanced above, be effectuated by TP polymers, and it should therefore be possible to explain the phenomenon on the basis of the physical properties ascribed to these cellular elements.

In our discussion of filopodia, it was inferred that the centriole embodies in its structure the information required to form stellate complexes of microtubules. The fact that the centre of each aster invariably is occupied by a centriole suggests that this information may also be exploited for the formation of asters which, in contrast to the filopodial complexes present in sf-cells, are essentially spherical complexes of microtubules. Incidentally, the function of the centriole may be envisaged to consist in the formation of a stellate complex containing a moderate number of microtubules ending in the cortex. Granting this, and the adhesion between mucopolysaccharides and TP polymers, it follows that when microtubules subsequently are formed spontaneously in the cytoplasm, they will become aligned between those already present. When this activity is continued, a dense stellate complex may be formed.

Chromosomal microtubules may originate in and around the kinetochores (Schrader, 1941), and this 'source' of microtubules may possibly also be involved in the formation of an astral cell division apparatus. However, the origin of the microtubules is really an irrelevant question in this case; functionally no distinction is feasible, and even formally this is true, for any cytoplasmic object to which microtubules may adhere can, through spontaneous formation and alignment of the polymers, give rise to microtubular complexes.

The whole series of events leading to the formation of a mitotic apparatus can be predicted from our generalizations. It is presumed that the rigid microtubules are extended at random from the astral centres, and that they are able to develop a tension after attachment to certain specified places (cortex, centre and chromosomes). If they do not become attached, or if the points of attachment yield to the tension, the microtubules will collapse, being replaced by new ones. As a consequence of this activity, the shape of the cell and of the mitotic apparatus, and the position of the latter in the former, will gradually be modified, corresponding at each moment to an equilibrium dictated by the points of attachment and the force developed by the linear TP polymers. It was pointed out above that if the microtubules originating from one centre may become uniformly attached to the cell membrane in a spherical cell, the ultimate result of their combined action will be that the centriole is placed at or near the centre of the cell. With two asters present, an equilibrium obtains in which both centrioles are located excentrically, owing to the fact that microtubules connect the two centres. Acting through the astral centres, the forces involved in the establishment of an equilibrium may be represented by resultants, located in the line connecting the two centres, the *spindle axis*, pulling towards the centre of the cell and towards the periphery, respectively. Simplifying somewhat, it may be considered that the former represent the chromosomal, the latter the cortical, microtubules.

The hyaline nature of the central parts of the spindle is accounted for by the adhesion between the hyaloplasmic mucopolysaccharide and the microtubules. It should be pointed out that the relation obtaining between microtubules and mucopolysaccharide in the dividing cell is different from that found in sf-cells, where bundles of microtubules are kept loosely together in the filopodia, and in cf-cells, where they are firmly glued together.

The saltatory movements displayed by particles present in the outer regions of the spindle, and the 'bubbling' of surfaces to which microtubules are attached, are accounted for by the dynamics of the polymerization discussed above. It may also be mentioned that a normal cell division often is preceded by formation of microtubules in the vicinity of the centrioles which, incidentally, most often lie near the nucleus. It has frequently been suggested that 'astral rays' may be instrumental in breaking down the nuclear membrane (cf. Figure 47). As mentioned above, the present notions imply that they may do this by converting it into microtubules. The dynamic nature of the mitotic apparatus is borne out by the observations of Inoué (1964).

Karyokinesis

The main task of karyokinesis is to ensure an equal distribution of the chromosomes between the daughter cells. This is accomplished by the mitotic apparatus after it has established itself in the cell. Before the chromosomes are separated they become located in a plane perpendicular to the spindle axis at equal distances from the two centres, i.e. in the median plane of the mitotic apparatus. This remarkable fact is the more fascinating when it is observed that a pair of chromosomes may leave the metaphase plate, but return to this location after some time, showing that the position of the chromosomes corresponds to a dynamic equilibrium. It is easy to explain this phenomenon on the basis of the properties ascribed to the microtubules. Thus, when equilibrium obtains, the force developed by each aster is the same, and since, for statistical reasons, it may be presumed that the same number of microtubules is linked with each pair of chromosomes, it follows that the latter must be located at equal distances from each centre. As a consequence of the continuous formation of microtubules, it may happen that the number coming from one side exceeds that of their opposites. The ensuing unequal force distribution leads to displacement of the chromosomes, but this adjustment entails that the microtubules in numerical excess become shorter, and a new equilibrium is soon established. When sooner or later equalization of numbers takes place, the chromosome pair will return to the metaphase plate. It may be anticipated that the amplitude of these movements will decrease, as the number of chromosomal microtubules becomes larger. As shown in Figure 50, this point is borne out by actual observations. With the increase in the number of microtubules, the forces residing in the mitotic apparatus will augment correspondingly. If the number of cortical microtubules grows faster than that of chromosomal ones, the distance between the two centres must be extended, since the only means by which the equilibrium can be preserved is that the latter become elongated.

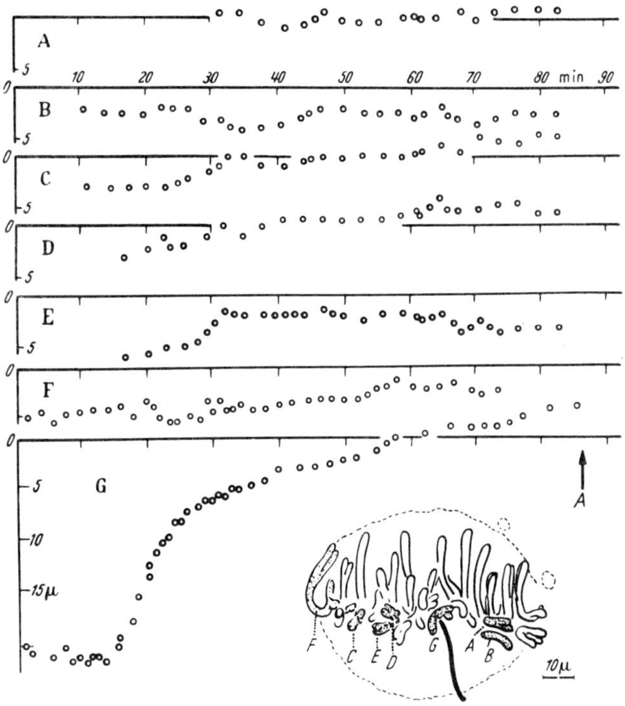

Figure 50. Oscillations of the kinetochores during metaphase in *Haemanthus* endosperm. The points represent successive positions of a kinetochore, as plotted from cinematographic records. (Reproduced from A. Bajer and J. Molé-Bajer, *Chromosoma*, 7 (1956), by permission of Springer, Berlin)

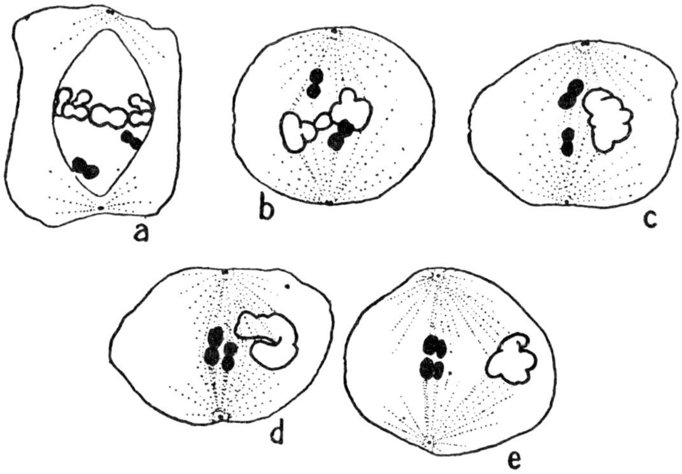

Figure 51. Lateral displacement of autosomes during spermatogenesis in *Brachystetus*. (a)–(e), successive stages. (Reproduced from F. Schrader (1953), *Mitosis*, by permission of Columbia University Press, New York)

At some stage the force acting on the chromosomes will be strong enough to overcome those keeping them together at the kinetochores. Owing to the fact that the pulling force is exerted through the centres, it may be expected that all chromosomes are torn apart simultaneously. The sudden disturbance of the equilibrium will cause some recoil, entailing among other things that the centres will move apart. Many microtubules will collapse because they are no longer under tension; this holds particularly for those attached to the chromosomes,, but new ones formed in their place will act to pull the chromosomes towards the centres.

The mechanism of karyokinesis which has been formulated here on the basis of the properties ascribed to the microtubules embraces features involved in hypotheses advanced by Östergren, Rashevsky, Schrader and many earlier authors (cf. Schrader, 1953). As pointed out by this author, a number of objections may be raised against each of these hypotheses. As far as I can judge, none of them apply to the mechanism proposed here. To mention but one example, it seems possible to account for the elimination of the autosomes in the spermatogenesis of certain Hemiptera (Figure 51) on the basis of the rigidity of the microtubules, provided they do not become attached to the chromosomes.

Cytokinesis

Cytokinesis, cell division proper, is one of the most important biological processes, the mechanism of which has been a constant challenge to the curiosity and imagination of biologists. Many theories have been advanced to explain this phenomenon; for discussion of previous work and new ideas the reader is referred to the reviews by J. C. Dan (1948), Ris (1955), Swann and Mitchison (1958), Wolpert (1960) and Roberts (1961).

The process of cell division may be associated with dramatic form changes but, as known from the cells of most plants and many animals, it may take place with slight or no modification of the external shape. In either case an increase in the surface area must occur in association with cytokinesis, and this phenomenon may be the smallest common denominator in all cell divisions, irrespective of possible variations in the mechanism involved. In the present context the attention will be focused on this aspect; the form changes will be dealt with only to the extent required for the general understanding of the mechanism.

Perusal of the literature shows that five different elements have been proposed to take part in or be responsible for cell division, viz., cortex (plasma membrane), cortical gel (ectoplasm), cortical microfilaments, astral rays (cortical microtubules) and certain small vesicles appearing in the cleavage plane after separation of the chromosomes has occurred. We shall in the following sections discuss separately the possible contribution of the cortical and the cytoplasmic elements.

Cortex

Cell division is invariably associated with enlargement of the surface area of the cell. Under these circumstances the crucial question is not whether the cortex is involved, but rather whether it is actively or passively engaged. Several authors (e.g. Swann and Mitchison, 1958; Wolpert, 1960; Rappaport, 1965) have advanced the contention that the cortex is actively involved. Their views differ in some respects, but they have in common the belief that in a dividing cell the cortical tension is higher in the *median* (equatorial) region than elsewhere, and that the division furrow results from the cortical 'contraction' which must take place in order to equalize the tension. This suggestion apparently implies that the cortex, presumably the plasma membrane, is elastic. According to the ideas propounded above, this contention is not correct, since the tension cannot be changed by stretching or slackening. As this statement may not carry sufficient weight to discourage adherents to the hypothesis, we may try to consider its consequences. Thus, if an elastic tension prevails, being higher in the median region, then a surface movement towards the median may be anticipated, because equilibration of the tension must involve decrease of the surface area here. So far the proposed mechanism is corroborated by actual observations, but it is difficult to grasp how this event can be followed by furrow formation, since this entails an increase in surface area. If the prevailing tension was elastic, the stretching occurring in the cleavage furrow should cause a very large increase in tension. This could never occur spontaneously, but would rather require the supply of energy, a possibility for which none of the suggested mechanisms provides.

Some theories consider the ectoplasm to be responsible for the division furrow, either through 'contraction' (W. H. Lewis, 1951; Marsland and Landau, 1954) or through a combination of 'contraction' and growth (Selman and Waddington, 1955). We shall later have an opportunity to show that there may be a substantial truth in the suggestion that cortical growth is involved, but here we shall first consider whether it is possible to attribute an active role to the ectoplasm.

It will be recalled that cortical attachment of microfilaments is an essential aspect of the formation of ectoplasm. When this fact is combined with the contention that linear

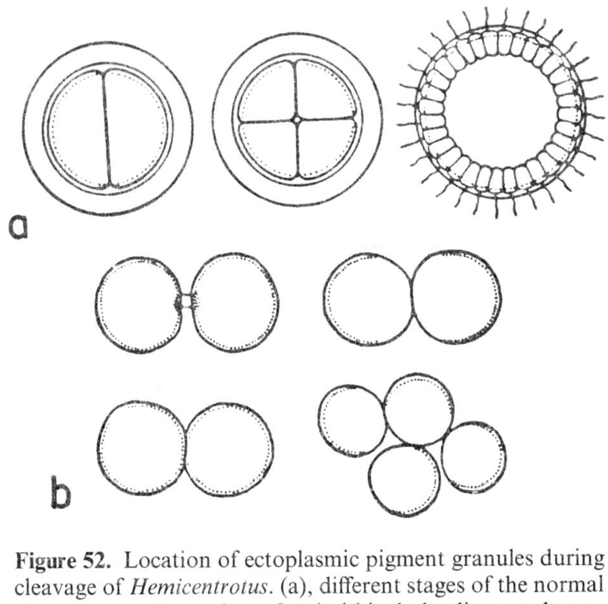

Figure 52. Location of ectoplasmic pigment granules during cleavage of *Hemicentrotus*. (a), different stages of the normal cleavage when the egg is confined within the hyaline membrane. The pigment is seen to remain at the outside; (b), cleavage after removal of the hyaline membrane. (Reproduced from K. Dan, *Internat. Rev. Cytol.*, **9** (1960), after Motomura, by permission of Academic Press, New York)

TP polymers become attached only to free surfaces, it is possible to conclude that no ectoplasm can be present beneath the incipient cell partitions inside a dividing cell, since these surfaces are not free. This conclusion is borne out by observations on the sea urchin *Hemicentrotus* (Motomura, cf. K. Dan, 1960), in which the cortical pigment granules, known to reside in the ectoplasm, are found to remain on the outer surface during cleavage and blastula formation when the embryo develops inside the normal envelopes (Figure 52a). In a calcium-free medium the egg membrane and hyaline membrane are removed, and the blastomeres are not pressed tightly together. This means that the entire surface is free, and consequently accessible to attachment of microfilaments and formation of ectoplasm. From Figure 52(b) it is seen that this expectation is realized, and also that when the blastomeres subsequently make intimate contact the ectoplasm is withdrawn. These observations strongly argue against the possibility that ectoplasm present at the margin of the furrow is responsible for the growth of the partitions separating the daughter cells, since ectoplasm simply cannot subsist on their surfaces.

Attempts have been made to measure the physical properties of the two cortical elements. The 'stiffness' determinations of Mitchison and Swann (1955) on the developing egg of *Psammechinus* show that the cortical resistance towards deformation increases many times during metaphase and anaphase, returning to approximately the original value at the end of cleavage (Figure 53). This observation has been confirmed by Selman

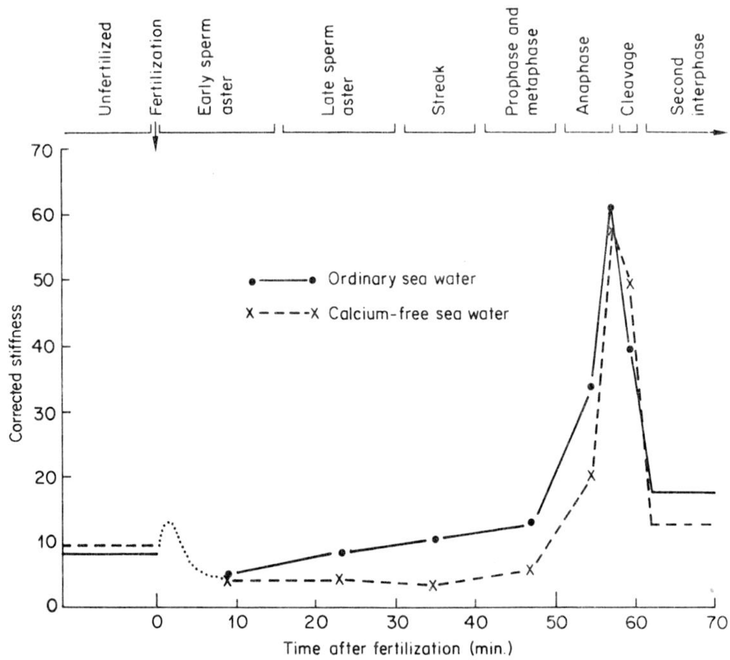

Figure 53. 'Stiffness' determinations on eggs of *Psammechinus* during first interphase and division. (Reproduced from J. M. Mitchison and M. M. Swann, *J. Exptl. Biol.*, **32** (1955), by permission of The Company of Biologists, Cambridge)

and Waddington (1955) in the egg of *Triturus*; it may be pointed out that the changes coincide completely with the rounding up observed in many cells just before division begins. This phenomenon was suggested to be the outcome of the activity of the cortical microtubules. It seems that a force which results in changing the shape of the cell must automatically oppose further changes. Since the 'stiffness' determinations actually measure the resistance towards deformation, it therefore is likely that the results obtained register a cytoplasmic rather than a cortical phenomenon. Attempts to demonstrate local variations in 'stiffness' were unsuccessful, except those of Wolpert (1966), which are very difficult to evaluate.

As pointed out by Wolpert (1960), the determinations of the 'gel strength' of the ectoplasm by Zimmerman, Landau and Marsland (1957) are hard to interpret, and the variations observed are not easily correlated with the process of cell division. These results therefore do not support the supposition, also unlikely for other reasons, that the ectoplasm is responsible for cell division.

Electron microscopical studies have recently revealed the presence of bands of ectoplasmic microfilaments in the furrow region of dividing eggs, oriented parallel to the plane of division. In eggs of Cnidaria their diameters were reported to lie in the range of 3–8 nm (Schroeder, 1968; Szollozi, 1968); in *Triturus* values of 8–10 nm were observed (Selman and Perry, 1970).

It has been suggested by the authors quoted that this ring of microfilaments is actively engaged in the formation of the contraction furrow. This idea may conform with those advanced above if the ends of the individual polymers are attached directly or indirectly to the plasma membrane, and if they prevail in the dynamic state of formation and collapse proposed to be typical of linear TP polymers. The possibility of this mechanism does not, of course, in any way entail that it is actually realized.

Admitting that the presence of microfilaments in the division furrow is an important feature which deserves close attention, I venture to suggest that their main function may be the stabilization rather than the formation of the division furrow. The discussion of this point will be resumed in a subsequent section.

Cytoplasm

According to Mazia (1961), the participation of astral rays, i.e. cortical microtubules, in the process of karyokinesis has generally been refuted. Contrariwise, their involvement in cytokinesis has been suggested by many authors in the past (e.g. Van Beneden, Boveri and Heidenhain, cf. Wilson, 1928) and in the present century (e.g. Gray, 1924; K. Dan, 1943; Rappaport, 1965).

Among the circumstances supporting this view the most convincing is that the furrow normally cuts right through the median plane between the two asters representing the remainder of the mitotic apparatus after karyokinesis has occurred. If the asters are displaced by experimental means, the division plane follows suit (cf. Rappaport, 1965). This spatial correlation demonstrates beyond discussion that the asters determine the orientation of the furrow. The observations by R. Chambers (cf. Chambers and Chambers, 1961) that division in cells and cell fragments deprived of asters generally is irregular and abortive also show that the asters participate in the process, even if they evidently are not the only factor involved.

Claims to the effect that asters cannot be of crucial import for the mechanism of cytokinesis have often been advanced. Thus it has been observed that cell division may be completed after the asters are obliterated by various means, such as solation with colchicine (Swann and Mitchison, 1953; Hiramoto, 1958), destruction by micromanipulation (cf. Chambers and Chambers, 1961) and removal with a pipette (Hiramoto, 1956). Common to all these results is the fact that division is suppressed if the asters are obliterated before the anaphase. Findings like these certainly support the contention advanced by Swann and Mitchison (1958) that no theory of cell division is acceptable which requires the presence of asters throughout the process. Granting this, it must on the other hand be pointed out that these results do not rule out that asters are instrumental in the initiation of cytokinesis, nor do they bear upon the role they may have in determining the plane of division under normal circumstances. The observations on these points therefore sustain the proposition that the asters are actively involved in the mechanism of cell division. The discussion of their possible contribution will be postponed to the following section. Participation of the spindle has generally been rejected; this standpoint is in agreement with the fact that the spindle has disappeared before cytokinesis begins.

The acceptance of the asters as collaborators in cytokinesis does not account for the formation of new cell surfaces. Since, as shown above, the cortex cannot be made responsible for this phenomenon, it follows that it must be accomplished by a cytoplasmic factor, residing in the median plane. In plants, where no form changes occur during cell division, it has been found that vesicles accumulate in the presumptive division

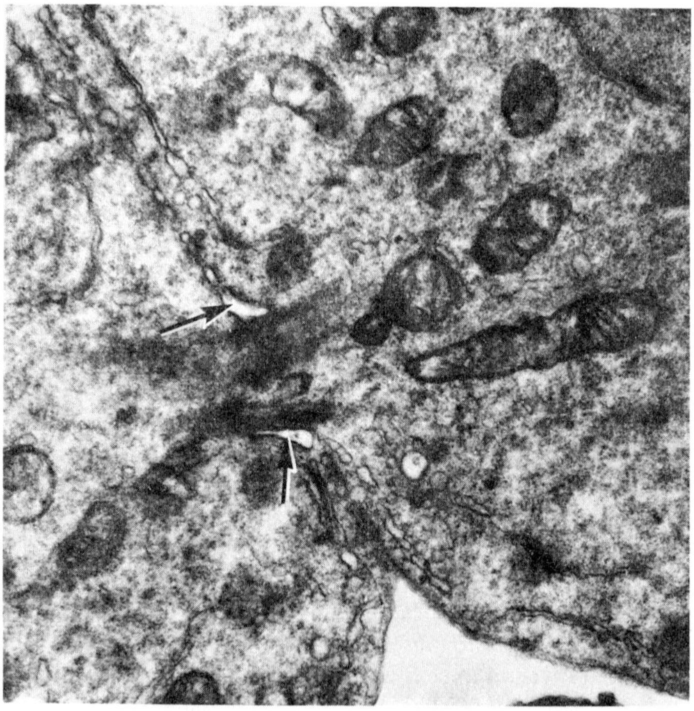

Figure 54. Cleavage region of an erythroblast showing the presence of cytoplasmic vesicles in the division plane. On each side of the midbody is situated an elongated vesicle (arrows) larger than the others. (Reproduced from R. C. Buck and J. M. Tisdale, *J. Cell Biol.*, **13** (1962), by permission of The Rockefeller University Press, New York)

plane, and it has been proposed that the new cell surfaces (plasma membranes) are formed by fusion of these vesicles (cf. Mazia, 1961). This suggestion presumes that the vesicles are built up by TP monomers. The occurrence of this mechanism in animal cells was suggested by Motomura (1950), and subsequently a number of electron microscopic studies have borne out the presence of vesicles in eggs (e.g. *Mytilus*, Humphreys, 1964) as well as in somatic cells (Buck and Tisdale, 1962), findings which suggest that this mechanism of cell surface formation is much more common than hitherto anticipated (Figure 54).

The mechanism of plasma membrane formation through fusion of membranous vesicles poses some problems. The first concerns the origin of the vesicles, which may be present before the division begins or arise *de novo*. The former possibility really does not constitute an answer, for sooner or later new vesicles must be formed. If they are traced back to the endoplasmic reticulum, as suggested by Porter (1956a), they must form spontaneously when the latter is broken down prior to cell division. However, if the vesicles are made up of TP polymers, their spontaneous formation is easily visualized, since this phenomenon may occur *in vitro* from membrane (TP) monomers in the presence of lipids (Green and Goldberger, 1967). The vesicles may be present in the cytoplasm as individual entities, or as multivesicular bodies, the latter being large membranous vesicles containing a number of small ones in a hyaline matrix. These formations, first detected by their metachromatical staining, are known to become attached to the astral microtubules (Pasteels and Mulnard, 1957; Rebhun, 1964). The second problem concerns the fusion of the vesicles to form large membranes; although a very likely proposition, it does not seem to have ever been demonstrated to occur either *in vivo* or *in vitro*.

The preceding discussion leads to the contention that the essential aspect of cytokinesis, the formation of new plasma membrane, is a cytoplasmic phenomenon, involving asters and cytoplasmic vesicles. The mechanism by which these components cooperate to achieve this end will be discussed in the next section, where we shall also consider a case of cell division for which a slightly different mechanism must be envisaged.

Overall mechanism

The process of cell division is initiated by the formation of two stellate complexes of microtubules, the asters. Most cells must be subjected to a number of changes before this phenomenon can take place. Thus it appears that cells in interphase normally contain one centriole; duplication of this structure is therefore a prerequisite for the formation of two asters. Lobocytes do not contain microtubules; thus the special factors responsible for their formation must be made available. In filocytes microtubules are present, but they do not prevail as asters, so in this case also changes must occur before the process of cell division can take place. The mechanisms responsible for these changes are completely unknown.

With two asters present in the cell an *amphiaster* may be formed, owing to the presence of central microtubules connecting the two centres. If chromosomes are present they will be incorporated, with the resulting formation of a mitotic apparatus, but cell division may occur in the presence of an 'achromatic' amphiaster; the chromosomes are thus no essential part of the process. It may be pointed out, however, that the collapse of the mitotic apparatus which occurs in association with karyokinesis, and the ensuing increase in the distance between the centres, probably normally plays a role in the overall mechanism.

The spatial distribution of the astral cortical microtubules is bilaterally symmetrical around the presumptive division plane. As far as I can judge, the observed influence of the asters on the determination of this plane must somehow be correlated with this fact. In certain theories of cell division it is claimed that the cortical microtubules (astral rays) do not become attached to the cortex, while in others this event is considered an important aspect of the mechanism. I have already vouched for the latter alternative, showing that on this assumption it is possible to account for the rounding up of the cell and for karyokinesis in a simple way.

If this contention is accepted, it follows that the density of cortical microtubules is particularly high in the median plane. This circumstance has already been invoked to account for the elongation of the cell usually observed, and it may even account for the formation of the incipient division furrow. This interpretation is supported by the formation of wrinkles at this location in many eggs (Figure 55). Selman and Perry (1970)

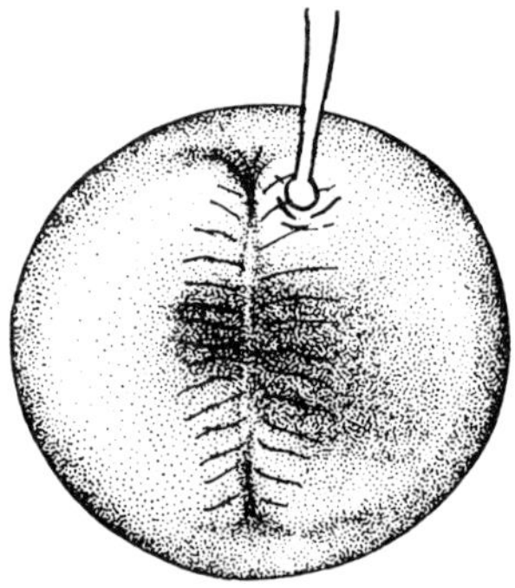

Figure 55. Surface wrinkles perpendicular to the first cleavage plane in a sturgeon egg. (Reproduced from A. I. Zotin, *Acta Biol. Cracov., Zool. Ser.*, **5** (1962), by permission of Academy of Science, Krakow)

suggest that these wrinkles are caused by the ring of microfilaments beneath the furrow, but there seems to be a number of reasons which argue against this proposal. Thus, although the microfilaments may be envisaged to be involved in the formation of the furrow, there seems to be no justification for presuming that the tension varies from point to point along the ring, and without this condition fulfilled, it is not possible to account for the wrinkles. Contrariwise, in the case of cortical microtubules, it would be unlikely that they were evenly distributed along the furrow, and the tension developed by them therefore might vary sufficiently to generate the observed wrinkles. The fact that these surface irregularities disappear upon treatment with colchicine (Zotin, 1962) corroborates this contention. Irrespective of the possible function of the microfilaments, they clearly outline the plane of the division. However, since this is ultimately determined by the asters, it seems to follow that the latter, and thus presumably the cortical microtubules in the presumptive division plane, are responsible for the changes required for the alignment of the ectoplasmic microfilaments.

It was pointed out above that cytoplasmic vesicles are found in the presumptive division plane. It does not seem too difficult to explain this phenomenon, for when the chromosomes have become separated, the astral microtubules directed towards the median planes will be free, i.e. unattached, and according to the properties ascribed to them it may be anticipated that they become engaged in a lively activity of extension and withdrawal. If, in the cytoplasm, vesicles are present which may become attached to the linear polymers, then they will be pushed or pulled about by the latter. The overall result of this activity will be that they accumulate in a plane at equal distances from the centres, since this location must correspond to the 'equilibrium position' of the microtubules under these conditions.

It is easy enough to visualize the fusion of these cytoplasmic vesicles, but the formation of new cell surface begins at the incipient furrow that, for some time at least, is covered by ectoplasm, which may prevent contact between plasma membrane and vesicles. Thus, even if the suggested function of the cytoplasmic vesicles is accepted, it appears that they cannot be involved right from the beginning. This is not a serious objection, however, for if, as suggested, the plasma membrane is made up of TP polymers, there will always be plenty of material around for the purpose of membrane formation. It follows that this argument is of particular importance in cases where apparently no cytoplasmic vesicles are present in the division plane (cf. Selman and Perry, 1970).

As a general rule it may be anticipated that in small cells and eggs, in which the mitotic apparatus occupies the whole volume, the formation of new cell surface may occur more or less simultaneously in the entire division plane, or else that the cortical growth is radially symmetrical. When the mitotic apparatus is excentrically located, as in eggs that are large or yolk-laden, or both, plasma membrane formation begins at the side closest to the mitotic apparatus, whence it gradually grows through the egg to the opposite side. This situation was described by Selman and Waddington (1955) in the egg of *Triturus*. These authors made some remarkable observations concerning the mechanism of furrow formation which are best presented in their own words: 'In the case of almost every egg there were clear signs of changes occurring in the subcortical cytoplasm in positions below and ahead of the furrow, whose progress could be seen on the surface of the egg. The modified cytoplasm lay only in the plane of the cleavage furrow, midway between the daughter nuclei produced by the mitosis and perpendicular to a line joining the two nuclei [Figure 56]. The modifications in the appearance of the cytoplasm which lay ahead of the course of the furrow [Figure 57] consisted first in a displacement of yolk granules, which in a section produced a line containing fewer or no yolk granules; secondly, in local accumulation of small pigment granules along this same line; and thirdly, in fibres which could be seen to lie across and perpendicular to the plane of modified cytoplasm' (p. 722). 'In several instances, moreover, the pigmentation in the modified region formed two lines of granules at a separation of about 4 μm, and this suggests that the new layer of gel may already itself be double' (p. 724).

These observations, which lately have been extended by the electron microscopical studies of Selman and Perry (1970), largely confirm the views advanced here. Thus, the 'fibres' turn out to be bundles of microtubules, and as they may be expected to accumulate hyaloplasm, the 'displacement of yolk' is also accounted for. The 'accumulation of small pigment granules' may be due to the adhesion of these bodies to TP polymers. That 'two lines of granules' are formed might be explained by the presence of the membranous vesicles, but as stated above, this particular feature is absent in the newt's

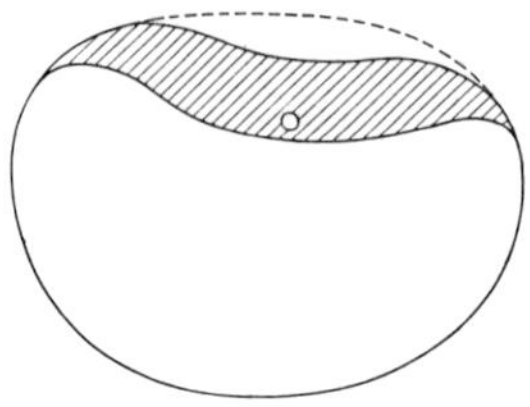

Figure 56. Diagram of a section through a *Triturus* egg in the plane of the first cleavage division, 8–10 minutes after the beginning of the division. The area between the dashed line and the upper surface of the egg indicates the extent of 'dipping-in' in the furrow region. The shaded region indicates the extent of the sheet of modified cytoplasm ahead and beneath the externally visible furrow. (Reproduced from G. G. Selman and C. H. Waddington, *J. Exptl. Biol.*, **32** (1955), by permission of The Company of Biologists, Cambridge)

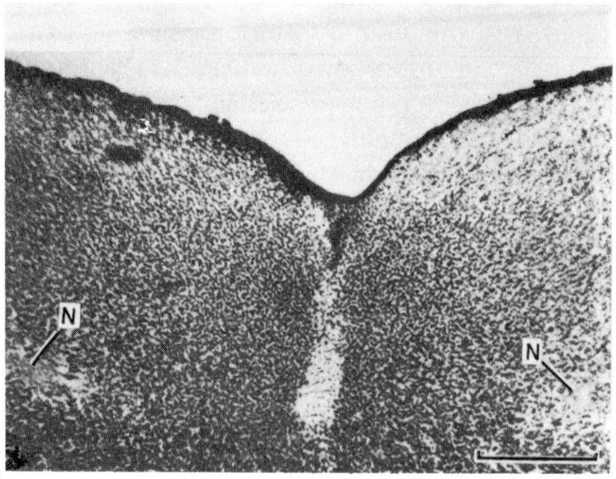

(a)

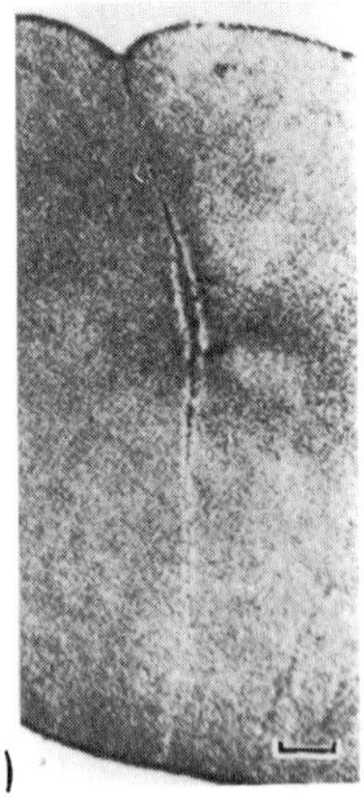

(b)

Figure 57. Cell division in *Triturus*. (a), vertical section perpendicular to the first division plane, about 8 minutes after the beginning of division. Fibres may be seen in the hyaline region. N, nuclei; (b), section cut in a plane about 10° from the horizontal and transverse to the furrow, showing the subcortical signs of the furrow extending right across the egg. A line of pigment granules is seen, and a double layer of hyaloplasm. The scale is 100 μm. (Reproduced from G. G. Selman and C. H. Waddington, *J. Exptl. Biol.*, **32** (1955), by permission of The Company of Biologists, Cambridge)

egg. Incidentally, the fact that the new surfaces are only weakly pigmented as compared to the old ones, even when the former are freely exposed to the external medium (cf. Selman and Waddington, 1955), seems to rule out that furrow formation involves ectoplasmic growth, since the pigment is located largely in the ectoplasm. The sparse pigmentation observed may rather be due to pigment granules derived from the cytoplasm.

It may be mentioned that if the formation of new surface is only a matter of ingrowth from the cortex, involving a kind of constriction, as envisaged by Selman and Waddington, then it is not easy to account for the incomplete cleavage occurring in discoidal cleavage. When the presence of astral microtubules is required, a plausible mechanism is ready at hand.

So far no specific function has been ascribed to the band of ectoplasmic filaments. It will be realized, however, that once formed these elements define the division furrow independently of the asters. If, as suggested by Selman and Perry (1970), they are attached to the cell surface, they will unquestionably be able to resist deformation of the dividing egg. They may, in other words, function to stabilize the division furrow, a task which may be of particular importance in large cells such as eggs (Figure 58). By this explanation it also becomes possible to understand that cell division may in some instances continue after the obliteration of the asters, the microtubules of which would otherwise be the only means capable of generating the physical force required to preserve the deformation of the egg during the process of division.

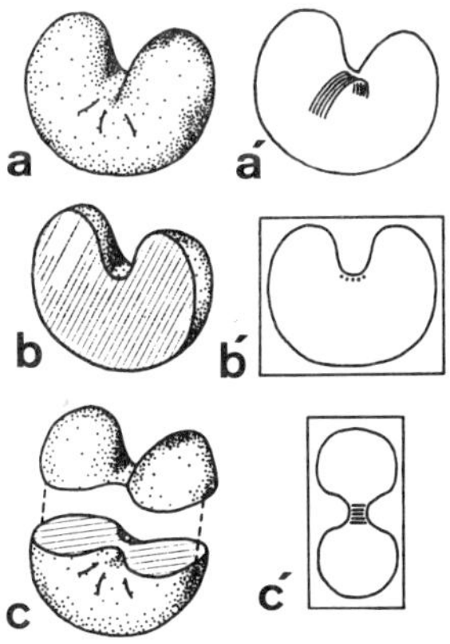

Figure 58. 'Contractile ring' filaments in dividing eggs of *Stomatoca*. (a), stereogram of a whole egg; (a′), schematic drawing showing the position, orientation and alignment of the filaments as observed in electron micrographs; (b), stereogram of an egg sectioned through the median plane; (b′), schematic drawing showing cross-sectional filament profiles; (c), stereogram of an egg sectioned latitudinally through the uppermost part of the division furrow; (c′), schematic drawing showing filaments cut lengthwise. (Reproduced from T. E. Schroeder, *Exptl. Cell Res.*, **53** (1968), by permission of Academic Press, New York)

The ring of microfilaments may be of special importance for the stabilization of the division furrow when the latter is formed through unilateral ingrowth, as, for instance, in amphibian eggs. That it is not a necessary participant in radially symmetrical divisions is demonstrated by its absence from the egg of *Dendraster* (Schroeder, 1968).

An interesting case of cell division has been observed by Dan and Dan (1947) in the cleaving egg of *Spirocodon* (Hydrozoa) which, similarly to that observed in some Echinodermata and Ctenophora, passes unilaterally from the animal to the vegetal pole. 'Spindle fibres', presumably representing central microtubules, stretch between

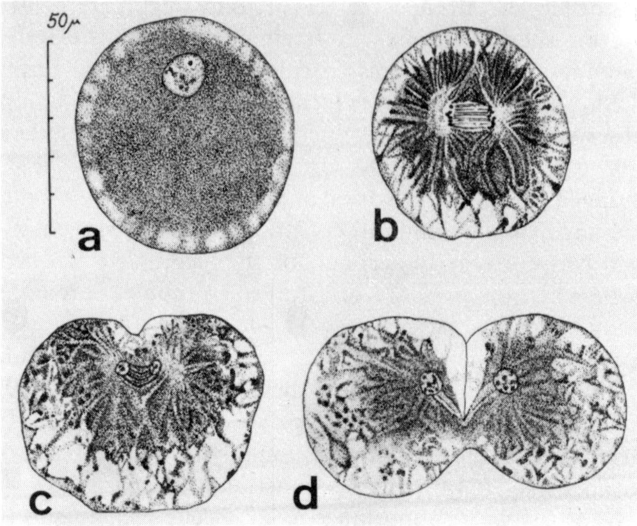

Figure 59. Cell division in *Spirocodon*. (a), unfertilized egg; (b), telophase with chromosomes and asters; (c), anaphase with asters, a 'bending' spindle and a shallow furrow; (d), later stage with a V-shaped spindle. (Reproduced from K. Dan and J. C. Dan, *Biol. Bull.*, **93** (1947), by permission of Marine Biological Laboratory, Massachusetts)

the nuclei, and as the cortex grows in from the animal pole, the direction of these microtubules is changed (Figure 59). This phenomenon has been interpreted to represent a 'rotation of the asters' and 'bending of the spindle' by Dan and Dan, but a different interpretation is possible. Thus, the reason for the 'rotation' may very simply be that microtubules do not become attached to the plasma membranes in the cleavage furrow, because the surfaces are not free, and consequently only those directed towards the intercellular connection will prevail. The rotation and the bending may therefore be only apparent, depending on the disappearance of microtubules from a certain spatial sector. This uneven distribution of linear polymers will, of course, be of some influence on the shape of the presumptive daughter cells. I believe that these observations are of particular importance in demonstrating the presence of astral microtubules exactly in the region where they are supposed to exert their function. Incidentally, it is seen in Figure 60 that when the furrow sides are free, the so-called phenomenon of 'spinning'

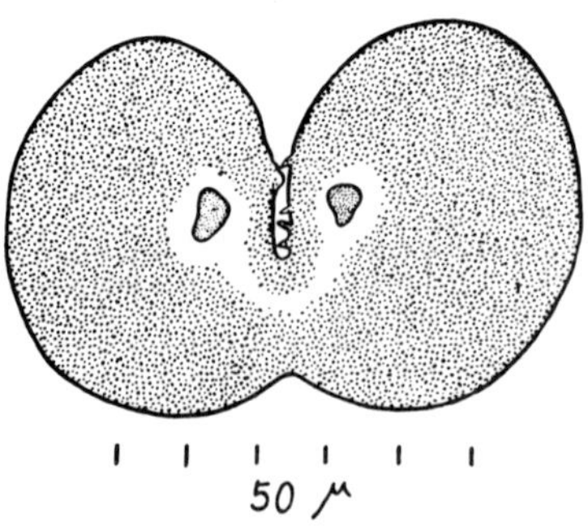

Figure 60. *Spirocodon* egg, showing 'spinning' activity on the free sides of the division furrow. (Reproduced from K. Dan and J. C. Dan, *Biol. Bull.*, **93** (1947), by permission of Marine Biological Laboratory, Massachusetts)

occurs (Andrews, 1897). We may, in agreement with present notions, assume that astral microtubules, which under these conditions may become attached to the cell cortex, are responsible for this activity.

Several theories of cell division have concentrated on the situation obtaining when the presumptive daughter cells round up individually during the process of cytokinesis, thus giving rise to a dumbbell-shaped complex, as observed in *Hemicentrotus* eggs in a calcium-free medium by Motomura (Figure 52b). Although this mechanism may be an exception rather than a rule, it is of course important to account for it. If we consider an egg just before karyokinesis it is seen that a large number of cortical microtubules cross the median plane (Figure 61). The resulting high density in this region has the

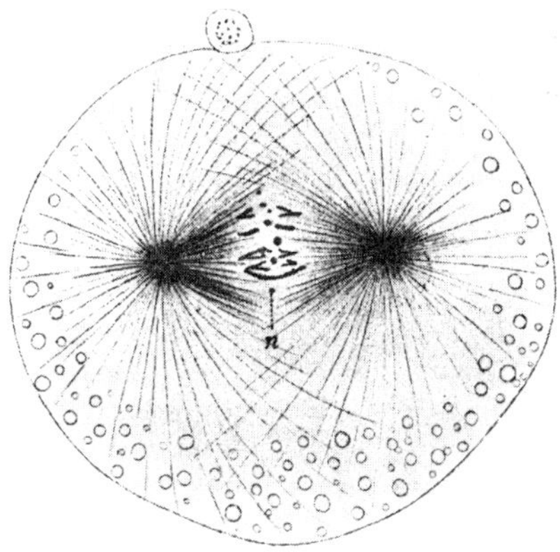

Figure 61. Dividing egg of an annelid, showing cortical microtubules in the median plane. (Reproduced from A. D. Mead, *J. Morphol.*, **14** (1898), by permission of The Wistar Press, Philadelphia)

consequence that the force residing there is higher than elsewhere in the cortex, which may be anticipated to lead to a reduction of the median cross-sectional area. Since the total volume is constant, this can occur only if associated with a simultaneous elongation of the cell, but this event is opposed by the force deployed by the spindle. This circumstance explains the fact that cell elongation usually takes place only after the separation of the chromosomes. If the conditions permit it, the crossed microtubules, or rather their descendants, will pull in the median region in the form of a constriction, and subsequently each aster behaves as an independent entity, organizing the part of the cell it occupies as a sphere. As shown in Figure 62, the number of 'spheres' which may be formed in this way is not limited to two. As this process approaches completion, the two daughter cells may be connected by a rather narrow stalk, but it is questionable whether the microtubules can ever accomplish severance. Recalling that ectoplasm accumulates around the stalk (Motomura, Figure 52b), the most likely mechanism is that microfilaments may cross the opening and pull the edges together.

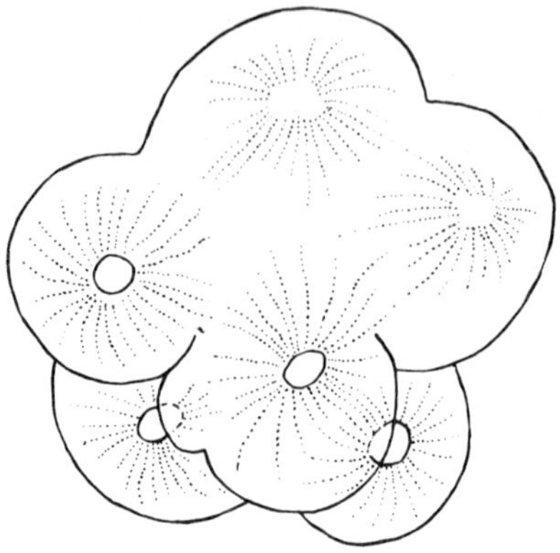

Figure 62. Nuclear asters and cytasters with partial furrows in parthenogenetically activated egg of *Lytechinus*. (Reproduced from R. Chambers and E. L. Chambers (1961), *Explorations into the Nature of the Living Cell*, by permission of Harvard University Press, Massachusetts)

It is evident that in this type of cell division the microtubules are directly responsible for the changes in shape which lead to cytokinesis. It may be anticipated that if the microtubules are suppressed, for instance by colchicine, the cell will round up, since the 'constricted form of the partly divided egg is not a least surface figure and can only exist by reason of internal structure and applied force' (Rappaport, 1966, p. 5).

It is obvious that in this mechanism the new plasma membranes are not formed in the cytoplasm. Observations on the movements of particles attached to the surface of dividing eggs (cf. Dan and Dan, 1947) convey the impression that in the sea urchin the formation of new surface occurs by stretching of the old one (Hiramoto, 1958). As seen from Figure 63, an initial increase in surface area at the poles is at first coupled with condensation in the median region; later this area is also enlarged. According to the views adopted here the surface, i.e. the plasma membrane, cannot be either stretched or condensed; in the former case it must be presumed that new TP monomers pass from the ectoplasm to the plasma membrane, in the latter the movement will be in the opposite direction. It will be realized that it is impossible from observations on superficial particles to distinguish between this mechanism and surface stretching.

From this survey it is seen that all the factors which by various authors have been alleged to be instrumental in cell division indeed are or may be involved, but that the cytoplasmic components on the whole are active participants, whereas the cortical ones generally are passive.

The process of cell division is a very intriguing phenomenon which has no counterpart in the inanimate world. The various physical models proposed, involving for instance suspensions of oil drops, can at best account for cytokinesis in small cells, and it is completely impossible to devise a physical model in which elastic forces may simulate the event of cell division. The mechanism advocated here is really an enormous step backwards in time to the early fibrillar theories of mitosis. The reason why it has been

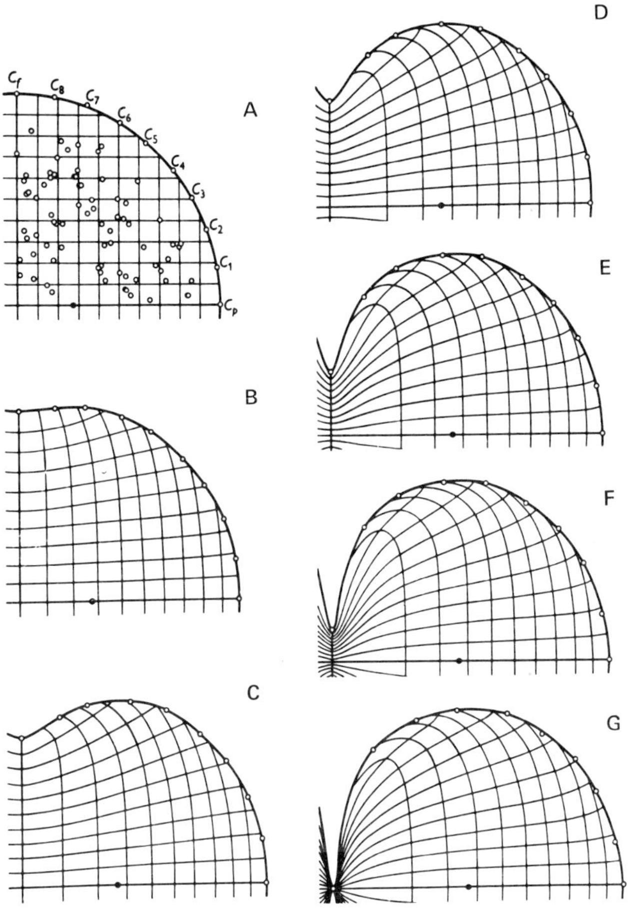

Figure 63. Diagram indicating surface movements during the first cleavage division in *Clypeaster*. One-quarter of the egg is shown. The movements of the white circles on the cell surface C_p, C_1–C_8, C_f in (A) can be followed in the subsequent stages (B)–(G). The black circles indicate the position of the astral centre. (Reproduced from Y. Hiramoto, *J. Exptl. Biol.*, **35** (1958), by permission of The Company of Biologists, Cambridge)

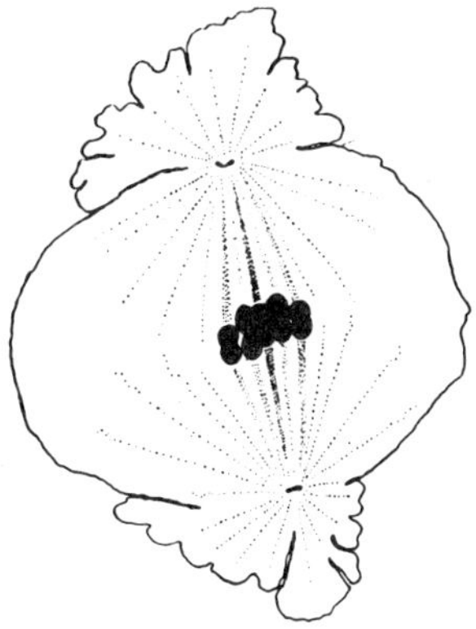

Figure 64. Metaphase in a large cell from *Arvelius*. The unusual cell shape may be accounted for on the basis of the properties ascribed here to the mitotic apparatus, if it is further assumed that the cortical tension is very low. (Reproduced from F. Schrader (1953), *Mitosis*, by permission of Columbia University Press, New York)

impossible to exploit this idea successfully before is primarily that the dynamic nature of the microtubules was unknown. The particular advantage of the present hypothesis is that it can account for a number of diverse phenomena associated with cell division, some of which will be discussed later. To quote just one example here, it seems that no other theory of cell division can account for the cell shape depicted in Figure 64. On the basis of the dual role of the microtubules, as both rigid and tensible structures, it is not difficult to explain this phenomenon.

MEIOSIS

Meiotic cell divisions occur during spermatogenesis as well as in the maturation process discussed earlier. From a karyokinetic point of view the processes are equivalent, but as regards cytokinesis, the event encountered in eggs is of particular interest, and in the present context only this phenomenon will be dealt with.

Generally, a cell divides into two daughter cells of approximately equal size, and as appears from the discussion of mitosis this outcome may be related to the fact that the position of the mitotic apparatus in the cell is bilaterally symmetrical. As pointed out, this situation presumes that cytoplasm and cortex are homogeneous with respect to the properties responsible for formation and attachment of microtubules. If inhomogeneities prevail, the metaphase plate will be displaced from the median plane of the cell, and since the former determines the location of the new cell partitions, an unequal division will ensue. As will be shown later, it is easy to account for the various cases of unequal divisions observed during the cleavage of most eggs within the framework of the mechanism of mitosis outlined above, provided such inhomogeneities are allowed for.

Considerable differences in size between two daughter cells may be brought about in this way, but they are always modest compared to the ratios usually obtaining between the mature egg and the polar bodies. Although it certainly appears desirable to presume that the mechanism of cytokinesis involved in meiosis is principally the same as that involved in mitosis, it is nonetheless evident that the existence of both must depend upon some fundamental differences. One such is that the meiotic apparatus often, but not always, is anastral, another that it is always confined to a restricted region near the animal pole. We shall in the following sections discuss these particular features, and try to account for their contribution to the overall mechanism of meiosis.

The meiotic apparatus

In contrast to the mitotic apparatus, the meiotic apparatus may be either astral or anastral. Under these circumstances it is possible to propose that an anastral meiotic apparatus may be a sufficient, but not a necessary, condition for a division leading to the formation of polar bodies, whereas an astral meiotic apparatus neither is necessary nor sufficient for achieving this purpose. For this reason the attention in the present section will be focused on the nature of the anastral meiotic apparatus.

It was observed by King (1901) that isolated kinetochores may give rise to aster formation. Pollister and Pollister (1943) showed that in atypical spermatogenesis in certain viviparid snails chromosome degeneration occurs, and it was found that the ensuing spermatids contained an excess of flagella, corresponding in number to the lost chromosomes. These records are among many that support the close relationship between kinetochores and centrioles originally suggested by C. D. Darlington and by Schrader (cf. Schrader, 1953).

This circumstance evidently suggests that in the formation of an anastral meiotic apparatus the role of the absent centrioles may be taken over by the kinetochores. According to the views advanced above this function does not comprise interference with the formation of microtubules, since this is a phenomenon occurring spontaneously in the cytoplasm. Rather, the kinetochores may provide sites where the microtubules can become attached. The shape of the microtubular complex thus formed may depend on limitations of the directions in which the microtubules may radiate. If no restrictions obtain, a stellate complex may be formed, as observed not only with centrioles and isolated kinetochores, but also with other cytoplasmic particles, which may give rise to cytasters. In contrast it seems that, as long as the kinetochores are associated with the chromosomes, microtubules may be extended only in a limited spatial sector. This function of the kinetochores is corroborated by the fact that the second meiotic apparatus generally is smaller than the first one, as may be anticipated as a consequence of the reduction of the number of kinetochores to one-half.

Accepting this mechanism, and taking into account the rigidity of the microtubules and the ability of mucopolysaccharides to keep the polymers together, it is possible to understand the intranuclear formation of what, without any risk of ambiguity, may be called *chromosomal spindles* in *Acroschismus* (Figure 65, Hughes-Schrader, 1924). From

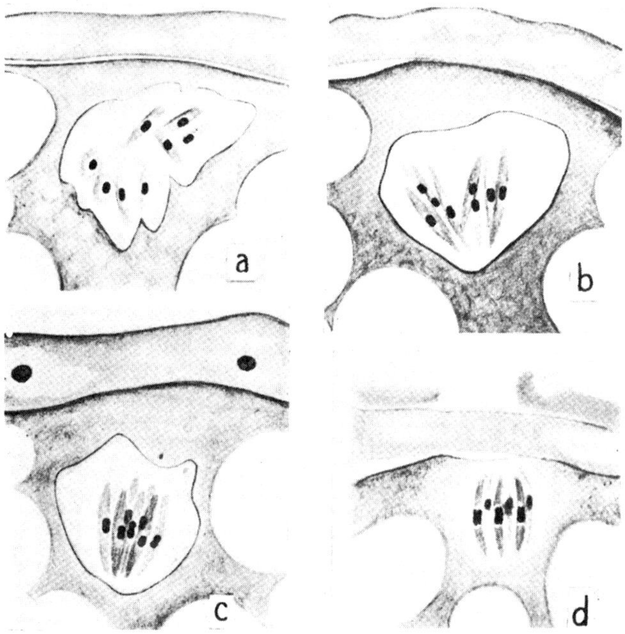

Figure 65. Intranuclear formation of the first meiotic apparatus in *Acroschismus*. (a) (d), successive stages showing how the chromosal spindles become aligned to form a barrel-shaped anastral meiotic apparatus. (Reproduced from S. Hughes-Schrader, *J. Morphol.*, **39** (1924), by permission of The Wistar Press, Philadelphia)

the figure it is seen how these spindles are gradually aligned to form a multipolar or barrel-shaped anastral meiotic apparatus, which becomes attached to the animal pole after the rupture of the germinal vesicle. The number of poles may be reduced when the spindles are brought together (Figure 49), an outcome in which mucopolysaccharides may be involved. Owing to their tapering shape, the chromosomal spindles and the bipolar anastral meiotic apparatus remind one of the spindle in the normal astral mitotic apparatus. It has already been pointed out that in the latter case the presence of centrioles is implied by definition, and that the tapering stems from the fact that all the microtubules originate in the centres. As regards the shape of the former structures no explanation exists, but the fact that the contour of the chromosomal spindles is so similar to that of the spindle-shaped cf-cell suggests that no special mechanism is required. It will be realized that only chromosomal microtubules are present in the chromosomal spindles, but when these join to form a meiotic apparatus, continuous fibres may occur, being in this case made up of *polar* rather than central microtubules.

In an astral cell division apparatus microtubules originating in the centres are arranged in stellate complexes, and apart from the central and chromosomal microtubules engaged in the formation of the spindle, the remaining ones are attached to the cortex. It is typical of an anastral meiotic apparatus that the microtubules pass from the poles towards the chromosomes and the opposite poles, but never in other directions. This feature, which is easy to understand on the basis of the mechanism of formation outlined above, entails that cortical microtubules tending to increase the distance between the poles cannot exist. If, as we may presume, cortical microtubules are present, then they are directed from the inner pole, and possibly from the chromosomes, towards that region of the cortex where the outer pole is attached. This particular feature may be of decisive importance for polar body formation.

Location of the meiotic apparatus

The meiotic apparatus, whether astral or anastral, is always located at the animal pole, at right angles to the cortex. Considering first the orientation, it is readily seen that this represents the equilibrium state if the force deployed by the microtubules from the inner pole is the same in all directions. As regards the location, we shall deal first with the astral meiotic apparatus. It is immediately apparent that an excentric location is in contrast with the behaviour predicted for asters and with that observed when the mitotic cleavage divisions begin after fertilization. In some cases, albeit not always, the asters involved in the latter case are derived from the sperm centriole, those in the meiotic divisions from that of the egg, and the difference thus possibly might depend upon the properties of the centrioles. This proposal is unlikely because, as envisaged here, the centriole only acts to supply the special information required for the formation of a stellate complex of microtubules, the actual establishment of the aster being a cytoplasmic function. Actually the asters in eggs of this type are often located centrally to begin with, but sooner or later they end up at the animal pole (fertilization type 2). This rather suggests that the region around the animal pole in the unfertilized or the recently fertilized egg offers particularly favourable conditions for attachment of microtubules (Figure 66), a situation that is reversed before the cleavage divisions begin.

Various changes are known to occur in the cortex in association with fertilization, among which are the ectoplasmic translocations accomplished by cortical movements and the elimination of the cortical vesicles. The former generally entail that the ectoplasm, originally more or less evenly distributed, becomes accumulated in certain regions. On the assumption that ectoplasm interferes with microtubule attachment, the observed changes may agree with prediction. It may appear to be in disagreement with earlier arguments that ectoplasm, the existence of which depends upon attachment of linear TP polymers to the cortex, should prevent the attachment of microtubules. However, it does not seem unlikely that the microfilaments may impede the establishment of contact between microtubules and plasma membrane, and it may also be recalled that in many eggs particulate bodies, notably pigment granules, are accumulated in the ectoplasm. It is possible that these may reduce the chances of the microtubules getting in touch with the plasma membrane. As we shall discuss in a later section, this mechanism

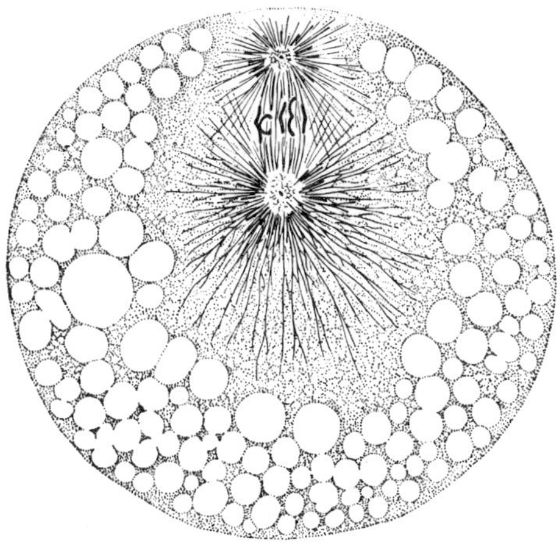

Figure 66. Astral meiotic apparatus in *Thalassema*. The location of the yolk granules suggests that they may be involved in preventing attachment of microtubules from the proximal aster to the cell cortex. (Reproduced from E. B. Wilson, *Biol. Lectures* (Wood's Holl), **1898** (1899))

may account for the action of some polarities. The cortical vesicles may act in a similar way, in which case the change in microtubular behaviour observed upon fertilization may be correlated with the fact that the vesicles either collapse or enter the cytoplasm in association with this event.

Whatever the exact basis of the peculiar properties of the cortex responsible for the localization of the meiotic apparatus, it seems that the latter feature is sufficient to account for the disparity in size between the polar bodies and the egg. This contention is supported by the fact that very large polar bodies may be formed if, in eggs with an astral meiotic apparatus, the latter by centrifugation is forced towards the centre of the egg (cf. Conklin, 1917). This observation, and also the formation of polar lobes in association with the maturation division, suggests that microtubules may become attached all over the inside of the cortex, even if the animal pole is the preferred site. Thus, in the presence of an astral meiotic apparatus, cortical differences appear to be a necessary requirement for meiosis.

As regards the question of the function of the anastral counterpart in ensuring polar body formation it seems that, once formed, the microtubules will sooner or later establish contact between it and the plasma membrane, and transport the meiotic apparatus to the cortex where, for reasons outlined above, it will be oriented perpendicular to the latter. However, in the absence of microtubules directed towards the interior of the egg, the meiotic apparatus can never be pulled towards the interior of the egg, and it may be concluded that the excentric location is the equilibrium condition for an anastral meiotic apparatus. This obviously implies that no particular requirements must be made with respect to the cortical properties; in principle, polar body formation may occur at any point on the surface of the egg, when an anastral meiotic apparatus is involved. This circumstance does not prevent the location of the animal pole being predetermined.

Overall mechanism

In mitotic cell division the separation of the chromosomes is achieved by the tension of the cortical microtubules which, acting through the centres, create oppositely directed forces at the kinetochores. In meiosis the situation is different, for here the force exerted by the cortical microtubules from the proximal pole acts to shorten rather than to lengthen the meiotic apparatus. Under such circumstances these elements can hardly be involved in chromosome separation. The only agents which may be made responsible for this operation are the chromosomal microtubules. These are fixed at the poles of the meiotic apparatus, and whatever action they exert on the chromosomes it must in the last analysis be balanced at their polar attachment. To explain the chromosomal separation it must therefore be envisaged that all the polar (extrachromsomal) microtubules form a cytoskeleton which is sturdy enough to counteract the pull exerted on the kinetochores by the other polymers. It will be realized that the force developed by the chromosomal microtubules in part depends upon the length of the meiotic apparatus; growth of the latter may thus be instrumental in chromosome separation. This mechanism may thus involve a certain 'pushing' action by the polar microtubules.

Allowing for the rigidity of the meiotic apparatus composed of microtubules, it follows that as the number of microtubules between cortex and meiotic apparatus increases, a pushing force develops, acting at the proximal pole of the meiotic apparatus. From observations it seems that this force may be strong enough to reduce the length of the meiotic apparatus. However, sooner or later the cortex yields to the pressure, resulting in the formation of a protrusion, containing half of the meiotic apparatus.

As regards the formation of the partition between polar body and egg, no records of the participation of cytoplasmic vesicles seem to exist. Most likely the cortical microtubules from both poles of the meiotic apparatus may contribute to the formation of new surface by pulling the plasma membrane towards the animal pole.

CELL DIVISION RULES

It was realized at an early period that certain regular features distinguish the process of cell division. On the basis of these observations a number of cleavage rules—or, more pretentiously, cleavage laws—were established. Since the word 'cleavage' in the present context is reserved for the initial subdivision of the fertilized egg, the expression *cell division rules* will be used here.

The first cell division rules were stated by Sachs:

'[I] Cells typically tend to divide into equal parts.

[II] Each new plane of division tends to intersect the preceding one at right angles' (1877, quoted from Wilson, 1928, p. 982).

According to Thompson (1942), the ideas embodied in the second rule were first formulated by Schwendener in 1860 and by Hofmeister in 1863. To these were added those of O. Hertwig:

'[III] The typical position of the nucleus (and hence of the mitotic figure) tends towards the centre of its sphere of influence, i.e. of the protoplasmic mass in which it lies.

[IV] The axis of the spindle typically lies in the longest axis of the protoplasmic mass, and division therefore tends to cut this axis transversely' (1881, quoted from Wilson, 1928, p. 985).

Lastly, one rule was formulated by Balfour (1885, p. 95), as follows:

'[V] The velocity of the segmentation in any part of the ovum is roughly speaking proportional to the concentration of protoplasm there; and . . . the size of the segments is inversely proportional to the concentration of the protoplasm.'

If an attempt is made to classify these rules with respect to their implications, it is seen that I is concerned with the size of the daughter cells arising from a division and II and IV with the orientation of the division planes, whereas III really does not state anything about cell divisions unless it is tacitly assumed that the position of the nucleus—the addition concerning the mitotic figure was kindly made by Wilson—determines the plane of division. If this is granted, the outcome predicted by Hertwig's first rule (III) coincides with that of Sachs' first rule (I). Thus, as it turns out, Hertwig's rules essentially are explanatory generalizations of those of Sachs.

As regards (V), it tries to account for the difference in size between the animal and vegetal cells in the amphibian embryo by postulating that the latter are large because they divide more slowly than the animal ones. It does not explain the existence of unequal cell divisions, for instance the third one in the amphibian embryo, which gives rise to two groups of four cells of different size. It is therefore questionable whether it may rightly be considered a cell division rule. As far as its factual content is concerned, its generality is challenged by the rule propounded by Kofoid (1894), based upon observations of cleavage in various Mollusca, according to which 'the greater the amount of the yolk the greater seems to be the tendency of the cells of the yolk-laden quartet to divide before those of the smaller quartets' (p. 196). Even in the specific context for which it was established its validity has been disputed. Thus it has been claimed by Sirikami (1958, quoted by K. Dan, 1960) that although there is a temporal animal–vegetal division 'gradient' during cleavage in the egg of *Bufo*, there is no distinct difference in the rate of division between animal and vegetal cells. However, in view of the slight disparity in size between the two cell-types at the 8 cell stage (Figure 73) as compared with later stages (Figure 83), it seems difficult to generalize this finding. It rather appears unavoidable to admit that differences in division rate exist which may be correlated with the relative content of yolk, but no general principle can be established as regards the influence of this factor.

Thus, rules I and III state that the daughter cells generally are of equal size. Rule IV does not predict anything about the size relation between the daughter cells; it may be valid in conjunction with either equal or unequal divisions. Indeed, the principle advanced may also be of importance for the orientation of successive division planes, so it may also be useful together with II.

Many exceptions to the four first cell division rules are known, and it was therefore emphasized by Wilson (1928, p. 985) that 'the rule of Sachs and Hertwig must not be pushed too far'. Certainly this statement implies that their predictive value is very limited, and they are thus not the kind of generalizations which belong to the domain of theoretical biology. We shall in the following section discuss the validity of the cell division rules on the basis of the concepts advanced above, considering separately the question of the size of the daughter cells and the orientation of successive division planes.

Size of daughter cells

In the present section we shall deal with 'equal' and 'unequal' cell divisions. These adjectives generally relate to the size of the cells, but it must be pointed out that a cell division should not be classified as equal unless the daughter cells are identical, not merely of equal size. Conversely, in an unequal division the daughter cells must be different with respect to at least one property, although usually several differences probably obtain, among which commonly is the size. However, when more than one differential prevails, their influences on the cell size may interfere, with the result that the disparity in size is imperceptible. This does not exclude that the cells are different, as may be revealed by their subsequent fate. If other influences can be excluded, it may be warranted, in compliance with the causal postulate, to infer that the cells were different in spite of the similarity in size.

It has been proposed above that the division plane coincides with the median plane of the spindle and that the location of the latter in the cell depends upon the forces generated by the astral microtubules. If a dividing cell is uniform with respect to those cortical and cytoplasmic properties that interfere with attachment and formation of microtubules, respectively, and unaffected by external deforming influences, it will assume a spherical shape. The midpoint of the spindle must under these circumstances lie in the centre of the cell, and any median plane may become the division plane. If an apical polarity is present, concerned with properties of the kind mentioned, the spindle may under exceptional circumstances lie in the apical axis, but in general the equilibrium condition for the mitotic apparatus is fulfilled when the spindle lies in an arbitrary median plane through the apical axis, perpendicular to the latter and at a distance from the respective poles that is determined by the properties of the polarities. The division furrow will lie in a median plane. If a median polarity obtains, the mitotic apparatus will be in equilibrium when situated at right angles to the median plane thus defined; the division plane will coincide with the latter. The mitotic apparatus may also lie in the median plane, and since the asters by definition must be of different size in this case, the division plane cannot be a median plane.

Polarity may also be impressed upon a cell by external forces influencing its shape. If by this action the cell becomes cylindrical, one median plane is defined perpendicular to, and many through, the long axis. Since the force developed by the microtubules is proportional to their length, it follows that the only stable equilibrium for the mitotic apparatus obtains when the spindle lies in the long axis, as predicted by Hertwig's second rule. It has been pointed out by Thompson (1942) that adherence to the latter implies that the areas of the new cell partitions are as small as possible. Hertwig's rule may thus be regarded as a special case of the rule of minimum surface area, which again is a special case of the law of minimum free energy. Although the physical basis of this rule unquestionably relates to the length of the mitotic apparatus, it seems to be practical to formulate the principle with reference to the area of the cell partitions.

The cases dealt with here may be summarized in two rules of cell division, the first concerned with equal, the second with unequal, divisions.

The first rule may be phrased as follows:

(1) A cell will divide into two identical daughter cells if spherical symmetry obtains with respect to the factors influencing microtubular activity. The same outcome is generally observed when the symmetry relations are radial or bilateral. If different median planes exist, the location of the division plane will be determined by the rule of minimum surface area.

This statement embraces cell division rules I, III and IV.

If no symmetry obtains, it follows that the two asters must differ with respect to size, and the daughter cells will consequently be unequal. As we have discussed above, the same obtains when the spindle lies in the symmetry axis or plane. In certain cells a median plane may obtain, but the mitotic apparatus thus determined may not ensure adherence to the minimum surface area; under these circumstances the latter must prevail and two unequal daughter cells may be formed. An illustration of this situation may be found in the third cleavage division in many eggs, where the spindle axis generally is parallel to the animal–vegetal axis. The second rule of cell division may be stated in the following way:

(2) A cell will divide into two different daughter cells if no symmetry obtains with respect to the factors influencing microtubular activity. The same outcome is observed if, in case of radial and bilateral symmetry, the spindle lies in the symmetry axis or plane, respectively, and whenever an equal division leads to a violation of the rule of minimum surface area.

Cell division rule IV is a special case of this statement. This second rule is of particular interest in the present context because the polarities prevailing in eggs generally establish symmetry relations of the kind which, as envisaged in the rule, affect the outcome of a cell division.

The extremely unequal cell division observed in association with the formation of the polar bodies is not covered by the second rule. This case may be represented by a third rule:

(3) A meiotic apparatus will be formed if, for reasons inherent in the cortex or in the cell division apparatus, the microtubules from both centres or poles become attached

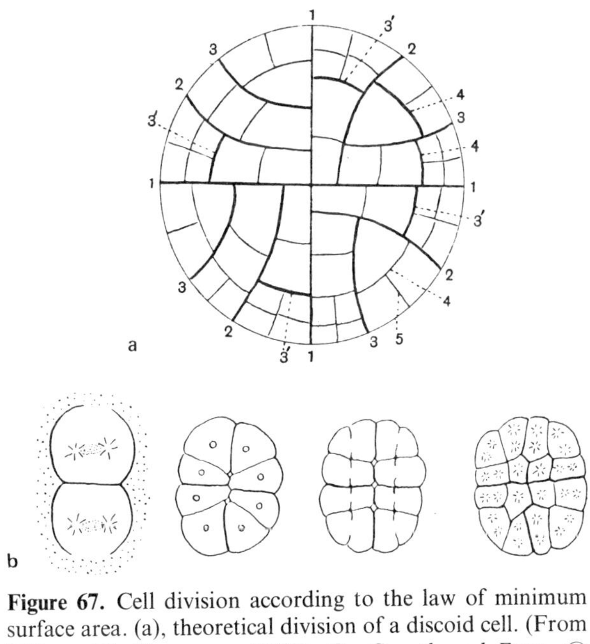

Figure 67. Cell division according to the law of minimum surface area. (a), theoretical division of a discoid cell. (From D'Arcy W. Thompson (1942), *On Growth and Form*, © Cambridge University Press). (b), cleavage in the blastodisc of *Serranus*. (Reproduced from H. E. Ziegler (1902), *Lehrbuch der vergleichenden Entwicklungsgeschichte*, after H. V. Wilson, by permission of Gustav Fischer, Jena)

to the same, rather than to opposite, regions of the egg cortex. The cell division which takes place under such circumstances, by definition located at the animal pole, is distinguished by a great disparity in size between the daughter cells.

Orientation of successive division planes

Sachs' second rule, stating that successive division planes intersect at right angles, was established on the basis of two-dimensional aggregates. To gain generality, it should hold also for three-dimensional aggregates. When this point can be tested in Metazoa, during the three first cleavage divisions, it is found to be valid in most eggs. However, after the third division the cells generally arrange themselves in two-dimensional aggregates, either forming the wall in a coeloblastula, or the external cell layer in a stereoblastula-gastrula. The continuation of divisions in three dimensions is so rare that it may be neglected in the following discussion.

It was shown by Thompson (1942) that in case no other form changes occur than those directly associated with the subdivision of the aggregate, adherence to Sachs' second rule is required to comply with the law of minimum surface area (cf. Figure 67). It may thus be envisaged that it is the shape of the daughter cells which determines the orientation of the following spindles.

However, it is known that the same change of axes occurs in those cases where the cells round up between division, suggesting that this interpretation cannot claim general validity. Examples of this kind are mostly observed when, in many instances of radial and spiral cleavage, replication of the centrioles occurs after completion of karyokinesis, in which case the new centrioles are pulled apart at right angles to the axis of the preceding mitotic apparatus. It is seen that this spatial relation determines a new mitotic apparatus at right angles to the preceding one (Figure 68). Thus, if no forces displace the

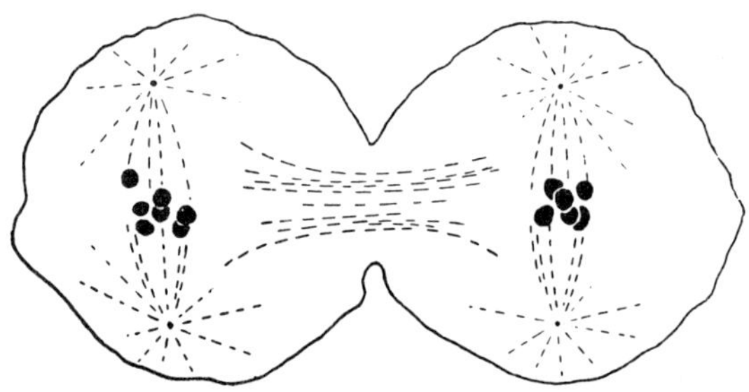

Figure 68. Separation of centrioles perpendicular to the division plane. Second spermatocyte division in two sister cells of *Pachylis*. (Reproduced from F. Schrader (1953), *Mitosis*, by permission of Columbia University Press, New York)

centrioles, their mechanism of replication—or rather separation—suffices to explain the orientation predicted by Sachs' second rule. If we look into the mechanism responsible for this separation it will be realized that the longest microtubules pulling in opposite directions often are found in a plane parallel to that of division. The perpendicular orientation of the cleavage planes may thus be ensured if the centrioles preserve their location when the cells round up.

In certain cases the daughter cells are known to lie in one row, successive division planes obviously being parallel. This phenomenon, which might be called 'one-dimensional' division, is known from *Parascaris*, where it is exhibited by the P_1 blastomere and its early descendants, and from the division of the teloblasts found in various embryos, particularly in those of Annelida. It is worth noticing that the formation of micromeres in many eggs, notably those with spiral cleavage, bears certain similarities to this one-dimensional division pattern. Nothing is known about the mechanism involved, but it seems likely that in *Parascaris* a cortical polarity obtains with the effect that the asters preserve their attachment to the same cortical location through successive divisions. In teloblastic divisions an accumulation of hyaloplasm occurs at the end where the daughter cells are budded off; presumably we are here dealing with a cytoplasmic polarity (Figure 69). Parallelism between successive division planes may also obtain if the cell is so long that this orientation is dictated by the law of minimum surface area (Figure 70).

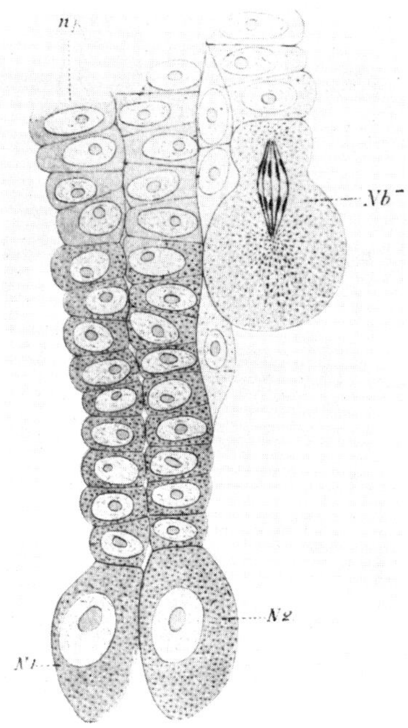

Figure 69. Teloblast division in *Allobophora*. n, nephro-myoblasts; Nb, neuroblast in division. (Reproduced from E. B. Wilson, *J. Morphol.*, **3** (1889))

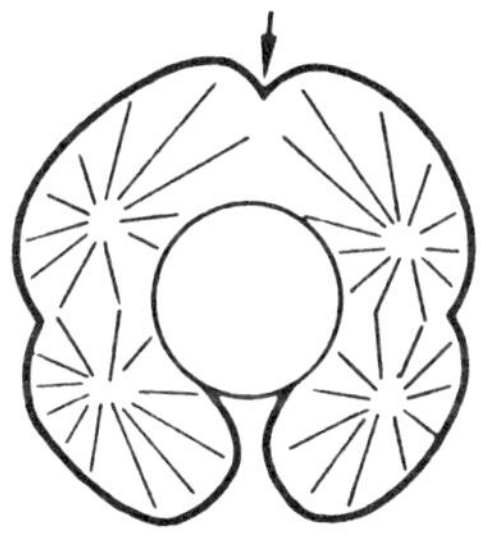

Figure 70. One-dimensional division in an elongated, binucleate cell. The middle furrow (arrow) is formed in the absence of a spindle, but not of asters. (Reproduced from R. Rappaport, *J. Theoret. Biol.*, **9** (1965), by permission of Academic Press, New York)

In the preceding section it was shown that adherence to this law under certain circumstances may lead to deviations from the size differential predictable from the symmetry relations imposed by the polarities. The present discussion shows that the perpendicular orientation between successive division planes may be replaced by parallel orientation in the presence of polarities influencing the formation and attachment of microtubules. We may summarize these considerations in a fourth rule of cell division as follows: (4) Successive division planes are oriented so as to comply with the law of minimum surface area. This entails that in general they intersect at right angles, but in certain special cases they may become parallel. The same result may ensue in the presence of factors interfering with the microtubular activity.

Interactions

It was pointed out in the introduction that epigenetic events represent the outcome of interactions between the embryogenetic substrate and the processes which owe their existence to the predetermined mechanisms. Among these cell division is the first to enter into action, and the present section deals with the interactions between the substrate and this embryogenetic process.

EMBRYOGENETIC SUBSTRATE AND CELL DIVISION

The development preceding gastrulation is generally considered to consist of two phases, the initial subdivision of the egg into a number of smaller cells and blastula formation or *blastulation*. For the first phase the names 'cleavage' or 'segmentation' are usually employed. Both are somewhat ambiguous: the first is often used to designate cell division in general, the other in the meaning of metameric segmentation. In the present context *cleavage* has been adopted, being used only in this specific sense. The following discussion represents an attempt to outline the nature of the interactions between the embryogenetic substrate and the cell divisions occurring during cleavage. This restriction does not imply that interactions do not occur during later embryonic stages, but they are too complex to be subjected to analysis at the present time.

The process of cleavage exhibits some unusual features which can be explained only with reference to the properties of the substrate. The first of these concerns the mitotic activity. During the early stages of development all cells or, to be more general, all nuclei in the embryo, undergo division. Although the number of dividing cells may go on increasing for the better part of the developmental period, the fraction of mitotically active cells rapidly diminishes. It appears that as a general rule gastrulation marks the stage where the transition occurs, as shown for example by the observation that the mitotic index drops from 55 to 8 per cent in *Bufo* during gastrulation (Bragg, 1938). The most remarkable aspect of the early mitotic activity is probably that the initial divisions are synchronous during a number of mitotic cycles. This phenomenon is very unusual, being met with nowhere else in Nature. The first part of this section will be devoted to a discussion of the synchronous cell divisions.

It was set out in the second and fourth rules of cell division that some of the polarities residing in the egg may influence the course of cell division in such a way that the ensuing daughter cells are different with respect to one or more properties, whereas others may affect the orientation of the cells. Depending upon the nature of the polarities, the early blastomeres become grouped in a number of different arrangements, called *cleavage patterns*. We shall here survey some of the most general cleavage patterns, and subsequently discuss the mechanisms by which they are established through influence of the polarities on the process of cell division.

Apart from these effects of the substrate on the process of cell division it should be mentioned that Driesch (1898) observed that in various hybrids of Echinoidea the rate of cleavage, and the number of primary mesenchyme cells, conform to the characteristics of the maternal species, thus being determined by properties of the egg. The former point has been corroborated by A. R. Moore (cf. 1945). This question will not be dealt with further in the present context.

Synchronous cell division

Normally, mitotic activity is separated by long interphases during which the cell 'grows' in preparation for the following division. Under these circumstances it is easily seen that even small individual differences with respect to external and internal conditions may suffice to establish asynchrony. It is therefore to be anticipated that 'relatively synchronized divisions are encountered more frequently in cells that do not grow between divisions, where the preparations for divisions are not so dependent on external nutrition and massive syntheses' (Mazia, 1961, p. 375).

The mitoses occurring in early embryos are distinguished not only by a very brief interphase but by an unusually long duration of the mitotic process (Figure 71). Hence

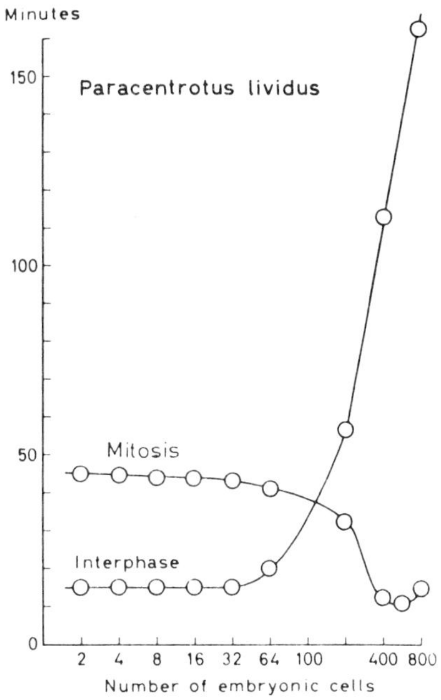

Figure 71. Changes in the mean duration of interphase and mitosis (prophase to telophase) during the early development of the sea urchin embryo. (From I. Agrell (1964), in *Synchrony in Cell Division and Growth*, John Wiley, New York)

the chances of observing all the cells simultaneously in the act of division are greatly enhanced. Whereas an increase in the interphase causes a decrease in the mitotic index (MI), which corresponds to an actual retardation of the process of cell division, a corresponding change in the mitotic phase will give rise to an increase in MI which does not represent a real acceleration of the mitotic activity. This circumstance indicates that the remarkable thing about these divisions may be not so much that they are synchronous, but rather that the normal ratio between the durations of mitosis and interphase is upset.

In order to approach an explanation it may first be proposed that synchrony depends upon cytoplasmic rather than upon nuclear factors. The influence of the latter is supposedly ruled out by the demonstration that the transition from synchronous to asynchronous cell divisions is unaffected by X-ray induced inactivation of the nuclei (Neyfakh and Rott, 1958, quoted by Dettlaff, 1964). If a cytoplasmic factor somehow furthers synchronous divisions, the fact that this activity sooner or later comes to an end suggests that the agent becomes exhausted, either by degradation or by utilization. The former possibility cannot be definitely excluded, but the latter is more likely. Accepting it, and presuming that the factor is homogeneously distributed in the cytoplasm and is used in equal amounts at each division, it follows that small cells, or rather nuclei exposed to limited volumes of cytoplasm, should reach sooner the state of exhaustion, involving the elongation of the interphase and, consequently, transition to asynchronous cell division.

This is borne out by many observations; thus it may be mentioned that in the sea urchin embryo the micromeres very often lag behind the remaining cells. When they have divided into eight cells, the two quartets are of different size, and here the outer larger micromeres always divide ahead of the inner smaller ones (Agrell, 1964). By microphotometric measurements, McMaster (1955) has shown that in *Lytechinus* the Feulgen reaction in the interphase nuclei reaches maximum intensity first in the macro-

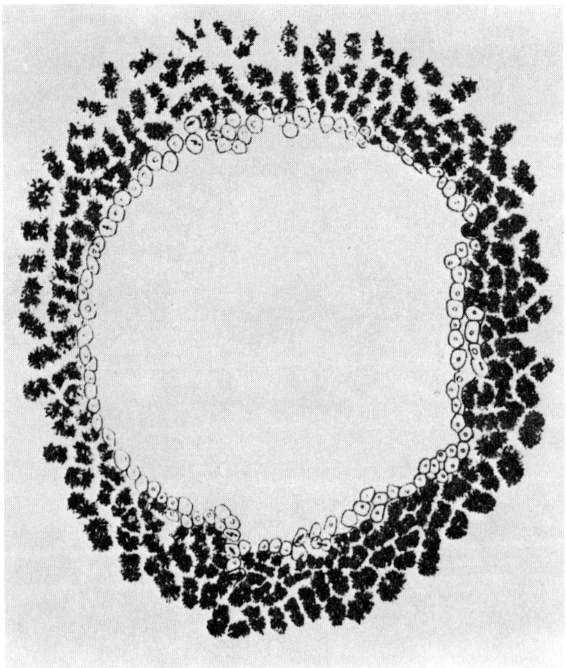

Figure 72. Division synchrony in the yolk syncytium during the thirteenth cell cycle in *Belone*. The inner part of the blastoderm is not drawn. (From F. Kopsch, *Internat. Monatschr. Anat. Physiol.*, **18** (1901))

meres, subsequently in the mesomeres and last in the micromeres. These results suggest that the preparation of the nuclei for the following division depends upon the volume of the cells. In the egg of *Belone*, a fish, synchrony obtains for eight division cycles, but in the peripheral syncytium where the nuclei are not confined by cell boundaries, the synchrony continues for at least three more cycles (Kopsch, 1901, Figure 72). Similar observations have been made in Cephalopoda (cf. Agrell, 1964). The data presented here thus support the contention that a substance is present in the cytoplasm of the fertilized egg which is of importance for the course and the rate of cell division, and which gradually becomes exhausted.

As regards the nature of this substance, Mazia (1961) has suggested that it may be the protein of the mitotic apparatus, i.e. TP monomers according to the present terminology. This suggestion is somewhat difficult to reconcile with the exhaustion phenomenon, since the total volume remains unchanged during cleavage, implying that although the number of mitotic apparatuses increases, the space they must span is correspondingly reduced.

Hoff-Jørgensen and Zeuthen (1952) published determinations of DNA in *Rana* made with a microbiological essay, specific for deoxyribonucleosides. They made the remarkable observations that up to the blastula stage the amount of 'DNA' remained constant, at a value of $6{\cdot}5 \times 10^4$ pg, whereas the amount per spermatozoon was about 12 pg, the material found in the cell thus corresponding to about 2200 diploid nuclei. Hoff-Jørgensen and Zeuthen figured out that this quantity cannot possibly be accommodated in the egg pronucleus, and were able to confirm this point by demonstrating that most of it is present in the cytoplasm. No cell counts were made, but it can be estimated that at the time the 'DNA' content increases, the cell number is of the same order as the number of nuclei mentioned. It was therefore natural to propose that the deoxyriboside-containing material is a precursor for the DNA incorporated in the cleavage nuclei, its function being to ensure that cleavage 'rapidly brings the embryo to the state . . . that it can begin its own synthetic activities' (Zeuthen, 1951, p. 65).

It has been found that the time at which asynchrony sets in in Amphibia (cf. Schönmann, 1938) corresponds to the stage at which the net increase in DNA content begins, and a similar correlation has been found in various Echinoidea (Hoff-Jørgensen, 1954; Agrell, 1964), and in Insecta (Durand, 1961). These observations indicate that the typical features of the synchronous divisions, long mitotic phase and brief interphase, depend upon the presence of the deoxyriboside-containing material. The results of Grant (1960) showing that the sensitivity of DNA synthesis towards folic acid analogues is very low during cleavage, as compared to later stages, suggest that a change in the mechanism of DNA synthesis occurs, and the time of transition corresponds well with that at which the material becomes exhausted. Findings of this kind lend strong support to Zeuthen's contention that it is used for the synthesis of DNA.

Many authors have published determinations of DNA in embryos, notably in Echinoidea and Amphibia. From my reviews of these data (1966b; 1966c) it appears that the results obtained are quite disparate, some estimates being quasi-astronomical when calculated as numbers of diploid nuclei. However, careful scrutiny shows that values of this kind are obtained when unspecific methods are employed; the more specific ones agree in giving values lying between 10 and 50 pg 'DNA' per embryo in sea urchins, and below 0·1 μg in Amphibia, corresponding to 4–5 divisions in the former, and to 11–13 (2000–8000 cells) in the latter case. Grant (1958) has shown that part of the deoxyriboside-containing material is low molecular (acid-soluble), in agreement with the assertions of Hoff-Jørgensen and Zeuthen (1952), and also that the acid-insoluble fraction increases during cleavage, in perfect agreement with the suggested function as reserve material. Kuriki and Okazaki (1959) found the total amount of 'deoxyribosides' to be 2–3 times larger than that reported by other authors using the microbiological method, and of this approximately 75 per cent was acid-soluble, mainly deoxynucleotides. This fraction remains constant during development, and its function is therefore difficult to ascertain.

To the molecular biologist the presence of DNA in the cytoplasm, except in mitochondria, is heresy, violating a dogma of this science. Attempts to reject and disprove the assertions just discussed have therefore not been lacking. The most thorough work in this spirit is that of Dawid (1965), who applied the conventional methods for DNA isolation to the cytoplasm of amphibian eggs. Dawid arrived at the conclusion that the material thus isolated is mitochondrial DNA, but since the yield was only a few per cent of the amount found in the analyses mentioned above, it follows that these endeavours do not settle the problem. It therefore seems possible to maintain that in various eggs a material is present, distinguished by containing deoxyribosides, without therefore being high-molecular DNA. Nothing definite is known about this material; it may consist of oligonucleotides rich in guanine (cf. Durand, 1961; Agrell, 1964). This material disappears after a certain number of division cycles, but as long as it is present it ensures that cell division occurs with a very brief interphase period.

It has been known for a long time that the transition from synchronous to asynchronous cell divisions is associated with the formation of nucleoli, absent in synchronously dividing cells (Richards and Porter, 1935; Schönmann, 1938). This phenomenon suggests that new synthetic activities are initiated at this stage, and Dettlaff (1964) has summarized substantial evidence suggesting that only in asynchronously dividing cells have the nuclei assumed their specific function. The onset of this activity entails that the presumptive fate of the cells is determined, or at least that the first step towards this end has been made. The correlation between the various phenomena is most clearly shown in determinate eggs, in which the fate of the blastomeres is determined at a very early stage. In *Lymnaea* it has been found that nucleoli appear at the 16 cell stage (Raven, 1966), and Neyfakh (1961, quoted by Dettlaff, 1964) has shown by X-ray irradiation that nuclear function begins at the 12–16 cell stage. Unfortunately, the cell lineage has not been published in detail for any member of this genus, but in the closely related *Physa* asynchronous divisions begin at the 16 cell stage (Wierzejski, 1905). Likewise, no DNA analyses have been made on *Lymnaea*, but in *Ilyanassa* Collier and McCann-Collier (1962) have found that the content corresponds to about 32 times the diploid amount, and that synthesis begins at the 25 cell stage. All these data agree very well to bear out the suggestion advanced by Dettlaff (1964) that the transition from synchronous to asynchronous cell division signals the onset of the determination process. If we go a step further and presume that asynchrony appears when the reserves for DNA synthesis are exhausted, then it follows that the stores should be small in determinate and large in indeterminate eggs, a point which is corroborated by most of the published DNA analyses (cf. J. Williams, 1965). If this supposition is correct, then it obviously follows that the main function of the deoxyriboside-containing material, rather than to further the synthetic activities of the cells, is to delay these until a certain number of cells is present.

Cleavage patterns

The cleavage patterns owe their existence to the fact that the cells or blastomeres in different parts of the embryo vary with respect to size and to their reciprocal spatial orientation and to the circumstance that in certain cases only part of the egg is involved in the process of cell division. It is evident that this course of events must be referred to the presence of factors which, as envisaged by the second and fourth rules of cell division, interfere with the formation and attachment of microtubules. The factors thus engaged are the polarities discussed above.

It is possible to distinguish two classes of cleavage patterns, viz., *holoblastic* ones, in which the whole mass of the egg takes part in the process of cell division, and *meroblastic* ones, in which this activity is confined to the superficial layers. To some degree a correlation obtains between egg size and class affiliation, meroblastic cleavage being found mainly in large, holoblastic in small, eggs, but deviations from this rule are common. Each of these classes contains a number of cleavage patterns, generally characterized by the symmetry pattern they impose upon the cleaving embryo. In the present context we shall discuss the most important cleavage patterns, establishing in a formal way the number and the nature of the polarities required for their genesis.

In order to do that, it will be necessary to formulate some rules concerning the effect of the polarities upon the process of cell division. It will be assumed that the polarities act in an all-or-none fashion, which implies that if a cell division plane cuts the polarity vector perpendicularly, the polarity will continue to prevail in one of the daughter cells and be absent in the other one. Contrariwise, if a division plane is parallel to the direction of the vector, the original polarity will prevail in both daughter cells. Finally, if the division plane intersects the polarity vector obliquely, it will be presumed that both daughter cells contain a polarity which, however, is attenuated or modified relative to the original one, and generally different in the two cells.

A comparison with the very comprehensive survey of cleavage patterns published by Siewing (1969) will show that I have neglected quite a few of the more complicated cases. This does not imply that the relations between cleavage patterns and polarities set out in the following sections are valid only for simple cleavage patterns, but it must be admitted that considerable ingenuity may be required to account for the causal connections in some of the more intricate instances observed.

Holoblastic cleavage patterns

On the basis of the symmetry relations displayed by the various holoblastic cleavage patterns we shall distinguish four groups, characterized by *radial*, *biradial* and *bilateral symmetry*, and *bilateral asymmetry*, respectively.

Radial cleavage

In the absence of any polarity all cells will be identical, and spherical symmetry must consequently obtain. As pointed out, this situation is never realized in Nature. If one polarity prevails it will, probably without exception, be an apical polarity. As a consequence of the latter the cells must vary with respect to either size or composition, or both, along the animal–vegetal axis. Usually the two first divisions are meridional and the third one latitudinal. Under such circumstances the presence of an apical polarity cannot be demonstrated before the 8 cell stage, in which there must be two groups of four cells each, differing with respect to size or other properties. The two groups of cells are generally characterized by the adjectives 'animal' and 'vegetal', respectively. This phenomenon is extremely common, being encountered in almost all eggs with holoblastic cleavage (Figure 73). The formal representation of the cooperation of the

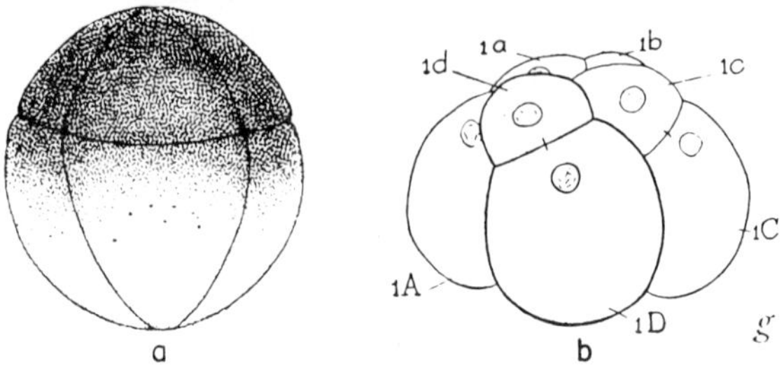

Figure 73. The 8 cell stage, showing the influence of an apical polarity on the cleavage pattern. (a), *Rana*; (b), *Trochus*. (Reproduced from T. H. Morgan (1927), *Experimental Embryology*, (b) after Robert, by permission of Columbia University Press, New York)

polarity in the establishment of the cleavage pattern is shown in Figure 74. If two apical polarities are present, and if the rule of perpendicularity with respect to the division planes is obeyed, it may be anticipated that two tiers of eight cells will be found at the 16 cell stage, the members of which contain either one or the other of the original

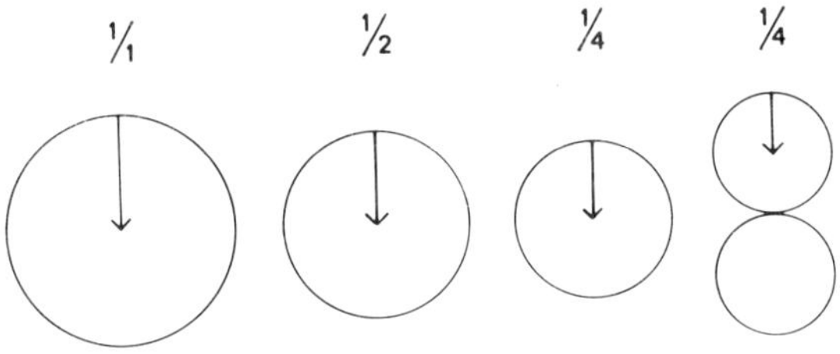

Figure 74. Formal representation of the influence of an apical polarity upon the cleavage pattern. The fractions indicate the part of the whole embryo represented by the respective diagrams. In the 8 cell stage the two cells have been made of equal size, thus indicating that the polarity must not affect the size even if this usually is the case

polarities. The following division plane must be latitudinal, leading to the formation of four groups of eight cells (Figure 75), of which two, the animal and the vegetal cells, contain the original polarities, whereas the intermediate sixteen cells contain no polarities. Whether or not these two tiers of cells are identical cannot be decided *a priori*.

If one of the apical polarities affects the orientation of the mitotic apparatuses, a more complicated cleavage pattern may ensue. An example of this is found in Echinoidea, where the fourth cleavage leads to the formation of two tiers of four cells in the vegetal, and one tier of eight cells in the animal, hemisphere (Figure 76d). In the following division the latter divide latitudinally and the former longitudinally, thus giving rise to

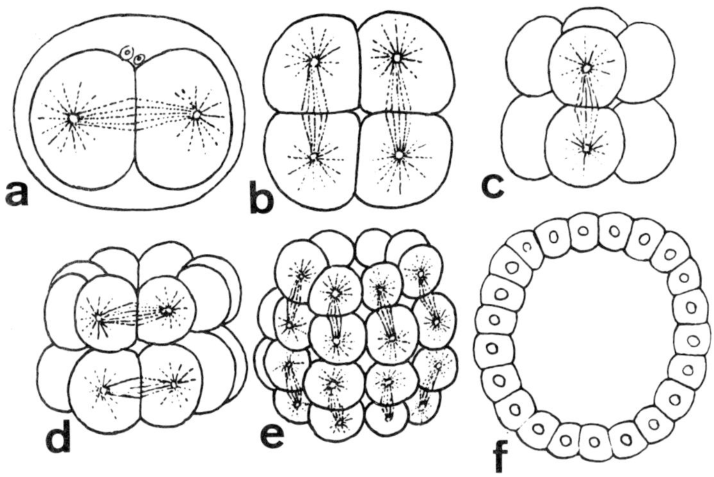

Figure 75. Cleavage in *Synapta*. (a)–(e), 2, 4, 8, 16 and 32 cell stages; (f), blastula. (Reproduced from E. Korschelt and K. Heider (1936), *Vergleichende Entwicklungsgeschichte der Tiere*, by permission of Gustav Fischer, Jena (after Selenka))

four groups of different cells (Figures 76e and 77). These may in principle correspond to the apical plate, epidermal, archenteron and mesenchyme cells, respectively. It must be pointed out that the segregation as far as the presumptive fate is concerned is not completed at the 32 cell stage. After the following division the endoderm has become separated from the ectoderm, but apparently the presumptive apical plate cells and those of the animal ectoderm still form one population. If the polarities act in an all-or-

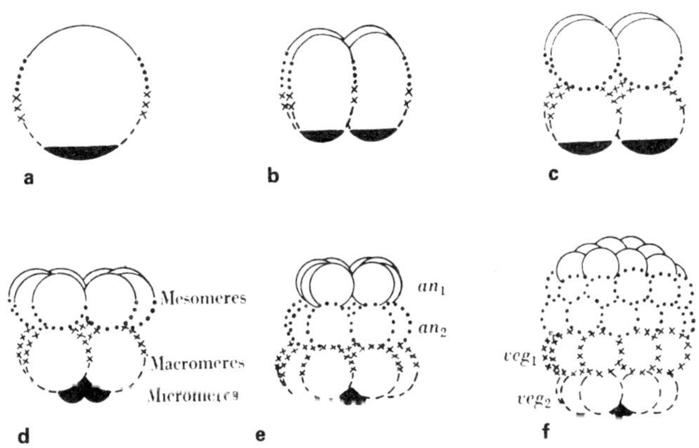

Figure 76. Cleavage in *Paracentrotus*. an_1, upper tier of animal blastomeres (———); an_2, lower tier of animal blastomeres (. . . .); veg_1, upper tier of vegetal blastomeres (×××); veg_2, lower tier of vegetal blastomeres (---), micromeres (black). (a), uncleaved egg; (b), 4 cell stage; (c), 8 cell stage; (d), 16 cell stage; (e), 32 cell stage; (f), 64 cell stage. (Reproduced from S. Hörstadius, *Pubbl. Staz. Zool. Napoli*, **14** (1935), by permission of Stazione Zoologica di Napoli, Naples)

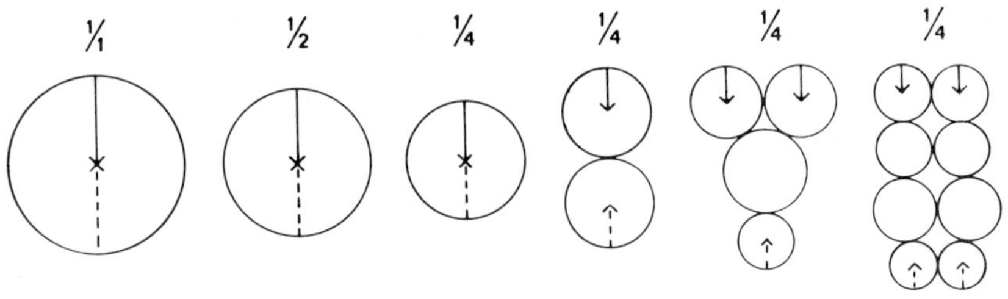

Figure 77. Formal representation of the influence of two apical polarities on the cleavage pattern, when the successive cleavage planes are orientated as in the sea urchin egg. The fractions indicate the part of the whole embryo represented by the respective diagrams

none fashion, two should give at most four different cell types, and it is therefore questionable whether further subdivision of the cells containing the polarities will give rise to new kinds of cells. Rather, it may be envisaged that during the immediately following divisions of the animal cells, one of the daughter cells is a presumptive epidermal cell, while the fate of the other is undetermined, but that after a certain number of divisions the cells around the animal pole are all presumptive apical plate cells. As far as the four vegetal cells in the 16 cell stage are concerned, there is reason to believe that all their descendants are mesenchyme cells.

If neither appearance nor immediate presumptive fate of the cells in each of the tiers of cells discussed above indicates that they are different, a radial cleavage pattern is said to obtain. When applied to the 8 cell stage, this name may be somewhat awkward, and Wilson (1928) therefore used the expression ‘orthoradial’. This term is hardly of more general service than the traditional one, and the latter is therefore employed here. From the preceding discussion it appears that the necessary and sufficient conditions for the generation of a radial cleavage pattern are that the three first division planes are perpendicular to each other and that at least one apical and no median polarity is present. It is not necessary to pay regard to the other kinds of polarity, for a dextro–sinistral polarity cannot exist in the absence of a median one, and a normal polarity does not interfere with the symmetry relations.

In radial cleavage the cells in adjacent tiers usually lie on top of each other. Provided that adhesive forces obtain between the blastomeres, this arrangement cannot correspond to a state of minimum free energy, because the surface contact evidently must increase if one tier of cells lies in the furrows between those of the other one. It may therefore be presumed that force must be deployed by some supracellular agent, generally the hyaline membrane, to prevent the cells from assuming the organization pattern corresponding to the state of minimum free energy. When the cells in radially cleaving eggs behave according to the prediction of physics, as for instance in some Cnidaria (Nyholm, 1943), it thus rather reveals the absence of a hyaline membrane than any more sophisticated mechanisms. It will be realized that the cells may arrive at their position by clockwise or counter-clockwise movements, the chances presumably being equal for either possibility. Under these circumstances it is misleading to call this phenomenon ‘pseudo-spiral cleavage’ (Siewing, 1969), since the typical feature of spiral cleavage is the circumstance that the direction of the movement is predetermined, not that the cells by their location happen to comply with the law of minimum free energy.

Biradial cleavage

It appears that biradial cleavage patterns usually arise when the three first division planes do not stand at right angles to each other.

The simplest example of this type is encountered in the 'spiral' cleavage with only two micromeres, as observed in certain Acoela like *Polychoerus* (Figure 78). Here two

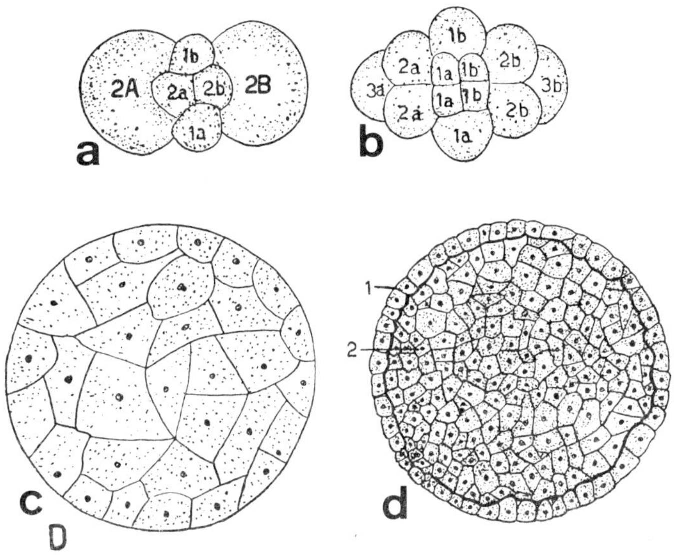

Figure 78. Development of *Polychoerus*. (a), 6 cell stage seen from above; (b), 16 cell stage seen from above; (c), early stereogastrula; (d), late stereogastrula, surface cells separated as ectoderm. 1, ectoderm; 2, meso- and endoderm. (Reproduced from L. H. Hyman, (1951), *The Invertebrates*, Vol. 2, by permission of McGraw-Hill, New York (after Gardiner))

median planes are defined, one through the centres of the macromeres and one between them, perpendicular to the former. No definite differences have been demonstrated between the macromeres, and there is therefore no reason to believe that a median polarity obtains. This point is borne out among other things by the fact that the stereogastrula to all appearances is radially symmetrical (Figure 78c and d). The fact that the adult animal is bilaterally symmetrical does not violate this conclusion, since this organization may be arrived at without the presence of a median polarity, as will be shown later. It is quite obvious that the particular symmetry relations obtaining in this case stem from the fact that the second mitotic spindle does not lie in the same plane as the first one.

The first eight blastomeres in Ctenophora lie in two rows of four cells each (Figure 79), a situation which is accounted for by the fact that the third division planes are parallel to the first one. It is known that the mitotic apparatuses during these early divisions are located at the animal pole, and the deviation from the usual three-dimensional orientation between the three first cleavage divisions indicates that the astral microtubules cannot penetrate the yolk-laden cytoplasm. This interpretation is corroborated by the fact that in the subsequent divisions the mitotic activity is confined to the ectoplasm, thus giving rise to a large number of hyaline micromeres (cf. Figure 104). Through the eight macromeres two median planes are defined, one separating two rows, the other two groups of four cells (Figure 79c). Since the outer cells in each row have been found to differ from the inner ones with respect to their presumptive fate, it must be presumed that polarities are present in the four blastomeres which by division give rise to the eight cells depicted in Figure 79. Symmetry obtains, however, around the median plane

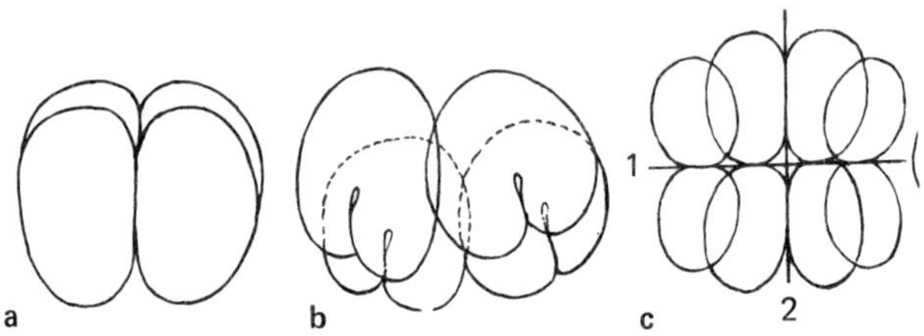

Figure 79. Macromere formation in *Ctenophora*. (a), 4 cell stage; (b), third division; (c), 8 cell stage, with the two median planes indicated (1 and 2). (Reproduced from E. Korschelt and K. Heider (1936), *Vergleichende Entwicklungsgeschichte der Tiere*, by permission of Gustav Fischer, Jena)

separating the two groups of four cells, and formally the present case may therefore be represented by two identical and oppositely directed polarities in the fertilized egg, lying in the presumptive median plane of the two rows of cells (Figure 80). In *Poly-*

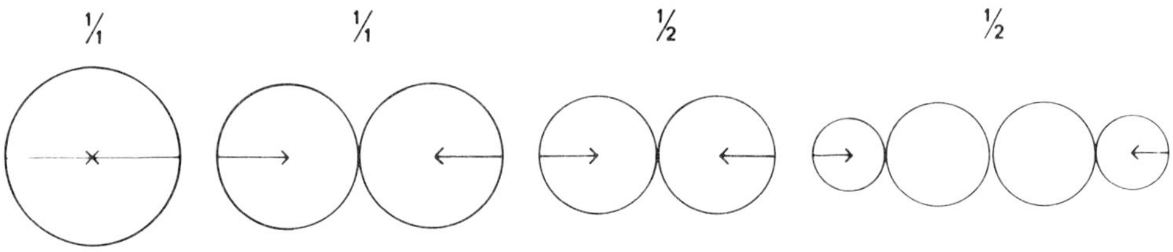

Figure 80. Formal representation of the influence of two identical and oppositely directed median polarities on the cleavage pattern. The fractions indicate the part of the whole embryo represented by the respective diagrams

choerus, where two median planes, but no polarities, obtain, the biradial cleavage pattern is of no consequence for subsequent development. It seems likely that the presence of two polarities of the kind suggested to exist in Ctenophora is a minimum requirement for establishing permanent effects.

Bilaterally symmetrical cleavage

The presence of one polarized median plane confers bilateral symmetry on the egg and the embryo, and a median plane of this type may be determined by two separate vectors, representing either one apical and one median polarity, or two of the latter type, provided they do not lie on the same axis. Since a bilateral cleavage pattern presupposes bilateral symmetry, it may be inferred that the conditions establishing the latter are the necessary and sufficient conditions for the former to occur.

Considering first the case where an apical and a median polarity are involved, a number of different cases may be distinguished. The simplest ones are those where the first cleavage coincides with and the second is perpendicular to the median plane

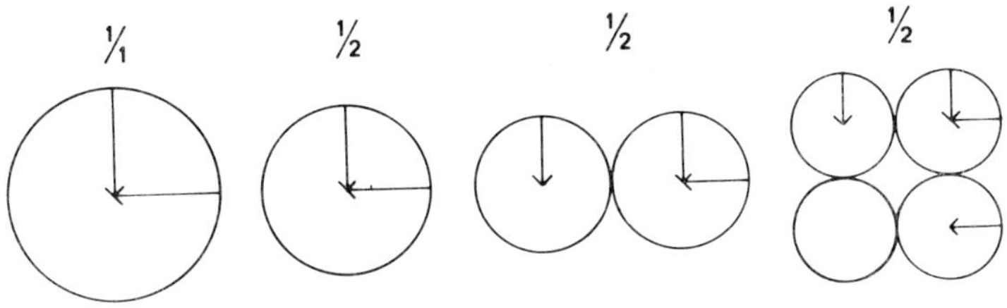

Figure 81. Formal representation of the influence of an apical and a median polarity on the cleavage pattern, when the first division plane lies in the median plane. The fractions indicate the part of the whole embryo represented by the respective diagrams

(Figure 81) and where the order is reversed (Figure 82). It is possible that these patterns are encountered in Nature, but representative examples do not appear to be abundant in the literature.

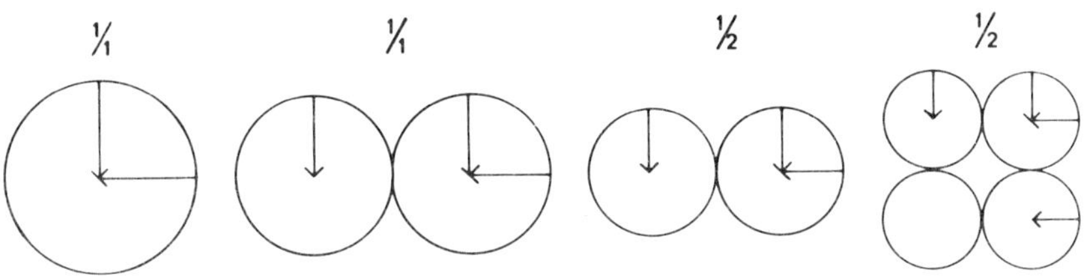

Figure 82. Formal representation of the influence of an apical and a median polarity on the cleavage patterns, when the first division plane is perpendicular to the median plane. The fractions indicate the part of the whole embryo represented by the respective diagrams

The actually observed bilateral cleavage patterns are more complicated. An interesting case is found in the Amphibia, where the formation of the grey crescent signals the establishment of the median plane of the embryo. Great pains have been taken to

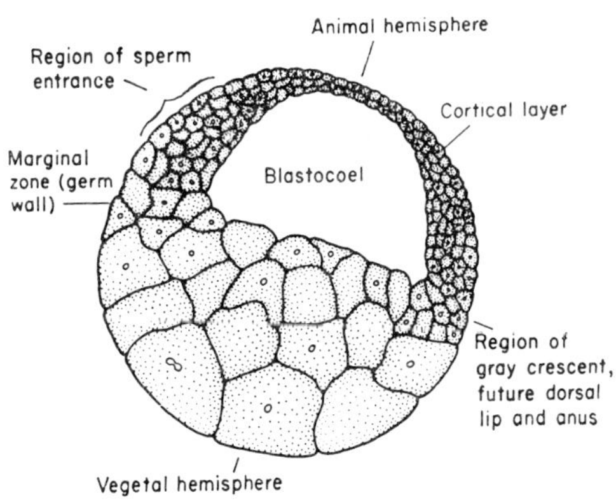

Figure 83. Section of a frog blastula. (Reproduced from R. Rugh (1951), *The Frog Its Reproduction and Development*, by permission of McGraw-Hill, New York)

demonstrate the coincidence of this with the first cleavage plane, but although a certain preference may be discerned, it nevertheless holds that any angle between the two is compatible with normal development (cf. Ancel and Vintemberger, 1948). This circumstance entails that the effect of the polarity cannot be anticipated to be displayed at the 8 cell stage. From this it cannot be concluded that the median polarity does not interact with the mechanism of cell division, and thus with the cleavage pattern. The opposite relation clearly obtains, for it has been observed many times that the cells at the dorsal side consistently are smaller than those on the ventral side (Figure 83).

Another particular case is found in the spirally cleaving eggs, in which the first cleavage division results in the formation of two unequal blastomeres, called AB and CD. In some cases a median polarity exists before the division, the plane of which may thus be envisaged to be perpendicular to the median plane. In other cases the cell division proper is instrumental in the conversion of a vegetal–apical into a median polarity. During the second cleavage division AB divides into the macromeres A and B which are essentially identical as far as their subsequent fate is concerned. The division of CD is unequal, giving the macromere C which is similar to A and B, and D which is different from the former (Figure 84). If a median plane exists in the fertilized egg it coincides with that determined by the blastomeres AB and CD, passing through their centres. A new median plane is determined after the second division, passing through the centres of B and D. This plane does not coincide with the ultimate plane of symmetry, but there is a constant relation between them within a species; however, between species and still more between higher taxa considerable variation obtains. The influence of the apical polarity is exerted only at the third division, when the micromeres 1a–1d are separated from the macromeres (Figure 84).

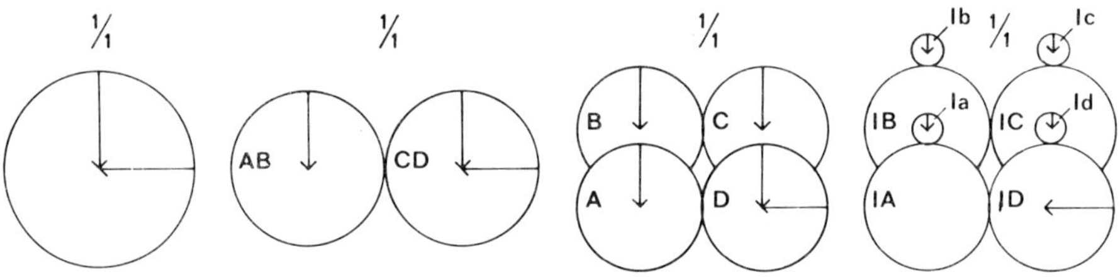

Figure 84. Formal representation of the influence of an apical and a median polarity on the cleavage pattern, when the median plane passes through the centres of the two first blastomeres. The fractions indicate the part of the whole embryo represented by the respective diagrams. The names of the blastomeres are indicated according to the accepted terminology for spiral cleavage

An example where the presence of two median polarities may be inferred is found in Ascidiacea. In this case the polarities are visibly displayed by the grey and the yellow crescents (Figure 46). In *Styela* the first division plane coincides with the median plane. According to the rules established for oblique polarities, the following two divisions, one longitudinal and one latitudinal, should give rise to four different cells in each half of the embryo (Figure 85). Actual observations seem to bring out this point (Figure 86). It must be pointed out that the principle illustrated in Figure 85 is a completely theoretical construction. Nothing prevents an apical polarity existing in *Styela*, and the fact that the animal cells are smaller than the vegetal ones (Figure 86) rather suggests that this is the case.

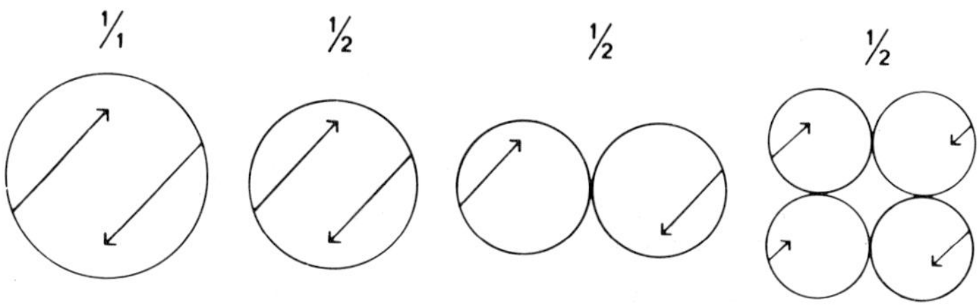

Figure 85. Formal representation of the influence of two different median polarities on the cleavage pattern. The fractions indicate the part of the whole embryo represented by the respective diagrams

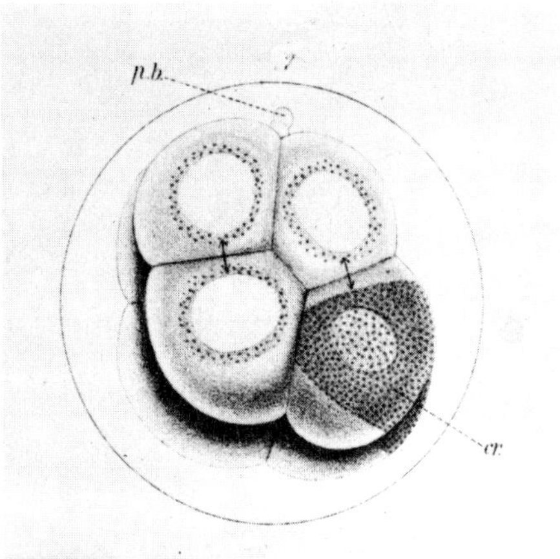

Figure 86. Cleavage in *Styela*, 8 cell stage in lateral view. Most of the yellow ectoplasm is located in the lower right blastomeres, but some is found around nuclei in the remaining ones. The animal cells are clearly smaller than the vegetal ones. cr, yellow crescent; p.b., polar body. (Reproduced from E. G. Conklin, *J. Acad. Nat. Sci. Philadelphia*, [2], **3** (1905))

Bilateral symmetry of a very special type is encountered in *Parascaris*, where an apical polarity is responsible for the fact that the spindles of the second cleavage divisions are oriented perpendicularly to each other. The resulting T-shaped aggregate determines a median plane passing through the centres of all four blastomeres (Figure 87). If A and B are identical, a second median plane prevails, going through the centres of P_2 and EMSt at right angles to the first one, in which case the aggregate is biradially symmetrical. The cell P_2 subsequently establishes contact with one of the animal blastomeres, thereby giving rise to a lozenge-shaped aggregate. Through this event the surface contact is increased. Theoretically, P_2 may move to the right or the left in the median plane, but if biradial symmetry obtains, these two possibilities are indistinguishable. In the lozenge-shaped aggregate the main axes are determined, the dorso–ventral one passing from A

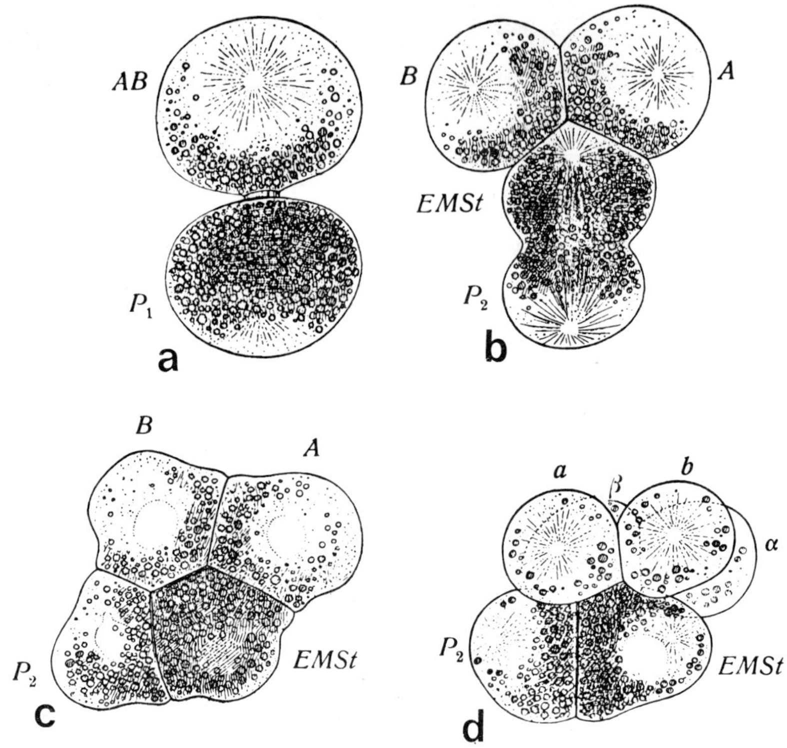

Figure 87. Cleavage in *Parascaris*. (a), 2 cell stage; (b), T-shaped 4 cell stage; (c), lozenge-shaped 4 cell stage; (d), 6 cell stage. The letters indicate the names of the blastomeres. (Reproduced from Th. Boveri (1910), in *Festschrift zum sechzigsten Geburtstag Richard Hertwig*, by permission of Gustav Fischer, Jena)

and B to P_2 and EMSt, and the antero–posterior one from EMSt to P_2. It will be realized that according to the principle that polarities cannot arise spontaneously, it is necessary to imply that two apical polarities exist. Of these the animal–apical one is responsible for the presumptive dorso–ventral axis, whereas the vegetal–apical one through the movement of P_2 is converted into a median polarity, determining the antero–posterior axis.

Bilaterally asymmetrical cleavage

Bilateral asymmetry is established in the presence of a dextro–sinistral polarity which may be represented by a vector perpendicular to a median plane determined either by an apical and a median or by two median polarities. Several of the examples discussed in the preceding section, for instance *Branchiostoma* and Amphibia, actually are distinguished by a dextro–sinistral polarity, as shown by an asymmetry in the adult body. It has not, however, been possible to demonstrate any effect of this polarity upon the cleavage pattern.

The situation is different in eggs with spiral cleavage, in which the first four micromeres come to lie in the furrows between the macromeres, for here the orientation between corresponding pairs of macromeres and micromeres is predetermined by a dextro–sinistral polarity. It was shown above that the macromeres determine a median polarity and that, provided one of the micromeres differs from the others, the latter will also establish a polarity. In this way two 'median' planes come into being, intersecting in the axis of the apical polarity at an angle of 45° (cf. Figure 88).

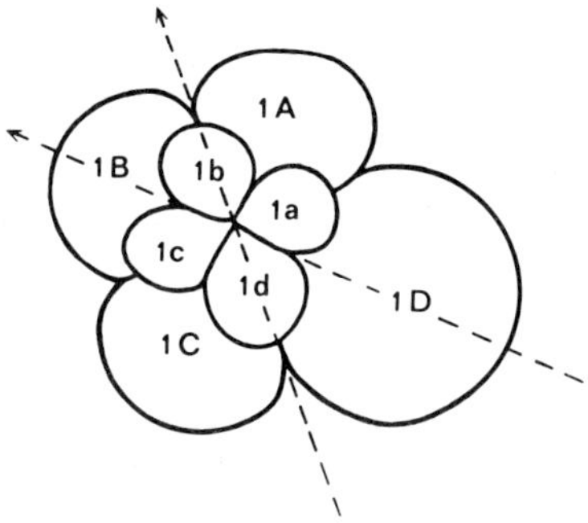

Figure 88. Establishment of a dextro–sinistral polarity through the presence of two median polarities. The names of the blastomeres are indicated according to the accepted terminology for spiral cleavage

In case the four first micromeres are identical, two median planes may be determined if the four second micromeres are different. In this instance the planes coincide initially, but as a consequence of subsequent division, this relation may be changed, thereby giving rise to a dextro–sinistral polarity. From what is known about the morphology of animals with spiral cleavage, it is obvious that coincidence between the planes is the rule rather than the exception; in actual fact it is only among Gastropoda and Bivalvia that asymmetry is found in the adult organism. A formal representation of the cleavage in an egg possessing a dextro–sinistral polarity is shown in Figure 89. It is seen that

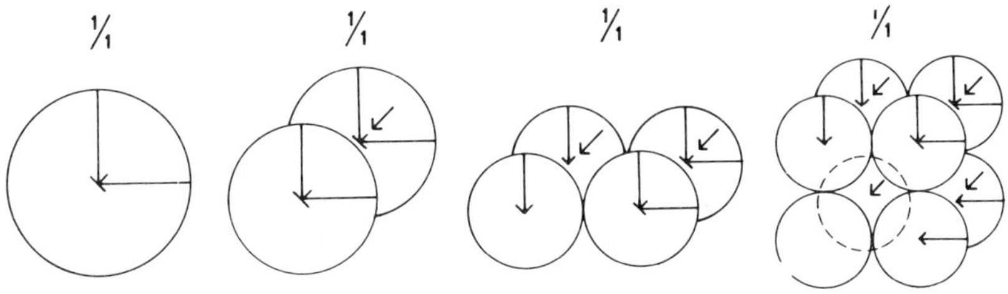

Figure 89. Formal representation of the influence of an apical, a median and a dextro–sinistral polarity on the cleavage pattern. The fractions indicate the part of the whole embryo represented by the respective diagrams

potentially eight different cells may be formed, but it is doubtful whether this possibility is ever exploited.

It was observed by zur Strassen (1896) that asymmetrical cleavage obtains in *Parascaris*, as demonstrated by the fact that at the 6 cell stage the four animal blastomeres may be arranged in two different ways, one being the mirror image of the other (Figure 90). This peculiar cleavage pattern presumably depends upon the fact that the four cells are of different size, but if the two divisions through which they arise were equal, it might

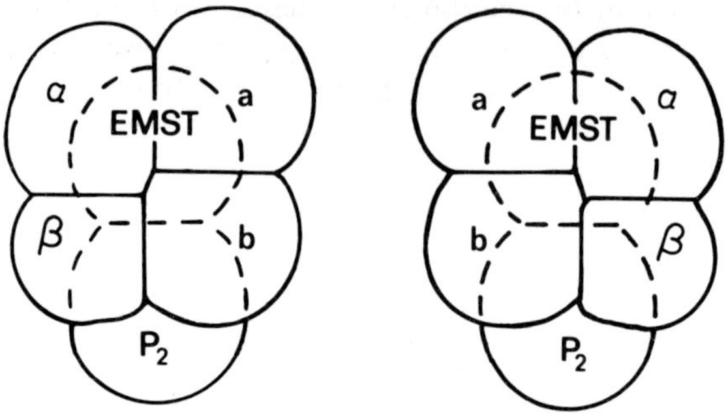

Figure 90. Bilateral asymmetry in *Parascaris*. (Reproduced from R. Bonfig, *Z. Wiss. Zool.*, **124** (1925), by permission of Akademische Verlag, Leipzig)

be anticipated that each of the two arrangements would appear with equal frequency. In point of fact one pattern was found to occur 30–40 times more often than the other one, suggesting the presence of a dextro–sinistral polarity. Bonfig (1925) resumed the study of cleavage in *Parascaris*, but unfortunately he was unable to settle the crucial problem, the material basis of the polarity. Interestingly enough, zur Strassen observed that among 125 adult animals, the nucleus of the excretory system was located to the right in four cases and to the left in the remaining ones, indicating that the arrangement of the animal blastomeres is of morphogenetic consequence.

Meroblastic cleavage patterns

Two types of meroblastic cleavage are known, their distinction being based upon the fraction of the egg surface which is involved in the process of cleavage. If the surface is only partially engaged, the cleavage pattern is called *discoidal*, otherwise it is termed *superficial*. Both patterns thus are really superficial; it might actually be more correct to characterize them as total and partial (or polar) superficial cleavage.

Superficial cleavage is usual, but not ubiquitous, in Arthropoda, but is found also in other phyla, e.g. Cnidaria (cf. Siewing, 1969). Discoidal cleavage is found in Cephalopoda and in some classes of Vertebrata, and occasionally in other parts of the animal kingdom. Transitions between holoblastic and meroblastic cleavage are known in many cases, thus in certain Crustacea the cleavage may initially be holoblastic. Likewise, in the lower orders of Neopterygii, e.g. in *Amia* and *Lepisosteus*, the cleavage pattern is transitional between holoblastic radial and discoidal cleavage.

Superficial cleavage

In the typical superficial cleavage the pronuclei fuse in the interior of the egg, and begin their multiplication while being gradually displaced towards the periphery, where they come to be located in the ectoplasmic layer (Figure 41). Cell formation only begins when the nuclei have arrived at the cortex. In a spherical and otherwise uniform egg, it might be anticipated that the cortical density of nuclei and cells would be uniform, amounting to a normal polarity and a spherical cleavage pattern. This situation does not obtain in the spherical eggs of Crustacea; rather, the nuclei tend to accumulate preferably at the animal pole, suggesting the presence of an animal–apical polarity.

This point is borne out by the observation by Manton (1928) on *Hemimysis*, in which the ectoplasmic layer is substantially thicker at the animal than at the vegetal pole. This polarity thus gives rise to a radially symmetrical 'cleavage pattern', based upon the density of the otherwise identical nuclei. The presence of a median polarity is demonstrated by the later formation of a germinal disc at the presumptive ventral side (Figure 91).

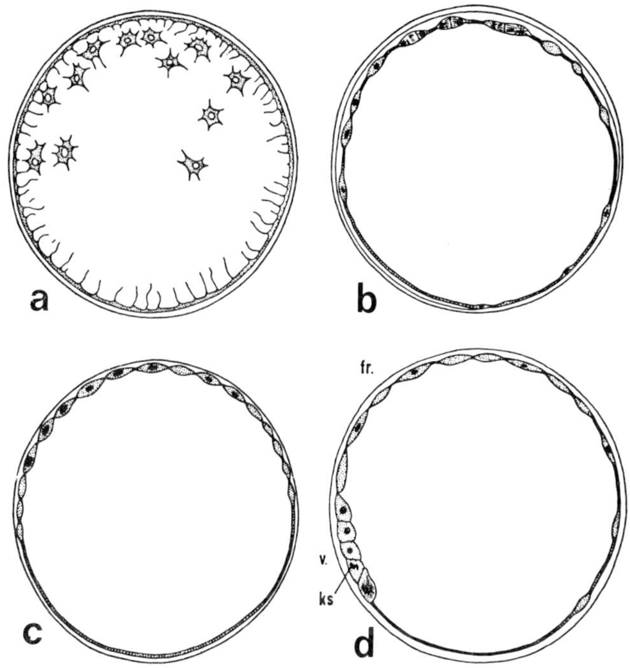

Figure 91. Superficial cleavage in *Hemimysis*. (a), 12 nuclei all projected in the same plane; (b) and (c), 64 and 128 cell stages, respectively, showing high density of cells at the animal pole; (d), formation of the germinal disc (ks) at the presumptive ventral side (v). fr, presumptive anterior end. (Reproduced from R. Siewing (1969), *Lehrbuch der vergleichenden Entwicklunggeschichte*, by permission of Verlag Paul Parey, Hamburg (after Manton))

It is evident that in superficial cleavage the normal polarity represented by the ectoplasm is used as an aspect of the mechanism of embryogenesis. The conversion of this polarity into apical and median polarities which occurs in so many other eggs must accordingly be suppressed. Under these circumstances the expedient of impressing one or both of the mentioned polarities upon the egg by means of a rigid envelope may be of great importance, a point that will be discussed later.

Discoidal cleavage

In discoidal cleavage the blastodisc is always situated at the animal pole, a fact that generally may be a consequence of the accumulation of ectoplasm at this location occurring during fertilization, thus involving the conversion of a normal to an apical polarity. If the egg is spherically or radially symmetrical at the outset, no median polarity can be present in the blastodisc, and the determination of the plane of bilateral symmetry may in that case be a matter of chance. This situation probably is normal in Pisces, as suggested by the fact that investigations of several species have failed to reveal any constant relation between the first division plane and the embryonic axis (Clapp, 1899; Oppenheimer, 1936).

In Cephalopoda the shape of the egg ensures that the blastodisc is bilaterally symmetrical (Figure 42). Here the first division plane coincides with the median plane, and the next one is at right angles to the former (Figure 92). This sequence seems to violate

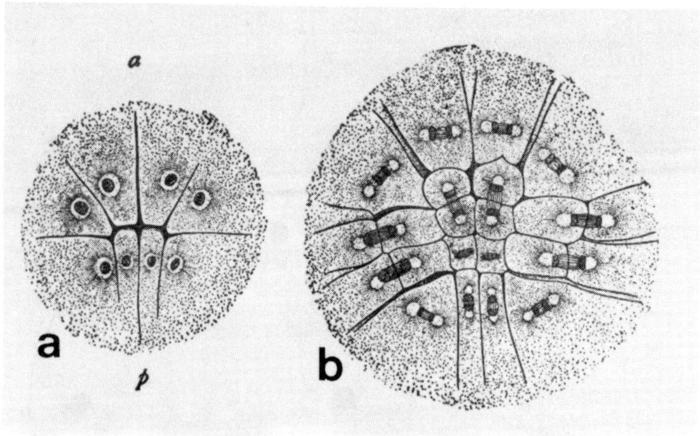

Figure 92. Bilateral cleavage of the blastodisc in *Loligo*. (a) and (b), successive stages. a–p, median plane. (Reproduced from E. B. Wilson (1928), *The Cell in Development and Heredity*, after Watasé, The Macmillan Company, New York)

the first rule of cell division, since the blastodisc is longer in the presumptive antero–posterior direction.

It is a typical feature of discoidal cleavage that the partitions formed during the earliest cell divisions cannot proceed into the yolk-laden part of the egg, presumably because the asters cannot penetrate here. As a consequence the blastodisc is subdivided by vertical cell walls, while remaining in contact with the remainder of the egg at the lower end. Only when the 'blastomeres' become so narrow that a mitotic apparatus perpendicular to the surface of the blastodisc obeys the rules of cell division will separation occur, with the simultaneous formation of yolk nuclei (Figure 93).

Polarities and cleavage patterns

We have in the preceding section discussed the relation between the polarities and the cleavage patterns in a purely formal way. It is the purpose of the present section to illustrate, by means of selected examples, the mechanisms through which the polarities exert the actions instrumental in the establishment of the cleavage patterns. The latter are also influenced by factors like the hyaline membrane, when present, and intercellular adhesion, but effects of this kind will be largely neglected in the present context.

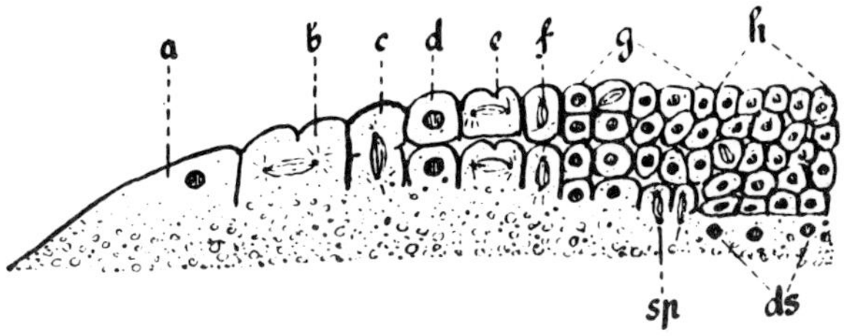

Figure 93. Germinal disc in an avian egg, showing vertically oriented mitotic apparatuses (c, f and sp). ds, yolk nuclei. (Reproduced from E. Korschelt and K. Heider (1936), *Verglechende Entwicklungsgeschichte der Tiere*, after O. Hertwig, by permission of Gustav Fischer, Jena)

Since the cleavage patterns arise through cell divisions, and since their distinguishing features are the relative size and orientation of the blastomeres, it follows that the action of the polarities in general must obey the second and the fourth rules of cell division. We shall discuss the effect of each of the four types of polarities separately.

Apical polarities

In most cases the polarities reside in either egg envelopes, cortex or cytoplasm. Since the mechanisms through which they exert their action vary considerably with the nature of the material substrate of the polarity, it is expedient to subdivide the treatment with respect to this point.

Egg envelopes

An elongated egg envelope imposes upon the egg an apical polarity which, however, cannot influence the cell size, but only the plane of division. From the first rule of cell division it follows that, in elongated cells where no other polarities obtain, the partition will cut perpendicularly through the longest axis at its midpoint. The first cleavage in

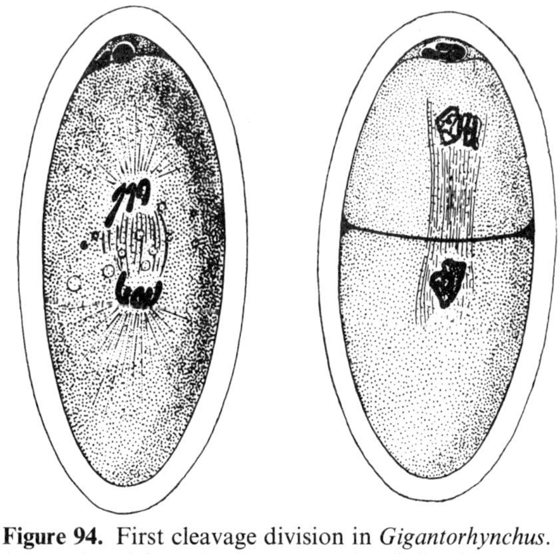

Figure 94. First cleavage division in *Gigantorhynchus*. (Reproduced from A. Meyer, *Zool. Jahrb. Abt. Anat.*, **50** (1928), by permission of Gustav Fischer, Jena)

certain Acanthocephala may possibly be accounted for by this mechanism (Figure 94). The first division in the eggs of Nematoda (Figure 87) is also perpendicular to the animal–vegetal axis, but in this case the orientation of the subsequent divisions of P_1 suggests that other factors may be responsible for the initial attachment of an aster to the vegetal pole.

Apical polarities are in many cases established through translocations of ectoplasm occurring in association with maturation and fertilization. In superficially cleaving eggs the ectoplasmic movements are suppressed, and the elongated shape of insect eggs may therefore represent an alternative method for establishing an apical polarity. The mechanism through which the effect of this polarity is exerted is not well known, except for the fact that it imposes a definite shape upon the embryo.

Cortex

When an apical polarity is present before fertilization, it is most often determined by the cytoplasmic distribution of yolk, as in telolecithal eggs. Contrariwise, the apical polarity is usually determined by cortical properties in alecithal and isolecithal eggs. From what we know about the mechanism of cell division it is evident that the cortex can interfere with this process by influencing the attachment of microtubules to the plasma membrane, and we have already seen that various observations may be accounted for on the assumption that the chances of microtubule attachment decrease with the amount of ectoplasm present in a certain cortical area.

The ectoplasm in the sea urchin egg is concentrated in a ring stretching from the equator towards the vegetal pole (cf. Figure 28). Presuming that the suggested mechanism is operative, the consequences of the distribution of ectoplasm for the attachment of microtubules are shown in Figure 95(a). It appears that even if no polymers obtain in the

1/1 1/4 1/8

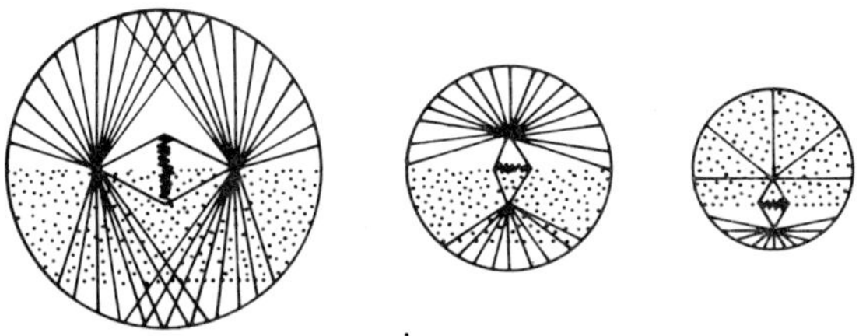

a b c

Figure 95. Representation of the cleavage divisions in the sea urchin embryo, based on the assumption that the ectoplasmic accumulation in the vegetal hemisphere interferes with the attachment of microtubules to the cortex. The fractions indicate the part of the whole embryo represented by the respective diagrams. (a), first division, (b), third division, (c), fourth division in the vegetal blastomere

region covered with ectoplasm, which may not be true, the microtubules anchored at the polar areas may suffice to ensure that the spindles are situated approximately in the equatorial plane. In the third division the shape of the cell dictates that the mitotic apparatuses are perpendicular to this plane. Once more, the ectoplasm may not seriously influence the cleavage pattern, because the attachment of the longest microtubules is unaffected (Figure 95b). If the division plane lies within the ectoplasmic region, interference with the microtubular activity may be anticipated. Interesting things begin to happen at the fourth division. In the four animal blastomeres the division plane is perpendicular to that of the third one, and the daughter cells are of equal size, suggesting that no polarities interfere with either size or spindle orientation. In the vegetal blastomeres the division is unequal, and the plane essentially parallel to the preceding one. In this case, therefore, polarities must prevail and from assumptions advanced above it follows that there is considerable inequality with respect to the number of microtubules directed towards the equator and towards the vegetal pole, respectively (cf. Figure 95c). The direction of the spindle suggests that one of the descendants of the aster originally attached to the vegetal region preserves this affiliation, thereby preventing the rotation of the spindle otherwise to be expected.

Cortical apical polarities most often exist in the form of either animal or vegetal polar plasm, accumulated as a result of cortical movements associated with maturation and fertilization. In some eggs two apical polarities are present, but since both cannot be efficiently exploited for morphogenetic purposes, it is usual for the vegetal one to be converted into a median polarity. As will be shown later, this conversion may be accomplished either through cell division or through morphogenetic processes.

Confining ourselves to the case where the polar plasm determines an apical polarity, thus that generally present at the animal pole, we shall presume that the accumulated ectoplasm, consisting mainly of hyaloplasm and microfilaments, constitutes an ideal matrix for microtubule formation. This proposal is not, it will be realized, contradictory to the suggested impediment of microtubule attachment by the ectoplasm. However, as formation of microtubules by definition is a cytoplasmic phenomenon, it follows that the polarity has been transferred from the cortex to the cytoplasm, thus coming into the scope of the next section.

Cytoplasm

In telolecithal eggs there is a difference in cytoplasmic composition along the apical axis to the effect that the yolk concentration is much higher in the vegetal, hyaloplasm in the animal, region. In many eggs, for instance those of Amphibia, this distribution is parallelled by a cortical polarity, based on the fact that the ectoplasm, as indicated by pigment accumulation, does not extend all the way to the vegetal pole. If, as contended, the ectoplasm interferes with attachment of cortical microtubules, it might be anticipated that small cells would be formed at the vegetal pole. As it is, the vegetal cells formed at the third division are somewhat larger than the animal ones (cf. Figure 73), suggesting that other factors than the cortical polarity affect the process of cell division. It may be inferred that it is the formation and extension rather than the attachment of microtubules which are impeded, and thus that a cytoplasmic polarity is involved. It is suggested that the relatively low concentration of hyaloplasm in the vegetal hemisphere is part of the mechanism responsible for the observed cleavage pattern, but unquestionably the yolk granules form a physical barrier which reduces the chances for penetration of microtubules through the cytoplasm, thereby impeding establishment of contact with the plasma membrane. This latter aspect is similar to the proposed effect of the ectoplasmic pigment granules. The circumstance that the difference in size between the animal and vegetal cells arising from the third division mostly is rather slight suggests that the effects of the two polarities are of the same order.

Accumulation of hyaloplasm at the animal pole may, by enhancing microtubule formation, function as a cytoplasmic polarity. The hyaloplasm may originate either from ectoplasm, transported to the pole through cortical movement, or from the cytoplasm, whence it is extracted by the microtubules of the meiotic apparatus. It is evident that the latter must be astral in order to carry out the suggested function. Through the accumulation of hyaloplasm at the animal pole, the material substrate of the cytoplasmic polarity is confined to a narrow region, entailing that the disparity between the size of the animal and the vegetal cells generally is much greater than in the Amphibia, as evidenced by observations on spirally cleaving eggs, in which this mechanism is prevalent. It may be noted that discoidal cleavage represents an extreme example of this mechanism.

Median polarities

There is good reason to presume that, except for those determined by egg envelopes, median polarities never exist before fertilization. This means that the determination of the median plane in the embryo must occur either through a chance phenomenon or through an 'external' directing influence which, without exception, may be traced to the fertilizing spermatozoon. In certain cases the median polarity may exist before the first division takes place, in other cases it is established through the division process proper. We shall here discuss the median polarities with reference to their material substrate.

Egg envelopes

The elongated shape confers an apical polarity, corresponding to the presumptive antero–posterior axis, upon the eggs of Insecta. If the curvature of the envelope varies along the circumference, a median polarity may also be determined. In either case the ventral side is distinguished by a greater number of nuclei and cells than the opposite dorsal side. If the wall of the egg envelope is straight, the preferential accumulation of nuclei at the ventral side presumably depends upon a preformed cortical polarity. If, on the other hand, the shape of the egg is bilaterally symmetrical, it may be envisaged that the polarity thus established can influence the direction of the nuclear movements.

In order to appraise this possibility it is necessary to consider in detail the mechanism responsible for the translocation of the nuclei. It is evident that, according to the views advanced above, astral microtubules, suspended between the nuclei and the cortex, afford the only possible means by which this phenomenon may be executed. This mechanism was, indirectly at least, propounded by Manton (1928). To account for the karyokinetic movements during the nuclear mitoses it is necessary to presume that internuclear central microtubules replace part of the cortical microtubules normally present in dividing cells. This proposed association between the individual nuclei must influence their approach towards the cortex. It will be realized that if any kind of self-propelling 'migration' were involved, the nuclei ought to spread spherically from the original location, whereas in reality the nuclei directed towards the poles progress much faster than the other ones (Figure 41). This difference follows from the contention that long microtubules exert a stronger force than short ones.

In non-cylindrical eggs it is always the most convex, or the least concave, surface, as seen from the inside, that becomes the ventral surface accommodating the largest number of nuclei. If this circumstance depends upon the cortical properties, the polarity derived from the shape will be superfluous. However, if the movement of the nuclei is random, then the observed accumulation may be due to the fact that from a point at equal distance from the two surfaces, the microtubules spanning a certain angle will be longer at the least concave surface (Figure 96). The force deployed by the microtubules

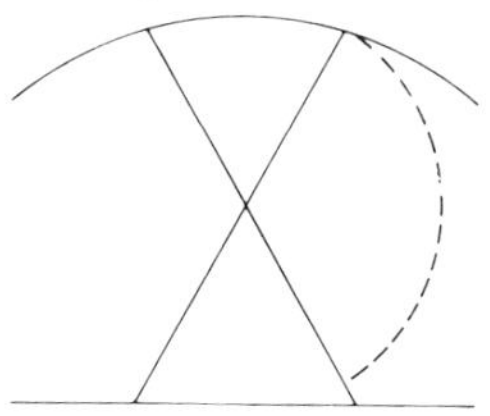

Figure 96. Diagram showing that from a point at equal distance from two surfaces, lines spanning a certain angle will be longer in the direction of the least concave surface

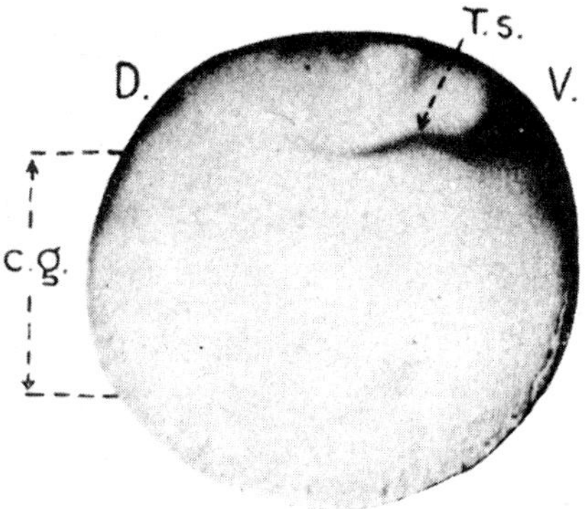

Figure 97. Section through a fertilized egg of *Rana*, showing accumulation of pigment at the ventral side (V.) and formation of the grey crescent (c.g.) at the dorsal side (D.). T.s., copulation path of the spermatozoon. (Reproduced from P. Ancel and P. Vintemberger, *Bull. Biol., Suppl.* **31** (1948), by permission of Bulletin Biologique, Paris)

being proportional to their length, it may be anticipated that more nuclei will be pulled towards the presumptive ventral side than in the opposite direction. To the extent that the cell distribution at the surface of the egg may be called a cleavage pattern, one may thus contend that the envelope acts to establish a bilaterally symmetrical cleavage pattern and a median polarity in the egg.

In Cephalopoda the shape of the envelope, and thus of the egg, has the consequence that the germinal disc assumes a bilaterally symmetrical shape, which subsequently is conveyed upon the cleavage pattern and upon the embryo.

Cortex

Median cortical polarities arise as the result of cortical movements initiated by the fertilizing spermatozoon. This circumstance implies that the fertilization must occur outside the poles and that the polarity bears a definite relation to the point of fertilization; generally the median plane determined by the latter will also be the plane of bilateral symmetry. Since cortical stimulation leads to accumulation of ectoplasm around the point of stimulus application, it may be anticipated that a cortical median polarity is distinguished by ectoplasmic condensation at the side where fertilization takes place, and a corresponding attenuation at the opposite side.

A typical example of a polarity of this kind is found in Amphibia, where the median polarity usually is visualized by the grey crescent, the formation of which involves an elevation of the pigment border at the presumptive dorsal side. In sectioned eggs pigment accumulation may be observed around the point of sperm entrance (Figure 97). Many attempts have been made to establish a correlation between the median plane thus determined and the plane of the first cell division. Various observations show a slight predominance for coincidence, but normal development is ensured even when this does not obtain.

According to the principles outlined above it may be predicted that the polarity may interfere with attachment of microtubules to the plasma membrane to the extent that this phenomenon will be enhanced at the dorsal side. As a consequence of this it may be expected that the dorsal cells are smaller than the ventral ones. It may not always be easy to demonstrate this effect during early cleavage; among other things it seems that for a clear-cut demonstration it is an inevitable requirement that the first division occurs in the median plane. However, reports supporting the prediction have been published, thus, as shown in Figure 98, the second cleavage division may be displaced towards the

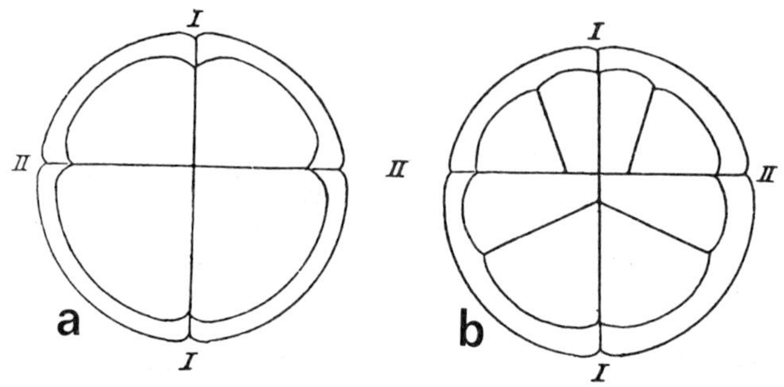

Figure 98. The division planes during the early cleavage stages in *Rana*. I and II, first and second division planes, respectively. (Reproduced from W. Schleip (1929), *Die Determination der Primitiventwicklung*, after Roux, by permission of Akademische Verlag, Leipzig)

dorsal side, and it has even been observed that the third division plane may be oblique, being closer to the animal pole at the dorsal than at the ventral side (cf. Schleip, 1929a). According to this author differences in the rate of cell division between the two sides may amplify the original differential and, as pointed out previously, there is general agreement concerning the existence of a distinct disparity with respect to cell size at later developmental stages (cf. Figure 83).

Many observations suggest that cytoplasmic factors are involved in the determination of the median polarity in Amphibia (cf. Ancel and Vintemberger, 1948). However, various attempts to change the polarity through cytoplasmic dislocation have confirmed that the cortical factors are the most influential (cf. Penners and Schleip, 1928; Pasteels, 1964).

Cytoplasm

The example just discussed shows that a cortical polarity may exert its function upon the process of cell division during an extended part of the period of cleavage. The situation is different as concerns cytoplasmic polarities, for in order to function in an all-or-none fashion the polarity must give rise to two cells differing with respect to cytoplasmic composition. It therefore follows that cytoplasmic polarities exert their effect during either the first or the second cell division.

We shall in the following discussion distinguish two cases, viz., one where an apical polarity, either before or in association with the first cell division, becomes a median polarity, and another where the orientation of the polarity is determined in relation to the point of sperm entrance.

Polarity conversion. Two different mechanisms can be distinguished which lead to the conversion of an apical to a median polarity. In the first one the polarity is established before the division begins, and in that case the agents responsible for the polarity direct the orientation of the division plane. In the second one the establishment of the polarity is intimately associated with the process of cell division.

The classical example of the first kind of polarity conversion is found in Ascidiacea as shown by the description of the development of *Styela* by Conklin (1905). We have already observed how the sperm aster transports the yellow ectoplasm from the vegetal pole towards the equator at one side of the egg, thereby establishing a median polarity. The aster must eventually end up in the centre of the egg, where equilibrium obtains.

As a consequence of the astral movements, a substantial amount of the ectoplasm is transferred to the cytoplasm where, as we shall subsequently see, it acts by promoting the process of microtubule formation. This circumstance justifies its being classified as a cytoplasmic polarity. Hence, the original cortical normal polarity established by the ectoplasm is converted first into a vegetal–apical polarity, which may be transitional between a cortical and a cytoplasmic polarity, and subsequently into a cytoplasmic median polarity.

If the formation of an amphiaster were delayed until the sperm aster had reached the centre, the cytoplasmic polarity might influence the direction of the mitotic apparatus in such a way that the spindle was located in the median plane. As it is, the replication of the centriole occurs while the aster is still close to the cortex, with the consequence that the spindle comes to lie perpendicular to the median plane. Since this orientation is not later changed, it follows that the first division plane divides the egg into two equal parts.

Since hyaloplasm is a favourable substrate for the formation of linear TP polymers it may be anticipated that the density of microtubules is higher at the dorsal side where the ectoplasm has been transferred into the cytoplasm, with the result that the second cleavage division is unequal, the smaller cells being those containing the yellow hyaloplasm. Conklin (1905) does not present a lateral view of the 4 cell stage which might bring out this point, and in the 8 cell stage the situation is complicated by the possible interference of an apical polarity. However, in the 16 cell stage it is clearly seen that the dorsal–vegetal cells, being particularly rich in hyaloplasm, are the smallest cells in the embryo (Figure 99).

An animal–apical polarity may also be converted to a median polarity through astral activity. This, at least, is the way I interpret the observations by Lillie (1901) on *Unio*. During the meiotic divisions the pronuclei remain at a location near the animal pole, and when maturation is over, the egg pronucleus is transported towards the side of the egg opposite the sperm pronucleus. Hyaloplasm is accumulated in front of the pronucleus, suggesting that microtubules are at work (Figure 45). It is not clear whether the egg or the sperm centriole is responsible for these movements, but the former possibility seems most likely.

As the egg pronucleus approaches the lateral cell cortex, the sperm pronucleus is moved towards its female counterpart, and after breakdown of their membranes a mitotic apparatus is formed, distinguished by the fact that one aster, being much closer to the cortex, is much smaller than the other one. This state of equilibrium can be

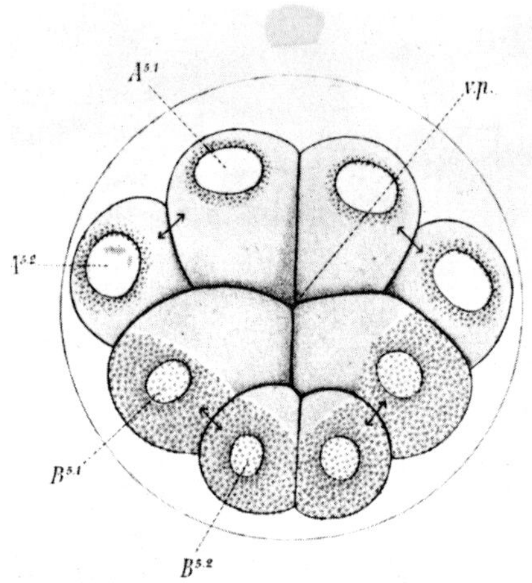

Figure 99. Cleavage in *Styela*. View of 16 cell stage from the vegetal pole. (Reproduced from E. G. Conklin, *J. Acad. Nat. Sci. Philadelphia*, [2], **13** (1905))

accounted for on the assumption that the smaller aster is located in the region containing the hyaloplasm from the animal pole. As a consequence of this distribution of cytoplasm, the daughter cells are of unequal size, determining a median polarity passing through their centres. The orientation of the polarity seems to coincide with the plane determined by the migration of the male pronucleus towards the animal pole. This correlation may be accounted for on the assumption that the sperm pronucleus shields off a certain spatial section from extension of microtubules, with the consequence that the female pronucleus migrates in the opposite direction, thereby establishing the median polarity.

The hyaloplasm which is transported by the egg aster must to a large extent be polar plasm, i.e. ectoplasm located at the animal pole during the meiotic division. It is obvious that if, as suggested above, the presence of this substance is of importance for the size differential between the macromeres and the micromeres (the animal–apical polarity), then the circumstance that part of this material is used for the creation of a median polarity must by necessity attenuate the apical polarity. It may be that the relatively large size of the micromeres in *Unio* and other Bivalvia can be accounted for in this way.

The examples discussed so far show that astral activity may accomplish a bilaterally symmetrical distribution of cytoplasm before the first cell division has taken place. We shall now discuss some cases in which the cytoplasmic distribution is radially symmetrical, but where the first division proceeds in such a way that the daughter cells are unequal, thereby giving rise to a bilateral polarity. It appears that this unequal cell division may occur in two distinctly different ways, characterized by the relative size of the two asters.

If these are equal, then the presence of a polar lobe is required which, at the end of the division, fuses with one of the otherwise equal blastomeres. The formation of a lobe at the vegetal pole presumes a vegetal–apical polarity. The latter may be formed by accumulation of ectoplasm if fertilization occurs at the vegetal pole, but apparently the polarity often already exists in the oocyte, as described for *Dentalium* by Wilson (1904) and for *Chaetopterus* by Lillie (1906), cf. Figures 31 and 100. It is often assumed that the cortex somehow is involved in the formation of the polar lobe, but it seems rather difficult to devise a satisfactory mechanism by this means. Noting that the lobe is formed

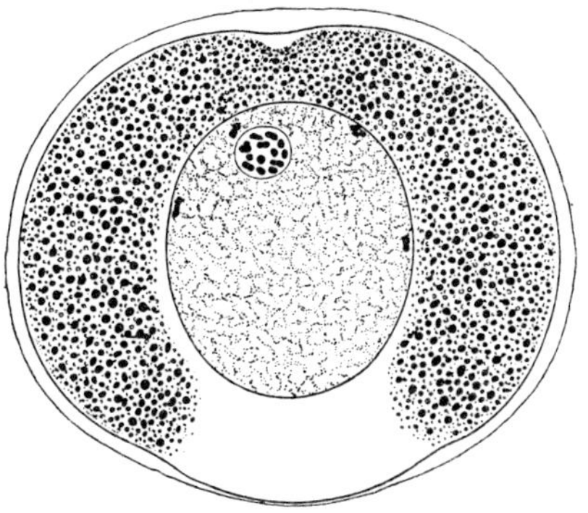

Figure 100. Unfertilized egg of *Dentalium*, showing a vegetal–apical polarity. (Reproduced from E. B. Wilson, *J. Exptl. Zool.*, **1** (1904), by permission of The Wistar Press, Philadelphia)

in association with the rounding up of the cell preceding cytokinesis, and recalling that this phenomenon must be ascribed to the activity of the asters, it may be more likely that the polar lobe owes its origin to the fact that microtubules cannot penetrate the ectoplasm located around the vegetal pole. If this is the case, cortical microtubules attached at the circumference of this region will act to pull the edges together to form a more or less narrow stalk, and the new cell partition will fuse with the contracted cortex at one narrow region, leaving the polar lobe in connection with one blastomere (Figure 101).

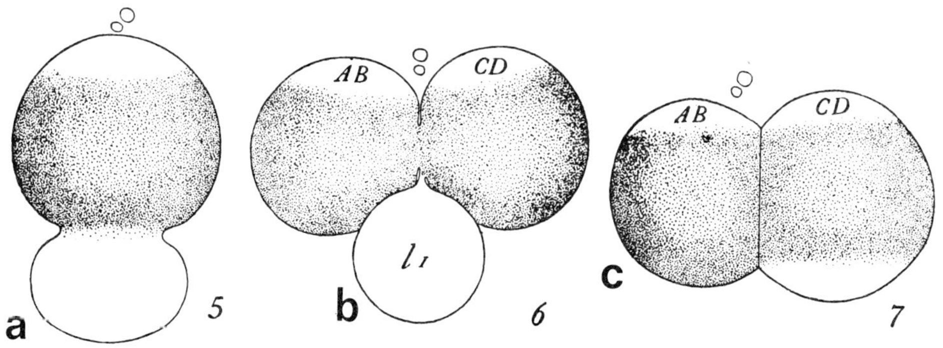

Figure 101. Polar lobe formation in *Dentalium* during the first cleavage division, resulting in the formation of two unequal blastomeres. (Reproduced from E. B. Wilson, *J. Exptl. Zool.*, **1** (1904), by permission of The Wistar Press, Philadelphia)

In this way a bilateral polarity is created, which may be further accentuated by polar lobe formation in the second cleavage division. Occasionally, for instance in *Dentalium*, an indication of a polar lobe may be observed during the maturation divisions. On the basis of the interpretation given above, it must be assumed that microtubules from the meiotic apparatus are involved. Evidently, this phenomenon can be anticipated only when the latter is astral, a prediction that seems to be borne out by factual observations.

Polar lobe formation requires that the asters span most of the cell volume, but in many eggs, notably large, yolk-laden ones, the asters are confined to the animal region. In that case, exemplified by *Crepidula* (Conklin, 1897), it cannot be expected that cortical microtubules can reach the vegetal cortex in sufficient number to form a polar lobe. This does not exclude that the division becomes unequal, for as the new cortex growing towards the vegetal region reaches the ectoplasm preventing microtubular activity, it may be forced to bend and follow the boundary of the vegetal–apical polarity. The effect will be exactly the same as when a polar lobe is formed.

Polar lobe formation is mostly found in Mollusca, but, as mentioned, it occurs also in Annelida. In this phylum, however, the mechanism involving two unequal asters is much more common. As discussed above, polar plasm is usually present at both poles in Annelida. If these act to prevent the attachment of cortical microtubules, the consequence will be that one aster comes to control more, the other less than half, of the egg volume. Under these circumstances the division plane will be situated to the same side of both poles, giving rise to one large (CD) and one small (AB) cell. The polar plasms are present in the former and finally end up in the D blastomere. This mechanism is indicated by the observations by Lillie (1906) on *Chaetopterus* (Figure 102).

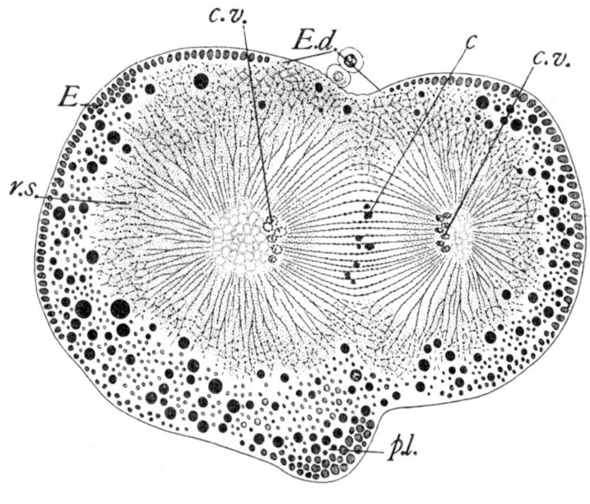

Figure 102. First cleavage division in *Chaetopterus*, exhibiting two asters of unequal size. c, chromatin; c.v., chromosomal vesicles; E, ectoplasm; E.d., zone at animal pole free of ectoplasm; p.l., polar lobe. (Reproduced from F. R. Lillie, *J. Exptl. Zool.*, **3** (1906), by permission of The Wistar Press, Philadelphia)

Determination by the sperm entrance point. As appears from the preceding section, it is usual that an apical polarity, created as a cortical response to the penetration of a spermatozoon at the vegetal pole, is converted into a median polarity. We may now proceed to consider the situation where the sperm, entering at a point outside the poles, gives rise to the establishment of a median polarity.

Under the specified conditions, the fertilizing sperm confers bilateral symmetry upon the egg. If one of the two asters of the ensuing mitotic apparatus preserves the contact with the cortex beneath the entrance point the original median plane may be preserved.

If hyaloplasm is accumulated by the sperm aster, it may be anticipated that the pole of the mitotic apparatus turned towards the entrance point is distinguished by the formation of a large number of short cortical microtubules, whereas those from the other pole must span a large part of the egg volume before they reach the opposite cortex. An equilibrium may therefore be established between two asters differing with respect to the number and length of the cortical microtubules, resulting in an unequal division, the smallest all being closest to the point of sperm entrance.

As it is, the sperm aster generally migrates along the inside of the cortex before the mitotic apparatus is formed. If the direction of movement is fortuitous, there must be an equal chance that the sperm aster moves to the left or to the right. If it remains stationary, the division plane will cut the egg surface at equal distances from the point of sperm entrance, but as it moves, for instance to the right, the left-side partition will approach the entrance point. Thus, the relation between division plane and point of fertilization depends upon the movement of the sperm aster. The latter phenomenon being subjected to the statistical laws, it follows that the distances covered are distributed normally. Morgan and Tyler (1930; 1938) have investigated the correlation between the first division plane and the point of sperm entrance in *Cumingia*, finding coincidence in 70 per cent, the angle being less than 45° in 13 per cent and between 45° and 90° in 8 per cent. This result emphasizes the statistical nature of the event, suggesting that the mean migration distance must be close to half the external circumference of the smaller cell, which is always closest to the point of sperm entrance.

The two blastomeres being unequal, the division plane does not pass through the animal pole, but is displaced towards the side where the smallest blastomere is formed. Considering only the 77 cases of coincidence of entrance point and division plane, Morgan and Tyler actually found that the small blastomere (AB) was to the right in 40 cases and to the left in 37 cases (Figure 103). The interesting observation was made that in the second cleavage division CD was always divided so that the smaller cell C was closest to the point of sperm entrance, with the consequence that the arrangement of

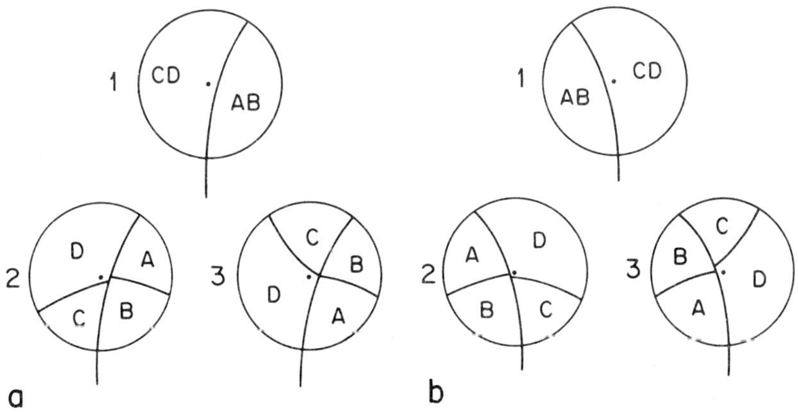

Figure 103. Orientation of the second division planes in *Cumingia* in those cases where the first one passes through the point of sperm entrance. (a), the first plane to the right; (b), the first plane to the left of the animal pole. a_2, a_3, b_2 and b_3 are four theoretical possibilities for the second planes of which only a_2 and b_2 become realized. (Reproduced from T. H. Morgan and A. Tyler, *Biol. Bull.*, **58** (1930), by permission of Marine Biological Laboratory, Massachusetts)

the blastomeres in half of the cases was clockwise, in the others counter-clockwise (Figure 103). This phenomenon, which of course has no influence on the median polarity, passing at this stage through the centres of B and D, strongly suggests that conditions prevail near the point of sperm entrance which in some way favour the formation of cortical microtubules.

The relation between division plane and point of sperm entrance was also investigated in *Chaetopterus*. Coincidence was found in 41–45 per cent (Morgan and Tyler, 1930; 1938), but among the remaining eggs the sperm entrance point was at the AB side in 69–76 per cent of the cases. This result is accounted for if the relative migration of the sperm aster is less extensive than in *Cumingia*. In *Chaetopterus* the sperm does not influence the orientation of the division of CD, an observation which may be related to the particular mechanism responsible for the unequal division in this case (cf. Figure 102).

That the median cytoplasmic polarities influence the cleavage pattern is demonstrated by the fact that the relative sizes of the blastomeres in the D quadrant generally deviate from their counterparts in the other quandrants.

Dextro–sinistral polarities

It was pointed out above that bilateral asymmetry is exhibited by animals in many different taxa, and that this phenomenon must be referred to a dextro–sinistral polarity. However, it is only in the Gastropoda that a relation between cleavage pattern and adult body structure has been established. In this particular case it was found by Crampton (1894) and Kofoid (1894) that the dextro–sinistral polarity determines the relative orientation between corresponding pairs of micromeres and macromeres in the third cleavage division. In the succeeding divisions of both macromeres and micromeres the orientation of the mitotic apparatus is alternately dexiotropic and levotropic, in accordance with the fourth rule of cell division.

A precondition for the utilization of the polarity is that the descendants of the odd and the even micromeres, being separated by an angle of 45°, each determine a 'median' plane. As far as the latter are concerned, the fact that 4d, in contrast to its counterparts in the other quadrants, gives rise to mesoderm ensures the establishment of a median polarity. If the descendants of the odd quartets exhibit a radially symmetrical cleavage pattern they will not determine a median plane. It will be realized that an essential difference between a trochophore and a veliger larva is that the velum of the latter is bilaterally symmetrical, and it may therefore be envisaged that the veliger larva is a necessary, albeit not a sufficient, condition for the existence of bilateral asymmetry in the adult. The shell is presumably another necessary factor, since 'detorsion' may occur in its absence.

It is possible that in spirally cleaving eggs the dextro–sinistral polarity may reside in the egg membrane as a 'spiral' vector. If this is true, it is established before the median polarity, but for reasons outlined earlier, it is not defined as a dextro–sinistral polarity before the median one exists. There is no reason to believe that polarities of this kind can be determined by either cortex or cytoplasm.

Normal polarities

A normal polarity usually prevails in the form of ectoplasm accumulated beneath the plasma membrane. We have seen that most often it is converted to apical or median polarities, except in Arthropoda, where it is of immediate importance for the 'cleavage pattern' distinguishing this phylum. Another instance where the normal polarity, as represented by the ectoplasm, influences the cleavage pattern appears to be in Ctenophora, where all the ectoplasm gradually ends up in the micromeres (Figure 104). It may

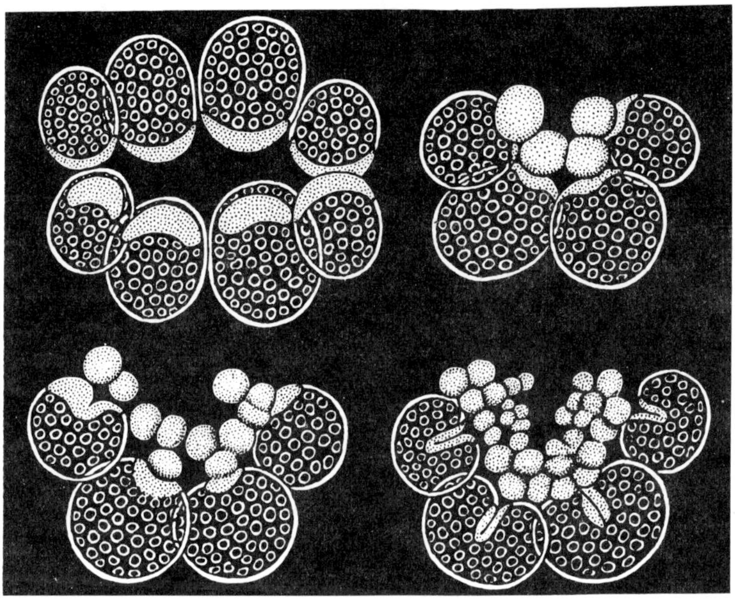

Figure 104. Formation of ectoplasm-containing micromeres in *Beroë*. (Reproduced by arrangement with Holt, Rinehart and Winston, Inc., New York, from *Principles of Development* (1939), by P. Weiss, after Spek, © Holt, Rinehart and Winston, New York)

finally be mentioned that in the embryos of Vertebrata, where paratangential cleavages are common in the blastula, the resulting difference in distribution and fate of the cells may be considered to be the effect of a normal polarity.

CELL DIVERSIFICATION

The developing embryo is continuously subjected to changes at various levels of organization. Thus, on the cellular level, new cell-types gradually make their appearance and these, together with various extracellular elements elaborated by some of the cells, contribute to innovations on higher levels, resulting in the formation of tissues and organs. The word 'differentiation' is often used comprehensively to denote all these phenomena, sometimes in association with specifying prefixes like 'cyto-', 'histo-', etc. Although it cannot be disputed that in the broadest sense of this word these processes are differentiations, the complexity of the problems nevertheless requires better terminological discrimination. In the present context we shall use 'cell diversification' to denote all changes on the cellular level, whereas events involving interaction between the cells and extracellular elements are called 'morphogenetic processes'.

The reason for the adoption of the former name rather than accepting the current 'differentiation' is that, as outlined in the introduction, it is important to distinguish between two kinds of variations on the cellular level, viz., those leading to the formation of representatives of either of the eight cell orders, and those imposing further changes upon the latter without interfering with the cell-order affiliation. As proposed, the former processes will be called 'cell transformations', whereas the expression 'cell differentiation' is reserved for the latter type of change, in compliance with the most widely endorsed sense of the word. There is good reason to believe that the same kind of mechanism is responsible for cell transformations and cell differentiations, thus warranting their being classed together. There are, however, two circumstances which justify transformations being treated as a separate group, the first being that they are concerned with the transition between cell orders, the second that, as will be shown later, they exhibit some properties which clearly set them apart from the cell differentiations.

The existence of different cell-types may confidently be attributed to modifications in the synthesis of proteins carried out by the cells. We can therefore define the phenomenon of cell diversification as a process through which changes in the prevailing pattern of protein synthesis are initiated or, paraphrasing Jacob and Monod: 'We shall consider that two cells are [diversified] with respect to each other if, while they harbor the same genome, the pattern of proteins which they synthesize is different' (1963, p. 31). This activity represents the second of the predetermined mechanisms, as defined in the introduction. The predetermination comprises two aspects, the genetic information concerning the total repertoire of proteins which may be synthesized by the organism in question, embodied in the nuclear DNA, and a mechanism of nucleo-cytoplasmic interaction allowing for selection among these.

The proteins distinguishing a certain differentiation pattern are often elaborated in such large quantities that there may be a tendency to overlook the fact that their number generally is small compared to the total array of proteins produced by the cell. This point has been brought out by Holtzer and Abbott (1968), who stress that from a qualitative point of view the disparity as regards protein synthesis may be slight between an undifferentiated and a differentiated cell. Thus, both kinds synthesize a substantial number of proteins necessary to sustain the life of the cell, over and above which the latter further elaborates at least one and probably several proteins, which may be classified as *differentiation products*. It must be emphasized that the class of differentiation products includes compounds other than proteins, for very often the latter are enzymes which catalyse the synthesis of polysaccharides, lipids, pigments, etc. The differentiation products are called 'luxury' or 'unessential' molecules by Holtzer and Abbott (1968). From a cellular point of view these adjectives are very generous, since elaboration of such substances often is associated with a suppression of the reproductive capacity, suggesting that under some circumstances 'obnoxious' or 'lethal' might be more appropriate epithets. Cell transformations do not appear to interfere with the faculty of cell division, so in this case the appellation proposed by Holtzer and Abbott may be appropriate.

The synthesis of the 'essential' proteins is an important phenomenon, which embodies problems of genetic, molecular biological and cell physiological interest. From the epigenetic point of view these events may be taken for granted, just as much as the supply of energy and similar phenomena. In contrast, the diversification products, which may be unessential or even lethal from a cellular point of view, are nonetheless indispensable for the formation and maintenance of the metazoan body. The phenomenon of diversification forms the basis of the ontogenesis and phylogenesis of the Metazoa, because the diversification products, to some degree in contrast to the essential molecules, largely are of importance for the whole body, being involved in morphogenesis (e.g. extracellular elements), regulation mechanisms (e.g. hormones), food uptake (e.g. digestive enzymes), etc. For this reason the phenomenon of cell diversification is of central significance for the study of embryogenesis.

The mechanism of cell diversification also involves aspects of genetics and molecular biology. In the present book it is my intention to steer clear of these disciplines as far as possible, but in this chapter it will be necessary to deal with certain problems on the molecular level. Among these, the mechanism of nucleo-cytoplasmic interaction is of prime importance. It is obvious that the realization of disparate cell-types in different regions of the embryo implies that dissimilar cytoplasmic conditions must exist at some stage in the said regions. This situation cannot obtain without there being some initial differences which, as proposed above, may be classified as polarities. It was seen in the preceding chapter that the latter may influence the process of cell division, and here it will be shown that they may also affect that of cell diversification, directly or indirectly, using the process of cell division as an intermediary. The interaction thus ensuing is responsible for the spatio–temporal distribution of the diversification patterns.

This question is the most significant part of the present topic, but before it can be treated properly it will be necessary to discuss the general aspects of the mechanisms involved in cell diversification and to present an outline of the most usual diversification patterns.

Mechanisms

Cell diversification involves that among the proteins elaborated by individual cells certain of them specifically characterize a particular *diversification pattern*. Thus, it may be inferred that the information passed on from the nucleus to the cytoplasm varies from one cell-type to another. This phenomenon may be explained in two different ways: either the genomes of the cells are different, owing to a mechanism of *differential replication* of DNA, or they are identical, in which case *differential transcription* of DNA must take place. The first of these mechanisms, encountered in the phenomenon of *chromatin diminution*, is exceedingly rare. Consequently, cell diversification normally must involve differential transcription of DNA, the outcome of which, for reasons given below, will be called *induced protein synthesis*. In either case it is necessary to presume that the particular behaviour of DNA reflects differences in the cytoplasmic composition of the respective cells. In other words, both differential replication and differential transcription must depend upon nucleo-cytoplasmic interaction.

The gradual appearance of new cell-types in the embryo implies that on several occasions during embryogenesis the descendants of an individual cell belong to separate diversification patterns, a circumstance which requires that the daughter cells differ with respect to the contents of their cytoplasm. This situation can prevail only if the division giving rise to the two cells is unequal, or if they have been subjected to different environmental conditions. These two phenomena, unequal cell divisions and the influence of extrinsic factors, notably surrounding cells, form the basis of the mechanism of cell diversification. We shall in the following pages discuss the role of chromatin diminution and of induced protein synthesis in this event.

CHROMATIN DIMINUTION

Chromatin diminution was first observed in *Parascaris* by Boveri (1899), but it also occurs in other Nematoda and, less commonly, in Insecta. The phenomenon described by Boveri involves that, in the cell closest to the animal pole, part of the chromosomal material is spilled into the cytoplasm during the second cleavage division. This incident is repeated three more times in one of the cells arising from the other of the first two blastomeres. At the 32 cell stage consequently only two cells have the full gene complement, whereas the remaining ones have undergone chromatin diminution. It has been found that the former are the primordial germ cells, the latter presumptive somatic cells. There can be no doubt that this phenomenon involves some kind of diversification of import for the further destiny of the cells. Being the first step of diversification, it should concern a cell transformation according to the present notions. Since furthermore it appears to be irreversible, it ought to be the lf-transformation. The consequence of this supposition, that all the somatic cells are filocytes, is by no means unacceptable, but the available evidence cannot decide this question. Since the bodies of the adult animals contain a number of different cell-types, it follows that the subsequent steps of diversification must be accounted for by another mechanism.

In certain Diptera it has been found that the cytoplasm located at the vegetal pole, distinguished by a high content of nucleic acids, is instrumental in preventing chromatin diminution in the presumptive germ cells. In this case it is thus clearly shown that differential DNA replication involves nucleo-cytoplasmic interaction. The observations of Boveri (1910) suggest that a similar mechanism exists in *Parascaris* (cf. Figure 87).

INDUCED PROTEIN SYNTHESIS

Cell diversification is unknown in microorganisms, all members of a culture being essentially identical. It is known, however, that in certain cases the full potential of the genome as regards protein synthesis is realized only under special conditions. The best-known example of this kind is provided by *Eschericia coli*, in which it has been found that the synthesis of β-galactosidase and some other enzymes takes place solely in the presence of lactose or one of a few related substances (cf. Monod, Cohen-Bazire and Cohn, 1951). This phenomenon, called 'induced synthesis of enzymes' or, in more general terms, 'induced synthesis of proteins', is important because it demonstrates that selective transcription of the genome is possible and that the range of activity may be affected by external factors. Since the influence of substances added to the medium must be exerted via the cytoplasm, induced protein synthesis may be regarded as an example of nucleo-cytoplasmic interaction. A wealth of observations have been made concerning the biochemical and genetic aspects of the induction of β-galactosidase synthesis in *E.coli*. This work culminated in the advancement of the operon hypothesis (Jacob and Monod, 1961; 1963).

The salient points of this hypothesis are as follows: the amino acid sequence of a protein is determined by a particular deoxynucleotide segment or *structural* gene. The primary product of the latter is an RNA copy of the gene, the messenger RNA, which carries the structural information to the cytoplasm where the protein is synthesized. An *operator* gene is proposed to control the transcription of one or several adjacent structural genes, which together with the operator constitutes an *operon*. The function of the operator is in turn controlled by another determinant, called a *regulator* gene. The latter forms a cytoplasmic product, somewhat ambiguously called a *repressor*, which has the property of being able to react with a low-molecular substance called an *effector*. This reaction, specific with respect to the repressor (R) and the effector (F), may be represented by $R+F \rightleftharpoons RF$. In some systems, called *inducible*, only the R form is active in blocking the transcription; thus, in the presence of the effector the repressor is inactivated. In other systems, called *repressible*, only the RF form is active, hence the function of the particular operon is suppressed by the effector. An example of the former type is represented by the enzymes involved in the metabolism of lactose in *E.coli* referred to above. When this bacterium is grown in the presence of tryptophane, the synthesis of the enzymes normally responsible for the production of this amino acid is suppressed; they therefore constitute a repressible system. It appears that inducing effectors are generally substrates of the induced enzymes, whereas repressing effectors are products of the repressed enzymes (cf. Jacob and Monod, 1963). In the latter case the control of the protein synthesis thus acts as a negative feedback system.

Studies of the phenomenon of induced protein synthesis have led to the establishment of two classes of proteins, called 'constitutive' and 'inducible', respectively. Jacob and Monod (1963) identified the former with the *non-repressible* proteins, the synthesis of which always proceeds at maximum rate. These two classes may not coincide exactly, but the proposed definition is preferable on the basis of the operon hypothesis. It would therefore be natural and correct to include the remaining proteins in the class of *repressible* proteins. As we have just seen, the latter comprises two subclasses, called 'inducible' and 'repressible', respectively. This terminology is obviously not very consistent; as a matter of fact it may be advantageous to use the names *R-repressible* and *RF-repressible* for the two groups. This does not necessarily imply that we must give up the notion of induction of protein synthesis. As will become evident later, there are good reasons for preserving it in the present context.

The derepression observed in the induced protein synthesis in microorganisms is characterized by the fact that this phenomenon is dependent upon the presence of an inducing effector in the external medium. This phenomenon may be called *transient derepression*. It is suggested that in Metazoa another mechanisms exists, in which brief exposure to an effector may ensure a lasting, if not necessarily irreversible, activation of a particular operon. It is furthermore proposed that this phenomenon, which may be called *permanent derepression*, is involved in the process of cell diversification in Metazoa. Although there is no direct evidence in support of this view, it will be argued that the stabilization may be achieved through a mechanism of positive feedback.

It is not implied that all the proteins distinguishing a particular diversification pattern are permanently derepressed. On the contrary, the fact that the composition of the intestinal secretions may be adjusted to that of the food certainly suggests that the synthesis of the digestive enzymes, which is part of the differentiation pattern represented by the intestinal cells, involves a mechanism of transient derepression. According to this proposition it appears that, so far as the proteins are concerned, the essential molecules of Holtzer and Abbott (1968) must constitute the class of non-repressible proteins, whereas the luxury molecules may be identified with the repressible ones, taken in the widest sense of the word. A survey of the various classes is presented in Table 2.

Table 2
Classification of proteins with respect to the mechanism through which their synthesis is controlled

Non-repressible (constitutive)		
Repressible	RF-repressible (negative feedback) (repressible)	
	R-repressible (inducible)	transient derepression
		permanent derepression (positive feedback)

Even at an early stage workers in this field realized that the induced synthesis of proteins may offer a clue to the mechanism of embryonic differentiation (Monod, 1947; Spiegelman, 1948). This idea has also been adopted by Jacob and Monod (1963), who suggested that the operon hypothesis may be applicable to this phenomenon. Various authors have argued that for a number of reasons it may be unwarranted to extrapolate in this way from microorganisms to animals, and alternative mechanisms have been advanced to account for the mechanism of cell diversification in Metazoa (Britten and Davidson, 1969; Tomkins, Gelehrter, Granner, Martin, Samuels and Thompson, 1969). As it happens, our ignorance regarding this question is so great that it is impossible to apply either of the hypotheses in the present context since none, or at best very few, of their predictions can be tested by experimental observations. The fact that in this chapter I adopt some notions from the operon hypothesis is a matter of expediency, which has little significance as far as the molecular mechanisms are concerned.

The phenomenon of differential transcription of DNA through nucleo-cytoplasmic interaction formally corresponds to the induced synthesis of proteins observed in microorganism. The cytoplasmic agents involved in this interaction correspond to the effectors of the operon hypothesis. For reasons which will soon become obvious, they will in the present embryological context be called *inductors*. There is no doubt that cell diversification pattern involves the synthesis of several proteins. This circumstance either implies that a specific inductor is required for each protein, or that an inductor may initiate the activity of several genes. The latter possibility being the simplest, it is reasonable to presume that the genes encoding the proteins distinguishing a specific diversification pattern are located in one or possibly a few cooperating operons. The latter eventuality is tacitly assumed from now on when the word 'operon' is used.

The gist of the present discussion is the proposition that the phenomenon of embryonic induction involves nothing but the induction of new diversification patterns through a mechanism of permanent derepression, the stabilization being accomplished by positive feedback. It may possibly be argued at this point that many changes in cellular properties occur spontaneously, without the cells being exposed to the influence of external stimuli. It is true that *selfdiversifications* of this kind take place, but it is proposed that the fundamental mechanism is the same as that prevailing in the presence of an external source of induction, the difference being that in selfdiversifying cells the inductors arise within the cells proper. It may thus be possible to distinguish between *endogenous* and *exogenous* induction processes, both of which will be discussed presently. However, first we shall examine the phenomenon of *competence*, as revealed by the fact that only under certain conditions are cells responsive to inductive stimulation.

Competence

The concept of competence was first introduced in the embryological literature by Waddington (1932) to denote the ability of embryonic tissues to react to inductive stimuli. In the present context competence is used to designate a property possessed by individual cells, involving the faculty to undergo a certain step of cell diversification. This definition thus comprises both spontaneous and induced processes of diversification, and it furthermore holds that a cell may have lost competence with respect to one pattern of diversification and yet be competent as to other patterns. The justification for using an already established term in this way is that the old meaning of 'competence' appears to be a special case of the phenomenon as defined here.

It has been found that in some embryos the cells may during early stages be refractory to inductive stimulation (cf. Needham, 1942). This situation changes with time, but subsequently the cells may again reach a stage of inertness relative to certain specific embryonic inductors. Embryonic cells thus may exist in two temporarily distinct states of incompetence. Supposing that they are also different in other respects, it may be appropriate to introduce the names *precompetence* and *postcompetence* for these phases (cf. Figure 105). We shall here discuss some of the factors responsible for the acquisition and loss of competence displayed by embryonic cells.

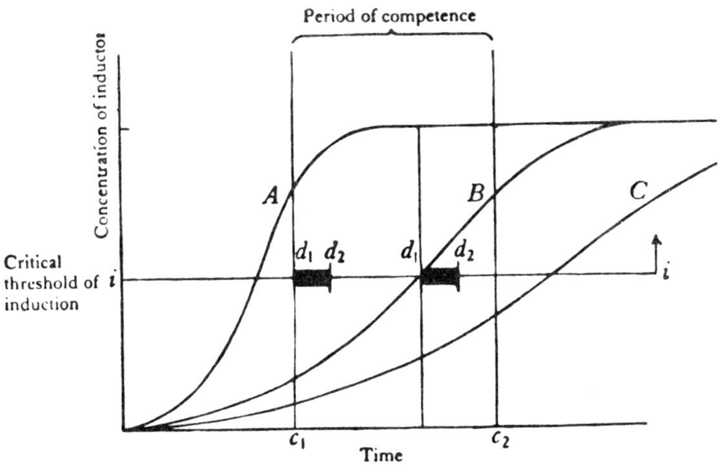

Figure 105. A hypothetical scheme illustrating the correlation between competence and inductor concentration. In A the inductor has reached the critical level before competence is achieved, and the response ('determination', d_1–d_2) is initiated as soon as competence begins. In B the threshold concentration is reached during the period of competence, with the consequence that a response obtains. In C the inductor concentration remains too low during the period of competence. (Reproduced from J. Needham (1942), *Biochemistry and Morphogenesis*, by permission of Cambridge University Press, London)

As proposed, the first step of diversification to which cells are subjected generally is one of cell transformation. The appearance of new cell-types is thus an overt demonstration of competence in the cells concerned. Among the cell orders which appear during early embryogenesis, the sf-cells occupy a prominent position, partly because their appearance facilitates the demonstration of their presence, and partly because they are instrumental in the execution of certain important embryogenetic processes. Thus, this cell-type is responsible for carrying out gastrulation in many eggs, notably those of Echinoidea and Amphibia. The suppression of the lf-transformation must in such cases lead to a developmental block at the blastula stage.

Cell diversification involves protein synthesis, and the precondition for competence is therefore the presence of a functional genome. As a matter of fact it has been shown repeatedly that whenever the genome is absent or otherwise impaired in the animal classes mentioned above, embryogenesis comes to an end before the stage of gastrulation. These findings also demonstrate that in the protein synthesis taking place during the initial phases of development transcription of DNA cannot be an essential aspect.

It may be of some interest to compare these notions with experimental observations on RNA synthesis during the early embryonic stages. Studies of this question have been made on many different species, but we shall confine ourselves to results obtained on Amphibia. It has been found in incorporation experiments that no new ribosomal RNA (rRNA) is produced before gastrulation and that, as expected, this activity is intimately associated with the appearance of nucleoli (Brown, 1964). This event appears to be of little embryogenetic consequence, for in an anucleolate mutation no rRNA is ever formed, yet development proceeds to an early swimming stage (Brown and Gurdon, 1964). It may therefore be inferred that synthesis of rRNA has nothing to do with the acquisition of competence. The observation is interesting in another way, because the formation of rRNA signifies that in the presence of appropriate genes, transcription of their DNA begins at gastrulation.

In contrast, soluble RNA (sRNA) becomes labelled before gastrulation, notably in the late cleavage stages (Brown, 1964). As regards sRNA, it appears that the incorporation is confined to the terminal bases (Brown and Littna, 1966), a phenomenon which does not involve transcription of DNA. The synthesis of messenger RNA (mRNA) is very slight before gastrulation, and appears to be qualitatively different from that occurring subsequently (cf. Denis, 1968). The significance of the last observation is somewhat ambiguous, but in spite of this it seems justified to suggest that replication of nuclear DNA is an activity that is gradually assumed by the embryonic cells, beginning first, it may be presumed, in the particularly active dorsal cells at a late blastula stage, whence it spreads to the cells in other parts of the embryo. In our discussion of the synchronous cell divisions it was pointed out that many findings tend to show that the embryonic nuclei do not assume their normal function before this mitotic pattern comes to an end (cf. Dettlaff, 1964). Without postulating that the transition from synchrony to asynchrony can directly be equated with acquisition of competence, I would suggest that the former is a prerequisite for the latter.

It appears that the induction of cell diversification may spread from cell to cell in an aggregate, a phenomenon which is called *contact induction*. The fact that several cell-types can cooperate in the formation of various tissues and organs indicates that the cells involved have reached a state of postcompetence, in which they are no longer susceptible to contact induction. A well-known example of postcompetence in embryonic cells is found in the ventral ectoderm of the amphibian embryo, which is known to become refractory to the influence of dorsal lip transplants at a certain stage of development (cf. Holtfreter, 1968). In this particular case the postcompetence may be concerned with the lf-transformation. We shall not discuss this question further, except to point out that it may be regarded as an illustration of the exclusivity principle of cell diversification, which is dealt with in the following section.

Induction

We shall here bring together various observations on the induction of cell diversification patterns in embryos and in isolated cells in an attempt to outline the characteristics of the phenomenon. The discussion will begin with a description of various general features. The generation of cells with different diversification patterns may be traced back either to the occurrence of unequal cell divisions or to the fact that initially identical cells become exposed to different environmental influences. If the differences between two cells, originating from an unequal cell division, are of consequence for cell diversification, it must follow that the cells proper or some of their descendants will come to differ with respect to the diversification pattern they assume. This mechanism, which must involve spontaneous formation of different inductors, may be called 'endogenous' induction. Through the process of contact induction a cell can impress its diversification pattern upon neighbouring cells, and presumably other external influences may exert a similar effect. For this event we shall use the designation 'exogenous' induction. Both types of induction will be discussed here, followed by an examination of the classical antagonism between cell division and cell diversification.

General features

As shown by P. Weiss (1939), cell differentiation or, in the present terminology, cell diversification, is distinguished by some outstanding features, viz., *genetic limitation*, *exclusivity*, *discreteness* and *progressiveness*. The first property may be explained by the fact that protein synthesis is involved, and the third one if, as proposed, each step involves one operon, corresponding to all the proteins belonging to a specific diversification pattern. The second trait obviously implies that only one operon, or certain particular combinations of these, can be active at the same time. In this respect the operons of the inducible synthesis patterns must deviate from those of the constitutive proteins, since presumably many operons are responsible for their synthesis. Finally, as regards progressiveness, it is not possible directly to account for this phenomenon, but there are two aspects which deserve mentioning. The first one is that diversifications in many cases appear to be irreversible, which explains the lack of retrogression. Secondly, with respect to the progression from one to a following diversification pattern, it may be suggested that the cytoplasmic status established by each of these may involve the elaboration of factors required for inducing the activity of new operons.

When protein synthesis is controlled by negative feedback, involving that the reaction products in the end of a metabolic sequence inhibit the synthesis of enzymes engaged in the initial steps (cf. Walker, 1965), it is possible to keep the concentrations of the enzymes and metabolites within certain limits. This situation does not obtain as regards the diversification products, which may be elaborated in large quantities and deposited inside or outside the cell. It has been pointed out by Jacob and Monod (1963) that this phenomenon is incompatible with a mechanism of negative feedback; if anything, the process might rather be presumed to involve positive feedback. Nothing is known about this point, but it may be mentioned that it is not necessarily all the proteins of the diversification pattern which are produced in 'excess'; possibly this holds only for the final products, and in that case the elimination of the latter by excretion, precipitation, etc., may suffice to explain the continued production. It is nonetheless important to keep in mind that differences on the molecular level must be invoked to explain the existence of both non-repressible and repressible proteins.

Rutter, Ball, Bradshaw, Clark and Sanders (1967) have investigated the differentiation of the exocrine cells in the pancreas of the rat embryo. Through determinations of the amounts of specific proteins, these authors could distinguish three different states of differentiation. Initially, the undifferentiated, presumptive exocrine cells contain no discernible amounts of specific proteins, but at 12–14 days an increase occurs, reaching a level corresponding to a state of *protodifferentiation*. After 19 days this is followed by a substantial increase (10^3–10^5 times) leading to the differentiated state (Figure 106).

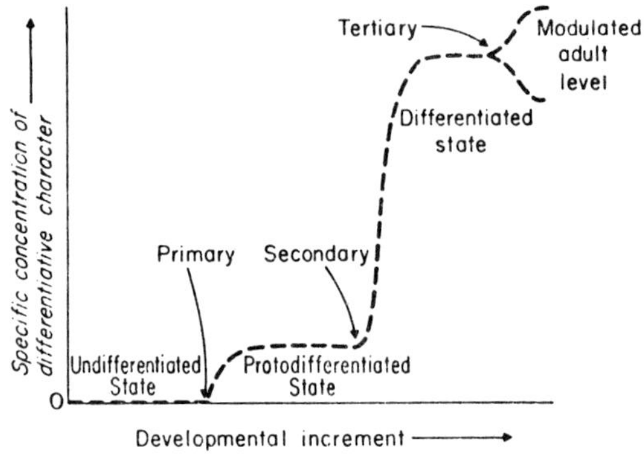

Figure 106. Phases of differentiation. (Reproduced from W. J. Rutter, W. D. Ball, W. S. Bradshaw, W. R. Clark and T. C. Sanders, 'Levels of Regulation in Cytodifferentiation', *Exptl. Biol. Med.*, **1**, pp. 110–124 (S. Karger, Basel/New York, 1967))

This succession of events is associated with distinct morphological changes, and especially noteworthy is the fact that 'growth', i.e. cell division, has ceased in the state of *differentiation* proper.

It seems that the 'primary regulatory event' of these authors corresponds to the inductive phenomenon. Subsequently, synthesis of the specific proteins may occur, and if attainment of the differentiated state was only a matter of accumulation of the particular inductor, a gradual increase might be anticipated to take place. This is not the case; rather, the differentiation products are kept at a constant level for several days, after which the phase of differentiation is initiated with a very steep increase. Since this occurs at the time when the cells have ceased to grow, it may be that we are facing a case of the antagonism between cell division and cell differentiation.

As pointed out by Rutter *et al.* (1967), the concentration of the various differentiation products presumably may oscillate within certain limits (cf. Figure 106), a phenomenon which is called 'modulation' by these authors. This expression appears to enjoy great popularity, as evidenced by the fact that it has been applied to at least three different phenomena. Introduced originally by P. Weiss (1939), it was meant to indicate the transition between activity and inactivity of differentiated cells. As an illustration,

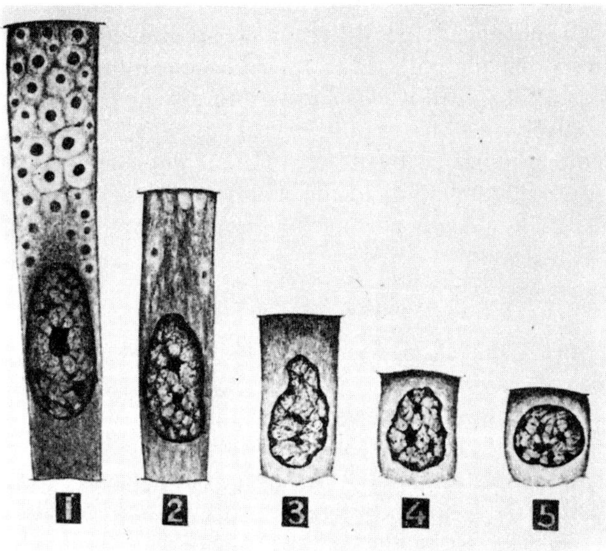

Figure 107. Inactivation of the cells in the seminal vesicles of the rat following castration. 1, Control; 2, 24 hours castrate; 3, 2 days castrate; 4, 20 days castrate; 5, 90 days castrate. (Reproduced from C. R. Moore, W. Hughes and T. F. Gallagher, *Am. J. Anat.*, **45** (1930), by permission of The Wistar Press, Philadelphia)

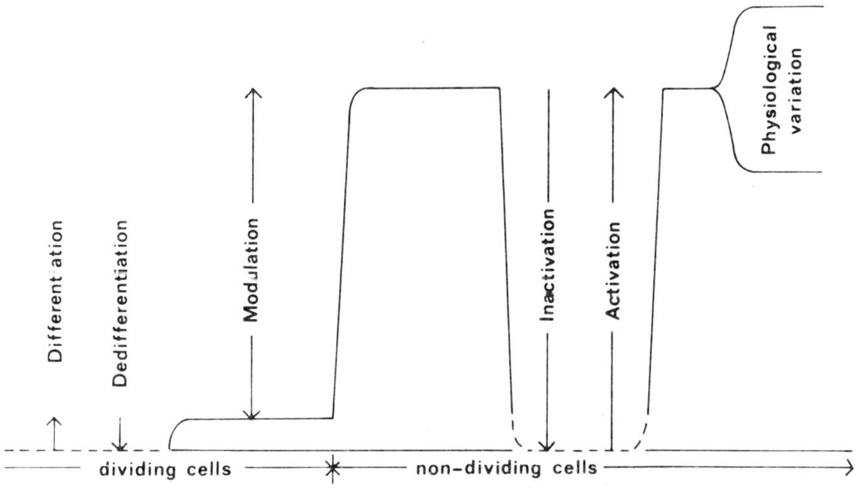

Figure 108. Suggested terminology concerning the changes in differential activity to which cells may be exposed

Weiss quoted the epithelial cells of the seminal vesicles, the function of which requires the presence of sex hormones (Figure 107). As we shall presently discuss, it is used by various authors for the oscillation between cell division and cell differentiation observed in cultured cells. This variation presumably corresponds to transition between the states of protodifferentiation and differentiation, but since the inactive cells in the example presented by Weiss do not engage in cell division, we are clearly dealing with two different phenomena. It is very difficult to devise an apposite name for the former event, possibly it may be most expedient to call it *modulation*, and use the expressions *activation* and *inactivation*, respectively, for the case described by Weiss. It seems appropriate to classify the oscillations indicated by Rutter *et al.* as *physiological variations*, and the transition from protodifferentiation to the undifferentiated state as *dedifferentiation*. The relations between the various phenomena discussed here are shown in Figure 108.

Endogenous induction

Theoretically, an isolated cell may gradually assume new diversification patterns, a circumstance which, if environmental influences may be excluded, must imply that through the metabolism of the cell new inductors are made available. In the embryo a number of different diversification patterns arise spontaneously in separate regions and, as stressed earlier, this event may be referred to the occurrence of unequal divisions. It was shown above that the latter in general depend upon a polarity, and it therefore follows that these are also involved in cell diversification. This fact raises the question whether the polarities are concerned with the cytoplasmic distribution of inductors. There appear to be several reasons for repudiating this proposal. First, it seems impossible to account for an inhomogeneous distribution of low-molecular substances and, at least according to the operon hypothesis, inductors are small molecules. Second, we have seen that the polarities are engaged in the control of microtubular activities. It is, of course, not inadmissible to presume they operate in more than one way, but it is certainly against the principle of Occam's razor. However, the third and most serious argument against the unequal distribution of inductors is that it implies either that all the descendants of a cell arising from an unequal division must assume the same diversification pattern, which is patently not true, or else that a continuous segregation of inductors must occur in subsequent cell divisions, which seems extremely unlikely. It therefore appears necessary to conclude that the inductors responsible for the various diversification patterns arise spontaneously in the individual cells, as suggested above to occur in isolated cells.

This discussion leads naturally to a consideration of the so-called 'morphogenetic substances', which repeatedly have been claimed to be present in various eggs. It has been proposed in the introduction that morphogenetic processes are the outcome of interactions between cells and extracellular elements, the nature of which has been outlined above. It may be justified to call the latter morphogenetic substances, but it is well established that none of these are present in the egg. The only type of compounds of morphogenetic importance, besides mucopolysaccharides and tensible proteins, that might be envisaged to occur in the egg would be inductors, since these in the last analysis are responsible for the formation of the various cell-types, which partly are themselves engaged in morphogenetic processes, and partly elaborate the extracellular elements. This latter possibility having been excluded above, it seems warranted to infer that morphogenetic substances do not exist, and as far as I can judge, this conjecture is corroborated by the fact that none of the innumerable attempts to demonstrate their existence has been crowned with success.

The spontaneous production of an inductor in a competent cell must, when a certain threshold concentration is exceeded, by definition lead to the establishment of a new diversification pattern. This endogenous induction, which above was called 'selfdiversification', may be concerned with either a *selftransformation* or a *selfdifferentiation*, both of which will be discussed in the following pages.

Selftransformation

Cell transformations are the first diversifications to occur, and are therefore easy to observe during early embryogenesis. All three transformations may be recognized in a number of different embryos, but among these the lf-transformation is for many reasons the most important and also the most readily distinguishable; the discussion therefore will be confined mainly to this process.

In Invertebrata the lf-transformation generally occurs at several separate locations. In Echinoidea the most obvious region is at the ventral pole where the mesenchymal sf-cells are formed. As we shall see later, there is good reason to believe that the apical plate cells at the animal pole are cf-cells. In spirally cleaving eggs, mesoderm cells, i.e.

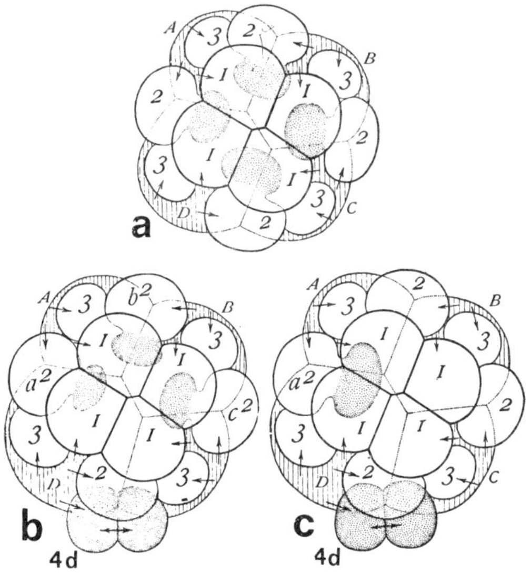

Figure 109. Sources of mesoderm (filocytes) in spirally cleaving eggs. (a), *Leptoplana*, small cells separated from the second quartet in all four quadrants. (b), *Crepidula*, small cells from the second quartet of the quadrants A–C, plus 4d. The latter is not particularly small, but the filocytes are formed by the budding off of small cells through a kind of teloblastic cleavage. (c), *Unio*, one large cell from 2a plus 4d. In both cases teloblastic cleavage appears to be involved. (Reproduced from E. B. Wilson, *Biol. Lectures* (Wood's Holl), **1898** (1899))

sf- or cf-cells, are formed from all four quartets. In many cases the differentiation patterns assumed by the cells from the D quartet differ from those of the filocytes from the remaining quartets. In Vertebrata the spontaneous occurrence of the lf-transformation appears to be a unique phenomenon, occurring only at one place in the embryos. It is known that in Amphibia this process is confined to the dorsal lip of the blastopore, and a site corresponding to this is probably common to all vertebrates, suggesting that the median polarity is responsible for the spontaneous occurrence of the lf-transformation.

A number of findings have been made which may throw some light on the conditions which favour this transformation. Thus, to begin with it may be pointed out that one of these pertains to the size of the cells, as is particularly well established in Invertebrata, where 'mesenchyme' cells generally arise from cells distinguished by their small size. This phenomenon is especially well known from the sea urchin embryo, but similar observations have repeatedly been made on spirally cleaving eggs (Figure 109). It was suggested above that small cell size may be responsible for the early cessation of the synchronous cell divisions. This circumstance alone cannot, however, be the causative factor, for in that case all cells should become filocytes once their mitotic activity is slowed down. Presumably, small cells are distinguished by some cytoplasmic trait which may lead to the formation of the inductor initiating the lf-transformation.

It was mentioned above that in Amphibia the lf-transformation occurs spontaneously at the dorsal side. The work of Ancel and Vintemberger (1948) has definitely established that the median polarity decisive for this cell class distribution is usually established at fertilization, and we also know that the dorsal cells generally are distinguished from the ventral ones by their smaller size. This property is thus somehow involved in, or associated with, the selftransformation, just as in the examples discussed above, but it is not the absolute size that is crucial, for the cells at the dorsal lip which become sf-cells are much larger than, for instance, the ectodermal cells near the animal pole.

In their studies of Schultze's inversion experiments, Penners and Schleip (1928) observed that the distribution of yolk and light cytoplasm at the surface of the egg is of great importance for the formation of a blastopore. The determination established at fertilization was rather hard to upset, for two-thirds of all observed blastopores would be located in the region of the grey crescent, but the remaining ones might form at any part of the surface, and in any direction relative to the original plane of symmetry. However, in almost every case invagination occurred at the margin of a white, i.e. a yolk-laden area of the surface (Figure 110). These observations suggest that a second condition favouring the spontaneous lf-transformation is related to the cytoplasmic composition; in cells identical in this respect the smallest ones will transform, in cells of different composition the likelihood for transformation increases with the yolk:cytoplasm ratio.

This is as far as we can approach the present problem with the aid of observations on intact embryos, but there are some additional observations on embryonic explants which bear on this question. It is well established that in Amphibia the ectoderm does not form a neural plate unless it is underlain by the filocytes formed at the dorsal lip of the blastopore, or maybe rather by cells induced by them. For this and other reasons, e.g. the shape and behaviour of the neural plate and neural crest cells, it has been proposed elsewhere by the author (cf. 1967a) that neural induction in fact involves the induction of the lf-transformation.

Numerous implantation and explantation experiments had led to the generalization

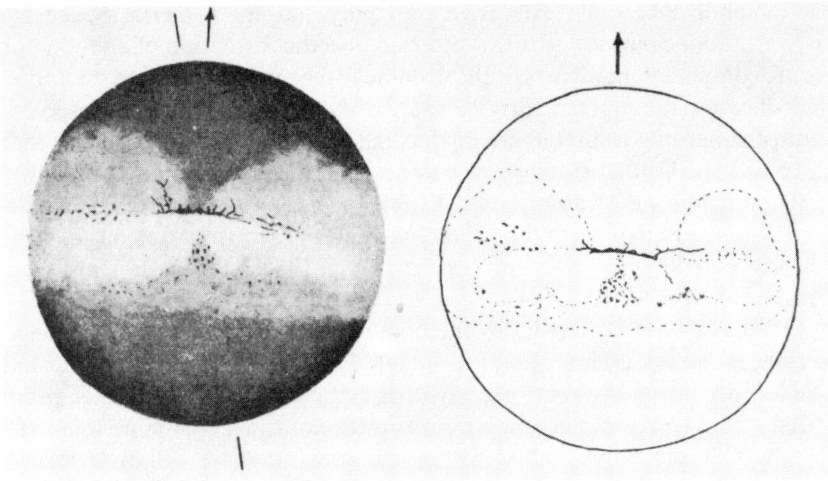

Figure 110. Invagination and cell ingression in an inverted egg of *Rana*, seen from the vegetal pole. The pigment accumulation at various points of the surface, mostly along the white margin, represents ingression of small numbers of sf-cells. The arrows indicate the median plane, the lines the plane of the first division. (Reproduced from A. Penners and W. Schleip, *Z. Wiss. Zool.*, **130** (1928), by permission of Akademische Verlag, Leipzig)

that physical contact with a normal or an artificial inductor is an obligatory precondition for 'neuralization'. This contention was overruled by Barth (1941), who was able to show that under certain circumstances isolated dorsal ectoderm may spontaneously differentiate into nerve cells. Holtfreter (1944b; 1945; 1947) analysed these, at the time, very unorthodox observations, and came to the striking conclusion that when explanted ectoderm is treated badly enough to provoke 'sublethal cytolysis', nerve cell differentiation may occur in the absence of an inductor. Expressed in terms conforming with the views presented here, Holtfreter's findings show that the spontaneous lf-transformation, and thus the formation of the appropriate inductor, is provoked by sublethal cytolysis.

We have now established three conditions which further the spontaneous lf-transformation, viz., small cells, yolk-laden cells and sublethal cytolysis. It seems a difficult task to find the common denominator which can be made responsible for the initiation of the transformation. We are here facing the problem of explaining how diversification, involving qualitative differences between cells, can arise in a population derived from one original cell. A clue to the solution may possibly be found in observations on the induced enzyme synthesis in *E.coli*. In the presence of glucose and lactose this organism will grow without synthesizing any β-galactosidase. The growth of the culture ceases when the glucose disappears, but immediately the synthesis of RNA and enzyme protein begins (1956a; 1956b), leading gradually to a resumption of the growth on the basis of the glucose moiety of lactose. In this case the milieu is the same for all the cells, and they consequently all behave in the same way. In a developing embryo the milieu is intracellular, which allows for the possibility that it may vary from one cell to another. It seems natural to presume that the activities going on in the early blastomeres are determined and restricted by the reserves present in the cytoplasm, and that they will continue these as long as the resources allow for it. If, in the course of time, some strategic metabolite becomes exhausted they may, like the bacteria, be led or forced along new metabolic pathways. If the latter involve the formation of an inductor and if the cells are competent, new parts of the genome will be mobilized, with the consequence that diversification takes place. If the substance, the loss of which is proposed to initiate the lf-transformation, is preferentially located in the hyaline cytoplasm and small enough to be lost through cytolysis, then clearly it may be possible to account for the three different conditions leading to the lf-transformation as discussed above.

The *exhaustion* mechanism outlined here, an analogy to the situation observed in *E.coli*, has been advocated by me on several occasions (e.g. 1965a; 1965b; 1967a). A variant of this mechanism, which may be termed *selective utilization*, was proposed by Ten Cate many years ago (1952). Referring to the high content of aromatic amino acids in the vegetal cells of the sea urchin egg (cf. Lindahl, 1936), Ten Cate pointed out that these might be left over from the purported synthesis of collagen by these cells, provided of course that stored yolk proteins contain a complete assortment of amino acids. If, among the substances thus accumulated, some are potential inductors, cell diversification may ensue. Although these two mechanisms differ in essential aspects, no distinction will be made between them in what follows, thus the word 'exhaustion' may imply either possibility.

That the lf-transformation is somehow associated with the metabolism is suggested by two separate observations. The first one is that in hybrid embryos of Amphibia, in which the metabolic rate is substantially lower than in control embryos (Barth, 1946), the dorsal lip is formed closer to the vegetal pole than usual (Figure 111), i.e. in cells distinguished by a higher yolk:cytoplasm ratio than those normally transforming into

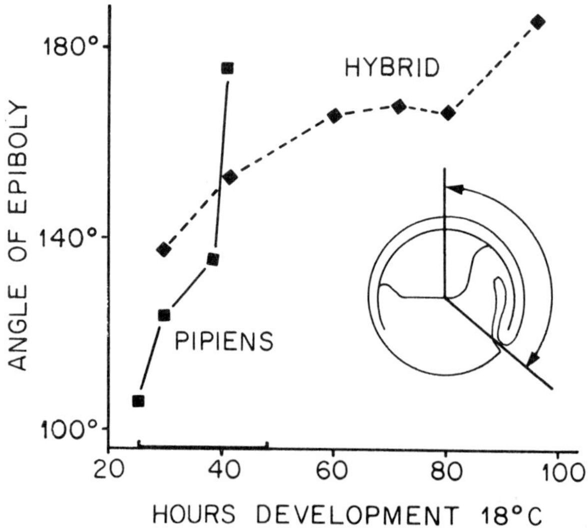

Figure 111. The angle of epiboly, as defined in the figure, in normal and hybrid frog embryos. It is seen that the invagination begins about 30° closer to the vegetal pole in the hybrids, as compared with the controls. (Reproduced from J. R. Gregg (1957), in *The Beginning of Embryonic Development*, after Moore, by permission of American Association for the Advancement of Science)

sf-cells (J. A. Moore, 1946). The other finding is that the median polarity, as evidenced by blastopore formation, may be inverted by unilateral restriction of the oxygen supply (Løvtrup and Pigon, 1958). The interpretation of the latter observation is somewhat complicated by the fact that anaerobiosis suppresses the synthesis of RNA (Wilde and Crawford, 1968), but this finding is not incompatible with the view propounded here. The basic idea behind this dynamic determination is thus that quantitative differences through the mechanism of exhaustion or selective utilization may lead to qualitative differences between initially identical cells, thereby creating the conditions required for spontaneous cell transformation through nucleo-cytoplasmic interaction.

Very little is known about the mechanism responsible for the sc-transformation. The fact that the cells exposed to the external medium always become colligocytes, whereas many cells in the interior are solocytes, may indicate that exposure to the external medium favours the sc-transformation. In most embryos the formation of colligocytes is almost simultaneous in the superficial cells, suggesting that if any polarity is involved, then it must be the normal polarity with respect to the environment, and that contact induction from cell to cell is not instrumental in this case. These findings support to some extent the contention that contact with the medium may further the transformation, but do not exclude that interior cells may become colligocytes, presumably through contact induction. Unfortunately, studies on cultured cells do not furnish any clue to the problem, but the fact that solocytes may be cultivated certainly demonstrates that the media employed in no way promote the sc-transformation.

Still less can be said about the nv-transformation. This event appears to occur spontaneously in both solocytes and colligocytes, when they are situated in such a way that part of their surface is exposed to a fluid medium.

Selfdifferentiation

With the exception of the presumptive germ cells, all cells will sooner or later undergo differentiation. This statement is valid irrespective of whether the cells have been subjected to transformation or not. Within each order the cells may assume a number of differentiation patterns. Thus, confining ourselves to the sf-cell, it is known that in various Invertebrata, where the lf-transformation largely seems to be spontaneous, the sf-cells may become nerve, muscle and pigment cells as well as mucopolysaccharide-producing mesenchyme cells and skeleton-forming cells.

Still more impressive is the range of differentiation patterns exhibited by the neural crest cells, in which the lf-transformation is induced. Thus, they may give rise to 'pigment cells of different kinds; Schwann's sheath cells; meninges of the brain; most if not all of the cells present in the spinal and autonomic ganglia; the adrenal medulla; tooth papilla; large amounts of mesenchyme (the so-called mesectoderm); and, last but not least, cartilages' (Holtfreter, 1968, p. 13). Upon this impressive list a few comments are required, because it does not enumerate differentiation patterns, but rather morphological cell-types. It is possible to reduce the number significantly by a rational analysis, even if it may still remain quite respectable. To begin with, there are at least two types of pigment cells, melanocytes and xanthocytes. The ganglion and the adrenal cells are closely related differentiation patterns, and it is possible that the sheath cells also belong to this group, or maybe rather to the glia cells, a cell-type overlooked by Holtfreter. Their particular behaviour may be accounted for on the assumption that they are colligocytes, being adhesive to both the axonal and their own surfaces. Fourthly, we have the mucopolysaccharide-producing cells, to which may belong not only the 'mesenchyme', but also those responsible for the formation of the *arachnoidea*. We shall assign to one group the collagen-producing cells, in which we may include the membrane-producing cells, responsible for the formation of *pia mater*, and also the chondrocytes. It will be shown later that these two kinds of cells belong to separate differentiation patterns, but that does not change the fact that collagen in both cases is a characteristic differentiation product. The classification of the cells responsible for the formation of tooth papilla is uncertain; probably they are collagenocytes.

The fact that the neural crest cells may assume such a variety of differentiation patterns presupposes that they may exist in a number of different states with respect to cytoplasmic composition, each characterized by the presence of a specific inductor. In trying to explain how this situation comes about it seems obvious that we may exclude external influences, for all these cells arise from the same region of the embryo and it is therefore preposterous to presume that they have been subjected to different environmental conditions. There appear to be only two possible ways to account for their origination. According to the first one all the cells may, in the course of their normal metabolism, successively pass through the various states mentioned above. The differentiation pattern assumed by the cells may thus depend upon the time at which the differentiation is initiated or upon the metabolic activities which have taken place prior to this event. It will be realized that this explanation, which assumes that we are dealing with self-differentiations, represents a special case of the exhaustion principle. The other possibility is that there exists a causal inductive relationship between the differentiation patterns, such that one type induces a second pattern in other cells. Since it is generally presumed that induction of the same pattern may also occur, this mechanism appears to require the existence of cell populations, which differ with respect to their response to inductive stimuli, and it therefore does not offer any specific advantage relative to the first one. It may nevertheless be worth while to test the proposition in the light of experimental evidence.

The neural crest cells are not well suited for this purpose because they arise at a relatively late embryogenetic stage, and we shall therefore use some other cell-types for the analysis. Among the earliest filocytes to arise are the chordocytes and the neurocytes, belonging to disparate differentiation patterns. Because of their peculiar shape it is difficult to substantiate the cell-class affiliation as regards the first cell-type. However, the facts that they contain microtubules (Eakin and Westfall, 1962) and that the chordocytes are the only agents of the 'chordamesoderm', which may be considered responsible for the induction of the lf-transformation leading to neurulation in Amphibia, do support the proposed classification (cf. 1967a). The chorda–spinal cord complex is known to induce muscle cell differentiation both *in vivo* and *in vitro*. The cell-class affiliation of the myoblasts is easily decided on account of their content of microtubules and their spindle shape. These three cell-types often occur in explants from amphibian embryos, and it is possible on the basis of the combinations in which they prevail to decide whether indeed an inductive relationship obtains between them.

In explants from the dorso–lateral regions of the gastrula representatives of these differentiation patterns are usually found (Holtfreter, 1938a; 1938b). Comparison between Figures 112 and 113 shows that this result does not conform at all with the presumptive destination as established in Vogt's fate maps (1929a), according to which the primordia of the chorda and the somites are localized in the dorso–ventral surface

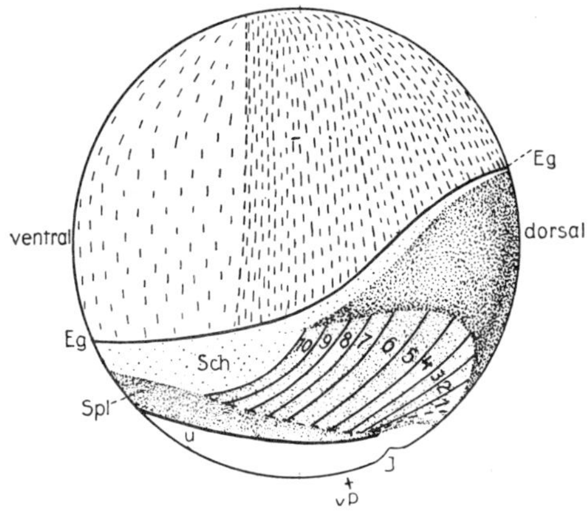

Figure 112. Fate map of the embryo of *Triton* at the beginning of gastrulation. Densely dotted area, presumptive chorda; Eg, limit of gastrulation; I, early blastopore; Sch, main part of the tail bud material; Spl, lateral mesoderm; U, later blastopore; VP, vegetal pole; 1–10, somites. (Reproduced from W. Vogt, *Arch. Entw.mech.*, **120** (1929), by permission of Springer, Berlin)

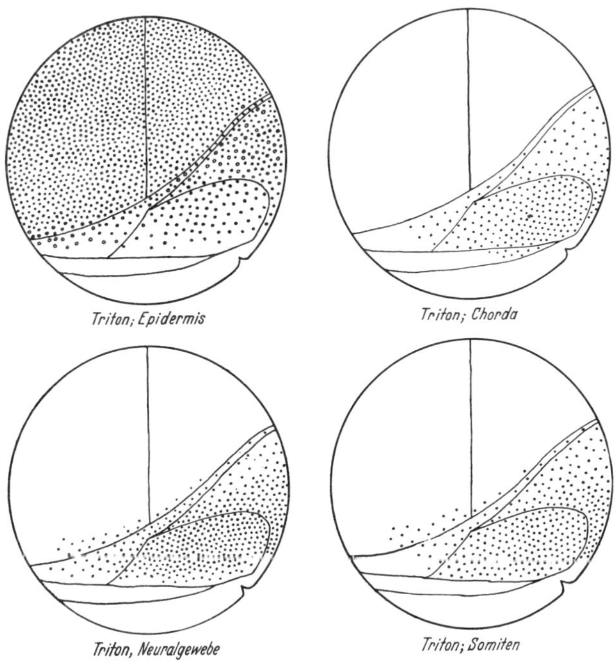

Figure 113. Distribution of differentiation patterns, as observed in explants from various parts of the ecto-mesoderm of a *Triton* gastrula. (Reproduced from S. Løvtrup, *Acta Zool.*, **47** (1966), after Holtfreter, by permission of AB Läromedelsförlagen, Stockholm)

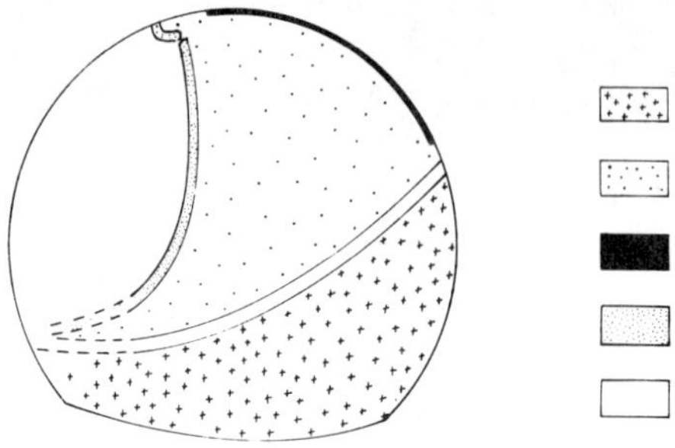

Figure 114. Revised fate map for the ectoderm of the gastrula of Caudata. (Reproduced from S. Løvtrup, *Acta Zool.*, **47** (1966), by permission of AB Läromedelsförlagen, Stockholm).
See Note 1, page 548.

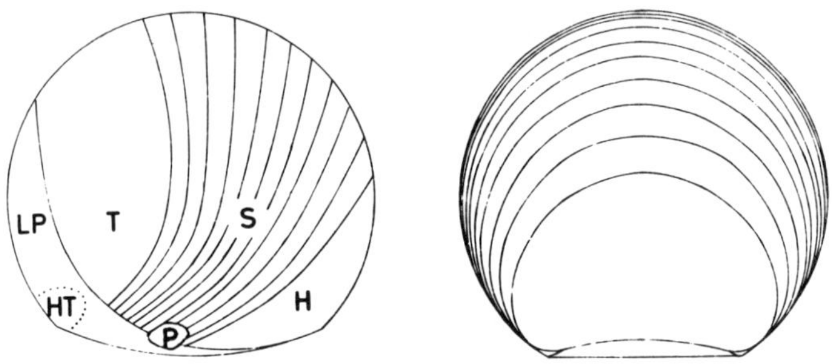

Figure 115. Theoretical fate map for the mesoderm of the gastrula of Caudata, based on the assumption that the germ layer is coherent, and not an agglomeration of free cells. H, head; HT, heart; LP, lateral plate; P, pronephros; S, somites; T, tail. (Reproduced from S. Løvtrup, *Acta Zool.*, **47** (1966), by permission of AB Läromedelsförlagen, Stockholm). See Note 2, page 548.

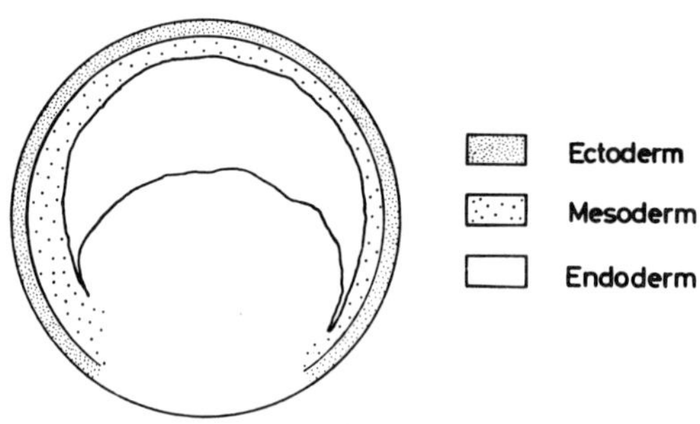

Figure 116. Median section showing the relative localization of the three germ layers in the gastrula of Caudata. (Reproduced from S. Løvtrup, *Acta Zool.*, **47** (1966), by permission of AB Läromedels-förlagen, Stockholm)

of the gastrula, the former around the dorsal median, and the latter in two tapering strips at the sides. Holtfreter (1938a) tried to account for these results by invoking two different processes, called 'selfdifferentiation' and 'regulation', respectively. The first of these is supposed to be involved when the differentiation corresponds to the presumptive fate (part of the cases where either chorda or muscle cells were formed), the second one when such agreement does not obtain (part of the chorda and muscle cells, and all nerve cells). As I have pointed out (1965a; 1966a) this interpretation cannot be upheld, for closer scrutiny of Vogt's own results reveals that the primordium of the notochord occupies the whole dorso–ventral surface but is confined to the superficial ectodermal cells, whereas the presumptive somite cells are the free cells in the interior (Figures 114, 115 and 116). These revised fate maps simplify matters to some extent; it is thus possible to modify Holtfreter's interpretation in such a way that only the nerve cell differentiation represents regulation, whereas the other cell-types will be cases of selfdifferentiation.

However, the introduction of the concept of regulation is not altogether satisfactory, because it involves that nerve cells may arise through two different mechanisms, by regulation in the dorso–lateral ectoderm and by induction in the remaining ectoderm. According to the views advanced above, it holds that whenever a cell is subjected to changes classified as differentiations, these must be evoked either by intrinsic (endogenous induction or selfdifferentiation) or by extrinsic factors (exogenous induction). This must hold also in the case of the explants discussed here, and if nerve cell formation is not selfdifferentiation, it must be induced by one of the other cell-types. The choice between these possibilities may be reached by establishing whether any causal relation obtains between the three differentiation patterns. It is important to point out before we start the analysis that these differentiations are by no means determined by factors inherent in the dorso–lateral ecto-mesoderm. This is evidenced by the fact that cells of indifferent 'ectoderm' which normally become cl-cells, if exposed to an appropriate inductor, will transform into filocytes; subsequently, they may differentiate in either of the three

Table 3

Differentiation patterns in explants of the dorsal blastoporal lip of *Triton* wrapped in ectoderm from *Bombinator*. The upper signs in each row represent the differentiation of the dorsal lip, the lower ones that of the ectoderm. From J. Holtfreter, 1936, *Arch. Entw.mech.*, **134**

Expt.	Chorda	Nerve	Muscle
1	+ −	+ −	+ +
6	+ +	+ −	+ +
8	+ −	+ +	+ +
11	− −	+ −	+ +
14	− −	+ −	− +
16	+ −	+ −	− +
17	+ +	+ −	− +
25	+ −	+ +	− +
26	+ +	+ +	+ +

directions (Holtfreter, 1936, cf. Table 3). The quotation marks above are meant to emphasize that these explants may have contained free mesodermal cells.

We may begin by investigating whether notochordal differentiations can be induced by either of the two other types. This proposition appears unlikely for two reasons: firstly because it stands in contradiction to current notions about the causal inductive relation between notochord (archenteron roof) and neural plate, secondly because the notochord seems to be the first organ to undergo differentiation, but none of these arguments is conclusive. However, a study of the explantation experiments of Holtfreter shows that chorda formation often is more frequent than muscle cell differentiation (Table 4). The occurrence of the former in an explant in which the latter is absent supports the contention that muscle cells do not induce chorda differentiation. I do not think this result will surprise any embryologist. The situation is a little more difficult with respect to the chorda–nerve cell relation, because the latter represents the most frequent differentiation pattern in the larger part of the explants. However, there are several cases in the extensive data published by Holtfreter in which the number of explants containing chorda cells exceeds those with neural differentiation (Table 4). Considering the conspicuous appearance of notochordal cells, it is unlikely that their presence would be overlooked, and the relatively few cases may therefore be considered conclusive. Consequently, chorda–cell formation must represent a case of selfdifferentiation.

Table 4

Cases of differentiation in explants where the formation of chorda cells is more frequent than that of nerve cells. Selected from J. Holtfreter, 1938, *Arch. Entw.mech.*, **138**

Species	Explant origin	Total	Number of explants Chorda	Nerve	Muscle
Triton	Dorsal lip	17	17	16	15
Triton	Lateral marginal zone	6	5	4	4
Triton	Ventral marginal zone	10	1	—	1
Ambystoma	Endoderm	7	4	—	3
Bombinator	Dorsal lip	16	12	9	13
Rana	Endoderm	4	2	—	3–4

That chorda cells may induce the differentiation of either of the other cell-types is a possibility which, as mentioned, would be in closer conformity with the temporal sequence in the normal embryo. However, this possibility is excluded by cases 11 and 14 in Table 3, and is also borne out by the results of Holtfreter on Caudata quoted above, thus in Figure 113 the areas of muscle and nerve cell differentiation are seen to extend much further towards the ventral side than that representing chorda differentiation. Many more observations supporting this point are found elsewhere in the literature.

At least one of the two remaining cell-types must therefore be a selfdifferentiation pattern, which in turn may induce the other. This possibility is suggested by the simultaneous occurrence of these two cell-types in many explants (cf. Figure 113). However, some of the results in Table 4 show that muscle differentiation often occurs in the absence of any nerve cells, thus excluding the likely contention that muscle cells are induced by nerve cells. The results in the first row of this table show that nerve cells may form without muscle cells being present; supplementary cases may be found in the quoted papers by

Holtfreter and elsewhere. Therefore, the less likely possibility that muscle cells induce the formation of nerve cells may also be excluded. That muscle differentiation is independent of either chorda or spinal cord has recently been demonstrated again by Muchmore (1958).

We can exclude that the conditions of explantation as such determine the mentioned differentiation patterns, for explanted ventral ectoderm will differentiate along these lines only if isolated in combination with dorsal lip. The results discussed thus suggest that inductive relations between differentiation patterns may not be a common phenomenon. It is, of course, unwarranted to generalize on the basis of the few observations dealt with here, but it seems that for filocytes the decisive factor in the determination of their fate may be the cell transformation. This event may be spontaneous or induced, but once it has occurred the cells will proceed to undergo selfdifferentiation, the pattern of which is determined by intrinsic factors. This mechanism thus corroborates the first possibility proposed above, according to which the cells may exist in different states corresponding to each of the differentiation patterns. That they may pass from one to another is suggested by the fact that nerve cells were found in the explants of dorso–lateral ectoderm, which normally forms only chordoblasts. This circumstance supports the suggestion that a mechanism of exhaustion is involved.

I have now on three occasions proposed that exhaustion may be of decisive importance for the determination of the activities and fates of embryonic cells, viz., in the transition from synchronous to asynchronous cell divisions and in the initiation of selftransformation and selfdifferentiation. Even if the mechanism as such is plausible, it may appear that I am going too far by thus applying it on repeated occasions. It must be observed, however, that as far as the cell divisions are concerned, the exhaustion was suggested to be concerned with a specific substance of direct importance for the synthesis of DNA. No reasonable objection can be raised against the contention that this material may occur together with, and become exhausted independently of, other substances which determine the presumptive diversification patterns.

As regards the latter, two possibilities obtain: the mechanisms for transformation and differentiation may be coupled or they may be autonomous, i.e. the exhaustion mechanism responsible for transformation is in no way involved in the one determining differentiation. The latter case implies that three separate groups of substances must independently be subjected to exhaustion. I have in previous papers suggested that, as far as the lf-transformation is concerned, coupling may obtain in the embryos of Amphibia (1967a). The fact that this transformation may occur spontaneously in many different cells in Invertebrata, leading to disparate differentiation patterns, contradicts this contention, for if exhaustion beyond a certain point is required for the lf-transformation to occur, then it may be anticipated that all cells assume the same differentiation pattern. It therefore rather seems that the mechanisms responsible for the lf-transformation and for the subsequent differentiations are autonomous. As regards the exhaustion mechanism, it should be mentioned that in cells richly endowed with reserves, the state of depletion should be a very unusual phenomenon. The circumstance that the lf-transformation occurs spontaneously at only one place in Amphibia, namely at the dorsal lip, from which it spreads to other regions through contact induction, may possibly be correlated with the fact that these eggs contain large amounts of reserves as compared with the eggs of Invertebrata in which selftransformation occurs at many locations.

In the scheme shown in Table 5, the concepts of precompetence, competence and postcompetence are combined with the idea of a succession of different states, arrived at through a mechanism involving exhaustion. It is shown how sl-cells, if exposed

Table 5
Selfdifferentiation with and without lf-induction

$$
\begin{array}{lcccccccc}
 & \text{Precompetence} & & \text{Competence} & & & & \text{Postcompetence} \\
 & & | & \textcircled{f} & & \textcircled{f} & & \textcircled{f} \;|\; & \textcircled{f} \\
\text{(a)} & sl_1 \rightarrow sl_2 \rightarrow \rightarrow sl_n & \dashrightarrow & sl_{n+1} & \rightarrow & sl_{n+2} \rightarrow \rightarrow & & sl_{n+m} & \dashrightarrow sl_{n+m+1} \\
 & & & \downarrow & & \downarrow & & \downarrow & \\
 & & & sf^{*}_{n+1} & & sf^{*}_{n+2} & & sf^{*}_{n+m} & \\
 & & & \downarrow & & \downarrow & & \downarrow & \\
 & & & sf^{**}_{n+1} & & sf^{**}_{n+2} & & sf^{**}_{n+m} & \\
 & & & \downarrow & & \downarrow & & \downarrow & \\
\hline
\text{(b)} & sl_1 \rightarrow sl_2 \rightarrow \rightarrow sl_n & \dashrightarrow & sl_{n+1} & \rightarrow & sl_{n+2} \rightarrow \rightarrow & & sl_{n+m} & \dashrightarrow sl_{n+m+1} \\
 & & & \downarrow & & \downarrow & & \downarrow & \\
 & & & sl^{*}_{n+1} & & sl^{*}_{n+2} & & sl^{*}_{n+m} & \\
 & & & \downarrow & & \downarrow & & \downarrow & \\
 & & & sl^{**}_{n+1} & & sl^{**}_{n+2} & & sl^{**}_{n+m} & \\
 & & & \downarrow & & \downarrow & & \downarrow & \\
\end{array}
$$

The indices 1 to n represent successive precompetent stages, n+1 to n+m successive competent stages and n+m+1 the stage of postcompetence. The asterisks denote successive differentiation patterns, and ⓕ the induction of the lf-transformation.

to an inductor of the lf-transformation during the phase of competence, may differentiate along separate lines (Table 5a). If they are not induced they will remain sl-cells, but as such they may of course also differentiate. If the state of postcompetence is concerned also with differentiation, the sl-cell must differentiate while being in one of the states indicated in Table 5(b). Postcompetence with respect to induction of the lf-transformation may be a consequence of the cells having begun differentiation as lobocytes which, according to the principle of exclusivity of diversification, may make them refractory to transformation induction.

It is unlikely that similar schemes exist for the other transformations, since these appear to be *accessory*, which means that they may be imposed upon already differentiated cells. This does not imply, of course, that competence always prevails, for if that was so, all cells would sooner or later be subjected to these transformation patterns.

Exogenous induction

It was observed at an early stage by the students of embryonic induction in Amphibia that two kinds of induction may be distinguished, those in which the inductor provokes the formation of its own kind, and those where no distinguishable similarity obtains. For these two phenomena we shall use the names *homotypic* and *heterotypic* induction, respectively (Grobstein, 1964). The existence of the heterotypic kind is implied in the interpretation of almost all experiments on embryonic induction. Thus, the formation of a secondary embryonic axis by an implanted presumptive notochord shows that a number of different organs and tissues arise as a consequence of this event. Even under less complicated conditions, for instance when the formation of kidney tubules is initiated in isolated kidney 'mesenchyme' by external factors (cf. Grobstein, 1964; Saxén, Koskimies, Lahti, Miettinen, Rapola and Wartiovaara, 1968), heterotypic induction is claimed to occur.

These phenomena cannot be correctly interpreted without a thorough analysis. First of all, it is necessary from the observed events to separate all the morphogenetic processes which may be initiated as a consequence of an induction, but cannot be induced themselves. However, even if we confine the analysis to the cellular level, very often many different cell-types may be observed. Thus, in the case of the formation of a secondary axis, it is possible to distinguish chordocytes, neurocytes, myocytes, mesenchyme cells, pigment cells, etc. As I have tried to show, the appearance of all these differentiation patterns may be accounted for on the assumption that the principal effect of the implanted, as well as of the normal, inductor is to ensure that the ectodermal and mesodermal cells in the dorsal region undergo the lf-transformation. The subsequent origination of a spectrum of filocytes, representing separate differentiation patterns, may be considered as due to selfdifferentiations, determined by the cytoplasmic composition of the cells at the time they begin their differentiation. This circumstance does not exclude, of course, that the cells may be subjected to other inductive influences, concerned for instance with the acquisition of accessory transformation patterns. The occurrence of this kind of event is suggested by the fact that some of the cell-types mentioned above are solocytes while others are colligocytes. If this interpretation is accepted it will be realized that many of the observations claimed to be heterotypic in reality are homotypic inductions.

It is understandable that during the early studies on embryonic induction, the really spectacular outcomes of induction experiments were so awe-inspiring that the analytic abilities of the observers were paralysed. It is a regrettable fact that this situation continued for many years; only recently have various authors begun to stress the necessity to distinguish between induction (diversification) and morphogenesis (cf. B. D. Davis, 1964; Holtfreter, 1968; Waddington, 1968). The fact that certain observations have been subjected to misinterpretations does not exclude that both homotypic and heterotypic inductions may exist, and since this question is of great interest from both theoretical and practical points of view, we shall discuss it in further detail.

Homotypic induction

Homotypic induction is simple to understand on the basis of a mechanism involving positive feedback, for in that case a diversified cell produces the inductor which serves to maintain the state in the cell proper. If the inductor can cross the boundaries to adjacent cells, the latter may undergo diversification provided they are competent. On the other hand, it is very difficult to envisage a satisfactory mechanism in the absence of positive feedback, for if no new inductor is produced by the cells that have responded to the induction there is no way to account for the spreading of the induction.

The notion of positive feedback goes a long way to account for tissue homogeneity, because the specific inductor must be present in an adequate concentration throughout the tissue. Under these circumstances all competent cells will respond to the influence of the inductor, whereas non-competent cells may occur as different cell-types, as demonstrated by many mixed tissues, migrate to other locations or succumb, possibly as a consequence of a mechanism of *inhibition* like that envisaged by Rose (1955).

It has been found that purified protein fractions may function as neural inductors (cf. T. Yamada, 1961; Tiedemann, 1963). To reconcile these observations with the views advocated here it must be presumed that the preparation contains the factors required for the synthesis of the lf-inductor. It cannot be taken for granted that these proteins normally are involved in the induction process, since the same result may ensue if the inductor molecule proper passes from the inducing to the reacting cells. It has been shown that induction stimulates synthesis of mRNA in the neural ectoderm (Tiedemann, 1966). This finding is in agreement with the contention that all steps of diversification are associated with protein synthesis.

One of the most remarkable findings in the field of embryonic induction is that a bewildering diversity of agents may act as inductors in the amphibian embryo (cf. Spemann, 1958). It seems possible that many of these inductors, particularly the inorganic ones, act by provoking cytolysis, but it cannot be excluded that in some cases a veritable exogenous induction of the lf-transformation obtains. Since supposedly most of the active inductors contain filocytes, there is no justification for assuming that heterotypic induction takes place.

As emphasized so often by Holtfreter (cf. 1968), the outcome of an inductive event depends largely on the potential of the reactive material. This is also borne out by the fact that inducers often include representatives of many differentiation patterns, supposedly containing their special inductors, yet without any visible effect upon the induction process. These odd inductors often do not result in the establishment of a secondary embryonic axis although, according to the interpretation suggested here, they are efficient as inductors of the lf-transformation. This shortcoming may be due to the fact that, in contrast to natural indicators, they are unable to participate in the morphogenetic events that normally accompany induction.

Heterotypic induction

From the preceding section it appears that many cases registered as heterotypic inductions upon closer scrutiny turn out to be in reality homotypic inductions of the lf-transformation, followed by selfdifferentiation. According to the present notions, heterotypic induction must involve that the metabolic activities associated with a certain differentiation pattern, besides the inductor required for its own maintenance, also elaborate an inductor for another pattern. That the cells in question do not react to the second inductor may be explained by the principle of exclusivity. However, if surrounding cells respond by either homotypic or heterotypic induction, it is evidently necessary to presume the existence of two different populations of cells, distinguished by their reactivity towards exogenous conductors.

Some of the early differentiation patterns may be accounted for by selfdifferentiation, but both observation and reason suggest that other mechanisms must also be invoked to explain the gradual appearance of new cell-types during embryogenesis. These can hardly arise by any other route than successive modifications of the initial differentiation patterns. It may be imagined that a given differentiation spontaneously progresses to new ones. However, since the number of differentiation patterns increases with time, branching must occur at various stages, and this circumstance shows that external factors must be somehow involved. It is possible that the latter act by influencing the metabolism of the cells, either by supply of some active substance other than the inductor or by changes in the physical environment, which subsequently leads to the formation by the cell proper of an inductor of a new differentiation pattern. Whether this mechanism can account for the phenomenon of heterotypic induction remains to be seen.

It has been proposed above that the synthesis of differentiation products is controlled by a mechanism of positive feedback, in contrast to the RF-repressible proteins where negative feedback is supposed to be involved. Undoubtedly this contention constitutes a great oversimplification, but one which successfully accounts for a variety of phenomena. The precondition for the mechanism of negative feedback is that the inductors required for the synthesis of the proteins always are present in sufficient quantity to warrant the activation of the genome. This possibility obviously cannot obtain in the case of the proteins distinguishing the differentiation patterns, but that does not exclude that negative feedback may be involved. However, under these circumstances the differentiation patterns would be characterized by some specific features. Thus, selfdifferentiation would be impossible, for as soon as a threshold concentration of inductor was formed, the induced metabolism would efficiently prevent further production of inducer. It therefore seems that such a differentiation pattern would be preserved only if the inductor was continuously supplied from an exogenous source. Secondly, the mechanism of positive feedback ensures that a differentiation pattern may spread from cell to cell through contact induction. Contrariwise, if negative feedback prevails, a differentiated cell may inhibit neighbouring cells from assuming the same differentiation pattern.

It is known that the activity of differentiated cells, as regards the synthesis of differentiation products, may be subjected to variations. One of these is the suppression observed in dividing cells, a problem we shall deal with later. Another case is observed in cells affected by hormones. As the latter are known to influence cells belonging to disparate diversification patterns, the effect cannot be concerned with induction, but rather with activation and inactivation. As an example we may mention the sex hormones, which influence the activity of many different cell-types. It is likely that an *activator* for certain differentiation patterns may be an *inactivator* for others (cf. Lasnitzki, 1965).

An interesting case in question is the effect of vitamin A on keratinizing epithelia (Fell and Mellanby, 1953). It has been found that when cultured in the presence of this vitamin, many normally keratinizing epithelia will come to consist of a mixture of ciliated and mucous cells, and conversely, in the absence of viatmin A, glandular epithelia may become squamous, keratinized epithelia. Keratin production is undoubtedly a phylogenetically younger differentiation pattern than mucus production. In view of this fact the action of the vitamin as a suppressor or inactivator of the latter differentiation pattern becomes most astonishing. It is possible to presume that mucus production does not represent a true differentiation pattern, but rather corresponds to a basal cell activity occurring even in keratinized cells but which is relatively unobtrusive in the presence of keratin production. The effect of vitamin A may in that case to some extent be explained by the fact that it stimulates the mitotic activity (cf. Lasnitzki, 1965), but the response appears to be too specific for this to be the only explanation.

We cannot leave this subject without a brief discussion of the 'heterotypic' induction through which collagen-producing cells are induced to become chondrocytes. It has been found that the necessary accessory differentiation pattern involving at least the synthesis of chondroitin sulphate, may be induced either by the notochord and the ventral spinal cord (Holtzer and Detwiler, 1953; cf. Holtzer, 1961) or by the endoderm of head (Hörstadius and Sellman, 1946; cf. Hörstadius, 1950). It is extremely difficult to imagine the common feature in these tissues which can be made responsible for the induction, certainly there is no reason to presume that they themselves are engaged in the synthesis of chondroitin sulphate. Possibly there are present in all of them mucoitin sulphate-producing colligocytes, but this does not seem to furnish a clue, since the inductive effect is quite specific, other colligocyte aggregates being inactive. The situation becomes further complicated by the claim that in the notochord it is the sheath cells, not the chordoblasts, which possess the inductive power (Cooper, 1965), because the former, as far as can be judged, are themselves nothing but collagen-producing sf-cells.

From the preceding discussion it appears that it is necessary to presume that external factors take part in the determination of some of the differentiation patterns which arise during development. According to the induction mechanism adopted here this can be achieved only by ensuring that the proper inductor is made available to the cells in question. If this occurs through direct supply of the inductor, the mechanism is identical with that of homotypic induction, and the fact that the inducing cells do not themselves possess the differentiation does not, in my opinion, warrant the introduction of the notion of heterotypic induction. This expression ought to be reserved for a mechanism involving other factors than the inductor. If this standpoint is accepted, it must be admitted that the existence of heterotypic induction has yet to be demonstrated.

Cell division and cell diversification

It is a classical embryological observation that cells actively engaged in proliferation generally do not differentiate, and this apparent antagonism between the two phenomena has been brought out also by studies on cells in culture. It has often been claimed that complete incompatibility prevails, but recently it has become evident that cultivated cells may be engaged in both activities simultaneously, although the synthesis of differentiation products occurs at reduced rates under these conditions (cf. Cahn, 1968; Holtzer and Abbott, 1968; J. R. Whittaker, 1968).

Dividing cells are, in contrast to differentiated cells, engaged in growth, involving synthesis of large amounts of non-repressible proteins. Under these circumstances it seems plausible that the drastic reduction of differentiation products observed in proliferating cells is merely a matter of competition of mRNA on the translation level, as suggested by Whittaker (1968). This author showed that various predictions of this hypothesis are substantiated by observations on cultured pigment cells. According to this idea the cells thus simply return to the level of protodifferentiation, as defined by Rutter *et al.* (1967). Whittaker, and others with him, uses the word 'modulation' to designate the change between differentiated and growing cells. This usage has been adopted here, even if the phenomenon is different from that to which the name was originally applied by P. Weiss (1939).

A problem of great importance is whether the phenomenon of differentiation is reversible, i.e. whether dedifferentiation can occur. It has been observed that, by appropriate changes, the proliferation of cultured cells may be suppressed, with ensuing assumption of the synthesis of differentiation products. This finding is in accord with the suggestion that the growing cells are in the state of protodifferentiation in which the mechanism of positive feedback will assure that inductor is present, although at a low level. The possibility exists that the activity may be reduced to the extent that the inductor is no longer available in sufficient quantity to allow the resumption of the differentiated activity under otherwise suitable conditions. Abbott and Holtzer (1968) observed that cultured chondrocytes under certain circumstances may lose the power to return to the differentiated state. These cells apparently have become dedifferentiated, and the fact that this situation is not changed in the presence of functional chondrocytes suggests that they have also lost the competence to respond to inducing stimuli.

In this connection mention must be made of the most important studies of Nameroff and Holtzer (1967) on the synthesis of mucopolysaccharides in proliferating and non-proliferating chondrocytes. As we shall discuss later, this cell-type is an sf-cell which through differentiation has acquired the ability to synthesize collagen and chondroitin sulphate. The mucopolysaccharide normally characterizing the sf-cell is hyaluronic acid, so that as regards mucopolysaccharide metabolism the differentiation involves at least two modifications, sulphatation and synthesis of galactosamine. It was found that in non-proliferating cells labelled sulphate and glucose were incorporated in a fraction behaving electrophoretically like chondroitin sulphate. After hydrolysis the isotope was found mainly in uronic acid and galactosamine, less than 5 per cent being recovered in the glucosamine fraction. When isolated chondrocytes were cultivated in monodisperse cultures a lively mitotic activity took place during the first five days. In this period no incorporation of sulphate occurred; as far as this activity is concerned the cells thus behaved as undifferentiated cells. The results relating to mucopolysaccharide synthesis were more ambiguous. Determination of incorporation of glucose showed label in the glucuronic acid, the glucosamine and the galactosamine fractions, the ratio between the latter being 1:1. The cells have evidently resumed the synthesis of the constitutive hyaluronic acid, without giving up the production of chondroitin.

The observed elaboration of chondroitin may partly relate to the fact that the cells prevail in the state of protodifferentiation, in which case they might be expected to carry out some sulphatation also. It must be borne in mind, however, that the results represent the outcome of five days' activity, during which time the cells gradually return to the metabolism distinguishing their differentiation pattern. This is very clearly shown by the analyses obtained from the second five-day period. Here two mucopolysaccharide fractions were obtained, one sulphated and one non-sulphated. In the former one the glucosamine:galactosamine ratio was 1:1, the same as in the non-sulphated fraction for the first period. This result is interesting insofar as it seems to show that both chondroitin and hyaluronic acid become sulphated. In the non-sulphated fraction, which may be presumed to be synthesized later than the sulphated one, the glucosamine: galactosamine ratio is 1:20, indicating a drastic reduction in hyaluronic acid relative to the differentiation product chondroitin sulphate. It is seen that these results rather conclusively demonstrate the antagonism between synthesis of non-repressible proteins and differentiation products.

We have thus far discussed only the relation between cell division and cell differentiation, and it therefore remains to consider the other type of diversification processes, the cell transformation. The fact that most or all cell orders are encountered in growing cell cultures (cf. Willmer, 1960) demonstrates that cell division does not interfere with the synthesis of the substances responsible for the preservation of the cell-class properties. If the mechanism of competition on the transcription level is accepted, this state of affairs suggests that the quantities of protein required for the maintenance of the transformation patterns are so slight that no competition obtains. This proposition is supported by the fact that there appears to be a substantial difference in the quantities of sulphated mucopolysaccharide synthesized by colligocytes, where this activity relates to a transformation pattern, and by chondrocytes, in which it represents a differentiation pattern. We shall return to this question presently. It may be noted that in flagellocytes the centriole supposedly is involved in the formation of the vibratile organelle. It goes without saying that in this respect the cell must be subjected to 'detransformation' before it can go through the process of cell division.

Diversification patterns

A considerable number of diversification patterns is found in adult organisms, particularly in animals of the high-ranking phyla like Mollusca, Arthropoda and Vertebrata. The circumstance that a diversification pattern is distinguished by a certain synthetic activity rather than by its participation in morphological structures entails, as pointed out above, that the number is smaller than that of histologically discernible cell-types. In this section we shall present a list of diversification patterns, mainly those found in the early embryo. Whenever possible we shall try to characterize the patterns in chemical terms. After this inventory of cell transformation and cell differentiations, some attention will be devoted to the question of the hierarchy of diversification patterns.

TRANSFORMATION PATTERNS

The eight cell orders were suggested to be derived from an original one on the assumption of the existence of three different processes of cell transformation. Characteristic of all of them is that, as far as can be judged, the transformation involves only very slight changes in the prevailing synthetic activity. Thus, the sc-transformation leading to colligocytes distinguished by the production of sulphated mucopolysaccharides may involve nothing more than sulphatation of the substances otherwise present in the solocytes. The two other cell transformations are characterized by the formation of microtubules, present either in vibratile organelles or in the cytoplasm. These structures are made up of TP monomers which are present also in amoebocytes, and so far the transformation patterns do not involve any novelty. According to the hypothesis of Oosawa and Higashi (1967), the elaboration of microtubules requires the presence of specific selector molecules. However, cytoplasmic microtubules are present in dividing amoebocytes, so the transformations can hardly involve the mobilization of an inactive operon, but maybe a change in its activity so that it operates continuously. As far as the nv-transformation is concerned, the function of the vibratile organelles requires the presence of substances otherwise acting as neurotransmitters, but other metabolic steps may be involved.

Since, as we shall presently discuss, the lf-transformation appears to be of much greater importance than the other ones, it may be presumed that it involves steps other than synthesis of the selector molecule. Next to nothing is known concerning this point, but it should be mentioned that the appearance of sf-cells in embryos of Echinoidea and Amphibia is associated with the synthesis of alkaline phosphatase (1966b; 1955).

There seems to be a significant difference between the sc- and the nv-transformations on the one hand, and the lf-transformation on the other. Thus it appears that the sl-cell is confined to a specific range of differentiation patterns, generally characterized by adherence to the principle of exclusivity. After subjection to the irreversible lf-transformation, the ensuing sf-cells may also differentiate along a number of separate lines, but these are different from those pursued by the sl-cells. Conversely, it seems that sf-cells belonging to at least some of the various differentiation patterns may assume either of the two other transformation patterns. These thus do not conform strictly to the principle of exclusivity and have therefore been characterized as 'accessory' diversification or transformation patterns. Certain observations on cell cultures and on embryos suggest that these transformations, in contrast to the lf-transformation, are reversible.

DIFFERENTIATION PATTERNS

The lf-transformation serves to create two groups of cells, the lobocytes and the filocytes, each of which are distinguished by a number of differentiation patterns that, with one possible exception, are specific to the particular groups. According to the proposition advanced above, a certain differentiation pattern may in principle be associated with any combination of the two other cell transformations, so that theoretically it might exist in four different forms. This possibility is probably never realized, but we shall in the following pages discuss examples showing that the same differentiation pattern may prevail as both solocytes and colligocytes, or solocytes and vibriocytes.

It has often been observed that the determination of the presumptive fate of embryonic cells occurs a long time before 'histodifferentiation', i.e. cell differentiation, sets in. It has therefore been suggested that two phases of differentiation exist, first it seems by Huxley (1924), who suggested the names 'chemical determination' and 'visible differentiation' for the respective events, but subsequently many other names have been suggested. If, as proposed here, lobocytes and filocytes differ with respect to the differentiation patterns they can assume, the lf-transformation amounts to a determination of the presumptive fate of the cells as far as differentiation is concerned. It is therefore likely that Huxley's phase of chemical determination corresponds to the lf-transformation. Clearly, that of visible differentiation may be identified with the transition between the states of protodifferentiation and differentiation as described by Rutter *et al.* (1967). It thus appears that between these two phases one more is intercalated, viz., the change from the undifferentiated to the protodifferentiated state, the process of differentiation proper, in which the activity distinguishing the differentiation pattern has begun, although this is generally difficult to visualize or otherwise establish.

Differentiation patterns are characterized by the chemical activities in which the cells are engaged, which often are revealed by the accumulation of differentiation products. It must be pointed out, however, that the visible substances may represent only the end of an extensive sequence of metabolic processes. Both solocytes and colligocytes may secrete their differentiation products, but only the former, notably the sf-cells, are able to elaborate a matrix in which they are themselves embedded. The colligocytes secrete globular proteins in the form of secretion granules. As appears from keratinizing cells, fibrous proteins are generally not secreted, but rather deposited intracellularly. Only with the death of the cell does the keratin become an extracellular element. Collagen, a fibrous protein, is secreted by living cells, but in the form of a non-fibrous precursor, tropocollagen.

Willmer (1960) has made an attempt to distinguish various *families* of differentiated cells. The present survey of the families of differentiation patterns in lobocytes and filocytes owes a lot to this work, even though I do not always accept the classification proposed by Willmer. It will be noticed that cell systematics exhibits the particular feature that it is the classes rather than the orders which are subdivided into families.

Lobocytes

Among the lobocytes we may distinguish three families, the members of which are solocytes, viz., the *gonocytes*, the *haemocytes* and the *osteoclasts*, and several families of colligocytes, of which we shall deal here only with *epitheliocytes* and *protein-* and *steroid-producing adenocytes*.

Gonocytes

The germ cells represent the original stock from which other cell-types are derived through diversification processes. To the extent that the latter are irreversible, it holds that some of the descendants of the egg, constituting the so-called *germ line*, must be protected against those conditions that promote diversification, including the phenomenon of chromatin diminution. The fact that germ cells generally arise from yolk-laden endodermal or mesodermal cells seems to support the contention that exhaustion of reserve material is involved in the mechanism of cell diversification.

The egg is distinctly different from the cell that originally left the germinal epithelium and which, we presume, is a direct descendant of the germ line. Yet, if the contentions advanced here are valid, the egg must at no time become engaged in the synthesis of anything but non-repressible proteins. In other words, if yolk and other storage proteins represent differentiation products, then they must be elaborated by differentiated cells. The yolk proteins may be produced by special yolk cells which become engulfed by the oocytes, or by follicle cells covering the surface of the latter, or they may be synthesized elsewhere in the body, for instance in the liver, and transported to the ovary by the blood. It thus appears that the egg may accumulate specialized proteins without becoming itself differentiated. As already pointed out, the behaviour and appearance of many eggs indicate that they are amoebocytes.

As far as the spermatozoon is concerned, the cells involved in its formation, spermatogonia, spermatocytes and spermatids, are, in conformance with anticipation, typical sl-cells. The sperm itself is a flagellocyte, and may be classified as a vsl-cell. This fact shows that it is subjected to cell transformation, and the sperm is therefore not completely shielded against cell diversification. Since little more than the pronucleus enters the egg, there seems to be no reason to expect that the transformation pattern of the spermatozoon can affect the egg.

Haemocytes

The blood cells comprise a spectrum of types, erythrocytes, three kinds of granulocytes, lymphocytes, monocytes and megakaryocytes. It is generally presumed that these differentiation patterns are genetically related (cf. Guyton, 1966). For various reasons it is believed that they may be derived from cells belonging to the reticulo-endothelial system. Willmer (1960) points out that the cells lining the sinusoids in various organs and tissues, e.g. liver and bone marrow, in their capacity to segregate vital dyes display typical amoeboid behaviour. Unfortunately this particular property is not very distinctive, for Kupfer's cells, belonging to the reticulo-endothelial system, are unquestionably phagocytic and yet the morphological difference between them and the remaining cells in the system is so typical that it warrants their classification as filocytes (Figure 117). It may be mentioned in passing that the network of cells in the liver sinusoids is so similar to the 'atypical' epidermis observed by Holtfreter (1934, cf. Figures 13 and 117) that it can hardly be questioned that the cells involved are cl-cells.

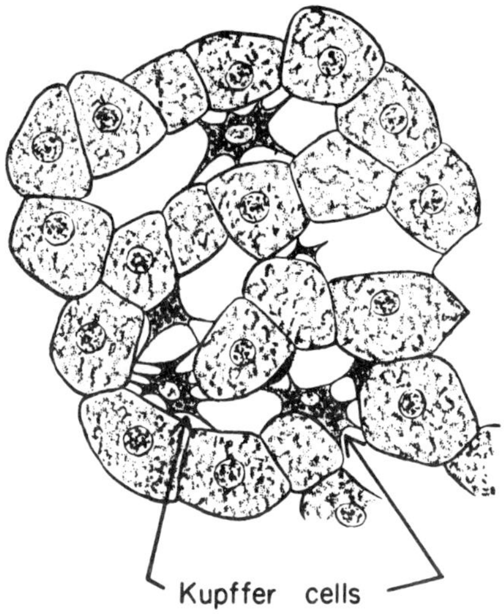

Figure 117. Kupfer's cells in liver sinusoids, showing phagocytosis of Indian ink particles. (Reproduced from A. C. Guyton (1971), *Textbook of Medical Physiology*, after Copenhauer and Johnson, by permission of W. B. Saunders Company, Philadelphia)

It appears that some of the various leucocytes under certain circumstances may assume a shape suggesting that they are sf-cells. Since cells of this order usually become immobilized through attachment, the free migration of the leucocytes in the tissues appears to be incompatible with this classification, unless they have become subjected to a change with respect to the adhesivity of their filopodia.

The erythrocytes contain microtubules, a filocyte feature. This circumstance may indicate that during their differentiation they become sf-cells, but as we have seen above, even lobocytes may under some conditions form microtubules. Lymphocytes are considered to participate in the formation of granulation tissue. Since an essential constituent of the latter is collagen, there is some reason to believe that the lymphocytes are transformed first to sf-cells, being subsequently differentiated into collagen producers. The examples discussed here show that it is difficult to classify the haemocytes; some may be filocytes and others may become transformed into this cell class.

Several of the blood cell-types are distinguished by their production of globular proteins, antibodies, etc., a feature which is characteristic for many lobocytes. The cells in question are mainly derived from the lymph glands and thymus. It is worth mentioning that these cells possibly constitute a reserve of competent cells which, after suitable diversification, may become somatic cells.

Monocytes *in vitro* contain acid phosphatase (Weiss and Fawcett, 1953), an enzyme characteristic of protozoan amoebae (Holter and Lowy, 1959). It is the only phosphatase present in the early embryo, while, as already mentioned, alkaline phosphatase appears together with the first filocytes.

Osteoclasts

Of these cells there is not much to say except that they emerge from bone tissue cultured *in vitro*, together with the osteoblasts which are typical sf-cells. The osteoclasts are large multinucleate cells, a typical amoebocyte feature (Hancox, 1946), distinguished by their possession of acid phosphatase.

Epitheliocytes

Epitheliocytes, i.e. cells participating in the formation of epithelia, constitute a group of families, each distinguished by a specific differentiation pattern, some of which will be discussed below.

In many embryos cl-cells form epithelia; as examples we may mention the ciliated cells in the pronephros and in the epidermis of vertebrate embryos. As shown in Figure 118, the shape of the cells in the epidermis of the amphibian embryo leaves no doubt

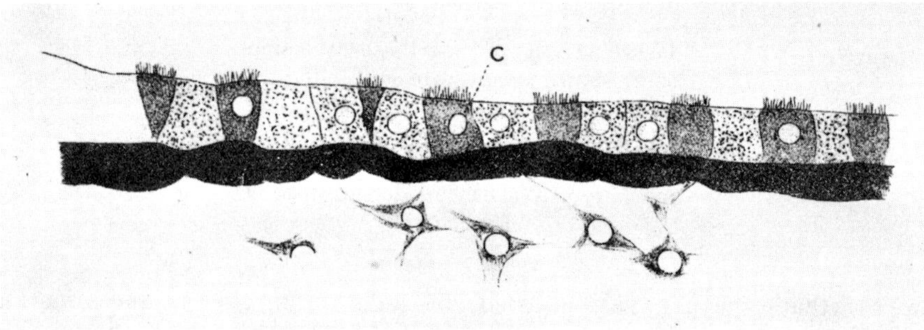

Figure 118. Section of the epidermis of a 6 mm tadpole of *Rana*. (Reproduced from R. Assheton, *Quart. J. Micro. Sci.*, **38** (1896), by permission of Clarendon Press, Oxford)

about their cell-class affiliation. However, there are great difficulties involved in determining the differentiation pattern represented by these cells. In many cases they may produce mucus, but whether they are involved in other kinds of synthetic activity is not known.

Sooner or later most epidermal cells assume the production of keratin or other scleroproteins. Generally, it seems that globular proteins are produced by lobocytes and fibrous ones by filocytes. In the present context it is of great importance whether this correlation has absolute validity, for in that case the keratin-producing cells must be cf-cells. The fact that, under the influence of vitamin A, they may become converted into mucopolysaccharide-producing cells (Fell and Mellanby, 1953) is no drawback to the argument, for this differentiation pattern is well known among filocytes. The cells apposed to the basement membrane are often elongated, but this cannot be used as a criterion, because tonofibrils may be responsible for this shape. The presence or absence of microtubules is not of much use either, since certain differentiated filocytes may cease to elaborate this type of polymer. Yet, it may be mentioned that in the ammocoete larva of *Petromyzon* microtubules are interspersed among the tonofibrils in the basal epidermal cells (Fawcett, 1966). It thus appears that a certain support obtains for the notion that keratin-producing cells are cf-cells. Similar considerations apply to chitin-producing cells. Although a definite classification of these cell-types thus cannot be made, they will in the present context be included among the filocytes, entailing that they properly belong to the following section.

It is possible that there are other differentiation patterns in this family, but our knowledge is too scanty to warrant further discussion.

Protein-producing adenocytes

As a common name for gland cells we shall use *adenocytes*. In the present family we may first deal with the hormone-producing cells of ectodermal origin, thus those present in the anterior lobe of the pituitary, representing several differentiation patterns, and those found in the thyroid and parathyroid glands. A corresponding endodermal group is found in the pancreas, including among others the insulin-producing islet cells.

Besides hormones, a number of other proteins are elaborated, mucoproteins by salivary glands, digestive enzymes by these and by the pancreas. The latter products are also formed in many cells in the walls of the digestive tract; it is possible that some of these are cl-cells. The liver cells are known to produce a rich assortment of proteins. It appears to be typical of this family, and possibly of the lobocytes, that the differentiation products are accumulated in secretion granules, which are transported out of the cells.

Steroid-producing adenocytes

In this case also it is possible to distinguish an ectodermal and an endodermal group. The former is represented by the adrenal cortex and various steroid hormone-producing sexual glands. The latter is represented by the liver, in which the bile salts are formed.

Filocytes

The filocytes exhibit a large number of easily distinguishable differentiation patterns, which even to a considerable extent may be characterized from a chemical point of view. We shall here deal with the following families: myxocytes, neurocytes, sclerocytes, myocytes, chromatocytes and protein-producing adenocytes. This list is not exhaustive, but it does comprise all the important differentiation patterns. As it turns out, it is not practical in this case to treat the solocytes and colligocytes separately.

Myxocytes

A great number of cell-types are distinguished by the production of mucopolysaccharides in quantities over and above those usually present in cells. Since it is rather inconvenient to use the expression 'mucopolysaccharide-producing cells' to characterize this family of differentiation patterns, we shall henceforth adopt the name *myxocytes* for this purpose.

The most prominent cell in this family is probably the mesenchyme cell, engaged in the elaboration of mesenchymal mucopolysaccharides, the main component of which is hyaluronic acid, at least in Vertebrata. Since the name 'mesenchyme cell' is very ambiguous, being used in connection with cells representing several orders and differentiation patterns, it may be preferable, if possibly not quite correct, to designate these cells as *myxoblasts*. This term is employed by Willmer (1960) as a name for the order of sf-cells, but it will be realized that the usage suggested here corresponds more closely to the etymological sense of the word. There are good reasons to believe that the mucopolysaccharide present in undifferentiated sf-cells is the same as that secreted by the myxoblast, and if that is true, the latter may represent a differentiation pattern involving a quantitative, but not necessarily a qualitative, change in the prevailing synthetic activities. In this context it may be mentioned that the endothelial cells in capillaries and other blood vessels appear to be sf-cells flattened against an adhesive substratum. We do not know what differentiation pattern these cells represent, but it is not excluded that they may be myxocytes. To this family also belong the mucous gland cells, represented for instance by the frontal gland cells in the amphibian embryo (Figure 119) and

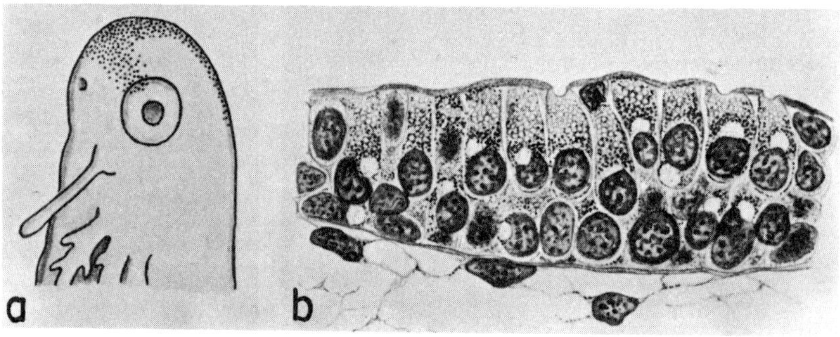

Figure 119. Frontal gland cells in the head epidermis of *Triton*. (a), distribution of the cells; (b), cross section through the cells. (Reproduced from J. Holtfreter, *Arch. Entw.mech.*, **127** (1933), by permission of Springer, Berlin)

by the goblet cells present in mucous membranes. The latter discharge their contents through the mechanism typical for colligocytes.

The notochordal cells and the proximal lens cells may also safely be included here, whereas the distal lens cells by their shape betray that they are cl-cells, although they exhibit the same differentiation pattern as the distal cells (Figure 120). These groups of cells exhibit their colligocyte affiliation by the fact that they participate in aggregate formation. The nature of the substance elaborated by chordoblasts and lens cells is not known, but the available evidence suggests it to be a sulphated mucopolysaccharide, but probably not chondroitin sulphate.

In this group we shall also include the trochoblasts in trochophore larvae and the comb plates in Ctenophora and other similar cells. Although the ciliation in these cells is of an unusual type, it probably nonetheless must be considered a transformation, rather than a differentiation, characteristic. Nothing is definitely known about the differentiation pattern of these cells, but since they are derived from small hyaline cells, and since, like other cells in this family, they are distinguished by their vacuolated cytoplasm, it seems reasonable to propose that they are myxocytes. This contention is

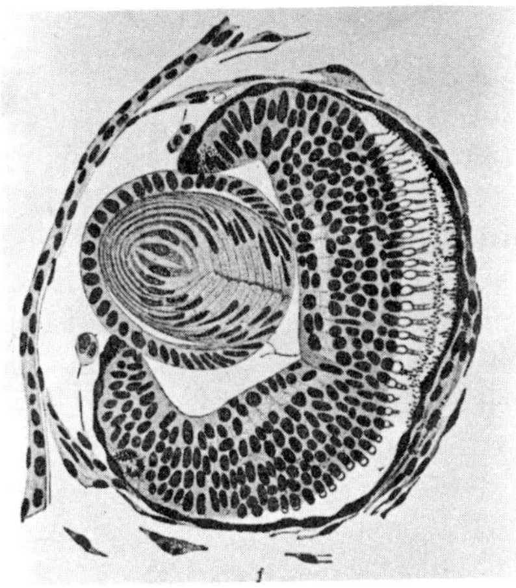

Figure 120. The lens in *Siredon*, showing the cubic distal and the elongated proximal lens cells. (Reproduced from C. Rabl, *Z. Wiss. Zool.*, **63** (1898), by permission of Akademische Verlag, Leipzig)

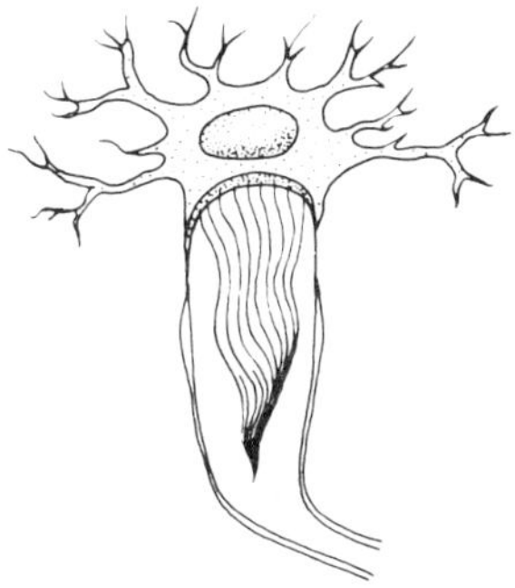

Figure 121. Protonephridium (flame bulb) from a cestode. (Reproduced from W. S. Hoar (1966), *General and Comparative Physiology*, after Goodrich, by permission of Prentice-Hall, New York)

supported by the circumstance that the origin of these cells is closely related to that of the myxoblasts in various spirally cleaving eggs, a subject to which we shall return at a later occasion. This relation suggests that they may be filocytes, and the intimate connection between this cell-type and various sf-cells in the sea urchin embryo corroborates this point, indicating that they may be classified as vcf-cells.

Much evidence intimates that among the protonephridial cells the flame bulbs are trochoblasts, the main body of which has been pulled into the body cavity. The difference between the two manifestations of this cell-type may be accounted for on the assumption that the protonephridial cells, in contrast to the trochoblasts, are exposed to a morphogenetic space, the mesenchyme-filled blastocoele, before membrane formation prevents 'invagination' of the cell (Figure 121).

It may finally be noted that the swelling and vacuolization of the test and the inner follicle cells in *Styela* (Tucker, 1942) suggest that they also belong to the family of myxocytes.

Neurocytes

It has been mentioned that sf-cells may span various cavities in the embryo with their filopodia, and if conduction of an electric potential is a generalized membrane property as much as the spreading of a tension–relaxation wave, it may be envisaged that the undifferentiated sf-cell can function as a primitive nerve cell, a contention that is supported by various embryological observations. It seems that this particular property has been further developed in the differentiation of the nerve cell. The most outstanding characteristic of this cell-type presumably is the ability to sustain an enormous outgrowth of one or two filopodia. That microtubules are involved in this activity is well documented (cf. Fawcett, 1966), and so far the differentiation pattern may be considered a quantitative modification of a cell-class property. Other features are known to distinguish the neuron, some of which may be specified in chemical terms.

Willmer (1960) has pointed out that filocytes are discriminated by their metabolism of aromatic amino acids. In this context may be mentioned the presence of noradrenaline in many nerve cells. The participation of 5-HT, the stimulant of microtubular activity, in the function of nerve cells is also worth noting, although it should be recalled that this substance is present even in lobocytes, at least during mitosis (Buznikow *et al.*, 1964). The existence of cholinergic neurons suggests that they are flagellocytes, thus vsf-cells, which have assumed the nerve-cell differentiation pattern. It is not unlikely that the antagonistic function of the autonomic nerve systems is based upon two different cell orders, a vsf-cell which is cholinergic and a nsf-cell which is adrenergic. From the comparative physiology it is known that the latter differentiation pattern is phylogenetically younger than the former one.

The endocrine cells of the adrenal medulla belong to this family, and probably also the polypeptide hormone-producing cells of the posterior lobe of the pituitary. The hormone secreted by the former, adrenaline, is methylated noradrenaline, and thus represents one further step in the metabolic conversion of tyrosine. The astrocytes and the oligodendroglia cells may constitute a special subfamily. These cell-types appear to be characterized by the presence of microfilaments rather than of microtubules, whereas neurons contain both kinds of linear polymers. Nothing is known about the physiological features that distinguish these cells from nerve cells. It may be mentioned that, according to Willmer (1960), microglial cells are amoebocytes (sl-cells). As a common name for the differentiation patterns in this family *neurocytes* is suggested.

Sclerocytes

In the family of *sclerocytes* we shall include all differentiation patterns involving the production of 'hard' tissues, thus all connective tissue cells and those responsible for the elaboration of epidermal coverings like keratin, chitin, etc.

The most important member of the family is unquestionably the collagen-producing cells represented in the animal kingdom from Protozoa to *Homo*. This cell-type does not appear to have a particular name; in the present context we shall, for convenience, use that of *collagenocyte*. It seems to be a characteristic feature that collagen cannot be secreted unless the cell is exposed to a matrix of sulphated mucopolysaccharide. If this condition does not prevail in a cell culture, potential collagen producers behave as ordinary 'fibroblasts'. The matrix may be supplied by the surface coat of colligocytes, which explains the formation of collagenous membranes, sheaths and septa on all surfaces of colligocyte aggregates in the embryo. Under these circumstances cartilage is never formed, a fact which presumably may be referred to the chemical nature of the matrix.

Contrariwise, if the sf-cells proper are induced to synthesize chondroitin sulphate, cartilage may be formed. I have on a previous occasion (1967a) suggested that the cells representing this differentiation pattern, the *chondrocytes*, are cf-cells, because they produce sulphated mucopolysaccharides, but the appearance of these cells, when actively growing in cell cultures, clearly demonstrates that they are sf-cells. It is therefore evident that synthesis of chondroitin sulphate represents a separate differentiation pattern, which may serve to convert collagenocytes into chondrocytes.

The cells producing elastin also belong to this family, as well as those responsible for the elaboration of keratin, chitin, etc. In chitin-producing cells a change in the mucopolysaccharide metabolism has resulted in the formation of an element useful for the construction of external body coverings. It may be pointed out that structural proteins are also involved (cf. Travis, 1960).

The process of mineralization, particularly calcification, generally requires a matrix consisting of collagen or a similar protein imbedded in mucopolysaccharide (cf. Moss, 1964). As appears from the discussion of Moss, biological mineralization comprises several different types. Of special interest is the intracellular calcification encountered in larval Echinoidea, a phenomenon studied by Okazaki (1960) and by Wolpert and Gustafson (1961a). These authors found that the spicules of the skeleton were deposited in an organic matrix formed by the fusion of filopodia extended by mesenchymal sf-cells. If indeed the structural component of the filopodia is microtubules, then it seems to follow that TP may take the place of collagen in the matrix, an observation that is the more interesting when the difference between the proteins as concerns their amino acid composition is taken into account.

This intracellular calcification seems to be an embryonic feature which among adult animals is preserved only in Holothuroidea and Asteroidea (cf. Hyman, 1955; Kühn, 1955). However, in all classes of Echinodermata collagen is present in the connective tissues of the body wall, both in non-calcified (Moss and Meehan, 1967) and in calcified structures (cf. Travis, François, Bonar and Glimcher, 1967). The latter point has been contested by Klein and Currey (1970), but in view of the convincing evidence presented by Travis *et al.* (1967), it seems that their discordant results primarily may be referred to an exceedingly low protein content in the organic phase (cf. Travis, 1970).

Calcification requires a special sclerocyte differentiation pattern which is responsible for the initial precipitation of the inorganic phase.

Myocytes

Linear TP polymers may develop a tension when suspended between two frames. This phenomenon may be exploited for the execution of mechanical work, as illustrated by the mesenchyme cells which pull the archenteron through the blastocoele in the sea urchin embryo (Gustafsson, 1961). This capacity has been further exploited in a special family of differentiation patterns, the muscle cells. It is a remarkable fact that one of the first steps in this line is the switch from microtubules in the myoblasts (Allen and Pepe, 1965) to microfilaments in the true muscle cells. This change does not, in my opinion, influence the cell-class affiliation, but it may be an indication that microfilaments are more efficient for the particular purpose of carrying out mechanical work.

In the simplest differentiation pattern in this family, the smooth muscle cell, only actin filaments are present. In skeletal and heart muscle cells, another protein polymer, myosin, has been added. As discussed above, this component may be instrumental in the phenomenon of muscle shortening. It is interesting to note that the two differentiation patterns mentioned are distinguished by containing different proportions of the two lactate dehydrogenase isozyme subunits (Markert, 1963).

In most invertebrates the muscle cells originate as sf-cells, which to a large extent settle in regular patterns in the body wall. Higher in the evolutionary scale the muscle cells are cf-cells, a circumstance which among other things allows for the formation of muscles rather than muscular layers.

As a common name for the various differentiation patterns in this family we shall use *myocytes*.

Chromatocytes

The family of pigment cells or *chromatocytes* comprises several differentiation patterns, distinguished by the accumulation of coloured substances in the cytoplasm. Of special interest are the integumentary chromatocytes, the *chromatophores*, of which a wide range exists, generally distinguished by the chemical nature of the enclosed pigment. Here we shall mention only three, viz., the *melanocytes* containing melanin, the *xanthocytes* containing carotenoids and the *guanocytes* or *iridocytes* containing guanine. The pigment movements observed in the usual melanophores depend, as already pointed out, on the fact that the melanin granules adhere to the microtubules in the filopodia of these cells. The various chromatocyte differentiation patterns do not appear to be generically related, in possible contrast to some or most members of the other families.

Protein-producing adenocytes

The uptake of food from the gut may occur either through phagocytosis or through resorption. In the former case the cells presumably are amoebocytes (sl-cells), in the latter they may be colligocytes. The shape of many of the cells in the digestive tract suggests that they are cf-cells. The circumstance that the intestinal wall is a rich source of alkaline phosphase also supports this contention. The presence of microvilli in the brush borders indicates, on the other hand, the presence of microfilaments. Whether this represents a special differentiation pattern, or whether cl-cells are also present, remains to be seen. In any case, the variety of products elaborated by the digestive tract, enzymes, hormones, etc., clearly demonstrate the existence of a number of differentiation patterns.

Against the background of the cell-order affiliations of gut cells suggested here it is of interest to note that ontogenetically the two types of gut cells differ insofar as the phagocytotic ones always arise directly from yolk-laden embryonic cells, whereas guts with resorbing cells may be traced back to special primordia, intimating that special epigenetic processes, most likely inductions, are involved in their origination. (Schlottke, 1934.)

HIERARCHY OF CELL DIVERSIFICATION PATTERNS

According to the views advanced in this chapter the lf-transformation is the first decisive step as regards the presumptive diversification of the cells in the early embryo. This transformation occurs spontaneously in certain cells and may spread to others through induction. After this event has taken place, two classes of cells prevail, the lobocytes and the filocytes, each of which may differentiate into patterns that largely are class-specific. We have seen that among the filocytes in the amphibian embryo at least three differentiation patterns arise through selfdifferentiation. In this particular case observational evidence suggests that the differentiation patterns arise from an initially homogeneous population of cells. Under these circumstances if was found necessary to postulate the existence of a mechanism of cytoplasmic exhaustion to account for the formation of different cell-types. Obviously, if this suggestion is correct, it must be possible to establish sequences of differentiation patterns such that the first one represents the least, and the last one the most, exhausted cells. We shall in the present section begin with an investigation of the possibilities for establishing sequences of this type.

In the larva, or in a corresponding stage in directly developing embryos, various diversification patterns are present, among which, at least in the higher animal phyla, most of the families listed above are represented by one or sometimes a few cell-types. The differentiation patterns in embryos generally are not identical with those present in adult organisms. This fact may be referred to the property of progressiveness ascribed by P. Weiss (1939) to the process of differentiation, and it may thus be presumed that the differentiation patterns are subjected to progression in the course of development. Insofar as the number of differentiation patterns simultaneously increases substantially, and since it is often possible to trace a group of patterns back to one of the early cell-types, it may be warranted to propose that ramification occurs at various occasions during ontogenesis. Very little is known about the two kinds of processes mentioned here, but for the sake of completeness they will be briefly discussed.

Sequences of selfdifferentiation patterns

The present discussion will be confined mainly to the selfdifferentiation patterns in the amphibian embryo, but the results arrived at will at a later occasion be shown to be of more general validity.

It was demonstrated by Holtfreter's experiments discussed above (1938a; 1938b) that in dorso–lateral explants from the gastrula the cells may through selfdifferentiation become chordocytes (i.e. myxocytes), neurocytes or myocytes, provided they are subjected to the lf-transformation, and the same result was obtained with cells from the ventral ectoderm (Tables 3 and 4).

According to the mechanism of nucleo-cytoplasmic interaction, these results evidently imply that the ecto-mesodermal cells can exist in at least three different states with respect to cytoplasmic composition, one corresponding to each of the differentiation patterns. It may furthermore be inferred that the myxocytes represent the last step in the sequence, since obviously the chordocytes normally assuming this differentiation pattern pass through states leading to neurocytes and myocytes before their fate is definitely established. It is in conformance with a mechanism of exhaustion that these cells, the most active ones in the dorsal region, which are almost exclusively responsible for the execution of the gastrulation movements (1965a; 1966a), will be the only ones to reach this last state. There is no direct evidence which can decide the relative positions of the two other selfdifferentiation patterns, but since the ectodermal cells apparently are much more active with respect to cell division, morphogenetic movements, etc., than the mesodermal ones, which in the normal embryo are the only ones to become myocytes, it may reasonably be presumed that the state leading to muscle cells comes before that representing nerve cells. The following *sequence* may thus tentatively be established: myocytes ⟶ neurocytes ⟶ myxocytes. (See Note 3, page 548.)

In the phenomenon of exhaustion two factors are potentially involved: time and activity. If no activity differential obtains, it may be anticipated that the cells differentiating first represent the first member of the succession, etc. On the other hand, if the cells differentiate at the same time, the least active cells ought to turn into the first differentiation pattern. In the above sequence the notochordal cells are the first to differentiate, and the muscle cells the last, in apparent contradiction with expectation. As regards this point, it will be seen that the time of differentiation may be referred to the differential obtaining with respect to the lf-induction; as is well known, the chordocytes induce this transformation in the presumptive neurocytes, and both cell-types may participate in the corresponding induction of the myocytes. This circumstance cannot explain the nature of the differentiation patterns assumed by the cells; it seems that in this particular case the activity differential exceeds the time differential to the extent that the temporal sequence is inverted.

If the suggested sequence is correct, and if Holtfreter's explants were homogeneous, then one might expect any combination of two and three cell-types except the one myoblast–chordoblast, since it would be unlikely to find cells in either of the two extreme states, and none in the middle one. Quantitatively, the situation may be formulated so that the frequency of neural differentiation can be lower than either one of the other two, but never lower than both of them. A survey of Holtfreter's results (1938a; 1938b) shows several cases where the nerve cells are absent and the other cell-types are present. This is brought out by Table 4, even if the listed results are selected ones, in which chordocytes are more frequent than neuroblasts. Thus, either the proposed succession is wrong, or the explants were not homogeneous.

There are, however, other results which may contribute to the elucidation of this question. In the previously quoted study of xenoplastic explants containing dorsal lips from one species wrapped in ectoderm from another species, Holtfreter (1936) reported the differentiation patterns for both tissues in each of the various explants. As appears from Table 6, several cases were recorded where, among the dorsal lip cells, notochordal

Table 6

Differentiation patterns in explants of dorsal lip wrapped in ectoderm from another species. From J. Holtfreter, 1936, *Arch. Entw.mech.*, **134**

	Number of explants				
	Total	Chorda	Nerve	Muscle	Chorda and muscle in absence of nerve
Dorsal lip *Triton*		14	13	17	3
Ectoderm	17				
Bombinator		3	17	4	0
Dorsal lip *Bombinator*		12	9	13	2
Ectoderm	16				
Triton		0	14	1	0

or muscle cells, or both, were present in the absence of nerve cells. However, the ectoderm, which may be considered more homogeneous and more indifferent than the dorsal lip with respect to the presumptive differentiation, behaved as predicted above.

A similar analysis of Muchmore's experiments with explanted somites (1957), which presumably also represent a more homogeneous material than the dorsal lip, shows that, except for one case, neural tissue represents the most frequent or the intermediate differentiation pattern (Table 7). These observations support the contention that the

Table 7
Differentiation patterns in explants of somites from *Ambystoma*. From W. B. Muchmore, 1957, *J. Exptl. Zool.*, **134**

		Number of explants			
	Somite plates	Total	'Chorda	Nerve	Muscle
Somite plates wrapped directly in ectoderm	1	127	4	35	11
	2	18	0	0	0
	4	14	1	6	2
Somite plates fused before wrapping in ectoderm	1	113	2	7	1
	2	73	3	20	65
	4	46	5	14	43
	6	30	10	6	27
	10	29	6	14	29

neuroblast represents an intermediate differentiation pattern, but it must be admitted that there are several uncertain factors involved in the interpretation of the explant results.

Part of this difficulty stems from the fact that the three cell-types originate in different cell layers. It may be more unambiguous, therefore, to compare the differentiation pattern of the free cells in the embryonic interior. I have proposed on a previous occasion that this population of cells in Caudata represents the source not only of the somites, but also of the sensory 'ectoderm' in the head region (1967a). The main differentiation patterns of these cells in the antero–posterior direction are the myxocytes of the lens placodes, corresponding to the chordocytes, the cells forming the auditory placodes and contributing to the various ganglia (cf. Yntema, 1943) which, we may presume, are neurocytes, and the myocytes of the somites. Here we thus meet the same succession as above, and again there seems to be a correlation between the activity of the cells (migration distance) and differentiation pattern. The nasal placode, consisting of neuroblasts, is formed in front of the lens. This phenomenon does not violate the generalization presented, because the placode is located in the outer ectodermal layer.

The differentiation patterns of the neural crest sf-cells were enumerated above. Among these we shall begin by discussing the following three: collagenocytes, ganglion cells (neurocytes) and myxocytes. In this case there is no obvious reason to assume that an activity differential obtains, since the presumptive neural crest cells are located closely together, and therefore the time factor may be expected to determine the differentiation patterns. This is corroborated by the fact that the last cells to leave the neural crest are myxoblasts, as shown by their participation in the formation of the dorsal fin. We are therefore justified in placing this cell-type last in the sequence. The formation of the *arachnoidea* on top of the *pia mater* also suggests that they appear later than the collagenocytes. Evidence presented earlier (1967a) indicates that the neuroblasts participating in the formation of the dorsal ganglia appear between these two differentiation patterns, and we can therefore establish the following succession:

collagenocytes ⟶ neurocytes ⟶ myxocytes

which, except for the first member, involves the same differentiation patterns as in the sequences established above.

Among the remaining differentiation patterns exhibited by the neural crest cells we may briefly consider the melanocytes. According to Wilde (1955), phenylalanine may be the specific inductor in this case. However that may be, it is worth noting that adrenaline and melanin are alternate products of the metabolism of phenylalanine, which may indicate that neurocytes and melanoblasts are closely related, but presumably the latter represent a branch from the succession of selfdifferentiation patterns.

It has been suggested that the intestinal cells represent a filocyte differentiation pattern, originating in cells richly endowed with yolk. This implies that if this pattern belongs to the selfdifferentiation sequence, it must be situated before those of myocytes and collagenocytes. This possibility is supported by the fact that in spirally cleaving eggs 4d may sometimes give rise to both myocytes and intestinal cells. Since the differentiation of the myocytes somehow is dependent upon the collagenocytes, the former ought to precede the latter. It is thus possible to establish the following sequence:

intestinal cells ⟶ myocytes ⟶ collagenocytes ⟶ neurocytes ⟶ myxocytes.

Since this series forms the basis for further differentiations it is proposed to call them the *primary* selfdifferentiation patterns.

This selfdifferentiation sequence probably represents a phylogenetic succession. Since, according to the exhaustion theory, the passage through the series represents a gradual limitation of differentiation possibilities, it is reasonable to postulate that the phylogenetic succession is represented by the sequence as read from the right to the left. This contention receives support from the following observations. While the production of mucopolysaccharide is known to occur in the Porifera, nerve cells first appear in Cnidaria, and cells specialized solely for the performance of mechanical work are found only in animals at still higher phylogenetic levels. It must be pointed out, however, that collagen has been observed in the two phyla mentioned, and even in Protozoa. 'Endoderm' is found in all metazoan phyla, but intestinal filocytes, distinguished by the secretion of digestive enzymes, are first encountered among the higher invertebrates. In passing, it may be mentioned that since presumably the nucleo-cytoplasmic interaction is independent of the confinement of a nucleus within the boundaries of a single cell it is to be expected that the nuclei present in the yolk sac of a discoidally cleaving egg may be induced to synthesize intestinal enzymes. Under these circumstances they may obviously be instrumental in the digestion of the yolk, the function traditionally ascribed to them.

Progression and ramification

The process of embryonic differentiation begins with the establishment of a number of primary selfdifferentiation patterns. Initially, the latter appear to be most conservative, thus in the sf-cells the two earliest ones in the succession established above are concerned with the synthesis and secretion of the mucopolysaccharide, which is otherwise deposited only in the cytoplasm and on the outer surface of the plasma membrane, and with a copious formation of cytoplasmic microtubules, respectively. The selfdifferentiation pattern represented by the collagenocyte appears to constitute an innovation from the point of view of synthetic activity, but it is interesting to note that the myocyte is distinguished first by the presence of microtubules, later by microfilaments, a feature which is known even in the lobocytes.

Some of these cell-types may remain relatively unchanged during the course of ontogenesis, e.g. the myxoblasts and the collagenocytes. Other cell-types may undergo successive changes, developing into differentiation patterns representing a considerably higher degree of specialization than that exhibited by the embryonic cells. To this group belong muscle and nerve cells in the first place.

It may be envisaged that in these cases the original cell-type spontaneously passes on to new differentiation patterns, possibly through the gradual realization of the information contained in the operon. Another alternative is that later differentiation patterns involve the activation of new operons, or the cooperation between several of these. In that case it must be presumed that the necessary inductors are produced as part of the activities of the preceding differentiation pattern. It seems likely that a *progression* of this kind may be responsible for the increase in complexity and perfection to which many differentiation patterns have been subjected during the course of phylogenesis.

However, it is only possible by this expedient to account for one cell-type for each of the primary selfdifferentiation patterns, but as we have observed above, there can be no doubt that the number of differentiation patterns increases during development, and that in many cases it is possible to trace several of these back to one of the original selfdifferentiation patterns. It is therefore evident that, as pointed out by Lillie (1929), a continuous branching of differentiation patterns must take place. For this process, Lillie employed the words 'segregation' and 'dichotomy'. The first of these will be used later in a different sense. The second implies a bifurcation, representing clearly the simplest possible mechanism, and therefore presumably in general corresponds to the actual situation. In the present context the less specific name *ramification* will be applied to this process.

Complete ignorance rules as far as the mechanism of this phenomenon is concerned. It is obvious, however, that, if the descendants of a given cell are going to acquire different properties, then either an unequal cell division must be involved, or else the daughter cells must be exposed to different external influences. During early development the former possibility is very likely to be realized, as will be shown in the following section. It may even be that inequalities between cells arising by chance can be of consequence for the subsequent differentiation. Yet, as the cells become smaller, this mechanism becomes rather improbable, but at the same time the complexity of the developing embryo, both as regards spatial organization and chemical activity, is so great that numerous possibilities for interaction obtain.

It has been suggested that each step of differentiation involves the activation of new genes, and it might thus be envisaged that sequential differentiation simply results in an addition to the repertoire of proteins synthesized by the cells. Little is known about this question, but it seems likely that suppression of certain synthetic activities may take place. As an example of this, it may be mentioned that the appearance of sf-cells in the embryos of Echinoidea and Amphibia is associated with the production of alkaline phosphatase (1966b; 1955). This enzyme is present in certain differentiated filocytes, for instance osteoblasts and cells of the digestive tract, but most often it is absent, suggesting that the assumption of certain differentiation patterns involves suppression of the synthesis of this enzyme.

It would be of great value to illustrate the process of ramification with examples based upon actual observations. It is a sad fact that, as far as I can judge, very little information has been published concerning this problem. If I venture to present here a scheme outlining the hierarchical order for the filocytes in the early amphibian embryo, then it must by necessity be guesswork to a large extent, even if I believe that some of the points may be borne out by the views advanced earlier in this chapter, and by additional evidence presented during the discussion. For the sake of simplicity, the scheme is based on the assumption that dichotomy obtains at each step.

As shown in Figure 122, the first step involves the transformation of an sl-cell into

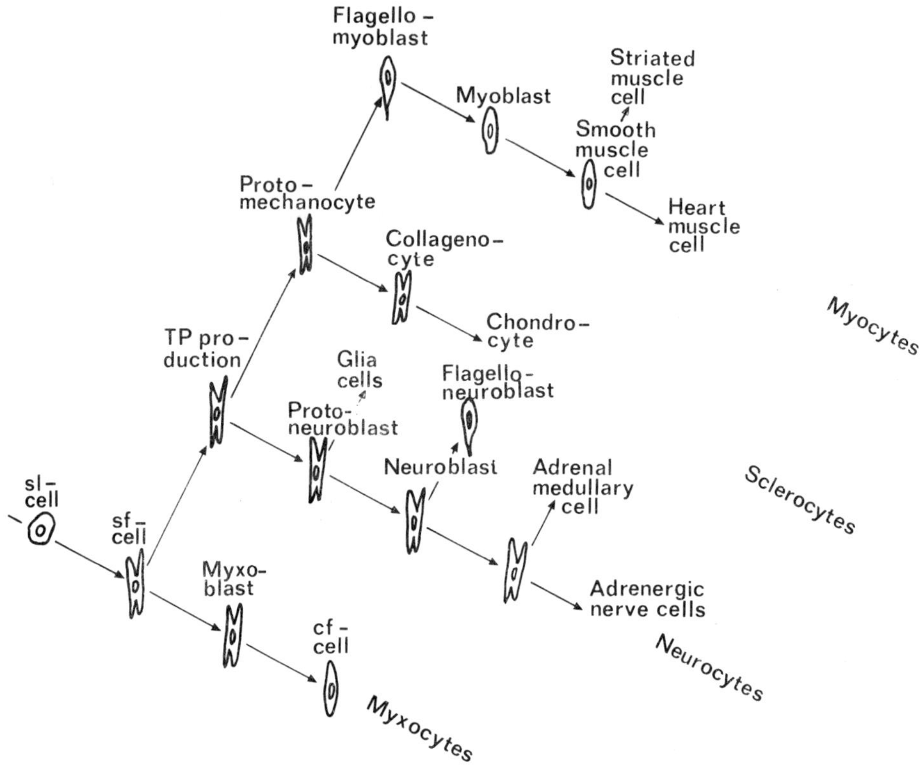

Figure 122. Hypothetic relationship between various filocyte differentiation patterns. Further details are found in the text.

what may be a generalized sf-cell, which is supposed to selfdifferentiate in two directions, one cell-type being specialized in the production of mucopolysaccharide, the other of TP monomers. The former, the myxocyte, may through the sc-transformation become a chordocyte, mucous gland cell, etc. The latter is shown to differentiate in two directions, a cell which is called a 'protoneuroblast', while the other, for lack of a better name, is termed a 'protomechanocyte'. It is suggested that the former either becomes a 'neuroblast' or diverges into a line leading to the glia cells. The neuroblast, in turn, is converted partly into a 'flagello-neuroblast', partly into a noradrenaline-producing cell. The name of the former cell-type suggests that the cells in question undergo the nv-transformation and become vibriocytes (vsf-cells). Whether or not such an intermediate stage occurs is not known, but the fact remains that the cholinergic nerve cells have assumed one aspect of the vibriocyte transformation pattern, viz., the production of acetylcholine, and it is suggested that the flagello-neuroblast is the ancestor of this type of nerve cell. The noradrenaline-producing cell splits up into the adrenaline-producing cell-type present in the adrenal medulla, and into the ancestor of all adrenergic nerve cells.

Returning to the protomechanocyte, it is proposed that the latter differentiates into a collagenocyte and a 'flagello-myoblast'. The former may, through induction, become a chondrocyte. The existence of the latter cell-type is suggested by the observation of flagella in myotomal cells (Figure 123). Certainly, these structures may be stereoflagella,

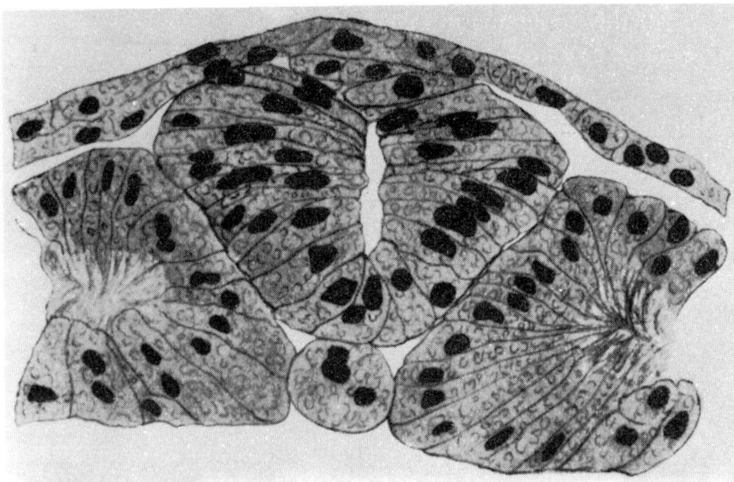

Figure 123. Embryo of *Triton*, sectioned through the middle of the trunk. It may be seen that the somite cells are flagellated, turning the flagella towards the central cavity. (Reproduced from F. E. Lehmann, *Arch. Entw.mech.*, **113** (1928), by permission of Springer, Berlin)

and the cells thus ncf-cells, but the formation of acetylcholine esterase during a certain stage of the development of the somites (Figure 124) intimates the presence of vibriocytes. In any case, the cells sooner or later undergo the sc-transformation, thereby presumably becoming 'myoblasts', which in turn become smooth, striated and heart muscle cells.

The scheme presented here accounts for the selfdifferentiation patterns whose succession was outlined above, except for the intestinal cells, and the relation between collagenocytes and myocytes. The close affiliation between the latter differentiation patterns is evidenced by the fact that all the sf-cells leaving the somites are collagen producers. It is interesting to note that the differentiation of muscle cells *in vitro* depends upon the presence of a substrate of collagen (Hauscka, 1968). This precondition is met when presumptive muscle cells settle on the basement membrane in various Invertebrata. However, the subsequent layers of muscle cells can hardly be affected by the collagen substrate, and the function of the latter must therefore be taken over by muscle cells. It therefore seems that we are not dealing with an inductive effect, but rather with an organizing one. This suggestion is supported by the fact that occasionally myocytes may differentiate in the absence of collagen. Muscle cell differentiation is associated with a change from microtubules to microfilaments in the cytoplasm. It is possible that

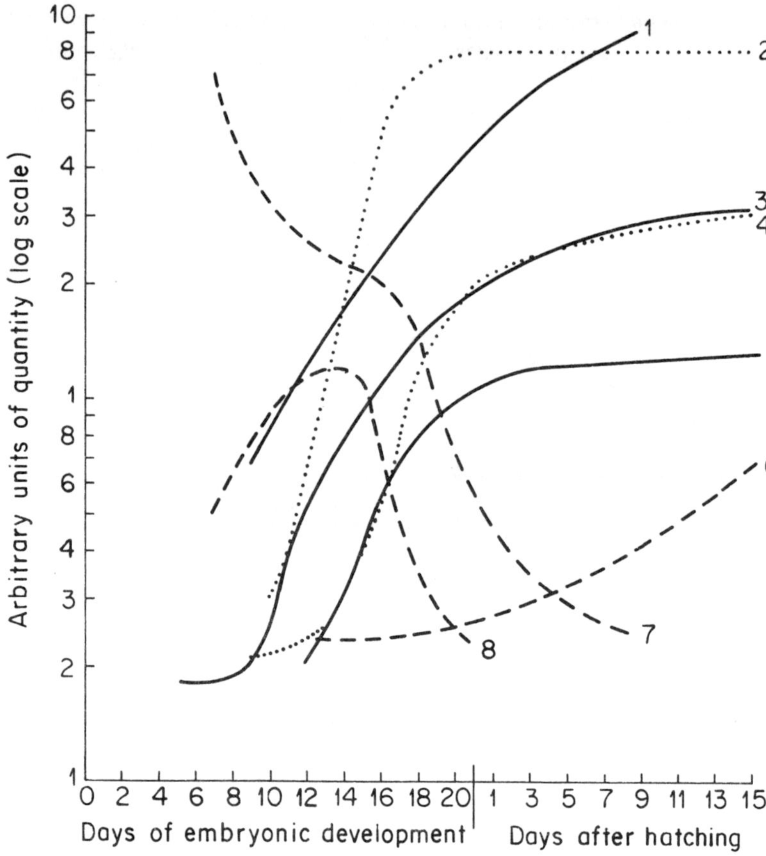

Figure 124. Biochemical changes in developing chick skeleton muscle. 1, myosin; 2, collagen; 3, glycogen; 4, actomyosin; 5, ATPase; 6, alkaline phosphatase; 7, per cent nuclei synthesizing DNA; 8, acetylcholine esterase. (Reproduced from S. D. Hauscka in H. Ursprung, *Stability of the Differentiated State* (1968), after Basil and Herrmann, Moog, Konigsberg, and Herrmann, Marchok and Herrmann, by permission of Springer, Berlin)

mechanical properties of the latter require an external orienting influence for the alignment distinguishing the muscle cell. It is seen that when the muscle cells are colligocytes this influence will never be missing, since collagen membranes are formed upon the surfaces of all colligocyte aggregates.

Interactions

The preceding discussion shows that to account for the phenomenon of cell diversification it is necessary to invoke a mechanism of nucleo-cytoplasmic interaction, which may involve that specific inductors activate a number of genes, whose joint activity serves to establish a particular diversification pattern. The stabilization of the patterns may occur through positive feedback, but competition apparently most often prevails between non-repressible and repressible proteins at the level of translation, with the result that the synthesis of the latter proceeds at a low level in dividing cells. It was also suggested that acquisition of competence, as well as spontaneous transformation and differentiation, in many cases may be explained on the basis of a mechanism of exhaustion of available reserves, by which quantitative differences may become converted to qualitative ones.

In the establishment of cells differing with respect to their cytoplasmic composition, which thus forms the basis for the process of cell diversification occurring during embryonic development, the embryonic substrate and the process of cell division are of great importance. In the following pages we shall try to outline some of the general principles involved in these interactions, and the ensuing distribution of diversification patterns in certain selected embryos.

EMBRYOGENETIC SUBSTRATE AND CELL DIVERSIFICATION

The embryogenetic substrate supplies both of the factors involved in nucleocytoplasmic interactions, the genome and the cytoplasm. From our knowledge of the mechanism of DNA replication, it may be presumed that the complete genome is present in every cell. This view, generally embraced today, forms the basis of the mechanism advocated here. Under these circumstances the importance of the genome for the process of cell diversification is restricted to its potential content of information as regards protein synthesis and to the role it plays in the mechanism of induction. The realization of disparate diversification patterns in separate regions of the embryo may, on the other hand, be referred to properties residing in the cytoplasm. The third cellular subunit, the cortex, may also interact with the process of cell diversification, but only through its influence upon cell division. The discussion of the contribution of this factor must therefore be postponed to the following section.

Genome

The main importance of the genome lies in the fact that it constitutes a means through which information concerning amino acid sequences may be stored. Thus, all the proteins which the cells may potentially synthesize are encoded in the form of DNA molecules. Protein synthesis may be controlled by nuclear factors, possibly functioning as regulator genes as proposed by Jacob and Monod (1961). The existence of such mechanisms by which the synthetic activity of the cell is regulated constitutes a further kind of information contained in the genome.

One of the most challenging aspects of the processes of cell diversification occurring in the course of embryonic development is that they constitute a gradual expansion of the synthetic repertoire. The emergence of new cell types, each endowed with the faculty to synthesize a circumscribed number of diversification products, shows that this build-up is under rigorous control. Since the DNA, and hence the genome, supposedly is the same in all cells it follows that the mechanism concerned must involve the derepression of specific operons in the cells representing the various diversification patterns.

The observed course of events suggests that all the genes encoding repressible proteins are masked in the nuclei of the egg and the early blastomeres, an inference corroborated by various kinds of observations (cf. Davidson, 1968). Actually, certain findings indicate that even the synthesis of essential proteins may be affected. Thus, although the cultivation of embryonic cells generally is a simple matter, nobody has apparently so far succeeded in ensuring the survival and proliferation of the early 'undiversified' cells in culture. That the apparent state of deficiency may be due to a restriction in the synthetic faculty is suggested by the recorded presence in the different embryos of various reserve material which may serve to bring the embryonic cells through the initial phase of incapacity. As proposed above, these stored substances may act to delay the onset of diversification, a phenomenon which seems to have been exploited with increasing success in the course of phylogenetic evolution.

The observed course of cell diversification leads to the obvious inference that this process is associated with parallel alterations in the nuclei in the various tissues of the body. Presuming these changes to be irreversible, this proposition was tested and corroborated through nuclear transplantation experiments by Briggs and King (cf. 1959). Gurdon's challenge (1964) of the notion of gradual nuclear modifications, although issuing in discordant results, cannot be regarded as conclusive since he chose nuclei from intestinal endodermal cells for the transplantation. According to the views advanced here the diversification of the yolk-laden endodermal cells probably is retarded relative to most other embryonic cells, a circumstance which, among other things, may warrant that they constitute the source of the germ line. The fact that the cells are ciliated, or maybe rather flagellated, at the time when they are employed can hardly be taken as evidence for their having been subjected to any irreversible steps of diversification. For this purpose it rather would be necessary to establish the presence of typical endodermal differentiation products, such as digestive enzymes, etc. As is well known (cf. 1955), proteins of this kind appear only during larval development, thus at a much later stage than that from which nuclei were isolated by Gurdon.

This brief discussion suffices to show that one of the most important problems in embryology is concerned with the discrete utilization of the information stored in the genome. Unfortunately, without exaggeration it seems warranted to state that hardly anything is known about this topic, but the fact that it is susceptible to molecular-biological analysis is promising as regards future progress. Indeed, one of the reasons why this field has remained undeveloped so far may be that we have been unable to formulate questions relevant for experimental corroboration; once this state has been reached the responsible mechanisms may be much simpler than they appear to be for the time being.

Cytoplasm

Cells differing with respect to cytoplasmic composition must obtain before disparate diversification patterns can be realized. As far as the initial substrate is concerned, the cytoplasmic polarities are the only factors which can be held to account for the establishment of the required differences. Whenever a cell partition cuts the axis or the plane of a cytoplasmic polarity, the potential basis for two separate lines of cell diversification is established. Even if the process of cell division is not actively engaged in the creation of the cytoplasmic differential, it is nonetheless a necessary prerequisite since, according to the principle of exclusivity, no cell can assume more than one diversification pattern. It may be mentioned in passing that observations contrary to this generalization have been published (cf. Wilde, 1961), but the very fact that the name 'confused cells' is applied to this phenomenon stresses that it is an abnormality.

It has been postulated that the polarities work in all-or-none fashion with respect to the differences they can establish between cells, implying that one polarity may be responsible for two, two for four, cell-types, etc. Since the number of cytoplasmic polarities hardly ever exceeds two, it appears that the maximum number of diversification patterns which can arise through their action must be four. The cytoplasmic polarities may conceivably be concerned with cell transformations, but most often they seem to determine the selfdifferentiation patterns assumed by the cells. This point may be illustrated by two examples, one referring to Invertebrata, the other to Vertebrata.

In spirally cleaving eggs, filocytes are known to arise in the A, B and C quadrants, mainly from the first quartet, but often also from the following ones. These cells usually differentiate into either neurocytes or myxocytes, even though other differentiation patterns may be represented. In contrast, the filocytes arising from the 4d micromere generally become myocytes, possibly also collagenocytes. In this instance it seems that we may charge the median polarity with the function of establishing the cytoplasmic difference required to allot two of the selfdifferentiation patterns in the sequence established above to one, and another pattern (or two) to a different group of cells. In the amphibian embryo this determination is achieved dynamically through exhaustion, whereas in the determinate eggs dealt with here the static mechanism involving cytoplasmic polarities is involved. The mechanism of exhaustion is corroborated by the fact that the last two members of the sequence originate from the relatively hyaline ventro–lateral micromeres, while 4d, being better provided with reserve materials, gives rise to the first member(s) of the sequence. Needless to say, the median polarity cannot alone ensure the formation of two different differentiation patterns among the descendants of the micromeres. Another mechanism must be invoked to answer for this phenomenon, the nature of which will be dealt with later.

Another example is observed in the amphibian egg. In contrast to the spirally cleaving eggs, the lf-transformation occurs only at one location, the dorsal lip, whence it spreads through induction. The median polarity is thus responsible for this step of diversification. The apical cytoplasmic polarity has a function similar to the one just described, ensuring that members of the classes of lobocytes and filocytes assume disparate differentiation patterns in the animal and the vegetal hemisphere. The decisive step as far as this polarity is concerned is the third cleavage division. As a consequence of this event the presumptive differentiation patterns of the cells are determined as rigidly as in determinate eggs, even if admittedly later cell divisions create an intermediate population of cells, the fate of which may depend upon environmental influences (cf. Table 4, where 'endodermal' cells are reported to become chorda and muscle cells). Thus the apical polarity is responsible for the fact that ecto-mesodermal filocytes become chorda, nerve and muscle cells, and that the endodermal ones differentiate into intestinal cells of various kinds. Likewise, ecto-mesodermal lobocytes form epithelia, glands, etc., whereas their endodermal counterparts become liver and pancreas cells, among other things.

The processes of cell diversification associated with embryogenesis presuppose that cell division takes place. This does not necessarily exclude that diversification may occur in the absence of cell division. In this connection we may take a look at the observations of Lillie (1902) on the activated egg of *Chaetopterus*, which was found to become ciliated and undergo certain other changes, although it does not divide. On this basis, Lillie claimed that cell differentiation may occur in the absence of cell division. Numerous authors have accepted this interpretation, which suggests that the observed phenomenon is of fundamental importance. In spite of this it has never led to fruitful experimental or theoretical advances, and some authors (e.g. K. Dan, 1960) have criticized this idea. If differentiation is equated with diversification, Lillie's view may be acceptable, for apparently what happens is that the treated egg cell becomes a ciliated cl-cell. The implications of Lillie's interpretation were more far-reaching, due partly to the fact that similarity obtains between the ciliated cell and a certain stage of the normally developing embryo, partly to the lack of a clear definition of the phenomenon of differentiation. From this point of view Dan's criticism is completely warranted.

CELL DIVISION AND CELL DIVERSIFICATION

Through the means of unequal divisions the process of cell division may be involved in the creation of populations of cells differing with respect to their diversification patterns. As stated in the second rule of cell division, unequal divisions may be traced back to factors interfering with microtubular activity, i.e. their formation or attachment. The latter phenomenon presupposes cortical inhomogeneities of the kind proposed as being established by the cortical polarities. If microtubules are more easily attached to one part of the cortex than elsewhere, a mitotic apparatus may be formed in which the density of cortical microtubules is uneven, with the result that the median plane of the spindle does not coincide with that of the cell, the daughter cells consequently being of unequal size. Since only cortical properties are invoked, the two may theoretically be of identical cytoplasmic contents.

However, microtubules tend to accumulate mucopolysaccharides and it is therefore to be anticipated that the smaller cell will be more hyaline and contain less yolk and other reserve materials. Since both their size and cytoplasmic composition may favour mitotic activity, small cells will often divide faster than large ones, even if we have noted above that exceptions exist to this rule. As a result, small cells may be depleted of essential metabolites earlier than their more richly endowed counterparts, a circumstance which, as discussed above, may ensure that they acquire competence and begin their diversification. Even the latter activity may be affected by the cytoplasmic composition. It is suggested that the mechanism outlined here is the one through which the cortical polarities affect the process of cell diversification.

An example of apical cortical polarities exerting effects of this kind is found in the sea urchin embryo, where the two polarities ensure that cells of differing size and presumptive diversification are distributed along the animal–vegetal axis (cf. Figure 76). In Amphibia the median polarity is largely cortical and, as we have seen, it leads to a differential in cell size between the dorsal and the ventral cells (Figure 98). In this case the influence of the polarity is concerned with the lf-transformation. Unequal divisions are also found in Ctenophora, in which all the ectoplasm gradually ends up in the micromeres (Figure 104). In this case it seems reasonable to postulate that it is the normal polarity represented by the ectoplasm which is responsible for the unequal divisions. Most of the micromeres are known to become epidermis and comb plates (Farfaglio, 1963; Reverberi and Ortolani, 1963). The latter are obviously nothing but modified trochoblasts, and if the views regarding differentiation patterns presented earlier in this chapter are correct, the difference between the fate of these two groups of cells may be a matter of cell transformation, the comb plates being cf-cells, while the epidermis is made up of cl-cells.

It has already been mentioned that the polarities do not suffice to account for the total number of diversification patterns present even in the early embryo. We have seen that the mechanism of exhaustion constitutes one means by which the number of possibilities may be increased, but there is reason to investigate whether unequal cell divisions cannot be involved in this process. For this purpose we shall consider the fate of the first micromeres in spirally cleaving eggs. These blastomeres divide into two cells called $1q^1$ and $1q^2$, where q stands for a, b, c and d. Usually the cells in the D quadrant differ with respect to their diversification from those of the other quadrants, an occurrence which evidently must be an influence of the median polarity; this particular point will not be considered here. The blastomeres $1q^2$ are generally, but not always, smaller than their sisters, they are often more hyaline, and their mitotic activity typically comes to an end very soon, after which in many species they develop into trochoblasts. In order to reach this destination it must be presumed that they spontaneously undergo the lf-transformations, and subsequently assume the selfdifferentiation pattern representing the most exhausted state in the sequence established above.

The descendants of $1q^1$ differentiate in two directions: some become epidermal cells, and the remaining ones participate in the formation of the apical plate and, after penetration into the blastocoele, of ganglia and other nervous structures, e.g. eyes. It appears that we are here facing the same subdivision into lobocytes and filocytes as occurs in the ectoderm of the amphibian embryo, and like these, the former become epidermal cells, often ciliated, whereas the latter turn into nerve cells. Among these the sensory cells of the apical plate may be cf-cells possessing a stereoflagellum, while the interior cells are sf-cells, at least at the time when they leave the surface. Nothing prevents their later undergoing the sc-transformation and participating in aggregate formation, as exemplified by eye vesicles. In various larvae of Mollusca myxoblasts are involved in the swelling required for the formation of the velum. The exact origin of these cells is not known, but if they are descendants of the first quartet it is worth recalling that they are sf-cells representing the same family of differentiation pattern as the trochoblasts.

This survey seems to bear out that the first division of the first set of micromeres leads to a cytoplasmic repartition which determines the presumptive differentiation of the progeny of the two daughter cells. When transformed into filocytes, the cells derived from $1q^1$ assume the intermediate, and those descending from $1q^2$ the last, of the selfdifferentiation patterns in the sequence established above. This outcome may be accounted for on the assumption that two apical polarities are present, one responsible for the macromere–micromere cleavage pattern, and another one answerable for the dichotomy just described. The situation thus appears to be completely analogous to that proposed to exist in the sea urchin embryo.

A problem of great importance is whether or not this second polarity existed in the fertilized egg. In those cases where only a prototroch is formed, the former possibility does not appear entirely unlikely. However, the fact that metatrochs and paratrochs are sometimes formed suggests that the phenomenon may be repeated in the following triplets or quartets, thus demanding the presence of several apical polarities. This proposition seems so improbable that it is necessary to look for other interpretations.

As it is, the only mechanism which at the time can be made responsible for the occurrence of unequal divisions is that of cell division proper. In this context it may be recalled that linear polymers, including microtubules, are supposed to be distinguished by becoming attached preferentially to free cell surfaces. If we look at the dividing first micromeres (Figure 125), it is seen that the spindles are oriented in such a way that the upper poles are directed towards the surface adhering to a neighbouring micromere, while the lower ones are directed towards a free surface. This circumstance must entail a higher density of microtubules and of hyaloplasm in the daughter cell $1q^2$, which as a consequence may be expected to be smaller than $1q^1$. This explanation may suffice to account for the subsequent difference between the descendants of the blastomeres as regards their selfdifferentiation pattern.

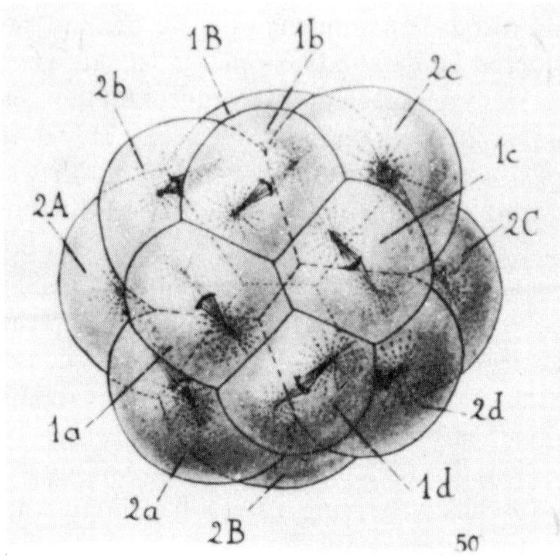

Figure 125. Division of the first quartet of micromeres in *Trochus*, showing that the cortical microtubules from one end of the mitotic apparatus become attached to the inside of cell surfaces in contact with a neighbouring micromere, those from the other end to that of a free surface. (Reproduced from A. Robert, *Arch. Zool. Exptl. Gén.*, **3** (1902), by permission of Centre National de la Recherche Scientifique, Paris)

Formally, it may be said of this occurrence that the spatial orientation of the blastomeres amounts to the creation of a new apical polarity. There is obviously no reason to doubt that a mechanism of this kind may operate more than once. As a matter of fact, it is not impossible that a similar principle is involved in the animal–apical polarity in the sea urchin egg. If this is so, we may be able to account for the *polar* or *apical dominance* observable in eggs of Echinoidea (Lindahl, 1941). This phenomenon may be demonstrated by removing the cells which normally form the apical plate. Under these circumstances the neighbouring cells will occupy the spatial location of the missing cells, and they will also assume their diversification pattern, contrary to their presumptive fate in the intact embryo (cf. Hörstadius, 1939).

In many spirally cleaving eggs mesenchyme cells are formed by paratangential divisions of the micromeres from one or more of the ventro–lateral quadrants, generally from the second quartet of micromeres (cf. Figure 109). This event generally takes place after the blastocoele has formed, a circumstance which explains the radial orientation of the spindle, since the free surfaces of the blastomeres now are facing towards the exterior and towards the blastocoele, respectively. Except for *Unio*, the inner daughter cell is always smaller than the one remaining in the blastula wall. This phenomenon, which may account for the differentiation pattern of the cells, suggests that the microtubules show preference, as far as attachment is concerned, for the surface exposed to the mucopolysaccharide-containing blastocoele. This contention is compatible with the suggestion made above regarding the affinity between filopodia and mucopolysaccharide-filled cavities. It is seen that in this case a normal polarity may be invoked.

The present survey shows that the normal course of the cleavage divisions is of direct importance for the process of diversification. If the cell divisions for some reason, experimental or natural as the case may be, proceed abnormally, it may be anticipated that the cell diversification also may be affected. This phenomenon, called 'cleavage-dependent inhibition of differentiation' by Lehmann (1956), has been observed on many different occasions.

DISTRIBUTION OF DIVERSIFICATION PATTERNS IN THE EARLY EMBRYO

We have now outlined the mechanisms which for the time being can be made responsible for the origination of the diversification patterns present in the early embryo. In the present section an attempt will be made to survey the distribution of the diversification patterns in Echinoidea, in Turbellaria, Annelida and Mollusca and in Amphibia, followed by a discussion of the relation between germ layers and diversification patterns.

Echinoidea

In the sea urchin embryo it is the two apical polarities which in the first place are responsible for the cell diversification. Considering first the vegetal polarity, it is evident that it is involved in the lf-transformation, since a large number of sf-cells arise around the vegetal pole. It is not known whether all the cells in question transform spontaneously, or whether only a few cells are subjected to this fate, acting in turn as inducers on neighbouring cells. The sf-cells are set free in two rounds, as the primary and secondary mesenchyme cells. The former are known to form the larval skeleton, the calcification occurring intracellularly in a matrix consisting of a syncytium of filopodia. There is no reason to presume that the differentiation pattern exhibited by these cells corresponds to that of the collagenocyte, and there is thus no way to decide their position in a sequence of selfdifferentiation patterns. The differentiation of the secondary mesenchyme cells is not known, but since they are very actively engaged in the process of gastrulation it might be envisaged that they reach a stage of maximum exhaustion and become myxoblasts. Against this possibility is the fact that oncotically active mucopolysaccharide is present in the blastocoele before these cells are formed; actually the observations of Monné and Slautterback (1950) and Monné and Hårde (1950a) suggest that this material is secreted by the sl-cells in the wall of the blastocoele. Another alternative is that these cells, or some of them at least, do not become exhausted to any significant degree, and become larval muscle cells.

As far as neurocytes are concerned it appears that two types may be distinguished, viz., the flagellated sensory cells (vsf-cells) located in the main flagellated ('ciliated') bands, and cells with filopodia (nsf-cells) suspended between the bands and the gut. Although the latter are not adrenergic, but rather are affected by 5-HT, it thus appears that even in this case a subdivision of the nervous system obtains (cf. Gustafson, 1969). The fact that these cells arise from the ectoderm is in agreement with the generalization that ectodermal cells subjected to the lf-transformation become neurocytes. The elongated cells with long flagella found in the apical plate have been suggested to be cf-cells, indicating that the animal–apical polarity is also concerned with the lf-transformation.

The cells between the two polar areas become ectoderm and endoderm. Possibly the endoderm consists of cf-cells, induced by contact with the mesenchyme cells, but the histological evidence is too scanty to decide anything on this point. It appears that, apart from the neurocytes discussed above, the whole region from the animal pole to the vegetal cells is subjected to the sc-transformation. A schematic representation of the distribution of the lf- and sc-transformations is shown in Figure 126.

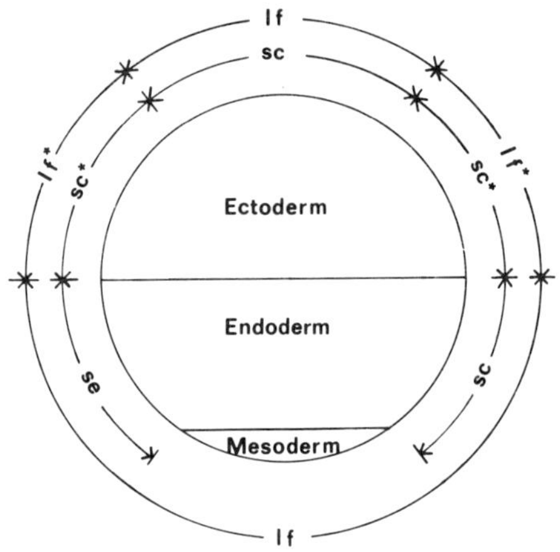

Figure 126. Cell-order distribution in the echinoid embryo. The cell transformations occurring in the various regions are indicated. sc*, except in the flagellated bands; lf*, only in the flagellated bands

Turbellaria, Annelida and Mollusca

We have already discussed several of the features distinguishing the distribution of cell diversification patterns in spirally cleaving eggs, but some points relating to the influence of the polarities need to be further elaborated.

The simplest situation obtains when no median polarity is present, as appears to be the case in certain Turbellaria, for instance *Leptoplana* (Wilson, 1899). Here the mesodermal cells are reported to derive from the second quartet, and the original apical cytoplasmic polarity may thus be responsible for the subdivision of the embryo into endodermal and ecto-mesodermal cells, respectively (cf. Figure 109a). When a median cytoplasmic polarity is present, various complications may arise. Thus, it is known that both in Annelida and Mollusca descendants from 4d may contribute to the endoderm, and in *Planocera* all endodermal cells come from this source, whereas the other micromeres of the fourth quartet and all the macromeres degenerate and are used as food (Surface, 1907). Under such circumstances the separation between ecto-mesoderm and endoderm is partly or wholly taken over by the median polarity, and in the latter case the function of the apical polarity is rather a matter of segregation between ecto-mesoderm and degeneration. This fate is not uncommon among large yolk-laden cells; it also occurs to some of the endodermal cells in Amphibia (cf. Vogt, 1909; Holtfreter, 1943a). The suppression of mitotic activity which always precedes the degeneration suggests that the cells have been more or less depleted of hyaloplasm, and that they consequently are no longer able to form a mitotic apparatus. The endodermal cells forming the intestine presumably are cf-cells, in the higher classes at least, and where a hepatopancreas is present there is reason to believe that it is formed by lobocytes, representing specific differentiation patterns.

It was observed that lobo- and filocytes, as well as solo- and colligocytes, arise from all or most of the micromeres, and that the lf-transformation occurs spontaneously on many occasions. As regards the differentiation patterns of the filocytes, it seems that neurocytes and myxocytes (myxoblasts and trochoblasts) arise from the first quartet in radially symmetrical Turbellaria, and from the ventro–lateral triplet in the remaining cases. The majority of the lobocytes derived from these micromeres become epithelial cells. In Annelida 2d is known to give rise to nerve cells, and it appears that in Bivalvia and Gastropoda the foot derives from the same source (cf. Lillie, 1895; Conklin, 1897), suggesting that epithelial cells, nerve cells and myxoblasts may all originate from this blastomere.

In the classical literature on cell lineage many authors have reported the formation of mesoderm from the second ventro–lateral triplet of micromeres (cf. Figure 109b) and even from the third one (*Thalassema*, Torrey, 1903). As regards the differentiation pattern of these cells the authors are far from unanimous. Lillie (1895, p. 38) originally suggested that 'only transitory organs, viz., the isolated muscle cells, which span the primary body cavity in various directions . . . and the adductor muscle of the larva, which is nothing but a bunch of these myocytes' are formed from this source. It will be seen that this description bears mainly on the cell order; the cells are obviously sf-cells, whereas the differentiation pattern remains uncertain, since myxocytes, neurocytes and collagenocytes may 'span the body cavity' as well. As far as the transience of the structures is concerned, Lillie has been challenged by many authors (e.g. Meyer, 1901; Treadwell, 1901; Torrey, 1903; Wierzejski, 1905), who all agree on the point that permanent muscles are formed from the ventro–lateral micromeres. In the dorsal quadrant myocytes are among the descendants of the blastomere 4d. Curiously enough, no information obtains with regard to collagenocytes. Yet, basement membranes are found in the animals belonging to the taxa discussed here, so this differentiation pattern must be present, arising probably in close association with the muscle cells. As far as the temporal relation is concerned, the collagenocytes presumably appear before the myocytes, since they are always located between the epidermis and the muscular layers.

In Mollusca the cells derived from 2d located nearest the animal pole do not spontaneously undergo the lf-transformation. It has been observed in *Lymnaea* and *Bithynia* that this transformation is induced in some of the descendants of 2d when they come in contact with the archenteron, as a consequence of which they form the shell gland (Figure 127, cf. Raven, 1966; Hess, 1956). This event must be referred to the competence

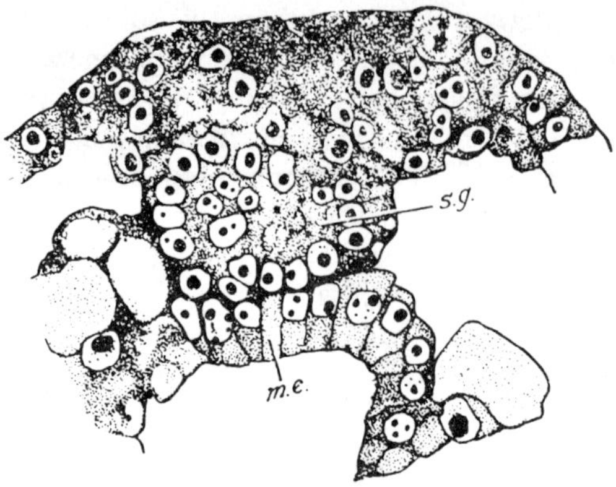

Figure 127. Induction of the shell gland in *Lymnaea* through contact between ectoderm and endoderm. m.e., small-celled midgut epithelium; s.g., shell gland. (From Chr. P. Raven (1966), *Morphogenesis: The Analysis of Molluscan Development*, Pergamon Press, Oxford)

of the cells in question, since in many embryos (e.g. *Crepidula*, cf. Conklin, 1897) all the ectodermal cells are intimately applied to the endodermal macromeres, and yet only this particular population of cells differentiates into shell gland cells. The latter are cf-cells which represent a differentiation pattern involving the production of a calcifiable matrix.

Various of the observations discussed in this section may be summarized diagrammatically. Beginning with a radially symmetrical egg, it is seen that the differentiation patterns assumed by the filocytes are arranged along the animal–vegetal axis according to the sequence established above (Figure 128a). We have seen that the apical cytoplasmic polarity involves an accumulation of hyaloplasm around the animal and yolk towards the vegetative pole. It is immediately obvious that if exhaustion is equated with absence of yolk, then this pattern of distribution is an excellent corroboration of the hypothesis, according to which exhaustion is responsible for the transition between the selfdifferentiation patterns. Incidentally, the fate suffered by many macromeres suggests that the very first 'step' in the sequence is degeneration.

When a median polarity is present, 2d may give rise to myxocytes and neurocytes, 4d to myocytes and possibly collagenocytes, and in those cases where the other micromeres of the fourth quartet degenerate, its descendants may even be responsible for the formation of the intestine. It needs little reflection to see that this outcome amounts to nothing but a tilting of the planes separating the various differentiation patterns such that they extend further downwards at the dorsal side (Figure 128b). This effect is easily understood when it is recalled that, in many cases at least, the median polarity is established by incorporating part of the polar plasms in D. At the animal end this will evidently have the effect that the 'apical' differentiation patterns (myxocytes and neurocytes) extend further down. At the vegetal end the presence of hyaline and polar plasm will reduce the chance of degeneration; both effects are thus predictable from the modified exhaustion hypothesis presented above. It is interesting to note that, as far as the filocyte differentiation patterns are concerned, there appears to be some overlapping

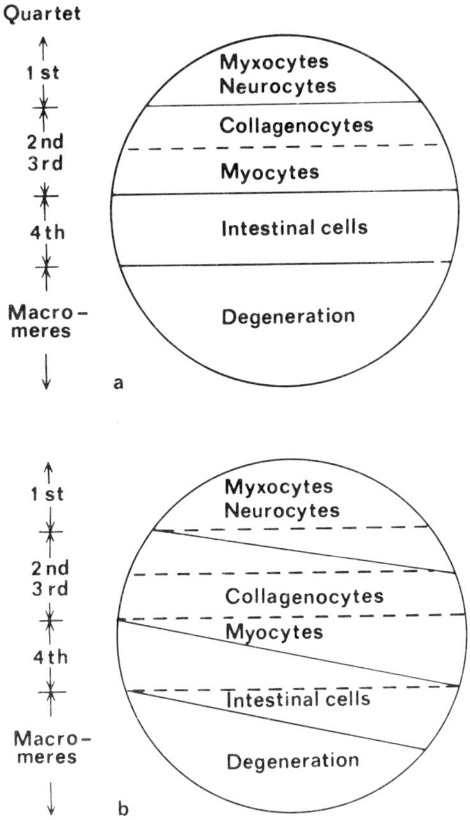

Figure 128. Polarities and differentiation patterns in spirally cleaving eggs. (a), diagram showing that in the presence of radial symmetry the distribution of the filocyte differentiation patterns corresponds rather closely to the sequence of selfdifferentiation patterns established for the amphibian embryo; (b), diagram indicating that a median polarity may involve nothing more than an inclination of the planes separating the various differentiation patterns

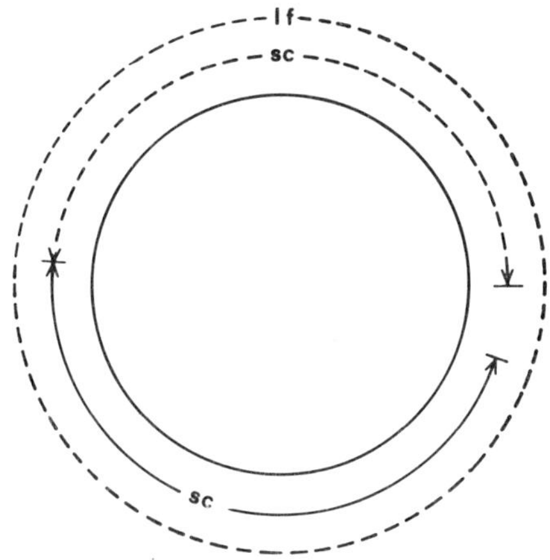

Figure 129. Cell-order distribution in an embryo with spiral cleavage, having a median polarity. The dashed lines imply that some but not all the cells in the particular region undergo the indicated cell transformation

among the first three or four ventro–lateral triplets. Clearly, four quartets of micromeres cannot be necessary for bringing the various differentiation patterns into existence, for in certain cases, e.g. *Clepsine* (Whitman, 1878), the mesoblasts are derived from 3d.

An outline of the cell-class distribution is given in Figure 129.

Amphibia

The cell differentiation patterns in Amphibia have already been discussed in some detail, and it therefore suffices to give a very cursory survey. The vegetal cells form the digestive tract and, as pointed out above, there is reason to believe that the gut is formed by cf-cells, whereas the liver and the pancreas are made up of endodermal cl-cells. The cytoplasmic apical polarity thus has essentially the same function as in the spirally cleaving eggs.

In the ecto-mesoderm the most decisive step is the lf-transformation, which occurs spontaneously at the dorsal lip, whence it spreads through induction. Most of the cells that do not undergo this change during the early stages become epithelial cells. Some of the ectodermal cl-cells may differentiate into myxocytes. The filocytes become myxocytes, neurocytes, collagenocytes or myocytes. Since the latter differentiation patterns are derived wholly or largely from the interior cells, it may be proposed that a normal polarity is involved in their determination, a function which, as we have seen, in the spirally cleaving eggs is exerted by the median polarity. In Amphibia random ingression of sf-cells occurs from the neural plate, but cells of this order mainly come from the neural crest. Most of the interior cells, except the myocytes, and the cl-cells forming the pronephros, do not undergo the sc-transformation suggesting that, as proposed above, the normal polarity is involved in this transformation. An illustration of the cell-order distribution patterns is given in Figure 130.

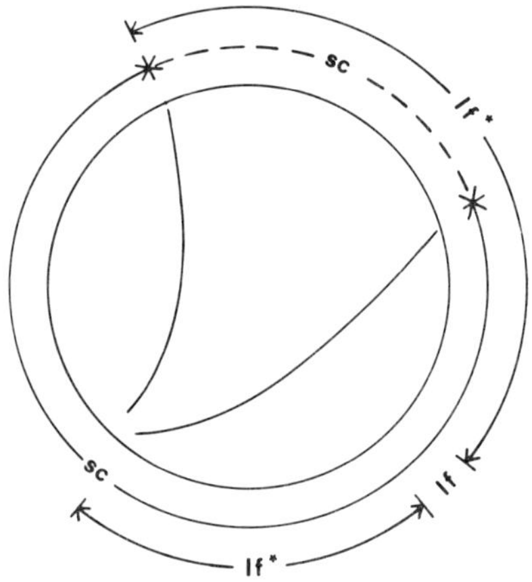

Figure 130. Cell-order distribution in an amphibian embryo. lf*, induced lf-transformation. The dashed line implies that some but not all of the cells in the particular region undergo the indicated cell transformation

Germ layers and cell diversification

The significance of the germ layers has been discussed almost from the time the theory was propounded and up to the present day (cf. Oppenheimer, 1940; de Beer, 1947). The two points which particularly have been contended and contested are that the segregation of the germ layers is associated with a determination of the presumptive fate of the cells, i.e. with the process of cell diversification, and that homologous organs are derived from the same germ layer. This question was analysed by de Beer (1947) who, showing that there are many exceptions to these contentions, was led to the conclusion that there 'is just sufficient constancy in the origin and the fates of the materials of which the germ-layers are composed to endow the ghost of the germ-layer theory with a provisional, descriptive, and limited didactic value' (p. 394) and that the germ-layer theory has 'a validity of its own [merely] as a generalization covering the structures of embryos' (p. 387).

The latter point is, of course, not without importance, since it obviously is valuable to have a terminology bearing upon the initial subdivision of the embryogenetic substrate. The following definition appears to comply with this standpoint: 'The term mesoderm is rather loosely applied to all cells, cell layers or cell masses which occur in the embryo between the ectoderm and the endoderm' (Hyman, 1951a, p. 18). It is evident, however, that the implications of this definition depend upon the embryonic stage to which it refers, and upon the definition of ectoderm and endoderm. Considering the amphibian embryo, it is known that during neurulation a number of membrane-bound sf-cells leave their station at the neural crest and migrate to the interior. The cells are considered to be mesodermal, and with reference to their supposedly ectodermal origin they are collectively called 'ecto-mesoderm'. Some of these cells settle beneath the epidermis, where they participate in the elaboration of the basement lamella. It would appear that their association with the ectoderm under these circumstances is as intimate as in their original location. The neural tube and various ectodermal and endodermal glands also sever their association with the external cells and penetrate into the interior, where they generally become more or less completely encapsulated in mesodermal tissue, yet these organs and tissues are referred to as ectodermal or endodermal, as the case may be.

The basis of this distinction is not obvious, but it may be suspected that it lies in the different methods of invasion, the mesoderm entering in the form of individual cells, the ecto- and endoderm as cell aggregates. That this inference is wrong is demonstrated by the fact that the ganglion cells originating in the neural plate and crest are classified as ectodermal, whereas coelomic cavities formed by the enterocoelous method from outpouchings of the archenteron are considered as mesoderm, not endoderm. In this case it is clearly either the diversification pattern or the nature of the morphological structure to the formation of which the cells contribute that is decisive for the classification. The latter possibility, which bears upon the question of homology, will not be dealt with here.

As far as the former is concerned, de Beer (1947) has already demonstrated several inconsistencies. In this context we shall first analyse the problem with reference to the cell-order hypothesis. Many mesodermal tissues are built up by filocytes, and there is the possibility that this feature distinguishes this germ layer from the other ones. However, we have already seen that the many intestinal cells undoubtedly are cf-cells, and the ectoderm also comprises filocytes, e.g. nerve cells. On the other hand, it cannot be questioned that some mesodermal cells are lobocytes, for instance the ciliated cells in the pronephric epithelia must be cl-cells, the osteoclasts sl-cells, etc. The germ layers therefore cannot be distinguished on this basis.

Since the differentiation patterns are subdivisions of the cell classes, it follows that a negative result also must obtain as far as they are concerned. We shall nevertheless investigate the distribution of the differentiation patterns on the basis of the sequence of selfdifferentiations for filocytes established above. It is seen that the first member of the succession, intestinal cells, belongs to the endoderm, the remaining ones to the 'ecto-mesoderm'. However, the second (myocytes), the third (collagenocytes) and some representatives of the fifth member (e.g. myxoblasts) are classified as mesodermal cells, whereas the fourth (neurocytes) and some cell-types in the fifth member (e.g. trochoblasts) are allotted to the ectoderm. This does not make much sense and rather corroborates the criticism voiced by de Beer (1947).

We may now briefly turn our attention to the two kinds of mesoderm called 'ecto-mesoderm' and 'endo-mesoderm'. In Amphibia, the former comprises collagenocytes, melanocytes and myxocytes, besides neurocytes which are considered to be ectodermal. The 'endo-mesoderm', or mesoderm, gives rise to myocytes, collagenocytes, neurocytes and myxocytes. The differentiation patterns are thus identical but for two members. Whether or not this warrants a terminological distinction may be a matter of dispute. In spirally cleaving eggs the differentiation patterns of the cells derived by these two mesodermal sources, represented by the ventro–lateral second and third triplets and by 4d, respectively, are not very well established, but as discussed above no definite differences have ever been established. Indeed, much evidence supports the opinion expressed by Treadwell that 'no hard-and-fast distinction can be made between the two kinds of mesoblast' (1901, p. 450). It therefore seems that the distinction between these two subdivisions, as well as between the ectoderm and the mesoderm, generally depends more upon intuitions and vague notions than on logical reasoning and actual fact, particularly as far as the diversification patterns are concerned.

In the present context I have not abandoned the germ-layer concept, but it is employed in a different sense from normal usage, namely with reference to the anatomy of the embryo, as proposed by de Beer (1947). Thus, to begin with the amphibian embryo, it is found that here the apical cytoplasmic polarity entails a clear-cut separation between the endoderm and the ecto-mesoderm. In the late blastula it is possible to distinguish two populations of cells in the ecto-mesoderm, one or two layers associated with the hyaline membrane, and the interior free cells (cf. Figure 116). The former is classified as ectoderm, the latter as mesoderm. As a consequence, the chorda and the neural plate, as well as the neural crests and the epidermis, are all classed as ectodermal tissues, whereas the mesoderm comprises muscles, connective tissues and kidneys, but also the nerve cells incorporated in cranial placodes, etc.

In the eggs of Invertebrata, where the blastula wall contains only one layer of cells, it is generally impossible at this stage to distinguish between ectoderm and mesoderm, but when an apical cytoplasmic polarity is present, it may act to separate the ecto-mesoderm from the endoderm. As discussed above, the situation is often complicated by a median polarity, since the dorsal quadrant may contain part or all of the endoderm. The cells which migrate to the interior may well, in agreement with the definition proposed by Hyman (1951a), be termed mesoderm, but in that case the nervous tissue must be included in this germ layer.

In Echinodermata the mesoderm will consist not only of the mesenchyme cells derived from the micromeres, but also of the coelomic pouches originating in the 'endoderm' and the various sf-cells which from the 'ectoderm' penetrate into the body cavity.

MORPHOGENETIC PROCESSES

In a previous chapter it was asserted that the material basis of development, the embryogenetic substrate, is a continuously changing entity. Starting with the fertilized egg, it is at later stages composed of an array of cellular and extracellular elements, characterized by various properties, some of which may be of importance for the morphogenetic processes taking place during embryogenesis. An attempt was made above to specify the properties in question, e.g. locomotion, adhesion, swelling, etc. These characteristics are in principle independent of the presence of other elements, and may be displayed by isolated cells and extracellular elements.

In the embryo, where various elements are in contact with each other, all sorts of interactions may occur. By this means the embryo is subjected to changes in shape, and these interactions may therefore appropriately be called 'morphogenetic processes'. Certain morphogenetic processes may take place outside the confines of the embryo, for instance in explants and cell cultures. The study of the activities occurring under such simplified conditions has contributed significantly to the analysis of the simpler morphogenetic mechanisms. It cannot be emphasized enough, however, that many normal morphogenetic events are suppressed when a part of the embryo is severed from the rest. It holds for morphogenesis more than for the other embryogenetic processes that its normal course can be studied only in the intact embryo.

The present chapter begins with an attempt to establish the nature of the most important of the morphogenetic mechanisms which, through cooperation, are instrumental in carrying out the morphogenetic processes occurring in embryos. Subsequently, we shall outline the relation of the latter to the preformed substrate, and to the processes of cell division and cell diversification, which are responsible for the formation of the substrate elements and for their spatial location in the embryo. I cannot end this introduction without mentioning the brilliant paper by Davenport (1895), in which a classification of elementary morphogenetic events was put forward. Many of the ideas advanced by Davenport coincide with those submitted by me.

Mechanisms

It may often be difficult to establish the contribution of the various substrate elements in the execution of morphogenetic processes. Here we shall nevertheless define as *morphogenetic mechanisms* those events which occur as the result of properties possessed either by cells or by extracellular elements. In general, the morphogenetic processes will thus be the outcome of cooperation between morphogenetic mechanisms.

ACTIVITIES INVOLVING CELLS

Morphogenetic activities carried out by cells may be referred to the various structures composed of tensible proteins and to the mucopolysaccharide-containing surface coat. The contribution of the former is largely a consequence of the power of locomotion which they confer on the cells. In this way they are responsible for a number of dynamic processes, but they also take part in certain static phenomena. The chemical nature of the surface coat first and foremost determines whether or not the cells may participate in aggregate formation, but in this case also the influence extends to other phenomena. The solocytes are largely involved in the various dynamic events and the colligocytes in aggregate formation, but this generalization does not cover all possibilities, and it is therefore necessary to discuss separately the processes involving adhesion and those involving tension.

Adhesion

In the present section we shall deal with two kinds of morphogenetic mechanisms in which adhesion plays a very important role, viz., cell aggregate formation and various morphogenetic mechanisms which represent the outcome of interaction between cell aggregates.

Cell aggregate formation

In the present context the concept of *cell aggregate* denotes a stable agglomeration of cells kept together by adhesive forces between the cells or between the latter and a common substrate, or both. Aggregated cells are not necessarily deprived of locomotive power, and it therefore follows that aggregation can occur only if the adhesive forces keeping the cells together are stronger than the locomotive forces tending to cause their dispersion. The proposed definition does not exclude that the balance between the two kinds of forces may change, with an ensuing segregation of the cells. It does, however, exclude incidental accumulations of cells. As an example of this we may mention the blastula wall in Amphibia, known to comprise several layers of cells. Except for those attached to the hyaline membrane, and in Salientia the subjacent cell layer also, these cells are free to move, and the multilayered condition therefore is not stable, even though adhesive forces do obtain between the cells. The lack of stability is clearly demonstrated when these cells are accumulated in the blastoporal region during gastrulation, for when the process of epiboly, which is instrumental in the accumulation of the cells, comes to a standstill, the cells immediately begin to scatter by migration.

The difference in behaviour between solocytes and colligocytes as regards aggregate formation is indicated by their names. It must be stressed, however, that this distinction holds only for the cells proper. Thus, only colligocytes can form aggregates by themselves, but both solocytes and colligocytes may form aggregates by adhering to a common substrate. Solocyte aggregates tend to be transient formations, either because the mobility of the cells leads to their derangement, or because the cells undergo transformation into colligocytes.

A distinction has been made earlier between two- and three-dimensional aggregates. A typical example of the former kind is observed when cl-cells are cultured on a plane surface, and the myotomes in embryos of Vertebrata may serve as an obvious example of a latter type. There are cases which are more difficult to classify, for instance the various vesicular and tubular structures encountered in the embryo. The lens and the auditory vesicle, as well as the neural tube and the pronephric ducts, are evidently three-dimensional structures, yet they are formed by unicellular layers of cells which, if they were spread out, would form typical two-dimensional aggregates. The situation is complicated by the fact that the dimensions of the aggregate may influence the classification. By this is meant that in those cases where the radius of curvature of the aggregate is relatively large, as for instance in the blastula wall, it seems more natural to consider the aggregate as two-dimensional than when the radius is small, as in the lens. This ambiguity is circumvented if a unicellular layer of cells is considered to constitute a two-dimensional aggregate, irrespective of the geometrical shape it may assume. By this expedient it appears to coincide with 'epithelium' in the widest sense of this word, including mesothelia and endothelia. In the literature there is a tendency to equate 'epithelial cells' with those classified here as colligocytes. It will be realized that this notion cannot be upheld.

In the following pages we shall discuss the formation and the properties of solocyte aggregates and of two- and three-dimensional colligocyte aggregates, as well as the mechanisms involved in aggregation.

Solocyte aggregates

Owing to the low intercellular adhesion, solocytes can only form two-dimensional aggregates. Turning first to the lobocytes, it appears that they never aggregate spontaneously, the reason presumably being that the phenomenon of contact inhibition does not obtain in this cell-type. This circumstance implies that solocyte aggregates can arise only in association with the process of cell division. The most common mechanism through which this occurs is found in eggs possessing a hyaline membrane. As cleavage goes on, all or part of the blastomeres will remain attached to the membrane, thus forming a two-dimensional aggregate. As known from many embryos, these cells may transform into flagellocytes, suggesting that they stop dividing, temporarily at least. In some cases they may also transform into sf-cells, and when this occurs, they will often leave their station, as exemplified by the primary mesenchyme cells in the sea urchin embryo and the neural crest cells in the amphibian embryo. Most frequently these cells, which are constantly exposed to the medium, transform into colligocytes, cl- or cf-cells, as the case may be.

In spirally cleaving eggs, and in other cases where a hyaline membrane is missing, micromeres or other small cells formed at the animal pole may overgrow the macromeres or the yolk. The cells involved in this activity form an aggregate similar to that discussed above, with the modification that the surface of the macromeres or the yolk has taken over the role of the membrane as the common substrate. It is seen that if, as proposed here, the micromeres are sl-cells, then it is necessary to presume that they are more adhesive than is normal for this cell-type. The fact that they never form cubic epithelia supports the contention that they are not cl-cells.

As discussed above, isolated sf-cells may flatten out on a suitable substrate, showing that cells of this kind may change their shape radically under the influence of adhesive forces. If, in the embryo, sf-cells are exposed to substrates to which their surfaces adhere, it may be anticipated that, obeying the law of minimum free energy, they will penetrate the substrate in order to establish maximum surface contact. If this for some reason is impossible, they must spread out on the adhesive surface. This explanation may account for the formation of various endothelia. The irregular outline exhibited by the cells in squamous epithelia of this type (Figure 131) suggests the presence of punctate adhesive contacts, which may be of the close or tight junction type (cf. Trinkaus, 1966; Trelstad, Hay and Revel, 1967). The initiation of the formation of the epithelia probably involves filopodial activity, resulting in immobilization of the cells, but the stabilization through elaboration of contact specializations appears to be one step over and above those responsible for contact inhibition. It must be mentioned that in certain endothelia real desmosomes are found (cf. Fawcett, 1966). Collagenocytes attached to the surface of colligocyte aggregates do not spread out to form epithelia. It is possible that the modification of the substrate accomplished by their own synthetic activity contributes to this changed behaviour. If this suggestion is correct it may be envisaged that the epithelia mentioned above are formed by myxoblasts.

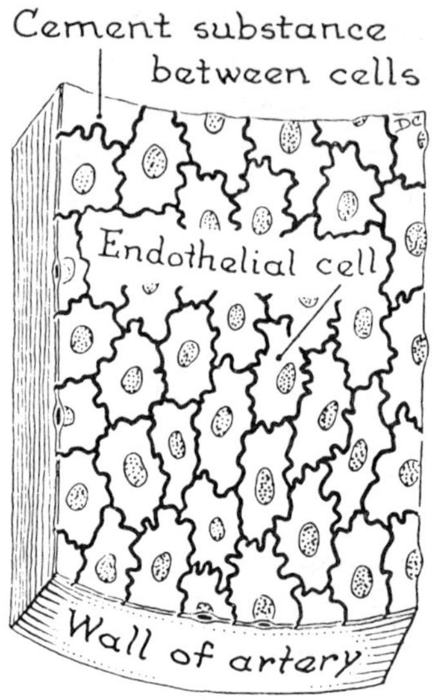

Figure 131. Diagram of the endothelial lining of the blood vessels, showing irregular outline of the endothelial cells. (Reproduced from A.W. Ham (1953), *Histology*, by permission of J. B. Lippincott, Philadelphia)

Many endothelia and mesothelia may be formed in the way that free sf-cells settle upon preexisting surfaces. This mechanism cannot be invoked to account for the formation of embryonic blood vessels, since these by necessity must arrive *de novo*, thus implying a mechanism of spontaneous aggregation. The precondition for this morphogenetic process seems to be that sf-cells are imbedded in a cavity containing a matrix of mucopolysaccharide to which they adhere strongly. Under these conditions the cells may migrate at random in these spaces. Provided a trapping mechanism like contact inhibition obtains, accumulation of cells may occur at locations representing the points of maximum statistical probability of cell encounter, presumably the central regions of the spaces. The accumulated cells subsequently form blood islands, i.e. vesicles consisting of flattened sf-cells and containing blood corpuscles. We have already discussed the fact that sf-cells exposed to a fluid medium will flatten out on a surface with appropriate adhesive characteristics. It seems that this phenomenon may account for the formation of the above-mentioned vesicles if fluid somehow accumulates between the cells. The subsequent fusion of the vesicles to form the blood vessels is too well known to require further comment. The process of fusion will be dealt with in a following section. According to the observations of Clark, Hitschler, Kirby-Smith, Rex and Smith (1931), capillaries in adult organisms are highly labile structures subject to constant remodelling, suggesting similarities with embryonic blood vessels. The participation of sf-cells in this case may explain the lability of these formations, as well as the low adhesion between the endothelial cells which to a large extent accounts for the high rate of permeation through the capillaries and for the fact that cells may pass through their walls. It may be mentioned that spontaneous formation of structures resembling capillaries has been observed in cultures of leukocytes by Hueper and Russel (1932).

Two-dimensional colligocyte aggregates

It has already been pointed out that the usual types of two-dimensional colligocyte aggregates, the cubic and columnar epithelia, presume that the adhesive properties of the substrate are identical with the intercellular adhesion. If the latter is stronger than the former, the cells will form a three-dimensional aggregate, if the opposite relation obtains, the cells will flatten out to establish maximum contact with the substrate. It may be anticipated that the first type of behaviour is more frequently displayed by the strongly adhesive colligocytes, the second by the weakly adhesive solocytes. It appears, however, that most often the colligocytes themselves secrete the material determining the adhesive properties of the substrate, thus ensuring the formation of the types of epithelium mentioned. As discussed previously, the existence of columnar epithelia requires an internal cytoskeleton. Generally, two-dimensional aggregates of colligocytes are composed of either cl- or cf-cells, mixed aggregates apparently being rare. In some cases, however, an aggregate of cf-cells may be located in a cell layer otherwise composed of cl-cells. In this case the cf-cell aggregate is often called a *placode*. Placode formation is a very general phenomenon; as examples we may mention the nasal and lens placodes in vertebrate, the apical plate and the shell gland in invertebrate embryos.

The two-dimensional colligocyte aggregate constituting the body wall typically arises through transformation of solocytes attached to the hyaline membrane or enclosing the macromeres. In Amphibia this transformation, as evidenced by the appearance of ciliated cells, occurs in the neurula stage, at the same time as the hyaline coat degenerates (Holtfreter, 1943a). A number of colligocyte aggregates arise through topological deformation of the body wall, thus for instance the digestive tract, the lens and otic vesicles in Amphibia, and the digestive tract, the coelomic sacs and the water–vascular system in Echinodermata. The shape of these secondary aggregates is determined as the resultant of intrinsic and extrinsic forces. Spreading of sl-cells on free, adhesive surfaces, and their subsequent transformation into colligocytes, is a theoretically possible mechanism of formation of two-dimensional aggregates. It may be that the coelomic walls are formed in this way in some embryos.

Isolated colligocytes may aggregate spontaneously under experimental conditions, and the same phenomenon can be observed in many embryos during their development. According to the law of minimum free energy the formation of compact spherical aggregates might be anticipated in the absence of extrinsic forces. This situation hardly ever obtains; most often vesicular aggregates arise under these conditions. An important, although not a necessary, aspect in the formation of vesicles is the elaboration of cilia or flagella. As we have seen previously, these cannot exist at surfaces where the cell has established adhesive contact with the surroundings. The frequently observed formation of two-dimensional aggregates with ciliated inner surface is a consequence of this particularity. Flagella have been observed in the somite vesicles of amphibian embryos (cf. Figure 123).

There is reason to believe that flagellated and ciliated surfaces are non-adhesive, and their contribution to the formation of vesicles is obvious so far. However, the establishment of free surfaces in the interior, to which the kinetosomes may adhere in the first place, is more difficult to explain, but presumably accumulation of fluid somehow occurs. The 'atypical' epidermis shown in Figure 13 suggests that formation of cilia has occurred at random, and that cavities obtain only where these are present. The aggregate shown in that figure is probably the closest approach to a solid spherical aggregate that may be observed. Figure 132(a) and (b) shows how the relative orientation between adhesive surfaces and aqueous milieu determines the polarization of the cells with respect

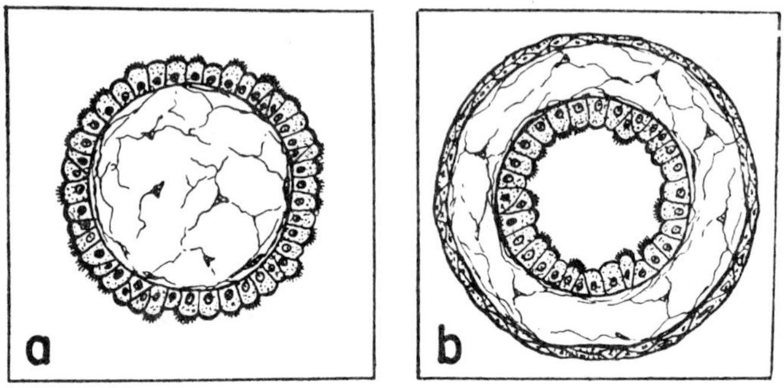

Figure 132. Vesicle formation in isolated cell aggregates. (a), vesicular cavity filled with mesenchyme, ciliation towards the external medium; (b), double vesicle, ciliation towards the central cavity. (Reproduced from J. Holtfreter, *Arch. Exptl. Zellforsch.*, **15** (1934), by permission of Gustav Fischer, Jena)

to ciliation. It is easily seen that if adhesive contact obtains at both sides of the aggregate, ciliation may be impeded. This may explain the lack of cilia in lens vesicles.

We have so far dealt with spherical aggregates which, as mentioned, must be the outcome of spontaneous aggregation when no extrinsic forces are involved. A deviation from this shape is exhibited by the somites which, although vesicular, are considerably deformed by the surrounding cell aggregates.

Under certain circumstances spontaneous aggregates may assume the form of tubes, a phenomenon that was first observed in cell cultures by Ebeling and Fischer (1922) and by Drew (1923). This shape cannot come into being without the participation of other morphogenetic mechanisms than aggregation. Taking again our examples from the amphibian embryo, two kinds of tubular cl-cell aggregates are encountered, straight ones as represented by the pronephric duct and tortuous ones as exemplified by the pronephric tubules. It seems likely that both arise through fusion of vesicles. The shape of the pronephric duct may be determined by its position between the lateral mesoderm and the somites. The tortuous form of the pronephric tubules suggests that they originate through random fusion of epithelial vesicles in a mucopolysaccharide matrix, thus through a mechanism similar to that envisaged for the formation of blood vessels. However, whereas irregularities in the latter may be obliterated when blood circulation begins, the original shape of the pronephric tubules appears to become stabilized by collagen membranes deposited at their surfaces. As pointed out above, this phenomenon requires the substrate of sulphated mucopolysaccharides which is present at the surface of cl-cells, but not at those of the sf-cells suggested to form the early blood vessels.

The salivary and mammary glands, as well as the lung and the pancreas, are tubular structures of much greater complexity than any of those discussed so far. The embryonic primordia of these organs may continue their development after explantation *in vitro*, and this circumstance has had the consequence that their morphogenesis has been subjected to intensive studies. Among the observations which have ensued from this work is that the morphogenetic events in question are always the outcome of cooperation between epithelial and mesenchymal cells. This fact has had the result that the processes are described or interpreted as inductive interactions between epithelium and mesenchyme. It is possible that inductions in the sense of the word adopted here do occur, but I contend that this terminology nevertheless obscures the essence of the problem. As stated repeatedly, morphogenesis is the result of spatio–temporal processes, in which the substrate elements are entities, the contribution of which depends upon their physical properties. The crucial question is therefore not the occurrence of inductions, but the nature of the participating agents.

As far as the two morphogenetic components are concerned, the major function of the epithelial cells, the colligocytes, according to the present nomenclature, is to form the wall of the tubular or glandular structures. It may be presumed that *in vivo* one of the most important functions of the mesenchyme is to supply, through the differential activity of myxoblasts, a three-dimensional substrate, a morphogenetic space, in which the morphogenetic activity can occur. In explants it is possible that this role is taken over by the culture clot. Further extracellular elements may be supplied by the mesenchyme, for instance collagen fibrils by collagenocytes. One further, and very important, contribution of the mesenchyme is the activity which may be deployed by the sf-cells present.

The studies on *in vitro* morphogenesis have shown that the mesenchymal, and to some extent even the epithelial, component is distinguished by its specificity. Thus, bronchial and mammary epithelia will, in the presence of salivary mesenchyme, branch in a very similar manner to that of the salivary gland. However, in the latter case features revealing the specificity of the epithelium may be observed (cf. Grobstein, 1953; Kratochwil, 1969; Spooner and Wessells, 1970).

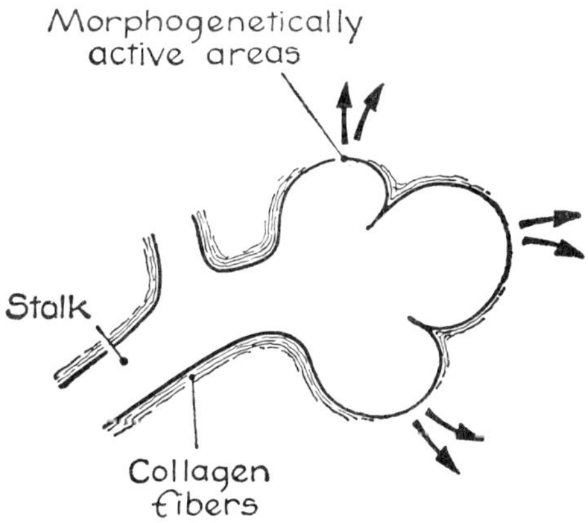

Figure 133. Possible relation between collagen distribution and morphogenetic activity in cultivated salivary epithelium. (Reproduced from C. Grobstein and J. Cohen, *Science*, **150**, 626–628 (29 October, 1965), by permission of American Association for the Advancement of Science)

That collagen, and thus collagenocytes, is an important agent in some of these morphogenetic processes has been demonstrated in various ways. Thus, Kallman and Grobstein (1965) have shown that from salivary mesenchyme, labelled with tritiated proline or glycine, radioactive material is released which may cross a filter membrane separating the mesenchyme from salivary epithelium. This material, which is instrumental in the formation of tubular glands, is accumulated at the epithelial surface with a distribution similar to that of ultrastructurally identifiable collagen. The importance of collagen was also demonstrated by the fact that the tubular outgrowth of salivary glands *in vitro* is suppressed by collagenase (Grobstein and Cohen, 1965). The absence of collagen at the morphogenetically active areas (Figure 133) indirectly suggests that the collagen membranes at the epithelial surfaces exert a stabilizing effect, while inhibiting the branching process proper. This latter point was confirmed by Wessells (1970), who observed that in mammalian lung development two kinds of mesenchyme are involved, a bronchial type which evokes or favours the branching of the lung buds and a tracheal type which inhibits this phenomenon, an outcome which may be correlated with deposition of collagen at the epithelial surfaces.

As far as the epithelial specificity is concerned, two factors seem to be obvious, viz., the rate of cell proliferation (cf. Wessells, 1970) and the time at which they produce the surface coat required for the deposition of the collagen fibrils. Over and above these possibilities a mechanism has been suggested by Wessells *et al.* (1971) involving the participation of microfilaments in the epithelial cells. The basis of this proposition is the observation by Baker and Schroeder (1967) that in the amphibian neurula microfilaments are present in the necks of the neural groove cells. In *Xenopus* these TP polymers are arranged in circular bundles, a circumstance which led these authors to the presumption that, acting like a 'purse string', they contrive to pull the neck of the cells together, thereby being instrumental in the formation of the groove.

We shall not discuss the morphogenetic aspects of neurulation on this occasion, but only mention some points which seem to contradict this supposition. For one thing, it was found by Baker and Schroeder (1967) that in *Hyla* similar microfilaments are arranged parallel to the groove lumen, an orientation which effectively seems to prevent their participation in the suggested operation. Secondly, the arrangement of the microfilaments in *Xenopus* appears to be very similar to that observable in dividing eggs which has been discussed in a previous chapter. In the latter instance a ring of cortical microfilaments was observed in the division plane (Schroeder, 1968; Szollozi, 1968; Selman and Perry, 1970), and its location was suggested to be the resultant of the dynamic activity of the microtubules emanating from the two asters. In the neural groove cells microtubules are also present, arranged perpendicular to the external surface (Baker and Schroeder, 1967) and their orientation may explain that of the microtubules. The presence of the latter at the free surfaces follows from the generalization concerning the preference of attachment of linear TP polymers. Furthermore, the most serious argument against the proposal of Baker and Schroeder (1967) is that a hemisphere constitutes the physical–geometrical limit to the extent of curvature which can be established in a two-dimensional cell aggregate by means of a constriction mechanism (cf. Gustafson and Wolpert, 1967). For the various reasons given here it seems unlikely that the microfilamentous ring is creatively involved in neural groove formation; possibly it exerts a stabilizing function, similar to that proposed to prevail during cell division.

In the salivary gland a ring of microfilaments is present at both ends of the epithelial cells, indicating that no differential obtains between the surface properties at the two ends of the cell. This circumstance would rather tend to exclude the involvement of a constriction mechanism, but the argument of Wessells *et al.* (1971) in favour of this alternative is based on the observation that cytochalasin causes simultaneously suppression of the branching of the gland and disappearance of the microfilaments mentioned. It seems plausible that the branches are stabilized initially, i.e. before the formation of collagen membranes, by the microfilament ring at the basal end of the epithelial cells. However, this inference neglects various other effects of cytochalasin, above all the repression of the undulating membranes, and thus the motility, of the mesenchymal sf-cells. It therefore cannot be excluded that the primary effect of cytochalasin is to block the morphogenetic contribution of these cells. This interpretation seems to be supported by some of the observations made by Wessells *et al.* (1971), for instance that colchicine also inhibits the morphogenetic activity, although in a different fashion. This is explained by the authors as an outcome of mitotic inhibition, but it may equally well ensue from the suppression of the formation of filopodia, so important for the activity of sf-cells.

The present discussion shows that the studies of *in vitro* morphogenesis have resulted in a remarkable progress in our understanding of this phenomenon, but also that many questions remain to be elucidated. However, the most striking advance is that in many cases explanations are now possible in simple physical terms, a circumstance which is very promising as regards the significance and progress of future research.

Three-dimensional colligocyte aggregates

Spontaneous three-dimensional aggregates most often become converted into two-dimensional ones. In Vertebrata a change in the opposite direction seems to take place when the cells of the somites assume the spindle shape of the myotomal cells, with the associated obliteration of the myocoeles (cf. Figure 14). The cell-type involved, the cf-cell, seems to be the only one capable of forming solid three-dimensional aggregates.

The parenchyme of various glands consists of cl-cells, but the tissues are never solid, being interwoven with ducts and sinusoids of various kinds, so that three-dimensional aggregates, when present, occur only in very restricted regions.

Epithelia containing several cell layers may be considered as three-dimensional aggregates. In this case each cell layer ought to act as a substrate for the following one, and consequently adherence to the law of minimum free energy appears to require that the shape of the cells is the same in all layers. If the upper cells are flattened, other forces than adhesive ones must be involved. The inference that a cytoskeleton must be present is confirmed by the presence of tonofibrils in various epidermal cells.

One more cell aggregate may be mentioned, the notochord. This does not conform to the definition of either two- or three-dimensional aggregates, indeed it is a question whether this structure should not be considered a one-dimensional aggregate. One circumstance which possibly can explain why the notochord is not tubular is that, when it is first formed, there are neither surrounding adhesive surfaces nor a mucopolysaccharide matrix available to ensure vesicle formation by polarization of the cells.

Aggregation mechanisms

Certain aspects of the mechanism of aggregation were discussed above, but there are a few more points which deserve attention. Aggregate formation by free cells involves migration, trapping or aggregation and sometimes also segregation. In membrane-bound cells the two first steps are substituted by cell divison. The remarkable precision with which aggregation of free cells occurs may easily give the impression that they are somehow directed towards their final goal. For this reason it has occasionally been envisaged that the cells are subjected to directed movements.

As far as this mechanism is concerned, two possibilities prevail, viz., chemotaxis and contact guidance. Chemotaxis is a well-established phenomenon, but among animal cells it is known only in leukocytes and monocytes (Harris, 1961), and it therefore seems to represent a case of specialization that can be of little importance for embryogenesis. Contact guidance (P. Weiss, 1934) is an occurrence which has been observed in cultures of fibroblasts and nerve cells. In the former case the direction of the migration of the cells, in the latter the outgrowth of axons, may be influenced by the ultrastructure of the substrate. In actual fact, it appears that the common denominator in these two cases is that the substrate affects the orientation of extended filopodia. If this is true, then only sf-cells can be involved, possibly without regard to the nature of the differentiation patterns. Since the substrate properties responsible for this effect are rather unspecific, it appears that, although contact guidance unquestionably is an important morphogenetic mechanism, it is not selective enough to account for the directed migration of special cell-types. It therefore seems necessary to conclude that cell movements are random in the embryo as well as in cultures. The random activities may involve migration of sl-cells on, or in the case of sf-cells exploration by filopodia of, available surfaces.

Aggregate formation implies that the migrating cells become trapped. Two different mechanisms have been invoked to account for this phenomenon. In the slightly adhesive and highly motile sf-cells contact inhibition, involving that filopodia from one cell become attached to the surface of another one, was suggested to account for the immobilization and the accumulation of the cells. In colligocytes the trapping is proposed to be the outcome of adhesive forces between cells and substrate or between the cells proper, and the active agent adhesive mucopolysaccharide present in the cellular surface coat and often also in the substrate. According to these contentions the mechanisms responsible are rather unspecific, being common for all sf-cells on one hand, and for all colligocytes on the other hand.

If thus migration is a random event, and aggregation relatively unspecific, some means of segregation seems to be necessary to account for the tissue homogeneity generally observed. Before we discuss this problem it must be pointed out that tissue homogeneity is relative, not absolute, for most organs and tissues contain a spectrum of different cell-types. As far as the segregation between colligocytes and solocytes is concerned, the differences in surface adhesivity and motility between these cell orders may suffice to account for this process. It is questionable whether membrane-bound sl-cells possess enough force to detach themselves, but the ingression of cells observed on various occasions during the development of spirally cleaving eggs suggests that this mechanism may occur with free cells. It must be pointed out that in this case overgrowth by colligocytes will have the same effect; often it may be difficult to decide which of the two phenomena actually takes place. As far as the sf-cells are concerned, it is well known that they may leave colligocyte aggregates by their own force, both when they are bound to a membrane (e.g. primary mesenchyme and neural crest cells) and when they are free (e.g. scleroblasts in the somites).

Steinberg (1963a; 1964) has shown that, according to the law of minimum free energy, segregation will occur between two populations of cells if these differ quantitatively with respect to surface adhesivity, provided random movements are allowed for. Unfortunately, the less adhesive cells will end up outside the more adhesive ones, a situation which does not seem to agree too well with normal morphogenetic events. It will be realized, however, that this mechanism accounts perfectly well for the segregation between solocytes and colligocytes just discussed.

As stressed by Steinberg, cell movements are essential for segregation. It appears that once a greater part of the cells have differentiated into colligocytes, potentially migrating cells may become trapped. This point is borne out by the observations made by Curtis (1961) on isolated and disaggregated ecto-mesodermal cells embedded in partly re-aggregated endodermal cells isolated from the *Xenopus* gastrula. It was found that the longer the time the latter were permitted to reaggregate before addition of the other cells, the less complete was the segregation of the various cell-types. From his results, Curtis concluded that there must exist a trapping mechanism which contributes to keep the cells in a certain position. The observations further indicated that the 'cell type to be outermost in the reaggregate changes its properties first, so that it becomes trappable. Random movements of such cells will bring them to the surface of the reaggregate where they will be trapped' (Curtis, 1961, p. 116). If the mechanism is a conversion of solocytes to colligocytes, the quotation supports the contention advanced here that this process is promoted by exposure to the external medium. A more detailed discussion of Curtis' results has been published elsewhere (1967b).

The expedients discussed so far do not allow for segregation between the two kinds of colligocytes, or between cells of the same class representing separate differentiation patterns. If more specific mechanisms do not obtain, it might therefore be anticipated that mixed tissues would prevail more frequently than is actually the case. Various experimental observations (e.g. Townes and Holtfreter, 1955; Curtis, 1961; Moscona, 1962; 1965; Steinberg, 1964) have been interpreted as support for the notion that tissue homogeneity is accomplished by segregation mechanisms. Thus it has been suggested that highly specific adhesive forces, for instance of the antigen–antibody type, obtain between cells, and that these forces may be both species and tissue specific (cf. Tyler, 1947; P. Weiss, 1959). According to Curtis (1966), evidence in favour of this view is scanty, but some of the observations on reaggregation of dispersed tissue cells indicate that a certain specificity obtains. It must be emphasized, however, that experiments of this kind do not apply to the conditions under which aggregate formation occurs in the embryo, since this generally takes place before the cells have assumed any specific differentiation patterns.

As a matter of fact, the cells may even be in the state of competence when they first aggregate, and under these circumstances it seems likely that the mechanism of induction, through which transformation and differentiation patterns may be impressed upon the cells, is primarily responsible for tissue homogeneity. It is possible, furthermore, that aberant cells by some inhibitory mechanism may be subject to degeneration, a question to which we shall return. However this may be, it seems justified to stress that, in the early embryo at least, the cell-types present are so few that the chances for mixed aggregates to be formed are rather slight.

Interactions between cell aggregates

Interactions between cell aggregates may be instrumental in morphogenetic processes.

In some cases only colligocyte aggregates are involved, but in other situations one of the reacting partners may be an aggregate of sl-cells. We shall consider these two cases separately.

Colligocytes

Under this heading we shall discuss partly the spreading of a colligocyte aggregate on a substrate and partly two types of aggregate fusion.

Epiauxesis. When an aggregate of cl-cells is located on a solid substrate, the shape of the cells will be dictated by the law of minimum free energy. When equilibrium obtains, no substantial enlargement of the area may be anticipated, even though the marginal cells possibly may flatten out somewhat. The situation changes, however, if the cells go on dividing, because each of the newly formed cells, in order to establish maximum surface contact, must preserve the contact with the surrounding cells and with the substrate. It is easily grasped that the consequence of repeated cell division will be that the thickness of the aggregate becomes reduced at the same time as the area enlarges. In contrast to the phenomenon of epiboly, which will be discussed presently, this kind of overgrowth is directly dependent upon concomitant cell division. In order to distinguish the two processes, the one dealt with here may be called *epiauxesis*. The mechanism of epiauxesis can account for the overgrowth of the macromeres by micromeres in various stereogastrulae, and also for blastula formation in the sea urchin egg according to the mechanism envisaged by Gustafson and Wolpert (Figure 134), even if in this case we are dealing with sl-cells attached to a hyaline membrane. It is probable that a similar phenomenon occurs in two-dimensional cf-cell aggregates.

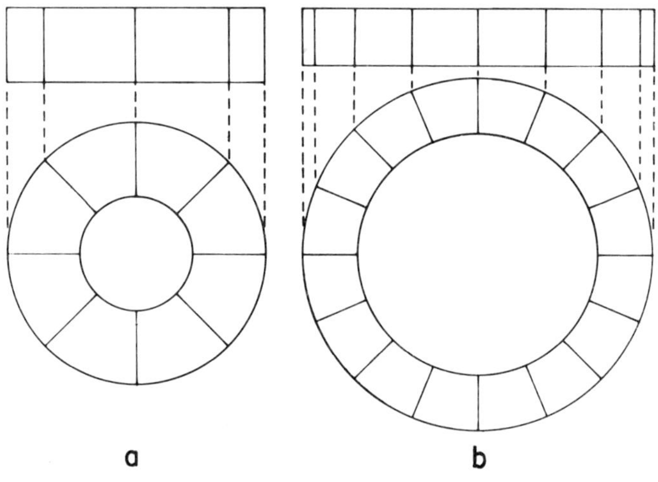

Figure 134. Schematic representation of epiauxesis during blastulation. (a), 8 cells arranged in an annular unicellular aggregate; (b), similar aggregate containing 16 cells. The following assumptions were made: (1) The total cell volume is constant; (2) The height of the cells is equal to the length (perpendicular to the plane of drawing) and to the mean width $2\pi/n \times R+r/2$, where n is the number of cells). This latter assumption is hardly correct since the cells are not colligocytes. (Based on views advanced by T. Gustafson and L. Wolpert, *Discovery*, **22** (1961))

As we have seen above, sf-cells, and under certain conditions even colligocytes, may form sqamous epithelia when exposed to an appropriate surface. In such flattened aggregates further cell divisions cannot be expected to contribute significantly to an increase in the area. Epiauxesis thus should not be possible in flattened epithelia, and if surface enlargement occurs in this case, it may therefore be presumed either that an active stretching of the aggregate takes place or that further material is added in the form of invading cells. The former of these possibilities may be classified as epiboly. The latter does not conform to the definition of epiauxesis given above but maybe a modification ought to be introduced to allow for this mechanism which, after all, is very similar to true epiauxesis. I have not found any examples of this sf-cell epiauxesis recorded in the literature.

Lamination and fusion. If two layers of colligocytes are apposed, they will adhere to each other, the cells possibly becoming oriented with non-adhesive ciliated surfaces away from the contact surface. This event may strongly influence the shape of the aggregate, as seen for instance when somites establish contact with body wall, neural tube, notochord, etc. (cf. Figure 123). This phenomenon, by which double-layered cell aggregates are formed, may be called *lamination.*

When layers of either cl- or cf-cells meet edge to edge, they may become united through the phenomenon of *fusion*. The closure of the endodermal gap in the amphibian embryo is an illustration of this event (cf. Figure 135). Examples of fusion between cl- and cf-cells seem to be rare, even if the two cell classes, as seen above, may form stable aggregates in concert.

In many embryos the mouth is formed after contact is established between ecto- and endoderm. One might in this case expect that lamination would take place, but it is possible that deformation of the aggregates occurs to the extent that they rather meet edge to edge. A similar mechanism is involved in the formation of neural tube, lenses and gill clefts in various embryos of Chordata. Once the colligocyte aggregates are covered by collagen membranes fusion can no longer take place.

Amoebocytes and colligocytes

In many embryos phenomena may be observed which involve rupture between cell aggregates or agglomerates, indicating that changes occur in the adhesivity of the surface. No general mechanism can account for these events, even if most often transformation of solocytes into colligocytes is the primary cause. We shall consider three separate processes, for which the names 'switchover', 'separation' and 'segmentation' are proposed.

Switchover. The phenomenon called *switchover* is responsible for the formation of the so-called 'endodermal gap' in amphibian embryos, as described in great detail for *Triton* by Vogt (1929a). When the presumptive notochordal cells during gastrulation migrate towards the anterior end, they establish intimate contact with the membrane-bound ectodermal cells in a narrow region around the dorsal median. Laterally, the notochordal primordium is united with endodermal cells through the hyaline membrane. The free mesodermal cells, which are present between the ectodermal and the endodermal cells, are pushed backwards by the latter towards the blastoporal lip. When the gastrulation movements slow down the mesodermal sl-cells begin to migrate towards the anterior end. Situated between the endoderm and ectoderm, they border laterally the notochordal primordium. As they advance, they adhere to the latter, which thereby

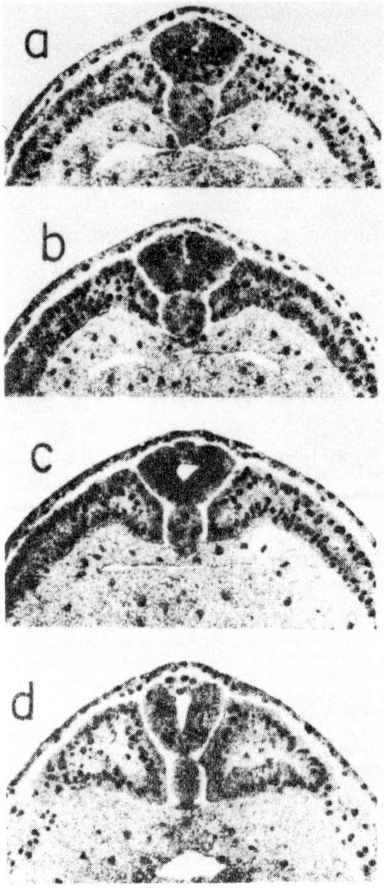

Figure 135. Transversal sections from early tailbud stage of *Pleurodeles*, showing various levels in a postero–anterior direction. Different stages of closure of the endodermal gap and formation of the hypochorda. (a), endodermal edges are still separated by a wedge of hypochordal cells; (b), the edges have met; (c)–(d), fusion has taken place. (Reproduced from W. Vogt, *Arch. Entw.mech.*, **120** (1929), by permission of Springer, Berlin)

switches its association from the endoderm to the mesoderm, thus giving rise to the endodermal gap.

This process comprises two aspects, one the weakening of the hyaline membrane which occurs when it is transferred to the interior (cf. Holtfreter, 1943a), and the other a differential in the adhesivity of the various sl-cells involved, the one between notochordal and mesodermal cells being stronger than that between the former and the endodermal cells. The mechanism has the consequence that the formation of the endodermal gap occurs in postero–anterior direction, a very unusual phenomenon since embryogenetic processes most often take the opposite course (Figure 135).

Separation. Another type of rupture occurs when a change takes place in the adhesive properties of an aggregate, as for instance in association with the transformation of solocytes, notably sl-cells, into colligocytes. The most obvious example of this phenomenon of *separation* is seen when the notochord and the dorsal mesoderm are separated in the amphibian embryo. The chordocytes transform much earlier than the mesodermal cells, apparently in the sequence sl→cl→cf (Figure 136, cf. Mookerjee, Deuchar and

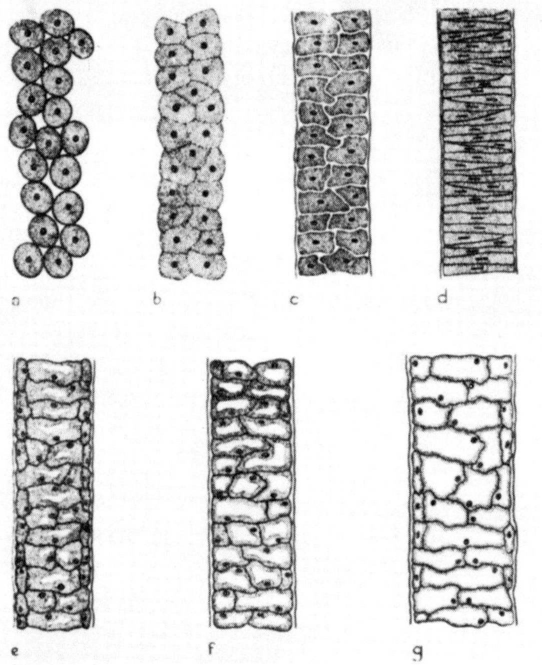

Figure 136. Successive stages of the diversification of the amphibian chorda. (a), late gastrula; (b), early neurula; (c), neural plate; (d), early tailbud; (e), middle tailbud; (f), late tailbud; (g), larva. The transformation from sl- to cl-cells, and from the latter to cf-cells (d), the deposition of the sheath and the differentiation, entailing swelling of the intracellular mucopolysaccharide, are clearly demonstrated. (Reproduced from S. Mookerjee, E. M. Deuchar and C. H. Waddington, *J. Embryol. Exptl. Morphol.*, **1** (1953), by permission of The Company of Biologists, Cambridge)

Waddington, 1953). It may be noted in passing that the change from wedge to 'pile of coins' shape to which the chordocytes are subjected after the formation of the sheath is predicted by the law of minimum total surface. Since the adhesion between amoebocytes and colligocytes is rather low, it follows that the association between the two cell groups is already reduced by the first step. Before the mesodermal cells begin their transformation, the first notochordal sheath has been formed, thus preventing fusion between the cell aggregates. No transformation gradient, in the sense defined in the next section, is involved in the process of separation, since the difference in time between the transformation of the notochordal and of the mesodermal cells into colligocytes is so great that it is impossible to speak about a dorso–lateral transformation gradient.

Segmentation. The concept of 'gradient' is of common use in embryological texts, covering, it seems, several distinct phenomena, among which is that of polarity as defined above. Another sense of the word represents the case where some kind of diversification spreads along one of the embryonic axes. This meaning is implied when the word is used here, thus denoting that diversification in some direction proceeds according to a *temporal gradient*. The mechanism may involve the spreading of an inductive stimulus, or the gradual acquisition of competence, followed by selfdiversification.

When transformation into colligocytes occurs along a gradient the transformed cells will form spherical aggregates which, assuming the shape of vesicles, may subsequently be subjected to deforming influences by the environment through the process of lamination. The vesicles will turn an adhesive surface towards the untransformed cells, and when these in turn become colligocytes, this surface may function as a substrate for lamination of the wall in the next vesicle. This mechanism may thus ensure the formation of a series of vesicles. Except for the gaps separating the aggregates, the existence of which would prevent interaction between the cells, the principle is excellently illustrated by Figure 137. This process of *segmentation* may account equally well for somite formation in Vertebrata and coelom formation in Annelida.

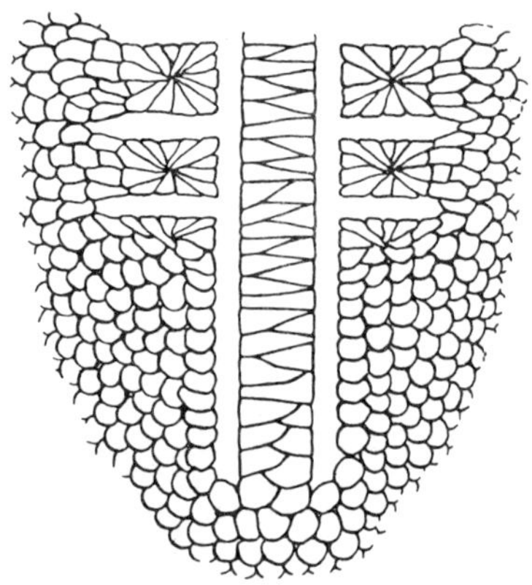

Figure 137. Diagrammatic illustration of mesoderm segmentation. (Reproduced from C. H. Waddington (1956), *Principles of Embryology*, by permission of George Allen and Unwin, London)

It appears that there is no fundamental difference between this mechanism and that invoked above for the formation of various tubular structures. It therefore seems necessary to assume that somehow the vesicles are prevented from fusing together. It is possible that lamination may temporarily stabilize the vesicles, but ultimately fusion is prevented by the formation of collagen membranes (septa) between the segments.

Another kind of segmentation, involving the formation of placodes, may ensue when the lf-transformation occurs in two-dimensional colligocyte aggregates, provided that the regions of cf-cells are separated by patches of cl-cells. Assuming that the transformation spreads from cell to cell through induction, the existence of placodes implies that the cl-cells were reached by the inductive stimulus after having lost their competence. Spreading of induction to more distant cells therefore presumes that each segment is separately subjected to inductive stimulation. In this case the segmentation obviously requires a 'gradient' with respect to competence. This mechanism may be invoked to account for the segmentation of the shell in Polyplacophora. The formation of the various placodes in the ectoderm of embryos of Vertebrata may also occur by this mechanism, even if they do not display segmentation in the morphological sense.

There are many other morphological phenomena which involve repetitions, thus representing some kind of segmentation. As examples we may mention tentacles, gill slits and gut diverticula in certain Platyhelminthes, and segmentation in Insecta. We shall presently discuss the mechanisms involved in some of these cases.

Tension

We have already dealt with the evidence indicating that the generation of tension in cells may be referred to a special type of protein, TP, which is localized in certain cytoplasmic and cortical structures. It was suggested that all dynamic activities displayed by single cells could be accounted for by the particular properties of TP polymers. Among these activities only those which entail interaction with the surroundings, i.e. pseudopodium formation and cortical movements, can be instrumental in morphogenetic mechanisms.

It follows from this restriction that we may expect the solocytes to be the foremost participants in the dynamic processes involved in embryogenesis, and we are therefore going to concentrate attention upon this cell class in the present section. It must be emphasized, however, that TP structures are present in the colligocytes, where they exert important functions, not the least in the form of cytoskeletons. Furthermore, it should not be forgotten that some kind of adhesion between cell and substrate is a precondition for locomotion. The morphogenetic contributions of sl-cells and sf-cells are remarkably different and we are consequently going to treat each cell order separately in the following discussion.

The sl-cell

It has been suggested that the egg and the early blastomeres are sl-cells, distinguished by their weak intercellular adhesion. Apparently, this contention is irreconcilable with the fact that the blastomeres stick together and participate in the formation of a blastula. As discussed above, this dilemma has been solved through the presence, in many eggs, of a syncytial membrane to which the sl-cells adhere strongly. In those cases where a hyaline membrane is not present, the blastomeres are often more adhesive than is typical of sl-cells, and furthermore the surfaces of the macromeres or of the yolk may possess adhesive characteristics serving to keep the blastomeres together.

In invertebrate embryos possessing a hyaline membrane all the cells of the blastula are attached to the latter; those cells which participate in the formation of internal structures detach themselves from the membrane and migrate to the interior, most often only after transformation into sf-cells. In vertebrate embryos a large number of free sl-cells are generally found in the interior of the blastula. The morphogenetic activities displayed by free and membrane-bound cells differ in many respects, as will be shown in the following sections.

Free cells

Free sl-cells exposed to a surface with appropriate adhesive characteristics will scatter by random migration. This is easily observed in cell cultures, but it also occurs in the amphibian embryo, where the free cells in the interior of the blastula tend to accumulate at the edges of the floor of the blastocoele. Gravity possibly influences the direction of movement in this case. During the epibolic gastrulation movements in Amphibia the free cells present between the ectoderm and the endoderm are to a large extent passively transported towards the blastopore. Once these movements come to an end, the accumulated mesodermal cells begin to migrate between the ectoderm and the endoderm. This migration occurs in the direction where free surface is available, i.e. towards the presumptive anterior region.

Membrane-bound cells

The hyaline membrane not only keeps together cells that would otherwise scatter because of their low intercellular adhesivity, but also 'amalgamates the peripheral cells into a plastic supercellular unit' (Holtfreter, 1943a, p. 266). In this way the membrane makes possible the translocation of the aggregate of membrane-bound cells as a unit. This morphogenetic mechanism, called 'epiboly', is the result of the motility of the individual cells bound to the membrane.

Epiboly. The movements of membrane-bound sl-cells were described by Holtfreter (1944a), who found that ectodermal cells united by a syncytial membrane will spread on a substrate of endodermal cells and ultimately enclose the latter completely (Figure 138). This phenomenon is satisfactorily explained on the assumption that the membrane-bound cells behave like free sl-cells, i.e. migrate at random on the available surface,

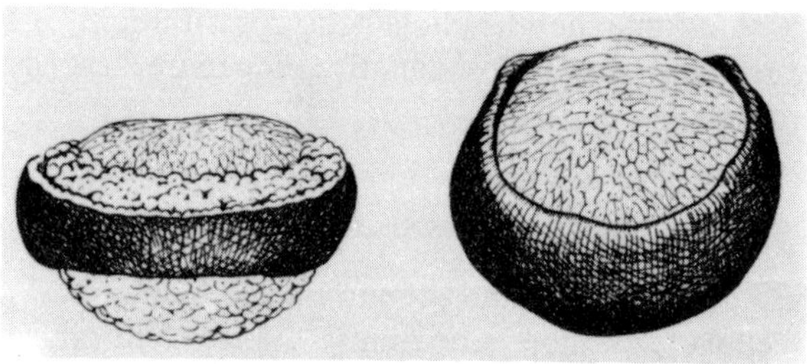

Figure 138. Epibolic movements in neurula of *Ambystoma*. After removal of the dorsal hemisphere and ventral epidermis the dorsal denuded cells are covered by coated endoderm, the ventral endoderm by epidermis. (Reproduced from J. Holtfreter, *J. Exptl. Zool.*, **95** (1944), by permission of The Wistar Press, Philadelphia)

which in this case is represented by the uncovered endodermal cells. The fact that the cells are bound together ensures that they can form an envelope around the endodermal aggregate.

If two layers of membrane-bound cells are apposed, they may migrate upon each other, and in opposite directions. This is exactly what happens when, during early gastrulation in Amphibia, the ectoderm is folded over the edge of the blastopore. The interior cells will migrate in the direction where free surface is available, thus towards the ventral side. In doing so, they will pull the exterior cells towards the dorsal blastoporal lip, and these may in turn by their own migration assist in pushing the interior cells ahead in the opposite direction. The explanation given here coincides with the view expressed in the following quotation: 'One can interpret ectodermal epiboly as due to the summative effect of the spreading tendency of each of the constituent cells of the cell sheet' (Townes and Holtfreter, 1955, p. 106). As previously described, Holtfreter (1948) observed two kinds of movements in sl-cells from the early amphibian embryo, one the normal polarized amoeboid movement known particularly well from Protozoa, and the other rotating movements in the cell membrane. It seems probable that the latter are responsible for the epibolic movements.

This phenomenon which, as we have seen, requires intimate contact between two layers of sl-cells, is generally called 'epiboly', or 'epiboly' and 'emboly', the latter referring to the activity of the interior cells. The latter distinction appears to be unnecessary, since evidently the two phenomena are separate aspects of one process. In the present context *epiboly* designates the spreading of a cell sheet as a consequence of the dynamic activity of some or all of the constituent cells. The distinguishing feature of this process is that it results from the motility of individual cells. Cell division is no integral part of the mechanism, as is clearly shown in the amphibian embryo, where DNA synthesis, and presumably also cell division, comes to a standstill for a brief period during gastrulation, just at the time when the epibolic activity is most extensive (cf. 1966c). Contrariwise, in the phenomenon of epiauxesis motile forces are not of great importance, the expansion of the aggregate being determined by the equilibrium established between the various adhesive forces involved, a situation that can be changed only if the cells are actively dividing.

The preceding discussion shows that sl-cells are involved in very few kinds of morphogenetic processes. In spite of this there is no doubt that their contribution is of great importance when, as we have seen, the whole, or most, of the process of gastrulation in certain taxa is carried out by cells of this class.

The sf-cell

The sf-cell exhibits two kinds of locomotion. Upon exposure to surfaces having suitable adhesive properties, the cell flattens out and moves by means of undulations in the membrane which is in contact with the substrate. When the cell is exposed to a space containing mucopolysaccharide, translocation may be accomplished by means of filopodia traversing the space. In either case sf-cells may display activities of morphogenetic importance. It appears that sf-cells are also involved in epiboly, and since claims have been made that both undulating membranes and filopodia may be instrumental in this event, it will be discussed under a separate heading.

Undulating membranes

The flattening of sf-cells required for the formation of undulating membranes may occur in both free and aggregated cells. Once this has taken place, the cells migrate about at random until further progress is inhibited. The phenomenon of contact inhibition may be responsible for immobilization of the cells, as a consequence of which the surface eventually becomes covered by a continuous cell layer. Various contact specializations may later contribute to the stabilization of the aggregate. The mesothelia and coelomic cavities in many animals probably arise through this mechanism.

Filopodia

Filopodia are responsible for a large number of morphogenetic processes. The mechanism involved may in all cases be referred to an interaction between the filopodia and the surfaces to which they become attached. This problem will be discussed before we turn to the processes in which filopodia are actively engaged. As for the sl-cell, we shall distinguish between free and membrane-bound cells.

Attachment of filopodia. Briefly summarized the properties of the filopodium are as follows: it is rigid when extended from the cell, the tip is adhesive, and if attachment occurs a tension develops which is proportional to the number of TP monomers. If the filopodium is not attached, or if relaxation occurs, it collapses. A precondition for this activity is generally that the cell is exposed to a space occupied by mucopolysaccharide.

The most detailed description of the mechanism of filopodium attachment has been published by Gustafson (1963). From observations on the process of gastrulation in the sea urchin embryo it was ascertained that filopodia extended from mesenchyme cells at the vegetal pole may establish contact with the inner surface of the blastocoele wall. Sometimes permanent attachment occurs, in other cases the filopodia are detached and withdrawn. The difference in the two cases is, according to Gustafson, that the adhesion between the surface and the filopodial tip varies, and if the force generated by the tension of the filopodium is stronger than the adhesive forces involved in the attachment, the latter will be severed and the pseudopodium withdrawn. Repetition of this event results in a random exploration of the available surface with respect to its adhesive properties.

These findings raised the question of the nature of the surfaces to which the tips may become permanently attached, a problem which was studied by Gustafson (1963) on embryos compressed under a paraffin drop; under these conditions the cells of the blastocoele wall are partially separated from each other, thus facilitating the observation of the attachment of the pseudopodia.

In order to discuss the results obtained it is necessary to describe and classify the four cell-types found in the sea urchin gastrula. The mesenchyme cells are easily identified as sf-cells (Figure 139a, b, c). The cells in the apical plate are elongated, flagellated cells, which do not separate under compression (Figure 139d). The latter feature suggests that they are colligocytes, their elongated shape that they are cf-cells. Among the cells located between the vegetal and animal regions two types may be distinguished. The first kind adhere strongly to the hyaline membrane, while the intercellular adhesion is weak (Figure 139 a, c, d). These cells may with some confidence be classified as sl-cells; the fact that they are occasionally flagellated suggests that they may transform into vsl-cells. The last cell-types, localized in the flagellated bands and at the level of the ring of primary mesenchyme cells, are united both at the inner and the outer surfaces. In the

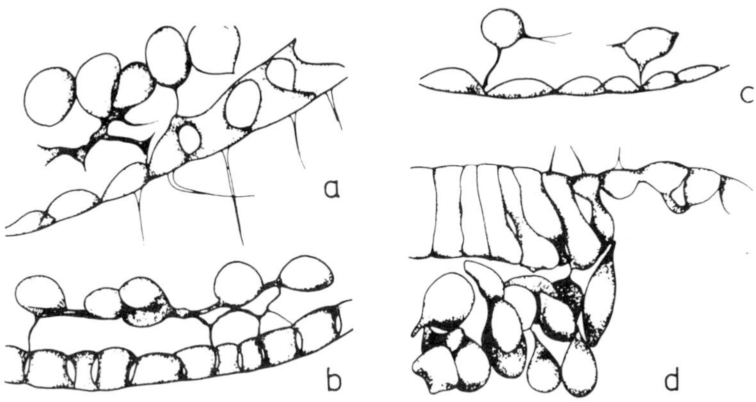

Figure 139. Cell-types in the sea urchin embryo. (a), sl-cells, some flagellated, those to the right kept together at the proximal side, presumably by the inner surface layer, and sf-cells; (b), sl-cells and sf-cells, the latter extending filopodia to the hyaline membrane; (c), sl-cells and sf-cells, filopodia attached to the contacts between the former; (d), sl-cells, sf-cells and cf-cells in the apical plate, to which no filopodia are attached. (Reproduced from T. Gustafson, *Exptl. Cell Res.*, **32** (1963), by permission of Academic Press, New York)

latter location the hyaline membrane may be assisted by desmosome-like structures (Balinsky, 1959; Wolpert and Mercer, 1963); in the former the only apparent agent holding the cells together seems to be the mucopolysaccharide-containing inner surface layer (Wolpert and Mercer, 1963; Immers, 1961; cf. Figure 24). When compressed, the intermediate cell walls become separated (Figure 139b). This feature suggests that the cells at the stage depicted are sl-cells, an interpretation that is supported by the fact that the inner contact may be broken, with the consequence that the cells assume the appearance of those classified as sl-cells above (Gustafson, 1963). The presence of desmosomes would be difficult to explain, because this otherwise appears to be a typical colligocyte feature. However, the junctions present in these cells are not true desmosomes; they may indeed be solocyte contact specializations similar to those found in certain sf-cells. It does not seem possible at the moment to account for these two modifications of sl-cells, particularly since there is no reason to believe that the inner surface layer is unevenly distributed.

Gustafson observed that filopodial tips in no case become attached to a cell surface. When free access obtains to the hyaline membrane, the filopodia may attach to it (Figure 139c). When the sl-cells are united at the inner surface, the filopodia may attach along the line of contact between the two cells (Figure 139a, b). The pseudopodial tips never adhere to the cells of the apical plate, not even at the lines of contact (Figure 139d).

The difference in behaviour of the filopodia towards the apical plate cells and towards the other cells clearly demands an explanation. A possible clue may be found in the results of Immers (1961) on the incorporation of $^{35}SO_4$. As shown in Figure 140, there is an extensive accumulation of the tracer, incorporated in sulphated mucopolysaccharides, at the proximal surfaces of the apical cells, as is to be expected if these cells are colligocytes. Presumably this covering prevents the filopodia from establishing contact with the cells in the apical plate. It must be emphasized, however, that the inner surface layer appears also to contain sulphated mucopolysaccharides and hence may act to

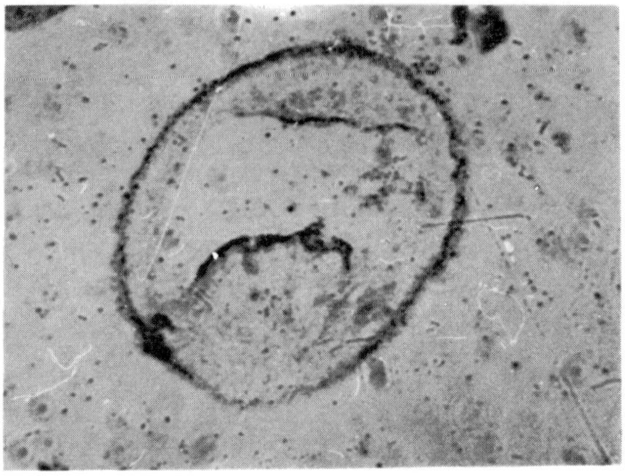

Figure 140. Incorporation of $^{35}SO_4$ in the sea urchin embryo. Early stage of invagination, showing sulphate uptake in the hyaline membrane, the blastocoelic gel above the invaginating archenteron and at the inner side of the apical plate. (Reproduced from J. Immers, *Exptl. Cell Res.*, **24** (1961), by permission of Academic Press, New York, and by courtesy of Dr. Janis Immers)

prevent the access of the filopodia to the cell surfaces, as suggested by Gustafson (1963). If it does, it may be inferred that it is a less efficient barrier than the material at the apical plate since it allows for attachment of filopodia at the lines of cell contact.

The preceding discussion shows that the morphogenetic activity exerted by sf-cells spanning mucopolysaccharide-filled cavities is subjected to constraint due to the fact that the attachment of filopodia is selective. It cannot be anticipated that the rules established through observations on sea urchin embryos are of general validity. On one point particularly it seems necessary to introduce an amendment. Thus, it was observed that no filopodia became attached to the only colligocytes supposedly present, the apical plate cells. In order to account for a wide variety of observations in the higher phyla it is necessary to postulate that usually filopodia may become attached to aggregates of colligocytes. Whether the attachment occurs only at points of contact or at the whole cell surface remains to be seen. This apparent difference may possibly be referred to the physico-chemical properties of the mucopolysaccharide layer covering the aggregates, but no definite explanation can be given for the time being.

Free cells. In this section we are going to deal with the situation where a population of sf-cells is present in space filled with mucopolysaccharide. In the embryo such spaces will always, except during the earliest stages, be surrounded by layers of colligocytes. Trapping of sf-cells may occur in the interior of the cavity, either through contact inhibition or through production of sulphated mucopolysaccharides. If this happens, cell aggregation may occur at a location determined by the statistical probability of cell encounters. This being highest in the centre of the space, it follows that when the latter is circular, the accumulated cells will form a spherical aggregate; when it is cylindrical, the aggregate will be elongated. This seems to be the mechanism responsible for the formation of blood vessels and chondrified structures. It will be realized that in the former case the first, in the latter case both, of the trapping mechanisms mentioned above may be involved.

Attached cells. We shall make no distinction between membrane-bound cells, which may be attached to the hyaline membrane at the time of their formation, and cells which, being free at some stage, become attached with their filopodia. If free sf-cells become exposed to surfaces containing points of attachment, the cells may become anchored there with their filopodia. As suggested by Gustafson (1963), the distribution of free cells will depend upon the density of the attachment points; if the latter is uneven, the cells will accumulate at those regions where the density is highest. In the sea urchin embryo the density of cell contacts is high on the inside of the blastula wall in those regions where the curvature is high. The latter phenomenon obtains in the ring bordering the ventral plate and at the bands of flagellated cells at the ventral side. The accumulation of mesenchyme cells at these locations can thus be explained as a statistical phenomenon on the basis of Gustafson's mechanism. It also corresponds to the observation of S. B. Carter (1965), according to which cells will move towards regions of maximum adhesivity.

Irrespective of the nature of the loci of attachment (hyaline membrane, cell contacts, etc.), a number of remarkable morphogenetic processes may take place whenever sf-cells are attached between different points of the same cell aggregate or between two different ones. It is possible to distinguish two kinds of events, which may be classified as *dynamic* and *static*, respectively. We shall describe some examples here, others will be discussed later in the book.

As shown by Dan and Okazaki (1956) and by Gustafson and Kinnander (1956), the archenteron of the sea urchin embryo is pulled through the blastocoele by filopodia originating from the secondary mesenchyme cells (Figure 141). The first filopodia are

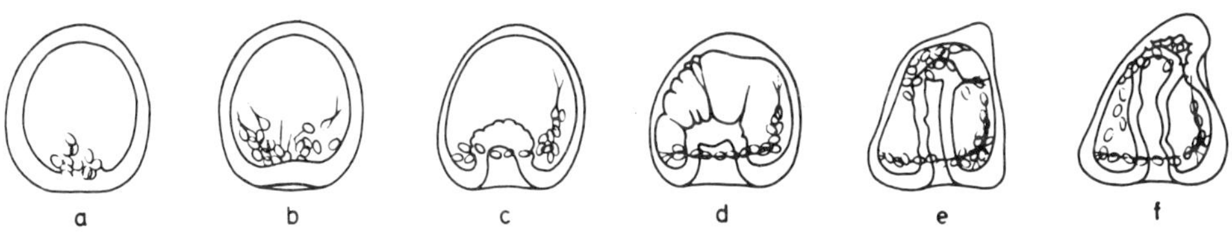

Figure 141. The early development of the sea urchin larva. (a), mesenchyme blastula, ingression of primary mesenchyme cells from the vegetal pole; (b) migration of the mesenchyme cells by means of filopodia; (c) arrangement of the mesenchyme cells in a ring around the archenteron and in two bands on the ventral side (one shown); (d), filopodia from secondary mesenchyme cells at the archenteron tip have become attached to the dorsal side of the blastula wall; (e), ingression of secondary mesenchyme cells, extension of filopodia from the archenteron to the ventral ectoderm, and incipient skeleton formation; (f), contact established between archenteron and ectoderm of the mouth region. (Reproduced from T. Gustafson (1964), in *Primitive Motile Systems in Cell Biology* (Eds. R. D. Allen and N. Kamiya), by permission of Academic Press, New York)

attached at the dorsal wall near the vegetal end, but gradually the filopodia advance towards the animal pole. Accepting that the tension is proportional to the length, it is understandable that the longer ones will prevail and pull the archenteron tip into the blastocoele. When the longest filopodia have reached the border of the apical plate, the end point of the movement, further advance of the archenteron tends to reduce the length of filopodia. In this connection the lability of these structures must be of importance; when the archenteron yields to the pull of the filopodia, the reduced tension leads to their collapse and withdrawal. The new and shorter ones subsequently formed may develop less tension than their predecessors, but they are still able to support the

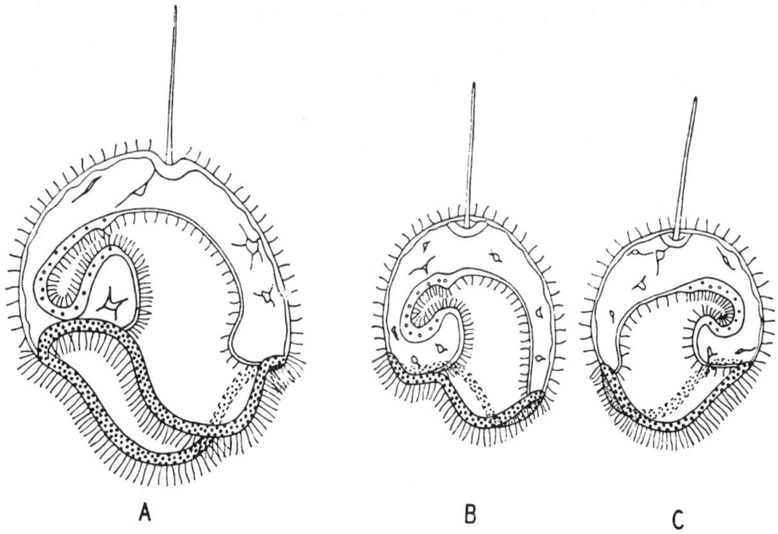

Figure 142. Pilidium larvae of *Cerebratulus*. (a), normal larva; (b), larvae from the isolated half blastomeres of one embryo. (Reproduced from S. Hörstadius, *Biol. Bull.*, **73** (1937), by permission of Marine Biological Laboratory, Massachusetts)

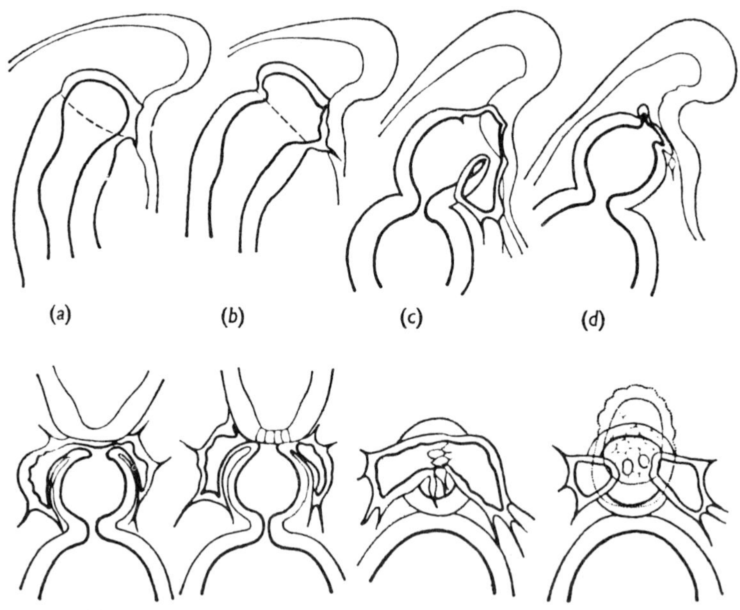

Figure 143. Coelom formation in the sea urchin embryo. (a)–(d), larvae in optical median section; (e) and (f), dorso–animal view; and (g) and (h), ventral view. The progressive constriction between coelom and oesophagus, and between the latter and the stomach, are shown. (Reproduced from T. Gustafson and L. Wolpert, *Exptl. Cell Res.*, **29** (1963), by permission of Academic Press, New York)

movement almost until the archenteron tip touches the body wall. In some cases, exemplified for instance by the pilidium larva, the archenteron does not advance so far into the blastocoele as in the sea urchin larva (Figure 142). It may be suggested that here the pressure in the blastocoele keeps equilibrium with the tensile forces of the filopodia. In the sea urchin embryo, the filopodia are subsequently instrumental in the approach between the archenteron and blastocoele walls which, through the process of fusion, culminates in the formation of the mouth (Figure 141). This kind of activity is also responsible for the formation of the coelom (Figure 143), for the closure of the endodermal gap in Amphibia (Vogt, 1929b, cf. Figure 135), and possibly also for the formation of the gill clefts (Løvtrup and Pigon, 1969).

The filopodia of sf-cells may also exert important morphogenetic functions while remaining in a fixed position. An example of this mechanism is observed in the dorsal fin in the amphibian embryo, in which a number of myxoblasts from the neural crest become trapped between the neural tube and the dorsal ectoderm. With the onset of their differentiation, the cavity in which they are enclosed begins to swell. One might anticipate the formation of semi- or quasicylindrical structures as a result of this activity, as readily observable under experimental conditions (Twitty and Bodenstein, 1941; cf. Figure 144). As established many years ago (Bytinski-Salz, 1931), the tapering shape of the dorsal fin owes its origin to the fact that sf-cells suspended between two sides of the fin serve to keep these together (Figure 145).

Epiboly. The phenomenon of epiboly is best known from amphibian, fish and bird embryos. The former case, already dealt with, was seen to be carried out by membrane-bound sl-cells. As far as the two latter are concerned, the appearance of the cells involved definitely excludes that they are sl-cells.

It was proposed above that extension of the area of a flattened aggregate of sf-cells might occur through a kind of epiauxesis involving the incorporation of new cells. There appear to be two circumstances which, without totally rejecting the participation of this mechanism, rather suggest that the so-called 'epiboly' in *Fundulus* conforms to the definition of this phenomenon advanced above. The first one is that the cells of the enveloping layer become more and more flattened during gastrulation (cf. Trinkaus, 1966). The second one is that the activity is very little affected by colchicine in concentrations that suppress cell division (Kessel, 1960). From this observation Trinkaus (1969) has drawn the conclusion that microtubules, the locus of attack for colchicine, cannot be involved in the epibolic movements. The validity of this reasoning may be questioned since the sensitivity towards colchicine appears to vary between different kinds of microtubules (cf. Stephens, 1968).

Epiboly in fish embryos has been studied by Devillers (1961) in *Salmo* and by Trinkaus (1966) in *Fundulus*. In the present context the discussion will be confined to the results obtained in the latter case. It has been established that the marginal cells of the enveloping layer show lively undulating membrane activity and that they are the only cells which adhere strongly to the substrate, the periblast. For this reason alone it appears justified to assert that they are responsible for the epibolic activity. The interesting observation was made that, probably as a result of tension, gaps may occur in the cell sheet. Under these circumstances the bordering cells may contrive to close the gap by means of undulating membranes or filopodia. Clearly this finding shows that both kinds of dynamic activity distinguishing sf-cells may occur, but that they become suppressed through cell contact.

It is worth noting that the cells of the enveloping layer cannot spread on the yolk

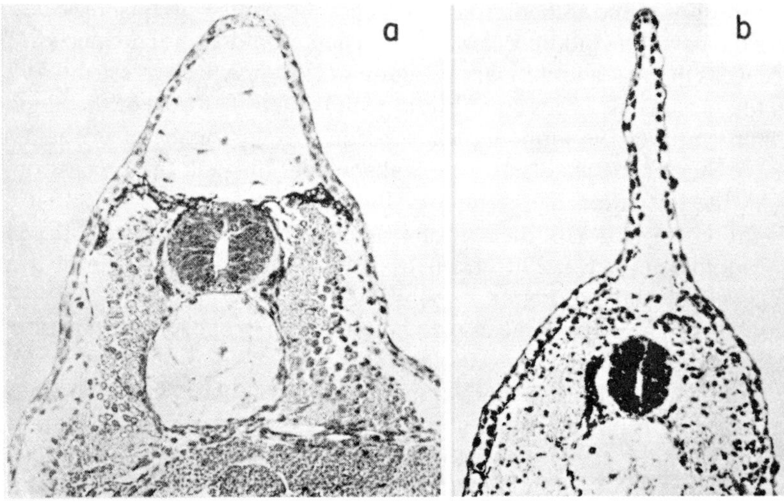

Figure 144. Dorsal fin in the amphibian larva. (a), abnormal fin obtained under experimental conditions; (b), normal fin. (Reproduced from V. C. Twitty and D. Bodenstein, *J. Exptl. Zool.*, **86** (1941), by permission of The Wistar Press, Philadelphia)

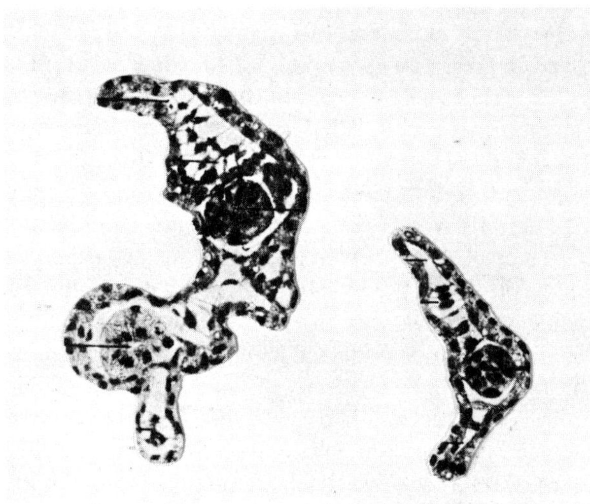

Figure 145. Cross sections through tail formation obtained under experimental conditions. The presence of sf-cells attached between the epidermal wall is indicated by lines. (Reproduced from H. Bytinski-Salz, *Arch. Entw.mech.*, **123** (1931), by permission of Springer, Berlin)

sphere surface beyond that underlain by the periblast. The latter is a syncytium present initially in the animal ectoplasm that has remained in the yolk sphere. Some time before the epiboly of the enveloping layer begins, this material starts to spread towards the vegetal pole in the thin ectoplasmic layer remaining in the yolk sphere. The mechanism involved in this nuclear migration is not known, but it is tempting to suggest that it is identical with the one proposed above as responsible for the nuclear translation in insect embryos, even though the participation of microtubules may be contra-indicated by the observations of Kessel (1960) referred to above. Electron microscopical evidence shows that the yolk sphere surface is substantially modified by the periblast, a fact which may account for the correlation between the epiboly of the enveloping layer and the spreading of the periblast (Lentz and Trinkaus, 1967).

If the epiboly involves stretching of the cells of the enveloping layer, it follows that these cells must be kept together by a force strong enough to withstand that responsible for the stretching. In this connection it may be mentioned that various kinds of contact specializations (close and tight junctions, desmosomes, etc.) gradually appear between the cells (Trinkaus and Lentz, 1967).

In the chick embryo also the marginal ectodermal cells of the blastoderm, constituting an enveloping layer, are actively engaged in the execution of the epibolic movements. However, some interesting dissimilarities prevail as compared with the phenomenon just described. Thus, for one thing it turns out that the inner surface of the egg membrane is the substrate on which the cells migrate (New, 1959). Secondly, it seems that in this case the activity is carried out by filopodia rather than by undulating membranes (Figure 146). Also in this case it was observed that the cells are kept together by contact specializations, desmosomes typically being rare before the third day of incubation (Bellairs, 1963). This change may, according to the propositions made above, indicate that the sc-transformation occurs in these external cells between the second and third day.

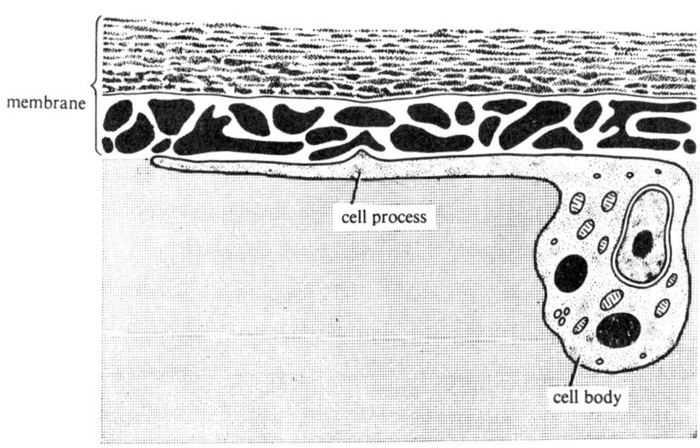

Figure 146. Diagram of a marginal cell in the chick blastoderm, showing the extension of a filoform cell process. (From R. Bellairs, *J. Embryol. Exptl. Morphol.*, **11** (1963), by permission of The Company of Biologists, Cambridge)

ACTIVITIES INVOLVING EXTRACELLULAR ELEMENTS

Extracellular elements, in the form of internal and external skeletons, cuticles and other coverings, as well as various forms of connective tissues, are of great importance for the shaping and the maintenance of the adult body. To the extent these elements are present in the embryo, they also exert such essentially static functions there, but over and above this contribution they are involved in a number of dynamic activities which largely are specific to the phase of embryogenesis. As development goes on, the number of extracellular elements gradually increases, but in a treatise on the early development the discussion may be limited to the four elements dealt with above, namely the hyaline membrane, oncotically active mucopolysaccharides, collagen fibrils and mineralized structures. In the following pages we shall discuss some of the mechanisms in which these elements are involved.

Hyaline membrane

The cohesion between the blastomeres is very weak in most embryos. One way to preserve the fixed spatial relation between the cells which is necessary for the orderly progression of embryogenesis consists in the transformation of the original sl-cells into colligocytes. This expedient involves a change from highly to slightly motile cells, a circumstance which greatly reduces the repertoire of possible embryogenetic processes. This disadvantage can be circumvented if amoebocytes, rather than being transformed, are kept together by the hyaline membrane in the form of a two-dimensional aggregate. This event is, in most embryos at least, the precondition for the formation of a coeloblastula and it is an absolute requirement for the execution of epibolic movements by sl-cells. Owing to the toughness of the hyaline membranes the cell aggregate may be subjected to more extensive topological transformations than are otherwise encountered in two-dimensional cell aggregates. Here we need only mention gastrulation and coelom formation in Echinoidea and gastrulation and neural tube formation in Amphibia. In the absence of a hyaline membrane the latter process cannot take place through folding of the neural plate. Rather, as observed in *Branchiostoma* and in Pisces, the neural plate is overgrown, through epiauxesis, by ectodermal cells. Subsequently, when the neural plate cells are enclosed in the body, surrounded by colligocyte aggregates at all sides, they reorganize themselves by lamination to form a tubular aggregate.

Oncotically active mucopolysaccharides

Binding of large quantities of water, associated with a volume increase by oncotic swelling, seems to be a property shared by all mucopolysaccharides, even if variations may obtain as regards the quantities absorbed. Oncotic and osmotic swelling bear many similarities, but the fact that a 2 per cent solution of hyaluronic acid binds almost 1000 times more water than may be anticipated from its molecular weight (Ogston, 1966) shows clearly that distinct differences obtain. The physico-chemical basis of this disparity remains to be elucidated.

Swelling has early been recognized as an important embryogenetic mechanism, having usually been considered as an osmotic phenomenon. It is only in recent years, which have witnessed important progress in our understanding of the properties of mucopolysaccharides, that the notion has become prevalent that the embryogenetic processes must be oncotic rather than osmotic events. From a morphogenetic point of view this difference is most important. Thus, as we have already discussed, a cavity filled with water-soaked mucopolysaccharide constitutes a *morphogenetic space*, open to the activities of filopodia extended by sf-cells. If, on the other hand, cells of this type are exposed to a fluid medium, they will generally flatten out on the substrate, and the formation of pseudopodia is suppressed. It is not only the filopodia of sf-cells, but the cells proper, which may penetrate into these morphogenetic spaces, and the same holds for sl-cells. This circumstance allows for the accumulation of the cells to aggregates of various forms. Two different functions may thus be ascribed to the oncotically active extracellular mucopolysaccharides: through their uptake of water they are responsible for volume changes which, being locally confined, may contribute massively to the intricate shaping of the embryonic body. The fact that morphogenetic spaces are created at the same time entails that various other cellular activities are made possible which may or may not influence the external shape.

Oncotic swelling

The earliest occasion on which oncotic swelling is morphogenetically active in the embryo is during the formation of the blastocoele. As another example of the morphogenetic effect of oncosis in an enclosed cavity we may mention the swelling of the anterior part of the neural tube, resulting in the formation of the brain in Vertebrata. This process requires the presence of many myxocytes associated with a modest number of collagenocytes, since the latter, by forming a membrane around the cell aggregate, serve to prevent the swelling. As discussed above, there is in the amphibian embryo an

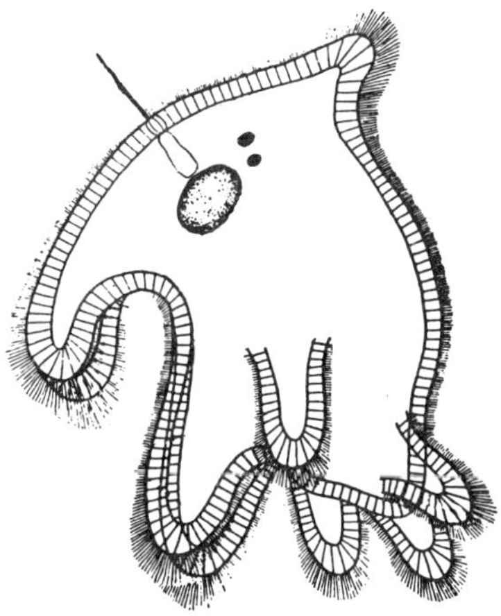

Figure 147. Müller's larva. (Reproduced from L. H. Hyman (1951), *The Invertebrates*, Vol. 2, after Kato, by permission of McGraw-Hill, New York)

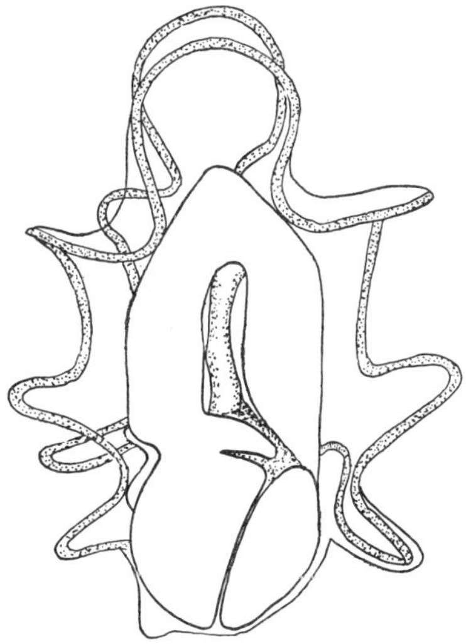

Figure 148. Bipinnaria larva of *Astropecten*, showing formation of lobes or arms along the ciliated bands (dotted). (Reproduced from L. H. Hyman (1955), *The Invertebrates*, Vol. 4, after Hörstadius, by permission of McGraw-Hill, New York)

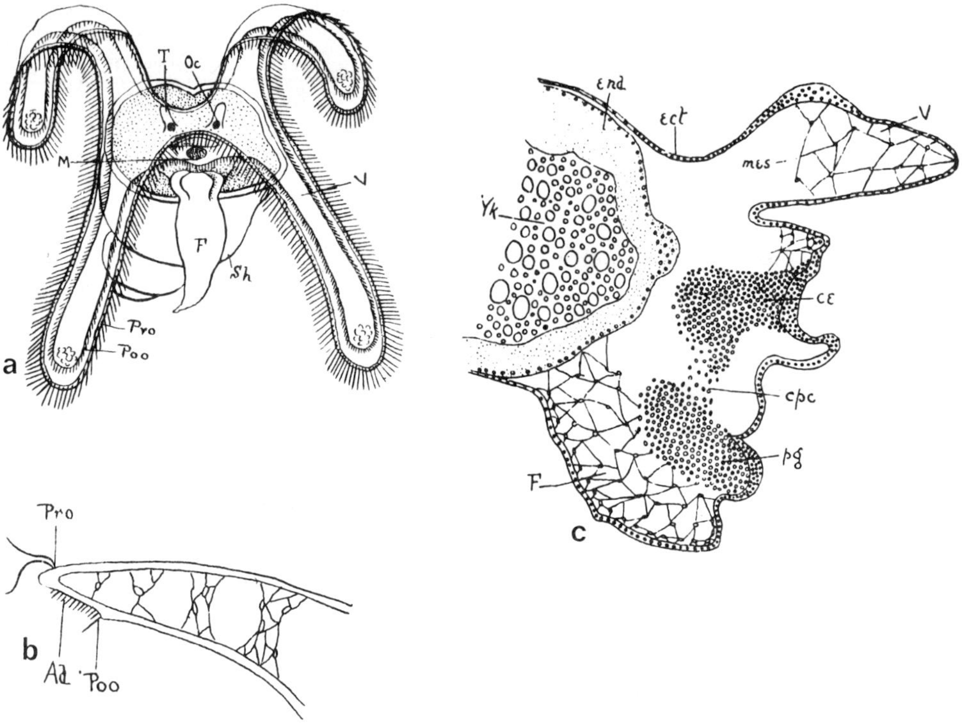

Figure 149. Morphogenetic activities of sf-cells in veliger larvae of Prosobranchia. (a), veliger larva; (b), transverse section through the velum; (c), sagital section through the cerebral and pedal ganglia of advanced embryo of *Fasciolaria*, showing that sf-cells also are involved in the formation of the foot. Ad, ad-oral cilia; ce, cerebral ganglion; cpc, cerebro-pedal commisure; ect, ectoderm; end, endoderm; F, foot; M, mouth; mes, mesoderm; Oc, eye; pg, pedal ganglion; Poo, post-oral cilia; Pro, pre-oral cilia; Sh, shell; T, tentacle; V, velum; Yk, yolk. (Reproduced from J. P. McMurrich, *Stud. Biol. Lab. Johns Hopkins Univ.*, **3** (1886))

antero–posterior variation in the distribution of these filocyte differentiation patterns corresponding with anticipation.

Apart from a few phyla, morphogenesis in the animal kingdom is characterized by outgrowths of various kinds. These may be larval structures like lobes in Müller's larva (Polycladida, Figure 147), arms in the bipinnaria larva (Asteroidea, Figure 148), velum in the veliger larva (Gastropoda, Figure 149) or tail and dorsal fins, balancers and gills, as observed for instance in Amphibia. They may also be adult structures like tentacles, fins, limbs, etc. It seems possible to propose that in all cases at least the initial stages in the formation of these anatomical entities depend upon the activity of myxoblasts, accumulated beneath the body wall. It is not possible to account for the mechanisms involved in all the cases mentioned, but some general principles still seem to obtain. Thus, it is quite typical that the larval lobes and arms, as well as the velum, are ciliated at the edges. It has been found that in the sea urchin embryo the skeleton-forming sf-cells accumulate beneath the 'ciliated' bands (Gustafson, 1961; Gustafson and Wolpert, 1967). If this rule has general validity for sf-cells, it follows that myxoblasts may also accumulate at these locations, which constitutes the first step towards the explanation of the morphogenetic mechanism.

It is easy enough to understand that a calcified skeleton may push out the body wall, as occurs in Echinoidea. It is more difficult to visualize how a plastic material like a mucopolysaccharide gel can accomplish a similar feat. The first prerequisite for this phenomenon to occur is that the pressure needed for the deformation of the ectodermal cell aggregate is counteracted by another pressure. The latter may be that of the body cavity, but another possibility is that the sf-cells with their filopodia span the area where the outpushing is to occur, a circumstance which may partly reduce the tension in the body wall and partly act to supply the necessary counterpressure. Once the initial bulging has been achieved, the sf-cells may assist in shaping the outgrowth by establishing connections between two sides of the ectoderm. The continued growth is not difficult to understand if, as suggested by Gustafson's observations, cells located at the edge or tip of the structure continue the activity dictated by their differentiation pattern.

The second aspect to account for is the fact that in some cases, e.g. arms and tentacles, the outgrowths are 'segmented', whereas in other cases, as represented by the velum in many Gastropoda, they are not. It may be suggested that 'segmentation' will prevail when a gradient obtains as regards the time at which the accumulated cells begin the elaboration of their differentiation products, but not if all the cells initiate their synthetic activity simultaneously. If this explanation is correct, there is one common feature, a diversification gradient, in this kind of 'segmentation', and that discussed previously in this chapter. As regards the proposed mechanism it may be pointed out that once an outpushing has become stabilized, for instance by sf-cells spanning its lumen, it will be impossible to develop enough force to push out the ectodermal aggregate at the base of the outgrowth, and this is obviously a requirement for the formation of an 'unsegmented' structure. I believe that the temporal sequence observed in the processes leading to the establishment of the morphological entities discussed here on the whole corroborates the suggestions advanced.

Turning now to the amphibian embryo, it will be realized that the cells responsible for the formation of the dorsal fin do not have to migrate to their seat of action; they are neural crest cells which become trapped between the ectoderm and the neural tube when the latter is formed. As regards the balancers and gills in the embryos of Caudata, Harrison (1921; 1925) clearly established that 'mesenchyme' cells are involved in their formation. The morphological difference between these structures, branching in gills, but not in balancers, may be explained by the fact that a basement lamella is present in the latter from their inception (Figure 150), whereas none is present in the developing gill.

In adult vertebrate structures like fins and limbs, it is evident that the morphogenetic role of the mucopolysaccharides is very soon overridden by other extracellular elements like collagen fibrils (cartilage) and calcified structures.

We have so far discussed the morphogenetic effect of the oncotic swelling of extracellular mucopolysaccharides. It may be mentioned that intracellular mucopolysaccharides after swelling confer upon the cells a certain mechanical strength which may be of morphogenetic significance. Thus, there can be no doubt that the biconic or flattened shape exhibited by many trochophore larvae may be referred to the shaping influence of the trochoblasts (Figure 151), and the same holds for the velum in many veliger larvae (Figure 149).

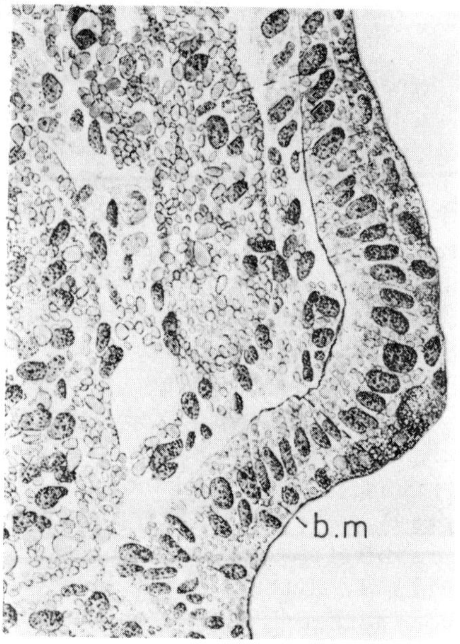

Figure 150. Section of an embryo of *Ambystoma*, showing an early stage of the balancer development. The presence of a basement lamella is clearly demonstrated. (Reproduced from R. G. Harrison, *J. Exptl. Zool.*, **41** (1925), by permission of The Wistar Press, Philadelphia)

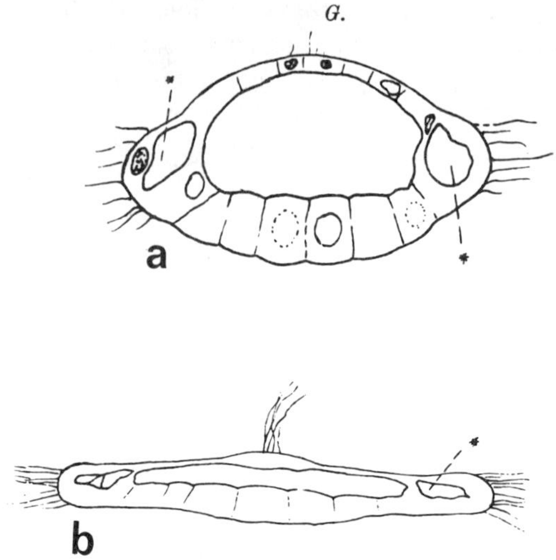

Figure 151. Blastula of *Polygordius*. The asterisks indicate the vacuolated cytoplasm of the swollen trochocytes. (a), 76 cell stage. The peculiar shape is best accounted for as a resultant of a pressure in the blastocoele and the rigidity of the ring of trochocytes; (b), 112 cell stage. The pressure in the blastocoele is evidently reduced relative to the force exerted by the trochocytes. (Reproduced from R. Woltereck, *Arch. Entw.mech.*, **18**, 1904, by permission of Springer, Berlin)

Cellular activities

The cellular activities going on in the morphogenetic spaces are, of course, of interest in this connection only insofar as they are made possible by the presence of mucopolysaccharide-filled cavities. We shall therefore not go into detail here, but give only a brief survey of the various processes which may be observed. Many examples have already been discussed, e.g. gastrulation and coelom formation in Echinoidea, stabilization of various outgrowths and shaping of the dorsal fin in Amphibia. In the acoelomate phyla the space between the digestive tube and the epidermis contains mucopolysaccharide. It is thus a morphogenetic space, and it may therefore be expected that signs of the activities of sf-cells and their differentiation products will be evident. As a matter of fact, the peculiar body forms found in many instances seem to depend upon activities of this kind (Figure 152). The diverticulation of the gut observed in many Tubellaria is another example of this activity (Figure 153). In contrast, the body cavity in the pseudocoelomate phyla contains a liquid medium, and in agreement with this the animals generally possess a simple worm-like shape and an undiverticulated gut.

A further morphogenetic process in which the participation of sf-cells may safely be invoked is the formation of the gastrovascular system in cydippid Ctenophora. Confining ourselves to the interradial and the meridional canals, it is seen that four of the former extend from the stomach into the body cavity, occupied by mucopolysaccharide.

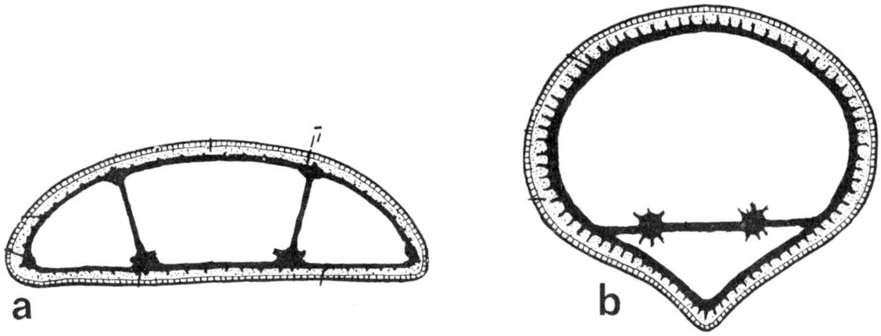

Figure 152. Body form in Turbellaria. (a), marine triclad; (b), land planarian. The dark lines indicate nervous connections, but it may be presumed that they are associated with connective tissue, the toughness of which contributes to the shaping of the body. (Reproduced from L. H. Hyman (1951), *The Invertebrates*, Vol. 2, by permission of McGraw-Hill, New York)

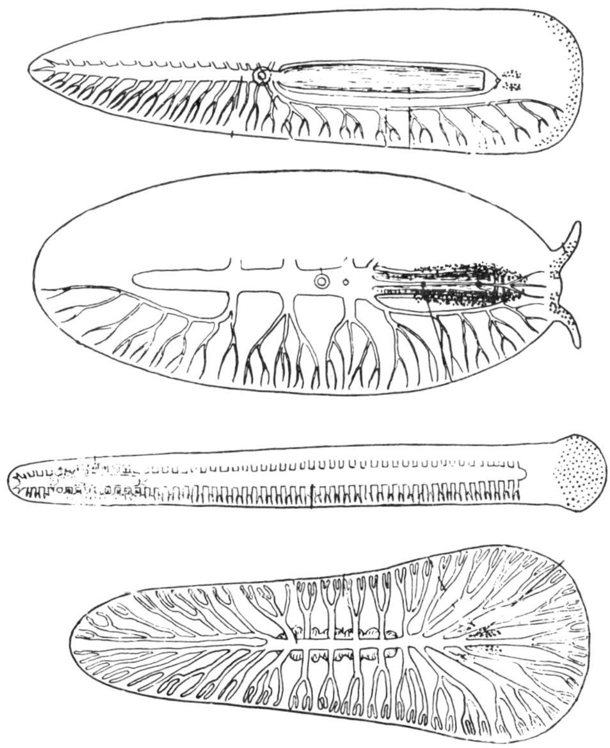

Figure 153. Gut diverticulation in various Polycladida. (Reproduced from L. H. Hyman (1951), *The Invertebrates*, Vol. 2, by permission of McGraw-Hill, New York)

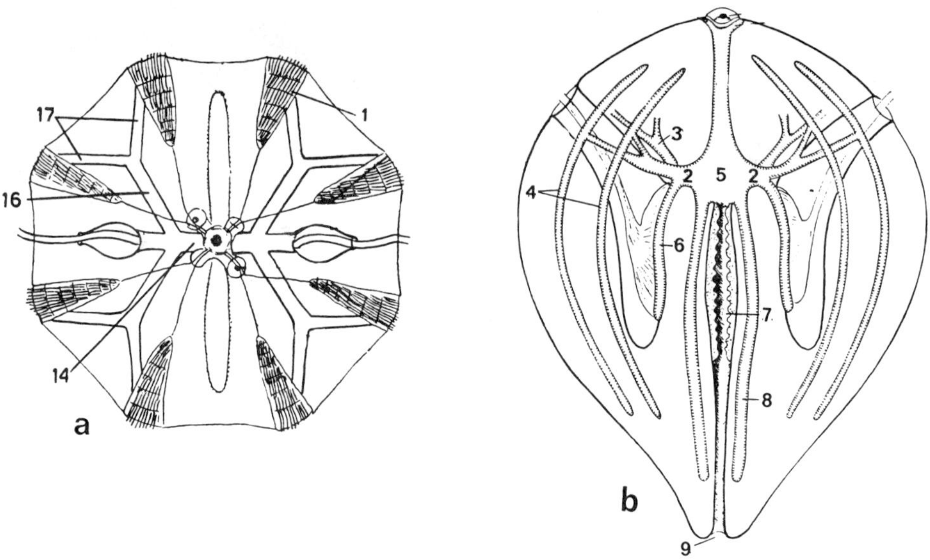

Figure 154. Cydippid ctenophores. (a), view from the aboral pole; (b), diagram of the digestive system. 1, comb row; 2, transverse canal; 3, interradial canal; 4, meridional canal; 5, stomach; 6, tentacular canal; 7, pharynx; 8, pharyngeal canal; 9, mouth. (Reproduced from L. H. Hyman (1940), *The Invertebrates*, Vol. 1, (b) after Chun, by permission of McGraw-Hill, New York)

Here they bifurcate, and end in the meridional canal, lying beneath the comb rows (cf. Figure 154a). Recalling that sf-cells show a particular preference for ciliated and flagellated cells, it may be assumed that they initially accomplish the formation of four outpouchings in the embryo, one for each of the comb rows then present. As the latter subsequently divide, the interradial canals follow suit, and with the growth of the animals the meridional canals are gradually elongated along the comb rows (Figure 154b). The morphology of this part of the gastrovascular system thus reveals in detail the course of the morphogenetic processes.

The morphogenetic space also allows for the accumulation and aggregation of sf-cells, with the ensuing formation of blood vessels, cartilage, etc., and of cl-cells, which, for instance, may give rise to pronephric tubules.

Collagen fibrils

Two separate differentiation patterns among the sclerocytes, the collagenocytes and the chondrocytes, respectively, are distinguished by their ability to produce collagen. The first step in this event involves the synthesis of tropocollagen molecules which, in the presence of a matrix containing sulphated mucopolysaccharides, may leave the cell and polymerize to form collagen fibrils. In the case of the collagenocytes, the matrix is generally supplied by colligocytes, with the result that two-dimensional entities, collagenous membranes, are formed at the surfaces of aggregates of this cell class. The chondrocytes produce their own matrix, chondroitin sulphate, a circumstance which has the consequence that, in contrast to the collagenocytes, they can produce three-dimensional structures, the substrate of which is cartilage.

Membranes

Although collagen is the protein most commonly engaged in membrane formation, its place can be taken by other fibrous proteins. Thus is has been observed by Pierce (1966) that various epithelia in the mouse embryo, in the absence of mesodermal cells, elaborate basement membrane *in vitro*, the proteins of which are immunologically different from collagen. Furthermore, the periodicity of collagen is lacking, and the chemical composition is unlike that of collagen, although hydroxyproline is present. The significance of these results remains to be elucidated.

Two kinds of collagen membrane exist, the basement membranes (basal or external lamina, cf. Fawcett, 1966) and the basement lamella. The former are present on the surfaces of colligocyte aggregates, the latter are found mainly in the dermis, notably in Vertebrata.

We shall discuss here first the formation of collagen membranes and subsequently their morphogenetic importance through participation in the stabilization and stretching of cell aggregates.

Formation

The origin of connective tissue membranes is a classical problem which in recent years has been subjected to renewed scrutiny. The traditional stand, and the one advocated above, is that these structures are the outcome of cooperation between different cell-types. It has been usual to describe this phenomenon as ectodermal–mesodermal or epithelio–mesenchymal interactions, but since it is easy to find examples which do not correspond to either of these expressions, it may be more appropriate, in accordance with the present terminology, to talk of colligocyte–collagenocyte interactions.

That cooperations of this kind do occur has been demonstrated in different contexts in the past; suffice it to mention here that the notochordal sheath is not formed in the absence of 'mesoblastic' cells (Holtfreter, 1939; Mookerjee, 1953). The formation of collagen fibrils from dermal fibroblasts has also been clearly demonstrated by Porter and Pappas (1959). On the whole, the evidence obtained recently does not contradict this view (cf. Edds, 1958), although some observations by Hay (1964) seem to indicate that even epithelial cells may produce collagen. On this point it should be noted that nothing appears to exclude that collagen-producing cf-cells exist, and these ought to be able to elaborate their own collagen membranes. It is not unlikely that the epidermal cells responsible for the collagenous cuticles in Nematoda and Annelida are of this type, but they may very well be found elsewhere.

The formation of ordinary basement membranes otherwise does not seem to pose any special problems. Their matrix is the layer of sulphated mucopolysaccharide proposed to cover the outside of colligocyte aggregates. Basement membranes are present solely at internal surfaces, these being the only ones accessible to the collagenocytes. The very fine collagen fibrils are apparently oriented at random (cf. Karrer, 1956; E. Yamada, 1960).

Departing from discovery of the orthogonal arrangement of the collagen fibrils in the basement lamella (Bytinski-Salz, 1938; Rosin, 1944), many studies have been devoted to this remarkable structure. It is now known that it consists of a number of plies or layers of fibrils, which may be up to about 50 nm in diameter. Each layer contains from one to several rows of parallel fibrils which are approximately perpendicular to those in the adjacent layers. In the outermost layer there is only one row of rather thin fibrils; the largest number of rows and the thickest fibrils are found in the intermediate layers (cf. Edds, 1964). Two problems have attracted particular attention, the location of the polymerization of the collagen fibrils, and the factors responsible for the crosswise arrangements of the layers. The circumstance that the innermost layer contains only one layer of thin fibrils has led to the view that this locus represents the place where the polymerization occurs. If the suggested cooperation between the two cell-types discussed above is correct, it entails that the tropocollagen molecules must reach the inner layer before they can participate in the formation of a new one. However, since these molecules can pass a membrane filter (Kallman and Grobstein, 1965), this proposition does not seem unlikely. That some molecules might be engaged on their way for the reinforcement of intermediate layers is also a natural consequence of this mechanism, which furthermore is supported by the isotope incorporation studies of Hay and Revel (1963). However, the orthogonal arrangement of the successive layers is not so easily explained. For one thing, it is known that the epidermal cells have some influence upon the orientation of the basement lamella (Porter, 1956b; Weiss and Ferris, 1956), an effect which obviously must be lost after the deposition of the first layer, whence it must be presumed that the latter and the succeeding ones take over the control, with the result that each new layer is perpendicular to the preceding one.

This problem has been completely revised through the recent advancement of the scindulene theory by Nadol, Gibbins and Porter (1969). In larvae and adults of *Fundulus*

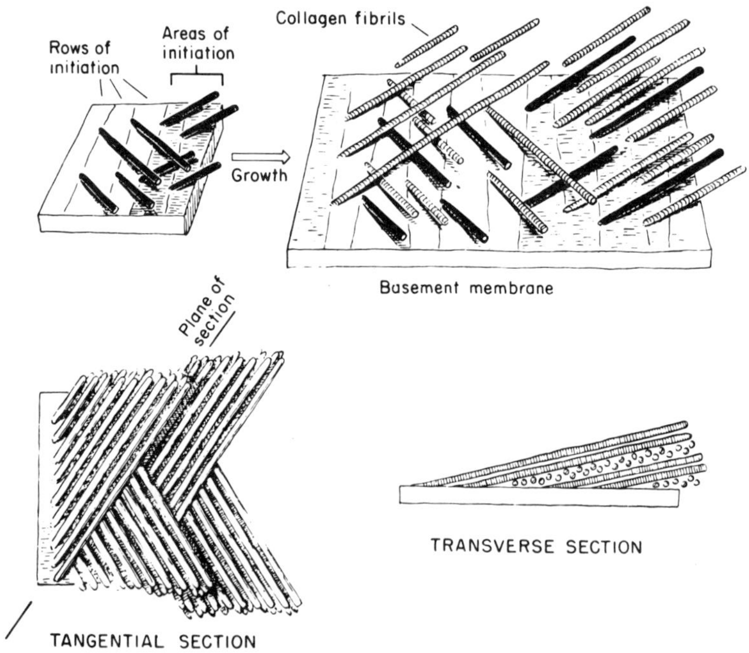

Figure 155. Basement lamella in *Fundulus*. The angle between the basement membrane and the collagen fibrils is exaggerated. (Reproduced from J. B. Nadol, J. R. Gibbins and K. R. Porter, *Develop. Biol.*, **20** (1969), by permission of Academic Press, New York)

these authors observed that, contrary to the prevailing opinion, the various layers are not parallel to the epithelial basement membrane; rather they form a slight angle, about 3°, with the consequence that the subsequent layers cover each other like shingles (Figure 155). One row of fibrils in each layer originates in the basement membrane, but with increasing distance from the point of attachment the average number of fibril rows increases. This finding is seen to concur well with the view that polymerization of collagen occurs at the inner end of the fibrils; in other words, no or very little transport of tropocollagen is required for the growth of the basement lamella.

The arrangement of the fibrils was found to be but quasi-orthogonal, the fibrils in two adjacent layers each forming an angle of 50–55° with the antero–posterior axis. As far as this arrangement is concerned, the model only requires that the first rows of fibrils are deposited crosswise successively along the axis. Since all the remaining ones are parallel to these, the latter may determine the orientation of the former, and the need for continuous switching is thus abolished. Nadol *et al.* (1969) are not very explicit about the mechanism by which the initial pattern is established, but they seem to suggest that the influence originates in the basement membrane. As a support for this contention they point out that the fine collagen fibrils in the latter also are arranged in an 'orthogonal' grid. It appears that this reasoning only carries the problem one step further back since, as we have seen, the fibrils in normal basement membranes are distributed at random. Furthermore, the fact that the same pattern is found in the admuscular lamella, which does not originate in the basement membrane, seems to suggest that a non-structural factor is involved, the most likely proposition being that some component of the dermal matrix is the responsible agent. In this context it is noteworthy that collagen fibrils under certain conditions, namely in the presence of 'a relatively large quantity of acid glycoprotein', may be precipitated spontaneously in an 'orthogonal' pattern (Gross, 1956, p. 265).

Nadol *et al.* (1969) point out a number of important and interesting implications of the scindulene model, and their enumeration is hardly exhaustive. The generality of these observations still remains to be established, but under any circumstances it is to be expected that they will turn out to be of great consequence for our understanding of various morphogenetic mechanisms.

In many animals successive layers of muscle cells are deposited upon the basement membrane or basement lamella of the body wall. It seems that the microstructure of the membrane, through the phenomenon of contact guidance, may act to orientate the muscle cells which, at the time they settle, are sf-cells. Since the epidermis apparently is responsible for the arrangement of the basement lamella, it follows that it also controls the alignment of the first layer of muscle cells. The fact that subsequent layers may be either perpendicular or diagonal to the preceding ones suggests that at least two different mechanisms may be involved in the orientation of muscle cells.

Cell-aggregate stabilization

It was observed by Holtfreter (1934) that if either ectoderm or endoderm from embryos of Amphibia is isolated free from mesoderm, the cells will first form an aggregate, kept together either by intercellular adhesion or by a hyaline membrane, but sooner or later the cells will disperse. The precondition for a stable epithelial organization of the cells, generally in the form of a vesicle, is the presence of a membrane, and the latter will be formed only if sf-cells are included in the isolate (Figure 132). It does not seem too preposterous to suggest that the membrane in question consists of collagen, produced by the sf-cells present.

The formation of collagenous basement membranes on the surface of colligocyte aggregates is a general phenomenon, observed in almost all animal phyla. Besides preventing dispersal of the cells, the membranes also may prevent the fusion of colligocyte aggregates. We have already seen that through this action membrane formation plays an important part in the mechanism of segmentation.

Stretching

Among Invertebrata a phenomenon of stretching must be involved in the morphogenesis of the elongated, worm-like animals found in several phyla. It will be realized that if the body wall at an early stage acquires a mechanical strength which opposes an increase of the cross-sectional area, then most of the growth must take place at one or both ends. The toughness need not be due to a basement membrane, a muscle layer would do just as well, but most often the former is present. It seems a sensible guess that this mechanism is at work in many cases, but whether other mechanisms are involved, or even possible, would require a detailed survey. In order to lend support to the suggested mechanism we may consider the correlation between presence of basement membrane and shape in the class Turbellaria. Great variation of body shape is encountered, from flattened, almost circular forms in certain Acoela to elongated, occasionally nearly cylindrical bodies in the other orders. Whereas in the latter a basement membrane is always present, it is lacking in many Acoela (cf. Hyman, 1951a).

In Chordata the stretching of the body is mainly the achievement of the notochord. This process is initiated by a swelling of the chordoblasts through uptake of water, a consequence of the fact that mucopolysaccharide, presumably a sulphated type (cf. Johnston and Comar, 1957), is accumulated in their cytoplasm (Figure 156). As the cells

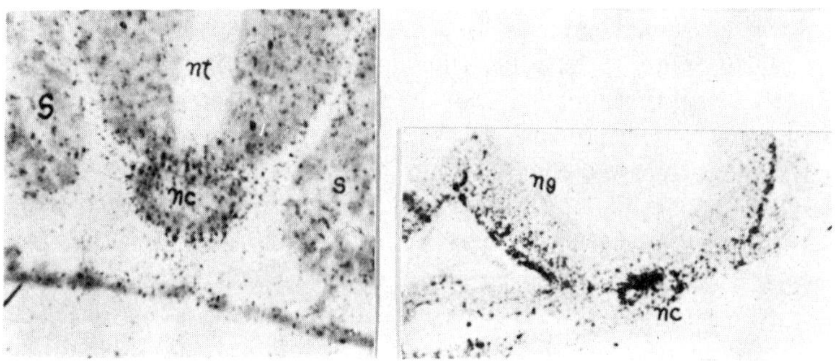

Figure 156. Incorporation of $^{35}SO_4$ in the chick embryo. nc, notochord; ng, neural groove; nt, neural tube; s, somites. (Reproduced from P. H. Johnston and C. L. Comar, *J. Biophys. Biochem. Cytol.*, **3** (1957), by permission of Rockefeller University Press, New York)

are confined within a collagenous sheath, the toughness of which only permits a moderate enlargement of the diameter, most of the volume change must be accommodated by stretching. This phenomenon is particularly extensive in the posterior region, where the notochordal sheath remains open and where uncommitted cells continue to differentiate. The morphogenetic outcome depends partly upon whether or not the notochord is anchored to the neural plate through sf-cells. In the former case (e.g. Amphibia), the stretching is almost exclusively directed towards the posterior end, and a large part of the neural plate will therefore become available for brain formation (Figure 157). If no

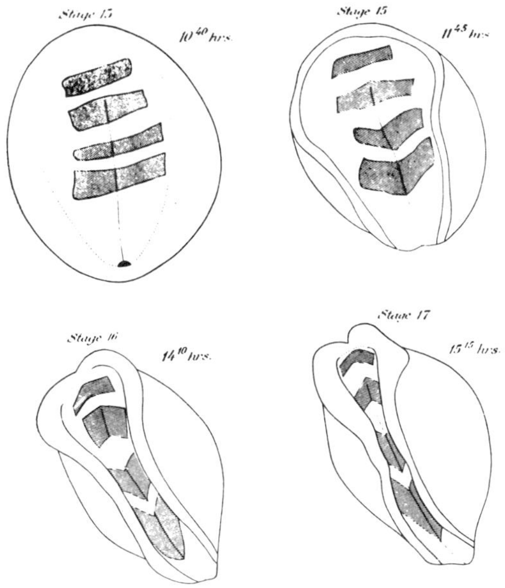

Figure 157. Stretching in the amphibian neural plate, as revealed by vital staining. (Reproduced from C.-O. Jacobson, *Zool. Bidrag, Uppsala*, **35** (1962), by permission of the author)

anchoring occurs, the notochord will extend also towards the anterior region, as in *Branchiostoma*, with the consequence that brain formation is more or less suppressed (cf. 1965a).

The notochordal stretching is normally responsible for tail formation, but under experimental conditions the latter has been observed to occur in the absence of a chorda, whose role has been taken over by the neural tube or even by myotomes (Bytinski-Salz, 1931). It is probably safe to assume that collagenocytes are present even in these cases, and that the membranes formed by them are instrumental in the observed stretching of the colligocyte aggregates. Other stretching phenomena, for instance that of the gut, may also be explained by this mechanism.

Cartilage

Owing to its mechanical properties, cartilage functions as an internal skeleton in many animals, but it is generally formed at relatively late stages of embryogenesis. For that reason it is not of great importance in the present context, and it therefore suffices to bring up a few salient points. As discussed in the previous chapter, cartilage is produced by sf-cells, which have acquired what may be considered two separate differentiation patterns, involving the synthesis of collagen and chondroitin sulphate, respectively. As sf-cells they may penetrate into morphogenetic spaces, where they will accumulate at statistically preferred locations. In the mechanism of accumulation the phenomenon of contact inhibition may be involved, but it is possible that the sulphated mucopolysaccharide also plays a role. The shape assumed by the cells after chondrification has begun suggests that the mechanical properties of the matrix have been changed to the extent that formation of filopodia is no longer possible.

Chondrified structures may be 'segmented', as shown by the vertebral column and the fins or limbs in various Vertebrata. In the former case the segmentation may be traced back to that of the somites, but the latter examples suggest that temporal gradients in the accumulation or differentiation of the chondrocytes may result in a process of 'segmentation'.

Calcified tissues

It holds for calcified tissues as for cartilage, that usually they are formed at a late ontogenetic stage, being therefore of little importance for embryogenesis. There are, however, some well-known cases where calcification, at least the initial phases of the process, begins so early as to influence extensively the larval shape, and for that reason a discussion of this morphogenetic element cannot be omitted.

Both calcified tissues and cartilage are employed in the formation of internal skeletons. The difference between the two elements as far as mechanical properties are concerned is that cartilage may possess some plasticity, whereas calcification leads to complete rigidity. Calcified structures are thus able to withstand much greater loads. Calcified tissues may also prevail in the form of exoskeletons like shells, etc.

It is a particular pleasure to deal with this subject, because it is one of the few biological topics for which a solid theoretical foundation has been created (cf. Moss, 1964). The general theory advanced by Moss will be briefly summarized in the first section. Subsequently, we shall discuss two specific examples of calcification, namely in the embryos of Echinoidea and Gastropoda.

General theory of biological calcification

The theory of Moss (1964) has been stated in the form of six postulates which are presented here, slightly condensed and with a few terminological changes:

(1) All biological calcifications are cellular processes, depending on cells belonging to the family of sclerocytes. Within this group various differentiation patterns may be distinguished.

(2) All biological calcifications consist of an organic matrix and an inorganic phase. The former is elaborated first, by sclerocytes, whereas the latter is formed in or on the preexistent organic phase.

(3) The organic phase has two components: a fibrous, organized phase and an amorphous phase; the latter is usually a viscous, highly hydrated, polyanionic colloid (sulphated mucopolysaccharide).

(4) The form of the calcified tissue is determined by the organic phase.

(5) Calcifiable matrix fibres are composed of complex polymers of proteins and carbohydrates. Classified by their principal components, it is possible to distinguish a collagen and a chitin group. In the former protein is predominant, in the latter carbohydrate. The chitin group is found only in Invertebrata, the collagen group in many phyla throughout the animal kingdom.

(6) Mineralization is not a simple chemical precipitation. Presumably the organic matrix, in providing unique spatial and electrochemical conditions, may initiate the nucleation of a particular type and species of crystal, and determine the primary crystallographic axes (cf. Glimcher, 1960).

A few comments are required. As regards the first postulate, it is seen that, according to the cell-class concept, scleroblasts are filocytes, being either sf- or cf-cells. That several distinct differentiation patterns of scleroblasts exist is demonstrated by the fact that different kinds of calcified tissue (e.g. bone, dentin and enamel) and uncalcified matrices may prevail simultaneously in an organism.

Moss (1964) points out that in certain lower invertebrates (Porifera and Echinodermata) cytoplasmic fibres may function as an intracellular matrix, in contrast to all higher phyla in which it is always extracellular.

Skeleton formation in Echinoidea

The morphogenetic importance of skeleton formation in the sea urchin embryo has been realized long ago (cf. von Ubisch, 1937; 1950). In recent years, however, studies by Okazaki (1960; 1965) and by Gustafson and Wolpert (1961a) have revealed the mechanism involved in fascinating detail.

These authors observed that the precipitation of the straight rods of the calcium carbonate skeleton occurs in cable-like syncytia formed by the filopodia of the primary mesenchyme cells. However, skeleton formation is first evidenced by the appearance of triradial crystals in the dense clusters of cells located where the ventro–lateral branches of mesenchyme cells meet those of the ring inside the lower part of the ectoderm (Figure 158). It appears that at this location a syncytium prevails in which the crystal is formed.

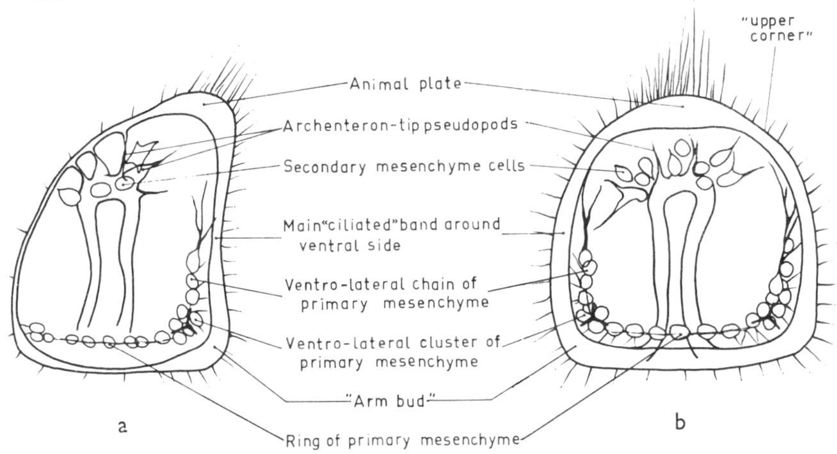

Figure 158. Schematic outline of an advanced gastrula of *Psammechinus*, showing the distribution of the primary mesenchyme cells and their relation to the main flagellated bands. (a), lateral view; (b), ventral view. (Reproduced from T. Gustafson, *Exptl. Cell Res.*, **32** (1963), by permission of Academic Press, New York)

The possibility that three linear syncytia meeting in one point are responsible for the crystal form is rejected by Gustafson and Wolpert (1961a), who assert that their shape has some 'crystallographic' basis. This contention is supported by the fact that the initial orientation of the crystals is fortuitous. Reorientation occurs, however, by the time the radii are 5µm long. This event is accomplished by the tension developed by filopodial syncytia originating from mesenchyme cells in the main ring and in the flagellated bands. These syncytia, it appears, become attached to the end of the crystal radii. As a consequence, one radius becomes orientated approximately parallel to a flagellated band, and one to the part of the ring lying between the bands. The direction of the third radius is hereby determined, and it is obvious that for purely geometrical reasons it cannot be parallel to the dorsal branch of the ring, even though a slight slanting of the crystal reveals the effect of the syncytium from this source (Figure 159).

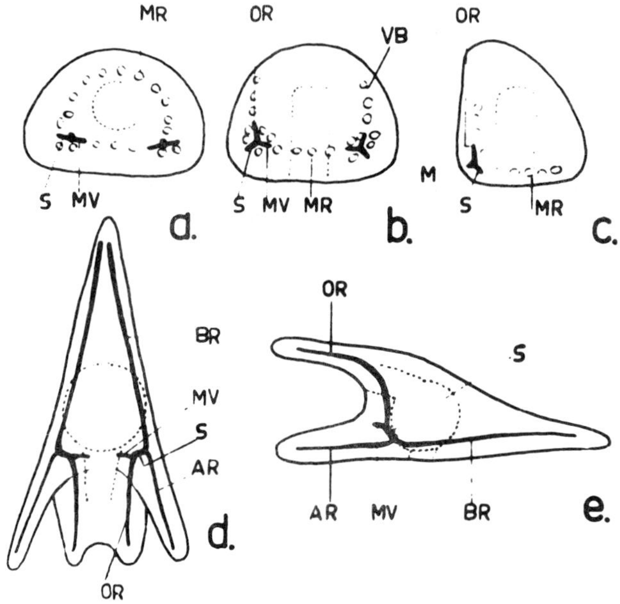

Figure 159. The main stages in the development of the skeleton in *Psammechinus*. (a), animal; (b), ventral; and (c), lateral view of the gastrula, showing the primary mesenchyme cells and the initial triradiate skeleton after reorientation. (d), animal and (e), lateral view of the pluteus larva. In (b) the three radii of the crystal are seen to correspond to the rudiments of the oral rod (OR), the mid-ventral rod (MV) and an anonymous segment (S). The latter bends backward in the pluteus, forming the body rod (BR). The anal rod (AR) branches off at the bending site. The two mid-ventral rods subsequently meet and fuse in the median plane. MR, main ring; VB, ventral branches of mesenchyme cells. (Reproduced from L. Wolpert and T. Gustafson, *Exptl. Cell Res.*, **25** (1961), by permission of Academic Press, New York)

In the subsequent accretional growth of the skeleton three different patterns may be distinguished: branching, bending and straight growth. According to the principles laid down by Okazaki (1960; 1965) and Wolpert and Gustafson (1961a), the two former phenomena are directed by the syncytial cables. Thus, whenever these are present ahead of the growing skeleton, their orientation determines that of the latter, otherwise it will grow straight, pushing out and remodelling the wall of the ectoderm. It appears that by applying these relatively simple principles it is possible to account for all the essential aspects of the anatomy of the pluteus larva, as will be shown in a later chapter. The importance of microtubules for skeleton formation is brought out by the fact that in *Psammechinus* a more complicated skeletal pattern develops when the filopodial activity is stimulated experimentally (von Ubisch, 1950; Gustafson, 1969).

Skeletal formation is suppressed in the presence of fluoride, in which case protuberances appear at the sites where arm formation would normally occur (Young, 1958). This shows that the distribution of the sf-cells is completely independent of their differentiation pattern, and that normal morphogenesis is dependent upon the latter. The protuberances may result from oncotic swelling, suggesting the presence of mucopolysaccharide-producing cells.

Gustafson and Wolpert (1961a) observed that a plug of cells is always attached to the tip of the skeleton in a growing arm, and that in their absence the growth process comes to a standstill. This demonstrates that the accretional crystal growth is under biological control. It was found that pigment cells may be present among the accumulated cells, indicating that there is only cell-order specificity with respect to adhesion to the crystalline skeleton.

Shell formation in Mollusca

Shell formation in Mollusca constitutes another example of the morphogenetic influence of calcification. What is of special interest in this connection is, of course, its effect upon the phenomenon of *torsion.* This process comprises two aspects, viz., the *ventral intestinal flexure*, by which the two ends of the digestive tract are approached, and the *lateral torsion*, which entails a 180° rotation of the shell and the viscera relative to the foot and the head (Figure 160).

Intestinal flexure occurs in Gastropoda, Scaphododa, Bivalvia and Cephalopoda, and may be referred to the growth of the mantle, with or without shell formation, around the body of the embryo. Since the growth mainly takes place at the posterior edge of the mantle, which is situated just behind the anal opening, it follows that the latter must be displaced anteriorly together with the mantle. The absence of intestinal flexure in Aplacophora, Monoplacophora and Polyplacophora may be accounted for by reference to the shape of mantle and shell, when present, in these classes.

The next step, the torsion proper, is known to occur in Gastropoda and Scaphopoda. This is an asymmetrical phenomenon, which requires a dextro–sinistral polarity. It has already been pointed out that the basis for this mechanism may be traced back to the spiral cleavage pattern, which explains its absence in Cephalopoda. It was shown that the descendants of the micromere quartets determine two different planes of bilateral symmetry, which formally corresponds to the existence of a dextro–sinistral polarity. The polarity established by the quartets of high order is in the first place determined by the unique diversification patterns of the descendants of the somatoblasts, particularly those of 4d. Most often they are sf-cells which enter the blastocoele and the possibility therefore obtains that their distribution may be directed by the already existing bilateral symmetry. Presumably this phenomenon may account for most of the cases where spiral cleavage is associated with bilateral symmetry.

A different situation obtains when a shell is involved. The shell gland is not derived from 4d, but rather from 2d. However, it has been shown that this gland, consisting

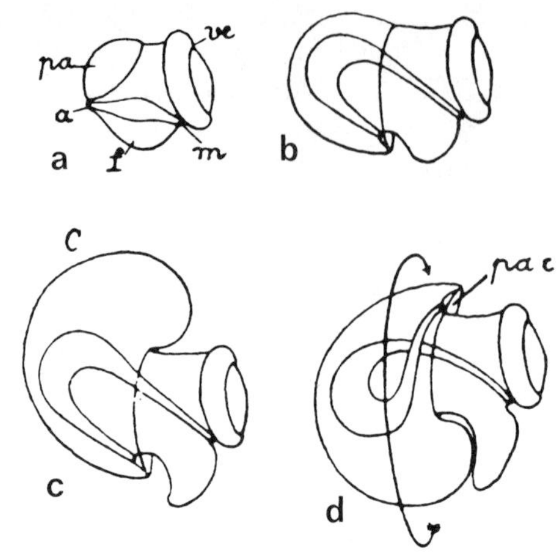

Figure 160. Torsion in Gastropoda. (a), embryo without, (b), embryo with, ventral intestinal flexure; (c), embryo with ventral flexure and exogastric shell; (d), embryo with lateral torsion and endogastric shell; the arrow indicates the direction of the lateral torsion. a, anus; f, foot; m, mouth; pa, mantle; pa.c, pallial cavity; ve, velum. (Reproduced from P. Pelseneer (1906), in *A Treatise on Zoology* (Ed. E. R. Lankester) after Robert, by permission of A. and C. Black, London)

presumably of cf-cells, is induced by endodermal cells (cf. Figure 127). Because of the asymmetrical location of the digestive tract relative to the symmetry plane determined by the first quartet cells, the shell gland will also be asymmetrical, and in this case there is no possibility of adjustment. As some muscles are attached to the shell, the internal anatomy likewise becomes asymmetrical, and there is some indication that this circumstance is part of the mechanism responsible for the torsion, certain muscles being instrumental in this process (cf. Eales, 1950; Crofts, 1955). It seems clear that if two shells are formed, as in Bivalvia, then the physical basis for torsion is lacking, even though asymmetry may obtain with respect to the size of the shells. The conclusion is thus that torsion requires that the growth of the shell gland leads to ventral flexure, and that the shell is asymmetrical relative to the median plane of the remaining body.

Interactions

The morphogenetic processes are executed by substrate elements which owe their origin to the activity of the embryogenetic processes of cell division and cell diversification. Intimate interactions thus obtain even in this case, which it is the task of analytical embryology to entangle. However, at this level the possibilities of reciprocal influence are so many that it is impossible, for the time being at least, to survey all combinations. The most which can be claimed for the following discussion is that it provides examples of modes of interaction.

EMBRYOGENETIC SUBSTRATE AND MORPHOGENETIC PROCESSES

As the various substrate elements arise through the activity of the processes of cell division and cell diversification, it is hardly correct to include them in a consideration of the influence of the substrate on the morphogenetic process. We shall therefore confine the discussion to the question of the effect of the initial embryogenetic substrate, i.e. the fertilized egg, on the morphogenetic processes.

The egg envelopes may influence the morphogenetic outcome by determining the shape of the embryo, a feature which is particularly striking in those instances where their shapes deviate from that of a sphere, but even in the latter case an influence may be exerted. The envelopes may also be instrumental in imposing polarities on the egg, either through the presence of a micropyle or by possessing bilateral symmetry.

The hyaline membrane is of morphogenetic importance insofar as this supracellular structure generally is the precondition for the formation of a coeloblastula. It is also indispensable for carrying out gastrulation whenever it occurs by invagination, as for instance in Echinodermata, and by epiboly, as in Amphibia.

The size of the egg is also of significance for the course of morphogenesis. This effect may be exerted on many different events, but we may once more use as an example the process of gastrulation. When eggs are small, gastrulation may occur through invagination executed by filopodia, as observed in the example just mentioned and in many other small eggs. If the egg size increases through deposition of yolk, a point is reached where the filopodia of the sf-cells are neither long nor strong enough to pull the endoderm through the blastocoele. In the amphibian egg the sf-cells are still responsible for the initiation of the process, by pulling membrane-bound cells at the dorsal lip into the interior of the egg where they establish contact with other membrane-bound cells. By this means the apposed layers of sl-cells can carry out the epibolic migrations through which gastrulation is accomplished. When the egg becomes still larger, as in Cephalopoda and in many Vertebrata, this mechanism is no longer possible, and gastrulation becomes to a large extent a matter of ingression of individual cells.

The polarities of the egg may also influence the morphogenetic processes. A comparison between the gastrulation in Echinodermata and in Amphibia may serve as an illustration, the difference in the geometrical location of the gastrulation process being directly correlated with the polarity responsible for the formation of the sf-cells which carry out or initiate the event.

CELL DIVISION AND MORPHOGENETIC PROCESSES

The contribution of cell division to morphogenesis comprises several aspects. Some of these are obvious and will only be mentioned in passing, viz., the subdivision of the original substrate which is the precondition for any morphological change, and the rise in level of organization which follows from this subdivision. There are two other aspects which must be dealt with in more detail, namely, the influence of the quantitative side of cell division, i.e. cell multiplication proper, and of the spatial location of the cells, as represented by the cleavage pattern, upon the morphogenetic processes.

Cell multiplication

There can be no doubt that growth is an important aspect of almost all morphogenetic processes, since they are generally associated with increases in mass. Unfortunately, much ambiguity prevails regarding the specific contribution of this phenomenon, owing to the fact that there is no general agreement concerning the definition of the concept of 'growth'. We shall begin the discussion by proposing a generally applicable definition. Subsequently, we shall analyse briefly the interaction between morphogenesis and cell division as well as the phenomenon of allometric or differential growth.

Growth

Growth in Metazoa is most easily measured by following the increase in weight, volume or length of the body. These changes may ensue as the result of an increase in cell number, in cell size or in the amount of extracellular elements. The third possibility definitely represents a case of cell differentiation. If, in the second case, we exclude the temporary enlargement preceding each cell division, it is probably correct to state that increase in cell size depends upon the elaboration and intracellular deposition of differentiation products, and thus this phenomenon is also in the first place a differentiation process.

Although the paramenters generally employed to estimate the process of growth may be convenient for practical purposes, it is thus quite obvious that they are very unsatisfactory from a theoretical point of view. This stems above all from the fact that for a thorough analysis of embryogenesis it is necessary to distinguish between growth and differentiation. Under such conditions it is evidently inconsequent to measure growth by means of parameters which are influenced by the process of differentiation. The only possibility of circumventing this difficulty is by equating growth with cell multiplication (1959d). Thus, a metazoan organism and, incidentally, cultures of cells and microorganisms as well, are subjected to growth as long as the number of cells increase. No allowance is made for reversibility; if cell degeneration occurs as part of the normal ontogenetic process or for other reasons, it is not classified as 'degrowth'. In the present context this occurrence will be treated as a differentiation phenomenon.

Regrettably enough, it is often difficult to apply this definition in practice, where it may be necessary to make use of traditional parameters. However, in small organisms, not the least in embryos, the cell number may often conveniently be estimated by determination of the total DNA content, as suggested by Davidson and Leslie (1950). Depending upon the method used for the analysis, the reserves for DNA synthesis found in many embryos may interfere with this approach, but this is not a serious drawback since *de novo* synthesis always begins at a relatively early stage of development.

That the process of growth, as defined here, is of great importance for 'growth' in the wider sense of this word is evidenced by the fact that the variation of body size in the animal kingdom exceeds the variation of cell size by many orders of magnitude, implying that the number of cells in large animals is vastly greater than that in small ones. Clearly, part of the mechanism answering for animal size involves some kind of control of growth through cell multiplication.

Growth and morphogenesis

As pointed out above, the very subdivision of the substance of the egg is an unconditional prerequisite for the occurrence of morphogenetic processes. This does not, however, imply that cell multiplication as such is instrumental in morphogenesis. This supposition has often been made in the past, but the idea that cellular attributes like motility and adhesion are involved was also expressed. Many of the mechanisms discussed in the present chapter corroborate the second alternative, and Townes and Holtfreter have taken sides in this issue to the extent that they stated that it 'is an anachronistic trend of many embryologists to speak of ingrowth, outgrowth or overgrowth when referring to gastrulation movements' (1955, p. 93). This view can be accepted for all those cases where gastrulation is accomplished through cellular motility, particularly since cells in the act of division generally temporarily lose this ability. However, as pointed out above, spreading of a cell layer may in some cases also take place when the slightly motile colligocytes are involved. This phenomenon, epiauxesis, which seems to account for gastrulation in many animals, is directly dependent upon the occurrence of cell division. It must thus be admitted that this event may play an important role in many morphogenetic processes, even though it unquestionably is less important than has been formerly believed.

Allometric growth

The rate of 'growth' may vary considerably in different parts of the body, e.g. limbs versus trunk. It has been found, however, that a certain mathematical relationship may obtain between the various growth processes (Thompson, 1942; J. S. Huxley, 1932). There is no doubt that the cell number is of importance for this so-called *allometric growth*, but it seems evident that differentiation processes, such as ossification, are of even greater influence, and it may therefore also be characterized as *differential growth*.

It was pointed out by Thompson (1942) that much of the variation of body shape observed within various classes and phyla may be accounted for as modifications of a basic form accomplished by differential growth. On the whole this activity occurs at rather late stages of ontogenesis, and it is therefore appropriate to postpone the further discussion of this question to a later context.

Cleavage patterns

A great many examples may be quoted to illustrate the influence of the cleavage pattern upon the morphogenetic processes. Since this question has been discussed in detail above, it may suffice to mention a few only. Thus, the biradial symmetry in Ctenophora may be related directly to the fact that eight macromeres are formed, lying in one plane. The bilateral asymmetry in Gastropoda ultimately depends upon the pattern of spiral cleavage. The loss of this cleavage pattern suffices to explain the absence of bilateral asymmetry in shell-bearing Cephalopoda. Finally, it was pointed out that the small number of cells formed from the first quartet in Bivalvia, as compared with other Mollusca, can be directly correlated with the primitive development of the head region in the former class.

CELL DIVERSIFICATION AND MORPHOGENETIC PROCESSES

Cell diversification and morphogenetic processes interact reciprocally in many different ways. A selection of examples which, it is hoped, is representative, will be discussed in the present section. The effects of cell transformation and cell differentiation are so different that special treatment is warranted. The morphogenetic significance of cell degeneration will also be included in the present section.

Cell transformation

We have already dealt with the participation of members of the different cell orders in the morphogenetic mechanisms which cooperate in the execution of the various morphogenetic processes. The cell transformations are important in this connection because they are responsible for making new cell-types available. Both the location and the time of transformation are of decisive importance.

The first point may be illustrated by two examples concerning gastrulation mechanisms. In the sea urchin embryo, the transformation of sl-cells to sf-cells begins at the vegetative pole and, as described above, this event forms the basis of the gastrulation. If the lf-transformation occurred at random everywhere in the wall of the blastula, the prevailing mechanism of gastrulation would not be possible. The fact that the gastrula becomes bilaterally symmetrical depends, as we have seen, on the fact that the filopodia cannot attach to the layer of sulphated mucopolysaccharide covering the inner surface of the apical plate cells, a circumstance which presumably is due to the transformation of these cells into colligocytes. This polarity conversion is thus the outcome of cooperation of two cell transformations and a morphogenetic process. As regards the importance of the time of transformation, it may be mentioned that if the apical cells do not transform in time, the sea urchin gastrula becomes radially symmetrical. This situation has been realized under certain experimental conditions.

In Amphibia the lf-transformation occurs at the dorsal margin. If it took place at either the vegetal or the animal pole, the gastrulation mechanism involving epibolic movements between two layers of sl-cells could not be realized. If the ectodermal cells in the amphibian embryo became transformed to colligocytes at an early stage, epiboly would be impossible. It may also be mentioned that transformation along a temporal gradient is the precondition for certain kinds of segmentation.

Interactions are of particular importance in eggs where induction of transformation is preponderant, i.e. in 'indeterminate' eggs. The classical example of this mechanism is, of course, the induction of the embryonic axis in the amphibian embryo. In this case a great number of interactions, inductive as well as morphogenetic, are initiated by the presumptive notochordal cells. As we have discussed above, the initial effect is the induction of the lf-transformation in the neural ectoderm, and this does not occur unless sl-cells from the dorsal lip through the epibolic movements are brought in contact with the cells in the dorsal median region of the neural ectoderm. The examples discussed here illustrate the reciprocity as regards interactions between cell transformations and morphogenetic processes.

Cell differentiation

The pattern, the time and the location of the differentiation process are important for the course of morphogenesis. As an example of the influence of the former we may mention the difference between the bipinnaria and the pluteus larva in Echinodermata, which may be referred to the fact that the differentiation pattern involving calcification is represented by some of the mesenchyme cells in the latter. As concerns Vertebrata, it is easy to visualize the importance of the differentiation patterns assumed by the chordocytes and the sheath cells, the former producing water-binding mucopolysaccharides, the latter collagen. Both of these are instrumental in the stretching of the embryonic body, which in turn is of importance for a great number of morphogenetic processes. It is true that some stretching may occur if the place of the chordocytes is taken over by neurocytes or myocytes, but these cells do not ensure the same degree of swelling. The sheath cells can hardly be replaced by cells of any other differentiation type. The influence of the various sclerocyte differentiation patterns on the morphogenesis of the larva and on the morphology in the adult is too well known to necessitate further comments.

As regards the temporal sequence of the differentiation patterns, attention may be directed towards the neural crest cells in Amphibia. It has been pointed out that the succession in which the various differentiation patterns appear (collagenocytes, neurocytes, myxocytes) corresponds to their place in the hierarchy. However, for the ectodermal filocytes (chordocytes, neurocytes) the order has been reversed. It requires little imagination to visualize the importance for morphogenesis of the fact that these differentiations occur in the proper temporal sequence.

As far as the spatial location is concerned, two examples may be quoted. First, it was seen that the asymmetrical position of the shell gland in Gastropoda is of decisive importance for the torsion and bilateral asymmetry of the body. Second, the large number of myxocytes in the anterior part of the neural tube is instrumental in the swelling associated with the formation of the brain in vertebrates.

The differentiation products discussed so far all belong to the group classified as structural, i.e. substances which directly participate in the morphogenetic processes. Among the remaining classes of differentiation products those with regulatory activity, such as hormones, may, through their impact upon the processes of cell division and cell diversification, profoundly affect the morphogenetic outcome. Mechanisms of this kind are of particular importance during late ontogenetic stages, and it may therefore be justified to omit a discussion of their function at the present stage. We must, however, take up this question in the last chapter of the book.

Morphogenetic processes may also affect cell differentiation. To the extent that this process is induced, morphogenetic processes may act to bring competent cells in contact. It has also been pointed out that the differentiation pattern assumed by the chordocytes, involving mucopolysaccharide production, may be accounted for, on the basis of the exhaustion hypothesis, by the lively activity displayed by these cells before they undergo differentiation.

Cell degeneration

Cell degeneration is an integral aspect of normal ontogenesis (cf. Glücksmann, 1951; Saunders and Fallon, 1966). A number of reasons may be brought forward to explain this event. First of all, as already discussed, it is common that large yolk-laden endodermal cells degenerate, their contents being taken up by the cells of the gut. In this case, part of the explanation may be that the cells gradually become exhausted with respect to hyaloplasm and lose their power to divide.

In some cases, cell degeneration is observed to coincide with the initiation of cell differentiation at some location. Glücksmann (1951) has suggested that the degenerating cells may be unable to assume the particular differentiation pattern represented by the surrounding cells. This is, of course, no explanation of the degeneration, since it might be anticipated that the cells could and would undergo a differentiation pattern of their own. That this does not occur must imply that one type of differentiated cells exerts an inhibitory effect on other types. The mechanism involved is entirely unknown, but it may be mentioned that it conforms to some extent to the ideas advocated by Rose (1955).

As a third possibility it may be envisaged that cells degenerate because they are not adequately supplied with blood. The appearance of holes in the central region of various membranes may very well be explained by this mechanism.

A fourth type of cell degeneration occurs at metamorphosis, during which a more or less extensive remodelling of the larval body occurs. In Insecta and Amphibia where this process has been thoroughly studied, it turns out that it is controlled by hormones, and that part of the mechanism involves an activation of lysosomal enzymes leading to cytolysis (cf. Weber, 1967; Whitten, 1968).

The attention has here been focused on the mechanism of cell degeneration. From a morphogenetic point of view the location of the degenerating cells is, of course, of particular importance. From the impressive list of examples on cell degeneration in vertebrate embryos compiled by Glücksmann (1951) it appears that this phenomenon is so common that it is impossible to establish any generalizations.

'That the entire organism in every stage of its development exercises a formative influence on all its parts appears to me an absolutely necessary hypothesis'.
(F. R. Lillie, 1901, p. 273)

EPIGENESIS AND ONTOGENESIS

Epigenesis is a spatio-temporally ordered sequence of causally, or maybe rather functionally (cf. Woodger, 1962), related events, which constitute interactions between the embryogenetic substrate and the embryogenetic processes. If, through the analytical approach, the nature and the contribution of these embryogenetic factors were established in complete detail, it should be possible for an embryologist, after exhaustive analysis of a fertilized egg, to predict the course of embryogenesis. Nothing like that is, of course, possible today.

A less ambitious undertaking would be to analyse the developing embryo at various stages in order to determine in each case the nature of the substrate elements and their spatial location, inferring by this means the epigenetic events responsible for the changes leading from one stage to the next one. I do not consider this an impossible task, even if some reservation must be made, because the analytic approach is far from being complete. Thus, the preceding part of this book is based on examples chosen from a relatively restricted range of taxa. It is far from certain that the embryogenetic factors thus established suffice to account for embryogenesis in all other animals; indeed, it is not unlikely that within the selected groups important mechanisms have been overlooked.

As stated by Lillie seventy years ago in the quotation heading this part of the book, interactions may occur at all levels in the developing organism. This circumstance has the consequence that embryogenesis in principle must be studied in whole embryos. Explantation, transplantation and other operative interferences may be important for the elucidation of embryogenetic mechanisms, but as soon as the normal organization of the embryo is disturbed one can no longer be sure that the various activities follow their usual course. Great caution is therefore required in the evaluation of the significance of results obtained by such expedients.

In transparent eggs the embryogenetic processes can be followed in the living state. Yet, in spite of the fact that thousands and thousands of embryologists have observed the developing sea urchin embryo under the microscope during most of the past hundred years, it is only in the last fifteen years that we have come to know about the morphogenetic mechanisms involved. Part of this achievement lies in the fact that time-lapse cinematography has been employed, since this method allows for the visual speeding-up of the process which is necessary to distinguish the separate steps (Dan and Okazaki,

1956; Gustafson and Kinnander, 1956). However, another part, and the crucial one, is due to the painstaking analysis to which the phenomenon has been subjected, particularly by Gustafson and his coworkers.

Unfortunately, most eggs are large and opaque, and only the superficial aspects of embryogenesis can therefore be studied on living material. Adherence to the demand that embryogenesis must be studied on whole embryos thus unavoidably seems to imply work with fixed and sectioned material, an expedient which may appear revolting to modern, experimentally-minded embryologists. It is my conviction, however, that if it is based upon proper knowledge of the nature and the function of the various epigenetic mechanisms, and is applied with discrimination, then it should be possible in a great many instances to reveal the factors responsible particularly for the morphological changes occurring between successive embryonic stages. Still, there is no doubt that very often 'serial sections of fixed material give only crude and often false information about the natural shape of the cells' (Holtfreter, 1943b), and this approach therefore cannot entirely replace work on living embryos.

As a matter of fact, as I have tried to show for the Amphibia (1965a; 1965b; 1966a), when satisfactory records have been published it may not even be necessary to make new observations; the method can be applied directly to the published material. Unfortunately, a perusal of the embryological literature shows that information on the development within many phyla and classes, among Vertebrata as well as among Invertebrata, is appallingly defective. A lot of work remains to be done before the science of comparative embryology is brought up to date.

An objection which has been raised against this approach is that it does not furnish 'proof' of the postulated mechanisms. In my opinion this argument stems from a misapprehension of the nature of the deductive method. To illustrate this point we may use an analogy from the kinetic theory of gases. As is known, it was postulated here, long before the existence of individual molecules was demonstrated, that such entities exist in all gases, and that they are distinguished by the possession of a momentum which they may transfer to a surface exposed to a gas, thereby exerting pressure on the surface in question. It never occurred to the physicists involved in the development of the theory that it was necessary to 'prove' this assertion by actually demonstrating that molecules with specified values of momentum did hit the surface. It would have been, and still is, impossible to furnish a verification of this kind.

Returning to embryology, we may consider the following contention: 'There exists a certain kind of cell, an sf-cell, distinguished by the possession of filopodia which, through the development of tension, may carry out mechanical work. If present in a cavity containing mucopolysaccharide, and bounded by colligocyte aggregates, sf-cells may, through attachment of the filopodia, become suspended between two different aggregate regions. Under such circumstances the tension of the filopodia will pull the aggregates together, a process that will go on until opposing forces equal the maximum force that may be generated by the filopodia. During the early phases of development this mechanism is the only one that can carry out morphogenetic processes of this kind.'

If this proposition is correct then it follows that, whenever shape transformations occur under conditions corresponding to those specified, we may infer that they are the outcome of sf-cell activities. On this point embryologists are in a better position than physicists, for in many cases it is indeed possible to demonstrate the presence of sf-cells at the strategic locations. However, even if special circumstances make it difficult or impossible to provide this kind of verification, it is nevertheless justifiable to trust that the processes are executed by sf-cells, unless complementary empirical evidence challenges the validity of the postulate. An example of this kind is found in the gill cleft formation in Amphibia (cf. Løvtrup and Pigon, 1969).

As I have tried to indicate, once the analytical work has been done, the synthetic approach is a 'do-it-yourself' job for anyone who has a suitable material and has confidence enough to believe that the task is worth while. However, in order to set an example, I shall here apply the method to some selected cases of embryogenesis, in which the attention will be concentrated upon the gastrulation and larva formation. Two other subjects will be treated in this context. In the first two chapters we shall discuss some problems concerning cleavage and blastulation, and in the two final ones we shall deal with various physiological and chemical changes which have been recorded in developing embryos.

CLEAVAGE

Scattered in the preceding chapters various points concerning the significance of the process of cleavage have been recorded, as well as of the mechanisms responsible for its occurrence. In the present context we shall bring together the most salient points and discuss certain questions which have so far been neglected. We shall also consider the interactions between cleavage patterns and polarities as reflected by the development of isolated blastomeres.

Implications of cleavage

The process of cleavage may exert its influence on three different phenomena, namely, on the determination of diversification patterns, on the distribution of diversification patterns and on the symmetry relations of the larval and adult body. Each of these effects will be considered in the following sections.

DETERMINATION OF DIVERSIFICATION PATTERNS

The impact of evolution is clearly revealed by a comparison of the typical invertebrate egg, having a diameter of 100 to 200 μm, with the large eggs common in the higher animals like Arthropoda, Cephalopoda and Vertebrata. Yet, in spite of this advance, it appears from the preceding discussion that, apart from some notable exceptions, Nature is usually, in each individual case of ontogenesis, facing the formidable task of creating differentials where none exist. We have already discussed the material basis of these differentials, the polarities, and we have seen that whether the latter are cytoplasmic or cortical they do interfere with the process of cell division. The outcome of this interaction is that cells differing with respect to size or cytoplasmic composition are distributed in the embryo in a regular pattern. The geometrical relations of this pattern serve to establish those of the larval or adult body.

The mechanism through which this outcome is ensured involves that the various cells assume different diversification patterns, this being the only means through which the process of cell division may give rise to permanent effects. In this context it may be appropriate to consider the phenomena of 'determinate' and 'indeterminate' cleavage.

It will recalled that the determination in question is concerned with the presumptive fate of the blastomeres with respect to cell diversification. As the names appear to indicate, certain embryos may be distinguished by the firmness with which the blastomeres adhere to a certain course of diversification, whereas in others the embryonic cells exhibit some degree of plasticity.

Actually, it may be more correct to postulate the existence of two kinds of determination, which may be designated *static* and *dynamic*, respectively. Since the determination is concerned with cell diversification, it is obvious that cytoplasmic differences must ultimately be involved. The early determination prevailing in the former case implies that it is based upon a cytoplasmic polarity, preexisting in the unfertilized egg or established during the first cleavage divisions. Contrariwise, in dynamic determination the diversification of the cells is determined either through unequal divisions, the occurrence of which requires the presence of a blastocoele or other epigenetically established conditions, or through exhaustion, caused by their participation in epigenetic processes of various kinds. Since it is feasible to interfere with epigenetic events of this kind, the possibility thus obtains in this latter case of altering the diversification of the cells.

It is important to emphasize that, according to the view advanced here, determinate and indeterminate *eggs* do not exist. In either type so labelled, the presumptive diversification of some cells is determined, wholly or partly, at the outset, while that of others is established during the course of early development. To illustrate this point, it may be mentioned that in the typically 'determinate' eggs of Mollusca, the differentiation patterns of the descendants of the first micromeres appear to depend in some cases upon the normal polarity established by the formation of the blastocoele. In the 'indeterminate' amphibian egg the apical cytoplasmic polarity is responsible for the disparity in differentiation exhibited by ecto-mesodermal and endodermal cells, respectively. I do not think that this statement is invalidated by the observation that the vegetal blastomeres may regulate to form an almost normal embryo (cf. Ruud, 1925), and that among explanted endodermal cells some nerve cells may obtain (Holtfreter, 1938a; 1938b). These findings rather suggest that the cytoplasmic polarity does not operate in an absolute all-or-none fashion. At any rate, it has never been proposed that ecto-mesodermal cells can differentiate into gut cells.

There is further evidence supporting the contention that no fundamental difference exists between determinate and indeterminate cleavage. Thus, our comparison between the selfdifferentiation patterns of filocytes in spirally cleaving and in amphibian eggs shows that the determination is concerned with the same sequence in both cases, and we may therefore be justified in presuming that the mechanisms involved also are the same.

It seems obvious that if the process of cleavage is to control the distribution of cell diversification patterns in the embryo through the process of unequal cell divisions, then the latter must proceed with great precision and regularity. It appears that this mechanism has reached its limits of exploitation in Annelida and in the lower classes of Mollusca. In Arthropoda, Cephalopoda and Vertebrata the cleavage patterns no longer exhibit the regularity observed in the lower taxa. In all these cases the diversification patterns of the cells are established through dynamic determination and induction. This combination seems to allow for much more flexibility and regulation, and thus also for much more complex epigenetic processes, than when static determination is involved.

We shall finish this section with a brief discussion of the principle of 'adaptation in cleavage' established by Lillie (1895). In this paper the following quotation is found (pp. 38–39): 'Almost every detail of the cleavage of the ovum of Unio can be shown to possess some differential significance. The first division in unequal. Why? Because the anlage of the immense shell-gland is found in one of the cells. The apical-pole cells divide very slowly and irregularly, lagging behind the other cells. Why? Because the formation of apical organs is delayed to a late stage of development . . . The peculiarities of the cleavage in Unio are but a reflection of the structure of the glochidium, the organization of which controls and moulds the nascent material'. That a correlation obtains is obvious, and to this extent Lillie is correct. However, he is clearly putting the cart before the horse when postulating that effects are antecedent in time to their causes. To take up only the examples quoted here, we have already suggested that the particular mechanism responsible for the establishment of the median polarity in Bivalvia has at the same time the effect that the first cleavage is unequal and that the division of the first micromeres, and thus the development of the head, is retarded.

DISTRIBUTION OF DIVERSIFICATION PATTERNS

Evidently, to the extent that cleavage is responsible for the establishment of the diversification patterns, the arrangement of the blastomeres, i.e. the cleavage pattern, may also influence the distribution of the various cell-types, except in those cases where the motility of the cells entails that they migrate from their original location.

As an example which is extremely common it may be mentioned that in many embryos, particularly in those with spiral cleavage, but in numerous other instances as well, the cells around the animal pole show high mitotic activity, giving rise to small cells some of which spontaneously undergo the lf-transformation and become myxocytes and neurocytes. Some of the former may become trochoblasts, and their particular significance will be dealt with presently. However, myxoblasts are often formed which, through the swelling they provoke, may give rise to the formation of a head vesicle in which various morphogenetic processes may take place. Among these are the formation of cerebral ganglia and similar structures. The concentration of nervous tissue to the anterior part of the body may thus in these cases be referred directly to the cleavage pattern. Another example is found in the teloblastic division of the descendants of 2d in certain Annelida which is responsible for the formation of the ventral nerve chords.

Numerous illustrations of this type of correlation may be quoted, but we shall discuss only one more cell-type, the cf-cell myxocyte, represented by the comb plates in Ctenophora and by trochoblasts and velar cells in embryos with spiral cleavage. As described in an earlier chapter, there are eight comb rows in larval and adult Ctenophora, one for each of the original macromeres. According to Morgan (1927), this number may be reduced experimentally in proportion to the macromeres, so that when one of the latter is left, only one comb row is formed, suggesting that each macromere supplies one comb row. However, the fact that initially four are formed, each of which subsequently divides into two, rather intimates that only half of the blastomeres are able to supply micromeres representing this differentiation pattern. This point has been borne out by Farfaglio (1963), Ortolani (1963) and Reverberi and Ortolani (1963), who found that only micromeres derived from the external macromeres become comb plates. It seems that in the embryo the latter arise in the furrows between the median and the external macromeres, whence they end up in positions near the medians of each of the eight macromeres, separated presumably by epidermal cells which also originate from the external macromeres. The fact that no separation occurs in isolated macromeres demonstrates that this phenomenon depends upon the presence of two blastomeres. It can therefore be stated that the morphological feature represented by the number and the distribution of the comb rows may be referred to the particular biradial cleavage pattern found in Ctenophora.

As a second example we shall consider the trochophore and veliger larvae in various Annelida and Mollusca. Scanning the list of cell lineages compiled by Robert (1902), it turns out that in ten cases the source of the trochoblasts and velar cells is indicated. In all Annelida (*Nereis*, Wilson, 1892; *Amphitrite*, Mead, 1897; *Capitella*, Eisig, 1898;

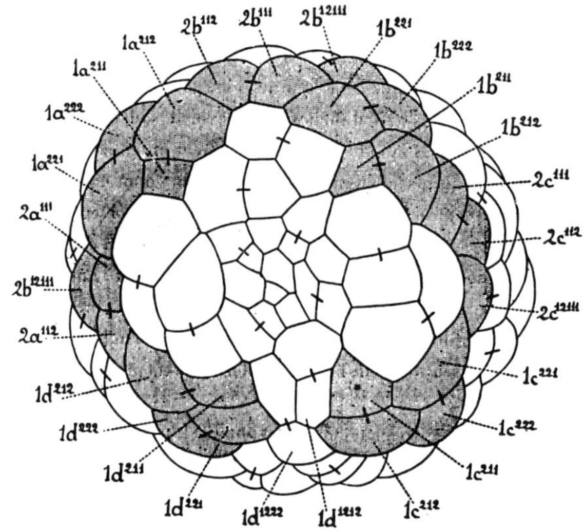

Figure 161. Cleavage in *Trochus*. The shaded cells become trochocytes. (Reproduced from A. Robert, *Arch. Zool. Exptl. Gén.*, **3**, 10 (1902), by permission of Centre National de la Recherche Scientifique, Paris)

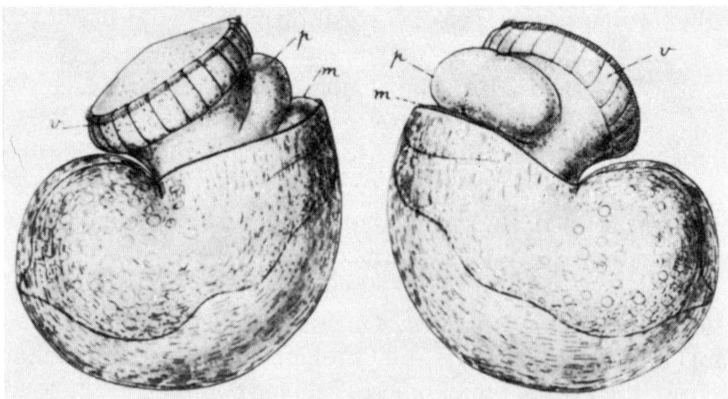

Figure 162. Veliger larva in *Trochus*. m, mantle; p, foot; v, velum. (Reproduced from A. Robert, *Arch. Zool. Exptl. Gén.*, **3**, 10 (1902), by permission of Centre National de la Recherche Scientifique, Paris)

Arenicola, Child, 1900; *Podarke*, Treadwell, 1901) the descendants of all four blastomeres $1q^2$ become trochoblasts, and the development passes through a trochophore stage with an essentially circular prototroch. Complete radial symmetry does not obtain, for generally blastomeres derived from 2a, 2b and 2c, but not from 2d, contribute to the prototroch. Among the Mollusca, $1q^2$ is the mother cell of trochoblasts in *Ischnochiton* (Heath, 1899), *Aplysia* (Carazzi, 1900) and *Trochus* (Robert, 1902), and in these forms the larva is either a trochophore or a veliger with an approximately circular velum (Figures 161 and 162). In *Crepidula* (Conklin, 1897) and in *Planorbis* (Holmes, 1900) only two or three quadrants of the first quartet contribute to the formation of ciliated velar cells, and both the cleavage patterns of the first quartet and the vela are markedly bilaterally symmetrical (cf. Figure 149). Generally, the trochoblasts and velar cells degenerate later, as a consequence of which these morphological features disappear.

SYMMETRY RELATIONS

The symmetry relations of the body may be referred to polarities prevailing in the fertilized egg. This statement implies that one polarity is required for the establishment of an antero–posterior axis, a second one for the dorso–ventral axis (or median plane) and a third one if bilateral asymmetry obtains. It has also been suggested that in a spirally cleaving egg a median polarity is defined by the centres of macromeres B and D, the direction of which bears some specific relation to that of the ultimate median plane, although they do not necessarily coincide.

Around the turn of the century there was much discussion about the relation between the presumptive median plane and the first cleavage planes, outlined by the contact lines between the macromeres, but it was shown at that time by Conklin (1897) that the median polarity is determined by descendants of the micromeres and not by the macromeres. Actually, the micromeres of the odd quartets lie in the cleavage furrows between the macromeres, and since those of the even ones lie between the others, it follows that 2b, 4b, 2d and 4d initially lie in the median plane determined by the corresponding macromeres. Since the two last mentioned micromeres generally are involved in the establishment of bilateral symmetry, coincidence between the initial and the ultimate median planes might be anticipated. That deviations occur is due to the fact that the aggregate of micromeres often is shifted back and forth in association with the budding of new micromeres, the degree of rotation depending upon the relative size of the macromeres and the micromeres (cf. Conklin, 1897; Robert, 1902).

It may seem astonishing that the macromeres do not affect the orientation of the median plane. However, in many cases the macromeres are of insignificant size at the end of cleavage (e.g. *Polygordius*, Woltereck, 1904), and in those cases where they retain a large size (e.g. *Crepidula*, Conklin, 1897), their main function is to supply food to the developing larva, their contribution to definite morphological entities being modest. This example illustrates a point mentioned above, viz., that in order to influence permanently symmetry relations or other properties of the larval or adult body the cleavage must determine the distribution of specific diversification patterns.

In both Annelida and Mollusca the cleavage of the first quartet proceeds with such precision that the blastomeres at a certain stage form an approximately orthoradial pattern (Figure 163). The circumstance that the annelidan and molluscan crosses shown in this figure are forming an angle of 45° to each other is, of course, a matter of convention which mainly refers to the relative cell size. However, the cells of the molluscan cross and the corresponding ones in Annelida (i.e. the non-cross cells) are the interesting ones, for they include the blastomeres 2d and 4d which, being different from the other cells of the cross, establish the median polarity. To the extent that the progeny of these blastomeres determine the bilateral symmetry of the trunk the two planes of symmetry thus coincide.

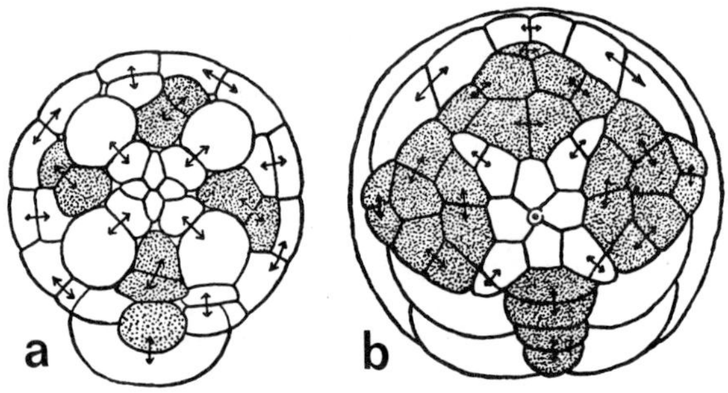

Figure 163. The cross in *Nereis* and *Crepidula*. (a), *Nereis*. The shaded cells represent the molluscan cross, except the most posterior one. The trochocytes lie at the margin of the egg. (b), *Crepidula*. The cross cells are shaded. The trochocytes lie around the margin. (Reproduced from E. G. Conklin, *J. Morphol.*, **13** (1897), (a) after Wilson, by permission of The Wistar Press, Philadelphia)

In Annelida, descendants of 4d generally are responsible for the formation of a series of vesicular cell aggregates consisting of cf-cell myocytes. This phenomenon is of great importance for the segmentation of the body, but does not seem to exert any influence on the determination of the dorso–ventral axis; if anything, it is rather incorporated as part of the antero–posterior axis. In those cases where some of the cells migrate into the blastocoele to form the mesodermal bands, the location of the latter obviously to a large extent must depend upon properties residing in the ectoderm, so even in this case the descendants of 4d do not determine the bilateral symmetry of the trunk. This function is, in many cases at least, carried out by the cells derived from 2d which form the somatic plate. When the latter spreads over the embryo, mainly through epiauxesis, it carries the neurocytes present at its margins to the ventral side of the body where they form the ventral cords, thereby establishing a distinct dorso–ventral axis. In Annelida, where there is no bilateral asymmetry, the three antero–lateral quadrants lose importance for the establishment of symmetry relations. This has been observed in *Bimastus*, where the A and B quadrants may be removed without serious embryogenetic consequences (Siewing, 1969).

In Mollusca the shell gland is derived from some of the descendants of 2d. As discussed above, the shell is instrumental in the establishment of the bilateral asymmetry. The precondition for the realization of the latter appears to be that the 'median' planes determined by the first and second quartets do not coincide, a stipulation the fulfilment of which generally is a direct consequence of spiral cleavage.

Development of isolated blastomeres

In embryological literature it is rather common to employ the word 'gradients' for the differentials here called 'polarities'. The reason for the present terminology is that, according to the contentions advanced above, the polarities are primarily involved in the determination of diversification patterns, and this being in general an all-or-none reaction, the notion of 'gradient' seems out of place. If the polarities were indeed gradients, for instance of a so-called 'morphogenetic substance', then the relative concentration might be the decisive factor, and it should therefore be possible to cut across the gradient and get complete regulation, as observed when a planarian worm is sectioned. As we shall discuss presently, there are many experimental observations which show that this possibility does not generally prevail.

In support of the gradient theory it may be argued that the sequence of selfdifferentiation patterns which are realized in the early amphibian embryo actually corresponds to a gradual exhaustion of strategic cytoplasmic components. I do not believe this view is acceptable, for the original cytoplasmic polarity apparently only ensures separation between the endodermal and the ecto-mesodermal selfdifferentiation patterns, the subsequent ramification being realized either through the mechanism of cell division or through activity or time differentials.

If we presume that the polarities act in the way suggested here, the development of isolated blastomeres will depend upon the number and nature of the polarities in a way that is theoretically predictable, as shown in Table 8. However, theory and practice do not always coincide, and although many of the points in the table are borne out by observations, there are certain findings which need further comment.

Table 8
Correlation between polarities and developmental potential of isolated blastomeres

Number and nature of polarities	Number of equivalent blastomeres	Number of blastomeres ensuring normal development	Symmetry
0 or 1 normal	2/2; 4/4; 8/8; etc.	2/2; 4/4; 8/8; etc.[1]	Spherical
1 or 2 apical	2/2; 4/4[2]	2/2; 4/4[1,2]	Radial
2 opposite identical median	2/2; 2 × 2/4; 2 × 4/8	0	Biradial
1 apical, 1–2 median	2/2[3]; 0/2[4]; 2 × 4[3]; 3/4[4,5]	2/2[3]; 1/2[4,6]; 2/4[3,6]; 1/4[4,6]	Bilateral
1 apical, 1–2 median, 1 dextro–sinistral	0	1/2[3,7]	Bilateral asymmetry

The raised numbers in the table denote: 1, as long as a certain critical minimum size is not reached; 2, if the third division is perpendicular to the animal–vegetal axis; 3, if the first division is in the median plane; 4, if the first division is perpendicular to the median plane; 5, the blastomeres not containing the median polarity may be identical; 6, the blastomere(s) containing the median polarity may develop; 7, the blastomere containing the dextro–sinistral polarity may develop.

In diploblastic animals the endoderm may in some cases arise through delamination or multipolar ingression, suggesting that either a normal polarity or chance determines the segregation between the two germ layers. Under these circumstances the possibility exists that no apical polarity is present, since the determination of the ultimate antero–posterior axis may be fortuitous or directed by environmental factors. Demonstration of this eventuality clearly demands that 1/8 blastomeres or still smaller ones may develop normally. The only phylum in which such an outcome may be anticipated is Cnidaria, since an apical polarity is generally quite evident in all other phyla. Actually, the only claims which have ever been made that smaller fragments than 1/4 blastomeres may develop normally are based on observations on embryos from this taxon (e.g. Wilson, 1903a).

Most often development is blocked when the fragments are so small; usually, 1/4 blastomeres are required to complete ontogenesis. This result may be interpreted in two ways: either an apical polarity prevails, or else the fragment, and consequently the number of cells, is too small for the required morphogenetic processes to occur. It is interesting to observe that 1/8 fragments of *Clytia* do form a normal planula larva (Figure 164), but tentacles never sprout, suggesting either that an essential differentiation pattern, e.g. myxoblasts, is missing, or that the volume or surface area of the larva cannot sustain the process of tentacle formation.

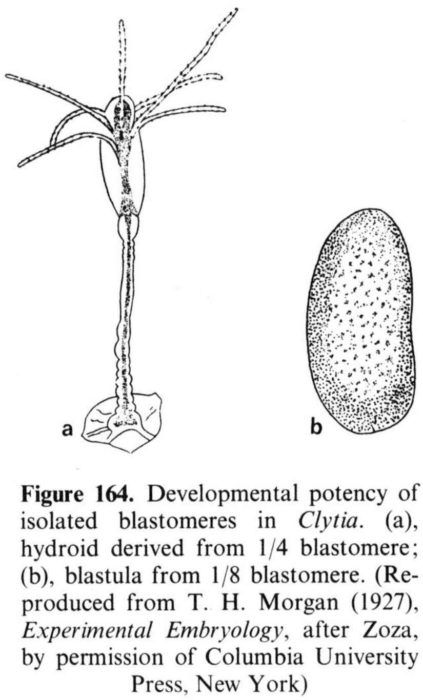

Figure 164. Developmental potency of isolated blastomeres in *Clytia.* (a), hydroid derived from 1/4 blastomere; (b), blastula from 1/8 blastomere. (Reproduced from T. H. Morgan (1927), *Experimental Embryology*, after Zoza, by permission of Columbia University Press, New York)

Turning to the second case, where only apical polarities are present, it is seen that each of the four blastomeres ought to develop normally. In *Cerebratulus*, Wilson (1903b) observed that some 1/4 blastomeres may develop into almost normal pilidium larvae (Figure 165), and this result has been borne out by other authors (cf. Morgan, 1927; Hörstadius, 1937). The number and nature of the polarities are not known in this case. The cleavage pattern does not reveal any median polarity (Wilson, 1930b), and the fact that mesoderm arises from only one or two quadrants (cf. Schleip, 1929a) does not necessarily demonstrate the existence of a polarity, for mesoderm formation may be a chance

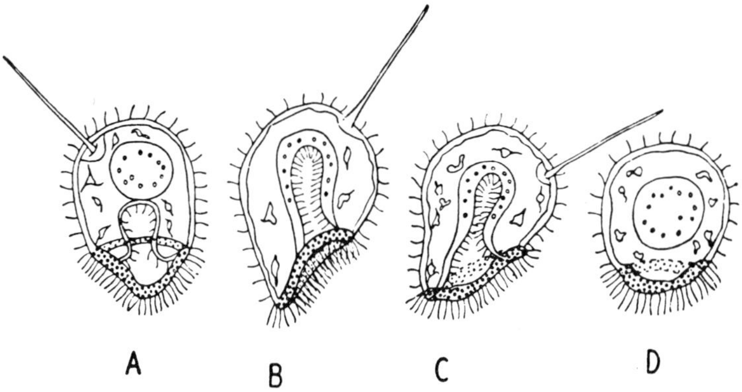

Figure 165. Developmental potency of isolated blastomeres in *Cerebratulus*. Four larvae derived from 1/4 blastomeres of one egg. (Reproduced from S. Hörstadius, *Biol. Bull.*, **73** (1937), by permission of Marine Biological Laboratory, Massachusetts)

phenomenon. However, if no median polarity prevails, we may infer that two apical polarities exist, one of which is epigenetically transformed into a median one, since the adult body possesses a dorso–ventral axis. If this situation obtains, with mesoderm formation being fortuitous, the first four blastomeres are equivalent, a circumstance which may account for the experimental observations.

As regards the sea urchin egg, various authors have claimed that the median polarity is predetermined (e.g. Runnström, 1926; Lindahl, 1932b; Hörstadius and Wolsky, 1936; Pease, 1942; Gustafson, 1952). Among these, Lindahl (1932b) and Pease (1942) have observed that the determination is labile, and that its orientation may be influenced experimentally. According to Hörstadius and Wolsky (1936), the first cleavage planes bear no relation to the presumptive dorso–ventral axis. It will be realized that if only apical polarities are present the two 1/2 blastomeres and the four 1/4 blastomeres should all possess the full developmental potential. If the observations made above are correct, the blastomeres should not be equivalent in this respect; rather, approximately half of the 1/2 blastomeres and one or two of the 1/4 blastomeres should be able to develop, the others not. Unfortunately, the literature is not conclusive on this point, because a record of the number of successful and unsuccessful cases is lacking (Driesch, 1892; 1897; cf. Hörstadius, 1939). However, as far as the 1/4 blastomeres are concerned, imperfect gastrulation has been observed in some cases, but apparently normal development never takes place. This result does not conform with the hypothesis that the polarities alone are responsible for the development, irrespective of the assumptions made concerning their nature. It seems natural to look for other causes for the failure, the most plausible being the one mentioned already, namely 'incompatibility between the size of the cells and the size of the sphere [blastula]. If, as seems to be the case, the cells are of the same size as those of a corresponding stage in the whole embryo, they are proportionately four times too large [and four times too few] in the smaller 1/4 embryo' (Morgan, 1927, p. 311). Apart from this weighty argument it must furthermore be recalled that even if the isolated 1/4 blastomere cleaves normally, the resulting cells must also be organized in a normal cleavage pattern, if embryogenesis is to follow its normal course. It does not seem unlikely that a mixing-up may occur. As far as I can judge, the question of a predetermined median polarity cannot be settled on the basis of the experiments discussed above.

Wilson (1893) found that 1/2 blastomeres of *Branchiostoma* may develop normally, and even 1/4 blastomeres gave rise to almost perfect embryos. Provided only half of the 1/4 blastomeres exhibit this capability, the observation is compatible with the presence of one or two apical polarities and one median polarity. Conklin repeated this work and came to the result that 2/4 blastomeres may develop, but only when separated from their counterparts along the first division plane which purportedly represents the median plane (cf. Conklin, 1933), whereas separation along the second division plane failed to give viable embryos.

Recently, this problem has been attacked anew by Tung, Wu and Tung (1958). These authors found that the two first blastomeres may give rise to two normal or two abnormal embryos, or one of each (Figure 166a). The same result was obtained when separation

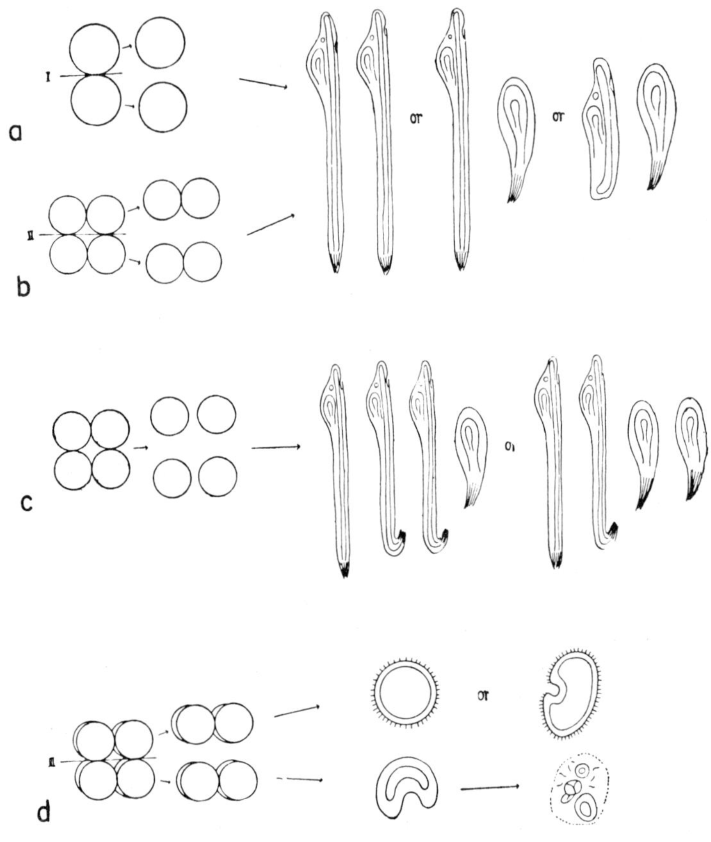

Figure 166. Developmental potency of isolated blastomeres in *Branchiostoma*. (a), development of 1/2 blastomeres isolated along the first division plane in the 2 cell stage; (b), development of 2/4 blastomeres isolated along the second division plane in the 4 cell stage; (c), development of 1/4 blastomeres in the 4 cell stage; (d), development of animal and vegetal blastomeres isolated along the third division plane in the 8 cell stage. (Reproduced from T. C. Tung, S. C. Wu and Y. F. Y. Tung, *Scientia Sinica*, **7** (1958))

occurred along the second division furrow, even though the frequency of normal development was significantly lower (Figure 166b). This result may to some extent be accounted for by the fact observed by Tung *et al.* (1958) that the first cleavage plane does not always coincide with the median plane. However, when 1/4 blastomeres were isolated it was found that quite often three out of four gave rise to approximately normal embryos (Figure 166c). This result is hardly compatible with a rigid fixation of the polarities at the 4 cell stage, a rather surprising outcome considering the fact that both the apical and the median polarities are cytoplasmic ones (cf. Conklin, 1932). These findings show that the Cephalochordata and Urochordata are much less similar, as regards the organization of the egg, than believed by Conklin (cf. 1924).

As an extension of the reported work, Tung *et al.* (1960) have followed the development of embryos in which one of the four tiers of eight cells at the 32 cell stage was removed. Among various combinations it was found that the two containing all the vegetal cells and half of the animal ones gave almost normal embryos, whereas absence of the vegetal blastomeres generally resulted in faulty development (Figure 167). These

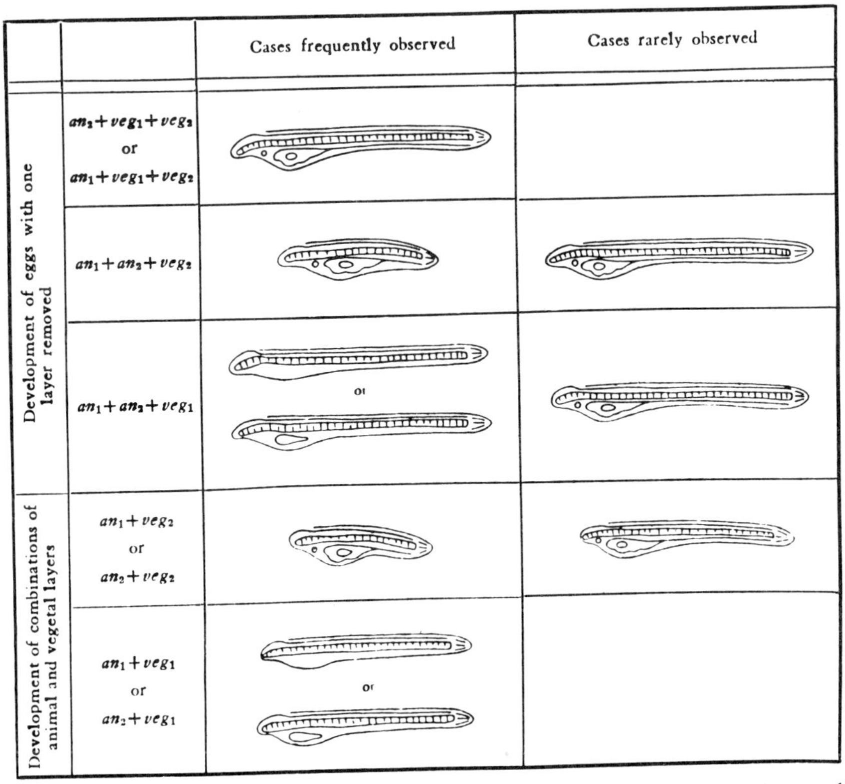

Figure 167. Developmental potency of various combinations of animal and vegetal tiers of blastomeres in the 16 cell stage in *Branchiostoma*. (Reproduced from T. C. Tung, S. C. Wu and Y. F. Y. Tung, *Scientia Sinica*, **9** (1960))

results corroborate to some extent the interpretation of the effect of the polarities given above.

In agreement with expectation it was observed by Ruud (1925) that when the first division plane lies in the median planes both blastomeres of *Triton* may give rise to a normal embryo, and the same result was obtained with the dorsal blastomere when the division furrow is perpendicular to the median plane. Normal development was rare in 1/4 blastomeres, but the fact that never more than two whole embryos were obtained from one egg is in agreement with the predictions in Table 8. Dalcq and Dollander (1948) and Dollander and Debry (1948) have reported that ventral half blastomeres occasionally may develop. These observations appear to be unique, but if the primary function of the median polarity is to initiate the lf-transformation it does not seem impossible that this event may occur if by chance the favourable conditions happen to prevail. The development of vegetal half embryos observed by Ruud (1925) has already been commented upon.

In spirally cleaving eggs the first division plane cuts across the median plane. According to Table 8, the dorsal blastomere CD may be expected to give rise to a normal embryo, a prediction that is corroborated by the observations on *Bimastus* mentioned above. Since apparently both body axes are determined by descendants of the D macromere in Annelida, this cell may even by itself ensure normal development. Corroboration of this prediction was claimed by Penners (1925), even though in this case the remaining blastomeres were not removed. Wilson's observations on *Dentalium* (1904) show that none of the first two or four blastomeres develop harmoniously, even though CD and D are relatively normal (Figure 168). It is interesting to note that the apical plate does not

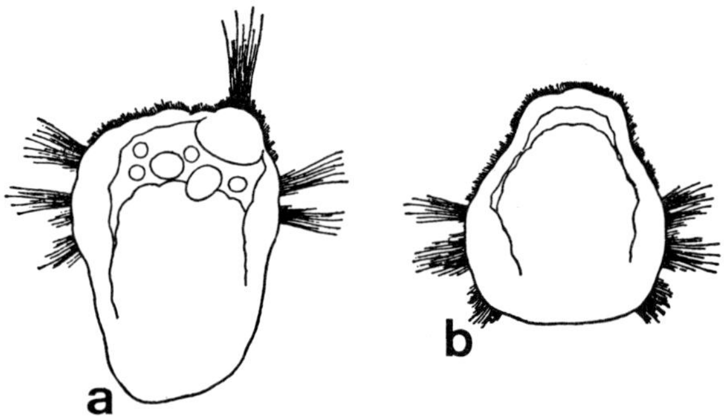

Figure 168. Developmental potency of isolated blastomeres in *Dentalium*. (a) and (b), twin larvae derived from the AB and CD quadrants, respectively, of the same egg. (Reproduced from E. B. Wilson, *J. Exptl. Zool.*, **1** (1904), by permission of The Wistar Press, Philadelphia)

form in A, B, C and AB, although they normally contribute to this structure. This finding seems to suggest that the lf-transformation in the animal pole cells may be induced from the D quadrant. The transformation does occur elsewhere in the blastomeres mentioned, as evidenced by the formation of trochoblasts. Since the organization of the veliger larva is so intimately associated with the cleavage pattern, it cannot be expected that isolated blastomeres from Gastropoda can sustain normal embryogenesis. This has been borne out by many observations (cf. Figure 169). The defective development may, formally at least, be referred to the absence of the dextro–sinistral polarity which, as we have seen, depends upon the cleavage pattern.

If the dextro–sinistral polarity acts in an all-or-none fashion, it may be anticipated that the epigenetic outcome of one of two blastomeres separated along the median plane

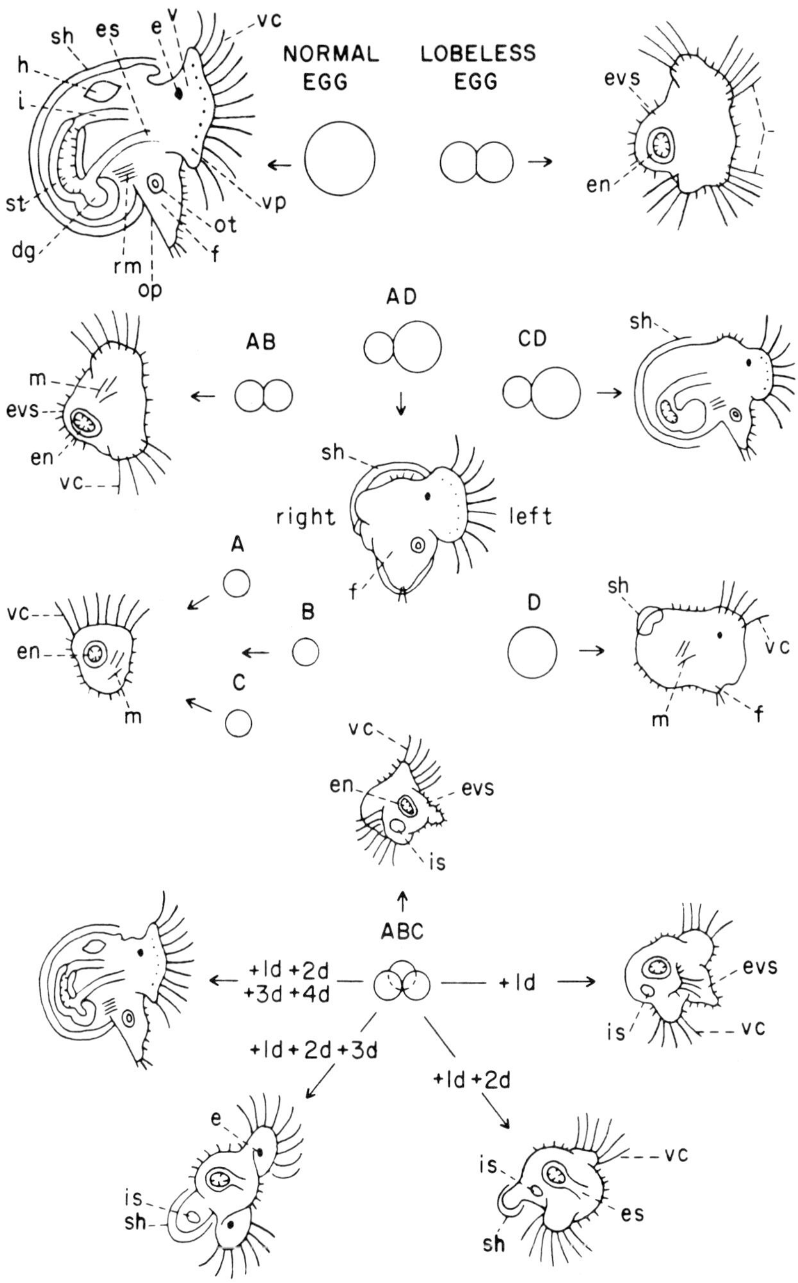

Figure 169. Developmental potency of isolated blastomeres in *Ilyanassa*. dg, digestive gland; e, eye; en, endoderm; es, oesophagus; evs, everted stomadaeum; f, foot; h, heart; i, intestine; is, internal shell fragment; m, muscle fibres; op, operculum; ot, otocyst; rm, retractor muscle; sh, shell; st, stomach; v, velum; vc, velar cilia; vp, velar pigment. (Reproduced from J. R. Collier (1965), in *The Biochemistry of Animal Development*, Vol. 1, after Clement, by permission of Academic Press, New York)

will retain the normal asymmetry, whereas that of the other one ought to be bilaterally symmetrical, or exhibit both kinds of asymmetry with equal frequency. The work of Tung *et al.* (1958) shows that two twin larvae of *Branchiostoma* both have the normal asymmetry. This result may be accounted for formally if the polarity is a gradient, or if it is a circular or spiral vector. Normal asymmetry also seems to prevail in embryos developing from isolated 1/2 blastomeres in Amphibia. It is interesting to note that when the blastula of the newt is separated in two by constriction in the median plane, the left half acquires normal asymmetry, while the right one shows the same frequency of this condition and *situs inversus* (Ekman, 1925). At this stage the polarity obviously works in an all-or-none fashion.

The comparison made here between the theoretically predictable and the actually observed development of isolated blastomeres show good agreement in some instances, but quite a few exceptions were noted. For the time being there seems to be no other way to account for this state of affairs than by recognizing that the biological reality is more complex than the theory.

BLASTULATION

In order to ensure the progress of embryogenesis, the cells originating from the process of cleavage must be arranged in an aggregate which may be subjected to further morphological changes. If, during this event, different kinds of cells, forerunners of the various diversification patterns, are formed, their relative location must also be preserved. The embryonic stage which serves these functions is the blastula. We shall in the present section discuss two aspects of blastulation, viz., the blastula types and the mechanism of blastocoele formation.

Blastula types

In the simplest possible case no polarities prevail, and all the cells will consequently be identical. If they are sl-cells, they will scatter unless they are kept together by a hyaline membrane or an egg envelope. If the cells do not adhere to this supracellular element more than to each other, a *stereoblastula* is formed, which is an agglomeration rather than an aggregate of cells. Examples approaching this kind of organization are found only among Cnidaria.

In all cases where a distinct blastula stage is found, the cells form a vesicular aggregate, a *coeloblastula*, enclosing a cavity, that is the blastocoele, which usually is filled with oncotically active mucopolysaccharides. Two different means may ensure the formation of this cell aggregate, either intercellular adhesion or a hyaline membrane. In the former case it would normally be necessary for the cells to be transformed into colligocytes, but the fact that blastula cells generally are flagellated rather than ciliated suggests that this possibility is seldom realized. Flagellocytes, vsl- or occasionally vsf-cells, may possibly acquire enough adhesion to form a stable aggregate, but most often a hyaline membrane is presumably involved in coeloblastula formation. Sometimes two other blastula types are distinguished, the *periblastula* and the *discoblastula*, found in eggs with superficial and discoidal cleavage, respectively. It seems justified to regard these as special cases of coeloblastulae, where in the former case the blastocoele is filled with yolk and in the second case the lower part of the blastula wall is replaced by the yolk sac.

A coeloblastula may comprise cells of unequal size, for instance small animal and large vegetal cells. If, in the case of such disparity of cell size, no hyaline membrane is present to limit the movements of the blastomeres, the small cells may spread on the surface of the large ones. In this case a kind of stereoblastula is formed, but since the cell migration results in separation between ecto-mesoderm and endoderm, it is usually, and appropriately, called a *stereogastrula*.

Mechanism of blastocoele formation

The creation of a blastocoele involves that water from the environment accumulates between the blastomeres of the cleaving embryo. This process has often been contended to occur through osmosis, but if this is correct the osmotically active substances must be high-molecular, in the sea urchin embryo at least, since molecules as large as those of sucrose may penetrate freely into the blastocoele in this case (A. R. Moore, 1945). In point of fact, as long as the blastomeres are sl-cells, they are not very adhesive, and during mitosis they actually separate from each other (cf. Figure 15), entailing that the molecules apparently may enter the blastocoele through the hyaline membrane and between the cells. This circumstance supports the suggestion advanced by Monné and Hårde (1950a), based on histochemical observations, that oncotically active mucopolysaccharides are responsible for the swelling of the blastocoele. The idea that uptake of water by a gelatinous material is involved in this event actually dates back to studies of *Branchiostoma* by Hatschek (1881) and Conklin (cf. 1932).

Wolpert and Gustafson (1916b) have pointed out that when all cells of the blastula adhere to the hyaline membrane, the area of the latter must increase gradually to accommodate the growing number of cells (cf. Figure 134). Support for the participation of this epiauxetic mechanism in Echinoidea is found in the fact that the volume of the blastocoele is enlarged in steps, in time with the cell divisions (K. Dan, 1960; Wolpert and Gustafson, 1961b; cf. Figure 170). This circumstance does not make the contribution of the muco-

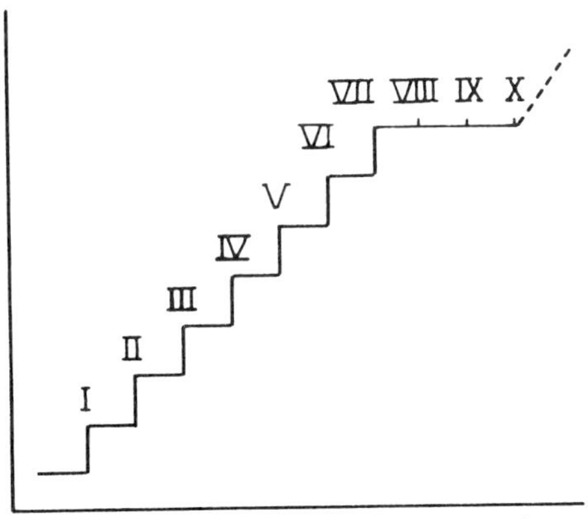

Figure 170. Stepwise enlargement of the blastocoele volume in *Mespilia* synchronously with the cleavage divisions. Abscissa, time; ordinate, volume. Cleavage division cycles indicated by Roman numerals. (Reproduced from K. Dan, *Internat. Rev. Cytol.*, **9** (1960), by permission of Academic Press, New York)

polysaccharides superfluous, for if the oncotic pressure does not follow suit with the enlargement of the blastocoele, water will either be squeezed out or the blastula will flatten under the influence of gravity.

Holtfreter (1943a) observed that in the amphibian embryo blastocoele formation may be suppressed in iso- or hypertonic salt solutions. This finding suggests that low-molecular substances cannot penetrate the wall of the blastula. It is hardly justifiable to infer from this finding that the hyaline membrane is impermable; a more likely explanation is that the cells are squeezed so tightly together that passage between them is excluded. The high tension in the hyaline membrane which must be invoked to explain this phenomenon entails that after a certain number of divisions the cells become elongated in a direction perpendicular to the membrane. As a consequence the mitotic apparatuses may be radially oriented, and the ensuing cleavage planes paratangential. In this way the blastula wall comes to comprise several layers of cells, in contrast to the situation prevailing among Invertebrata. If oncosis in involved in the formation of the blastocoele in this case, it is evident that the pressure is of modest magnitude since it may be counteracted by an isosmotic salt solution.

It is a surprising fact that the volume change associated with blastula formation may be referred exclusively to the blastocoele, the total volume of the cells remaining constant (Tuft, 1962). As discussed on an earlier occasion (1965c) it is likely that the tension at the cell surface is involved in the osmoregulation. From the equation $P=2T/R$, it appears that for constant tension T the internal pressure P increases as the radius R of the cells decreases with subsequent cell division. This circumstance may explain why small cells do not swell when exposed to strongly hypotonic solutions.

It many cases coeloblastulae and stereoblastulae occur together within the same taxon. This phenomenon is particularly well known among spirally cleaving eggs. Kofoid pointed out that a certain correlation obtains between the habitat of the eggs and the type of blastula, the stereoblastula being most usual in marine forms, and the coeloblastula being 'confined to eggs developing in *fresh water or moist situations*, and reaches its maximum manifestation in forms enclosed in thick encapsuled albumen, like Limax' (1895, p. 103). However, in eggs containing a large amount of yolk, as *Neritina*, Kofoid noted a tendency for suppression of blastocoele formation. It appears to me that these observations are most easily accounted for on the assumption that the tonicity of the surroundings of the egg is involved in blastocoele formation, but that this event may be prevented for purely geometrical reasons when the size ratio between the macromeres and micromeres is too large. In specific cases the morphogenetic potential of the two blastula forms differs significantly, but the examples just quoted show that no generalization is possible as regards this point, since the same type of adult organization may arise from either kind of blastula.

GASTRULATION

The significance of blastulation comprises three aspects: subdivision of the embryogenetic substrate into an array of cells, preparation for, and occasionally even the initiation of, the process of cell diversification, and the creation of a cell aggregate which, through the activity of morphogenetic processes, may be subjected to further morphological changes. Among these the first is the gastrulation by means of which a fraction of the total number of cells becomes enclosed by the remaining ones. This disparity in the fate of the cells must be referred to a polarity, for without any differential it would be a violation of the causal postulate that the cells assume different stations. The polarity may be concerned with the size of the cells or otherwise be associated with cell division, in which case the process of gastrulation may begin at a very early stage, or it may be a matter of diversification, notably cell transformation, implying that the process of gastrulation must await the commencement of these events.

We shall begin the present section with an outline of the various mechanisms of gastrulation, and subsequently discuss this phenomenon in some selected cases, viz., Echinoidea, Cephalochordata and Amphibia.

Mechanisms

Gastrulation is by most authors defined as, or equated with, segregation of the germ layers. It is possible to distinguish a number of different mechanisms involved in this event which in the embryological literature generally are designated 'ingression', 'delamination', 'invagination', 'involution' and 'epiboly', respectively. It is generally implied that the gastrulation mechanism in any particular case may be assigned to one of these various categories. However, as pointed out occasionally (e.g. Hyman, 1940), the gastrulation type is determined on the basis of the relative translocation of the two primary germ layers, the ectoderm and the endoderm, the means by which the mesoderm arrives at its final station being neglected. In view of the accepted definition of gastrulation this approach is hardly warranted. In reality, different processes are usually involved in the segregation of the various germ layers. If allowance is made for this fact, it becomes impossible to allot individual cases of gastrulation in triploblastic animals to one class. In the following discussion the mechanism of segregation of each germ layer will be treated separately under the appropriate heading. It must be stressed that the mesoderm segregation does not always come to an end during gastrulation.

Gastrulation being an epigenetic event, it follows that if the survey of epigenetic processes presented above is complete, it must include all those involved in the execution of gastrulation. Attempting to show that this is the case, we shall discuss the various gastrulation mechanisms under the headings: ingression, cell transformation, invagination, epiauxexis and epiboly.

INGRESSION

Ingression is a mechanism whereby individual cells leave the aggregate constituted by the blastocoele wall. Two kinds are usually distinguished, called *multipolar* and *unipolar* ingression, respectively. In spite of its name the former is more primitive than the latter, and we shall therefore begin with a discussion of this phenomenon.

Multipolar ingression implies that randomly distributed cells leave their station and enter the blastocoele. This behaviour is a direct indication that the cells in question are solocytes. If they are sl-cells, as supposedly are all the cells of the blastula, it may be envisaged that a differential obtains among the total population of cells, maybe with respect to the force with which they are attached, entailing that certain cells are freed and others are not. Subsequently the two groups of cells assume different diversification patterns, but there is no reason to presume that this must be referred to inherent differences at the time of segregation; the normal polarity established by the process of ingression may just as well be inferred to affect the diversification of the cells.

Ingression need not involve detachment of individual cells, but may also arise as a consequence of *paratangential* cell divisions. This phenomenon is usually called 'primary delamination', but there seems to be little sense in making a distinction on the basis of the way by which the cells enter the blastocoele, particularly since both may occur simultaneously. Hyman (1940) thus treats them both under the heading of 'delamination'; here they are brought together under the less exacting name 'ingression'. It is seen that with paratangential cleavage the normal polarity to which the cell is subjected may be employed for the execution of unequal cell divisions, thereby ensuring that the two populations of cells assume disparate diversification patterns (Figure 171).

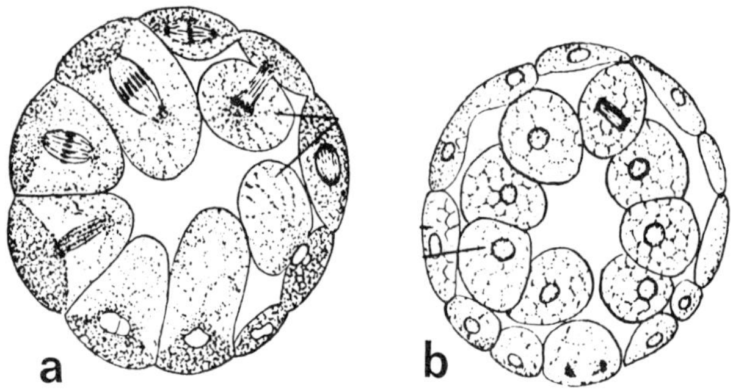

Figure 171. Gastrulation through ingression in Cnidaria. (a), endoderm formation through paratangential divisions in *Geryonia*. (b), completion of endoderm formation in *Liriope*. (Reproduced from L. H. Hyman (1940), *The Invertebrates*, Vol. 1, by permission of McGraw-Hill, New York)

For the segregation of the two primary germ layers, multipolar ingression obtains only in certain Cnidaria and possibly in various Vertebrata with discoidal cleavage. However, for the separation of the mesoderm from the two other germ layers this mechanism is very common. Suffice it to mention here the situation prevailing in spirally cleaving eggs, where mesodermal cells arise from all four quadrants and from several quartets, even if the process here does not occur quite at random.

Unipolar ingression differs from the mechanism just discussed only insofar as an apical polarity is involved, and ingression consequently occurs only from one region, which is always located around the vegetal pole. Unipolar ingression often occurs from flagellated blastulae, and it thus appears that the cells at the vegetal pole pass through the reversible transformation between vibriocytes and non-vibriocytes (cf. Figure 172). As a method of segregation between the primary germ layers this mechanism is known from Porifera and Cnidaria. In the latter case the cells remain in the interior, forming the endoderm, but in the former case they penetrate to the exterior, forming the ectoderm. It is interesting to observe that if the diversification of the cells was used in the classification of the germ layer, the outer layer in Porifera surely ought to be called mesoderm, since generally it is comprised of sf-cells assuming typical sclerocyte differentiation patterns.

The mesoderm is often segregated by unipolar ingression; as an example we may mention the Amphibia, in which the mesoderm is constituted by the interior cells in the animal hemisphere, arising through paratangential cell divisions.

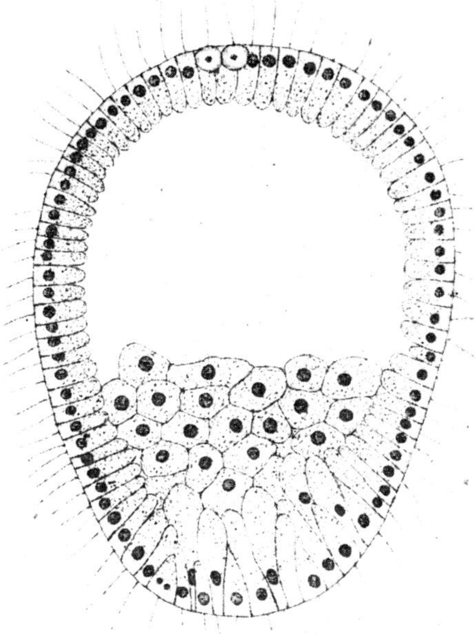

Figure 172. Gastrulation through ingression in *Octorchis*. Endoderm formation through unipolar ingression. (From E. Metschnikoff (1886), *Embryologische Studien an Medusen*, Hölder Verlag)

In some cases the cells are so firmly anchored in the blastula wall that only sf-cells have the force required to detach themselves. In this situation the lf-transformation is instrumental in the segregation, and the mechanism therefore will be discussed in the following section.

CELL TRANSFORMATION

Both the sc- and the lf-transformation may be involved in germ-layer segregation. As a mechanism for separating the primary germ layers, the former plays a role in stereoblastulae, where the transformation of the outer cells into colligocytes is the first sign of a distinction between the two germ layers (Figure 173). In those cases where sf-cells enter the blastocoele through multipolar ingression, the possibility exists that almost all cells become filocytes. In that case the random occurrence of the sc-transformation may ensure that some cells remain at the surface.

The lf-transformation is probably the most usual means employed for mesoderm segregation, particularly in those cases where all blastula cells are attached to a hyaline membrane. The most well-known example is found in the Echinoidea, where the mesodermal cells enter the blastocoele after transformation into sf-cells, partly at the

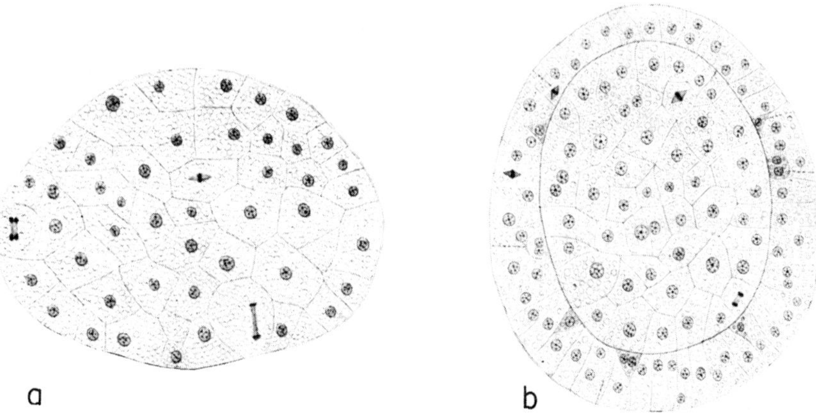

Figure 173. Gastrulation through cell transformation in *Clava*. Germ-layer separation through transformation of the peripheral cells into colligocytes. (a), stereoblastula; (b), stereogastrula. (Reproduced from K. Harm, *Z. Wiss. Zool.*, **73** (1903), by permission of Akademische Verlag, Leipzig)

vegetal pole, and partly at random locations in the ectoderm. Instances of this kind may probably be found throughout most of the animal kingdom.

INVAGINATION

The phenomenon of *invagination* is here defined in a more limited sense than is usually the case, namely, as a process of gastrulation where the endoderm is transported into the blastocoele through the activity of sf-cells. In the discussion of this and the following gastrulation mechanisms the segregation of the mesoderm is left aside, because the latter always occurs by one of the more primitive methods dealt with above.

As already stated, the work by Gustafson and Kinnander (1956) and by Dan and Okazaki (1956) has established that in the Echinoidea gastrulation is carried out by invagination as defined here. It is worth noting, however, that the idea that sf-cells are participating in the mechanism of gastrulation is anything but new, as evidenced by the following quotation: 'As concerns the process of gastrulation proper, then we obviously cannot by any means trace it back to purely mechanical factors. A glance at Figure 174, where the continuations extended spontaneously from the ecto- and endodermal cells simply form a forest, presumably supplies us with the best assurance of the activity of the blastomeres during the process of gastrulation' (Wierzejski, 1905, p. 639, author's translation). In this passage there is one expression, viz., 'purely mechanical factors', which is unacceptable, since surely mechanical work is executed in the process of gastrulation, even if the agents happen to be living cells, but otherwise the idea is in complete conformance with the views beginning to prevail among embryologists today. Similar ideas had been advanced by Schierholz (1889), but they were rebuffed by Lillie (1895, p. 57) with the statement: 'One cannot deny the originality of this view of the mechanics of development; neither can one accept it'. One may wonder about the status of present-day embryology if these correct notions had been accepted when they were advanced at the turn of the century.

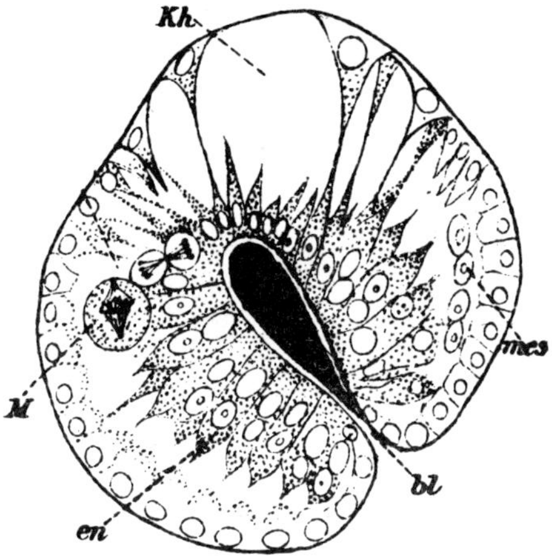

Figure 174. Gastrulation through invagination in *Physa*. Optical section of a gastrula. bl, blastopore; en, endoderm; Kh, blastocoele; M, mes, mesoderm. (Reproduced from A. Wierzejski, *Z. Wiss. Zool.*, **83** (1905), by permission of Akademische Verlag, Leipzig)

The details of the mechanism through which the filopodia execute the process of invagination have been discussed above, and no further comments are needed in this context. It suffices to point out that since there is a physical limit to the length and the strength of the filopodia, gastrulation through invagination can take place only in relatively small eggs. Once this reservation is made, it is probably correct to state that this is one of the most widespread mechanisms of gastrulation, found as it is in Echinodermata (Figures 158 and 178), Nemertina (Figure 143), Mollusca (Figures 174 and 175) and probably many other Invertebrata. As we shall discuss presently, it is probably also involved in the gastrulation of Cephalochordata, and it is active in the initiation of this event in Amphibia and presumably in other Vertebrata.

EPIAUXESIS

Four conditions must be fulfilled for epiauxesis to occur, namely, absence of a hyaline membrane, presence of a surface to which the blastomeres adhere and on which they may spread, a certain intercellular adhesion, even though the cells need not be colligocytes, and that they are mitotically active.

Since formation of a hyaline membrane usually occurs in association with the expulsion of the cortical granules, it follows that epiauxesis cannot be a very common mechanism of gastrulation, but it appears to be found in Ctenophora, in various spirally cleaving eggs, and probably in some Vertebrata. As far as the surface mentioned is concerned, it is always presented by the vegetal part of the egg, either in the form of large

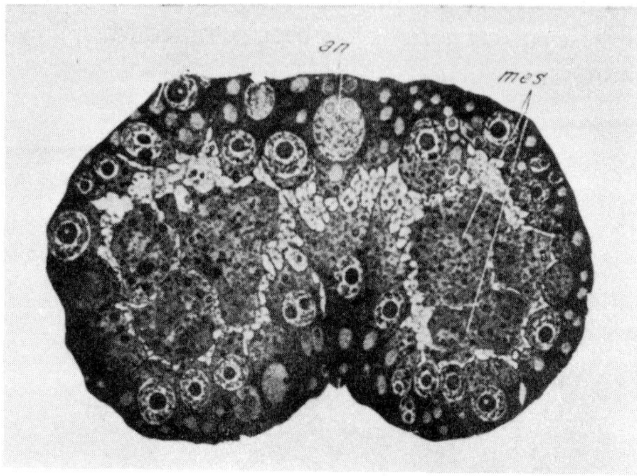

Figure 175. Gastrulation through invagination in *Lymnaea*. Section showing filoform pseudopodia stretching from endodermal cells towards the animal pole (an). mes, mesoderm. (Reproduced from Chr.P. Raven, *Morphogenesis: The Analysis of Molluscan Development* (1966), Pergamon Press, Oxford)

macromeres or of the yolk. Epiauxesis is particularly easily observed in yolk-laden eggs of Annelida and Mollusca (cf. Figure 176). It may be noted that the overgrowth in this case depends not only upon the subdivision of existing micromeres, but also upon the addition of new ones budding off from the macromeres. The flattening of the ectodermal cells that occurs during the latest stage depicted is not an aspect of the epiauxetic mechanism, but it may, of course, contribute to the execution of the process of gastrulation. Presumably this event is caused by a change in the adhesive properties of the surface of the macromeres. It has been pointed out above that a special kind of epiauxesis may be responsible for the overgrowth of the yolk in Pisces.

It should be mentioned that in some cases where gastrulation occurs through invagination, epiauxesis may be involved in later developmental stages, such as neurulation.

EPIBOLY

The passivity of the endoderm in the process of gastrulation has been borne out by the two preceding mechanisms. Invagination is carried out by the dynamic activity of mesodermal sf-cells, and epiauxesis is largely a consequence of the mitotic activity of the ectodermal cells. As discussed in a previous section, epiboly may be referred to the motility displayed by amoebocytes. This mechanism ensures that a large number of cells act in concert, which may be the reason why it is exploited in large eggs, for instance those of Amphibia, where invagination cannot do more than initiate the gastrulation. The process of epiboly is not confined to such cases, for as far as I can see the process of inversion occurring in certain Porifera must be executed by movements carried out by the non-flagellated cells which, judging from their appearance, are sf-cells (Figure 177).

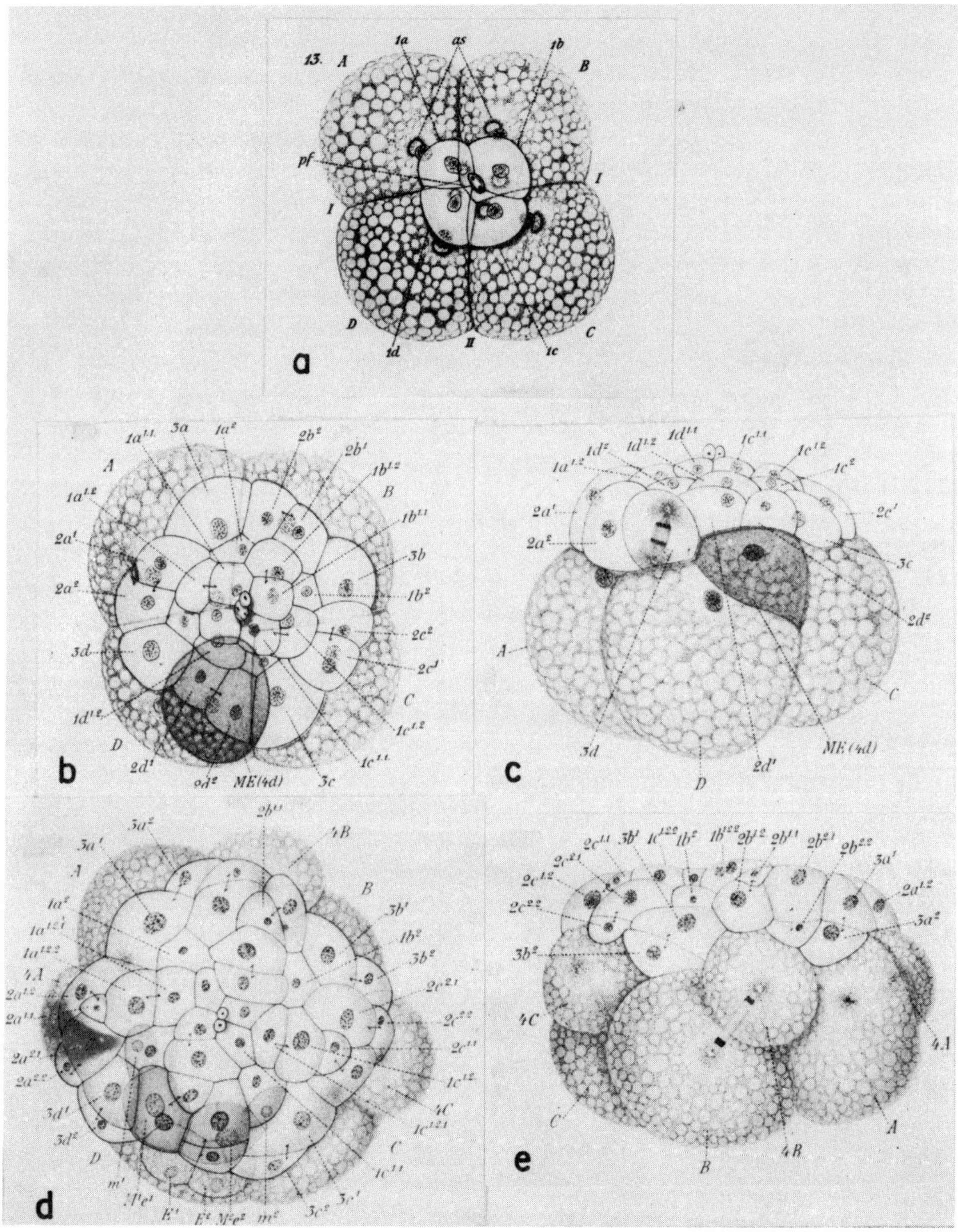

Figure 176. Gastrulation through epiauxesis in Crepidula. (a), 8 cell stage; (b), 29 cell stage; (c), lateral view of b; (d), 52 cell stage; (e), lateral view of d; (continued on p. 302)

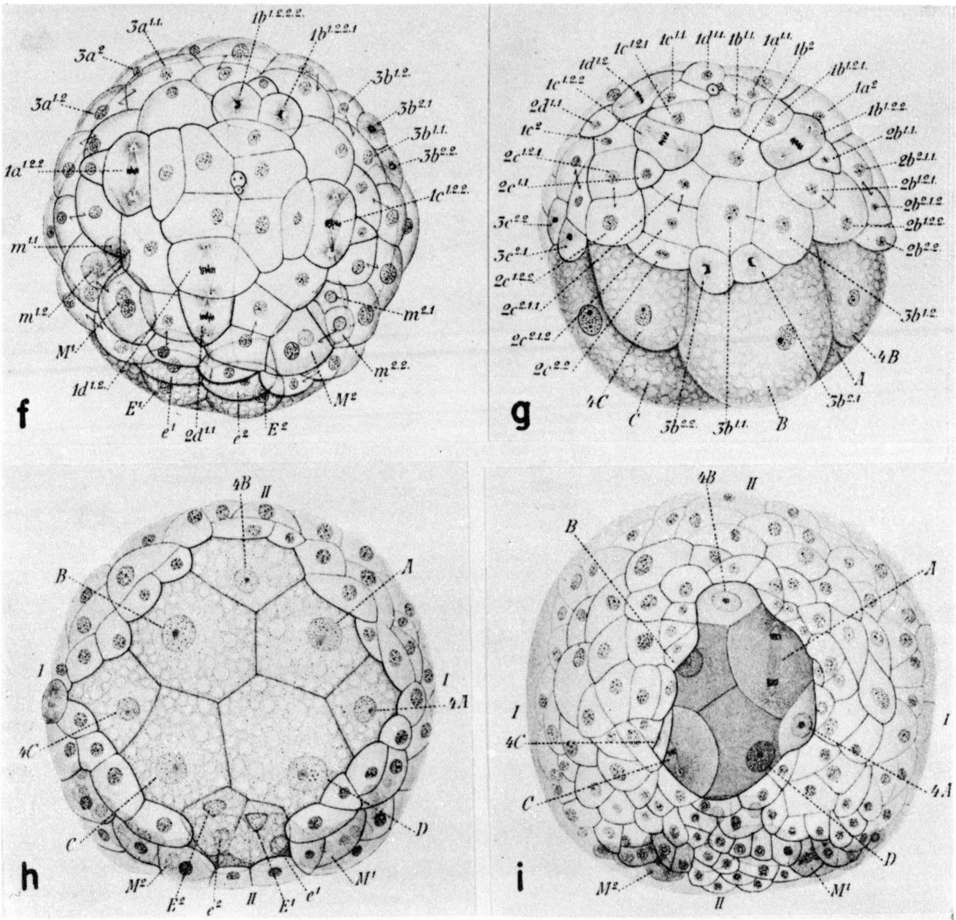

Figure 176. (*continued*) (f), 68 cell stage; (g), lateral view of f; (h) and (i), ventral views of two successive late stages showing the narrowing of the blastopore. (Reproduced from E. G. Conklin, *J. Morphol.*, **13** (1897), by permission of The Wistar Press, Philadelphia)

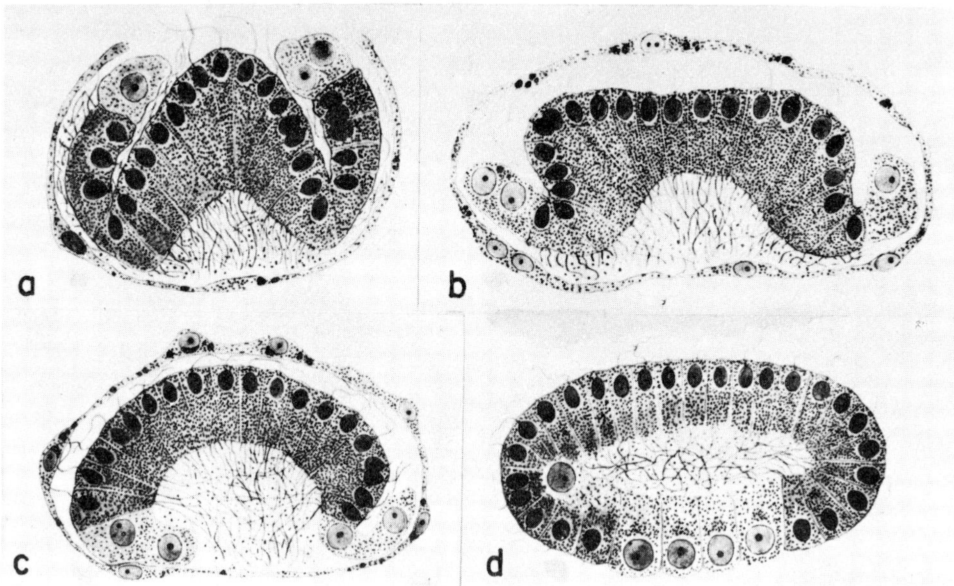

Figure 177. Gastrulation (inversion) through epiboly in *Grantia*. (a)–(d), successive stages of the inversion process. The active participation of the amoeboid non-flagellated cells is clearly indicated. (Reproduced from O. Duboscq and O. Tuzet, *Arch. Zool. Exptl. Gén.*, **79** (1937), by permission of Centre National de la Recherche Scientifique, Paris)

Echinoidea

The elucidation of the dynamic aspects of the mechanism of gastrulation in Echinoidea has initiated a new era in embryology. Being presumably the most thoroughly investigated embryogenetic process, this phenomenon has been used in various contexts in the preceding part of this book for the illustration of individual embryogenetic mechanisms. There are, however, certain aspects concerning the cooperation between the separate mechanisms which have been neglected so far. These will be taken up in the present section, but first I shall call attention to the figure published by Selenka in the last century which clearly shows that gastrulation in *Synapta* is executed by the same mechanism as that observed in sea urchins (Figure 178).

The mechanism of gastrulation in Echinoidea has been discussed in great detail by Gustafson and Wolpert (e.g. 1967). The present survey is based largely on their results and, except for some particular points which will be mentioned below, my interpretation coincides with the views advanced by them. We shall begin the discussion with the blastula, where at a certain stage a thickening of the wall occurs at the vegetal pole. Gradually a concavity arises, at the same time as a lively pulsatory activity may be observed at the proximal side of the vegetal cells. Gustafson and Wolpert (1967) have suggested that the thickening and the curvature of the cell sheet result from increased adhesion between the cells. This proposition is not acceptable, for, as shown above, if cells adhere more to each other than to the substrate, then the law of minimum free energy predicts that they will separate from the latter and form a spherical aggregate. Furthermore, elongation of the cells under the given circumstances cannot occur unless an

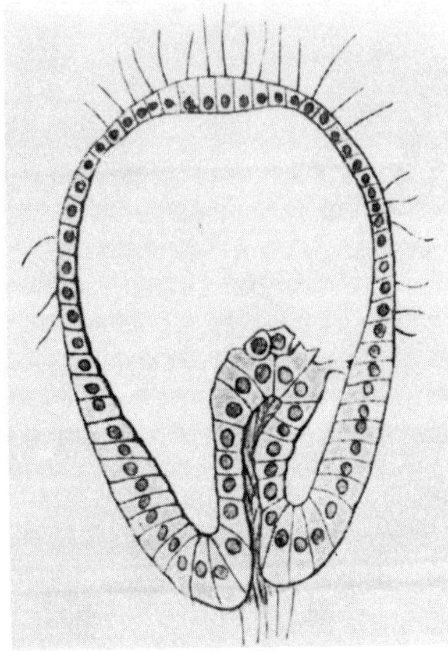

Figure 178. Gastrulation in *Synapta* through participation of filopodia from cells at the vegetal pole. (Reproduced from E. Korschelt and K. Heider (1936), *Vergleichende Entwicklungsgeschichte der Tiere*, after Selenka, by permission of Gustav Fischer, Jena)

internal cytoskeleton prevails. However, this precondition seems to be fulfilled, since the cells are about to be transformed into sf-cells, as revealed also by the pulsatory activity, which can be nothing but incipient filopodium formation. As sf-cells do not adhere to each other, it is obviously necessary to account for the origination of the concavity by other means than adhesive forces between the cell surfaces. On this point I have no particular suggestion except that the cells may be club-shaped, the thin ends being attached to the hyaline membrane. If this explanation is correct, it follows that the pressure in the blastocoele cannot be very high.

The first generation of sf-cells formed at the vegetal pole, the primary mesenchyme cells, become detached and enter the blastocoele, but the secondary mesenchyme cells, which subsequently arise, preserve their association with the hyaline membrane, thus ensuring that they may be instrumental in the execution of the process of gastrulation. Located at the presumptive tip of the archenteron, they extend filopodia that become attached to one side of the blastula wall, the presumptive dorsal side, along which they pull the endodermal cells towards the animal pole. The fact that this activity stops short of the apical plate was related above to the finding that the sl-cell filopodia cannot attach themselves to the inner surface of these cells (Gustafson, 1963).

There is a fixed spatial relation between the side at which this event takes place and the location where the stomodaeum is formed; they are always situated at diametrically opposite walls of the blastula. This implies that, from a purely geometrical point of view, predetermination of either the dorsal or the ventral side automatically entails fixation of the other one also. Functionally, however, the situation is different, for since the filopodial migration at the dorsal occurs earlier in time than any activity at the ventral side, it is seen that the location of the latter unconditionally is determined with the onset of gastrulation. To presume that the ventral side is predetermined therefore seems to involve enormous, and quite unnecessary, requirements as to the organization of the egg.

The possibility exists, of course, that a median polarity prevails in the fertilized egg, responsible for the determination of the dorsal side. However, the very fact that the bilateral symmetry is obliterated when the diversification of the apical plate cells is suppressed rather suggests that the animal–apical polarity is directly involved in the fixation of the symmetry plane. Hence it seems that support obtains for the contention that no median polarity prevails in the sea urchin egg, and that the attachment of the filopodia to one side of the blastula wall is a chance phenomenon by which the median plane becomes determined. As a consequence of this event the vegetal–apical polarity is epigenetically converted to a median one through a morphogenetic mechanism. This proposal does not exclude that the orientation of the median polarity may be influenced by experimental means.

Another aspect of the bilateral symmetry must be mentioned, namely, the varying thickness of the cell aggregate in the blastula wall. As evidenced in Figure 158, the dorsal wall is thin compared to the ventral one, but the greatest wall thickness is found in the presumptive flagellated bands. This bilaterally symmetrical pattern must either be predetermined or else be functionally related to those activities suggested as being responsible for the median polarity. As already pointed out, it cannot be referred to the adhesivity of the cells. As a matter of fact, it seems quite unlikely that the blastula wall, or rather the hyaline membrane to which many filopodia become attached, should be rigid enough to resist the pull exerted by the latter without deformation. And if this occurs, the direction of the forces predicts that cells must be accumulated towards the ventral side, just as is actually observed. In advancing this suggestion it is submitted that a certain elongation may ensue when cells are squeezed together. The same phenomenon is also revealed in a median view, where the stretching entails a displacement of the apical plate towards the ventral side (cf. Figure 158). In this context it is worth noting that

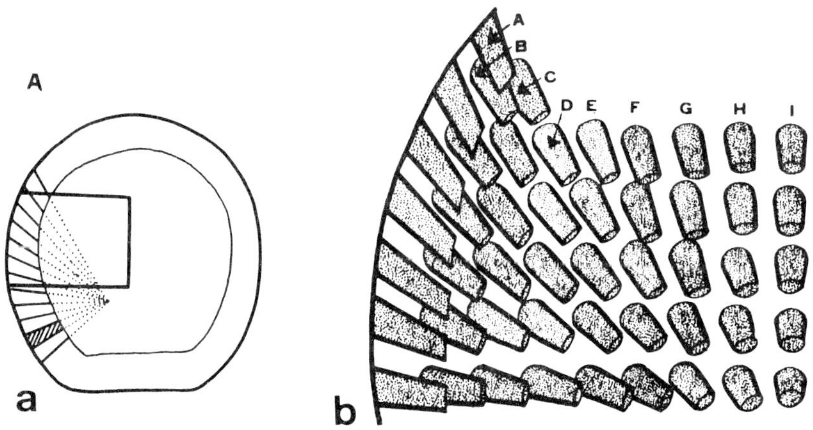

Figure 179. Orientation of the ectodermal cells in the sea urchin embryo. (a), whole embryo; (b), perspective of the enclosed part of (a). (Reproduced by permission from K. Okazaki, T. Fukushi and K. Dan, *Acta Embryol. Morphol. Exptl.*, **5** (1962))

Okazaki, Fukushi and Dan (1962) have observed fan-shaped optical patterns radiating from the mesenchyme cells, suggesting strains of the nature envisaged here (Figure 179).

For the sake of completeness it may be mentioned that, as pointed out by Gustafson and Wolpert (1967), the swelling of the blastocoele must also result in a thinning of the cell aggregate on the blastula wall, but the effect of this phenomenon should be more evenly distributed.

Cephalochordata

The development of *Branchiostoma* offers a challenge to the embryologist for at least two reasons, namely, the strategic phylogenetic position occupied by the subphylum to which it belongs and the fact that, in spite of several well-documented reports on its embryogenesis, no comprehensive explanation of the mechanisms involved has so far been provided. The present attempt to improve upon this situation is based mainly on Conklin's very careful studies (1932).

It is a typical feature of most observations on the development of *Branchiostoma* that the cells of the blastula are depicted as rather typical epitheliocytes, except at the ventral side, where the small cells located there are distinctly spherical, thus suggesting that they are sl-cells (ms, Figure 180a and b). Whether or not the remaining cells are

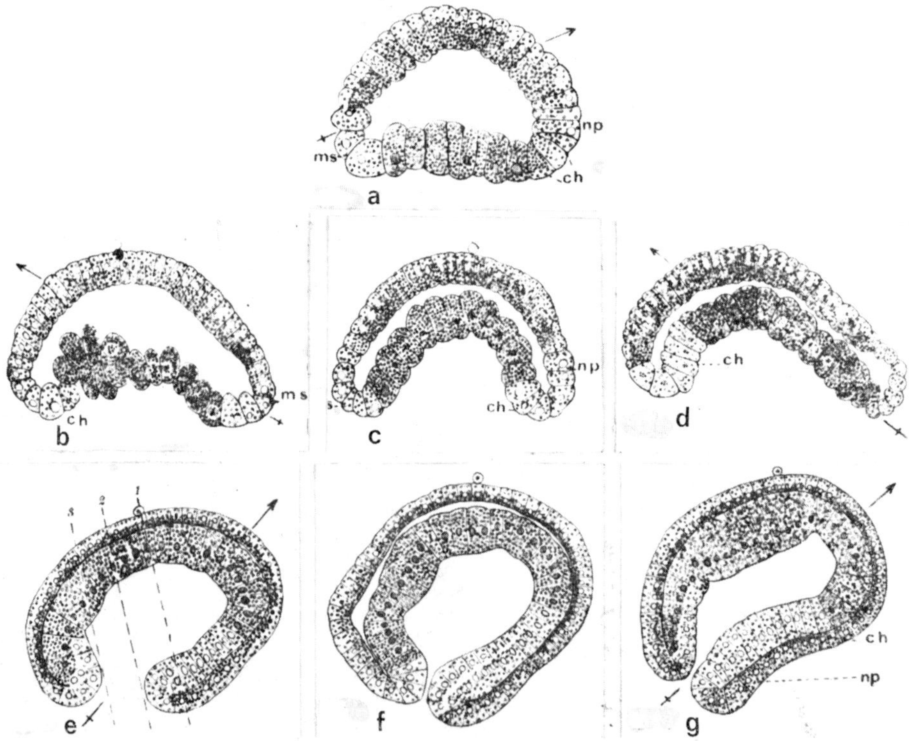

Figure 180. Gastrulation in *Branchiostoma*. (a)–(g), median or sagital sections of successive stages. Animal pole indicated by a polar body. ch, chorda; ms, mesoderm; np, neural plate. (Reproduced from E. G. Conklin, *J. Morphol.*, **54** (1932), by permission of The Wistar Press, Philadelphia)

colligocytes is impossible to decide, but they are clearly more adhesive than the ventral cells. Under any circumstances it must be inferred from the fact that the purported sl-cells remain in the aggregate rather than invade the blastocoele that they are kept together by a hyaline membrane. As shown by Conklin (1932), the blastula has the form of a pear, the narrow end being towards the ventral side. If the shape of the embryo is determined by the joint tension of a supracellular membrane and the intracellular adhesion, this deviation from a sphere is easily accounted for by the difference in cell adhesivity just mentioned, and it seems that the flattening of the aggregate of large endodermal cells may be explained in the same way (Figure 180a).

Subsequently, gastrulation is initiated with an arching of the endodermal cell aggregate into the blastocoele. Conklin (1932) mentions that Hatschek (1881) tried to account for this phenomenon by 'exosmosis', presumably involving that fluid is transported from the blastocoele to the surroundings. Without regard to the probability of this mechanism as such, it may be stated that if it occurs, the result should, in the first place, be a deformation of the part of the blastula wall possessing the lowest, and not, as is actually the case, the one with the highest, rigidity. Conklin himself suggests that the 'change in the shape and rate of division of the endoderm and ectoderm cells is apparently the principal cause of the infolding of the endoderm plate' (1932, p. 88). Concerning this mechanism it may be pointed out that if a vesicular aggregate is enclosed in a rather tight-fitting envelope, then continued cell divisions, entailing an increase of the surface area, may indeed have the outcome envisaged by Conklin. However, even this epiauxetic mechanism should result in folding at the weakest, not at the strongest, side. At any rate, there is a large perivitelline space between the egg and the egg membrane (Figure 36), which definitely serves to reject Conklin's proposition.

It therefore seems necessary to infer that the arching of the endodermal plate involves the deployment of some other kind of force, and on the basis of the gastrulation mechanisms outlined above, the most natural supposition appears to be that filopodia from sf-cells are engaged in this process, which may thus be an invagination. On this point the literature rather leaves one in the lurch, for nowhere is it possible to find a description of the participation of filopodia in this event. However, Conklin writes in a footnote (1932, p. 89): 'The early stages of invagination show many irregularities of form, especially in the invaginating area . . . Owing to these irregularities, many specimens are hard to orient and only the more regular ones are shown in the figures'. There is good reason to presume that these irregularities are the outcome of filopodial activity.

As it is, I have found only two illustrations which suggest that invagination takes place. One is published by Conklin (Figure 180b), and if this one reflects the actual event, it appears that the filopodia are formed by the endodermal cells proper, indicating partly that they are not colligocytes at this time, and partly that they undergo the lf-transformation which in a previous chapter has been proposed to be part of the normal diversification of intestinal cells. The other figure is found in MacBride (1910), and is reproduced in Figure 181. As it represents a rather late stage of gastrulation, it is not possible to infer anything about the initiation of this process, but the presence of filopodia between the germ layers is clearly shown in the drawing.

As the invagination comes to an end, a rather shallow gastrula is formed (Figure 180c and d), the opening of which gradually becomes closed at the same time as an elongation takes place (Figure 180e, f and g). Before we can discuss the mechanisms involved in this process and its consequences for the further development it is necessary to state certain

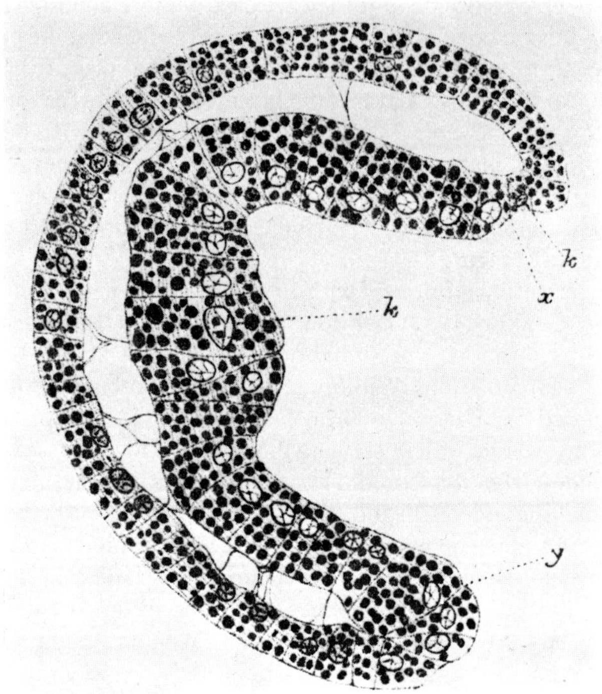

Figure 181. Filopodia between ectoderm and endoderm during gastrulation in *Branchiostoma*. (Reproduced from E. W. MacBride, *Quart. J. Micros. Sci.*, **54** (1910), by permission of the Clarendon Press, Oxford)

topological rules of gastrulation which greatly facilitate the visualization of the changes taking place. These rules, valid only when a hyaline membrane is present, are as follows:
(1) All the cells that cover the walls of the archenteron are, before gastrulation, situated below the limit of invagination (or gastrulation).
(2) When the blastula wall consists of one cell layer, all the cell-types present in the interior after gastrulation are before this event represented by distinct primordial areas in the surface.
(3) The distribution of primordial areas formed by the cells kept together by the hyaline membrane, i.e. generally all cells when the blastula wall contains one cell layer, and the outer cells when several layers obtain, is topologically equivalent before and after gastrulation.

From the two first rules we can infer that the primordial areas for the notochord, the mesoderm and the endoderm must in *Branchiostoma* be present in the blastula wall below the limit of gastrulation.

One of the remarkable discoveries made by Conklin (1932) during his studies of embryogenesis in *Branchiostoma* is that, whereas the presumptive notochord, as is usual, is located at the dorsal side, the mesoderm is represented by the small cells at the ventral side mentioned above; actually the primordium has the shape of a ventro-lateral crescent (Figure 182a and b). On the basis of Conklin's observations we can construct a fate map (Figure 183a). A schematic representation of the topology of the

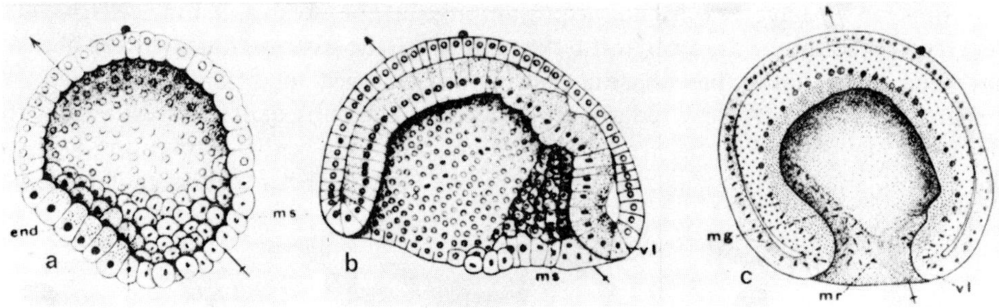

Figure 182. The origin of the mesoderm in *Branchiostoma*. (a), mesodermal crescent at the ventral lip; (b), mesodermal crescent at ventral lip after invagination; (c), shift of mesodermal crescent from ventral to dorsal side. end, endoderm; ms, mesoderm; vl, ventral lip. (Reproduced from E. G. Conklin, *J. Morphol.*, **54** (1932), by permission of The Wistar Press, Philadelphia)

invaginated tissues is shown in Figure 183 (b). The fate map produced on these considerations fits quite well with the one established by means of vital staining by Tung *et al.* (1962), as evidenced by Figure 184. The extension of the mesodermal crescents below the notochordal primordium shown in this figure is an improvement from a topological point of view, because it better accounts for the location of the mesoderm after gastrulation.

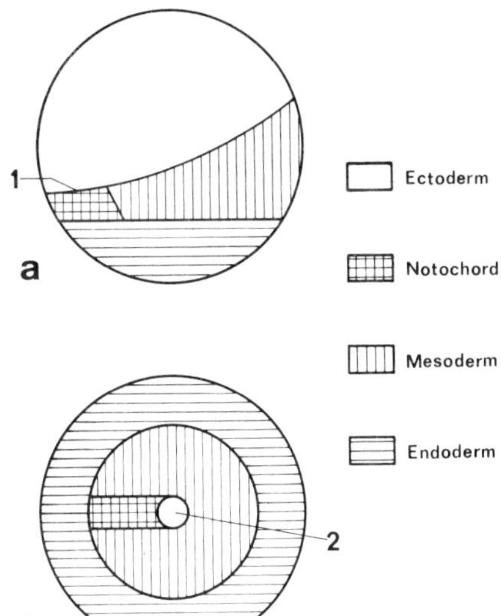

Figure 183. Topology of gastrulation in *Branchiostoma*. (a), fate map before gastrulation; (b), topological map of the archenteron, as it would appear after an approximately radially symmetrical process of gastrulation. The endoderm, which is supposed to have undergone the most extensive deformation, must be imagined to form a closed hemisphere behind the plane of the paper. 1, limit of gastrulation; 2, blastopore

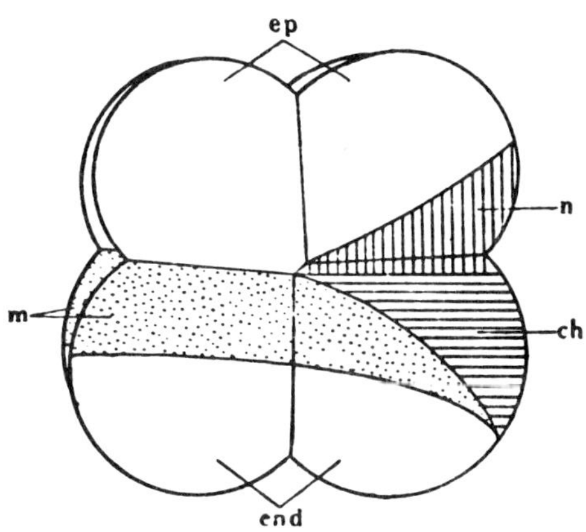

Figure 184. Fate map of *Branchiostoma* at the 32 cell stage. ch, notochord; end, endoderm; ep, epidermis; m, mesoderm; n, neural tube. (Reproduced from T. C. Tung, S. C. Wu and Y. F. Y. Tung, *Scientia Sinica*, **11** (1962))

If, thus far, we have established the presumptive fate of the invaginating cells, we may next try to account for the mechanism through which the gastrulation is carried out. We have seen above that the first phase of gastrulation is carried out through an invagination executed by sf-cells. A very remarkable feature of the subsequent process is that the closure of the blastopore occurs through extension of the ventro–lateral lips, whereas the dorsal lip is hardly engaged at all (cf. Figure 185). It seems that an explanation of

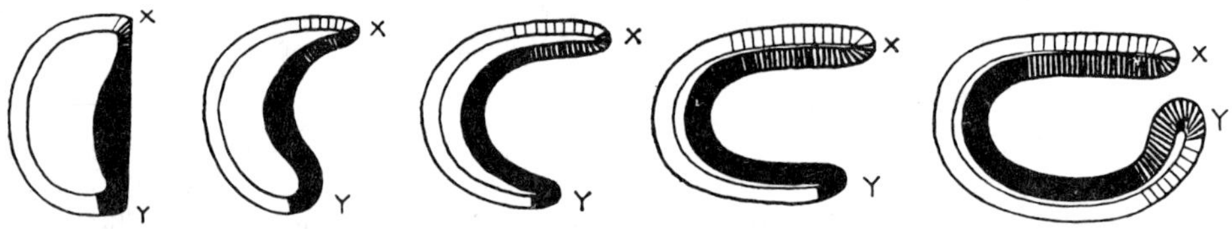

Figure 185. Gastrulation in *Branchiostoma*. Dark area, endoderm; white area, ectoderm. X and Y indicate the regions where extension of the germ layers occurs. (Reproduced from E. W. MacBride, *Quart. J. Micros. Sci.*, **54** (1910), by permission of the Clarendon Press, Oxford)

this state of affairs may be found in the cell-type distribution discussed above. For one thing, apart from the mesodermal cells, the other cells in the blastula appear to be colligocytes. If this holds for the presumptive chordocytes and neurocytes, it may be anticipated that the cell aggregates they form, after establishment of contact, become laminated. As a consequence these cells cannot contribute to morphogenesis except through epiauxesis, provided they undergo division.

Contrariwise, the cells in the mesodermal crescent, being sl-cells, expectedly engage in a lively epibolic activity as soon as they come in contact with the ectoderm, resulting in a stretching of the ventro–lateral sides of the blastopore. Owing to the shape of the mesodermal primordium these epibolic movements must be most extensive in the median plane, decreasing gradually from there at both sides. This distribution of the activity corresponds exactly to that required for the observed manner of blastopore closure. The proposed mechanism of epiboly is in my opinion corroborated by Figure 182 (b), from which it appears that the ectoderm cannot follow suit with the epibolic movement, which therefore leads to an accumulation of mesodermal cells.

Conklin (1932) observed lively mitotic activity in the mesoderm and explained the mechanism discussed here as epiauxetic, in accordance with prevailing notions. It must, of course, be accepted that the enlargement of the areas of the various cell aggregates of the embryo partly or wholly is a consequence of cell divisions, but it is impossible to accept the proposal of Conklin that the folding of the mesoderm in Figure 182 (b) is a result of epiauxesis, for if the mesodermal cells were colligocytes they ought to remain attached to the substratum. A final remark: expansion of the cell aggregates does not seem automatically to lead to constriction of the edges; some kind of tension is probably necessary to account for this event.

As shown in Figure 183 (b), the notochordal and mesodermal primordia may in a topological map be represented by a ring surrounding the blastopore. In actual fact it seems that very few mesodermal cells are left at the ventral lip, so a dorsalwards change in their location must occur in association with the epibolic movements, a change that is recorded by Conklin (1932) in Figure 182 (a) and (b). It is very difficult to visualize how this displacement comes about.

We may finish this section with a discussion of the distribution of the cell-diversification patterns. Since both the myocytes and the chordocytes are filocytes, it appears that all the equatorial cells belong to this cell class. Superficially, then, it may seem difficult to invoke the median polarity in the lf-transformation. It must be recalled, however, that the mesodermal cells remain sl-cells for a long time, their transformation occurring after they have lined up along the notochord. It therefore seems likely that the median, or rather the dorsal–median, polarity is responsible for the lf-transformation in the presumptive chordocytes, whence it spreads through induction, partly to the ectodermal neurocytes and partly to the mesodermal myocytes.

The shape of the mesodermal primordium can be explained only on the assumption that a ventral–median polarity obtains, an important function of which clearly is to delay the sc-transformation in the mesodermal cells. A cytoplasmic apical polarity seems to ensure the establishment of the particular differentiation pattern represented by the gut cells. Referring to the proposed mechanism of invagination, apparently filocytes are present in the endodermal plate at an early stage, suggesting that the lf-transformation takes place spontaneously in some of these large yolk-laden cells. This event would correspond to that occurring in the dorsal lip of Amphibia.

Conklin (1932) observed that the cytoplasmic translocations caused by the movements of the pronuclei are involved in the establishment of the median polarity. In particular, it was found that 'hyaloplasm' is accumulated at the ventral side. In agreement with the function in cell division ascribed to mucopolysaccharides, the cells in this region are consistently smaller than at the opposite side, as evidenced already at the 8 cell stage (Figure 186). The fact that these cells later become myocytes does not agree with the exhaustion

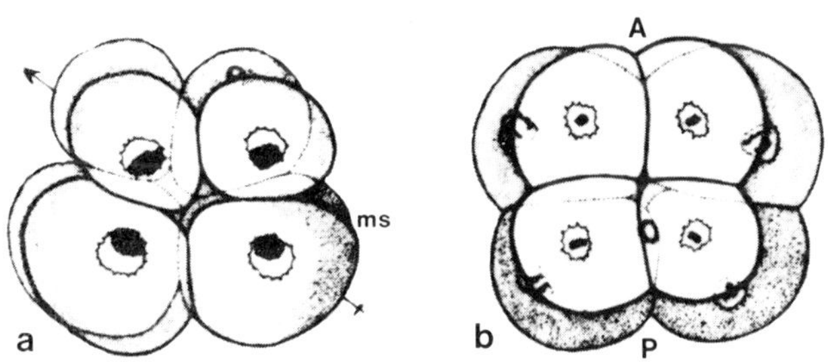

Figure 186. Cleavage in *Branchiostoma*. 8 cell stage showing that the animal cells are considerably smaller than the vegetal ones. (a), seen from the side; (b), seen from the animal pole. (Reproduced from E. G. Conklin, *J. Morphol.*, **54** (1932), by permission of The Wistar Press, Philadelphia)

hypothesis advanced above. Before these observations are used to falsify the latter it must be stressed that there are two median polarities, and it cannot be excluded without further study that at least the dorsal–median polarity conforms with the theoretical prediction.

According to the mechanism of gastrulation in *Branchiostoma* outlined here, the development is to a considerable extent 'indeterminate'. This viewpoint does not agree with the ideas advanced by Conklin (1932), but it is largely corroborated by the experimental observations of Tung *et al.* (1958; 1960).

Amphibia

Few biological phenomena have been more thoroughly studied than gastrulation in Amphibia. In the first quarter of the present century divergent opinions were voiced regarding various aspects of this process. In particular, it was held that gastrulation in Salientia and Caudata differs essentially in many respects. Vogt's vital staining experiments (1925; 1929a), designed to clarify this issue, did not really settle the dispute, for many of his interpretations were radically new. Assured of the superiority of his approach, Vogt felt that his results represented the unambiguously final answer to the problem of early amphibian morphogenesis. It is apparent that most workers in the field share this view, for almost all subsequent experimental work has been based on the fate maps established by Vogt. Dissident voices have been raised occasionally, but such views have either been rejected or ignored.

In some recent papers (1965a; 1966a) I have shown that careful scrutiny of Vogt's own published results leads to the conclusion that they do not support the claims advanced by Vogt. On the whole, it appears that the notions antedating Vogt's papers are correct. This is especially true with respect to the profound difference between the two orders, which appears to stem above all from the fact that in Salientia the ectoderm comprises two cell layers that, among other things, make it very difficult to account for its contribution to the various tissues and organs by vital staining. In the present context the discussion is confined mainly to Caudata, but certain points will be illustrated by reference to Salientia. Two developmental phases will be dealt with, *pregastrulation* and *gastrulation*, the latter of which may be equated with archenteron formation.

PREGASTRULATION

In the late blastula, shortly before the appearance of the blastopore, a certain pregastrulatory activity may be observed in both orders of Amphibia. Vogt (1929a) distinguished three different processes during this phase of development, namely, displacement of the marginal zone, 'contraction' of the vegetal area and inclination of the animal–vegetal axis (cf. Figure 187).

It appears that in the late blastula changes occur in the tension of the superficial layer of the embryo, resulting in an increase of the ratio between that prevailing at the vegetal and the animal regions, respectively. As a consequence, the pigment border moves towards the vegetal pole. It was ascertained by Vogt (1929a) that the entire vegetal area is involved in this 'contraction' by the demonstration that the area of stain marks placed all around the vegetal pole decreases during this process. Cells associated with the hyaline membrane will move together with the latter, hence a displacement of the marginal zone occurs, resulting in an accumulation of small cells from the animal hemisphere wall below the equator. This activity is more extensive in the dorsal than in the ventral side, and stain marks placed at the animal pole will therefore be shifted dorsally, an observation which was interpreted by Vogt as an inclination of the animal–vegetal axis. All three pregastrulation phenomena are obviously causally related.

We do not know anything about the mechanism responsible for these events. Nieuwkoop and Florschütz (1950) suggest that a 'contraction' of the surface layer in the vegetal zone is involved, but this proposal cannot account for the observed dorso–ventral difference. However, it is possible to distinguish three regions in the embryo, the dorsal,

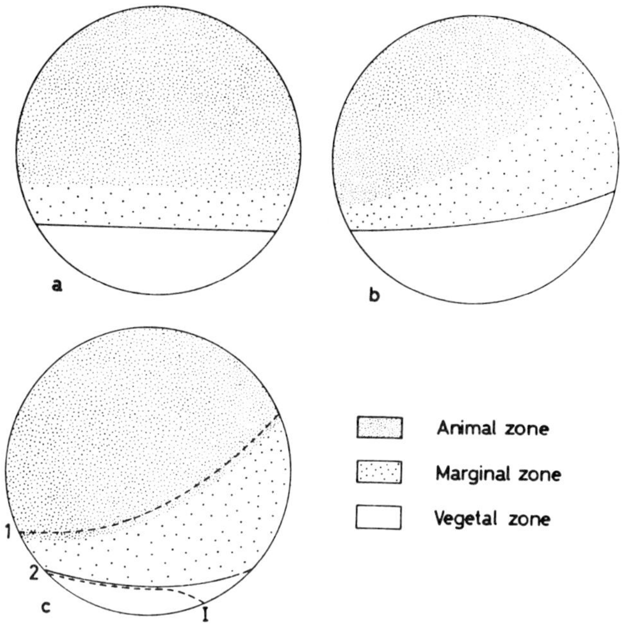

Figure 187. Pregastrular cortical changes in Caudata. (a), before fertilization; (b), immediately after fertilization; (c), late blastula to early gastrula. 1, limit of gastrulation; 2, limit of blastopore formation; I, site of initial invagination. (Reproduced from S. Løvtrup, *Acta Zool.*, **47** (1966), after Vogt, by permission of AB Läromedelsförlagen, Stockholm)

the ventral and the vegetal one, ordered with respect to increasing cell size (cf. Figure 83). If we assume that, as a general rule, the tension decreases when the superficial cell aggregate is subdivided into smaller and smaller cells, then the above ordering will also refer to the cortical tension, and hence a movement of the pigment border towards the vegetal pole might be anticipated, being more extensive at the dorsal than at the ventral side. The dorso–ventral difference in the thickness of the blastula wall, the dorsal being the thinner one (Schultze, 1895), would partly be explained by this mechanism. All the observable differences between the dorsal and the ventral side can thus be referred to one parameter, viz., the cell size. As discussed above, the initial median polarity may be responsible for the disparity in size between the dorsal and the ventral cells. It must be emphasized that the present discussion is concerned with the relative strength of the tensions; in absolute terms a gradual increase seems to occur during the blastula stages (cf. 1965c).

GASTRULATION

The process of gastrulation comprises two aspects, viz., certain morphological changes and the forces responsible for their execution. As far as the last point is concerned it has already been pointed out that the function of the median polarity involves that the lf-transformation occurs spontaneously at the dorsal side, first in the dorsal lip region. Being situated in the vegetal hemisphere, the sf-cells thus formed are quite large, and they

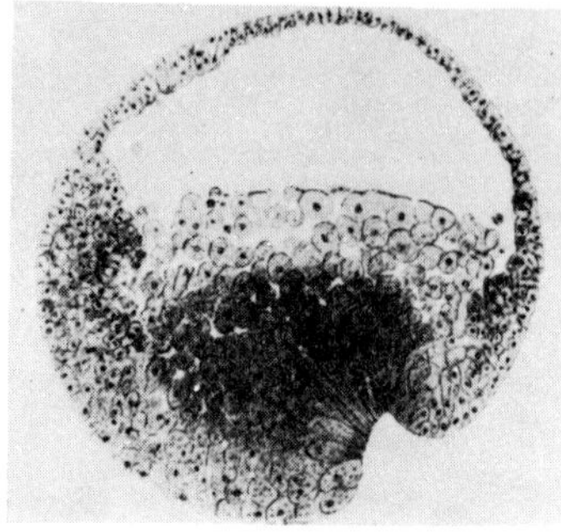

Figure 188. Section through the dorsal lip of *Triton*. The activity of the large sf-cells (bottle cells) in the dorsal lip is clearly indicated. (Reproduced from W. Vogt, *Arch. Entw.mech.*, **120** (1929), by permission of Springer, Berlin)

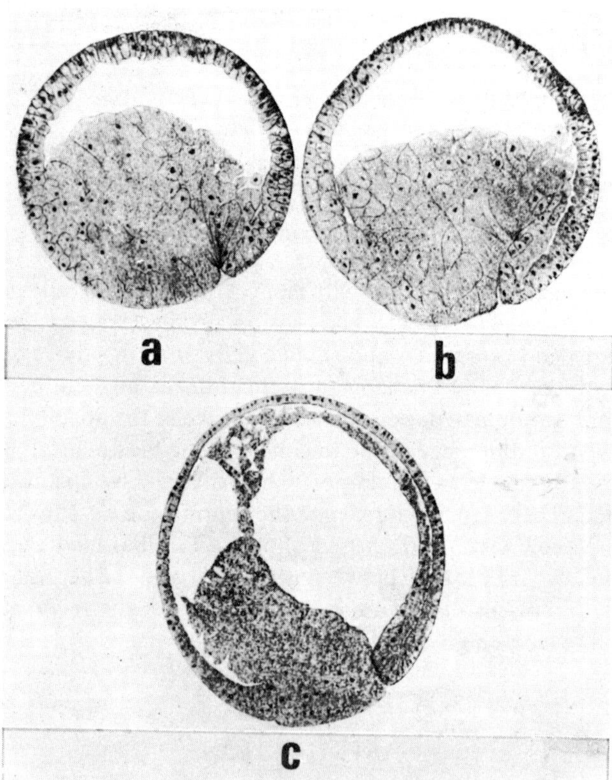

Figure 189. Gastrulation in *Pleurodeles*. Median sections through various stages: (a), early gastrulation; (b), crescent-formed blastopore; (c), circular blastopore. (Reproduced from W. Vogt, *Arch. Entw.mech.*, **120** (1929), by permission of Springer, Berlin)

are therefore able to pull the hyaline membrane into the interior of the egg (Figure 188). Gastrulation thus is initiated by a process of invagination, just as in *Branchiostoma*. It is likely that some of the filopodia, as in Echinoidea, establish contact with the dorsal wall, thereby pulling the tip of the incipient archenteron towards the animal pole. This point is not brought out clearly by any published figure I have seen, but it may be suggested by Figure 189 (a) and (b) for reasons which will become apparent presently. It is certain, however, that at some stage the invagination comes to an end and the rest of gastrulation is carried out by epibolic movements.

Epiboly requires the participation of layers of membrane-bound cells. The blastula wall in the animal hemisphere contains several layers of cells, but when the epibolic movements gather speed, two unicellular layers have established direct contact (Figure 189c). It may be argued that the thinning is only apparent, being a direct consequence of the stretching associated with the epibolic movements. This contention is not acceptable for the simple reason that the situation presented in Figure 189 (c) prevails only in a narrow region in the dorsal median; laterally there are still several cell layers between the endoderm and the membrane-bound cells (Figure 190). This state of affairs allows for

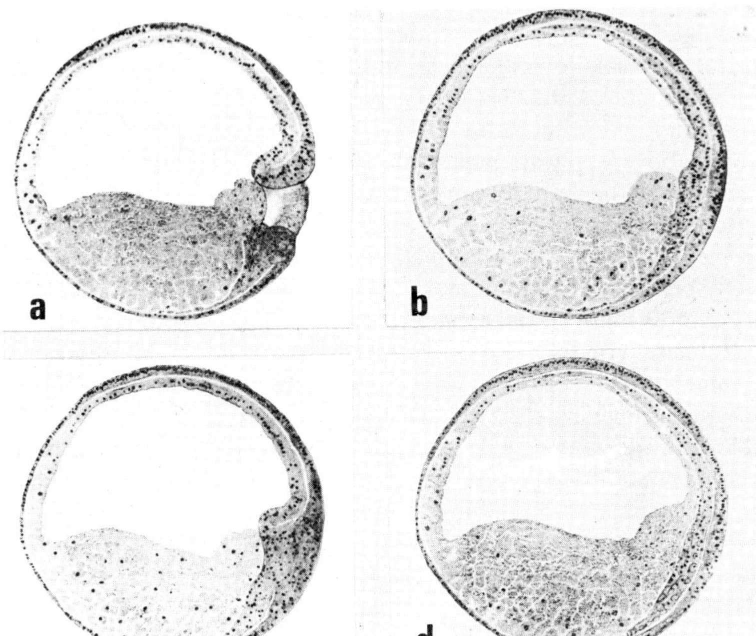

Figure 190. Sagittal sections of a late gastrula of *Pleurodeles*. (a), median section; (b), paramedian section; (c) and (d), still further lateral sections. (Reproduced from W. Vogt, *Arch. Entw.mech.*, **120** (1929), by permission of Springer, Berlin)

the conclusion that the epibolically active cells are located in the mid-dorsal region, and that consequently all other spatial displacements which occur are largely passive. Referring to Figure 189 (a) and (b) it will be realized that if these are median sections then there is some justification for the assumption that invagination is still going on at the stages depicted, since the blastula wall contains several layers of cells at this time.

According to the second rule of gastrulation advanced above, it is evident that since there is more than one cell layer in the blastula wall it is not necessary that all subsequently internal cell-types are present as superficial primordia below the line of gastrulation, but nothing prevents, of course, that they are. This latter stand was taken by Vogt (1929a) on the basis of his staining experiments, which were considered by him to demonstrate the presence of the mesoderm in two areas located laterally to the chordal primordium (cf. Figure 112). However, the entire extension of the archenteron is covered by endoderm, except in the dorsal median where the presumptive chordocytes are located, a fact suggesting that, as predicted by the first rule of gastrulation, only these two cell-types are present in the surface of the blastula before gastrulation.

In order to substantiate his interpretation Vogt (1929a) was forced to assume that a rupture occurs between mesoderm and endoderm as soon as they have turned around the blastoporal lips. The presence of the endodermal gap was taken as proof of this contention. Subsequently, the interior germ layers were supposed to form closed tubes by dorsal convergence in the endoderm and ventral convergence in the mesoderm.

I have previously (1966a) tried to show that the very detailed documentation published by Vogt (1929a) clearly demonstrates that gastrulation does not occur in the manner envisaged by Vogt. I base my view on various facts, among which an important one is that Vogt was unable to track down the location where the purported rupture occurs. This issue cannot be definitely settled before the mechanism of formation of the endodermal gap is discussed in a following section.

As mentioned above, application of the first two rules of gastrulation suggests that the superficial cells below the limit of gastrulation not belonging to the endoderm represent the notochordal primordium. From the third rule it may be inferred that, since after gastrulation the presumptive notochord stretches all the way to the blastopore, it must border directly on the endoderm almost to the ventral side. This conclusion has the consequence that the notochordal area covers the territories allotted by Vogt (1929a) to notochord and mesoderm (cf. Figures 112 and 114). It is not possible to decide exactly how far the notochordal primordium stretches laterally, but as I have pointed out (1966a), Vogt's own results show that it extends much further to the sides than suggested by his fate map. This is quite convincingly indicated in Figure 191, where it is seen that the material lying between the red and blue stain marks (Figure 191b), and possibly even some of that stained blue, has been condensed in the dorsal median forming the presumptive notochord (Figure 191c).

It is evident that, in agreement with the notions of Vogt (1929a), a very extensive dorsal convergence occurs. It must be pointed out, however, that only the presumptive notochordal cells are involved in this phenomenon. It is associated with a considerable stretching of the chordal primordium which may be referred to the fact that among the interior cells only the notochordal cells are engaged in the epibolic movements. As far as the endoderm is concerned, it is seen, by comparison of its area before and after gastrulation (cf. Figures 112 or 114 and 190), that a substantial expansion must occur. Referring to the fate of the notochordal cells, this process must involve a ventral convergence. None of the figures published by Vogt (1929a) shows this phenomenon in Caudata, but in Figure 192 (b) it is clearly borne out in *Bombinator*. The shape of the stain marks demonstrates the course of archenteron formation typical of Salientia: initially only elongation occurs, the widening of the cavity being confined to the very last stages of gastrulation (Figure 193). Vogt considered that only the horizontal parts of the marks in

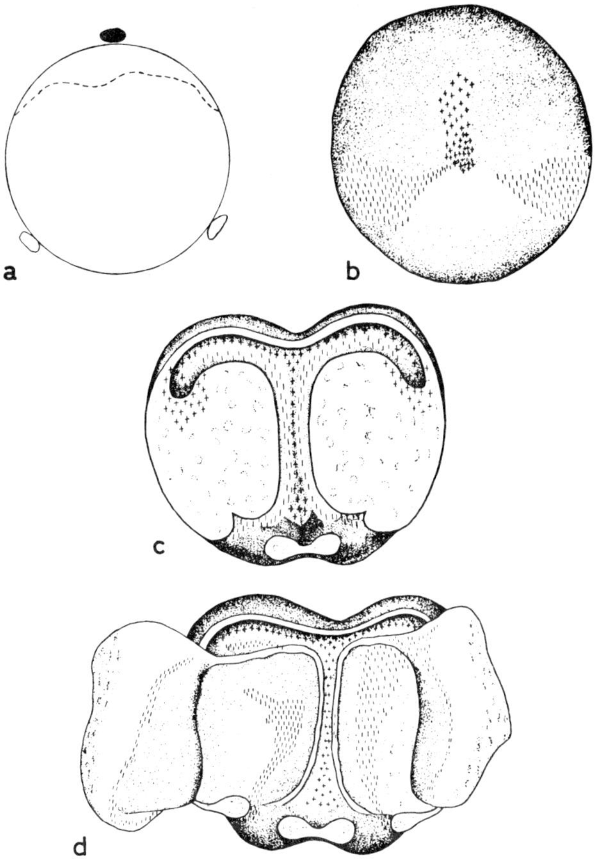

Figure 191. Gastrulation movements in *Triton*, as evidenced by vital staining. (a), location of the stain pads, one red at the dorsal and two blue at the ventro–lateral equatorial regions; (b), early gastrulation; (c), early neurula opened; (d), the endoderm has been lifted away from the mesoderm. Crosses indicate red, dashes blue staining. (Reproduced from S. Løvtrup, *Zool. Gothoburgensia*, **1** (1965), redrawn after Vogt)

Figure 192(d) belong to the endoderm, the vertical branches being the mesodermal marks shining through. Considering the relatively good agreement between the marks in Figure 192(b) and (c), it seems rather unlikely that mark 2, for instance, can have the shape indicated in Figure 192(d), when viewed through the endoderm, and presumably the situation would be still worse with respect to mark 1 which, for some unkown reason, has not been shown in Figure 192(c).

It remains to consider the free cells covering the inside of the blastula wall. It has already been made clear that these cells are pushed away from the dorsal region by the cells engaged in the epibolic movements. Everywhere else the free cells are intercalated between two cell aggregates, one which is pulled towards the blastopore, the other in the opposite direction. As a result the cells are transported backwards (Figure 194), being

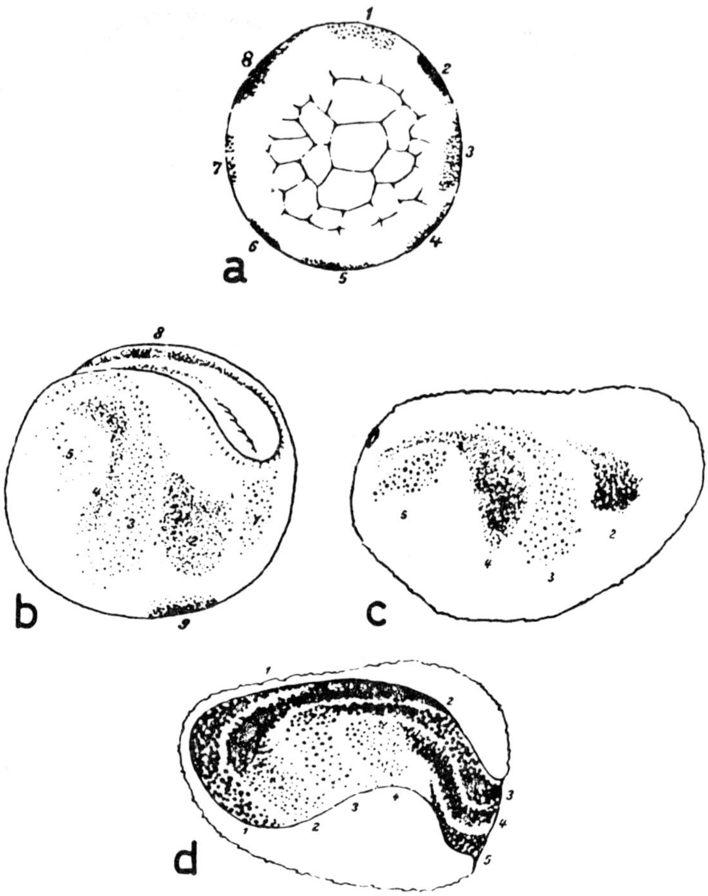

Figure 192. Gastrulation movements in *Bombinator*, as evidenced by vital staining. (a), location of the eight stain marks placed equatorially in the early blastula; (b), external view of the neural plate stage; (c), the embryo after removal of the ectoderm; (d), when opened in the median plane. (Reproduced from S. Løvtrup, *Zool. Gothoburgensia*, **1** (1965), redrawn after Vogt)

accumulated around the blastopore, mainly in the dorsal region. Since the mesoderm has been relegated from the superficial layers, there can be no doubt that it is represented by these cells which subsequently move towards the anterior end. Being free sl-cells, they may migrate independently by their own force. This notion is partly corroborated by the fact that their movements begin only when the epibolic activity has ceased, and partly by the circumstance that, as noticed by Vogt (1929a), the mesodermal marks tend to be more blurred than the other ones. Their independent displacement is borne out by Figures 191 and 192, even if the downward trend in the latter case indicates that they are influenced by the ventral convergence to which the endoderm is subjected.

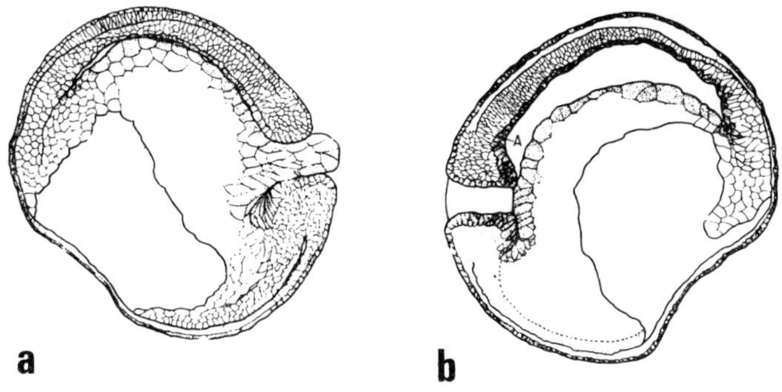

Figure 193. Archenteron formation in *Xenopus*. (a), median section from small yolk plug stage in which the archenteron is still a narrow slit; (b), median section from an early neurula, in which the widening of the archenteron through ventral convergence has been initiated. (Reproduced from P. D. Nieuwkoop and P. A. Florschütz, *Arch. Biol.*, **61** (1950), by permission of Archives de Biologie, Brussels)

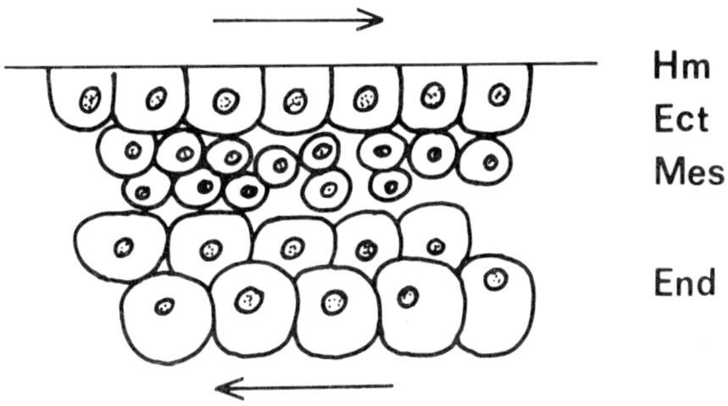

Figure 194. Diagram showing the location of the mesodermal cells between the ectoderm and the endoderm during epiboly in the amphibian gastrula. The arrows indicate the direction in which the outer germ layers are transported by the epibolically active cells around the dorsal median. From this sketch it is not obvious that the mesodermal cells will move to the right, i.e. towards the blastopore, as is actually observed. Ect, ectoderm; End, endoderm; Hm, hyaline membrane; Mes, mesoderm

LARVA FORMATION

In the present chapter we are going to deal with the most important aspects of the development following gastrulation in the three cases discussed in the preceding chapter. As far as Echinoidea are concerned, it will be possible to follow the development until the formation of the pluteus larva. In Cephalochordata and Amphibia the process of embryogenesis soon becomes very complicated. For this reason it is necessary to discontinue the discussion at a more or less arbitrary point. Here the neurula has been chosen as the last stage, a decision which, I believe, may be compatible with the chapter heading insofar as the neurula may be considered as an incipient larva.

Echinoidea

The most important aspect of larva formation in Echinoidea is the growth of the skeleton and the shaping of the gut. Both of these phenomena will be discussed here.

The basic principles of skeleton formation have been outlined above. It was seen that the primary mesenchyme cells are distributed according to a definite pattern on the blastula wall, namely, as a ring around the invaginating archenteron and two ventro–lateral chains. This distribution indicates a differential with respect to attachment of filopodia, and in agreement with the rule established by Gustafson (1963), the cells appear to accumulate where the density of cell contacts is highest. It may thus be conceded that the ectoderm acts as a template for congregation of the mesenchyme cells (cf. Gustafson and Wolpert, 1967), but it is important to note that this template is created through their own activity, the ring being situated where the blastula wall is curved as a consequence of invagination and the ventro–lateral chains where the cells are squeezed together as a result of the stretching of the dorsal wall.

In our preceding discussion of the work of Okazaki (1960; 1965) and of Gustafson and Wolpert (1961a), we have seen that skeleton formation begins with the formation of two small triradiate crystals, one in each of the clusters of cells located where the branches meet the ring. The crystals are subsequently oriented so that one of the radii is parallel with the ventral part of the ring and the other with one of the branches (Figure 159). These two radii will begin to grow straight, forming the midventral and the oral rods, respectively. When the latter reach the stomodaeal region they bend forwards, resume their straight growth and form the hood. This particular growth pattern is easily understood when it is recalled that the ventral ectoderm is curved inwards as a result of mouth formation, so that it becomes possible for the mesenchyme cells located in the ectoderm to extend syncytial filopodia to the tip of the skeleton, thereby changing its direction of growth. The two midventral rods will eventually fuse in the median plane. The third radius is initially almost perpendicular to the ring of mesenchyme cells (Figure 159). As the latter extend cables to the tip of the crystal, two branches are formed, the one directed forwards being the presumptive anal rod, the one pointing backwards the body rod. As these continue to grow straight the anal arms and the scheitel are formed. It is interesting to notice that the shaping of the latter involves a rotation of the two skeletal parts, obviously effected through interaction between the body wall and the body rods as the latter continue their growth. This rotation leads to a widening of the distance between the anal arms. The very slow growth rate of the midventral rods, the reason for which is unknown, is a precondition for this particular morphogenetic mechanism.

The invagination of the archenteron involves that filopodia are gradually attached closer and closer to the apical plate. It has been pointed out that this mechanism by no means involves a gradient of adhesivity as regards the filopodial tips, but simply that the force developed by the filopodia is proportional to their length. It therefore follows that when the archenteron tip approaches the apical plate, the filopodia which can be extended towards the ventral side are longer than any other that can be formed under prevailing circumstances, and the direction of invagination will therefore be subjected to a corresponding change. We shall not discuss in detail the mouth and coelom formation, shown in Figure 144, except to point out that as far as the latter is concerned, attachment of filopodia to the flagellated bands is of great importance (cf. Gustafson and Wolpert, 1963a; 1967).

Of outstanding interest, but as yet incompletely understood, is the subdivision of the endoderm into oesophagus, stomach and protodaeum. In contrast to the other parts, the stomach is thick-walled, supporting the view that it represents a particular diversification pattern. This circumstance may explain the absence from the stomach wall of the 'contractile' filopodia which form a peristaltic network around the oesophagus and presumably also around the proctodaeum, and possibly also the formation of sphincters between the endodermal subdivisions and between the proctodaeum and the ectoderm. However this may be, when the peristaltic activity is initiated, water is pumped from the oesophagus to the stomach, thereby dilating the latter. When the hydrostatic pressure has become large enough to open the pylorus sphincter, water is pressed out into the proctodacum and from here to the outside (Gustafson and Wolpert, 1967). We are obviously here witnessing a phenomenon that may be called *functional morphogenesis*, and which probably will turn out to be of great importance once we come to know more cases of ontogenesis in sufficient detail.

In larvae from various classes of Echinodermata, filopodia from cells in the flagellated bands establish contact with the oesophagus, coeloms, etc. (Metschnikoff, 1884; Semon, 1888; Gustafson, 1969). It has been surmised that these cells act as nerve cells, conducting impulses to the internal organs from the flagellated cells, which thus function as sensory cells. It is interesting to observe that here also we have two different cell classes represented among the neurocytes, some vibriocytes, vsf-cells, in the flagellated band, some non-vibriocytes, nsf-cells, the filopodia of which serve as axons. Gustafson (1969) has demonstrated that the former contain acetylcholine esterase, suggesting that they are cholinergic, whereas various kinds of evidence indicate that 5-HT is involved in the activity of the latter.

Gustafson (1969) advanced the proposal that the mesenchyme cells situated beneath the presumptive flagellated cells induce the latter to become nerve cells. This contention is in agreement with the generalization put forward here that whenever presumptive epidermal cells are subjected to the lf-transformation they become neurocytes. Another interesting discovery which has emerged from the various findings discussed here is that at an early stage it is impossible to distinguish between nerve cells and muscle cells. This strongly supports the close relationship between these differentiation patterns implied by the notion that they derive from a generalized type of sf-cells.

Besides the flagellocytes discussed here another flagellated cell-type exists in the larva, namely those in the apical plate. It has been contended above that the latter are cf-cells, possessing stereoflagella which are nothing but extensions of the microtubular cytoskeleton, and being covered by a dense layer of sulphated mucopolysaccharide which prevents the attachment of filopodia. Since it is unlikely that stereoflagella can be involved in sensory activities, it appears that these two properties exclude that the apical plate can function as a sensory organ in the pluteus larva. This suppression of the animal pole region as the main location of neural tissue seems to be the price that has had to be paid for the establishment of bilateral symmetry in this and related classes.

Cephalochordata

It was suggested in the preceding chapter that the major phase of gastrulation in *Branchiostoma* is the outcome of epibolic movements, executed by a ventro–lateral crescent of sl-cells. When this activity comes to an end, the endodermal cells are still united, presumably through the hyaline membrane, with the presumptive myotomal and notochordal cells. Soon afterwards separation occurs between the three cell populations, with the consequent formation of an endodermal gap. The latter is closed by fusion when the edges of the endodermal aggregate meet in the dorsal median.

In order to account for these observations according to the morphogenetic mechanism of separation discussed above, it is necessary to presume that both the notochordal and the endodermal cells are colligocytes, and the mesodermal cells solocytes. Actually, the appearance of the endodermal cells suggests that they undergo the sc-transformation at a very early stage, and as regards the chordoblasts their passivity during gastrulation has been explained by their being colligocytes. There had been a tendency in the previous literature to consider the separation to involve an 'evagination' of a colligocyte aggregate consisting of the mesodermal cells, and much theoretical speculation has been based on this supposition. If this was true, there should always be an evagination cavity in the centre of the mesoderm. Actually, this is observed occasionally, but not consistently. Careful scrutiny of Conklin's drawings reproduced in Figure 195 brings out this point.

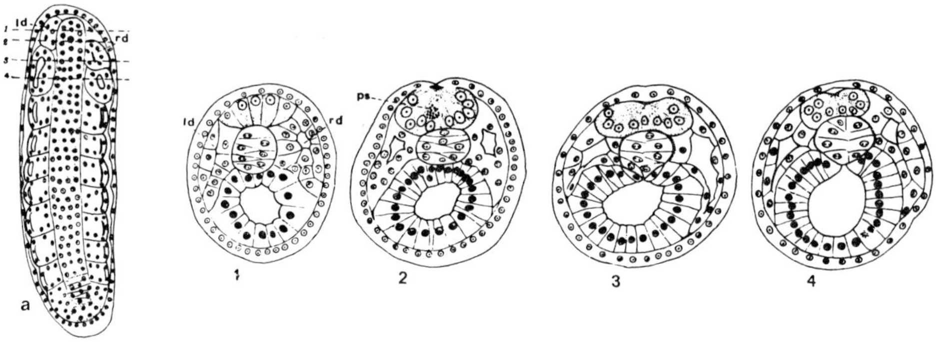

Figure 195. Mesoderm formation in *Branchiostoma.* (a), frontal section from an early larva; 1–4, transversal sections through the levels indicated in (a). ld and rd, left and right diverticula of the gut; ps, pigment spot. (Reproduced from E. G. Conklin, *J. Morphol.*, **54** (1932), by permission of The Wistar Press, Philadelphia)

The less schematical illustrations in Cerfontaine's paper (1906) are even more convincing; the irregular arrangement of the mesodermal cells definitely supports the notion that the cells in question are sl-cells (Figure 196). The sporadic presence of an evagination cavity may depend on the persistence of the hyaline membrane, which will impose a certain order on the cells. The phenomenon of separation shows that the membrane at some stage no longer unites the various cell groups, and it may therefore be justified to assume that it gradually disappears. This phenomenon may be the most important aspect of the mechanism of separation. When the mesodermal cells have left their association with the endodermal and notochordal cells, the latter might be anticipated to fuse when they meet. The only explanation I can mobilize to account for the fact that this does not occur (Figure 195) is that collagenocytes have begun to elaborate a sheath around the chorda, thereby preventing the fusion.

The function of the notochord for the subsequent stretching of the larval body has already been discussed. It may be mentioned in passing that no observations have ever accounted for the origin of the collagenocytes which form the sheath, but it may be presumed that these sf-cells arise among the mesodermal cells.

If the presumptive myotomal cells are sl-cells we have, as seen here, a means to explain the separation between the various cell aggregates in the interior of the embryo. However, even more important is the fact that by this means we have the clue to the

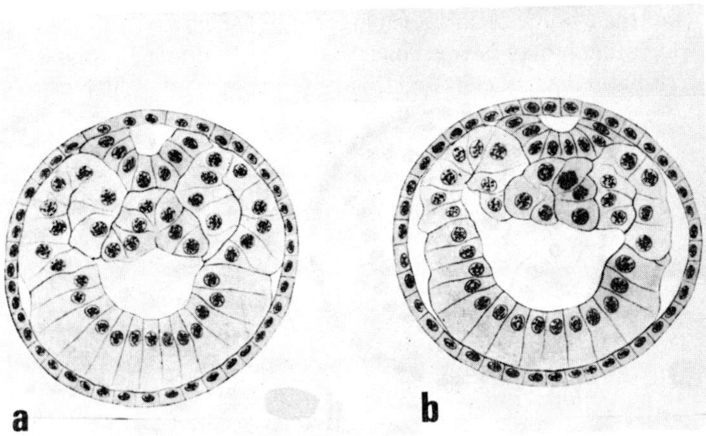

Figure 196. Mesoderm formation in *Branchiostoma*. (a), transverse section from a larva with nine pairs of somites; (b), transverse section near the posterior end close to the neurenteric canal. A virtual myocoele is indicated by heavy lines, but it is questionable whether this interpretation is justified. (Reproduced from P. Cerfontaine, *Arch. Biol.*, **22** (1906), by permission of Archives de Biologie, Brussels)

segmentation occurring in the mesodermal cells. If the latter evaginated as colligocyte aggregates, there would be every reason to anticipate that they formed tubular aggregates at each side of the chorda, just as the nerve cells do on top of the latter. Contrariwise, if they are solocytes, which transform into colligocytes along a temporal gradient, then the segmentation, involving the formation of vesicular aggregates through lamination and polarization, is easily understood.

We shall finally briefly discuss the events occurring in the ectoderm simultaneously with those just described. The first thing happening is that the cells above the notochord become transformed into filocytes, as indicated by their elongated shape. This process clearly begins in the posterior end, spreading gradually forwards (Figure 197). This

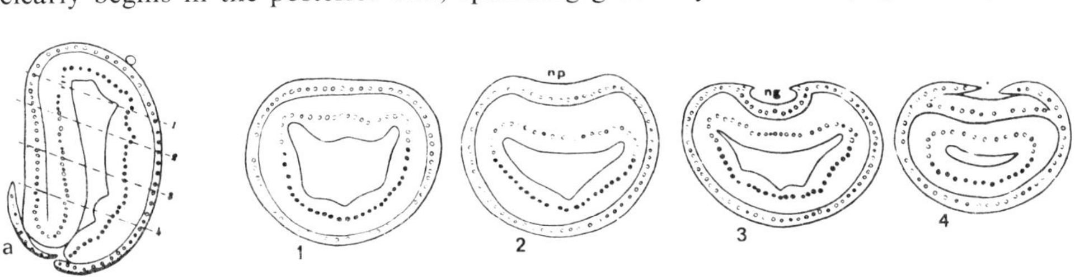

Figure 197. Neural plate formation in *Branchiostoma*. (a), median longitudinal section from early neurula; 1–4, transverse sections through the levels indicated in (a); ng, neural groove; np, neural plate. (Reproduced from E. G. Conklin, *J. Morphol.*, **54** (1932), by permission of The Wistar Press, Philadelphia)

course of events suggests that the transformation is induced by the subjacent cells. That neural induction occurs in *Branchiostoma* has been demonstrated by Tung *et al.* (1962b). As this process advances, ectodermal cl-cells begin to spread over the neural plate cells, presumably by epiauxesis. It is evident from the cavity formed between the two cell aggregates that they do not adhere to each other, a fact which reasonably may be referred to the impact of the hyaline membrane (Figure 198). This cavity has nothing to do with the neurocoele, rather it may be presumed that the membrane disappears after the transfer to the interior and that the cells first form a solid aggregate which through lamination

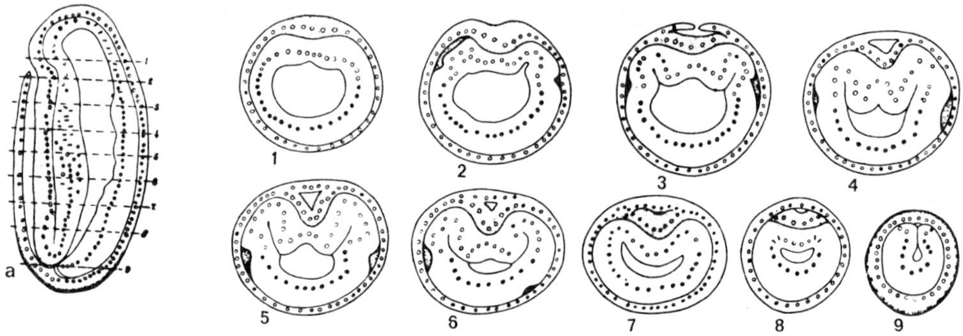

Figure 198. Neurulation through expiauxesis in *Branchiostoma.* (a), median longitudinal section from a neurula; 1–9, transverse sections through the levels indicated in (a). (Reproduced from E. G. Conklin, *J. Morphol.*, **54** (1932), by permission of The Wistar Press, Philadelphia)

subsequently becomes tubular. In the overgrowth of the neural plate, separation must occur between the cells in the neural plate and the cl-cells surrounding them. In this case the two kinds of colligocytes obviously do not form a stable joint aggregate.

Amphibia

In the present section two separate phenomena will be discussed, the process of *postgastrulation* (Vogt, 1929a) comprising all the activities taking place in the interior of the embryo, and *neurulation* proper, which is concerned with the induction and the closure of the neural tube and various associated events.

POSTGASTRULATION

Postgastrulation embraces the following processes: forward migration of the mesoderm, separation between the notochordal and endodermal primordia with the ensuing formation of the endodermal gap, separation between the notochordal and mesodermal cells, segmentation of the latter and closure of the endodermal gap.

At the end of gastrulation the free mesodermal cells accumulated around the blastopore begin to migrate forwards between the ectoderm and the endoderm, a phenomenon which is clearly shown in Figure 199. In the early neurula transversal sections show that these mesodermal cells are associated with the notochordal cells, whereas a cleft, the endodermal gap, separates the two edges of the endoderm (Figure 200). Sections made in the late gastrula demonstrate that this event progresses in an anterior direction and that it is intimately associated with the advancement of the mesodermal cells, i.e. it does not occur before these have established contact with the notochordal cells (Figure 201). It is proposed that, as discussed above, differences in intercellular adhesion between the various cell-types are instrumental in this process of switchover, but it must be pointed out that it is a precondition that the hyaline membrane no longer exerts its function of keeping the cells together.

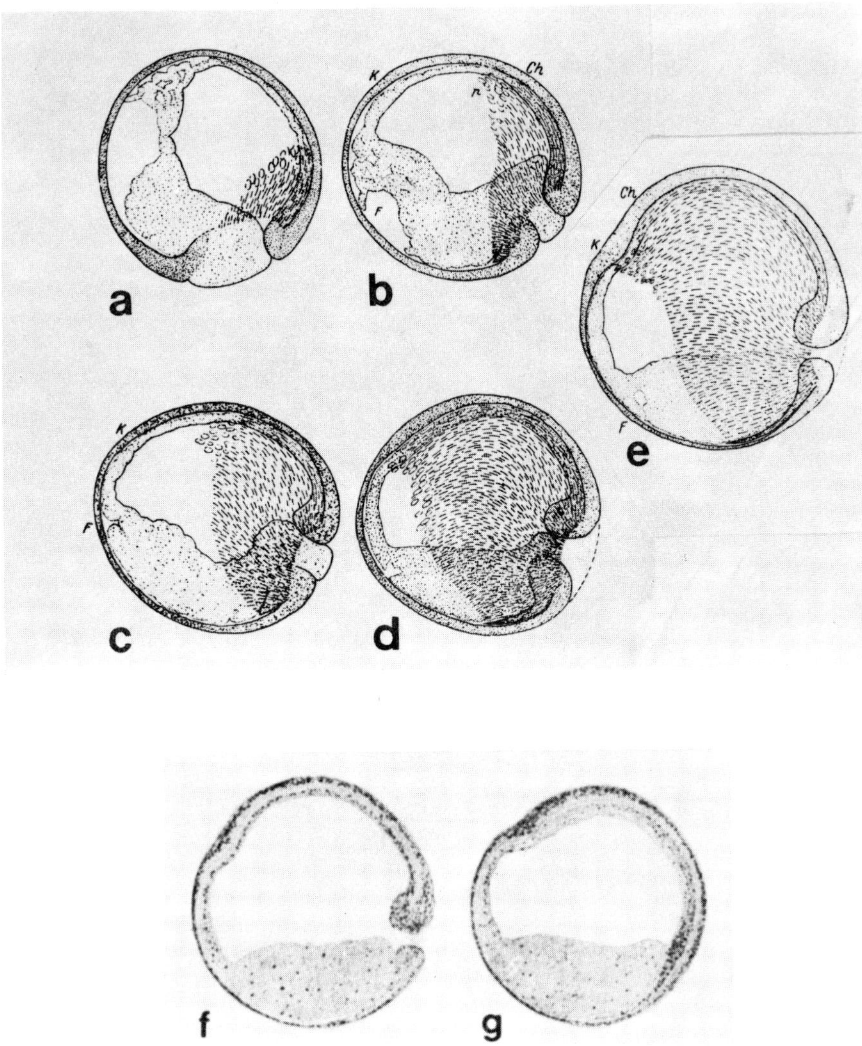

Figure 199. Mesoderm migration in *Pleurodeles*. (a)–(e), diagrams of median sections with reconstruction of the mesodermal layer; (f), median, and (g), paramedian, section from the stage represented by (e). Ch, presumptive end of the notochord; F, blastocoele; K, anterior end of the neural plate; p, prechordal mesoderm. (Reproduced from W. Vogt, *Arch. Entw.mech.*, **120** (1929), by permission of Springer, Berlin)

The proposal outlined here is supported by the fact that the anterior ends of the endodermal gap, the notochord and the lateral mesoderm all lie in the same transversal plane (cf. Vogt, 1929a). This substantiates not only that the notochord is separated from the endoderm, but also that the process is accomplished by the mesoderm. Vogt noted that, judging from the appearance of the cells, there is no difference between the notochordal cells and those lying anteriorly between the latter and the endoderm. This cell aggregate, the prechordal plate, is laterally bordered by mesodermal cells, but these cells

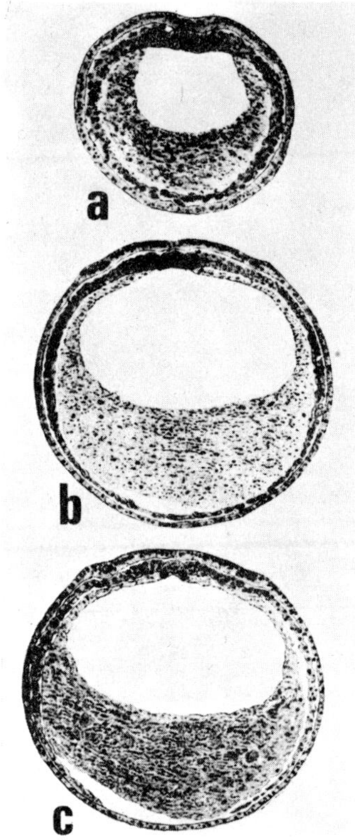

Figure 200. Endodermal gap in *Pleurodeles*. Transversal sections from an early neurula. (a), posterior; (b), the middle, and (c), the anterior, region. (Reproduced from W. Vogt, *Arch. Entw.mech.*, **120** (1929), by permission of Springer, Berlin)

advance in scattered formation ahead of the more coherent sheet of presumptive somite mesoderm (Figure 199e). The very dispersion of these cells seems sufficient to account for their failure to accomplish a separation between the endoderm and the prechordal plate cells.

Subsequently, separation occurs between the notochordal and the mesodermal cells. This phenomenon may be a consequence of the transformation of the former into colligocytes, and the ensuing deposition of a collagenous sheath (Figure 136). The formation of the somites must be based on the mechanism invoked to explain segmentation in *Branchiostoma*, i.e. a gradual transformation of the mesodermal sl-cells into cf-cells, leading to cellular vesicles, the walls of which become laminated with surrounding cell aggregates (Figure 137). Associated with the formation of the somites, an extensive accumulation of mesodermal cells occurs in the dorsum. Nothing is known about the

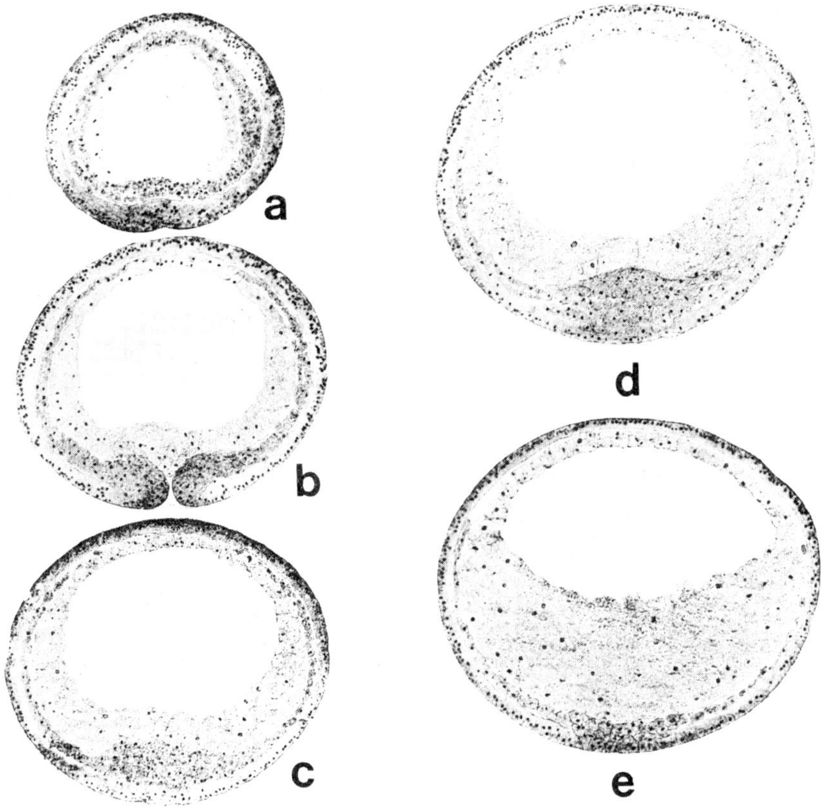

Figure 201. Formation of the endodermal gap in *Pleurodeles*. Frontal sections from a late gastrula: (a), dorsum; (b), blastopore; (c), ventral lip; (d) and (e), below the ventral lip. (Reproduced from W. Vogt, *Arch. Entw.mech.*, **120** (1929), by permission of Springer, Berlin)

causes of this phenomenon; possibly the transformation of the cells into colligocytes acts as a trapping mechanism.

It finally remains to account for the closure of the endodermal gap. The mechanism of this process was described by Vogt (1929b), who observed that cells leaving the sclerotome angle of the somites attach themselves to the free endodermal edges, bridging the gap and pulling the edges together. Once this task is executed, they leave the endoderm and form the hypochorda (Figure 135). Judging from their behaviour, the cells engaged in this operation are sf-cells. They later scatter without any trace, a phenomenon which also is understandable if the suggested classification is correct.

NEURULATION

In this section three separate phenomena will be discussed, namely, the changes on the cellular level accompanying neurulation, the formation of the neural tube, and the nature and the function of the sf-cells which during neurulation leave the neural plate and neural crests.

Neural plate cells

The initial step of neurulation is the induction of the lf-transformation in the presumptive neural plate cells. This fact is borne out by many different observations; suffice it to mention here that the shape and activities of neural plate and neural crest cells clearly show that they are filocytes. It is also certain that the only agent which can be made responsible for this event is the notochordal primordium (cf. 1967a), implying that the induction occurs only in a narrow strip around the dorsal medium, whence it subsequently spreads by contact induction from cell to cell in the neural plate, as suggested originally by Dalcq (1946). If the lf-induction can progress in this way among the neural plate cells and if the whole dorso–lateral region contains the notochordal primordium, then there does not seem to be any reason why the latter must be transferred into the interior in order to induce the neural plate. However, as pointed out on a previous occasion (1966a), it is, for purely topological reasons, necessary to postulate that the notochordal and neural plate areas are separated by a strip of presumptive tail epidermis (cf. Figure 114). It may thus be presumed that the lf-induction, starting at the dorsal lip, spreads in the dorso–lateral ectoderm, but that this process is so slow that the cells at some distance from the point of origin have lost their competence before they are reached by the wave of induction and that they consequently will become cl-cells and, as such, act as a barrier for the further advance of the induction. Under these circumstances it evidently becomes a necessity that the potential neural inductor circumvents this obstacle by migration. The mechanism outlined here clearly presupposes that there is a dorso–ventral gradient with respect to the time at which the cells reach the stage of postcompetence.

The neural plate is surrounded at all sides by presumptive epidermal cells, suggesting that once more the lf-induction is stopped by postcompetent cells. It must be emphasized that the limit of transformation by no means coincides with the one delineating the material to be incorporated in the neural tube. Thus the presence of the roof plate in the spinal cord suggests that the fusion occurs between cl-cells at the edges of the neural plate. This phenomenon is still more clearly shown in the diencephalon and the myelencephalon, the thin roofs of which indicate that they are not made up of filocytes. The mechanism involved in the latter case has never, as far as I know, been adequately accounted for, but it must presumably be related to the peculiarities in the closure of the anterior neural tube (Figure 202), and this in turn may be referred to the fact that the notochord does not reach into this region (cf. below). If the spread of the lf-induction is checked by the circumstance that the cells reach postcompetence along a dorso–ventral gradient, it might be anticipated that the border lines between the various primordia would be approximately great circles. As it is, they are extended towards the ventral side along the limit of invagination. It does not seem difficult to explain this fact with reference to the surface displacements occurring during gastrulation.

Even if a certain stretching occurs in the median direction as a consequence of the epibolic movements, it is nevertheless evident that a substantial accumulation of material occurs in the presumptive neural plate during gastrulation, particularly in the posterior part. It may be argued that the elongation of the cells results from their thus being squeezed together, i.e. through a mechanism similar to that stated to occur in the sea urchin embryo. This possibility seems to be substantiated by the observation by Lehmann (1929) that if very early neural plate is transplanted to the ventral region, they lose their typical shape. There is no justification, however, for the assumption that the cell shape may be explained by this fact alone. Rather, the elongation must primarily be due to the transformation of the cells into filocytes, as evidenced for instance by the fact that it may occur in isolated explants (Barth, 1941).

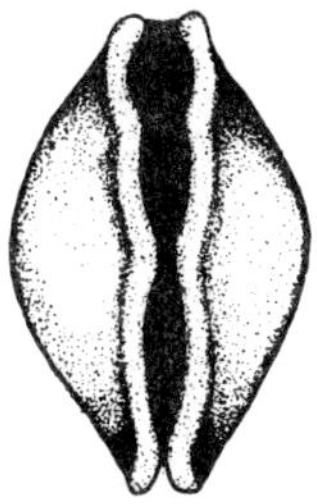

Figure 202. Neural tube closure in *Ambystoma*. The approach between the two sides varies at different levels along the neural fold. (Reproduced from R. Rugh (1962), *Experimental Embryology*, 3rd edn., after Harrison, by permission of the Burgess Publishing Company, Minneapolis

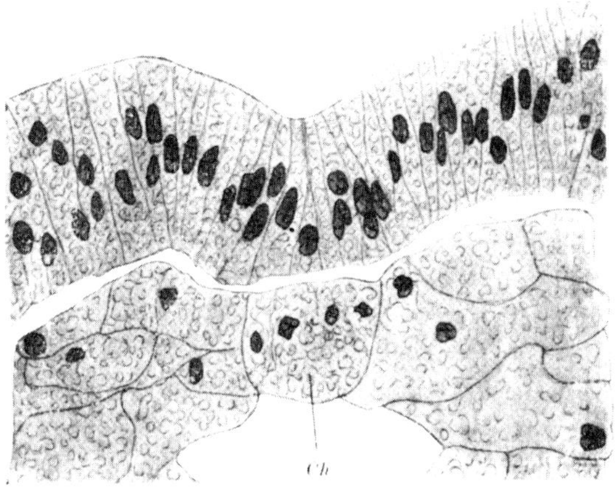

Figure 203. Transversal section through the neural plate of *Triton*. Ch, notochord. (Reproduced from F. E. Lehmann, *Arch. Entw.mech.*, **113** (1928), by permission of Springer, Berlin)

I have on previous occasions (1965a; 1966a; 1967a) put forward the view that the cells in the neural plate are cf-cells. This contention is certainly borne out by the shape of the cells (Figure 203), and also by the behaviour of the cell aggregate during the closure of the neural tube and in the participation of the latter in lamination processes with the surrounding cell aggregates (Figure 137), but above all by the circumstance that a collagen membrane, the *pia mater*, is deposited on the surface of the neural tube in a matrix of sulphated mucopolysaccharide (Figure 156). However, the presence of sf-cells at various locations in the neural plate is well documented and, as we have discussed on an earlier occasion, there is good evidence in support of the view that nerve cells are mainly flagellated and non-flagellated sf-cells. This point is brought out by many different observations; suffice it here to reproduce an illustration in which the behaviour and form of the cells in the neural tube clearly reveal their cell-order affiliation (Figure 204). The most satisfactory way to solve this apparent predicament is to assume either that the cells, when exposed to the external medium, undergo the reversible sc-transformation but revert to the solocyte form after their transfer to the interior, or that both solocytes and colligocytes are present in the neural tube.

In the early neural plate a system of pigmented lines appears, first in the median and later at the outlines of the neural plate. This pigment accumulation reflects the formation of shallow *grooves*; following the terminology of Baker (1927), a *median longitudinal* groove traverses the neural plate in the median plane, whereas two others, the *intermedial* and the *lateral* grooves, are outlining the neural crest more or less continuously around the neural plate (Figure 205). Pigment accumulation may also occur less regularly at other locations (cf. Lehmann, 1929). Concerning this phenomenon, Holtfreter (1943a, p. 271) stated that 'just as in the blastoporal region, the irregular [and the regular too,

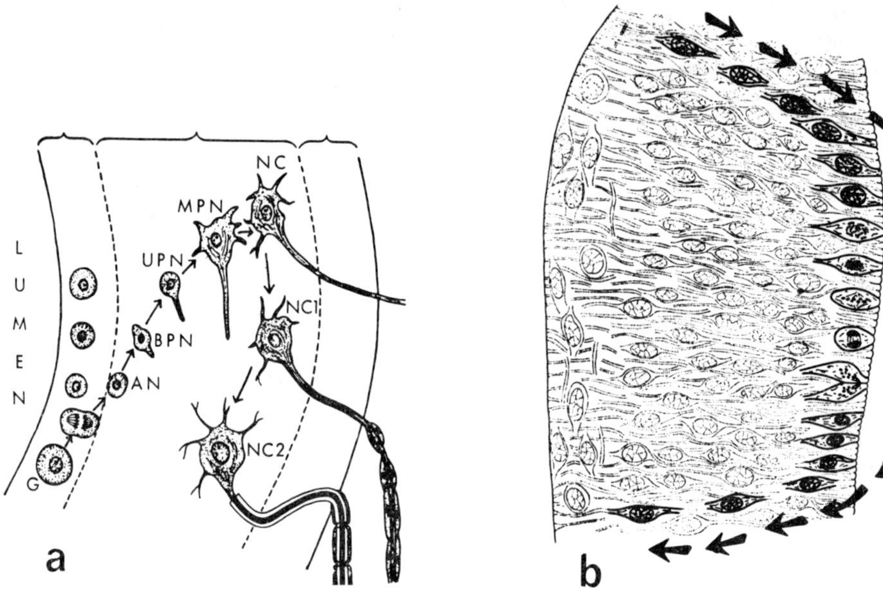

Figure 204. Cells in the neural tube. (a), schematic representation of diversification of a cell from the germinal layer into a nerve cell; (b), schematic representation of cells in the alar plate region of the neural tube wall of a 10 mm pig embryo. Cells with nuclei in various stages of karyokinesis are shown in dark outline. The arrows indicate a succession of mitotic stages in the cells which migrate towards the lumen during early prophase and away from it during late telophase. The participation of filopodia in these displacements may be expected and is indicated in the figure. G, germinal cell; differentiated nerve cells. (Reproduced from W. J. Hamilton, J.D. Boyd and H. W. Mossman, (1972), *Human Embryology—Prenatal Development of Form and Function*, by courtesy of W. Heffer, Cambridge and from F. C. Sauer, *J. Comp. Neurol.*, **62** (1935), by permission of The Wistar Press, Philadelphia)

presumably] accumulation of dark coating material in the medullary plate is the result of a preliminary ingression of a number of surface cells'. Since the formation of the blastopore may be referred to the sf-cells located there, and since this cell-type is the only one which is able to carry out activities of this kind, it is reasonable to infer that there are sf-cells present in the neural plate. The pigmented grooves suggest that the cells retain their association with the hyaline membrane, but also that they have filopodia attached somewhere in the interior.

In this context it is important to note that the extension of the median groove and the notochord coincide completely (Kingsbury, 1924; Baker, 1927), and since the formation of the former is suppressed in the absence of the latter (Lehmann, 1928), we are justified in proposing that the cells in the median groove are attached to the notochord. The sf-cells responsible for this activity do not have any name; apparently their existence generally has been overlooked. Since they presumably are responsible for neural keel formation in *Ambystoma jeffersonianum* (Baker, 1927, Figure 206), I have previously (1965a; 1966a; 1967a) adopted the name *neural keel* cells, even if a neural keel normally is not formed in either Amphibia or other Vertebrata. As far as the other grooves are concerned, it seems that their formation may be explained on the assumption that filopodia from the sf-cells involved, the *neural crest* cells, span the inside of the neural crests.

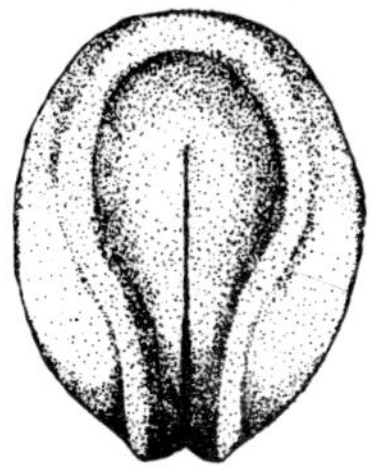

Figure 205. Neural plate stage of *Ambystoma*, showing the five pigmented streaks demonstrating the activity of the neural keel and neural crest cells. (Reproduced from R. Rugh (1962), *Experimental Embryology*, 3rd edn., after Harrison, by permission of the Burgess Publishing Company, Minneapolis)

We are here approaching an interesting problem, namely, whether the formation of the grooves is a sign that these particular locations are particularly rich in sf-cells, or whether it merely reflects that the cells residing there have special opportunities for the attachment of filopodia in their surroundings. The length of the median groove rather supports the latter alternative. Recalling the rules of attachment of filopodia in the sea urchin embryo, it may be envisaged that the neural keel cells become anchored to cell contacts in the notochordal cells which, as we have seen above, at an early stage transform into colligocytes (Figure 136). As far as the neural crest cells are concerned, the folding of the ectodermal cell aggregate may make the hyaline membrane at the opposite sides accessible to their filopodia. Thus it does not seem implausible that, at some stage, most or all of the neural plate cells are sf-cells before some of them, temporarily at least, become cf-cells. This proposition is corroborated by the fact that in the presumptive brain area the cells form a pseudostratified columnar epithelium, in which the lower cells remain attached to the hyaline membrane through slender processes (cf. Lehmann, 1928). Apparently the transformation succession in the neural plate is sl→sf→cf (→sf), whereas in the notochord it is sl→cl→cf (cf. Figure 136).

Neural tube formation

Neural tube formation involves two events, viz., that the neural tissue is transferred into the interior of the body and that it is arranged as a tubular aggregate. As revealed by the process of neurulation in *Branchiostoma*, all that is required for this purpose is that the neural plate is overgrown by the presumptive epidermis; subsequently the cells may

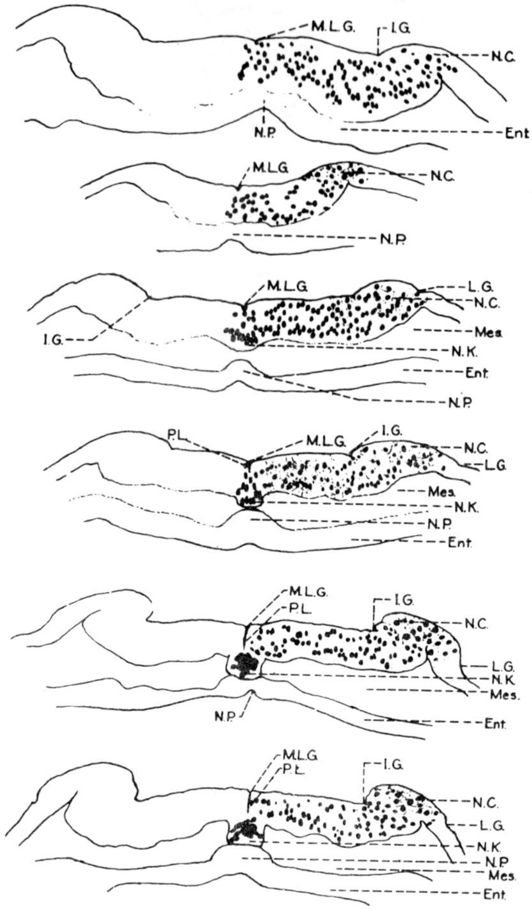

Figure 206. Successive stages in the development of the neural plate in *Ambystoma*, demonstrating the function of the neural keel. Ent., endoderm; I.G., intermedial groove; L.G., lateral groove; Mes., mesoderm; M.L.G., median longitudinal groove; N.C., neural crest; N.K., neural keel; N.P., notochord primordium. (Reproduced from R. C. Baker, *J. Comp. Neurol.*, **44** (1927), by permission of The Wistar Press, Philadelphia)

spontaneously form a tubular aggregate. A similar mechanism is observable in Pisces (cf. Nelsen, 1953), whereas in the other vertebrates the mechanism of neural tube formation is more complex. Using neural tube formation in Amphibia as illustration, we shall try to establish the differences obtaining, and their consequences.

There is no doubt that in either case the essential feature is a convergence of the ectoderm towards the dorsal midline where fusion occurs. In Amphibia this goes hand in hand with the dorsal convergence of the mesoderm mentioned previously, which poses the question whether these two processes are functionally related. It is possible that the mesoderm in its migration carries the ectoderm, a mechanism which obviously requires intimate contact with the germ layers. It may also be imagined that the ectoderm is the active partner, but in that case the mesoderm is necessary as a substrate. The latter alternative is supported by the observation by Selman (1955) that the force directed towards the dorsal median plane is reduced by incision in the lateral ectoderm, although neurulation is not suppressed by this expedient. It has been shown by Jacobson (1962) that the endo-mesodermal mantle must be intact for neurulation to occur, but this finding does not demonstrate the active participation of the mesoderm.

This question clearly is not settled, but the drawings shown in Figure 207 indicate that both factors are involved, the mesoderm mainly during the initial, and the ectoderm during the final, phases of neural tube closure. As far as the ectodermal activity is concerned, two possibilities obtain, viz., epiboly and epiauxesis. In the former case the ectodermal cells might be expected to push the mesodermal ones in a direction opposite to that of their own displacement, similarly to what happens during gastrulation. Since this prediction is not borne out, there is some reason to believe that the spreading occurs through epiauxesis, and hence that the ectodermal cells have transformed into colligocytes at that stage of development. This latter point may be supported by the fact that the formation of the epidermal basement membrane begins at about this time (Kemp, 1959).

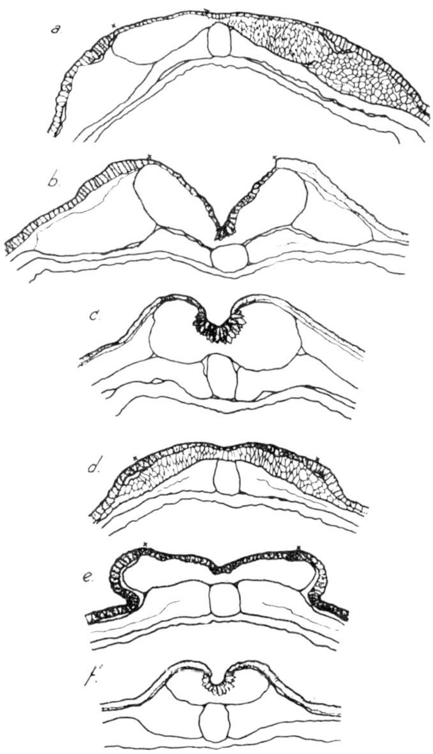

Figure 207. Neural tube formation in *Xenopus*. Successive stages in (a)–(c), the brain region and (d)–(f), the anterior trunk region. (Reproduced from P. D. Nieuwkoop and P. A. Florschütz, *Arch. Biol.*, **61** (1950), by permission of Archives de Biologie, Brussels)

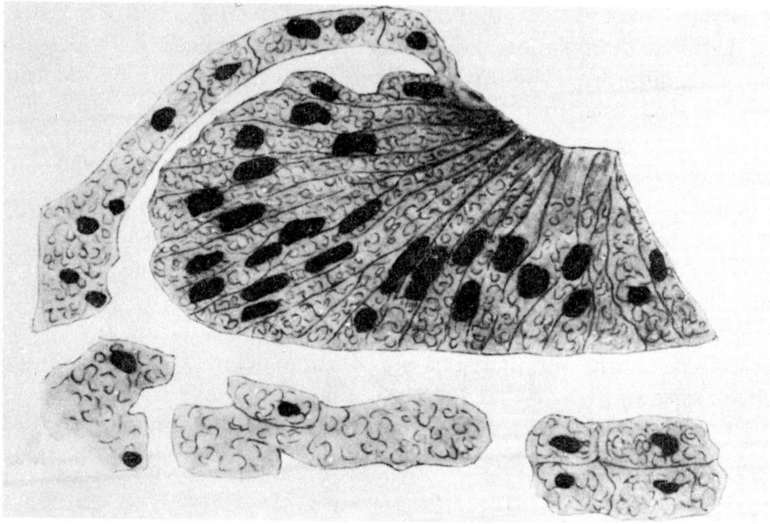

Figure 208. Transversal section through the neural crest of *Triton*. (Reproduced from F. E. Lehmann, *Arch. Entw.mech.*, **113** (1928), by permission of Springer, Berlin)

Besides the participants discussed here it is possible that the neural crest cells also are involved, pulling the border of the ectoderm over the edges of the neural plate (Figure 208). In this way they may ensure that cl-cells rather than cf-cells fuse in the dorsal median, thereby forming the roof plate. Through the mechanism outlined here the fusion leads directly to the formation of a tubular aggregate. The spontaneous formation occurring in Cephalochordata makes greater demands upon the intercellular adhesion of the neural cells; this may be one reason why dorsal ganglia are absent in this subphylum.

As already observed, the neural plate is anchored to the notochord through the filopodia of the neural keel cells. This seems to be a necessary, if not a sufficient, precondition for the hinge-like folding of the neural tube observed in many cases (Figure 207). This association is not indispensable for neural tube closure, since this event may take place even when the notochord is partially suppressed experimentally, but its function is evidenced by the fact that the neural plate bulges out in the region concerned (Lehmann, 1928). As just pointed out, *Branchiostoma* is distinguished by the absence of neural keel cells, a circumstance which entails that no attachment occurs between notochord and neural tube. It will be realized that this feature does not prevent elongation of the neural tube together with the rest of the body when the notochord becomes subjected to stretching. There is thus no doubt that the latter is instrumental in the lengthening of the neural tube, as envisaged by several authors (Lehmann and Ris, 1938; Nakamura, 1942; Hörstadius, 1944; Mookerjee, 1953), and that the attachment is of no special use in this context. However, this expedient seems nonetheless to be of utmost importance, for it ensures that the stretching of the neural tube is confined largely to the posterior part, thereby reserving an essential portion of the neural tube for brain formation (Figure 157).

Differentiation and function of sf-cells from the neural plate and neural crest

The present section is concerned with the differentiation patterns and the final location of the various sf-cells which are known to leave the neural plate and neural crest. This question, particularly the aspect concerning their contribution to the dorsal ganglia, has attracted much attention, and various methods have been employed to solve this problem (cf. Hörstadius, 1950). In the present context the discussion will be confined to the xenoplastic transplantation experiments of Raven (1936; 1937), because they, better than any other method, make it possible to trace the fate of the cells which migrate far from the site of the transplant. The work was performed with *Ambystoma* and *Triturus*, in which the differences in nuclear size and colour are so distinct that it is easy to determine the origin of a cell on this basis.

When a median piece of the neural plate is transplanted, it is found that most of the cells remain in the neural tube, but some leave and contribute to the formation of other structures (Raven, 1937). The mobility thus displayed suggests that the cells in question are sf-cells, or at least that they were so at the time they left the neural plate. Raven does not pay any particular attention to the neural keel cells, and the experimental set-up does not permit any discrimination between these and the neural plate cells. It may be justified to presume, however, that the results to a substantial extent pertain to the former. Some of these cells from the transplanted neural plate become Schwann cells. As already discussed, the differentiation pattern of these cells is unknown; their only distinctive feature is the adhesivity of their surfaces. Other cells from this origin contribute to the sympathetic ganglia, showing that they differentiate into neurocytes. In some of the results discussed by Hörstadius (1950) it was observed that neural plate (keel) cells may also end up in the spinal ganglia. If the migration of the cells is random, this finding appears very plausible, even though the relative location of the two types of ganglia rather favours the possibility that neural plate cells form the sympathetic, and neural crest cells the spinal, ganglia. Apart from these two differentiation patterns some melanocytes were found in the *pia* and *dura mater* of the spinal cord. It is a matter of surprise that no neural plate cells are present in the notochordal sheaths, the first of which is formed during the early stages of neurulation (Mookerjee *et al.*, 1953), long before the neural crest and the sclerotome cells have begun their differentiation. Since the chordocytes cannot themselves form such membranes (Holtfreter, 1939; Mookerjee, 1953), and since no other possible source prevails, it seems justified to postulate that the collagenocytes responsible for this achievement must be derived from the neural plate or keel.

Similar experiments with the neural crest (Raven, 1936) show that cells of this origin end up in the spinal and sympathetic ganglia (neurocytes), others in the *pia mater* and probably also in the epidermal basement lamellae (collagenocytes). Many of these cells also become myxoblasts, the main locations of which are in the dorsal fin and the head. Melanocytes from the neural crest are found partly randomly dispersed together with cells of other differentiation patterns, and partly in the dorsal epidermis in a pattern which may be species specific (Figure 209).

The differentiations exhibited by the various sf-cells discussed are members of the sequence of selfdifferentiation patterns established above, their order being:

collagenocytes→neurocytes→myxocytes.

The melanocytes may not be located in this succession but appear to be closely related to the neurocytes. According to the exhaustion hypothesis it may be anticipated that, if no

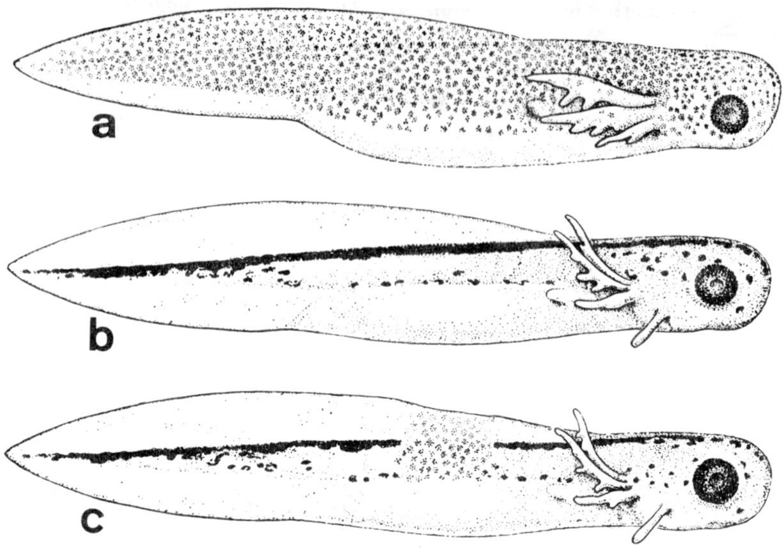

Figure 209. Pigmentation patterns in *Triturus*. (a), larva of *T. rivularis*; (b), larva of *T. torosus*; (c), larva of *T. torosus* bearing grafted segment of *T. rivularis* neural crest. The grafted melanocytes are seen to adhere to the donor pattern, except that they are unable to pass beyond the margin outlined by accumulated host pigment cells. (Reproduced from V. C. Twitty, *Growth, Suppl.*, **13** (1949))

other differences obtain, the cells differentiating first ought to be collagenocytes, followed by the neurocytes, the melanocytes and the myxocytes.

If this supposition is correct it may be of consequence for the spatial distribution of the various differentiation patterns. The collagenocytes, suggested to be the first to arise, are located in the *pia mater*, in the notochordal sheath and in the epidermal basement lamella. The source of the cells in the second membrane was proposed to be the neural plate or keel, whereas those in the *pia mater* were of mixed origin. As the somites and the epidermal cells transform into colligocytes they will adhere to each other, thus preventing or retarding the migration of the sf-cells. The cells forming after this occurrence therefore have more limited possibilities for migration. If, as anticipated, these cells mainly are neurocytes, it explains why this differentiation pattern is confined to the space between the somites and the neural tube, where they contribute to the formation of the spinal and the sympathetic ganglia. Before the last neural crests are ready to leave, the neural tube has already established contact with the epidermis. These cells which, as predictable from the selfdifferentiation sequence, are myxocytes, will be trapped in the narrow space around the dorsal median. When they assume their differential activity, the hyaluronic acid secreted by them will be instrumental in the morphogenesis of the dorsal fin. Some of these cells, escaping entrapment, may be responsible for the formation of the *arachnoidea*. The melanocytes are located dorsally, suggesting that the association between the epidermal and the coelomic walls prevents their access to the ventral part of the body. They migrate further than the neurocytes, which may indicate that they arise before the latter. On the other hand there are some indications that melanocytes have a remarkably high motility (Twitty, 1948; Gustafson and Wolpert, 1967), a property which possibly may account in part for their distribution.

The present discussion shows that in a crude way the temporal sequence of the self-differentiation patterns may account for their spatial distribution. However, if random movements are responsible for the migration of the cells, it can hardly be avoided that mixtures occur in many cases, as is so clearly brought out by the presence of melanocytes at many different locations. The consequences of this event may possibly be counteracted by homotypic induction. However, the failure of such adjustment to occur may not have serious repercussions, as long as the majority of the cells in a given location have the proper differentiation pattern.

PHYSIOLOGICAL EVENTS

As development goes on and various organs and tissues are formed, their functioning involves that new physiological activities are assumed and there is good reason to believe that in many cases they are directly engaged in the further progress of ontogenesis. It might be feasible to describe development just as well on the basis of these physiological changes as in terms of the morphological ones, which is the customary approach, even though the great similarities in physiological mechanisms found in the animal kingdom undoubtedly would allow for less discrimination. In the present context only one physiological problem will be dealt with, namely, the energy consumption and the nature of the substances utilized for this purpose. This subject has been intensely studied in Echinoidea and Amphibia and a few other of the favourite objects of experimental embryology, but the present discussion will be confined to the classes mentioned.

Echinoidea

Three problems concerning the energy metabolism will be treated here, the shape of the curves of oxygen uptake and their correlation with the epigenetic events, quantitative and qualitative aspects of the consumption of energy sources and the metabolic pathways through which the energy is made available.

OXYGEN UPTAKE

Studies on the respiration of the developing sea urchin embryo may be subdivided into two groups with respect to the methods employed. In the first group, relatively insensitive manometric methods have been used, requiring samples of several thousand embryos. This approach has been adopted by many authors, e.g. Gray (1927), Lindahl (1936; 1939b), Whiteley and Baltzer (1958) and Immers and Runnström (1960). The curves obtained exhibit a number of typical traits (Figure 210). Thus, shortly after fertilization the rate of oxygen uptake begins to rise, following an approximately exponential course. Around hatching this increase is discontinued, the rate of respiration remaining constant for several hours. In the mesenchyme blastula the curve rises again, at first steeply, later on more slowly. Similar results were obtained with the more delicate Cartesian diver method by Borei (1948). The other group comprises the work of Zeuthen (1951) and Scholander, Claff, Sveinsson and Scholander (1952), who used very sensitive methods in which the sample size varies between one and a few hundred embryos. The curves obtained by these authors are distinguished by cyclic variations in the rate of oxygen uptake, but in their general course they seldom resemble those just discussed. This

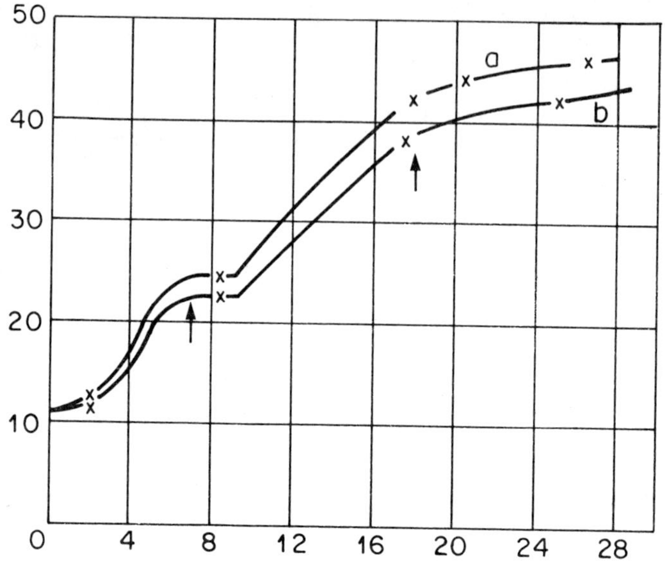

Figure 210. Oxygen uptake during the development of *Paracentrotus*. Curves a and b represent different clutches. First arrow, hatching; second arrow, end of gastrulation. Abscissa, hours; ordinate, µl O_2 per mg N and hour. (Reproduced from P. E. Lindahl, *Z. Vergl. Physiol.*, **27** (1939), by permission of Springer, Berlin)

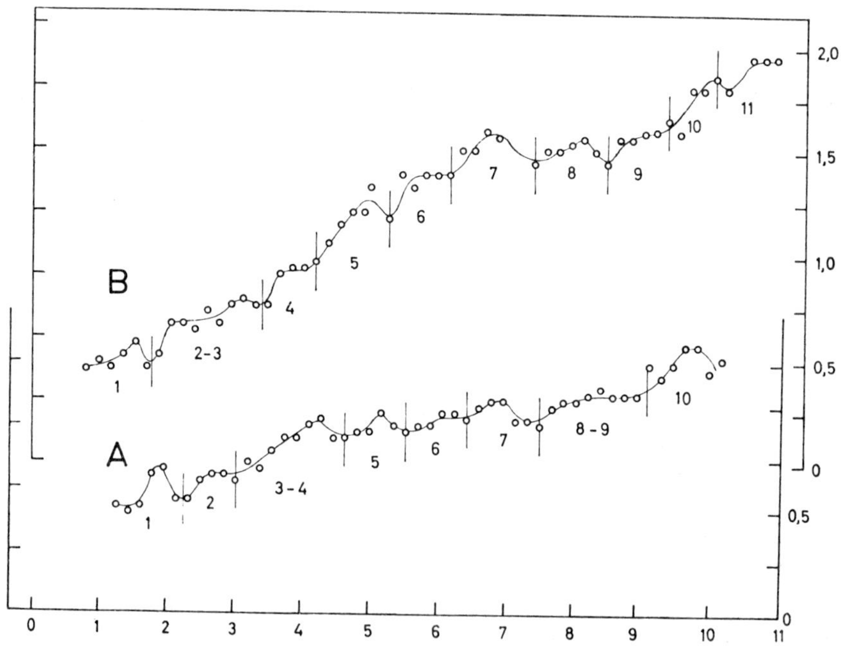

Figure 211. Oscillations in the oxygen uptake during the development of *Lytechinus*. Abscissa, hours; ordinate, µl O_2 per embryo and hour. (Reproduced from S. Løvtrup and R. M. Iverson, *Exptl. Cell Res.*, **55** (1969), by permission of Academic Press, New York)

disagreement may largely be accounted for by the fact that the time periods covered in the experiments of the second generally are shorter than in the first group.

A kind of reconciliation between the two groups was recently accomplished through studies on the oxygen consumption in *Lytechinus* (Løvtrup and Iverson, 1969) by means of the automatic diver balance (Larsson and Løvtrup, 1966). This method, which accommodates samples of 50 embryos, resulted in curves the general trend of which is similar to those recorded in the first, but exhibiting a number of oscillations reminiscent of the cyclic variations obtained in the second, group (Figure 211). Zeuthen (1951) and Scholander *et al.* (1952) have contended that the observed variations follow the same rhythm as the mitotic activity, and that they may represent the energy consumed in association with the cell divisions. This correlation was not borne out by the results obtained with the automatic diver balance, an outcome which maybe is not surprising, since cell division is only one out of three embryogenetic processes, and probably the most modest one as far as energy consumption is concerned.

It therefore seems likely that the various peaks and plateaux displayed by the curves in Figure 211 correspond to epigenetic events, executed mainly by the other two kinds of embryogenetic processes. As far as cell diversification is concerned, it is well known from visual inspection that separate groups of cells, for instance those located at the two poles, independently undergo transformation and differentiation. Multiphasic activities have furthermore been revealed on the chemical level (cf. Gustafson, 1965; Løvtrup, 1966b). Also, morphogenetic processes are known to occur in separate phases, as discussed in detail in preceding sections. Since these various events undoubtedly require

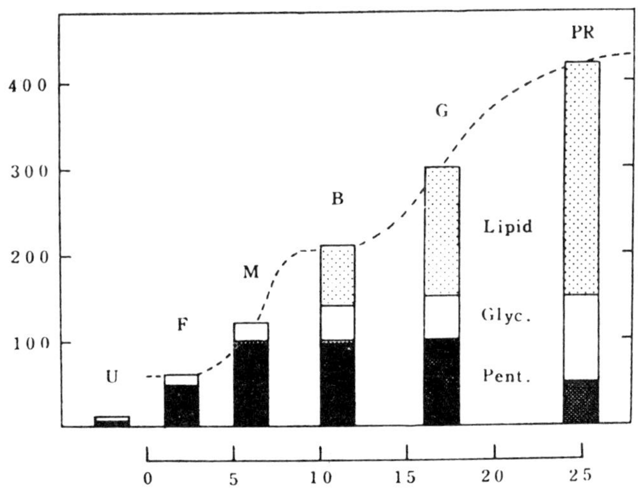

Figure 212. Utilization of energy sources during the development of *Anthocidaris*. Abscissa, hours; ordinate, μl O_2 per 10^6 embryos and hour. U, unfertilized eggs; F, fertilized egg; M, morula; B, swimming blastula; G, gastrula; PR, prism. Black zone (Pent.), carbohydrate utilization through the PP cycle. White zone (Glyc.), carbohydrate utilization involving glycolysis and mitochondrial oxidation. Dotted zone (Lipid), oxidation of lipids. Dashed line, change in the rate of oxygen uptake. (Reproduced from T. Gustafson (1965), in *Biochemistry of Animal Development*, Vol. 2, after Isono, by permission of Academic Press, New York)

supply of energy, it does not seem too far-fetched to propose that an essential part of the oxygen consumption consists of bursts of limited temporal duration, the maxima of which are large enough to influence the curves representing the total oxygen uptake.

ENERGY SOURCES

The fact that total nitrogen content remains constant in the sea urchin embryo (Gustafson and Hasselberg, 1951) suggests that proteins and other nitrogen-containing substances are not to any measurable extent utilized for the supply of energy. Earlier estimations of the changes in the content of lipids during development (Ephrussi and Rapkine, 1928; Hayes, 1938) have not given a clear-cut picture. The RQ determinations of Isono (1963) indicate that the oxidation of lipids begins in the blastula stage and increases substantially during the following developmental stages (Figure 212). From this figure it can be assessed that the combustion of lipids accounts for 45 per cent of the total oxygen consumption, whereas carbohydrate is the substrate for the remaining 55 per cent. The glucose is suggested to be utilized by two different pathways, the pentose phosphate (PP) cycle, which is quantitatively most important (35 per cent), particularly during the early stages, and oxidation through glycolysis and mitochondrial oxidation (20 per cent).

Although these results were obtained on a genus (*Anthocidaria*) different from those on which carbohydrate analyses have been made, it may still be of some interest to make a comparison. The data compiled in Table 9 show that considerable variation obtains

Table 9
The carbohydrate content of the sea urchin egg (pM per embryo)*

Authors	Total carbohydrate	Glycogen	Glucose
Paracentrotus			
Ephrussi and Rapkine (1928)	47·5	45	—
Zielinski (1939)	—	56·5	5·5
Örström and Lindberg (1940)	—	29	6·5
Arbacia			
Perlzweig and Barron (1928)	10·5	—	—
Hutchens *et al.* (1942)	22·5	10·5	—

*Conversion factors: *Paracentrotus*, 36 mg N per ml eggs; 0·245 g dry weight per ml eggs; 0·58 ml per 10^6 eggs (Örström and Lindberg, 1940). *Arbacia*, 25 mg protein per mg N; 5·9 mg N per 10^6 eggs (Hutchens *et al.*, 1942). pM, picomole.

with respect to the published results. It is beyond the scope of the present work to evaluate the methods employed, but it should be mentioned that Hutchens, Keltch, Krahl and Clowes (1942) found that the various steps involved in the glycogen determination give rise to a gradual loss of total acid-hydrolysable carbohydrate, suggesting that not all of the polysaccharide present is insoluble in ethanol–water mixtures or else that part of the reducing substances is not polysaccharide. According to Hutchens *et al.* (1942), the material in the glycogen fraction isolated from the sea urchin egg is similar to normal glycogen in most respects, but many results would be easier to explain on the assumption that there is another carbohydrate reserve with different chemical properties.

Örström and Lindberg (1940) observed that 18 per cent of the total carbohydrate—mainly glycogen—disappears in the first 10 minutes following fertilization, while Zielinski (1939) found a decline of 4 per cent during the first 2 hours of development. Ephrussi and Rapkine (1928) found a 7 per cent decrease in the glycogen content after 12 hours, and no traces left after 40 hours. In contrast, no changes were observed in total carbohydrate after 12 hours, and after 40 hours there was still left about 63 per cent (30 pM) of the original content. These results suggest that the carbohydrate reserves are mainly used for synthetic purposes, only about one-third being claimed for supply of energy. As a matter of fact it appears that an essential part of the glucose may be used for synthesis of deoxyribose. This is evidenced by the observation that there is a net increase of about 6000–7000 pg DNA during the development up to the pluteus stage (cf. below) which, assuming an average nucleotide molecular weight of 300, corresponds to about 20 pM deoxyribose.

Using the values of Isono quoted above, and various data on oxygen and substrate consumption, it is possible to make a balance sheet for the utilization of carbohydrates (Table 10). It is seen that if the glucose supposed to pass through the PP cycle is not

Table 10

Carbohydrate metabolism during the first 25 hours of sea urchin development*

	Oxygen consumption per embryo (pM)				Substrate consumption per embryo Glucose (pM)			Total	
	Total	Lipids (45%)	Glucose in Krebs cycle (20%)[1]	Glucose in PP (35%)	Krebs cycle[2]	PP cycle[3]	PP cycle with complete oxidation[2]	Calculated	Found
Paracentrotus	670[4]	300	135	235	22	115	39	61	18[5]
Arbacia	130[6]	59	26	45	4·3	22·5	8·3	12·6	8·5[6]

*Conversion factors: *Paracentrotus*, 36 mg N per ml eggs; 0·58 ml per 10^6 eggs (Örström and Lindberg, 1940). *Arbacia*, 25 mg N per g wet weight; 5·9 mg N per 10^6 eggs (Hutchens *et al.*, 1942).

The raised numbers in the table denote: 1, Isono (Figure 211); 2, 1 mole glucose requires 6 moles oxygen for complete oxidation; 3, 1 mole glucose requires 2 moles oxygen for conversion to acetate; 4, Lindahl (1939b); 5, Ephrussi and Rapkine (1928). The value is not corrected for the increase in ash content. If that is done, the glucose consumption becomes 12·5 pM; 6, Hutchens *et al.* (1942).

completely oxidized, then the carbohydrate consumption must be much higher than anything which can be accounted for by chemical analyses. Even if an intermediate is accumulated and subsequently oxidized by the mitochondria it is impossible to account for the oxygen consumption in *Paracentrotus*. Although some of the assumptions made in these calculations undoubtedly involve rather large errors, the discrepancy is still so large that it seems unavoidable to infer that there are present carbohydrate or other oxidizable reserves which are not determined by the analytical methods employed or else that the oxygen consumption values are too high. In this context it may be mentioned that the results obtained with *Lytechinus* (Figure 211) are of the same order as those observed by Lindahl (1939b) in *Paracentrotus*. For some unkown reason the agreement in *Arbacia* is much better, but it should be pointed out that Hutchens *et al.* (1942) were unable to register a loss in the content of carbohydrate corresponding to the oxygen consumption during the early developmental stages. The data presented here show that it is impossible to give a completely satisfactory answer to the apparently simple question about the nature of the energy sources in the development of the Echinoidea.

METABOLIC PATHWAYS

We have seen in the preceding section that glucose initially may be metabolized through the PP cycle, and that the ordinary pathway through glycolysis and mitochondrial oxidation only gradually enters into action. This observation clearly implies either that the appropriate enzymes are present, but put out of action through some unknown regulatory mechanism, or else that they are absent. We shall try to answer this problem on the basis of the available evidence. The question of the terminal oxidation will also be briefly discussed.

Tricarboxylic acid cycle

The presence of mitochondria in the early embryo has been decisively demonstrated. Various methods have been employed for this purpose: thus microscopical observation after vital staining by Gustafson and Lenique (1952; 1955), this approach, as well as counting in homogenates by Shaver (1956; 1957), and electron microscopy by Berg, Taylor and Humphreys (1962) and Berg and Long (1964). On certain points disagreement obtains between these authors, but this question has been dealt with in detail on a previous occasion (1966b), and it is therefore possible to confine the present discussion to the most important aspects of their findings.

As far as the appearance of the mitochondria is concerned, it seems that a gradual modification occurs, comprising size, cristal density and shape. The arrangement of the cristae seems to be very irregular in the early mitochondria. In the gastrula distinct differences obtain between mitochondria from animal and vegetal cells (Table 11). The

Table 11
Differentiation of mitochondria in the sea urchin embryo

		Relative[4] size	Relative[4] cristal density	Shape[1–4]	RMD[1,2]
Cleavage		1·00	1·00	Spherical or oval	Low
Gastrula	Animal	1·24	1·15	Spherical, rod-shaped?	High
	Vegetal[5]	~1·7–2·5	1·25	Rod-shaped	Medium

The raised numbers in the table denote: 1, Gustafson and Lenique (1952); 2, Gustafson and Lenique (1955); 3, Shaver (1956); 4, Berg and Long (1964); 5, Includes primary mesenchyme and invaginated cells.

number of mitochondria also increases; already before the 32 cell stage there is a definite rise in the relative mitochondrial density (RMD). This phase is of short duration, and for several hours there is no further change, but in the advanced blastula the RMD increases again, a process which lasts until the 25th hour, after which a decline takes place (Figure 213). The change in mitochondrial morphology may either imply that the mitochondria of the early type are subjected to changes or that they disintegrate and are replaced by particles of the more advanced type. The latter alternative appears to be correct according to the observations by Gustafson and Lenique (1952). It therefore seems probable that their results actually represent the sum of two separate curves as suggested in Figure 213.

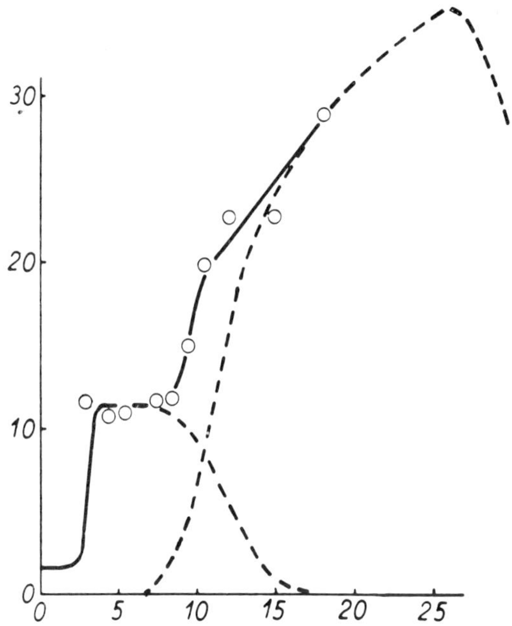

Figure 213. Relative mitochondrial density during the development of *Psammechinus*. Abscissa, hours; ordinate, relative mitochondrial density (RMD). The dashed lines represent a suggested resolution of the solid curve. (Reproduced from S. Løvtrup, *Bull. Swiss Acad. Med. Sci.*, **22** (1966), after Gustafson and Lenique, by permission of Swabe, Basel)

Several mitochondrial enzymes have been studied with regard to their changes during sea urchin development, e.g. malate and succinate dehydrogenases (Gustafson and Hasselberg, 1951), cytochrome oxidase (Deutsch and Gustafson, 1952) and cathepsin B, glutaminase I and ATPase (Gustafson and Hasselberg, 1951). For further enzymes the reader may refer to Løvtrup (1966b). The ATPase activity is presumed to represent two or more enzymes, out of which one is bound to the mitochondria (de Duve, Wattiaux and Baudhuin, 1962). It was found by Gustafson and Hasselberg that the activity of most mitochondrial enzymes exhibits a similar pattern of change, involving a large increase in activity during the mesenchyme blastula stage, which can be neatly correlated with the subsequently observed augmentation of mitochondrial number. However, certain irregularities obtain, suggesting that the mitochondria are subjected not only to multiplication, but also to differentiation, a conclusion in excellent agreement with the morphological observations just discussed.

If, during embryogenesis, different mitochondrial populations successively replace each other, this occurrence presumably must be reflected in the enzyme activity curves, at least in the form of phases with constant activity. The determinations on mitochondrial enzymes do not include sufficient points to reveal finer details in the curves, but it may be noted that a curve obtained by Gustafson and Hasselberg for succinate dehydrogenase (SDH) actually shows constant activity for several hours (Figure 214).

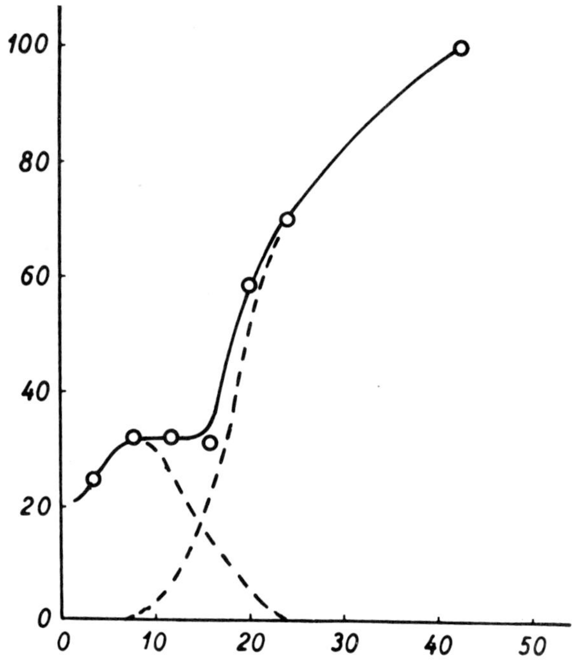

Figure 214. Succinate dehydrogenase activity during the development of *Psammechinus*. Abscissa, hours; ordinate, relative activity. The dashed lines represent a suggested resolution of the solid curve. (Reproduced from S. Løvtrup, *Bull. Swiss Acad. Med. Sci.*, **22** (1966), after Gustafson and Hasselberg, by permission of Swabe, Basel)

As indicated in the figure, this part of the curve may indeed be composed of a declining and an increasing component, representing two different mitochondrial populations. Comparison between Figures 213 and 214 suggests that the early mitochondria have a relatively high content of SDH.

The changes in cytochrome c oxidase activity present a very complicated picture (Figure 215). If we resolve curve 1, we get two curves which, with respect to time, correspond reasonably well with those for the changes in mitochondrial number and SDH activity. However, the early curve for cytochrome c oxidase is distinguished by a very high relative activity, when compared with the curves for RMD and SDH. From the work of Maggio (1959), we know that in isolated mitochondria there is, shortly after fertilization, an activation of cytochrome oxidase activity, amounting to 20–35 per cent, but that no further changes occur in the specific activity during the following three hours.

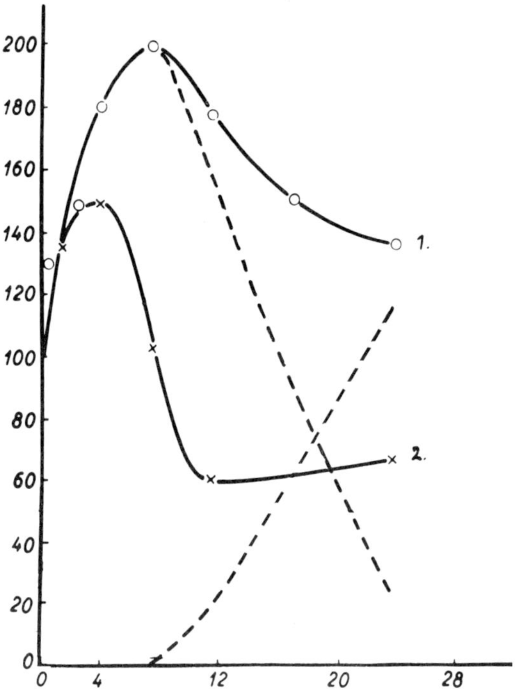

Figure 215. Cytochrome c oxidase activity during the development of *Psammechinus*. Abscissa, hours; ordinate, per cent of initial activity. Curves 1 and 2 represent different clutches. The dashed lines represent a suggested resolution of curve 1. (Reproduced from S. Løvtrup, *Bull. Swiss Acad. Med. Sci.*, **22** (1966), after Gustafson and Deutsch, by permission of Swabe, Basel)

It seems that the very steep increase in enzyme activity (Figure 215) at a time when there is no increase in mitochondrial number may be accounted for by these findings.

However, if the observed enzyme activity is bound to the mitochondria and if, as suggested by Maggio's results, the specific activity remains constant, then the increase in mitochondrial number and cytochrome oxidase activity should be proportional. This expectation is not fulfilled, for while the mitochondrial density increases at least fivefold in the early embryo, the rise in enzyme activity is at most 50 per cent after correction for the fertilization activation. This discrepancy indicates that a considerable part of the activity represents an extramitochondrial cytochrome oxidase. This question has been investigated by Hutchens, Kopac and Krahl (1942), who found that 67 per cent of the total activity in an egg homogenate is present in the supernatant and only 1 per cent in the 'mitochondrial' fraction. As only 80 per cent of the total enzyme content was accounted for, the possibility exists that the mitochondrial enzyme is more active in the unfractioned homogenate. These authors employed a manometric method, whereas Deutsch and Gustafson (1952) and Maggio (1959) used spectrophotometry. It cannot be excluded that different methods give somewhat divergent results, but nevertheless it seems justifiable to infer that there are two different cytochrome oxidases present in the early embryo, both of which begin to decline in the mesenchyme blastula.

The various curve resolutions presented above (Figures 213, 214 and 215) are not directly supported by recorded observations. However, the results of Immers and Runnström (1960) on the DNP stimulation of the oxygen uptake of intact embryos clearly show two separate curves. The minimum uniting them is located at the end of the phase of constant respiration (Figure 216). If, as these authors suggest, the curve of

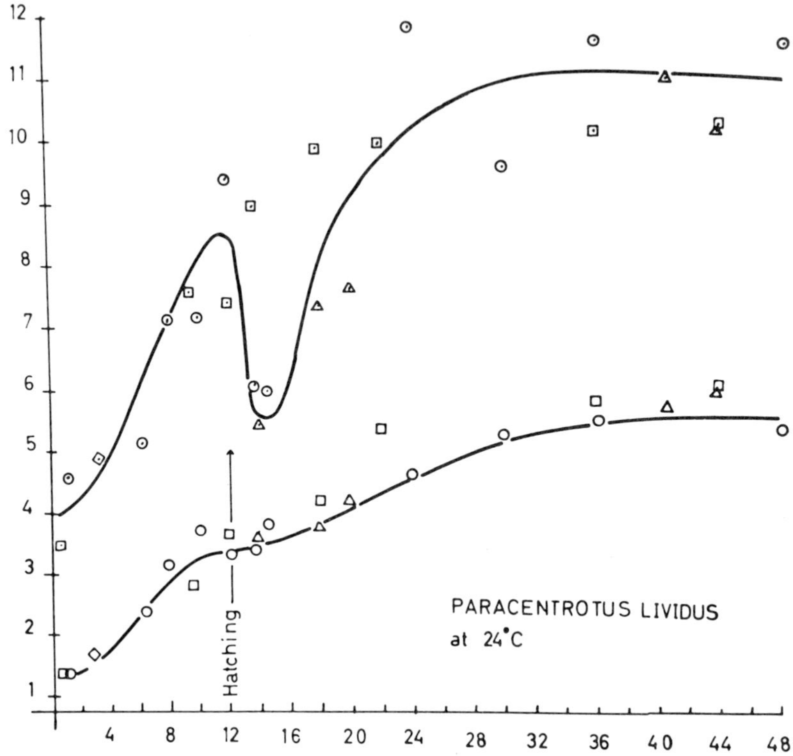

Figure 216. Respiratory control during the development of *Paracentrotus*. Abscissa, hours; ordinate, 0·1 µl per embryo and hour. Lower curve, normal respiratory rate; upper curve, oxygen consumption after addition of 2,4-dinitrophenol (5×10^{-5} M). (Reproduced from J. Immers and J. Runnström, *Developm. Biol.*, **2** (1960), by permission of Academic Press, New York)

DNP-stimulated oxygen uptake represents the maximum oxidative capacity in the embryo, then it seems absolutely unavoidable to conclude that two different mitochondrial—or extramitochondrial—oxidative systems occur in the sea urchin embryo, the first being gradually replaced by the second one during gastrulation.

As regards the remaining enzymes, cathepsin B, glutaminase I and ATPase, their activity increases in time with the second population of mitochondria. For the former two enzymes the relative increase is much higher than for the dehydrogenases, suggesting that the early mitochondria contain little or no activity of these enzymes (Figure 217). Conversely, the relative increase in ATPase is quite low (Figure 218), a phenomenon which also, of course, may bear upon the enzyme content of the first mitochondrial population. It is more likely, however, that this finding should be related to the fact that, as mentioned above, this enzyme is not located exclusively in the mitochondria. The

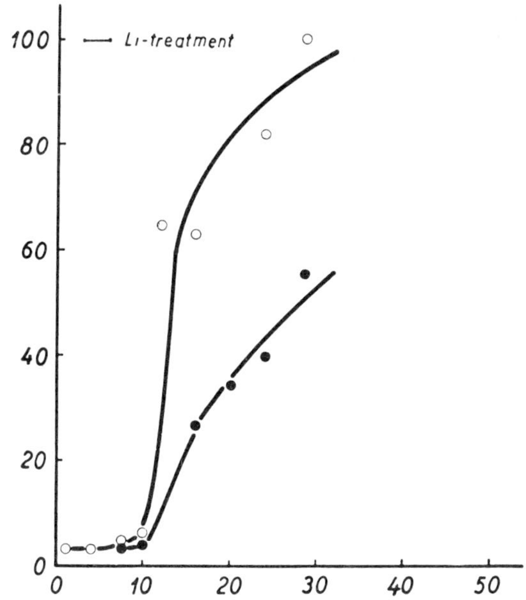

Figure 217. Cathepsin B activity during the development of *Psammechinus*. Abscissa, hours; ordinate, relative activity. (Reproduced from T. Gustafson and I. Hasselberg, *Exptl. Cell Res.*, **2** (1951), by permission of Academic Press, New York)

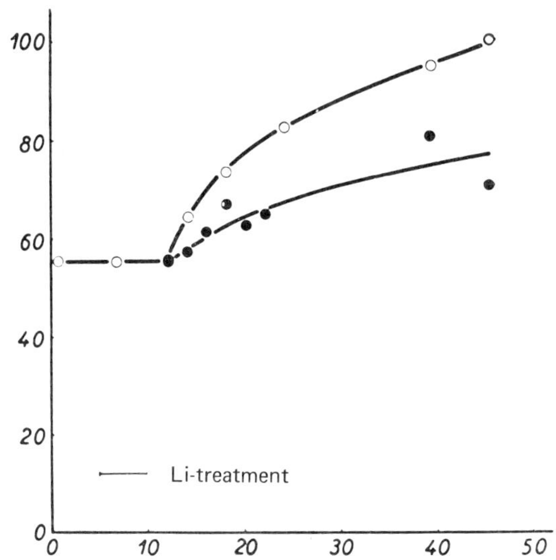

Figure 218. ATPase activity during the development of *Paracentrotus*. Abscissa, hours; ordinate, relative activity. (Reproduced from T. Gustafson and I. Hasselberg, *Exptl. Cell Res.*, **2** (1951), by permission of Academic Press, New York)

observed activity may be the sum of a constant, extramitochondrial component and an enzyme contained in the second generation of mitochondria.

The findings reported here strongly suggest that the mitochondria present in the early embryo are distinguished with respect to their contents of enzymes, but they do not convey any information as regards their activity. To answer this question we must turn to other kinds of experimental evidence. Crane and Keltch (1949) found that Krebs cycle intermediates, notably oxalacetate, are oxidized by cell-free egg homogenates. The rate of oxygen consumption was about 8 μM O_2 per ml egg and hour, as compared to a value of 20–40 for intact unfertilized eggs (Krahl, 1950). This result is corroborated by the observation by Cleland and Rothschild (1952b) that in homogenates a slight stimulation of the oxygen uptake occurs upon addition of Krebs cycle metabolites and that malonate and fluoracetate inhibit the endogenous O_2 consumption by 44 and 23 per cent, respectively. Oxidation of pyruvate was also observed; under optimal conditions stimulation after addition of this substance was 4 μM O_2 per ml egg and hour. These results seem to demonstrate that the mitochondria in the egg are active, but that their oxidative capacity is rather low when compared with the oxygen uptake of intact eggs.

Observations on homogenates do not necessarily reflect the situation prevailing in normal eggs and embryos, which may be better revealed by isotope incorporation studies. This approach has been employed by Hultin (1953a; 1953b; 1953c) who, using labelled NH_3, alanine, CO_2 and acetate, was led to the conclusion that mitochondrial activity is very low before the mesenchyme blastula stage, but rises very rapidly with the beginning of this developmental phase (Figure 219). Studies on the formation of CO_2 from labelled glucose have confirmed these results (Krahl, 1956; Bäckström, Hultin and Hultin, 1960).

The evidence presented here shows that the mitochondria in the early embryo are rather inefficient from a metabolic point of view, a circumstance which may suffice to account for the modest role played by mitochondrial oxidations during this phase of development.

Glycolysis

Until the presence of the PP cycle in the sea urchin embryo was indicated by Lindberg (1943), glycolysis was presumed to be the major pathway for glucose breakdown, even if lack of inhibition by iodoacetate (Runnström, 1933) suggested that certain peculiarities prevail. Since then much study has been centred around the question of whether both pathways exist in the embryo, or whether only the PP cycle is present. Thus, Krahl, Keltch, Walters and Clowes (1955a) have studied the hexokinase activity in *Arbacia*. It was found that glucose, 2-deoxyglucose, mannose and fructose are phosphorylated at appreciable rates, but glucose-6-phosphate (G-6-P) and fructose-6-phosphate are hardly attacked. This very low phosphofructokinase activity does not support the occurrence of glycolytic activity.

Studying the NADP reduction with various substrates, Krahl *et al.* (1955b) could establish the presence of glucose-6-phosphate dehydrogenase (G-6-PDH), 6-phosphogluconate dehydrogenase (6-PGDH), phosphoglucomutase, hexose isomerase and fructose-1, 6-diphosphatase. The first two enzymes are normal members of the PP cycle, the third and the fourth of the glycolytic pathway. When fructose-1, 6-diphosphate (HDP) was incubated with NAD, the coenzyme reduction proceeded at a rate of 1–8 per cent of that observed with NADP in the presence of G-6-P. The HDP–NAD reaction

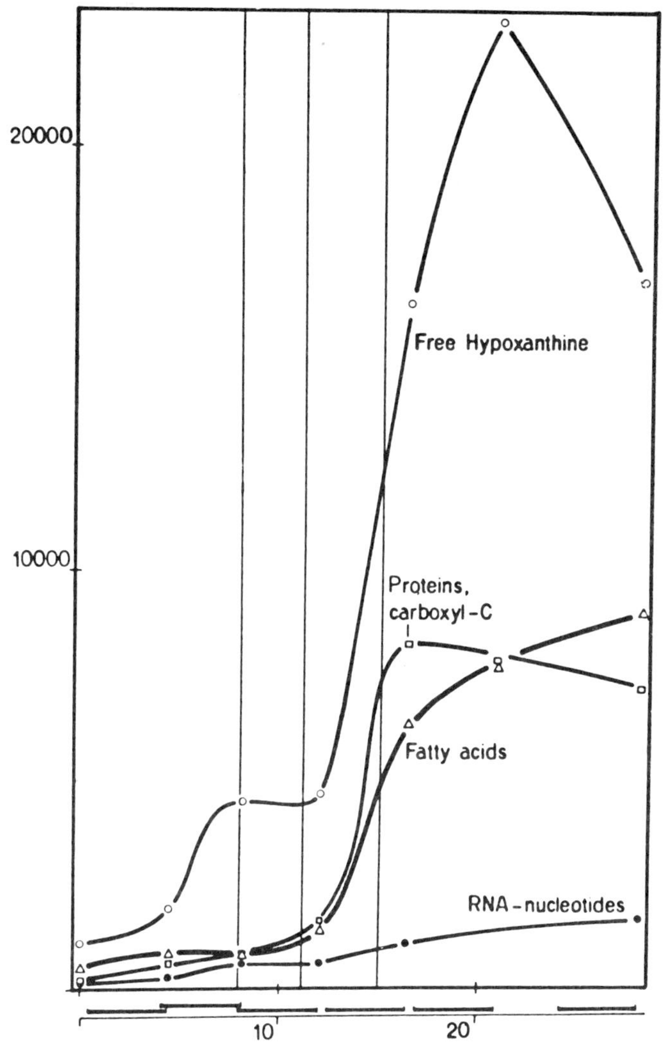

Figure 219. Incorporation of 1–[14]C-acetate during the development of *Psammechinus*. Abscissa, hours; ordinate, counts per minute. Vertical lines indicate hatching, appearance of mesenchyme cells and start of invagination, respectively. The periods of isotope treatment are indicated by horizontal bars at the bottom of the figure. (Reproduced from T. Hultin, *Ark. Kemi*, **6** (1953), by permission of Kungl. Vetenskapsakademin, Stockholm)

is supposed to involve two steps, splitting of the substrate by aldolase into two triosephosphates and oxidation of the latter by triosephosphate dehydrogenase (TPDH). In this procedure the observed rate must be that of the slowest reaction, yet even the slight activity observed is remarkable since Jandorf and Krahl (1942) could observe no TPDH in egg extracts.

The presence of this enzyme in sea urchin eggs was first questioned by Runnström

(1933), when he found that iodoacetate, an inhibitor of TPDH, does not interfere with the oxygen uptake of intact eggs. In later experiments it was observed that in *Arbacia* a higher concentration (0·03 M) does reduce the respiratory rate (Runnström, 1935), but since the treatment leads to abnormal development, the significance of this finding is difficult to evaluate. Runnström's original result was confirmed on homogenates by Lindberg and Ernster (1948). These authors also found that the oxygen uptake in the presence of HDP and NAD was the same as in the absence of substrate; an increase occurred after addition of iodoacetate. These observations hardly support the contention that either aldolase or TPDH is present in the sea urchin egg.

Perlzweig and Barron (1928) observed the formation of lactate in unfertilized and fertilized eggs. Cleland and Rothschild (1952a), ardent advocates of the glycolytic pathway in the present context, measured the lactate production in intact unfertilized eggs under anaerobic conditions. They established a rate of 1·5 μM per ml egg and hour, corresponding to an oxygen consumption of 4 μM for oxidation of the corresponding quantity of pyruvate. Substantially higher values were obtained in homogenates after addition of various cofactors.

It is evident that the interpretations of the observation presented here are contradictory. However, as far as I can judge, this stems from the assumption that lactate formation can occur only through glycolysis. This may be the case almost everywhere else, but that does not imply that other pathways exist in embryos. If this turns out to be so, their elucidation may be of great phylogenetic interest.

Pentose phosphate cycle

Many findings support the contention of Lindberg (1943) that the PP cycle is active in the sea urchin egg. Among these should be mentioned the finding that glucose and G-6-P, but not HDP, is oxidized by homogenates of *Strongylocentrotus* eggs, the rate being about 6 μM per ml egg and hour at 6° C, which may correspond to 20–30 μM at 20° C, and this oxidation is coupled with oxidative phosphorylation (Lindberg and Ernster, 1948). As already mentioned, Krahl *et al.* (1955b) observed that G-6-PDH and 6-PGDH are highly active in *Arbacia* eggs. Bäckström (1959; 1963), studying the changes in the activity of these enzymes during development in *Paracentrotus* and *Psammechinus*, has found that in the cleavage stages there is a substantial increase, but subsequently a distinct decline occurs; various oscillations were observed in the activity curves (Figure 220). These findings clearly suggest that the enzymes, and the corresponding metabolic conversions, are typical features of the early embryonic cells. Their activity in the intact embryo is borne out by the observations on CO_2 production from labelled glucose quoted above which indicate that a large fraction of this substance is metabolized by means of these enzymes (Krahl, 1956; Bäckström *et al.*, 1960). According to Isono (1962) and Isono and Ishida (1962), all the enzymes of the PP cycle are present, a circumstance which implies that triosephosphate may be formed. If TPDH was present, pyruvate might be formed from this intermediate, but as we have seen above, it has not been possible to demonstrate this enzyme so far.

Among the metabolic pathways discussed here, the evidence seems overwhelmingly to suggest that the PP cycle, or least an essential part of it, is active during the early stages of development in Echinoidea. One reason for the apparent reluctance to accept this notion may be the once-held belief that oxidation of NADPH, the coenzyme of the

PP cycle, in contrast to that of NADH, is not coupled with phosphorylation. Since this has been proved wrong, for mitochondria at least (Ernster and Navazio, 1957), there seems to be no obvious reason for rejecting the proposition that the PP cycle is responsible for the supply of energy in the sea urchin embryo. Even so, it appears that this question is far from being definitely settled, since many details remain obscure.

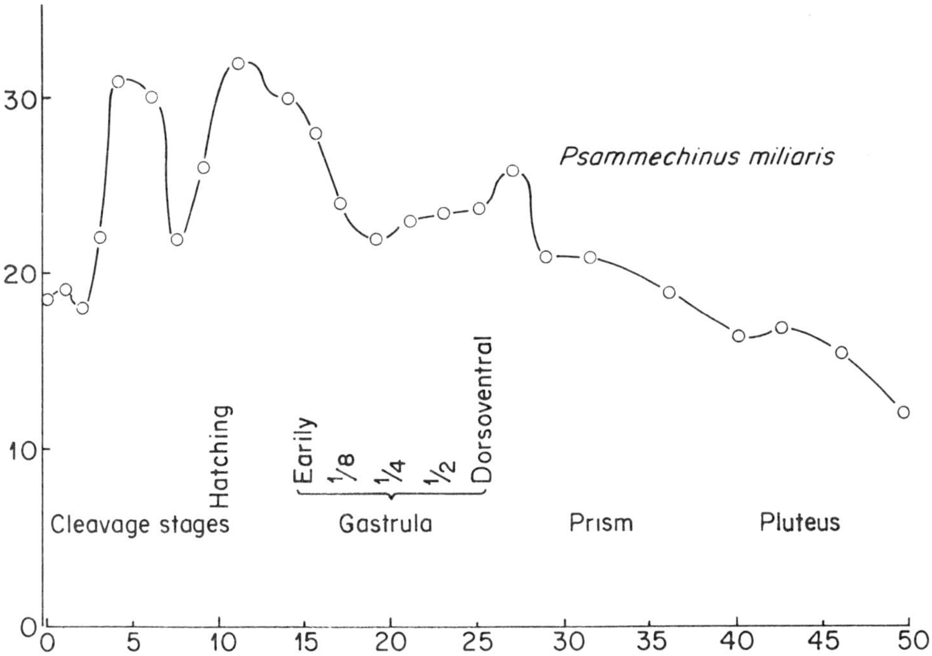

Figure 220. Glucose-6-phosphate dehydrogenase activity during the development of *Psammechinus*. Abscissa, hours; ordinate, arbitrary units. (Reproduced from S. Bäckström, *Exptl. Cell Res.*, **18** (1959), by permission of Academic Press, New York)

It has been known for a long time that the carbohydrate metabolism of various embryos differs from that encountered in adult tissues (cf. Needham, 1942). It is a remarkable fact that, in spite of the great strides made in the field of intermediary metabolism since then, the mechanism of glucose turnover in Echinoidea embryos still awaits its definitive elucidation.

Terminal oxidation

As discussed above, the sea urchin egg contains a very powerful cytochrome c oxidase. The presence of cytochrome c might consequently be anticipated, but in spite of numerous attempts (Ball and Meyerhof, 1940; Krahl, Keltch, Neubeck and Clowers, 1941; Rothschild, 1949; Borei, 1951; Yčas, 1954), only two cytochromes have been detected spectrophotometrically, belonging to the groups a and b, respectively. Maggio and Ghiretti-Magaldi (1958) reported the presence of cytochromes a, a_3, b and c in sea urchin mitochondria, but unfortunately the last one was lost during the preparation.

Cytochrome a_3 (=cytochrome oxidase or at least part of this enzyme) cannot, according to Rothschild (1949), be demonstrated spectrophotometrically in this material because certain pigments interfere with the determination. The approach used by Maggio and Ghiretti-Magaldi should circumvent this difficulty. In support of its presence Rothschild (1956) points out that cytochrome c oxidation occurs in homogenates (Krahl *et al.*, 1941), and the oxidation of dimethyl-p-phenylendiamine and similar substances has also been used as evidence on this point (cf. Runnström, 1956). Regarding substances of this type it was observed by Pappenheimer and Williams (1954) that they may be oxidized through a pathway not involving cytochrome c, and that this method therefore cannot provide an unambiguous measure of cytochrome oxidase. The results with cytochrome c oxidation are, of course, more clear-cut, but they only prove the presence of cytochrome a_3 on the presumption that this is the only terminal oxidase that can react with cytochrome c.

According to Keilin and Hartree (1938), cytochrome oxidase is a copper-containing enzyme. Krahl *et al.* (1941) tested the effects of two inhibitors of copper-containing enzymes, 8-hydroxychinoline and diethyldithiocarbamate, on the cytochrome c oxidation in *Arbacia*. None of these substances produced inhibition; on the contrary, a considerable stimulation was observed with the latter compound. This finding does not support the contention that the terminal oxidase is the same as that found in other animal tissues, and the failure to demonstrate the a_3 spectrum may therefore be a simple consequence of the fact that the enzyme is absent. This may not hold for the mitochondria, but as we have seen, they are only responsible for a small part of the total cytochrome oxidase activity. As a matter of fact, the values obtained for the latter do not seem to bear any correspondence to biological reality, as is shown by a comparison of the absolute values for oxygen uptake. Thus, in the unfertilized egg of *Arbacia* there is—assuming 0·1 mg N per mg dry weight (Hutchens, Kopac and Krahl, 1942)—an oxygen consumption of 4–5 ml per mg N and hour, the corresponding value for the fertilized egg is 20, whereas a dialysed enzyme preparation gave a value of 100 after addition of cytochrome c (Krahl *et al.*, 1941). The homogenates used by Hutchens *et al.* (1942) had an oxygen consumption of 2400 ml per mg N and hour.

The cytochrome b found in the sea urchin embryo has been proposed to be b_5 (Yčas, 1954), the microsomal cytochrome (cf. Strittmatter, 1961). This, together with $a+a_3$, are the main cytochromes present in the midgut of larvae of the *Cecropia* silkworm (Pappenheimer and Williams, 1954). These authors found that b_5 is an electron donor in the reduction of cytochrome c by NADH, but it also reacts with NADPH (Strittmatter, 1961). This last point appears to be of great significance, since the PP cycle dehydrogenases, including those demonstrated in the sea urchin embryo, i.e. G-6-PDH and 6-PGDH, utilize this coenzyme. Pappenheimer and Williams (1954) also showed that cytochrome b_5 is autoxidizable, in a reaction which is uninhibited by cyanide and carbon monoxide, and suggested that this enzyme represents the terminal oxidase in tissues deficient in cytochrome c. Some of the observations made with respect to the inhibitors mentioned (cf. Lindahl, 1939a; Krahl *et al.*, 1941; Robbie, 1946; Rothschild, 1949;

1956; Runnström, 1956) may possibly be explained with reference to the presence of cytochrome b_5, but other findings suggest that it cannot be the terminal oxidase; presumably even cytochrome a is involved.

From the discussion in this and the preceding section it appears that the metabolic pathways available for the early embryo are simple compared to those found in later stages. Similar pathways have been recorded in certain Protozoa, a fact which may indicate that the changes observed during embryogenesis represent a case of recapitulation on the biochemical level. As appears from various investigations, the enzymes required for an efficient metabolism of glucose, including mitochondrial oxidation, are encoded in the genome. Yet, the embryos must pass through a relatively extended part of their development before these enzymes are at their disposal. The significance of this state of affairs remains completely obscure.

Amphibia

Many investigations have been made on the metabolism of developing amphibian embryos. It is therefore possible to discuss this problem from the same angles as in the preceding chapter, namely, oxygen uptake, energy sources and metabolic pathways.

OXYGEN UPTAKE

I have in a series of papers (1953a; 1953b; 1959a; 1959b; Løvtrup and Werdinius, 1957) studied the changes in oxygen uptake associated with ontogenesis in Amphibia. In these a detailed discussion of the previous literature may be found, and it may hence be justified, for the sake of brevity, to discuss the problem on the basis of the results reported there. In the early studies it was found that in *Ambystoma* the respiration increases gradually up to the end of gastrulation, subsequently the respiration remains constant for several days, and in the early tailbud stage the increase in respiration is resumed, being interrupted by the second phase of constancy during larval development. Subsequently a maximum is reached at the end of prefeeding development. It was later found that in many cases two respiratory maxima may be observed (Figure 221). Work at different temperatures has clearly shown that, apart from the first constant level, the particular features of the respiratory curves do not coincide with specific developmental stages. The various phases therefore cannot be correlated with distinct epigenetic events, as suggested to be the case with the various phases observed in the sea urchin egg. There is, of course, no doubt that the general rise in oxygen uptake during development reflects an increase in energy-consuming processes, but so far the curves obtained have not permitted any resolution on this level. The reason for this failure may in part be due to the relatively insensitive methods employed, requiring 25–50 embryos in a sample, but also to the circumstance that the development lasts for such a long time that considerable differences may obtain between individuals of the same age.

When amphibian embryos develop at different temperatures, it turns out that the oxygen uptake associated with development up to a morphologically distinguishable stage is larger at low than at high temperatures (1953b; Ten Cate, 1953). This result may be related to the well-known fact that embryos reared at high temperatures are smaller than their counterparts developing in a cooler environment.

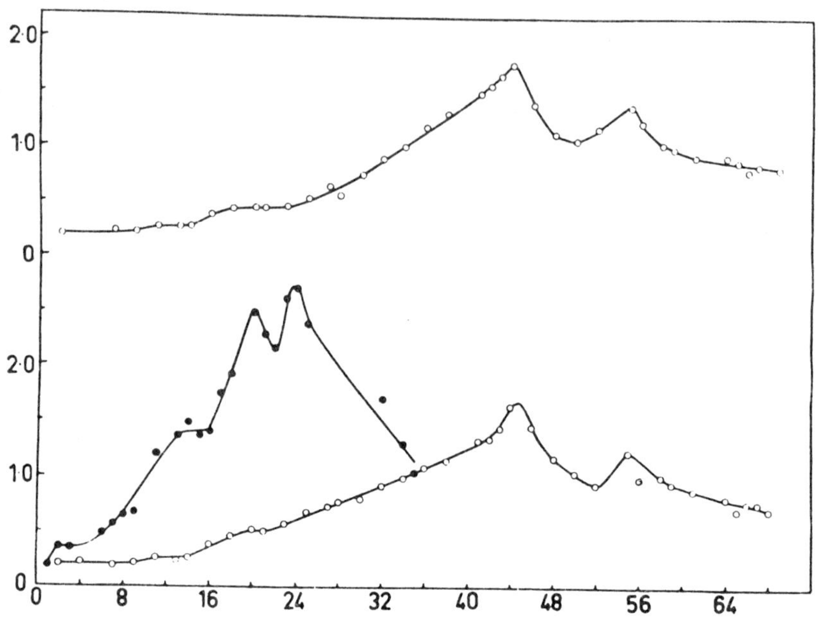

Figure 221. Oxygen uptake during development of *Ambystoma*. Respiratory curves with two horizontal phases and two maxima. Abscissa, days; ordinate, μl O_2 per embryo and hour. Filled circles, 18°C; open circles, 10°C. (Reproduced from S. Løvtrup, *J. Exptl. Zool.*, **140** (1959), by permission of The Wistar Press, Philadelphia)

ENERGY SOURCES

Many authors have attacked the problem of the utilization of energy sources in amphibian embryos by means of RQ determinations and chemical analyses (cf. Needham, 1942). The former approach involves the assumption of complete oxidation, and since there are weighty reasons to believe that during the early developmental stages accumulation of intermediates occurs, the reliability of the results obtained may be questioned. In the present context I shall review some of the most pertinent findings arrived at through chemical analyses. The investigations of Brachet and Needham (1935) and of Gregg (1948) on different species of *Rana* agree in showing that no glycogen disappears before gastrulation, while between one-third and half of the glycogen present at fertilization has disappeared at the time of hatching. Gregg's results furthermore show that the loss in total reducing carbohydrate is nearly identical with the decrease in glycogen, the latter thus apparently being the only carbohydrate energy source. Glycogen was found to amount to two-thirds of the total carbohydrate in the egg. In none of these studies were the analyses continued to the end of prefeeding development, but some older results (Fauré-Fremiet and Dragoiu, 1923) show that at this time only traces of glycogen are present. Combining this observation with those published by Gregg (1948), we may sketch the carbohydrate utilization in *Rana* in the following way: before gastrulation, no or very little consumption; one-third of the total carbohydrate, or half of the glycogen, is utilized during the embryonic stages, and the rest of the glycogen is combusted or otherwise utilized during the remaining part of the prefeeding larval development.

Relatively few determinations have been made of the consumption of lipids during amphibian embryogenesis. There is, however, rather good agreement between Bialaszewicz and Mincóvna (1921), who found no loss in fatty acids before hatching, whereas 50–60 per cent disappeared during the later stages, and Barthélemy and Bonnet (1926), who found the total loss to range from 25 to 55 per cent. Parnas and Krasinska (1921) confirmed the results of Bialaszewicz and Mincóvna by showing that no loss of lipids occurs before hatching, while Fauré-Fremiet and Dragoiu (1923) found a decrease of about 25 per cent.

The results concerning the utilization of protein disagree more than those relating to carbohydrate and lipid consumption. Bialaszewicz and Mincóvna (1921) found that at the time of hatching 9 per cent of the total nitrogen (TN) had disappeared, while Fauré-Fremiet and Dragoiu calculated that 12 per cent of the yolk protein had been consumed at this stage. Against these results stand those of Gortner (1914) on *Crytobranchus* and of Gregg and Ballentine (1946) on *Rana*, which showed only a slight, if any, change in TN from fertilization to hatching. The same was observed by Wills (1936) who, working with *Triturus*, *Ambystoma* and *Rana*, determined the TN of embryos and larvae from fertilization to metamorphosis. No significant changes were found before hatching. At the end of prefeeding development Bialaszewicz and Mincóvna (1921) recorded a loss of about 50 per cent TN, while Barthélemy and Bonnet (1926) found a decrease of 25 per cent at the time when the external gills disappear in *Rana*. Contrary to the authors who found losses in TN, Wills (1936) supplied food to the larvae, thereby presumably eliminating losses due to starvation, which may otherwise obtain after hatching. Under these circumstances the observable decreases in TN were slight, from 0 to 17 per cent, suggesting that most or all of the drop in TN observed by other authors may represent starvation. The various results thus unquestionably support the contention that protein may be utilized for energy supply through combustion, but the claim for TN constancy during the embryonic phase of development is also well supported.

The disadvantage of chemical analyses is that samples are taken continuously from a given population which, in order to comprise sufficient material, may consist of several clutches. Differences in developmental rate and in mortality must entail that a certain selection occurs which in turn may influence the results. This source of error is circumvented, theoretically at least, if it is possible to measure the consumption of energy sources without killing the embryos. As known from physiological textbooks an indirect approach exists, involving simultaneous determination of oxygen consumption, carbon dioxide production and nitrogen excretion. As already pointed out, RQ determinations imply complete oxidation of the energy sources, an assumption which probably is not always fulfilled. Another dependent parameter, which in principle may replace determinations of CO_2 production, is the weight. In terrestrial animals this approach is hampered by the fact that water losses may occur, but in aquatic animals it is possible to determine the submerged or reduced weight (RW) on a diver balance, either the Cartesian manometric type (Zeuthen, 1948) or the automatic electromagnetic version (Larsson and Løvtrup, 1966; Bergfors, Hansson-Mild and Løvtrup, 1970; Løvtrup, 1973a).

The RW of an object is the weight W minus the weight of the medium it displaces, $V \cdot \varphi_m$, where V is the volume of the object and φ_m the density of the medium (Linderström-Lang and Holter, 1940). Thus

$$RW = W - V \cdot \varphi_m = V(\varphi_o - \varphi_m) = W(1 - \frac{\varphi_m}{\varphi_o})$$

where φ_o is the density of the object. In an object of known composition the RW can be calculated from

$$RW = \Sigma_i W_i \, (1 - \frac{\varphi_m}{\varphi_i})$$

in which W_i and φ_i denote weight and density of a component i (Bottelier, Holter and Linderström-Lang, 1943). Consumption of carbohydrates and proteins which have densities higher than 1 will cause a fall in RW, whereas disappearance of lipids having densities below 1 will be reflected as an increase in RW. The use of RW determinations in combination with respiratory measurements for metabolic studies was introduced by Holter and Zeuthen (1948).

In order to determine the consumption of the three possible energy sources, the changes in three dependent variables must be known. In the present case it might be possible to include determinations of the nitrogen excretion. A very slight accumulation of nitrogen-containing substances has been registered in the surrounding medium during the early development of *Rana* (J. Brachet, 1939). However, this loss is small enough to warrant the assumption of TN constancy during early development. Under these circumstances there are only two independent variables and it is therefore possible to determine the nature of the energy sources consumed on the basis of measurements of ΔO_2 and ΔRW. At a relatively early larval stage the calculated consumption of carbohydrate approaches zero, and since the RW curves clearly show that combustion of lipids cannot account for the recorded oxygen consumption, it may be assumed, in agreement with the observations discussed above, that proteins are mobilized for the supply of energy during this part of development. It is thus possible to continue the calculations with the same two dependent parameters.

This approach has been employed by the author on several occasions (1953a; 1953b; 1959a; 1959b; Løvtrup and Werdinius, 1957). From the results in Figure 222 it is seen that the utilization of carbohydrate takes place in two phases, each displaying a distinct maximum. The consumption of lipids begins after four days and stays at a constant level between the seventh and the tenth day; subsequently it increases further, reaching a maximum after 14–15 days. The calculated combustion of protein begins around the fourteenth day and reaches a maximum after about 22 days. It will be noted that the two respiratory maxima coincide in time exactly with those for lipid and protein, respectively. The two constant phases, the latter of which is indicated only as a slight bend in the curve shown in Figure 222(a), coincide with the two maxima for carbohydrate utilization. It therefore seems that it is the availability of oxidizable substrates which is reflected by the various irregularities observed in the oxygen consumption curves.

From the curves in Figure 222(b) it is seen that the consumption of lipids begins when the first carbohydrate maximum is reached, and that the rise above the constant level starts at the time of the second maximum. These results suggest that carbohydrate is the preferred energy source, but that lipids are mobilized when the reserves of the former run short. The fact that this phenomenon occurs twice indicates that there are two different regions which enter into activity at distinctly separate times. These regions account essentially for all of the carbohydrate utilization, and since this substance is present in

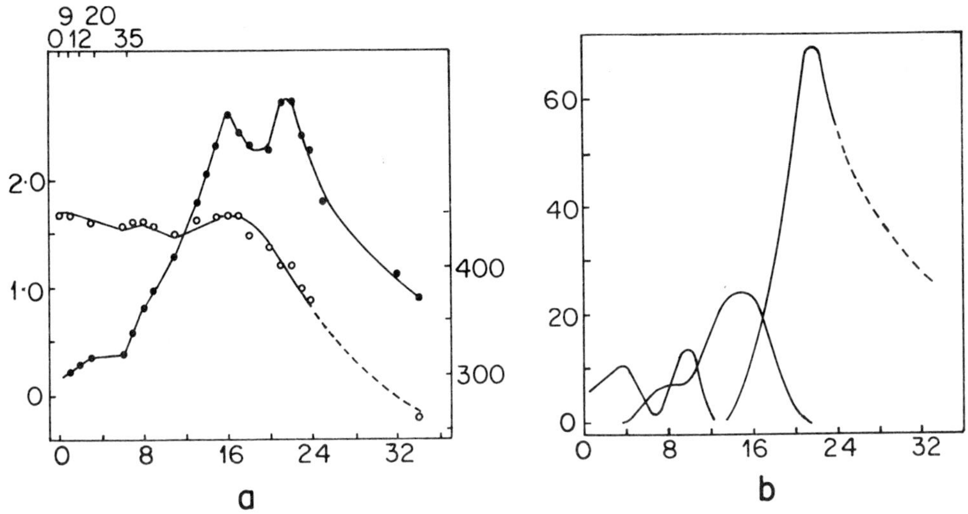

Figure 222. Consumption of energy sources during the development of *Ambystoma*. (a), changes in the rate of oxygen consumption (filled circles) and in reduced weight (open circles). The stippled part of the RW curve has been drawn on the basis of respiration curve. Lower abscissa, days; upper abscissa, stage of development; left ordinate, µl O_2 per embryo and hour; right ordinate, µg RW per embryo. (b), calculated consumption of energy sources. First curve, carbohydrate; second curve, lipid; third curve, protein. Abscissa, days; ordinate, µg per embryo and 24 hours. (Reproduced from S. Løvtrup, *J. Exptl. Zool.*, **140** (1959), by permission of The Wistar Press, Philadelphia)

very low amounts in the vegetal hemisphere (Heatley and Lindahl, 1937; Gregg and Løvtrup, 1950), the regions in question must be confined to the ecto-mesoderm. Although some overlapping must occur, it may be justified to infer that the first phase is mainly confined to the dorsal ectoderm, the second one to the ventral ectoderm and the mesoderm. Presumably, combustion of protein takes place in all parts of the embryos as the carbohydrate resources become exhausted.

When the results obtained are related to the amounts present at fertilization (1953a), it turns out that 60–70 per cent of the carbohydrate is consumed, an amount corresponding to the glycogen content. At the time of the respiratory maximum 41 per cent of the lipids (58 per cent of the neutral fat) and 16 per cent of the protein have been lost. This outcome is in very good agreement with the chemical analyses discussed above, and in combination these observations permit us to outline the utilization of energy sources in the following way: The energy source during very early development is not known. It may be carbohydrate, but in that case it is not completely oxidized, as shown partly by the fact that no change in total reducing carbohydrate can be registered (Gregg, 1948) and partly by the fact that very low RQ values are observed (Barth, 1946). During gastrulation oxidation of carbohydrate begins and it continues until all the glycogen is utilized. When this occurs, the combustion of lipids gathers speed. During larval development some protein is oxidized, but if the larvae are fed, the quantity thus consumed is rather slight. The natural succession of energy sources is thus: carbohydrate, lipid and protein.

It was mentioned above that at high temperatures the oxygen consumption associated with the development up to a given stage goes down. This does not imply that the efficiency of the process of development increases with temperature, for the losses in reserve materials are greater at high temperatures. Actually, it seems that under these conditions energy sources are lost before they are completely oxidized (1953a). This entails that a rather large amount of protein is utilized for the supply of energy, a fact which may explain the relatively small size of animals reared at high temperatures. It is interesting to note that the protein consumption is high also at low temperatures, apparently because of limitations in the rate at which lipids can be mobilized. It is therefore evident that there is an intermediate optimum temperature range, around 15°C for many species, where the protein consumption is minimal.

METABOLIC PATHWAYS

The metabolism of amphibian embryos, not least during the early developmental stages, has been subjected to numerous investigations. We shall here review the most important of the results thus obtained with respect to the Krebs cycle, the glycolytic pathway and the PP cycle. As will appear from this survey, the situation is similar to that observed with sea urchin embryos, which means that there are many problems awaiting solution.

Tricarboxylic acid cycle

It has been observed that the mitochondria present in the early embryo are small spherical bodies devoid of cristae. As development proceeds, the number of mitochondria increases, and profound morphological modifications may be observed, particularly in the cells of the neural plate, involving elongation of the particles and a very large increase in cristal density (Karasaki, 1959). Except that no mitochondrial degradation was observed, these findings are in excellent agreement with those reported by Gustafson and Lenique (1952; 1955) in the sea urchin embryo. It is interesting to speculate whether the changes in the mitochondrial morphology are a consequence of the cell transformation, i.e. whether it is a cell-class specific feature that the mitochondria in filocytes like muscle and nerve cells are tightly packed with cristae in contrast to those in lobocytes like liver cells. If this correlation is correct, then it must be a general rule that mitochondria in the sl-cells of early embryo are of the less elaborate type.

A number of mitochondrial enzymes have been followed during amphibian development, e.g. NADP-linked isocitrate dehydrogenase (ICDH), NAD- and NADP-cytochrome c reductase, NAD-linked malate dehydrogenase (MDH), succinate dehydrogenase (SDH) and cytochrome oxidase. The first four enzymes were studied by Wallace (1961), the two last plus the two reductases by Lang and Grant (1961).

The oxidation of isocitrate in the tricarboxylic acid cycle involves both NAD and NADP (Ernster and Navazio, 1956; 1957). The NAD-linked ICDH is found only in the mitochondria, while the other one may be located outside these particles. Since the former enzyme cannot be demonstrated in *Rana* embryos (Plaut and Sung, 1955), there is some reason to accept the assumption made by Wallace (1961) that the enzyme studied by him is localized in the mitochondria. It is seen from Figure 223 that the activity of this enzyme is negligible during the early stages of development, but that from gastrulation onwards it rises steeply.

As mentioned, the two cytochrome c reductases have been studied by both Lang and Grant (1961) and by Wallace (1961). From Figures 224 and 225 it appears that the results with NADP-cytochrome c reductase are in good agreement as far as the curve shape is concerned, and even the absolute values are comparable, for when recalculated, the

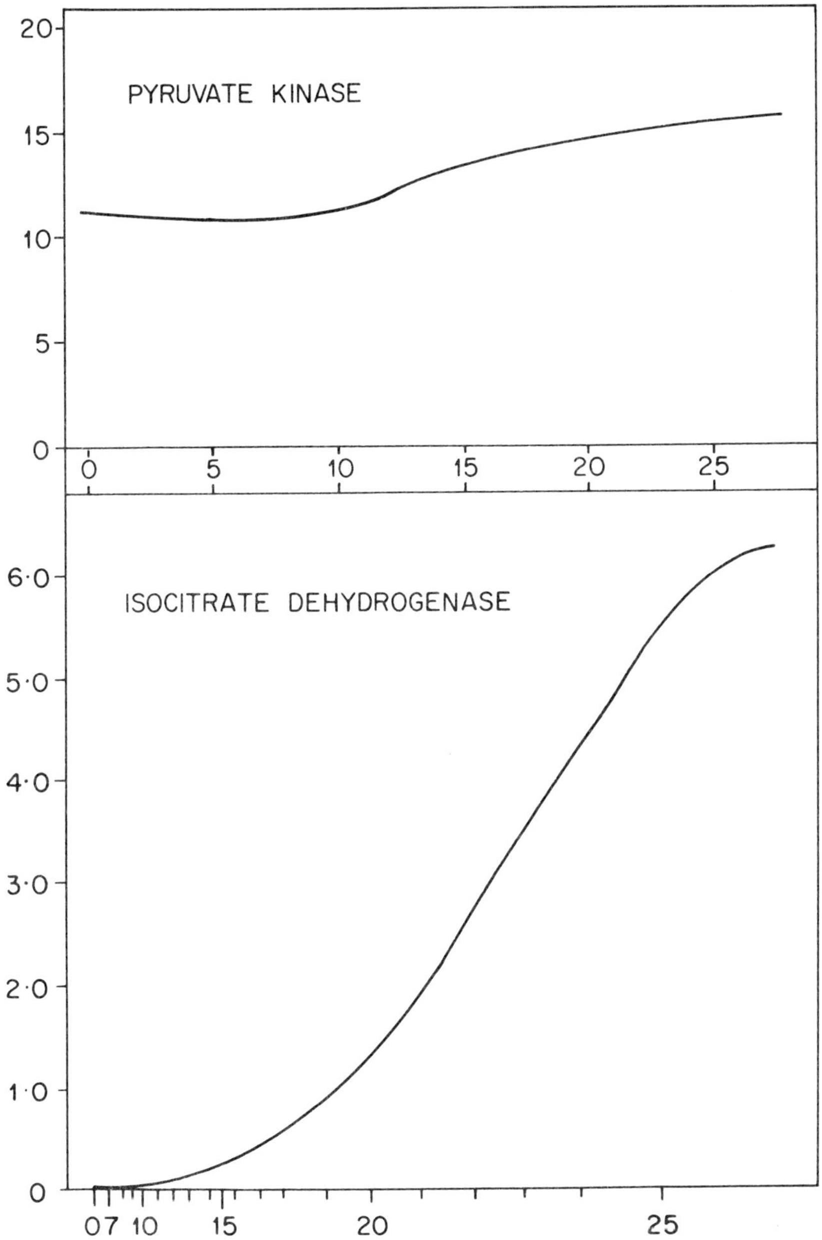

Figure 223. Pyruvate kinase and isocitrate dehydrogenase activities during the development of *Rana*. Lower abscissa, stage of development; upper abscissa, days; lower ordinate, μM NADP+ reduced per embryo and hour; upper ordinate, μM NADH oxidized per embryo and hour. (Reproduced from R. A. Wallace, *Developm. Biol.*, **3** (1961), by permission of Academic Press, New York)

constant level in Lang and Grant's results is found to correspond to about 0·017 μM NADP per hour and embryo, whereas Wallace found about 0·022. Two distinct phases are seen in Wallace's curve.

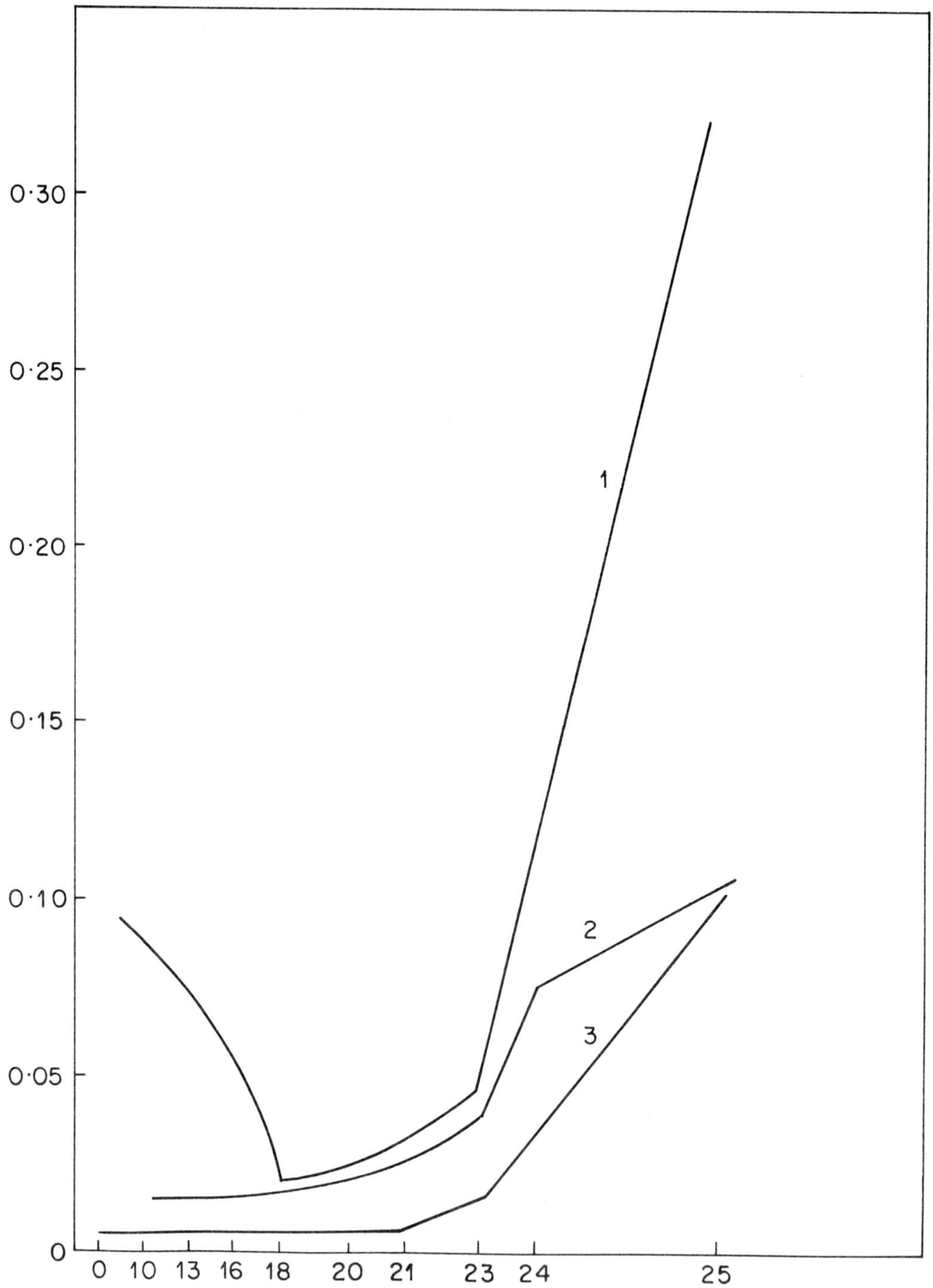

Figure 224. NAD-cytochrome c reductase (1), NADP-cytochrome c reductase (2) and succinate dehydrogenase (3) activities during the development of *Rana*. Abscissa, stage of development; ordinate, μM per embryo and hour. (Reproduced from C. A. Lang and P. Grant, *Proc. Nat. Acad. Sci.*, **47** (1961), by permission of the National Academy of Sciences, Washington)

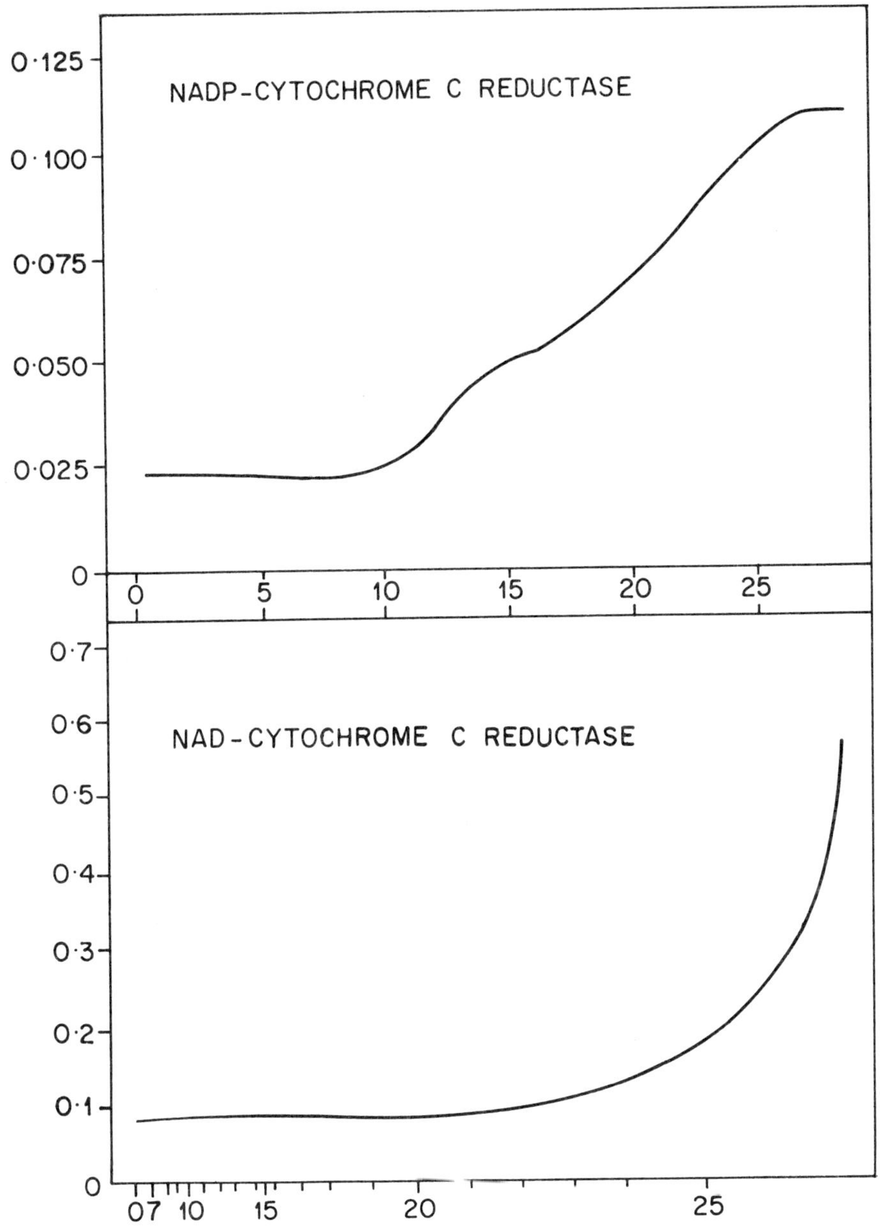

Figure 225. NADP-cytochrome c reductase and NAD-cytochrome c reductase activities during the development of *Rana*. Lower abscissa, stage of development; upper abscissa, days; lower ordinate, μM NADH oxidized per embryo and hour; upper ordinate, μM NADPH oxidized per embryo and hour. (Reproduced from R. A. Wallace, *Developm. Biol.*, **3** (1961), by permission of Academic Press, New York)

In contrast, the curves obtained for NAD-cytochrome c reductase are quite dissimilar, Wallace finding constant activity up to stage 23, whereas Lang and Grant observed a substantial decline up to stage 18, whence the activity increases again. This disagreement is remarkable, since the work was carried out on the same species and with rather similar methods; even the rearing temperatures were not much different in the two cases. If we compare the absolute values it is seen that both curves begin at approximately the same level (~ 0·1 μM per hour and embryo). The constant activity observed by Wallace thus lies above that measured by Lang and Grant. No attempt will be made to account for these divergent results, but it is presumed that the results of Lang and Grant correspond to a biological reality. If that is true, then it seems to follow that two different NAD-cytochrome c reductases are present in the developing embryos, one prevailing during the early stages, the activity of which is continuously reduced, and another one which begins to be synthesized at some uncertain stage, possibly at gastrulation.

The NAD-linked malate dehydrogenase is typically found both in and outside the mitochondria. The results of Wallace (1961) suggest that the activity of this enzyme is constant until stage 20. This observation seems odd if the various mitochondrial enzymes are synthesized simultaneously. However, it may be that the earliest part of the synthesis of new enzyme is masked by a decline in the original enzyme. This possibility is more likely to obtain when the relative increase in activity is low, as is the case with MDH (Figure 226).

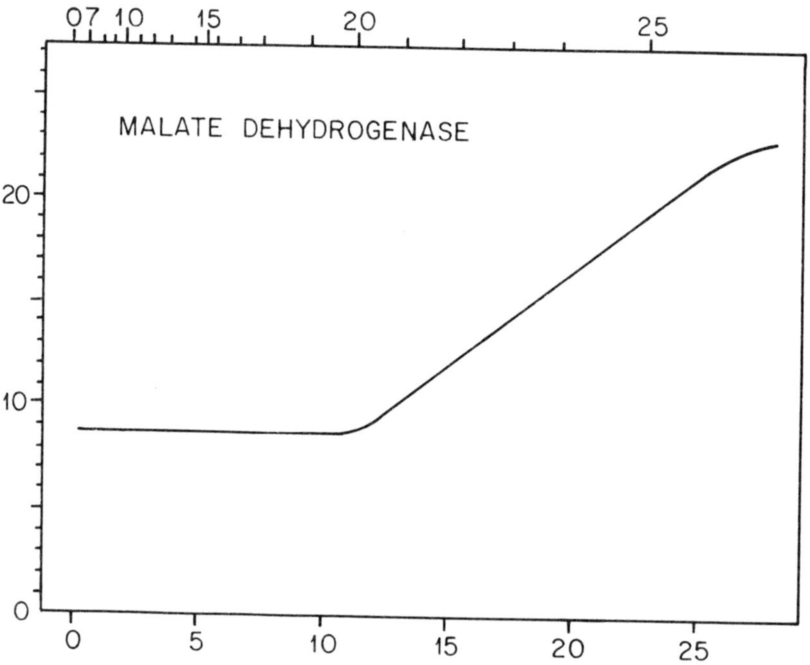

Figure 226. Malate dehydrogenase activity during the development of *Rana*. Lower abscissa, days; upper abscissa, stage of development; ordinate, μM NAD+ reduced per embryo and hour. (Reproduced from R. A. Wallace, *Developm. Biol.*, **3** (1961), by permission of Academic Press, New York)

Two enzymes remain to be discussed, the typical mitochondrial enzymes succinate dehydrogenase and cytochrome oxidase. These enzymes have been shown to be localized partly outside the mitochondria in chick embryos (cf. de Duve *et al.*, 1962). As discussed above, most of the cytochrome oxidase in the eggs of *Arbacia* is found in a non-sedimentable fraction, but in *Xenopus* embryos at least 60 per cent of the activity is recovered in the mitochondria (Boell and Weber, 1955). The results of Lang and Grant (1961) for succinate dehydrogenase (Figure 224) show constant activity during most of development; the final value at stage 25 corresponds to about 0·10 μM succinate or 0·05 μM O_2. The actual oxygen consumption at this stage is at least 0·10–0·15 μM O_2 per hour and embryo. The curve for cytochrome oxidase is practically identical with the one for NAD-cytochrome c reductase shown in Figure 224, suggesting that in this case also an enzyme present during the early development gradually becomes substituted by another one.

If the chemical data presented here are correlated with Karasaki's observations, it seems possible to infer that the synthesis of ICDH and the second phases of NAD-cytochrome c reductase and cytochrome oxidase activity may be referred to the new generation of mitochondria which are observed in the neurula. It may be presumed that these mitochondria also contain MDH and SDH and possibly NADP-cytochrome c reductase. The activities of these enzymes do increase during the later stages of development, but the shape of the activity curves does not indicate the appearance of the second generation of mitochondria. As far as the oxidative system in the first phase is concerned, we do not know whether it is mitochondrial or extramitochondrial. There is no doubt, however, that, just as in the sea urchin embryo, a degradation gradually occurs of the enzymes present in the early embryo.

The presence of two different oxidative systems is also indicated by results of Gregg (1960) shown in Figure 227, which demonstrate that the increase in oxygen uptake after

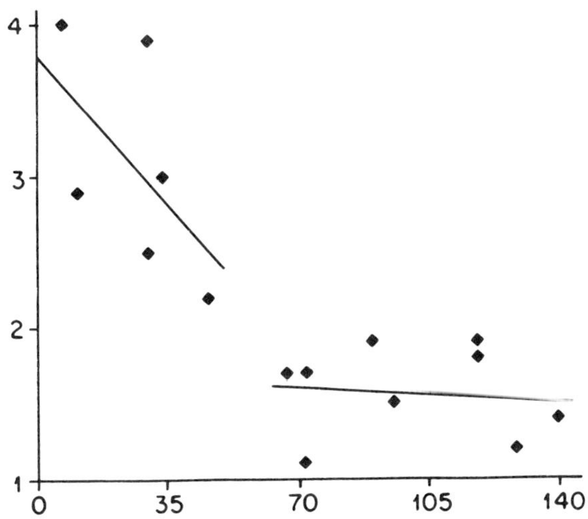

Figure 227. Respiratory control during the development of *Rana*. Abscissa, hours; ordinate, respiratory control quotient. (Reproduced from J. R. Gregg, *Biol. Bull.*, **119** (1960), by permission of Marine Biological Laboratory, Massachusetts)

addition of DNP is much higher before gastrulation than in later developmental stages, although it must be admitted that this outcome may be interpreted in other ways. The change in azide sensitivity observed by Spiegelman and Steinbach (1945) similarly suggests that a fundamental change in the properties of the oxidative system occurs around gastrulation (Figure 228). This modification of the oxidative metabolism may

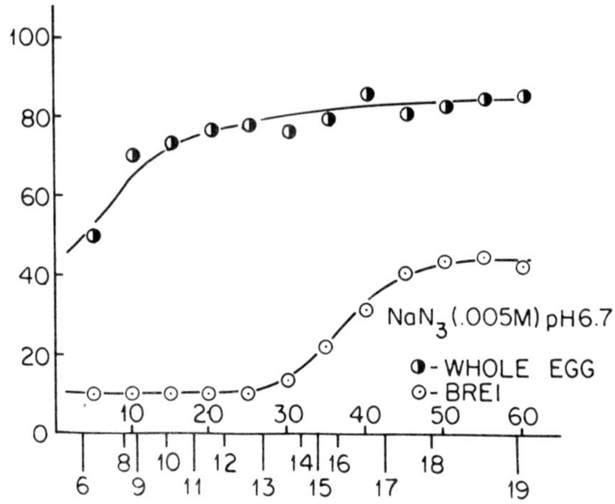

Figure 228. Azide sensitivity of the oxygen uptake of intact embryos and homogenates during the development of *Rana*. Lower abscissa, stage of development; upper abscissa, hours; ordinate, per cent inhibition of respiration. (Reproduced from S. Spiegelman and H. B. Steinbach, *Biol. Bull.*, **88** (1945), by permission of Marine Biological Laboratory, Massachusetts)

represent a case of biochemical recapitulation. This assertion will be valid even if the second generation of mitochondria is typical of the filocytes, since this cells class is phylogenetically younger than the lobocytes.

The results discussed so far indicate that among the tricarboxylic acid cycle enzymes, the activity of ICDH is very low during the early developmental stages. It seems almost impossible that a metabolic pathway involving this enzyme can be of importance under these circumstances. S. Cohen (1954) has published results concerning the incorporation of labelled CO_2 in *Rana*. In the blastula 50 per cent of the total activity was found in the soluble fraction, and of the latter 28 per cent was found in glutamate, 12 per cent in aspartate, and small amounts in other di- and tricarboxylic acids. In commenting upon these results, Wallace (1961) suggests that the dicarboxylic acid shuttle

$$\text{pyruvate} + CO_2 \rightarrow \text{malate} \rightarrow \text{oxalacetate} + CO_2 \rightarrow \text{citrate}$$

may account for the results of Cohen. The large amount of glutamate is rather disturbing, however, for if it is formed from α-ketoglutarate it seems absolutely necessary that ICDH is involved. In this context it may be mentioned that quantities of pyruvate and α-ketoglutarate accumulated in the early embryo are reduced during gastrulation and neurulation (Barbieri and Salomon, 1962). It may be worth pointing out that both the reductive carboxylation of pyruvate to malate and the reductive oxidation of α-ketoglutarate to glutamate require NADPH (cf. Horecker and Hiatt, 1958).

It is seen that, just as in the case of the sea urchin embryo, it is impossible to arrive at a clear-cut answer as regards the importance of the tricarboxylic acid cycle in the early amphibian embryo. It seems likely that active mitochondria are present, but they are

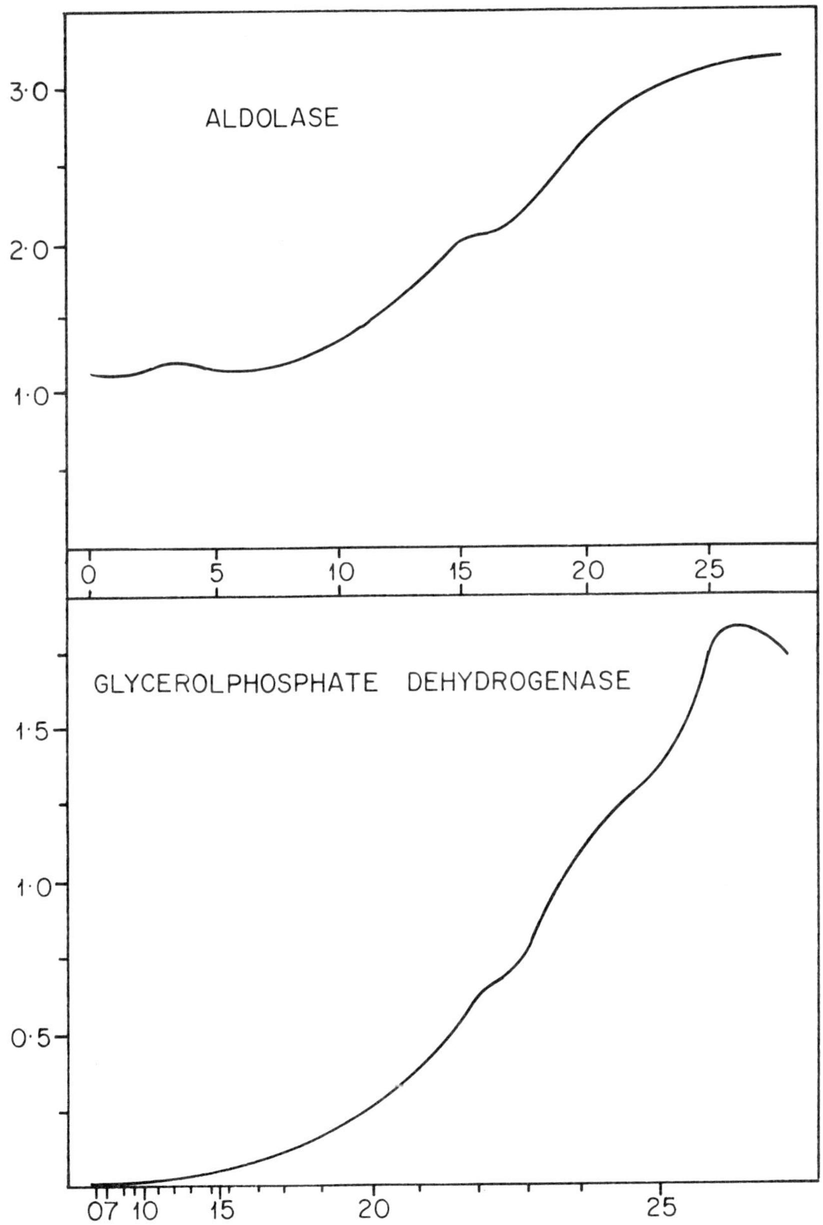

Figure 229. Aldolase and glycerol phosphate dehydrogenase activities during the development of *Rana*. Lower abscissa, stage of development; upper abscissa, days; ordinates, μM NADH oxidized per embryos and hour. (Reproduced from R. A. Wallace, *Developm. Biol.*, **3** (1961), by permission of Academic Press, New York)

deficient in at least one enzyme, and display other characteristics which set them apart from the mitochondria found in later embryonic stages and in adult tissues. According to Wallace (1961), the ICDH activity at the gastrula stage can account for only one-fiftieth of the prevailing oxygen consumption, and it is therefore questionable whether this enzyme, and thus the Krebs cycle, are involved to any significant extent in the supply of energy.

Glycolysis

Among the glycolytic enzymes the following have been studied by Wallace (1961): NAD-dependent glycerolphosphate dehydrogenase (GPDH), lactate dehydrogenase (LDH), pyruvate kinase and aldolase. LDH has also been investigated by Adams and Finnegan (1965). These enzymes are usually observed to be soluble, but it is interesting to note that during the first days of development in the chick embryo, LDH is located in a particulate fraction (cf. de Duve *et al.*, 1962).

The GPDH activity is extremely low during the early stages, but at stage 15 a very distinct increase is observed. The exact time at which the synthesis begins is difficult to decide. Subsequently there is a very extensive rise in activity which seems to occur in different phases (Figure 229). Wallace estimates the relative increase to reach about 170 during development, but it must be rather difficult to establish whether the enzyme is present at all during early development. At any rate it seems quite evident that it can be of little significance for the metabolism at this stage. However, GPDH is only an acces-

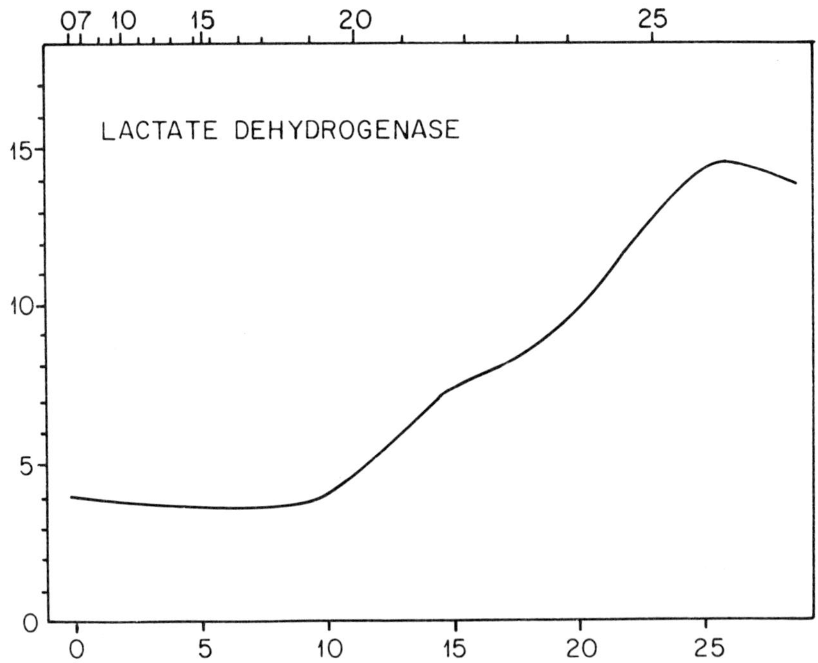

Figure 230. Lactate dehydrogenase activity during the development of *Rana*. Lower abscissa, days; upper abscissa, stage of development; ordinate, μM NAD^+ reduced per embryo and hour. (Reproduced from R. A. Wallace, *Developm. Biol.*, **3** (1961), by permission of Academic Press, New York)

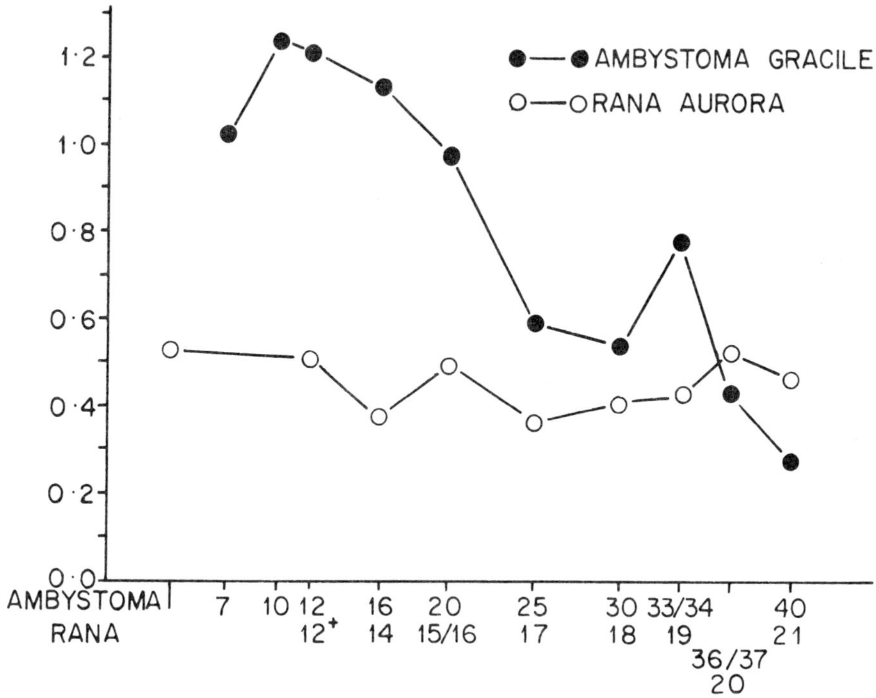

Figure 231. Lactate dehydrogenase activity during the development of *Ambystoma* and *Rana*. Abscissa, stage of development; ordinate, specific activity. (Reproduced from E. A. Adams and C. V. Finnegan, *J. Exptl. Zool.*, **158** (1965), by permission of The Wistar Press, Philadelphia)

sory glycolytic enzyme which is not involved in the transformation of glucose to pyruvate.

According to Wallace (1961), the activity of LDH oscillates slightly during early stages of development, but the real rise in activity begins around stage 18. The final relative increase is about fourfold, and two phases may be distinguished (Figure 230). The oscillations were also observed by Adams and Finnegan (1965); in *Ambystoma* a very marked decrease occurred, but no rise in activity was registered (Figure 231). The pyruvate kinase activity begins to increase at the same time as LDH; the total increase during development is hardly 50 per cent (Figure 223). The high initial activity may indicate that the formation of ATP from phosphopyruvate is an important part of the energy metabolism in the sl-cells.

The curve for aldolase is very similar to that of LDH, the maximum relative increase being about 3. Two distinct phases may be observed in the rising part of the curve (Figure 229). These observations rather clearly demonstrate that, possibly in contrast to the early sea urchin embryo, the glycolytic pathway is present in the amphibian embryo. Even if the production of lactic acid cannot be taken as a definite demonstration of glycolytic activity, this point seems corroborated by the repeated demonstration of lactic acid production under anaerobic conditions (Lennerstrand, 1933; Barth, 1946; A. I. Cohen, 1954; Gregg, 1962; Barbieri and Salomón, 1963). As shown in Figure 232, it may be possible, both in normal and hybrid embryos, to distinguish two phases in the capacity of lactate production, one of which begins to decline and the other to rise in the normal gastrula.

No change in total carbohydrate can be registered during early development (Gregg, 1948), nor can a stoichiometric relation be established between glycogen disappearance and lactate formation under either aerobic or anaerobic conditions (A. I. Cohen, 1954; Barbieri and Salomon, 1963).

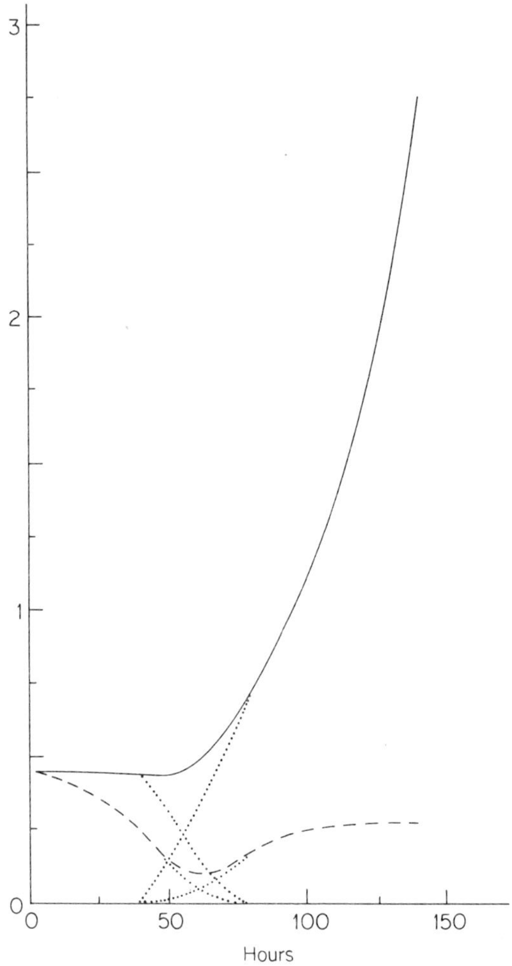

Figure 232. Anaerobic lactate production in normal (*R. pipiens*) and hybrid (*R. pipiens* ♀ × *R. sylvatica* ♂) embryos during the development of *Rana*. Unbroken curve, normal embryos; dashed curve, hybrid embryos. The dotted lines represent suggested resolutions of the curves. Abscissa, hours; ordinate, μg lactate per embryo and hour. (Reproduced from J. R. Gregg, *Biol. Bull.*, **123** (1962), by permission of Marine Biological Laboratory, Massachusetts)

It is interesting to note that although presumably the enzymes of the glycolytic cycle are encoded in the genome in both Echinoidea and Amphibia this pathway appears to be available in the early embryonic cells only in the latter case. The significance of this difference, which, we may presume, is an outcome of phylogenetic evolution, is completely unknown.

Pentose phosphate cycle

It has been found that enzymes of the PP cycle generally are highly active in embryonic tissues (cf. Herrmann and Tootle, 1964). The results of Wallace (1961) show that in the larval stages the activity of glucose-6-phosphate dehydrogenase (G-6-PDH) is of the same magnitude as aldolase and GPDH, but around 10 times lower than that of LDH and pyruvate kinase (Figure 233). The activity of 6-phosphogluconate dehydrogenase (6-PGDH) is more than 10 times lower than that of G-6-PDH. These results do not allow the conclusion that the PP cycle as such is present in the amphibian embryo, only that the initial stages of this pathway are catalysed. It is remarkable that during the initial stages the enzyme activities are very low, particularly that of 6-PGDH. In contrast to the Echinoidea, where the activities of these enzymes gradually diminish, there is a substantial rise in the amphibian embryo, beginning at gastrulation. Since this rise is in parallel with that of the mitochondrial enzymes, there is no reason to presume that the enzymes in this case are involved in the supply of energy. They may rather, through the generation of NADPH, be involved in various kinds of synthetic activities. As a number of lobocytes, e.g. liver cells, are active in the larva the rise in activity is compatible with the proposition that the PP cycle is typical of this cell class.

BIOCHEMICAL EVENTS

The synthesis of new cell constituents is an inherent aspect of cell division and cell diversification. As far as the first process is concerned, we have seen that during cleavage it may occur in the absence of any significant synthetic activity. Under these conditions the only biochemical parameters which unconditionally must change are DNA and presumably certain other nuclear compounds. As discussed above, the increase in DNA may not involve a change in the total content of deoxyribosides.

If cell division forms part of a process of 'pure growth', a doubling of all prevailing cytoplasmic components must precede each mitotic step, and it may thus be possible to follow the process of cell division by analysis of almost any substance, since the ratio between the various cellular components remains constant. This phenomenon is hardly ever encountered in Nature, except in cultures of microorganisms, and it is particularly unlikely to be observed in developing embryos, because the phenomenon of cell diversification is an essential aspect of embryogenesis. This process involves that the synthetic activity varies from one cell-type to another. Under these circumstances the ratios between the various parameters may be expected to change continuously during development (1959c). The synthetic activities, of course, comprise all kinds of substances, nucleic acids, proteins, mucopolysaccharides, lipids, etc. In the present context we shall deal only with one type of nucleic acid, viz., DNA, and among the proteins, which may be structural as well as catalytic ones, we shall discuss only the latter. As in the preceding chapter, we shall consider only enzyme patterns in whole embryos, and as there the discussion will be confined to Echinoidea and Amphibia.

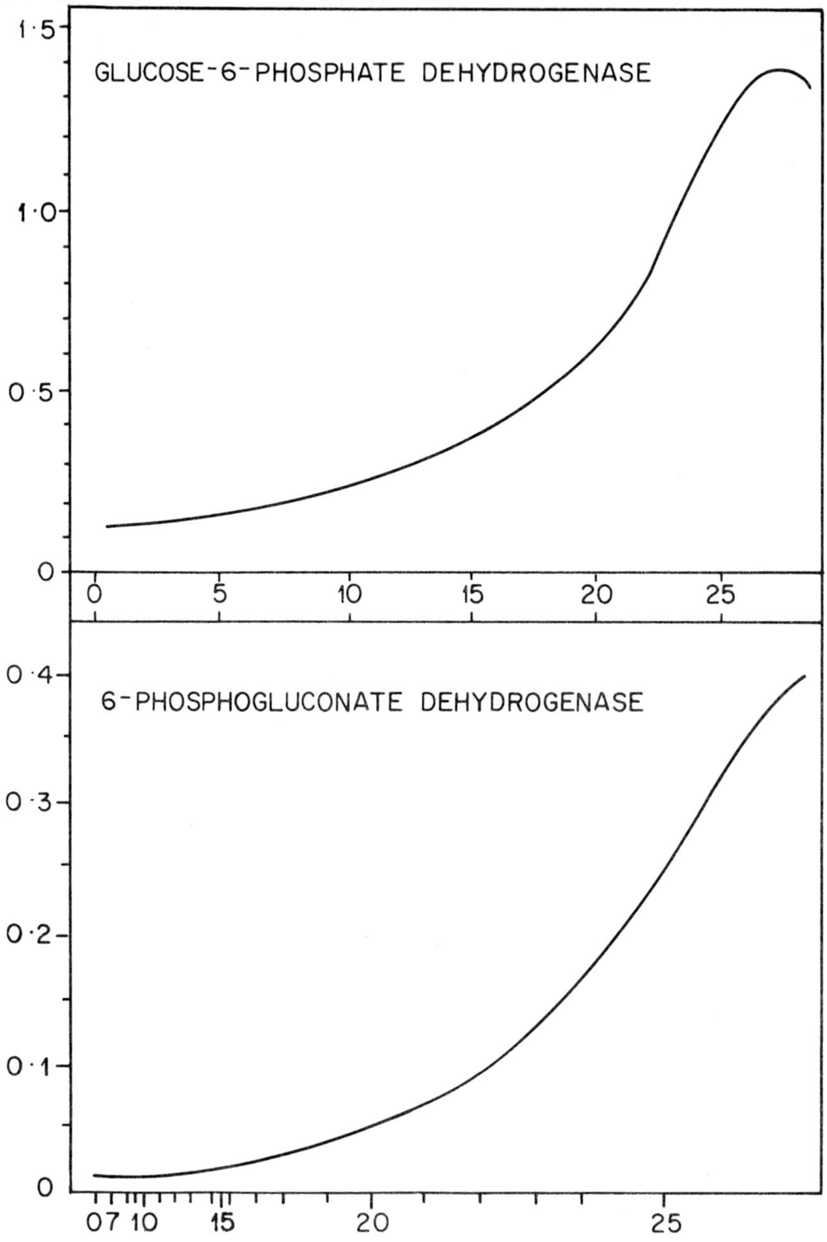

Figure 233. Glucose-6-phosphate dehydrogenase and 6-phosphogluconate dehydrogenase activities during the development of *Rana*. Lower abscissa, stage of development; upper abscissa, days; ordinate, μM NADP+ reduced per embryo and hour. (Reproduced from R. A. Wallace, *Developm. Biol.*, **3** (1961), by permission of Academic Press, New York)

Echinoidea

In the present section we shall try to trace the embryogenetic processes of cell division and cell diversification on the basis of records concerning chemical changes associated with sea urchin development.

CELL DIVISION

Since, with few exceptions, the amount of DNA per nucleus is constant, cell multiplication is easily followed by means of DNA determinations. The situation is more complicated as regards cell growth, i.e. the increase in the amount of cytoplasm associated with cell divsion, for it is not feasible to decide unambiguously which enzymes are truly representative of the cytoplasm, and in any case we are restricted by the fact that only a limited number of enzymes have been investigated.

DNA

Many reports have been published dealing with the DNA content of sea urchin gametes and embryos. In two cases the DNA content of the sperm has been found to 0·7 pg (*Arbacia*, Vendrely and Vendrely, 1949; *Paracentrotus*, Hoff-Jørgensen, 1954). The values observed for the pluteus stage are also essentially in accord, showing that the *Paracentrotus* larva contains about 5000–6000 pg DNA, corresponding to 3500–4000 cells (e.g. Whiteley and Baltzer, 1958; Baltzer and Chen, 1960). The *Arbacia* pluteus may have a slightly higher DNA content.

As regards the unfertilized egg, the agreement is less striking, the recorded values ranging between 12 and 1000 pg. As discussed previously (1966b), the results may be classed in two groups according to the specificity and sensitivity of the analytical methods, and when this is done it turns out that the lower values are the most reliable. Hence, the amount of 'DNA' present in the egg probably does not exceed that corresponding to 16–32 diploid nuclei. This material must represent a reserve which is utilized during the early mitoses.

A plateau or at least a hump in the DNA curves (Figure 234) has been observed in *Paracentrotus* by J. Brachet (1933), Elson, Gustafson and Chargaff (1952) and Baltzer and Chen (1960), and in *Arbacia* by Whiteley and Baltzer (1958). This feature occurs after 20–30 hours of development, and even if direct comparison is difficult, it seems certain that it cannot be correlated with the phase of constant respiration discussed earlier (Figure 210).

Enzymes

It is proposed here that the changes in the activity of lysosomal and certain supernatant enzymes may represent cytoplasmic growth. Obviously this choice is quite arbitrary, since the various cell-types may contain lysosomes differing with respect to their enzyme content and since a supernatant enzyme may be ubiquitous and yet be present in particularly large amounts in special kinds of differentiated cells. However, as will be observed from the following survey, in the sea urchin embryo this approach gives an unambiguous answer.

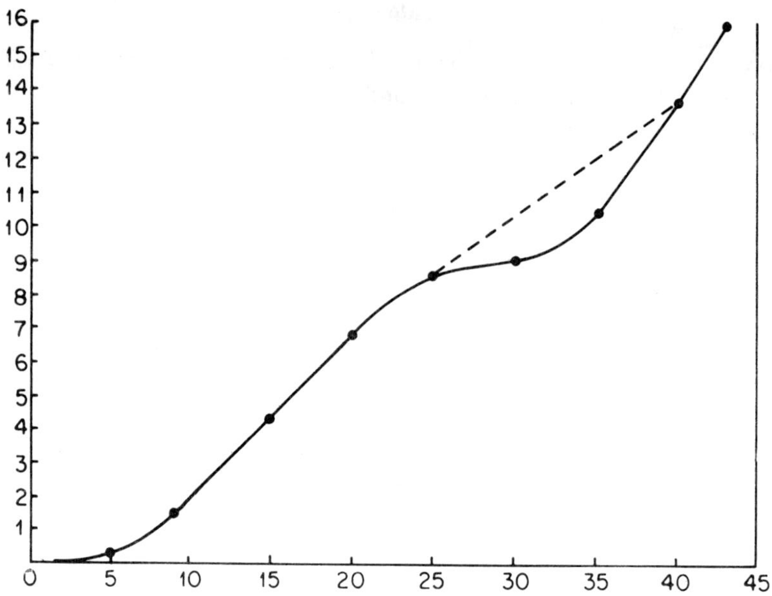

Figure 234. DNA synthesis during the development of *Paracentrotus*. Abscissa, hours; ordinate, mg DNA per g dry weight. (Reproduced from J. Brachet, *Arch. Biol.*, **44** (1933), by permission of Archives de Biologie, Brussels)

Among the enzymes which according to de Duve (1963) are located in the lysosomes, the following have been investigated by Gustafson and Hasselberg (1951): acid phosphatase, cathepsin D and arylsulphatase. It may be mentioned that the substrate affinity of the latter enzyme does not correspond exactly to that recorded for rat liver arylsulphatases, but the pH optimum suggests that it is a lysosomal enzyme (cf. 1966b). The activities of all these enzymes remain constant during development (Figure 235).

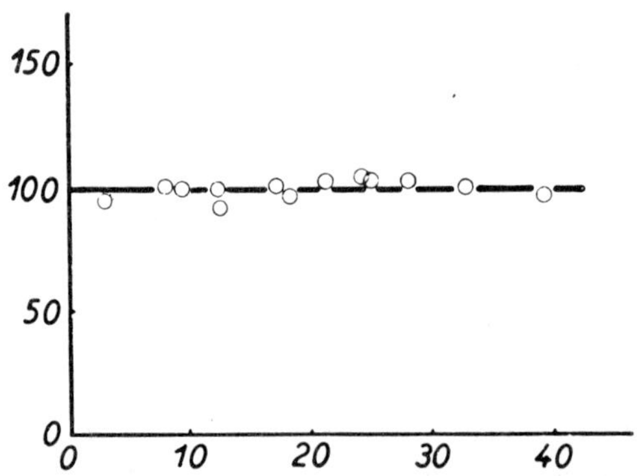

Figure 235. Cathepsin D activity during the development of *Psammechinus*. Abscissa, hours; ordinate, relative activity. (Reproduced from T. Gustafson and I. Hasselberg, *Exptl. Cell Res.*, **2** (1951), by permission of Academic Press, New York)

The supernatant enzymes which have been studied are: dipeptidase (Holter and Lindahl, 1940), adenosine deaminase, pyrophosphate (and hexametaphosphate) phosphohydrolase (Gustafson and Hasselberg, 1951). No increase occurs in the activity of any of these enzymes. Thus, if the proposed index of cytoplasmic growth is accepted, it follows that the development of the sea urchin embryo up to the pluteus stage is not accompanied by any measurable change in this parameter.

CELL DIVERSIFICATION

We shall here investigate separately those enzymatic changes which may be associated with cell transformation and with cell differentiation.

Cell transformation

The process of cell transformation comprises two aspects, a disappearance of sl-cells and a simultaneous formation of new cell-types. The first phenomenon must be associated with a reduction in the content of enzymes specific for the sl-cell. It has been suggested above that filocytes, at least in the early embryo, are distinguished by their content of alkaline phosphatase, and consequently it ought to be possible to use the changes in this enzyme as an index of the occurrence of the lf-transformation. The colligocytes are distinguished by their content of sulphated mucopolysaccharides, and analyses of the activity of the appropriate sulphotransferases might therefore function as an indicator of the sc-transformation. Unfortunately, no information obtains on this point. Finally, as regards the nv-transformation, it has been proposed that certain vibriocytes typically contain acetylcholine esterase, the formation of which therefore should coincide with their transformation.

Amoebocyte enzymes

It would, of course, be desirable to establish in advance a list of enzymes specific for sl-cells, and subsequently turn the attention to their developmental fate. The opposite approach employed here, viz., that the enzymes observed to decrease during embryogenesis are classed as sl-cell enzymes, is not logically unassailable but, owing to our ignorance, it must be accepted for the time being.

We have already seen that G-6-PDH and 6-PGDH decline steadily from the time of hatching (Bäckström, 1959; 1963), and it was suggested that the PP cycle may be involved in the supply of energy in the sl-cells in the sea urchin embryo (Figure 220). Two more enzymes have been observed to decrease in activity during development, viz., catalase (Deutsch and Gustafson, 1952) and 2′-deoxyribosyl 4-aminopyrimidine 5′-phosphate deaminase, dAPDase (Scarano, 1961). The reduction of the activity of these enzymes occurs in two more or less distinct phases (Figures 236 and 237). The second enzyme catalyses the conversion of deoxycytidylic acid deoxyuridylic acid and of deoxymethylcytidylic acid to thymidylic acid; the second of these reactions may be of importance for the synthesis of DNA. The timing of the two phases shown in Figure 237 does not correspond to that shown for DNA synthesis in *Paracentrotus* (Figure 234), but this discrepancy may possibly be correlated with the fact that the dAPDase has been studied in a species belonging to a different genus, *Sphaerechinus*, which apparently develops rather slowly.

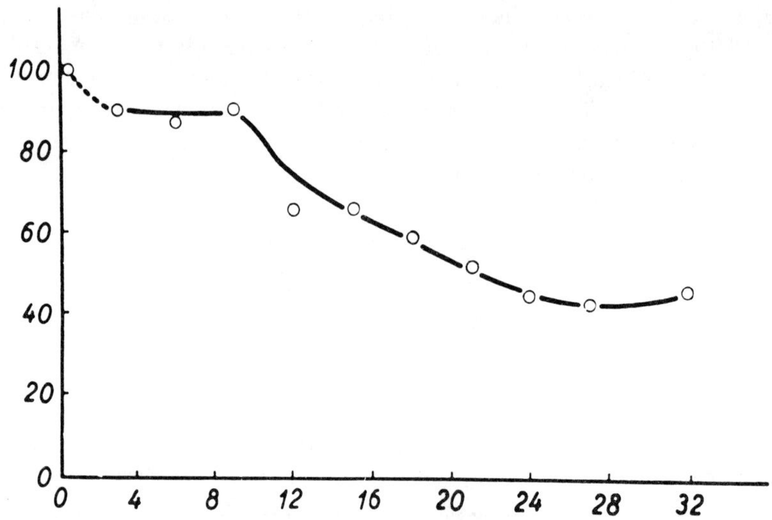

Figure 236. Catalase activity during the development of *Psammechinus*. Abscissa, hours; ordinate, relative activity. (Reproduced from H. F. Deutsch and T. Gustafson, *Ark. Kemi.*, **4** (1952), by permission of Kungl. Vetenskapsakademin, Stockholm)

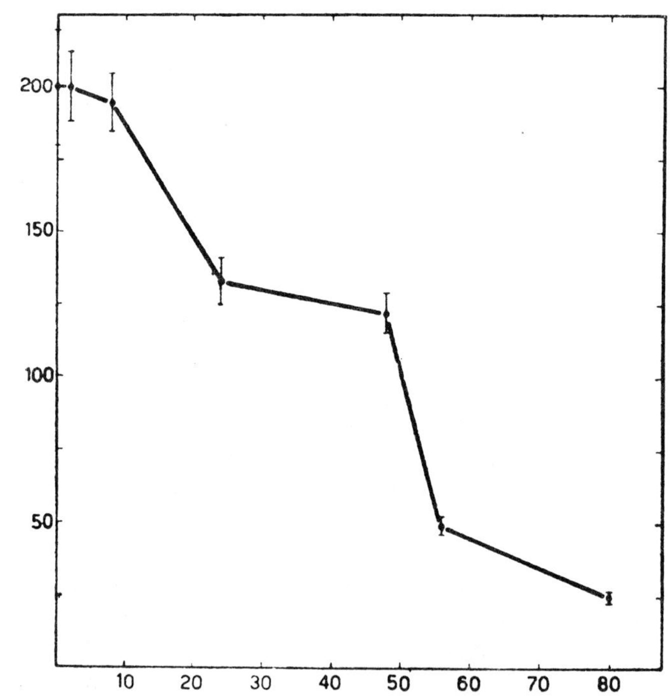

Figure 237. 2′-deoxyribosyl 4-aminopyrimidine 5′-phosphate deaminase activity during the development of *Sphaerechinus*. Abscissa, hours; ordinate, arbitrary units. (Reproduced from E. Scarano in *Symposium on the Germ Cells and the Earliest Stages of Development* (1961), by permission of Istituto Lombardi Accademia di Scienze e Lettere, Milan)

The lf-transformation

Possession of alkaline phosphatase (APh) may distinguish the filocytes, at least the first generations of this cell-type. Among differentiated filocytes certain are known to contain this enzyme in large amounts, whereas it is absent in other cases. As we shall presently see, this point is borne out by observations on the amphibian embryo, but determinations on Protozoa also seem to support this claim. Thus, in *Chaos* (Holter and Lowy, 1959) and in *Urostyla* (Pigon, 1962) which, judging from their shape must be characterized as lobocytes, it has been possible to demonstrate acid but not alkaline phosphatase. On the other hand, in *Trypanosomas*, which above has been suggested to be a vsf-cell, alkaline phosphatase is present (Harvey, 1949).

In the sea urchin embryo the situation is not clear-cut, for as shown in Figure 238,

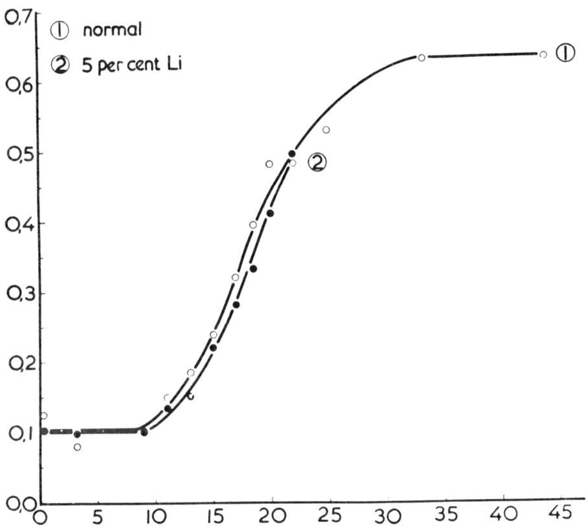

Figure 238. Alkaline phosphatase activity during the development of *Psammechinus*. Abscissa, hours; ordinate, arbitrary units. (Reproduced from T. Gustafson and I. Hasselberg, *Exptl. Cell Res.*, **1** (1950), by permission of Academic Press, New York)

there is APh activity in the unfertilized egg (Gustafson and Hasselberg, 1950). The moderate increase in relative activity during development shows that the amount of enzyme present at the beginning is quite substantial. Two arguments may be advanced in support of the contention that this observation need not violate the principle advanced here. The first one is that the activity at alkaline pH may represent residual activity of the acid phosphatase. As we have seen, this enzyme remains constant during development and since its activity in the gastrula stage is not much lower than that of APh (Figure 239), it follows that during the earlier stages it must be much more active than the latter enzyme. The second one is found in the results published by Hsiao and Fujii (1963), suggesting that the relative increase in activity is much lower with glycerol-2-phosphate as substrate than that found with phenylphosphate by Gustafson and Hasselberg.

This observation indicates that we are dealing with two separate enzymes with different substrate affinities, the one synthesized during development hydrolysing

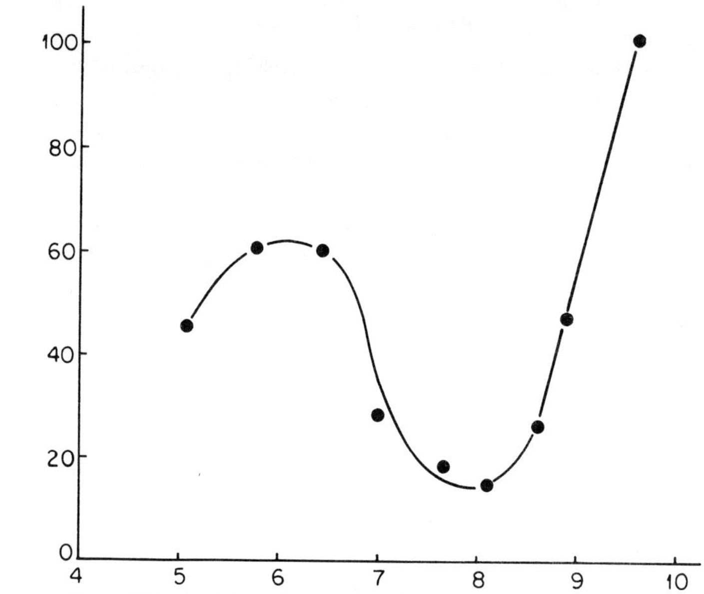

Figure 239. Activity–pH curve for the splitting of phenylphosphate by extracts of *Paracentrotus* gastrulae. Abscissa, pH; ordinate, relative activity. (Reproduced from T. Gustafson and I. Hasselberg, *Exptl. Cell Res.*, **2** (1951), by permission of Academic Press, New York)

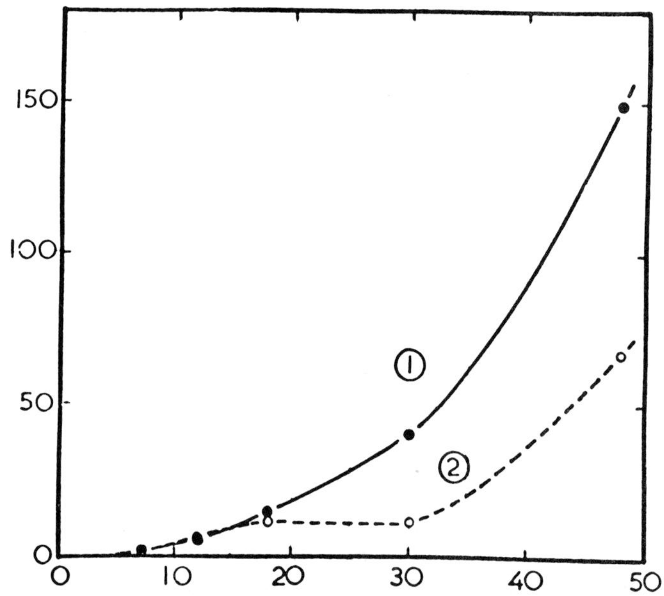

Figure 240. Acetylcholine esterase activity during the development of *Paracentrotus*. Abscissa, hours; ordinate, arbitrary units. 1, normal embryos; 2, vegetalized Li^+-treated embryos. (Reproduced from K.-B. Augustinsson and T. Gustafson, *J. Cell. Comp. Physiol.*, **34** (1949), by permission of The Wistar Press, Philadelphia)

phenylphosphate at a greater relative rate than the one present in the egg. However this may be, my proposition is supported by the fact that histochemical methods show the APh activity appearing after the mesenchyme blastula stage to be localized in the mesenchyme cells and in the gut (Evola-Maltese, 1957; Hsiao and Fujii, 1963), and there is no doubt that the increase in activity begins at the time when the first sf-cells appear.

The nv-transformation

Augustinsson and Gustafson (1949) have studied the activity of acetylcholine esterase (AChE), purportedly a vibriocyte attribute, during the development of *Paracentrotus*. No demonstrable enzyme activity could be recorded in the unfertilized egg, but subsequently a gradually rising curve was recorded (Figure 240). Tracing this curve back to zero activity, it can be estimated that the synthesis begins 4–5 hours after fertilization. On the basis of this result the authors concluded that there is a close correlation between the appearance of vibratile organelles and of AChE activity. It was found possible by means of treatment with Li^+ to distinguish three different phases of synthetic activity, of which the middle one is suppressed by this ion. The first phase begins, as mentioned, a few hours after fertilization and lasts to about 15–18 hours. This activity may represent the formation of flagella in the ectodermal cells. The second phase ends at 30–35 hours and may reflect the development of the ventral flagellated bands. The third phase was suggested by Augustinsson and Gustafson to correspond to the differentiation of the gut.

Cell differentiation

It is possible that some of the synthetic activity registered for the various enzymes discussed above represents cell differentiation, thus for instance the third phase of AChE synthesis. It does not seem possible to answer this question satisfactorily, part of the reason unquestionably being that the process of cell differentiation is not far advanced in the pluteus larva.

Amphibia

In this section we shall try to extract the information regarding cell division and cell diversification which is hidden in various chemical analyses performed on developing amphibian embryos. Since in many cases the same compounds have been analysed in the two classes of Amphibia dealt with here, it will be possible to make comparisons serving to illustrate the differences obtaining.

CELL DIVISION

The synthesis of DNA and of certain selected enzymes will be used as indices of cell division and growth, exactly as in the case of the Echinoidea.

DNA

Many authors have determined the DNA content during amphibian embryogenesis. On the whole, agreement obtains for the later stages, but the values obtained before

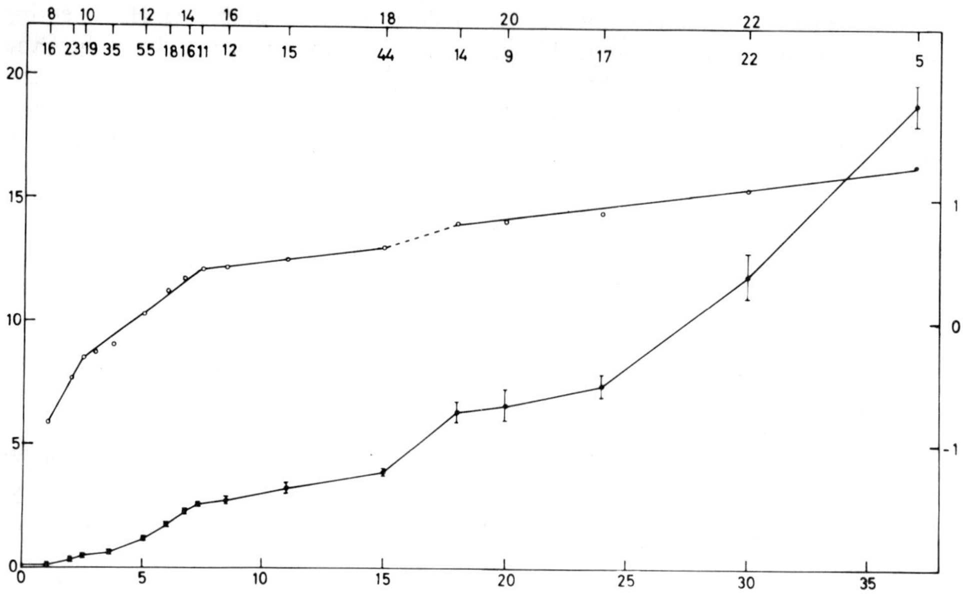

Figure 241. DNA synthesis during the development of *Rana*. Lower abscissa, days at 8°C; upper abscissa, developmental stage. Left ordinate, μg DNA per embryo; right ordinate, logarithm of μg DNA per embryo. The numbers above each experimental point indicate the number of determinations, the vertical bars the standard error of the mean. (Reproduced from S. Løvtrup, *Exptl. Cell Res.*, **43** (1966), by permission of Academic Press, New York)

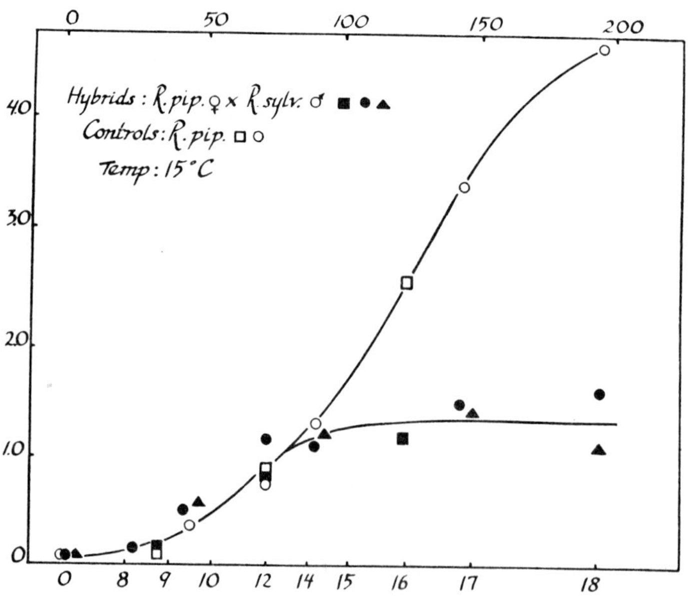

Figure 242. DNA content during the development of normal and hybrid embryos of *Rana*. Lower abscissa, stage of development in controls; upper abscissa, hours; ordinate, μg DNA per embryo. (Reproduced from J. R. Gregg and S. Løvtrup, *Biol. Bull.*, **108** (1955), by permission of Marine Biological Laboratory, Massachusetts)

gastrulation vary greatly. It appears that in this case also the methodological approach is decisive, the more sensitive and specific methods giving the lowest results (cf. 1966c). The most likely value for the deoxyriboside-containing material in the unfertilized egg lies in the range between 0·05 and 0·1 μg. As seen from Figure 241, the DNA synthesis occurs in four distinct phases, shown best maybe in the logarithmic plot. The first phase is particularly clearly demonstrated in hybrid embryos (Gregg and Løvtrup, 1955; 1960), in which synthesis of DNA ceases at stage 14–15 (Figure 242). The phases in DNA synthesis presumably are correlated with the phases in energy consumption described earlier, but since the results are obtained on different species it is not possible to bring out this point.

Enzymes

Among the enzymes which are relevant in the present context, three are supposed to be localized in the lysosomes, viz., acid phosphatase, deoxyribonuclease II and cathepsin D. The first of these has been studied by De Cesaris Coromaldi (1955), using glycerophosphate and phenylphosphate, and by myself (1955), using only the former substance as substrate. As far as the common substrate is concerned, the results are in good agreement, showing constancy until the beginning of larval development (Figure 243), after which a rapid increase sets in. Maximum activity in *Ambystoma* corresponds to an eightfold increase (1955). With phenylphosphate, De Cesaris Coromaldi found the increase to begin at the end of gastrulation, suggesting that these determinations are concerned with a different enzyme.

Deoxyribonuclease II was investigated by Coleman (1962), who found very little activity before the neurula stage; subsequently a very substantial rise was observed (Figure 244). Cathepsin D was determined by Urbani and De Cesaris Coromaldi (1954b; 1955) and by myself (1955). The results agree in showing the activity to increase in an early tailbud stage (Figure 245). The relative increase is ten–fifteenfold.

If these enzymes are found exclusively in the lysosomes, it seems possible to distinguish three populations of these particles. Those present in the egg contain acid phosphatase and cathepsin D, but hardly any deoxyribonuclease II. The synthesis of the two last mentioned enzymes which is initiated at the end of neurulation suggests the elaboration of lysosomes which contain no acid phosphatase. At the beginning of larval development a new phase of synthetic activity is indicated, but this time all three enzymes are involved. The break in the deoxyribonuclease II curve (Figure 244) corresponds quite well to the time at which acid phosphatase is being synthesized, and a similar phenomenon may be observed in the cathepsin D curves (1955). I am well aware that the results discussed here may be interpreted in other ways, but this question can only be answered by studies of the enzyme composition in isolated lysosomes.

Among the supernatant enzymes, the aminopeptidase acting on dipeptides has been studied by Urbani and De Cesaris Coromaldi (1954a; 1954b) and by Urbani, De Cesaris Coromaldi and Scollo-Lavizzari (1957). This enzyme and a tripeptide-splitting enzyme were followed by me during the development of *Ambystoma* (1953c; 1955). The results on dipeptide hydrolysis agree in showing that the synthesis begins in the tailbud stage. In *Ambystoma* (1953c) and in *R. dalmatina* no changes were observed before this stage, but in *R. esculenta* and in two species of *Bufo* a decline in activity is indicated during the early stages (Figure 246). The constant activities observed in the other cases may actually

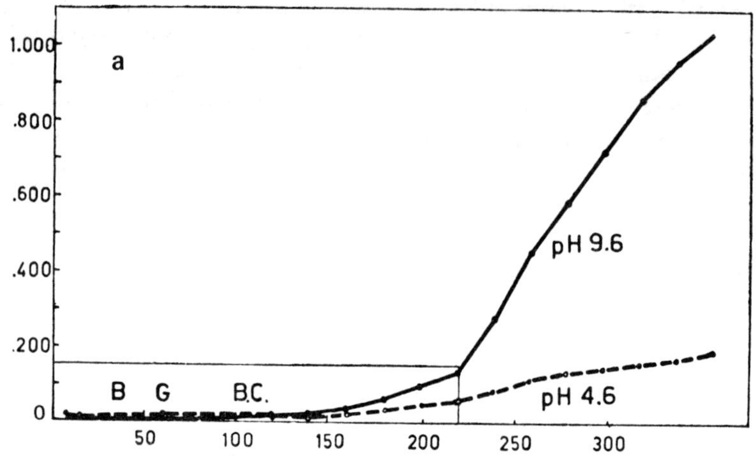

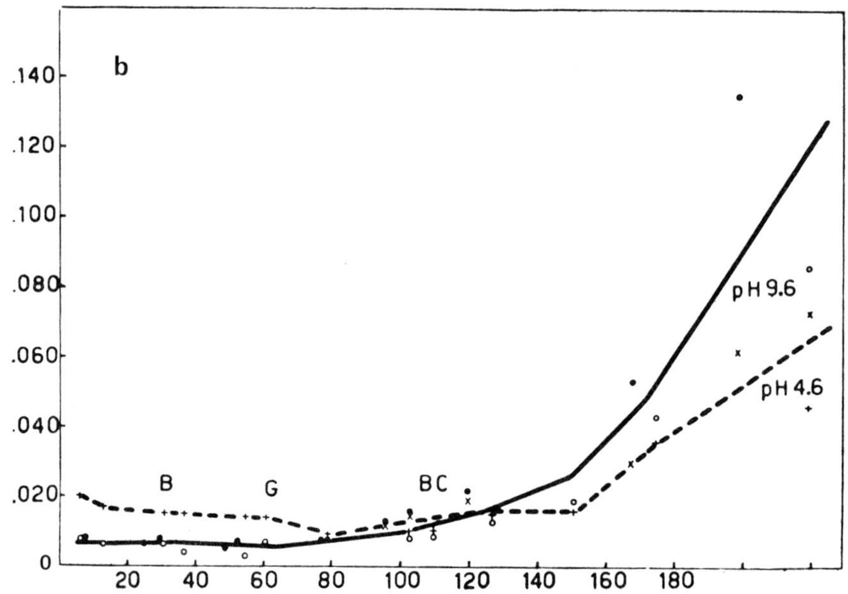

Figure 243. Acid and alkaline phosphatase activity during the development of *Bufo*, using glycerol phosphate as substrate. (b), enlargement of the framed part of (a). Abscissa, hours; ordinate, arbitrary units. B, blastula; G, gastrula; B.C., tailbud. (Reproduced from L. De Cesaris Coromaldi, *Ric. Sci.*, **25** (1955), by permission of Consiglio Nazionale delle Ricerche, Rome)

represent the summation of a rising and a declining enzyme curve. The total relative increase of this enzyme activity lies around two, in contrast to the tripeptide-splitting enzyme, which is subjected to a tenfold rise (1953c; 1955), suggesting that it is a separate enzyme.

If the enzymes discussed here are accepted as indices of the amount of cytoplasm, it follows that there is a distinct increase in this parameter during the embryonic develop-

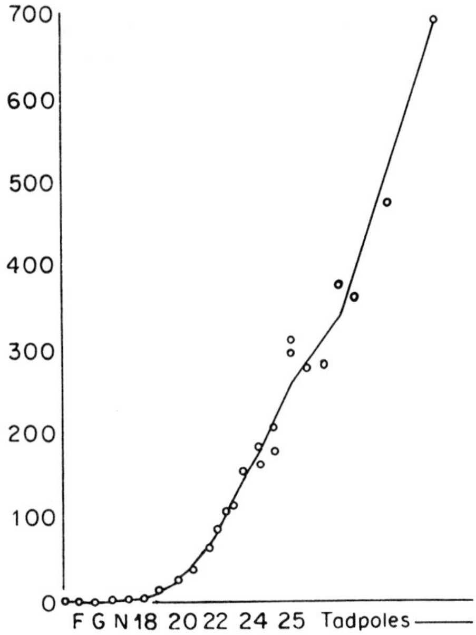

Figure 244. DNase II activity during the development of *Rana*. Abscissa, stage of development; ordinate, arbitrary units. (Reproduced from J. R. Coleman, *Developm. Biol.*, **5** (1962), by permission of Academic Press, New York)

ment of Amphibia in contrast to that observed in the Echinoidea. It seems reasonable to refer this disparity to the difference in yolk content in the two cases. Since the values for relative increase vary considerably, it is not possible to express the cytoplasmic growth in exact quantitative terms, but apparently 10 may serve as an approximate value.

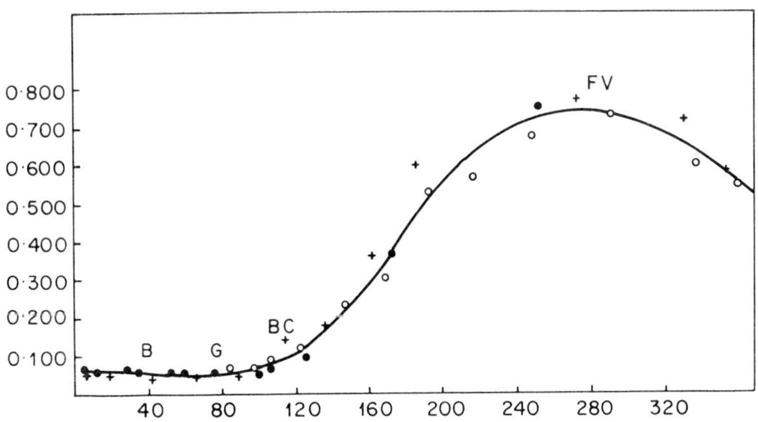

Figure 245. Cathepsin D activity during the development of *Bufo*. Abscissa, hours; ordinate, arbitrary units. FV, end of prefeeding development; other abbreviations as in Figure 243. (Reproduced from E. Urbani and L. De Cesaris Coromaldi, *Acad. Naz. Lincei* [8], **18** (1955), by permission of Accademia Nazionale dei Lincei, Rome)

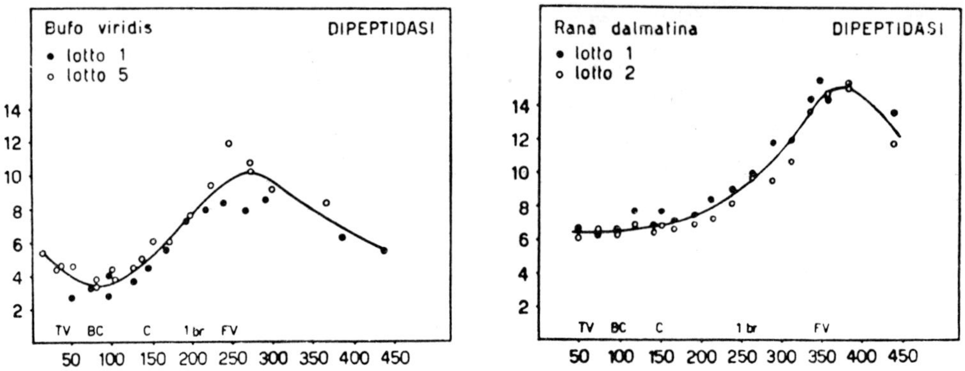

Figure 246. Dipeptidase activity during the development of *Bufo* and *Rana*. Abscissa, hours; ordinate, arbitrary units. 1 br, one gill covered; other abbreviations as in Figures 243 and 245. (Reproduced from E. Urbani, L. De Cesaris Coromaldi and G. S. Scollo-Lavizzari, *Accad. Naz. Lincei* [8], **23** (1957), by permission of Accademia Nazionale dei Lincei, Rome)

CELL DIVERSIFICATION

If the assumptions made in the discussion of the chemical diversification in Echinoidea are accepted, it is possible on the basis of published observations on Amphibia to arrive at a picture which exhibits many similarities to, but also clear differences from, that established for the former class.

Cell transformation

We shall discuss here those enzymes which with some reason may be anticipated to be typical of the order of sl-cells, as well as those which are suggested to represent the classes of filocytes and vibriocytes.

Amoebocyte enzymes

In the sea urchin embryo the enzymes observed to decrease during development were proposed to be typical for sl-cells. This decrease may be related to the fact that apparently the amount of cytoplasm does not change noticeably. In order to accommodate new cell-types with their specific enzyme pattern it is reasonable to anticipate degradation and disappearance of those enzymes which are no longer synthesized. The situation is different when a substantial cytoplasmic growth takes place, because under such circumstances the appearance of new cell-types does not require a decrease in the number of amoebocytes. We have seen that there is a very distinct dissimilarity as regards the PP cycle enzymes in the two cases investigated here: in Echinoidea there is a continuous decline in activity, whereas in Amphibia the activity is quite low at the outset, but is subjected to a considerable increase during development (cf. Figures 220 and 233). It must be noted, however, that the difference in initial concentration may be related to a difference in function, the PP cycle being involved in energy supply in the former case during the initial stages. The relatively large increase in G-6-PDH and 6-GPDH during amphibian development does not exclude that the enzymes are sl-cell enzymes, as suggested here, but the possibility exists that they represent cell differentiation rather than cell transformation.

Among the remaining enzymes which may be amoebocyte enzymes are amylase and lipase. The first of these was investigated by Urbani and Scollo-Lavizzari (1955), Urbani (1957) and myself (1955). The results obtained agree in showing that a substantial activity is present initially. The former authors found a very pronounced decrease in both *Rana* and *Bufo* during the development up to the larval stage (Figure 247). In

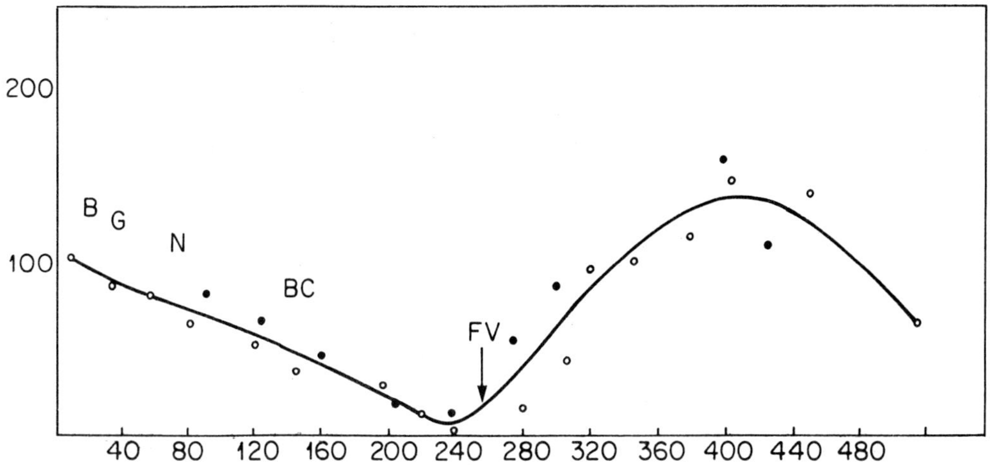

Figure 247. Amylase activity during the development of *Bufo*. Abscissa, hours; ordinate, arbitrary units. N, neurula; other abbreviations as in Figures 243 and 245. (Reproduced from E. Urbani and G. S. Scollo-Lavizzari, *Ric. Sci.*, **25** (1955), by permission of Consiglio Nazionale della Ricerche, Rome)

contrast, I found in *Ambystoma* a slight increase during the very early stages, subsequent to which the activity remained constant until the rise during larval development recorded also by Urbani and Scollo-Lavizzari (Figure 247). This disparity cannot be related to the circumstance that purportedly two different enzymes (α-and β-amylase, respectively) were estimated, for Scollo-Lavizzari (1956) found that both enzymes behave similarly in the toad. The observed differences thus seem to be species-specific traits, a point which is further borne out by results obtained on lipase (Urbani, 1957). As shown in Figure 248, the changes in activity are quite different in *R. esculenta* and *B. vulgaris*. The latter species seems to be quite exceptional, for the curves found for *R. dalmatina* and *B. viridis* are similar to that observed in *R. esculenta*.

The lf-transformation

Alkaline phosphatase, suggested to be specific for the filocytes, has been studied during amphibian development by Krugelis (1950), Krugelis, Nicholas and Vosgian (1952), De Cesaris Coromaldi (1955) and myself (1955; 1958; 1959c; 1961). Generally, a slight activity has been recorded during the very early stages, but since a rather potent acid phosphatase is present, the possibility exists, as already mentioned, that the observed hydrolysis represents residual activity of the latter enzyme at high pH values. It must be emphasized that, particularly in Salientia, this early activity is barely measurable.

The curves obtained by the various authors differ in several aspects. Krugelis (1950) and Krugelis *et al.* (1952) observed an exponential increase. With phenylphosphate as

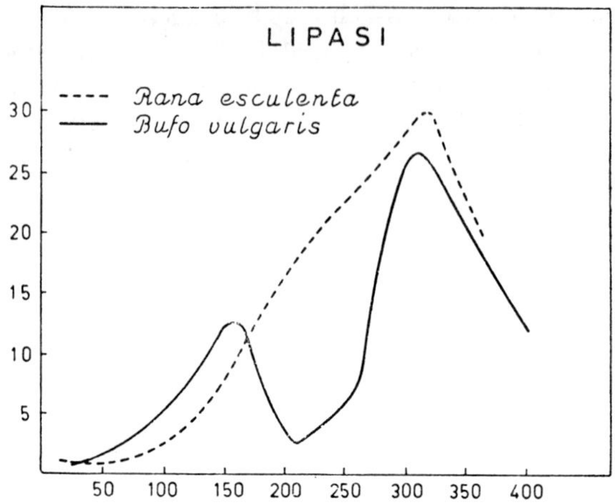

Figure 248. Lipase activity during the development of *Bufo* and *Rana*. Abscissa, hours; ordinate, arbitrary units. (Reproduced from E. Urbani, *Istituto Lombardo, Rend. Sci.* [B], **92** (1957), by permission of Istituto Lombardo Accademia di Scienze e Lettere, Milan)

substrate, De Cesaris Coromaldi (1955) found constant activity to the tailbud stage, but with glycerolphosphate a rise was already observed during gastrulation, thus exactly opposite to that observed for the acid phosphatase.

I have consistently found that, irrespective of the substrate, the synthesis occurs in phases, the first one during gastrulation, the last one during larval development (Figure 249). It will be noted, however, that the relative increase with glycerolphosphate is much lower than with phenylphosphate, suggesting that in this case too the enzyme initially present is different from that which is synthesized by the embryo. As we shall presently discuss, the last phase may unquestionably be correlated with the differentiation of the intestine. In the early work (1955) only one phase of synthesis was interposed between those just mentioned, but later it has been possible to demonstrate that there are altogether four phases (1961). The demonstration of the two intermediate ones is possible only at low rearing temperatures (Figure 250). It is interesting to note that the four phases of APh synthesis correspond in time with those observed for the synthesis of DNA (cf. Figures 241 and 250).

The contention that the formation of filocytes is accompanied by synthesis of APh is supported by the fact that the first wave of enzyme increase occurs at gastrulation when the first filocytes make their appearance at the dorsal side. The correlation between synthesis of DNA and APh may imply that, among the new cells formed in each phase of mitotic activity, a certain fraction become filocytes, but the situation is complicated by the fact that APh, as mentioned, may also represent a differentiation pattern. At any rate the coincidence is not complete, for the synthesis of DNA is less influenced by temperature than is that of APh (cf. 1959d; 1961). The results in the latter paper suggest that at low temperatures the formation of filocytes is somewhat retarded, as indicated both by the synthesis of APh and by morphogenetic peculiarities.

O'Day and Finnegan (1970) have investigated the APh activity in the embryos of two salamander species (*Ambystoma* and *Taricha*), using para-nitrophenyl phosphate

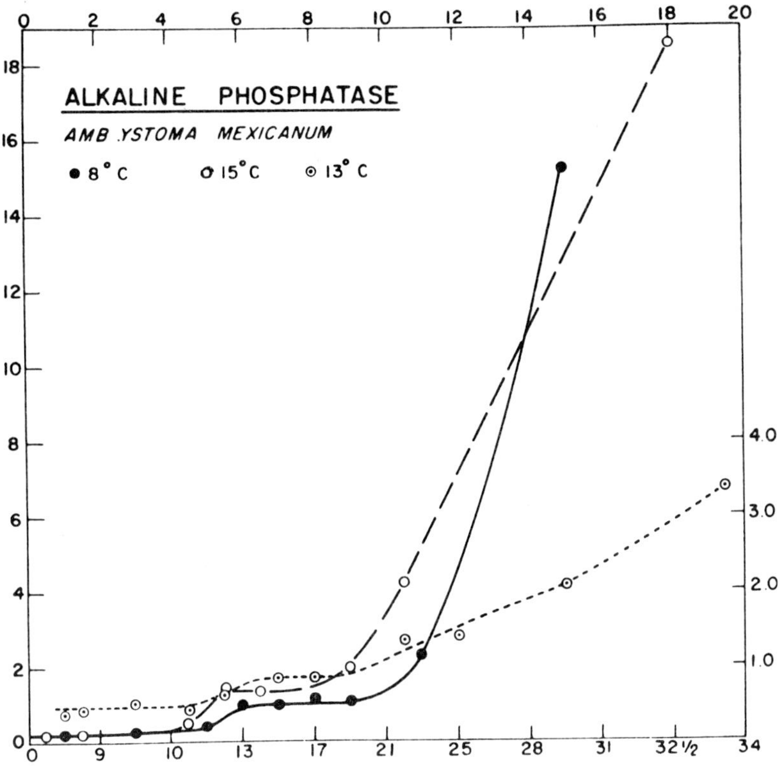

Figure 249. Alkaline phosphatase activity during the development of *Ambystoma*. Lower abscissa, stage of development; upper abscissa, days at 8°C; left ordinate, μg phenol per embryo and hour; right ordinate, μg P per embryo and hour. At 8 and 15°C phenylphosphate, at 13°C glycerolphosphate, was used as substrate. (Reproduced from S. Løvtrup, *C. R. Lab. Carlsberg, Sér. chim.*, **29** (1955), by permission of the Carlsberg Foundation, Copenhagen)

(PNPP) and glycerophosphate (GP) as substrates. On the basis of their results they distinguish a transitory, PNPP-preferring, and a permanent, GP-preferring, activity. Interpreting their data in conformance with the views advanced here the former activity may be correlated with the lf-transformation, whereas the latter, as pointed out by the above authors, reflects various differentiation phenomena.

The nv-transformation

The changes in acetylcholinesterase have been studied by several authors (Youngstrom, 1938; Sawyer, 1943; Ten Cate, Kooij and Zuidweg, 1951). None of the authors have concentrated on establishing whether or not there is any enzyme activity in the very early embryo, but this very circumstance suggests that, as in the sea urchin embryo, the initial activity is negligible. As is evidenced by the results of Ten Cate *et al.* (Figure 251), a rise in enzyme activity begins during neurulation. It is known from the work of Assheton (1896) that at this time the first ciliated cells begin to appear in the ectoderm around the neural plate (Figure 118), so clearly a correlation is suggested between these two phenomena.

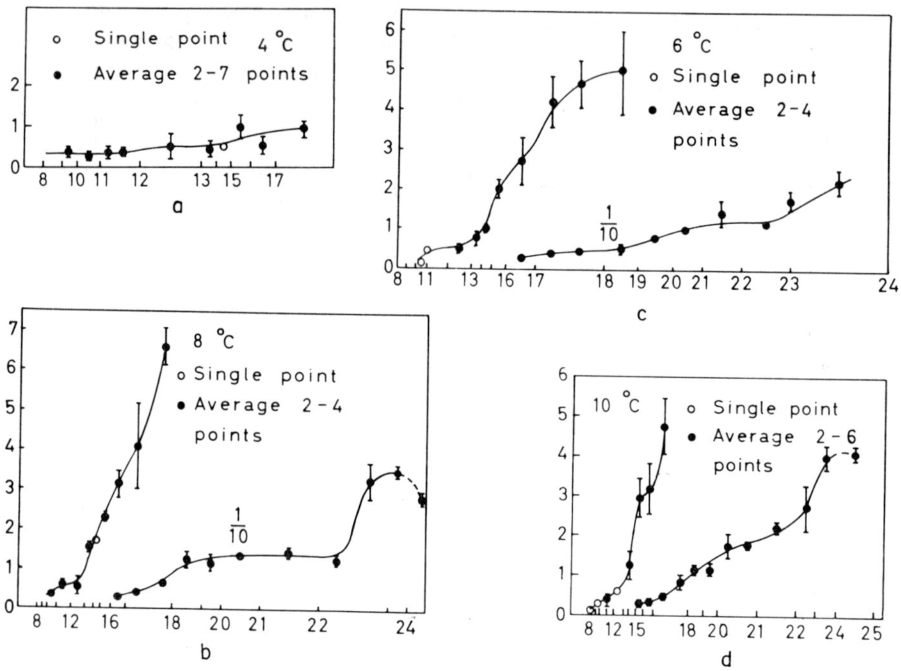

Figure 250. Alkaline phosphatase activity during the development of *Rana* at low temperatures. Abscissa, stage of development; ordinate, μg phenol per embryo and hour. In (b), (c) and (d) the ordinate for the later part of the curve has been increased by a factor of 10. (Reproduced from S. Løvtrup, *J. Exptl. Zool.*, **147** (1961), by permission of The Wistar Press, Philadelphia)

As I have redrawn Figure 251 the enzyme synthesis comes to a standstill around stages 20–21, after which it is resumed. It seems natural to suggest that the first phase represents the cell transformation mentioned, whereas the second one represents the differentiation of the nervous system, as particularly emphasized by Youngstrom (1938) and Sawyer (1943). Matters are obviously not as simple as that, however, for the nv-transformation may continue in various parts of the body during the later developmental stages also, for instance in the pronephros. It may be argued that the formation of the central nervous system begins at neurulation, and that thus the whole curve represents nervous differentiation. This cannot be accepted, partly because AChE is not specific for the nerve cells, and partly because what take place in the neurula are morphogenetic processes, not cell differentiation. Neural tube formation does not require that the neural plate cells are differentiated into neurocytes, but only that they are transformed into filocytes. As suggested above, it is only part of the total population of nerve cells which are flagellocytes (vsf-cells) and thus producers of AChE. The remaining ones are either ncf- or nsf-cells, among which the larger part presumably are adrenergic cells.

The order of appearance of the two purportedly cell-order specific enzymes (APh and AChE) is reversed in Amphibia as compared with the Echinoidea. This finding is, however, in complete agreement with morphological observations, for in the latter case flagellation of ectodermal cells occurs before the primary mesenchyme cells are formed. In Amphibia the ectodermal cells never become flagellated, and the formation of ciliated colligocytes does not begin before the stage of neurulation is reached.

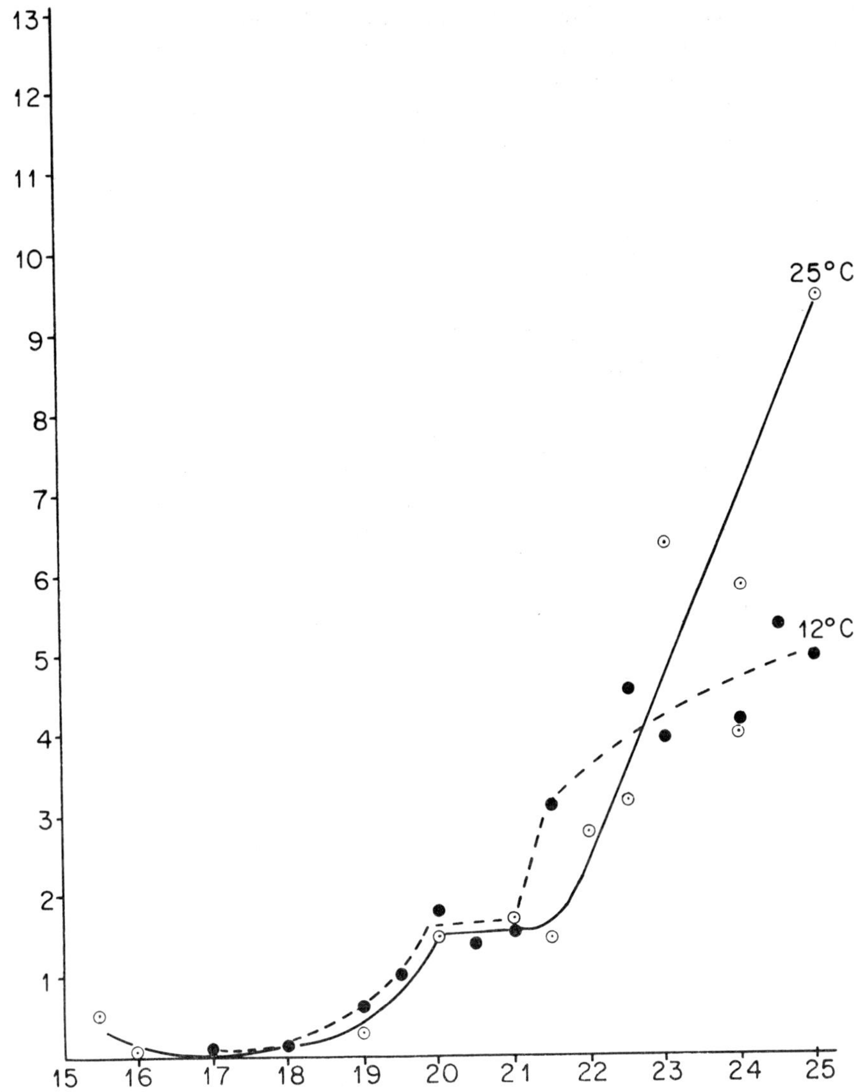

Figure 251. Acetyl choline esterase activity during the development of *Rana*. Abscissa, stage of development; ordinate, arbitrary units. (Reproduced from G. Ten Cate, J. S. Kooij and M. H. J. Zuidweg, *Proc. Konikl. Nederl. Akad. Wetensch.* [C], **54** (1951), by permission of Koninklijke Nederlandse Akademie van Wetenschappen, Amsterdam)

Cell differentiation

In our discussion of the synthesis of AChE it was suggested that the last phase of synthesis of this enzyme represents the differentiation of the nervous system. Apart from this case, it seems that it is primarily the differentiation of the endoderm which is reflected by the enzyme determinations made so far on developing embryos. The digestive tract contains a number of different enzymes, some of which are localized exclusively there, whereas others may be found in other tissues as well. Among the enzymes known to be present in the digestive tract, the following have been studied in quantitative investigations: trypsin, amylase, lipase, deoxyribonuclease and alkaline phosphatase.

My own results (1955) and those of Urbani (1957) show that the proteinase activity in alkaline media is negligible until late in prefeeding development (Figure 252). The

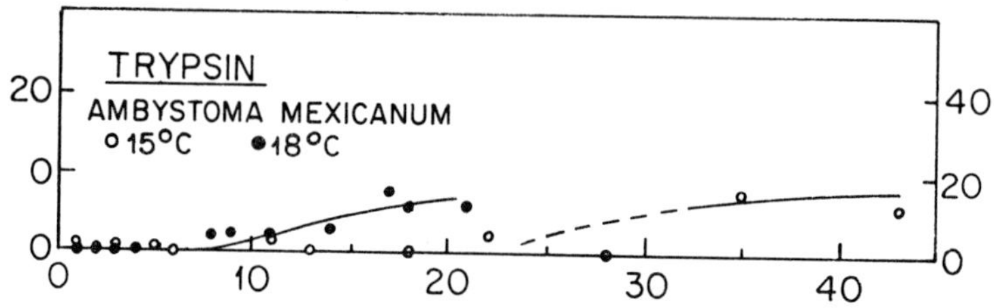

Figure 252. Trypsin activity during the development of *Ambystoma*. Abscissa, days; left ordinate, μg nitrocasein per embryo and hour; right ordinate, μg 'tyrosin' per embryo and hour. At 15° nitrocasein, at 18° casein, was used as substrate. (Reproduced from S. Løvtrup, *C. R. Lab. Carlsberg, Sér. chim.*, **29** (1955), by permission of the Carlsberg Foundation, Copenhagen)

exact stage at which the enzyme activity begins to rise has not been recorded, in the former case because of the difficulties involved in staging the axolotl embryo after that corresponding to Harrison stage 40 in *Ambystoma punctatum*.

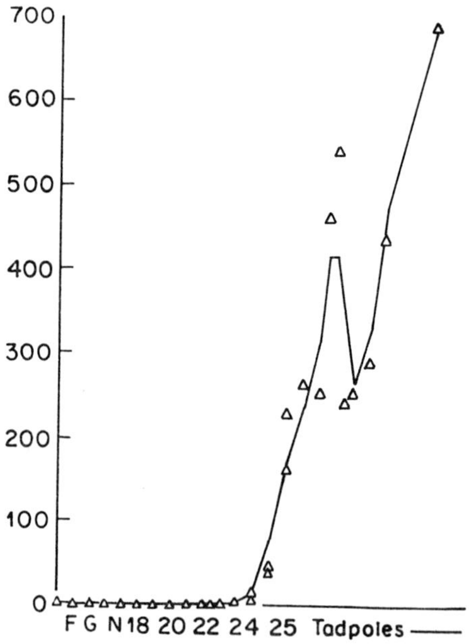

Figure 253. DNase I activity during the development of *Rana*. Abscissa, stage of development; ordinate, arbitrary units. (Reproduced from J. R. Coleman, *Developm. Biol.*, **5** (1962), by permission of Academic Press, New York)

However, in a study of the differentiation of the digestive tract of the latter species, Dorris (1935) has found in qualitative tests that trypsin, as well as amylase and pepsin, activity appears between stages 42 and 43.

As already discussed, amylase is present in the early amphibian embryo (1955; Urbani and Scollo-Lavizzari, 1955), in agreement with the circumstance that amylase is not exclusively found in the gut, but occurs in other tissues as well. In both the quoted papers a marked increase in enzyme activity was observed during the latest stages of development (Figure 247), coincident with the rise of trypsin activity. It can therefore hardly be doubted that this phase of enzyme synthesis reflects the differentiation of the gut.

The changes in lipase activity have been studied in various species by Urbani and his collaborators (cf. Urbani, 1957). A gradual increase in activity was observed except in *B. vulgaris*, where two distinct maxima were found (Figure 248). The stage at which synthesis of lipase is resumed corresponds pretty closely to the one at which synthesis of trypsin and amylase begins, suggesting that the second phase represents the differentiation of the gut. There is no reason to doubt that this phenomenon is associated with lipase synthesis in the other species also, even if the curves do not permit a clear-cut resolution.

The deoxyribonuclease active at alkaline or neutral pH is a pancreatic enzyme, but it has been found in other cells also, bound to the mitochondria. Coleman's results (1962) suggest that the enzyme is synthesized in two phases. Possibly the first of these represents the mitochondrial enzyme, but the latter coincides completely in time with the differentiation of the intestine (Figure 253).

The last enzyme to be treated in this group is alkaline phosphatase. This enzyme is found in various cells and tissues, but it occurs in particularly large amounts in the intestinal wall. As already discussed, the synthesis of this enzyme occurs in several phases during development, and since the last phase coincides approximately with the synthesis of the other intestinal enzymes described here, it is obviously justified to propose that it represents differentiation of the gut.

'The orderly minded orthodox biological world closed its ranks against [the] suggestion that revolutionary processes may happen. It became accepted that the only respectable doctrine is that evolution never involves anything but step-by-step Fabian gradualism, plodding along a weary way similar to that by which the annual milk yield of dairy cows or egg yield of hens is slowly improved—the occurrence of a little allopolyploidy or rearrangement of chromosomes by two or three breaks could be admitted, but would only push the basic philosophy from Bourgeois-Liberal to right-wing Social-Democrat'.
(C. H. Waddington, 1969, pp. 123–124)

EPIGENESIS AND PHYLOGENESIS

The ultimate aim of biology is to account for the mechanisms responsible for the origin of life. From a human point of view, the most important aspect of this problem is the evolution of animals, notably the Metazoa. In order to fathom intellectually a process of such bewildering complexity as metazoan phylogenesis it is an absolute necessity to operate within a firmly established hypothetical superstructure, since this measure clearly provides the sole means capable of bridling the imaginativeness of those engaged in this pursuit. The only evolutionary theory prevailing today is that of neo-Darwinism, which essentially is a merger between the Mendelian hypothesis of inheritance and the Darwinian hypothesis of evolution. From this origin of neo-Darwinism it follows that genetics, recently backed up strongly by molecular biology, has more or less monopolized the topic of phylogenetic evolution.

Yet, in spite of its spectacular exploits and its apparent hegemony over the minds of most biologists, neo-Darwinism does not stand unchallenged. A number of authorities, representing a variety of scientific disciplines, have time and again suggested or demonstrated shortcomings of various kinds inherent in the hypothesis. Later on we shall have the opportunity to deal with objections raised from other angles, but in the present context the attention will be focused on the criticism raised from an embryological point of view. The majority of embryologists appear to have accepted neo-Darwinism rather uncritically, but dissenting opinions have been voiced by several authors. It is the general consensus among these critics that a full understanding of the workings of evolution will not be attained before we know something about the role of the genes in the process of epigenesis through which the phenotype is brought into existence.

Contending that the preceding chapters of the present book have shed some light on this problem, I propose to undertake a renewed analysis of the question of phylogenetic evolution. I do not claim that this endeavour leads to the solution of any specific questions, but it appears to me that it does serve to bring some order into the whole complex of issues involved in this subject. We shall approach this task by considering separately the theories of *Darwinism*, of *neo-Darwinism* and of something which in the present context will be called the *complete theory*.

DARWINISM

The first theory of phylogenetic evolution which gained general, if not unanimous, acceptance was advanced by Darwin (1859) in his celebrated book. Various kinds of attack were levelled against Darwin's book from its first publication, much of which was of an emotional character, but even scientific arguments were advanced. As far as the substance of the theory is concerned no dissent obtains any longer; hardly anybody questions today the fact the evolution has taken place. However, with respect to details the situation is different, for criticism on this level has never ceased completely. To be sure, it abated in the wake of the advancement of the neo-Darwinian theory, but it is interesting to note that the improved insight into genetic mechanisms, which in recent years has been afforded through the rise of molecular biology, has led to a renewed challenge of certain aspects of the theory of evolution. We shall have the opportunity to discuss this kind of criticism on several occasions in the present chapter.

Darwinism has also been attacked by several authors from a more formal logical angle. This stand may be represented by the statement of von Bertalanffy (1952, p. 89) that 'a lover of paradox could say that the main objection to the selection theory is that it cannot be disproved'. Since Darwinism is, after all, a highly successful scientific theory, objections of this kind must be considered very seriously, and it may therefore be justified to spend some time on this question. The criticism takes various forms, but usually it is concerned with the expression 'the survival of the fittest' in which, so the argument goes, 'fittest' means 'fittest to survive', entailing that 'the whole "theory" becomes a tautology, a complicated way of saying simply that what survives survives' (Grene, 1969, p. 66). Being tautologous or vacuous, the renowned statement cannot be subjected to an objective test of verification since it cannot be refuted. However, even if the famous slogan unquestionably embodies the gist of Darwin's theory, it seems unjustified to use it as a basis of attack. For that purpose it is clearly necessary to go to the propositions advanced by Darwin.

Unfortunately, Darwin originated the tradition, subsequently followed by most writers on the topic of evolution, of crowding the text with observations on natural history to the extent that it is difficult to trace the logic of the argument. Indeed, interpreters of the theory do not always agree as regards the number of basic assumptions, but generally it is suggested that there are three or four.

In my opinion the Darwinian theory is based upon the following six premises:

(1) *Reproduction*. Ancestral (or parent) populations give rise to descendant (or offspring) population. Through his reading of Malthus' book Darwin came to the conviction that the reproductive potential of an ancestral population exceeds its own number, as shown for instance by the following quotation: 'Hence we may confidently assert, that all plants and animals are tending to increase at a geometrical ratio, that all would most rapidly stock every station in which they could any how exist' (Darwin, 1859, p. 65). As regards the numerical relations between succeeding generations we shall in the present context talk of *progression* if the number of offspring exceeds that of the parents, *regression* if the inverse situation obtains, and *survival* if the numbers are equal.

(2) *Inheritance*. The various properties exhibited by the offspring are similar to those of the parents, indicating that the process of reproduction involves a mechanism through which parental characteristics are transferred to the offspring. This was expressed by Darwin in the following way: 'Like produces like' (1859, p. 12) and 'the inheritance of every character whatever [is] the rule' (1859, p. 13).

(3) *Variation*. Owing to his particular focus of interest, Darwin paid most attention to intraspecific variation, particularly in domesticated plants and animals, thus: 'When we

look to the individuals of the same variety or sub-variety of our older cultivated plants and animals, one of the first points which strikes us, is, that they generally differ much more from each other, than do the individuals of any one species or variety in a state of nature' (1859, p. 7). Implicit in this premise is, of course, the total amount of variation that has been realized by extinct and still living organisms, and also the potential variation which may materialize in the future.

(4) *Mutation.* The general tenet of the fourth premise is that descendants may differ from their parents with respect to one or more properties and that, occasionally at least, these properties may be innovations which never before have occurred in that particular type of organism which they represent. This idea is implicit in Darwin's notion of 'descent with modification', discussed on various occasions in his book. In passing it may be noted that Darwin devoted a whole chapter to a discussion of the difficulties involved in accounting for this phenomenon which, in modern terminology, is called 'mutation'.

(5) *Competition.* Populations of certain organisms are competitive successful, i.e. they may progress in the face of competition with members of other populations. The most fundamental concepts in Darwin's theory are 'struggle for existence' and 'natural selection'. Darwin envisaged that a perpetual competition occurs between living beings and that the outcome of this struggle is the selective progression of more at the cost of less 'fit' individuals. As an illustration the following citation may serve: 'If we wished in imagination to give the [organism] the power of increasing in number, we should have to give it some advantage over its competitors' (1859, p. 78).

(6) *Environment.* The participation of the environment in the process of evolution involves two aspects. For one thing, the occurrence of competition clearly demands that the available space is limited. Hence, it may be contended that usually the various ecological niches, to employ present-day terminology, contain the maximum possible number of individuals. Or Darwin: 'Each area is already fully stocked with inhabitants' (1859, p. 69). Clearly, this statement is not generally valid, for the colonization of new elements in the course of time, as evidenced by the fossil record, shows that empty niches have existed occasionally at least. The importance of this fact is that under these conditions a population may undergo progression without the otherwise concomitant struggle for existence; hence selection may at least temporarily be non-operative.

The other role assumed by the environment refers to the circumstance that the competitive success of an organism to a large extent depends upon its adjustment to the inanimate surroundings; the better it knows how to cope with the latter, the better, presumably, it will hold its own in competition with other living beings.

The formulation of the premises above is hardly such as to allow for a formal derivation of conclusions, but nevertheless it appears that the following statements may be derived:

(1) *Isolation.* Any population of living organisms may progress and survive in an empty niche. This inference, or at least a somewhat similar one, was made by Darwin: 'Isolation . . . may sometimes be of importance in the production of new species' (1859, p. 105).

(2) *Selection.* Competitively successful organisms may progress and survive in an occupied niche. The population of less successful organisms present will be extinguished. Darwin expressed himself thus: 'As each . . . favoured form increases in number, so

will the less favoured forms decrease and become rare. Rarity, as geology tells us, is the precursor to extinction' (1859, p. 109).

(3) *Evolution*. The bearing of isolation on evolution is not clear, but from the fact that selection occurs it follows that in the course of time a continuous replacement of forms occurs in the organic world, i.e. that evolution takes place. This point was expressed by Darwin in the following way: 'As new forms are continually and slowly being produced, unless we believe that the number of specific forms goes on perpetually and almost indefinitely increasing, numbers inevitably must become extinct' (1859, p. 109). The present conclusion has some consequence for the criticism quoted above, for if the statement 'what survives survives' may be a triviality, the inverted expression 'what does not survive becomes extinct' clearly is an important and necessary conclusion in a theory of evolution.

(4) *Progress*. Whatever the exact significance of 'competitive successful', it obviously implies improvement of some kind, thus warranting that to the extent selection is involved, the direction of phylogenetic evolution must be one of progress towards forms that are better and better adjusted to their environment. In the words of Darwin: 'The inhabitants of each successive period in the world's history have beaten their predecessors in the race for life, and are, in so far, higher in the scale of nature' (1859, p. 345).

(5) *Divergence*. The continued operation of selection in space and time must result in a multitude of forms which, owing to their genetic relations, exhibit similarities and dissimilarities that can be used for their systematic classification. If fossils are incorporated in the system, a gradual change should be apparent. Thus Darwin: 'We can understand how it is that all the forms of life, ancient and recent, make together one grand system; for all are connected by generation' (1859, p. 344) and 'it is evident that the fauna [and flora] of any great period in the earth's history will be intermediate in general character, between that which preceded and that which succeeded it' (1859, pp. 333–334).

(6) *Direction*. The classification mentioned above has resulted in the erection of a hierarchy of taxa, distinguished by the fact that a number of taxa at any level in the hierarchy are included in a taxon at the next higher level, etc. It is reasonable to presume that the direction of evolution somehow is reflected in the arrangement of the taxa, hence it may be inferred that evolution has proceeded either from the apex to the base of the hierarchy or the other way round. Because of his preoccupation with intraspecific variation, which led him to the conviction that varieties are 'incipient species', Darwin rather was forced to presume that divergence on the species level led to the creation of genera, etc., as exemplified by the following quotation which is one among many that might illustrate this point: 'Thus it is, as I believe, that two or more genera are produced by descent, with modification, from two or more species of the same genus. And the two or more parent-species are supposed to have descended from some one species of an earlier genus' (1859, pp. 123–124).

It seems justified to state that the preceding account of the logics of Darwinism is a generous rendering. Yet, even a more critical analysis demonstrates that Darwin's reasoning holds together logically, the most important conclusion being that evolution has in fact occurred (Edman, Hermeren and Løvtrup, 1973). This being granted it can nonetheless be postulated that, considering the scope and the implications of the theory, the basis of the premises upon which it rests is slender indeed. This is borne out particularly by latter-day attempts to axiomatize the theory of evolution (cf. Woodger, 1952; M. B. Williams, 1970).

As mentioned above, it has been claimed that Darwin's theory cannot be falsified because it leads to a tautologous conclusion. If the present analysis is correct this view is wrong since it is possible to derive no less than six conclusions, none of which is tautologous, and hence the possibility obtains to refute the theory by comparing the predictions stated in the conclusions with empirical observations.

With respect to the first conclusion, the often observed survival of various relics supports, in my opinion, the effect of isolation as stated by the first conclusion. Admittedly, this stand implies different notions about the effects of mutation (modification) and selection than those nurtured by Darwin. The following four conclusions are concerned with the 'evolution of form in time and space' (Croizat, 1958) and may be tested through confrontation with, especially, morphological, palaeontological and biogeographical observations. The sixth conclusion, finally, is testable on the basis of the systematic classification of living beings, as will be shown later in this chapter.

Various kinds of palaeontological and biogeographical evidence were known to Darwin, and since these data are incompatible with the rival theory, Special Creation, they were mobilized in support of his theory. For independent testing of the theory, facts over and above those known to Darwin must be required since, after all, the latter were part of his inspiration. However, in the century since the publication of his book an enormous amount of evidence has been accumulated which is compatible with the conclusions stated above and incompatible with any other known theory concerned with the origin of life. It is certainly possible to envisage observations that would have refuted the theory, but such have never materialized.

It must be admitted, however, that these six conclusions represent the essence of what can be derived from the premises stated above. Although they are of considerable importance, the theory is still of limited value, because it does not lead to more detailed predictions which partly might contribute to our understanding of the mechanism of phylogenetic evolution and partly might be tested through confrontation with empirical observations. In order to achieve this goal it is necessary to complement the Darwinian theory with further hypothetical statements, specifying the nature of the agents involved.

It may be historically incorrect, but logically defensible, to state that an attempt to this end was made by Lamarck in proposing that new animal types arise through the inheritance of acquired characters. This hypothesis contains an assertion concerning the mechanism of inheritance in living organisms. Provided that present-day genetical studies are concerned with hereditary properties of the kind that have been instrumental in evolution, the hypothesis may be subjected to falsification through genetic experiments. As is well known, geneticists, assuming the stated provision, have been able to refute Lamarckism.

Darwin did also, in his book and elsewhere, advance some suggestions as to the genetic and epigenetic mechanisms responsible for phylogenetic evolution. Owing to want of concrete knowledge, these ideas were rather vague, and they have not stood up to the test of time.

However, Darwin's main contribution was to convince a large part of his contemporaries that life on earth has evolved through natural selection, thus ensuring that all subsequent biological research was performed and interpreted against this background. Therefore, I do not believe that the present outline of the implications of Darwinism, omitting as it does all references to specific details, does injustice to its creator.

NEO-DARWINISM

If the mentioned refutation of Lamarckism is warranted, the obvious and consequent next step is to introduce into the evolution theory the theoretical framework established by genetics. This step was made by the neo-Darwinists* when they effectuated an alliance between the theories of Darwin and Mendel. The outstanding feature of neo-Darwinism, as developed by Fisher (1930), Wright (1931) and Haldane (1932), was its formulation in mathematical terms. As pointed out in the introduction to the present book, there is no doubt that this expedient may be of great importance for the advancement of a branch of science because incorporation within the deductive system of mathematics enhances the facility with which predictions can be made. It is important to stress that the mathematical formulation as such does not imply any corroboration of a theory. The merit of such endeavours surely depends primarily on the extent to which the inherent variables and parameters have biological significance.

Any theory of phylogenetic evolution must specify the mechanisms involved in each of the steps represented by the six basic premises of Darwinism outlined above. It therefore follows that an appreciation of the neo-Darwinian theory is most easily obtained if we establish the nature of the postulates it makes with respect to these premises.

Actually, neo-Darwinism only advances three propositions, one of which is concerned with the first three, one with the fourth and one with the last two premises. The first proposition, relating to reproduction, inheritance and variations, asserts that in an organism *all* relevant properties, and thus also all variations, are determined by Mendelian particulate genes. In this assumption the process of reproduction is dealt with as successive generations of genotypes, in which the gene frequency is determined by the currently accepted modification of Mendel's original scheme.

The neo-Darwinian theory further postulates that innovations can come about *only* through mutations that, with respect to their scope and nature, are identical with those observable in intraspecific crossings. As regards the first aspect, the salient point is that the effect of an individual gene is quite slight, entailing that large modifications can be obtained only through accumulation of the effects of many genes. As to the nature of mutations, genetic experiments have disclosed that for a given gene location a number of alleles obtain and that conversion between the alleles may occur through mutation. The frequency of this event, which may be different for the appearance and disappearance of a particular gene, and from one gene to another, the mutation rate, is one of the parameters inherent in the mathematical theory. It will be noticed that new mutations, i.e. such that do not form part and parcel of the genetic complement of a given population, are not accounted for. This does not imply that such events are excluded by the theory, but once they have occurred they must behave according to the scheme and change back and forth regularly.

Premises 5 and 6, dealing with the favourable variations and the environmental effects, respectively, are taken into account through the introduction of selection coefficients, which stand for the relative advantage or disadvantage of a mutation. The values of these parameters are usually rather low, in conformance with the Darwinian

*There seems to be some discord regarding the meaning of 'neo-Darwinism'. Most commonly, and also in the present context, this word is used to designate the synthesis between the Mendelian and Darwinian theories. Some of the adherents of this theory contend that 'neo-Darwinism' should be reserved for an earlier phase in the history of Darwinism, the name 'the synthetic theory' being more appropriate for the later version.

and the neo-Darwinian postulate (cf. Fisher, 1932) that the effect of any one mutation is slight.

In a complete theory of evolution, the mode of contribution of each of the steps or agents outlined by the six basic postulates must be separately particularized. As far as this kind of specification is concerned, neo-Darwinism, being based entirely on genetic principles, is narrower in its scope than Darwin was in his book, although, of course, more correct as regards the genetic aspects of evolution. However, this does not change the indisputable fact that neo-Darwinism primarily is 'a genetical theory of population changes' (Kuhlenbeck, 1967, Vol. 1, p. 123). As an evolutionary theory it is nevertheless incomplete, 'an extrapolation the boldness of which is made acceptable by the impressiveness of its basic conception' (von Bertalanffy, 1952, p. 86).

Extrapolation is an acknowledged scientific expedient which often leads to right answers, but it is a dangerous approach unless it is employed with discrimination. It is possible that, by sheer luck, the neo-Darwinian theory is correct. This point can be tested by comparing its predictions with empirical observations. As a theory of population genetics, it may make some prognostications about the outcome of contemporaneous population-genetic experimentation, but it is important to stress that corroboration in this case does not bear upon its validity as a theory of evolution. This question can be answered only through reference to observational evidence concerned with the actual process of evolution. Being, as it is, a theory built upon a rather narrow foundation, it might be expected that, if obvious support is not obtained through this expedient, the advocates of the theory would look for modifications likely to improve the relation between prediction and observation.

We shall return to the question of the evolutionary prognostications of neo-Darwinism in the following section, in which an attempt will be made to establish a complete theory of evolution.

THE COMPLETE THEORY

In 'neo-Darwinism' the prefix 'neo-' is used to indicate that it deviates from the original Darwinian theory through the introduction of the specifications mentioned in the preceding section. As follows from the above discussion, amendments over and above these are required in a purportedly complete theory. Lest misunderstandings arise, I want to emphasize that in the present context 'complete' is used to mean 'comprehensive', not 'perfect'.

In the present chapter we shall first set out specifications regarding the six Darwinian premises, based upon relevant practical and theoretical considerations. Secondly, some of the conclusions which can thus be derived will be outlined and compared with our actual knowledge about the process of evolution, and finally a comparison is made between the predictive values of the neo-Darwinian and the complete theory.

Before we begin this quest it should be pointed out that asexually and parthenogenetically reproducing organisms are excluded from the neo-Darwinian theory, if not therefore necessarily from participation in phylogenetic evolution. Evidently, the complete theory must account for the latter event also. Unless otherwise stated, the following discussion is considered to be generally valid, when the amendments necessitated by particular manners of propagation are made.

Premises

The six Darwinian premises deal with all those aspects of life which are involved to ensure 'survival'. The quotation marks indicate that the word is used here rather in the same broad sense as in the famous slogan. The first four premises are concerned with reproduction, inheritance, variation and mutation, properties presumed to distinguish all living beings. The last two premises deal with the process of competition, the means through which certain properties are selected under special conditions and with the role of the environment in this event. It will be shown in the following pages that a lot of information has been accumulated about each of the phenomena or agents dealt with in these postulates which cannot be ignored in the establishment of the complete theory of phylogenetic evolution.

REPRODUCTION

To the neo-Darwinist the individual is equated with its genotype. This is a myopic view, for surely it is the organism proper and not the genotype which develops, eats, fights, reproduces and dies, and which thus is exposed to the process of selection (cf. Waddington, 1957; 1969). If we were to grant one-to-one relationship between genotype and adult organism, implying that ontogenesis is completely specified by the genotype, it might possibly be warranted to adopt the neo-Darwinian view when dealing with organisms developing along closely similar lines. However, as pointed out in the introduction, phylogenesis is the outcome of successive changes in the process of ontogenesis, and we therefore cannot hope ever to understand how genes may accomplish the former phenomenon unless we know how they carry out the latter. It is for this reason that the study of epigenesis is so important also for the theory of evolution.

Epigenesis is one aspect of the process of reproduction. The series of events involved in the latter, which collectively may be called a *reproductive cycle*, comprises two phases, located in successive generations. The first phase begins when, in an embryo, the precursors of the presumptive gametes become separated in the germ line and ends when, after maturity has been reached, an egg and a spermatozoon have united to form a fertilized egg. No unambiguous name appears to exist for this phase, and it will therefore be called here *progenesis*. The second phase begins with the fertilized egg and ends with the formation of a sexually mature adult. The process through which this change is achieved is that of epigenesis.

Progenesis

A number of features in the fertilized egg contribute to the outcome of epigenesis. Among these the nuclear genome assumes an outstanding position, but many other traits such as egg envelopes, cell-type of organization, size, shape, content and distribution of yolk and various other cytoplasmic components also play important roles. The egg supplies an overwhelming part of the material of the embryogenetic substrate, the contribution of the sperm generally being confined to half of the nuclear genome and often also a centriole and some mitochondria. However, the sperm exerts a very important function through its participation in the final establishment of the organization of the egg, usually occurring during fertilization and most often involving the institution of the polarities which, as we have seen, in turn are causally related to the subsequent epigenetic events.

Since all the mentioned properties of the gametes, including those of the sperm ensuring that fertilization occurs, are established during the progenetic phase, it follows that the events occurring [illegible] through their influence upon epigenesis, m[illegible]ed in other words, phylogenetic chan[illegible] outcome of progenesis.

The prospect o[illegible] upon the type of reproduction (ase[illegible]is, bisexual reproduction). This asp[illegible]ng of the gametes, belongs to the pha[illegible]well surveyed in the neo-Darwinian lit[illegible]nediate importance in the present co[illegible]

Epigenesis

The whole of the theory of epigenesis outlined above enters at this stage as a premise in the theory of evolution. Since phylogenesis represents the creation of new types of ontogeny it follows that modifications of the mechanisms or processes responsible for the latter constitute the only means available to Nature for the establishment of innovations of this kind. Hence we may infer that the course of phylogenetic evolution must be explainable through changes in the epigenetic processes carried out by the predetermined mechanisms, cell division, cell diversification and morphogenetic processes. Besides these questions we shall also devote a subsection to the problem of causality and epigenesis.

Cell division

The existence of a variety of embryonic cleavage patterns shows that this embryogenetic mechanism may be subjected to modifications, and we have seen that the only agent which at the time of cleavage is available for this purpose is the embryogenetic substrate. It was also possible to demonstrate that the cleavage pattern is of epigenetic importance insofar as there is a causal relationship between the symmetry of the cleavage pattern and that of the embryonic and adult body, stated in the generalization that the order of symmetry, as expressed in the number of polarities required for its determination, can never be higher in the latter than in the former. Not only the symmetry of the cleavage pattern, but also the general type, holoblastic, meroblastic, etc., may in many cases be of decisive influence upon morphogenetic processes.

Cell division, and thus cell multiplication, is the method of reproduction in microorganisms, a process limited only by space and available resources. Uncontrolled cell division is entirely incompatible with the maintenance of the organization of the metazoan body, and apparently this process is quite strictly controlled in the latter. The mechanisms responsible for this kind of regulation are not known, but no doubt the antagonism between cell division and cell diversification represents a phylogenetically early expedient. In higher animals more complex regulatory mechanisms, involving hormones, are known to occur. Since cell size varies within a relatively narrow range in

Metazoa, it follows that increase in size must depend upon interference with the process of cell multiplication.

Evolutionary changes may be, and have been, obtained through participation of the process of cell division, but apparently the possibilities of variation are relatively limited, and the effects are therefore usually of a general nature.

Cell diversification

The participation in diverse epigenetic processes of the various cell-types arising through cell transformation has been discussed above. Suffice it to mention that increase in the number of cell orders represented in the embryo potentially constitutes a means through which phylogenetic innovations can be achieved. It may be recalled that in some cases cell differentiation products (e.g. oncotically active mucopolysaccharides) are necessary for the cells to make their epigenetic contribution.

According to the function of the diversification products three subdivisions were distinguished above, namely metabolic ones (e.g. enzymes involved in digestion, intermediary metabolism and excretion), regulatory ones (e.g. hormones) and structural ones (e.g. intracellular structural protein polymers and extracellular substrate elements). Certain differentiation products are coloured (haemoglobin, melanin, aminopterins, etc.). In the first case the function is well established, and there is no doubt that it belongs to the first group. Melanin may have a physiological function (e.g. radiation protection), but for the aminopterins and many other pigments no other function is known than the one associated with the fact that they are coloured. When deposited in the superficial layers of the body these pigments may serve for protection, recognition and other ecological purposes. It may therefore be justified to establish a fourth group, comprising those *pigmented* differentiation products the main function of which is linked with their colour. It would seem reasonable in view of this definition to exclude haemoglobin from, but include melanin in, this class.

All four kinds of substances have been of importance for phylogenetic evolution. The properties form and pattern, the ones most attended to by systematists, evidently are materialized through the participation of structural and pigmented differentiation products. Certain hormones are, besides other functions they may have, known to be of great importance for morphogenesis (e.g. growth hormone, sexual hormones, thyroxin and parathormone). Effects of this kind are known mainly from Vertebrata and the higher Invertebrata in which they are confined to relatively late ontogenetic stages. It seems ascertained that some phylogenetic innovations have been instigated by hormones, clearly a most efficient mechanism since these agents often have very general and comprehensive actions which, in some instances at least, may affect the processes of cell division and cell diversification.

Other regulatory and metabolic differentiation products are concerned with the function of the body. On this level a considerable advance has occurred in the course of evolution, but the variation is much less than with respect to form and pattern, a particular type of function generally characterizing large groups of animals. It must be emphasized, however, that a certain group of metabolic differentiation products, the enzymes responsible for syntheses of various kinds, are responsible for the elaboration of all the other differentiation products.

During the early developmental stages the process of cell division, working in cooperation with the embryogenetic substrate, is the master of the process of cell diversification, determining the nature, location and time of appearance of the various diversification patterns. This process is subjected to constraints of various kinds, the fixed sequence of selfdifferentiation patterns probably being the most important one. With increasing complexity of the embryonic body, thus particularly in higher animals, other means of controlling cell diversification appear, among which the interaction between cells belonging to different diversification patterns presumably plays a decisive role.

Altogether it thus appears that through the mechanism of cell diversification Nature possesses a very powerful instrument for the inception of phylogenetic changes.

Morphogenetic processes

According to present-day molecular biological concepts, the genome is a set of discrete lumps of information, mainly concerned with the amino acid sequence in particular polypeptides. The total sum of this information is supposed to constitute a blueprint for an organism, and phylogenetic evolution is envisaged to be the outcome of the accumulation of changes in the composition of the coded proteins. It seems that in general the neo-Darwinists have adopted this proposition, but certain physicists and mathematicians have recently become worried about its implications.

For one thing, the theoretically unlimited number of mutations seems to be a disquieting phenomenon (Eden, 1967), but worse still is the fact that a link, currently neglected, is required to bridge the realm represented by the genetical information and that of the living organisms (Schützenberger, 1967). In a similar vein, Eden (1967, p. 11) has suggested that a 'way to look at the genotype is as a generative algorithm . . . , a sort of carefully spelled out and fool-proof recipe for producing a living organism . . .'. In view of the ideas and the evidence presented in the preceding pages of the present book it appears that these 'outsiders', but also certain biologists, for instance Waddington (cf. 1968; 1969), have a much more realistic notion about the relation between genome and morphogenesis than the one represented by the blueprint hypothesis.

As it is, morphogenesis is an exercise in spatial geometry or topology extended in time (cf. Kuhlenbeck, 1967), a physical process carried out by physical means. The latter are provided by the various substrate elements, cells and extracellular substances, and the physical forces exerted by them. We have seen that in widely different taxa the same, and rather few, elements are responsible for elaborating quite different body forms, a result explainable primarily by changes in the pattern of interaction between the substrate elements in time and space. Over and above this means it is true that phylogenetic innovations of morphogenesis have arisen through the introduction of new types of structural differentiation products, but even so the total number of existing structural elements is exceedingly small.

The creation of new forms does not imply, as is more or less inherent in the blueprint hypothesis, that new structural materials are introduced, but only that the old ones are moulded in different forms. On closer consideration, it appears that this is the same way human beings proceed in their creative and constructive efforts, and that any other approach, were it required for morphogenetic innovation, would imply an enormously complicated mechanism of phylogenetic evolution. There is no doubt that the views advocated here amount to a significant simplification of this process. In this context it may also be pointed out that the secret of phylogenetic progression does not seem to lie in the elaboration of a more and more complicated starting point, as represented by the fertilized egg, but rather in establishing the preconditions for a more and more complicated and extended series of causally connected epigenetic (morphogenetic) events.

Accordingly, very considerable constraints must prevail between the genotype and the outcome of morphogenesis. The number of substrate elements is limited, and even if variations occur in the primary structure of a certain protein (e.g. collagen) throughout the animal kingdom, it is certain that whatever mutations arise, they can be tolerated only as long as the basic mechanical properties of the protein fibrils are conserved. As to the modes of interaction, it likewise seems incontestable that their number must be restricted, and it therefore follows as an unavoidable conclusion that the total number of body forms or body plans realizable by Nature must be extremely limited. We shall return to this question presently.

Causality and epigenesis

It was postulated by von Baer (1828; 1837) that each developmental stage is a necessary condition for the realization of the subsequent one. Presuming that the 'essential nature' of the parents is somehow involved in the control of embryonic development, von Baer could not accept that it is also a sufficient condition. This step was taken by His (1874), who proposed that each stage is the cause of, i.e. the necessary and sufficient precondition for, the following one.

This idea was later adopted by Roux (cf. 1895) and by many others after him. It was professed with great eloquence and clarity by A. Brachet, as shown in the following quotation (1921, p. 63): 'The truth is that embryonic development consists of a long series of successive transformations, of which the antecedent ones act as immediate causes of the subsequent ones; each of them, in a certain sense, creates something new and provokes, in the same stroke, the activity of factors which up to then had had no occasion to intervene. They must thus be realized in a determinate order, since they are causally linked together' (author's translation). The present status of this epigenetic hypothesis is difficult to ascertain; undoubtedly it is accepted by most biologists, but it seems that the implications are very often ignored. It must be emphasized, however, that it is by no means implicit in the currently dominating blueprint hypothesis. The causal relationship between successive developmental stages is the main thesis of the present book.

The fact that epigenesis is not a process which is programmed in detail, but rather, like a computer operation, prevails in the form of a set of data and some operational instructions, is of great import for our understanding of the course of phylogenetic evolution. As pointed out in the introduction, small variations in the data (embryogenetic substrate, including the genome) or in the operational instructions (embryogenetic mechanisms) may change the course of the whole operation, and so will interference from the environment. If, for any of these reasons, the realization of one stage, S_i, in the succession of steps in a complex epigenetic operation

$$S_1, S_2, \ldots S_i, S_j, S_k, S_l, \ldots$$

is suppressed, then the otherwise predictable sequence will not materialize; either the total operation will be blocked, or a new outcome

$$S_j', S_k', S_l', \ldots$$

will ensue.

This result is completely different from that which would follow if the whole process was programmed in the genome, for in that case a mutation might be anticipated to entail a minor change in an otherwise normal epigenetic process. The mutations studied by geneticists are mostly of this kind because they are concerned with small intraspecific variations. Under these circumstances one may get the impression that the ontogenetic process is programmed, and this fact undoubtedly in part accounts for the prevalence of the blueprint notion.

The important consequence of the view advocated here is that a large phylogenetic change determining, say, the origin of a new taxon of high rank need not arise as a result of a long series of accumulated mutations. If a change, even a slight one, occurs sufficiently early in the sequence of operational steps, it may have very far-reaching effects. Thus, it is not at all necessary, nor even warranted, to presume that a new protein or differentiation pattern is engaged for, as we have seen, the same very limited number of differentiation patterns are involved during the early embryonic stages in widely diverging animal taxa.

Another implication of the computer analogy is that whenever the state S_i is reached, we may expect the sequence

$$S_j, S_k, S_l, \ldots$$

to follow. Obviously, S_i need not be approached through the sequence presented above; the same outcome might ensue from

$$S'_1, S'_2, \ldots S_i.$$

When, in this context, the two states designated by S_i are characterized as being identical, it is to be understood that this applies only to those properties which are causally involved in the subsequent states. Thus, two embryos may exhibit differences in a number of respects, chemical as well as physical, which sooner or later will influence the epigenetic processes. Yet, if they reach the same stage as defined here, their development will proceed along the same sequence of causally related steps until some of the differential factors are able to exert their influence.

Morphogenetic processes are carried out by relatively simple agents, even though they may through interaction give rise to organisms of great complexity. Nevertheless, since they are causally linked, it ought to be feasible to establish the general principles for their cooperation. Once this is accomplished, we should be able to predict, step by step, the actually realized epigenetic events. With sufficient insight it may even be possible, from knowledge about the ontogenetic end product, the adult, to make inferences about the epigenetic factors involved in its creation, an achievement which might be of great value in palaeontological studies.

INHERITANCE

In ordinary language the principal sense of 'inheritance' is 'that which is transmissible to an heir, or to offspring' (Webster's Collegiate Dictionary, 1939). I do not believe that any valid arguments can be raised against adopting this explication as a definition of 'biological inheritance'. Since what is transmitted from generation to generation, in sexual reproduction at least, is a fertilized egg, it is immediately obvious that this, and only this, entity can be the vector of biological inheritance. It is reasonable to presume that the various parts of the fertilized egg contribute differently to epigenesis. However, according to the evidence presented in earlier chapters, it must be granted that every subunit, from the egg envelope, when present, to the nuclear genome, contributes its share to the overall result. This stand is at variance with Mendelism, today the universally acclaimed theory of biological inheritance, which postulates that the only material of hereditary importance transmitted from parents to offspring is the nuclear genome, half of which comes from each parent. The disparity between these viewpoints is obvious, and is particularly striking because the Mendelian theory rests on a solid basis of experimental evidence.

Yet, the discordance is only apparent and stems from the fact that, owing to limitations inherent in their experimental approach, geneticists can study solely modifications arising as a result of genetic crossing. This method thus amounts to an establishment of various gene combinations, and accordingly the results obtained bear exclusively upon the action of nuclear genes. Owing to fertility barriers, geneticists mostly are confined to working with intraspecific crossings, and the gene effects observed generally impose only slight modifications on the animals. For this reason it appears that, unless the changes leading to transspecific evolution arise through the concerted action of such intraspecific gene mutations, this aspect of evolution may not be susceptible to genetic analysis. We shall return to this question later.

The extranuclear parts of the fertilized egg are not subjected to the usual type of Mendelian inheritance. This does not imply that proteins, and thus genes, are not involved in their establishment, but the mechanisms through which they are inherited are such that they cannot at all, or only with difficulty, be subjected to genetic analysis. As examples of the first kind we may mention the cell-type of organization and the centriole, the former transmitted through maternal, the latter generally through paternal, inheritance. The second kind may be represented by properties or features which are the outcome of activities in the maternal body, thus, presumably, egg envelopes and possibly also the direction of shell coiling in *Lymnaea*. In the latter case the mechanism of inheritance was proposed by Boycott and Diver (1923) and by Sturtevant (1923) to be Mendelian with one generation delay. I venture to propose that the situation prevailing in *Lymnaea* may represent a lucky incidence involving, as it does, only a geometrical modification in an otherwise viable organism. Probably the various extranuclear factors in the embryogenetic substrate exert functions which are so essential to the normal course of epigenesis that very slight variations may have deleterious effects. This circumstance must seriously hamper the study of their inheritance.

It is to be anticipated that the limitations of the Mendelian approach outlined above must somehow be reflected in the conceptual framework of the theory. A closer analysis of the terms *genotype* and *phenotype* shows that this is indeed the case. It is not easy to find explicit definitions of these notions, but, among the better ones, the following may be quoted: 'The whole hereditary material of an individual, the whole aggregate of its genes, [is] called a "genotype", and the form of body such a genotype produces is called a "phenotype"' (G. S. Carter, 1951, p. 51). These definitions show that the genotype is identical with the nuclear genome, and it is readily seen that if, as is often the case, the phenotype is identified with the adult body, then the statement quoted is invalid, since many other features are required for supporting the epigenetic process leading to the adult. If, however, 'form of body' refers only to such differences in body characters as are due to differences between various genotypes, then the above definition becomes meaningful. I do not think this point is generally realized, even if geneticists to some extent admit the existence of extranuclear hereditary factors.

Paraphrasing the above definition, and including ideas expressed by Waddington on several occasions (e.g. 1957; 1969), as well as names coined by him, it is possible to formulate the following definition: The whole hereditary material of an individual is called the *epigenotype*, and the form of body such an epigenotype produces through an epigenetic process is called the *epiphenotype*. A special clause referring to the influence of the environment should be added, but this is not important in the present context. It is quite clear that the epigenotype is identical with the embryogenetic substrate, i.e. the fertilized egg, while the epiphenotype is the adult animal *in toto*, and not just certain selected aspects of the latter. I believe it is very important for the understanding of phylogenetic evolution to realize that the concepts as generally employed by geneticists have a more limited scope than the epigenetic ones, particularly because they usually are implied to have the wider sense.

As a general conclusion of the present section it may be stated that the fertilized egg, the embryogenetic substrate, comprises some factors which are, and others which are not, transmitted according to the predictions of Mendelism. The geneticist studies the epigenetic effects of modifying the former, i.e. the nuclear genome. Contrariwise, the epigeneticist accepts the genome as a given entity while investigating, largely, the epigenetic outcome of alterations in the extranuclear parts of the embryogenetic substrate.

There is no doubt that, as far as the amount of epigenetic information is concerned, the greatest part is confined to the genome, and it therefore cannot be questioned that nuclear mutations have been of utmost import for evolution. However, the extranuclear factors are also of ample significance, particularly for the early developmental stages when the body plan is laid down, and it must therefore be concluded that neo-Darwinism cannot possibly claim to account for all aspects of inheritance of evolutionary concern; but nonetheless it obviously holds that the modern version of the Mendelian theory enters as a premise in the theory of phylogenetic evolution.

VARIATION

Biological variation in animals covers an enormous range, from the differences exhibited between members of separate phyla, or even subkingdoms, to individual variations. Irrespective of the mechanisms through which such modifications arise, the third Darwinian premise may be said to embrace the whole scope of biological variation.

In the present section it will be our task to establish the general nature of this diversity. Actually, even a cursory survey of the animal kingdom shows that at least four different parameters may be subjected to modification, viz., *form*, *pattern*, *function* and *behaviour*. The last one of these factors will be neglected throughout the following discussion. However, even with this substantial limitation it would be impossible to present a comprehensive review of the topic of biological variation. In the present context the functional aspect will be confined to a discussion of variations in the proteins synthesized by different animals, while the variations in form and pattern will be treated in a section dealing with the nature of systematic characters.

Proteins

We have observed that proteins may be subdivided into two classes, the non-repressible proteins, supposedly synthesized by all cells, and the repressible ones, which in turn may be separated into many subclasses, each of which distinguishes a particular cell diversification pattern.

Non-repressible proteins

For all we know, the genetic code, as well as the various steps involved in the synthesis of proteins, are essentially the same in all living organisms. That biosynthetic pathways also have remained largely unchanged during the eons of tellurian life is beautifully shown by the observations of Vogel (1965) on the biosynthesis of lysine. As appears from Figure 254, all living organisms can be subdivided into three groups with respect to this

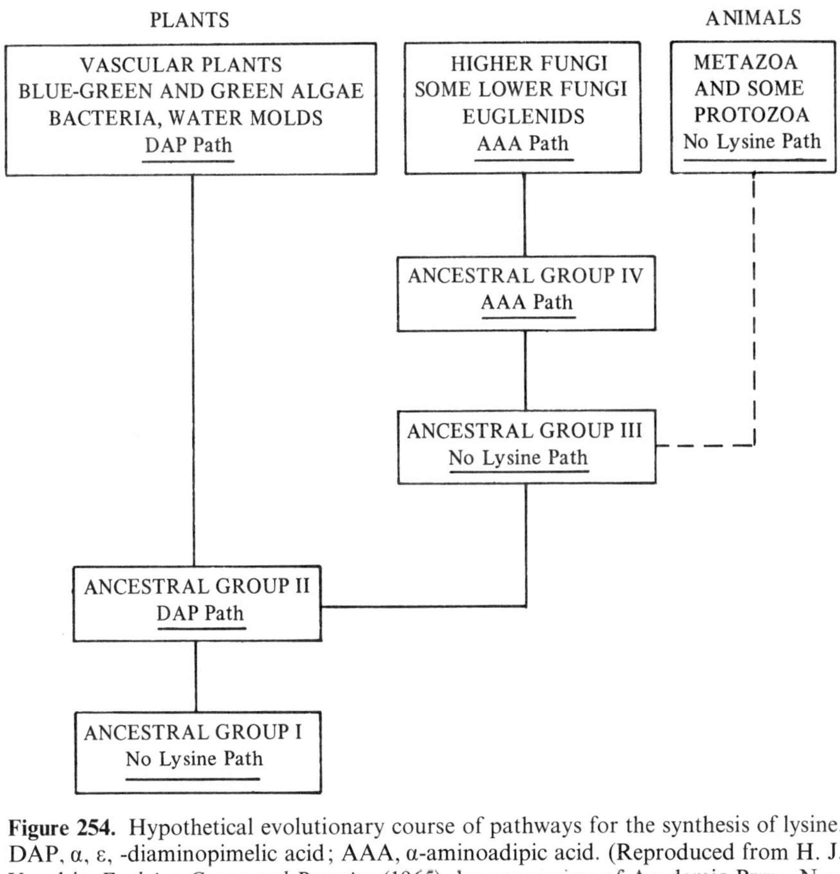

Figure 254. Hypothetical evolutionary course of pathways for the synthesis of lysine. DAP, α, ε, -diaminopimelic acid; AAA, α-aminoadipic acid. (Reproduced from H. J. Vogel in *Evolving Genes and Proteins* (1965), by permission of Academic Press, New York)

point, those that synthesize the amino acid via α-ε-diaminopimelic acid (DAP), those in which α-aminoadipic (AAA) is a key intermediate, and those that cannot synthesize lysine. These data strongly suggest that a pathway for the biosynthesis of lysine has been set up only twice during several milliards of years.

As far as energy-supplying processes are concerned, the three most important ones, glycolysis and the PP and tricarboxylic acid cycles, are present in almost all living organisms. The observations of Kaplan (1965) on vertebrate dehydrogenases show that the proteins involved may differ in their physical and catalytic properties, but that this variation follows a pattern which may be correlated with the known course of evolution. Studies of the characteristics of fructose diphosphate aldolase in a great variety of living organisms have made it possible to demonstrate that all investigated enzymes belong to either one of two distinct classes (Rutter, 1965), and also that other enzymes catalysing aldolase-type reactions can be similarly classified. This finding certainly intimates a remarkable conservatism as regards evolutionary innovation.

It appears that all existing intracellular structural proteins belong to a class which has been called 'tensible proteins'. As far as amino acid composition and other properties are concerned, the proteins obtained from different sources exhibit many similarities, but also, admittedly, some dissimilarities. The points raised here, which might be supplemented by numerous examples, suggest that although modifications have occurred in the properties and functions of the non-repressible proteins in the course of evolution, as well as acquisition of new ones and loss of others, the functions exerted by this class of proteins are remarkably similar in essentially all living cells. If all we knew about evolution were based on modifications of the non-repressible proteins, our notions about this event would be very vague indeed. Thus, although it cannot be doubted that the comparative biochemistry of the non-repressible proteins may in the future become of great importance for the elucidation of tricky phylogenetic problems, it still seems possible to assert with confidence that the variations responsible for the major aspects of phylogenetic evolution should not be looked for at this level.

Repressible proteins

The cooperation of different cell-types, representing disparate diversification patterns, is a distinguishing mark of the Metazoa. Therefore, the phylogenetic step from Protozoa to Metazoa must have involved at least one diversification process. It is equally certain that in ascending phylogenetic series there is a gradual increase in the number of diversification patterns. The latter are distinguished by their chemical activity, involving the synthesis of repressible proteins. These are themselves diversification products, but they may also be involved in the elaboration of non-proteinaceous substances thus labelled. In either case the various diversification, and in particular differentiation, products exert functions of general importance for the whole body, in contrast to the non-repressible proteins, whose range of action generally is confined to the cellular level. This circumstance suggests that we may look among the diversification patterns for some of those inheritable variations which have been instrumental in phylogenetic evolution.

Several comprehensive works on comparative biochemistry have appeared in the last decade, a scrutiny of which shows that, although the phylogenetic variation among the differentiation products, and hence among the repressible proteins, clearly exceeds that distinguishing the non-repressible proteins, it is nonetheless typical that many substances are almost ubiquitous in the animal kingdom, e.g. collagen, melanin, 5-HT. Evidently, evolution on the chemical level is a slow and cumbersome process compared with the strides made on the morphological level. This state of affairs has prompted Medawar (1967, p. 48) to advance the following statement: 'In a certain important sense all chemical evolution in living organisms stopped millions of years before even our faintest and most distant records of life began. So far as I know, no new kind of chemical compound has come into being over a period of evolution that began long before animals became differentiated from plants. . . . I have no views on the processes of evolution that brought new kinds of chemical compounds into existence, but I should not be surprised to find them very different from the forms of evolution that have been in progress since'.

With the reservations stipulated above, it is possible to endorse this statement wholeheartedly. As a consequence, it appears that every attempt may be rejected to explain evolution on the morphological level solely with reference to modifications on the chemical level.

Taxonomic characters

The characters used for the taxonomic classification of the members of the animal kingdom may largely, if not exclusively, be characterized as variations in form and pattern.

It is not possible to make a clear-cut distinction between morphological and non-morphological variation, but as a general rule it may be stated that significant differences in form are observed only on the higher systematic levels, while in the lower taxa morphological variation amounts at most to trivial modifications, and in many cases it may be confined to changes in pattern.

It is obviously out of the question to present even a cursory survey of the present subject; it can be illustrated only with a few examples. The large and small variations will be discussed under separate headings.

Archetypes

Comparative morphology, or 'formanalysis' (Kuhlenbeck, 1967), has established the existence of a number of distinct body plans in the animal kingdom. The gross aspects of this generalization were presumably known in prehistoric times, but its use for classificatory purposes was first established by Cuvier (1816). The value of this typological approach has been advocated or acknowledged by many authors (e.g. Owen, 1847; Gegenbaur, 1870; Jacobshagen, 1927; Naef, 1931; Kälin, 1945; Zangerl, 1948; D. D. Davis, 1949; Schindewolf, 1950; Kuhlenbeck, 1967; Waddington, 1967). It has been repeatedly criticized, particularly by adherents of the neo-Darwinian theory (e.g. Simpson, 1961; Mayr, 1963), as being an attempt to return to idealistic morphology, but as far as I can judge, this is a misrepresentation of the aim and purpose of formanalysis.

The concept of homology plays an important role in typology. As a general rule it may be stated that the homologous characters obtaining between the various phyla are few and rather trivial, but within each phylum this phenomenon is of great classificatory value. The phyletic body plans apparently represent rather unique evolutionary achievements; the relatively restricted number of animal phyla suggests that very little variation has been possible on this level.

This circumstance seems to warrant that Owen's (1847) notion of *archetype* should be adopted to stand for the general body plan of a phylum. Within each phylum the lower taxa, down to a level which may vary from case to case, represent homological modifications of the archetype, generally through *substitution* of one character for another one, but sometimes through *addition* of new characters. The concept of 'archetype' stresses

Table 12
Class characters in living Reptilia and Aves

	Reptilia	Aves
Feathers	Absent	Present
Preen gland	Absent	Present
Pneumatized bones	Absent	Present
Vertebral articulation	Amphicoelous or procoelous	Heterocoelous
Thoracic, lumbar, sacral and caudal vertebrae	Separate	Fused to varying degrees
Interclavicle	Present	Absent
Expansion of sternum	Absent	Present
Radius and ulna	Short	Long
Tibia and fibula	Present	Tibia fused with proximal tarsals, fibula reduced
Carpals and metacarpals, tarsals and metatarsals	Separate	Fused to varying degrees
Digits in front and hind limbs when present	5	3–4
Terrestrial locomotion	Quadrupedal or apedal	Bipedal
Brain case and brain	Small	Large
Skull sutures	Distinct	Much ankylosis
Quadrate and palatal elements	Fixed	Movable
Premaxilla	Small	Large
Teeth	Present, except Testudines	Absent
Bill or beak	Absent, except Testudines	Present
Syrinx	Absent	Present
Respiratory system	Blind-ending air passageways	Open interconnecting air passageways, air sacs
Systemic aortae	Left and right	Right
Tympanic membrane	Superficial	Deep

the fact that all members of a phylum possess a common body plan which, we may presume, has arisen through morphogenetic mechanisms which are the same in all individuals, and different from those responsible for the ontogenesis of animals belonging to other phyla.

We shall deal later with the morphogenetic elements involved in the formation of some of the most important archetypical body plans. Here we shall confine ourselves to the discussion of the phenomenon of homologous modifications in the two vertebrate classes, Reptilia and Aves. The members of these taxa embody a number of features which demonstrate that they possess the archetype of the subphylum Vertebrata. The traits in question comprise an axial skeleton, a central nervous system, a pituitary body, laterally segmented body musculature, a liver, etc. They furthermore exhibit features showing them to belong to the 'branch' Gnathostomata, thus jaws, paired appendages, various endocrine glands, etc. It will be noted that apart from the jaw bones, which represent substitutions for branchial arches, all the other characters mentioned are additions as compared with the Agnatha. Other features mark out their affiliation with the 'grade' Tetrapoda, e.g. limbs, substitutions for the pectoral and pelvic fins in the remaining Gnathostoma, lungs, corresponding to additions or to substitutions for a swimbladder or for lungs (in the lung fishes), etc. Finally, the class Reptilia is distinguished by various traits, among which a very typical one is the keratinous dermal scutes, covering the whole body. Class Aves is characterized by a number of features, which represent modifications of properties on all the levels discussed above. Thus, a tail formed by the axial skeleton is typical of all vertebrates except Salientia and Primates, but in the birds the caudal vertebrae have fused to form the coccyx. The tooth-bearing jaw bones of the Gnathostomata have been replaced by a beak, the scutes of the Reptilia largely by feathers, etc. (cf. Table 12).

When we go further down in the taxonomic classification of birds, we shall find that the ordinal, and usually also the familial, characters amount to further modifications of the particular avian version of the vertebrate archetype. These changes, which usually are associated with particular ways of living, may concern the presence or absence of wings, modifications in the shape of the feet, the beak and the wings, etc. This list of ordinal characters is not complete, but I believe that it is representative. The characters representing the lower taxa may sometimes be of the kind just discussed, but more often they are minor variations with respect to size and pattern, a subject which will be treated in the following section.

The substance of the present discussion is that the morphological innovations involved in phylogenetic evolution constitute distinct and discrete variations which, among other things, allow for the distinction of archetypes and modifications of the latter, and thus for taxonomic classification of animals.

Size and pattern

As a general rule, to which exceptions may be found, all members of any lower taxon have the same general body form. The minor variations obtaining between such taxa may be classified as being either quantitative or qualitative.

To the former class belong modifications in the size of the body or in parts thereof. Thus, in birds the changes may involve the length of the beak and of the feet, the length and width of the wings, the length of some or all of the tail feathers, etc.

The qualitative variations, or changes in pattern, are generally concerned with pigmentation, for instance of the feathers in birds. In many cases it would be possible to characterize an order or a family by a general colour scheme, modifications in which very frequently are used as distinctive characters on the generic and specific level. Variations obtain within the confines of a species which sometimes are used for the erection of subspecies, but most often they are accepted as subtaxonomic variations.

MUTATIONS

'The most basic weakness of [Darwin's] concept of evolution was the lack of an adequate theory of the production of variations upon which natural selection acted' (Provine, 1971, p. 130). Indeed, as appears from the historical surveys in the book quoted and in that of Rádl (1930), much of the often very heated controversy that has gone on about the theory of evolution was centred around this problem. In point of fact, in spite of the hegemony of neo-Darwinism, dissenting voices are still raised today, so in a way it is justified to state that this issue has not yet been settled.

On one point it seems that agreement obtains, namely, that whatever the nature of these variations, or rather their cause, they are called 'mutations'. It deserves mention that this word has been employed in different senses in the course of time. Thus, around the turn of the century 'mutation' was equated more or less with 'sport' as used by Darwin (1859), signifying large and suddenly appearing inheritable changes. As used today, influenced by genetics and cytogenetics, it may be defined as 'any inheritable modification in the genotype that may affect the phenotype', but under the impact of recent molecular biological discoveries there is a tendency to delete the stipulation about the effect.

Without disputing the expediency of this measure it seems necessary in the complete theory to adopt a different stand and let the implication of 'mutation' be 'any inheritable modification of the epigenotype that may affect the epiphenotype'. This circumstance notwithstanding, it holds, of course, that the theoretical superstructures of the disciplines mentioned are parts of the premise concerning the occurrence and significance of mutations.

The present section is introduced by a brief survey of the general nature of mutations, followed by discussions of their various effects.

Nature

It is possible to distinguish three kinds of nuclear mutations, viz: '(1) gene-mutations, i.e. changes or losses of single genes; (2) chromosome-mutations, i.e. structural changes in single or several chromosomes (deletions, duplications, inversions, translocations); and (3) genome-mutations, i.e. changes in the number of single, several or all chromosomes (haplo- or polysomy, heteroploidy, polyploidy)' (Timofeeff-Ressovsky, 1940, p. 76).

With the discovery of the genetic code it has become possible to formulate very clear notions about the nature and function of a certain class of genes, the so-called *template* genes, and thus of their mutations. The template genes determine the amino acid sequence, i.e. the primary structure, of the proteins that potentially can be synthesized by the organism. The information required for this purpose is specified by the sequence of bases in the DNA molecules which are the genes. Any changes in the base sequence is regarded as a mutation. Most common are the point mutations, involving that one base is replaced by another one, but others are known which result in different kinds of change in the DNA molecule (cf. Jukes, 1966).

Owing to the degeneracy of the genetic code, many point mutations will not entail coding for a different amino acid (cf. Sonneborn, 1965; King and Jukes, 1969). Even if allowance must be made for the possibility that the specificity of the transfer RNA's with respect to the various codons may ensue in modifications at the translation level in some instances (Caskey, Beaudet and Nirenberg, 1968), the actual existence of 'silent' or 'non-operative' mutations seems beyond doubt. Hence, obviously, a number of 'mutations', in the modern sense of this word, do not have an effect that can be recorded by any other means than base succession analyses in the DNA or the messenger RNA molecules.

Template genes also exist for ribosomal and transfer RNA; it may be anticipated that mutations can also occur in this instance, but about the phenotypic effect of these nothing seems to be known.

We have had above an opportunity to discuss the control of protein synthesis in *E.coli*, in which it has been found that various *control* genes are active. The existence of these kind of genes in Metazoa can be inferred with certainty, and clearly mutations leading to changes in their mode of action may be anticipated to be of great effect. Whether or not point mutations would suffice for this purpose is unknown.

Apart from losses through deletions, chromosome mutations cannot entail any changes in the repertoire of proteins which the organism in question is capable of synthesizing. They might therefore be expected to be of no consequence. This prediction has been borne out in some cases, but in others clear-cut effects have been obtained. Results of this kind are interpreted to reveal a position effect, the activity of the genes supposedly being dependent upon their location in the genome (cf. Dobzhansky, 1951). There is some reason to presume that it is the interaction between template and control genes which is effected under these conditions. The mechanism of chromosome mutations, very different from that of gene mutations, has been amply elucidated through cytogenetical studies.

A substantial number of genome mutations have been described, not the least in recent years. With respect to their effects these mutations cover the whole range from being lethal to having no discernible effect at all. The mechanisms responsible for this type of mutation are known in many cases.

No general rules can be established for the mutation rate of the three kinds of mutation, but by and large it appears that point mutations are the most, genome mutations the least, frequent ones.

According to the definition of 'mutation' proposed above it follows that modifications in extranuclear factors which are of epigenetic consequence must fall under this heading. Apart from their cellular localization, they differ from usual mutation in that they are inherited in a non-Mendelian fashion. Nothing precludes, of course, that this kind of mutation in many cases, if not necessarily always, may be traced back to nuclear mutations. The effects of such extranuclear mutations expectedly concern, in the first place, nucleo-cytoplasmic interactions and morphogenetic processes.

Effects

The information prevailing about gene action on the molecular level allows for outlining at least some of the potential effects of mutations of template and control genes. On the basis of the epigenetic theory propounded here it is even possible to sketch the principles which must be involved when mutation effects are manifested on higher levels of organization. We shall take up these questions in the following pages, as well as the problems concerning the size and the quality of mutations and the constraints of various kinds to which they are subjected.

Template genes

The mutations affecting the template genes are principally gene mutations, notably point mutations. Although the latter supposedly are random, their effects are not, as many mutations do not lead to amino acid substitutions. If, and when, such a replacement occurs, the properties and functions of the protein in question may be influenced to an extent which varies with the nature and the site of the amino acids involved.

This question has in recent years been elucidated by studies on the amino acid composition in haemoglobins from a wide variety of vertebrates (cf. Zuckerkandl and Pauling, 1965). It has been found that altogether only eight out of more than a hundred amino acids have not been subjected to evolutionary replacements. It appears that the amino acids at these *invariant* sites are intimately associated with the physiological function of the molecule. Among the variable sites, certain have undergone *conservative* substitutions, involving amino acids closely related in molecular properties, whereas at the remaining sites widely different amino acids may be found. According to Zuckerkandl and Pauling the residues at the invariant and conservative sites are mainly concerned with the elaboration and stability of the tertiary structure of the haemoglobin molecule. The fraction of *variant* sites varies from one protein to another, as evidenced by the observation that in cytochrome c as much as 50 per cent of the residues are invariant (Margoliash and Smith, 1965).

Some interesting inferences may be drawn from these findings. Thus, on the assumption that mutations are random, it follows that all operative ones concerning the invariant, and some of those involving the conservative, sites have been weeded out by selection. In this way the latter expedient obviously operates for the preservation of an existing function. As regards all the apparently haphazard replacements which have occurred in the variant sites, the neo-Darwinian theory postulates that their survival implies that they are somehow to advantage for the function of haemoglobin in the milieu where the various animals subsist. In this context should be mentioned, however, the suggestion by King and Jukes (1969) that mutations involving amino acid substitutions may become established through 'non-Darwinian selection'.

However this may be, it is not too difficult to understand the cooperation between mutation and selection in the preservation and the improvement of the function exerted by an existing molecule. The situation becomes considerably more complicated when we turn to the question of evolutionary novelties on the chemical level, i.e. proteins capable of executing hitherto non-existing functions. As regards this point, it was suggested above that such innovations have occurred, even if we, with Medawar (1967), must admit that metazoan evolution is remarkably defective in this respect.

Clearly, the process of selection cannot become involved before the molecule has first assumed its function, and without this agent it is very difficult to grasp how mutations, thus undirected, can ensure the generation of molecules with new properties. At any rate, it is certain that this mechanism cannot involve existing functional genes, for as we have seen, selection in this case works for preservation. The phenomenon of *gene duplication* or *replication* (E. B. Lewis, 1951; Ohno, 1970; Britten and Davidson, 1971) has paved the way towards an understanding of how innovations may come about. The possible existence of 'idle', or at least redundant, genes, exempt from the selectional demands as regards specific function, certain radically increases the likelihood that non-repressible, but particularly repressible, proteins with new properties may arise fortuitously. This proposal stresses the interesting and logically compelling inference that divergence on the molecular level requires 'isolation' just as much as does taxonomic divergence.

The existence of this evolutionary mechanism is strongly supported by the demonstration that many of the proteins secreted by endodermal, and to some extent ectodermal, glands exhibit distinct similarities with respect to the composition and sequence of amino acids (Adelson, 1971). It will be realized, however, that without the guiding influence of selection the rate of progress must be very slow, and the frequency of success extremely low, compared to those changes observed in genetic experiments. There can therefore be no doubt that the first origin of 'new' proteins must belong to the 'unique events' in Wright's classification of factors of evolution (1955).

Control genes

Control genes may be concerned with the regulation of both non-repressible and repressible proteins. However, because of the enormous importance of cell diversification for the evolution of form and function in Metazoa it may be inferred that mutations concerning the control of the latter must have been of particular phylogenetic significance.

It may be anticipated that control gene mutations can affect the number of proteins belonging to a particular diversification pattern, as well as the rate at which one or more of the proteins are synthesized. Mutations relating to the mechanisms of nucleo-cytoplasmic interactions are also imaginable.

Of still more interest are mutations leading to new diversification patterns. We have seen that some of the simplest diversification patterns may involve primarily a copious synthesis of normal cell constituents, suggesting that in such cases the modification is concerned with the transcription of genes for elsewhere non-repressible proteins. Most repressible proteins do, however, represent new functions, and their origination may therefore be referred to the mechanism discussed in the previous section. The establishment of the peculiar nucleo-cytoplasmic interaction that is a prerequisite for the functioning of a new differentiation pattern may be presumed to have occurred very few times during the course of phylogenetic evolution, as substantiated by the fact that, although their exact number is unknown, it is certain that even in higher organisms the number of differentiation patterns is rather restricted. It thus seems safe to postulate that the origination of new diversification patterns is also a unique evolutionary event.

A particular type of control gene is represented by those responsible for the shaping of the structural differentiation products, for instance keratin. As is known, this substance may be deposited as skin or in the form of scutes, feathers or hairs. The primary structure of keratin undoubtedly varies from one taxon to another, and there is some indication that disparities of this kind exist between the 'soft' and the 'hard' types (cf. Mercer, 1961), but since different configurations may occur in the same animal, it may safely be assumed that the morphological form cannot be referred solely to the template genes for keratin. It therefore seems necessary to infer the presence of a particular kind of 'morphogenetic control gene', concerned with the shape rather than the composition of a particular substance.

Epigenetic processes

A number of proteins are involved in the elaboration of the epigenetic substrate before and after fertilization. Some of these are structural proteins, others metabolic differentiation products, of which some in turn may elaborate structural differentiation products. This circumstance entails that mutations in template genes may have substantial effects upon epigenesis.

A similar situation obtains with respect to the control genes which are responsible for the time of onset and the rate of production of various proteins and, in combination with other embryogenetic processes, notably morphogenetic ones, may also influence the location of where these activities occur. It is obvious that, taking into account the causal connections between the successive stages of development, even mutations which entail only slight changes in the activity in control genes may through this *epigenetic amplification* give rise to extensive alterations in the final epigenetic product, the adult organism.

Mutations affecting the production of regulatory differentiation products constitute a further possibility, and a potentially very powerful one, for interference with epigenetic processes. For instance, a mutation affecting the rate of production, specificity or any other activity parameter of a hormone affecting an embryogenetic process may be expected, under favourable, or unfavourable, conditions, as the case may be, to exert profound effects.

Size

The question of the size or magnitude of the effect of the individual mutations which have been responsible for the execution of phylogenesis is the most controversial issue in evolutionary discourse.

Darwin (1859) started out by postulating that intraspecific variations through accumulation could lead to taxonomic divergence. Already among the early 'Darwinists' doubts were raised about the plausibility of this contention, a stand that was taken by T. H. Huxley, Galton and Bateson, to mention only English dissenters (cf. Provine, 1971). The neo-Darwinists, Fisher (1930) in particular, claimed that, under certain simplifying assumptions, mutations of the size and the frequency observable in genetic experiments would, under the influence of natural selection, be accumulated at a rate sufficiently high to account for the occurrence of phylogenetic evolution.

According to the neo-Darwinian notion the genotype is, in a particular environment, alone responsible for the generation of the adult organism, the (epi)phenotype. Consequently the latter is a kind of mosiac work to which each gene contributes its share, and the effect of each of the randomly occurring mutations is to modify the present status a tiny bit for better or worse, or not at all. Being conceived on the basis of observable mutations this idea is obviously well substantiated. It must be recalled, however, that the effect of mutations of this kind amounts to little more than slight modifications of the epiphenotype; and for this reason a number of arguments may be, and have been, raised against this conception.

For one thing, the observable mutations have not, so far, led to transspecific evolution, and in general it holds that they rarely, if ever, are concerned with the changes in form which obviously must have been involved in the generation of the various high-ranking taxa. Some few observations of transspecific mutations have been recorded in plants, but none of these have led to any spectacular morphological changes.

Secondly, certain properties do not allow for their origination through accumulation. To this category must be assigned 'quantum evolution' (Simpson, 1953a), i.e. modifications involving numerical changes (number of segments, vertebrae, appendages, digits, heart chambers, etc.) where, for obvious reasons, relatively large changes must be allowed for, even by adherents of neo-Darwinism. However, this objection applies in many other cases as well; thus, for instance, it is not easy to envisage how a functional circulatory system can arise through a process involving numerous small steps.

Thirdly, it is argued, even in the cases where the gradual origination of a property, say, a limb, is possible, it is unwarranted to invoke natural selection before the new function has been assumed. Thus, the small bud that would represent an incipient limb, rather than a benefit, might be a nuisance against which the selective forces would act. Under such circumstances phylogenetic advance could be ensured only by presuming directed evolution, 'orthogenesis', a proposition which contradicts the notion of random mutation.

As mentioned above, the existence of large mutations was propounded already at the time of Darwin and, although their number may have dwindled, there have throughout the history of the evolution theory been advocates of this view. Typically, the latter are to be found among biologists with morphological leanings, palaeontologists, biogeographers, systematists, embryologists, etc. As outstanding recent representatives of this group may be mentioned Goldschmidt (1940), Schmalhausen (1947), Dalcq (1949, 1954), de Beer (1951), Schindewolf (1950), Waddington (1957; 1969), Vandel (1936) and Brien (1969). It is most encouraging, in view of the stand adopted by me, to learn that the current progress within the field of molecular biology has led some students to the conviction that large-step mutations may be necessary, or at least possible, evolutionary events (cf. Ohno, 1970; Britten and Davidson, 1970).

Most, if not all, of these authors do not contend that large mutations are different from other ones on the molecular level, but only with regard to their effects, and in general there is agreement that the effects of mutations of this kind are amplified through interference with the course of ontogenesis. In relation to the classical antithesis, they thus advocate an epigenetic point of view, whereas the neo-Darwinian theory, considering the difficulty with which new mutations are accounted for, rather may be said to represent preformation on the genetic level. It must be mentioned, however, that even neo-Darwinians like Haldane (1932b) and J. S. Huxley (1932) have acknowledged the possibility of mutational interference with epigenetic processes.

The reason for the dispute on this question is evident; none of the parties can supply conclusive empirical evidence for their standpoint, and both can provide cogent reasons for this state of affairs. Thus, the neo-Darwinians claim that no case of artificial selection has been continued for sufficiently long time to warrant the creation of a new species, let alone taxa of high categories. The adherents of larger-scale mutations, on the other hand, contend that transspecific mutations for obvious reasons must have a frequency so extremely low that the chances of observing such an event are infinitesimal.

This controversy has gone on now for more than a century and it is hardly possible to produce any new arguments. Yet, for the sake of the following discussion it is necessary to deal with a few points, particularly regarding the classification of mutations with respect to the size of their effects. Joining the opponents of the Darwinian stand, I suggest that for convenience three classes may be distinguished, to which the adjectives *large*, *intermediate* and *small*, respectively, will be applied. It must be stressed that this subdivision is an artifact; either the mutations form one, or more, continuous spectra, or else it should be possible to erect more groups.

I suggest that on the morphological level this gradation may correspond approximately to the three kinds of variations discussed above, viz., archetypical modifications, changes in size and pattern, and intraspecific variations, respectively. I submit that, within certain limits, there is inverse proportionality between the effect and the frequency of these mutations, whence it follows that those observed in genetic studies mostly are small, albeit even intermediate ones may occur.

Some morphological modifications require the *cooperation* of several diversification patterns, and thus of several mutations. It may be noted that the word 'cooperation' here has an implication which is radically different from 'accumulation'. As an example illustrating this proposition we may mention the notochord. This structure is elaborated through the cooperation of cells representing two diversification patterns, viz., collagenocytes and chordocytes. The former are cells of the order nsf-cells, belonging to the family of sclerocytes; the latter are ncf-cells and myxocytes. The differentiation patterns represented by these two cell-types are common among Invertebrata, so in this respect the first appearance of a notochord was not at all an innovation. However, myxocytes usually are present as sf-cells, myxoblasts, and one of the modifications required for the genesis of the first chorda therefore was that certain myxocytes underwent the sc-transformation. Since the class of colligocytes undoubtedly was present in the incipient chordate, this conversion was not a chemical novelty either. I do not claim that the changes mentioned here are the only ones necessary for the formation of a notochord—the location and time of appearance of the cell-types are, of course, also important—but the main point is that through the proper cooperation of elements the properties of which are trivial, a phylogenetically revolutionizing result may ensue. It goes without saying that events of this kind are extremely rare, otherwise invertebrate embryologists ought to observe occasional chordate monsters, if not necessarily hopeful ones, in their cultures.

As regards the chemical level, it is beyond dispute that large mutations, represented by the origination of substances capable of executing new types of function, have occurred, but as observed above, a remarkable conservatism prevails in this domain. This low rate may be accounted for if large mutations of this kind require the cooperation of two or more mutations in a redundant gene before the new function can be assumed. Amino acid substitutions entailing the loss or the serious impairment of the function of a given protein may also belong to this group. To the intermediate mutations may be reckoned the specialization of a particular function, as exemplified by the various endodermal proteinases described by Adelson (1971). Finally, to the small mutations may belong amino acid substitutions leading to slight modifications in the efficiency of the activity exerted by a protein.

It is often argued against the large-scale mutations that they have never been observed to occur. It seems warranted to question whether this is correct; rather, it seems to me that the 'sports' observed in plants, and discussed by Darwin (1859), belong to this category, and similar observations have been made in animals. Thus, Goldschmidt (1940) pointed out that the various races of the domesticated dog exhibit a number of features which, were these animals encountered in Nature, would have warranted their subdivision into taxonomic categories well above the species level. The fact that the responsible mutations can survive only through the protection of man shows that they may be of little advantage under natural conditions, but this does not alter the fact that the mutations exist. Some of them, at least, must be classed as intermediate mutations, in the sense defined above, displaying quite tangible effects. This, indeed, seems to be a necessity, for if the establishment of the various races had involved the accumulation of hundreds of imperceptible mutations, directed not by natural, but by human, selection, then it seems most unlikely that they would ever have come into existence, particularly since man seems to be inferior to Nature as regards the evaluation of small mutations (cf. Simpson, 1953a).

It has been argued against Goldschmidt that all dogs are races of one single species, as evidenced from the observable racial interfertility. Actually, the establishment of fertility barriers seems to be primarily a consequence of isolation; granting that a Great Dane and an Alsatian may give fertile offspring today, the situation might well be different if the two races were kept isolated for ten or a hundred thousand years.

Needless to say, these observations on dogs are in no way unique. Extensive variations, based on single mutations, have been observed in mice (cf. Grüneberg, 1943), and similar findings have been recorded in other mammals. Likewise, Cole (1967) has found that the formation of clawed digits on the wings of chickens is controlled by one dominant gene. Villee (1942) has pointed out that certain mutations in Arthropoda entail changes which usually are characters typical of taxa of quite high categories.

As a last illustration of this point we may quote the distribution of primitive and advanced characters in two families of whales, Balaenidae and Balaenoptera (Abel, 1929). From Table 13 it is seen that features of the two categories are distributed for-

Table 13

Distribution of primitive and advanced characters in two families of whalebone whales. Framed features are primitive, non-framed ones are advanced. From O. Abel, 1929

	Balanidae	Balaenopteridae
Fingers	Five	Four
Throat plaits	None	Many
Dorsal fin	Absent	Present
Tibia rudiment	Present	Absent
Rostrum	Narrow and bent	Broad and flat
Cervical vertebrae	Fused	Free
Radius and ulna	Short	Long
Coronoid processes of mandibles	Reduced or absent	Small, but well developed

tuitously between the two families. The animals of these two taxa live under essentially identical environmental conditions, and if the modifications of the listed characters were the outcome of selection of random mutations occurring at reasonably high frequencies it might be anticipated that the evolution would have been parallel in the two families. Actually, in those cases where interaction with the environment is possible (e.g. fingers, dorsal fin, rostrum, radius and ulna) the advanced condition should be expected in both groups, and in those cases where this mechanism cannot be invoked (e.g. fibia rudiment and possibly cervical vertebrae) the preservation of the primitive state would be the most likely event. The actually observed situation seems explicable only on the assumption that intermediate mutations, occurring at random and at very low frequencies, are responsible for these morphological characters.

Quality

Among the mutations available to genetical experimentation, whether spontaneous or artificially provoked, it is possible to distinguish three classes, *positive*, *neutral* and *negative*, with respect to the advantage they confer upon the epiphenotype. The word 'advantage' may be rather ambiguous when used without further qualification, but the discussion of its implications must be postponed to the following section. In any case, since the mutations generally are small, the advantages or disadvantages are slight, and the viability of the animals is not impaired. As regards the three kinds of mutations, it is possible that the neutral ones are the most common; general experience has shown that the positive are less frequent than the negative ones (cf. Fisher, 1932).

Sometimes mutations are observed which have a more thorough influence upon the epiphenotype, but always in a negative direction. This circumstance led Fisher (1930) to the categorical statement that all large mutations are deleterious. Concerning this point it may be noted with regard to the intermediate and large mutations, supposedly involving substantial changes of form and function, that the likelihood of their being neutral obviously becomes increasingly smaller as the extent of their effect enlarges. Allowing for the much lower frequency of these types of mutations, and the ratio between positive and negative mutations, which may well be still lower in this case, it becomes clear that the chances of observing a large mutation are slight, but those of discovering a large positive mutation are infinitesimal. This inference seems to account for, if not to endorse, Fisher's generalization.

Among the negative mutations particular attention must be paid to those involving losses. From the preceding discussion it appears that a point mutation in a template gene may suffice to suppress completely the function of a particular protein, and other kinds of mutation are known that may lead to the loss of a certain gene. In cases where the activity carried out by the gene product is of vital importance, a loss mutation may hence be lethal. It must be pointed out, however, that from the comparative biochemistry several cases are known where the absence of a special protein distinguishes a whole taxon, suggesting the incidence of a loss mutation with no particular effect.

Various epigenetic processes may depend upon the activities of several genes. It goes without saying that phenomena of this kind must have required much evolutionary experimentation for their establishment. However, one mutation, impairing or deleting one of the participants in such a process, may suffice to abolish the compound phenomenon. This example illustrates a way in which negative mutations may incur losses, and emphasizes that in evolution, as elsewhere, destruction is much easier than construction.

Constraints

The generally acclaimed contention that nuclear mutations are responsible for phylogenetic evolution has been accepted above with only a very slight reservation. It was furthermore asserted that the mutations may bring forth innovations which affect the function or the form of the epiphenotype. The latter type of alteration may arise through regulatory interferences which do not involve the creation of new chemical compounds.

Considering the established nature of the mutations, it appears that the number of possible gene mutations is infinitely great, and even if chromosome and genome mutations are less frequent events, it is nevertheless likely that many variations may take place even in these cases. The potential for mutational modifications inherent in the genome is therefore enormous. Evidently, various forms of constraints must prevail between the occurrence of a mutation and the establishment of a demonstrable effect.

The potential outcome of point mutations is astronomical: thus, with twenty different amino acids it is possible to form about 10^{325} different polypeptide chains with 250 residues (Eden, 1967). Nothing of the sort is, of course, realizable, and the first constraint to counteract this, apart from the physical impossibility, is the selection for preservation of function. Whenever the coding for a protein with some important function has been established, its continued presence is zealously guarded, any mutation impairing its activity being rejected. Only if by chance it loses its functional importance may its loss by mutation be sustained. As regards innovations, a constraint may be found in the fact that only very few of the many possible modifications of an existing protein can reasonably be anticipated to convey a new function upon the molecule. Kilkson (1968), in a paper on the theory of macromolecular structure of phage particles, has discussed the mutational constraints on this level of organization.

Many of the mutations affecting morphogenesis presumably do not exert any noticeable effect, but when they do, they may either upset the ensuing epigenetic processes or initiate a new chain of causally connected events which give rise to an improved epiphenotype. Considering the number of elements involved in morphogenesis, and their possible modes of interaction, the likelihood of a happy outcome is extremely slight. A further restriction as far as morphological evolution is concerned is found in the fact that morphogenetic processes perforce must obey the physical laws, a subject that is the central theme in Thompson's famous book (1942).

It may finally be mentioned that the various constraints may be altered as evolution proceeds. Thus, it is highly possible that mutations, which under certain conditions do not lead to any significant modification or entail any particular advantage, may be extremely successful, or unsuccessful, in a new, or at least a different, setting, as represented for instance by an archetypical body plan or an environmental parameter.

COMPETITION

The phenomenon of competition, or struggle for existence, is in Darwin's theory intimately associated with those of fitness and natural selection, being the instrument through which organisms distinguished by fitness are selected, i.e. progress, at the cost of less successful ones. Selection, one of the conclusions of the theory, must be dealt with later, and consequently the present section may be devoted to a discussion of fitness and competition.

The implication of fitness is competitive success, hence the former may in principle be measured by the latter, an cxpedient which has been adopted in the Darwinian theories and which, as we have seen, with some justification may be said to involve a tautology. Clearly, in order to predict the course of evolution it will be necessary to establish criteria of fitness that are independent of the competitive success or survival of the organisms in question. Nobody has so far been able to do this, but in the following section it will be shown that a first step in this direction may be taken by introducing the concept of *adaptability*.

One of the section heads in Darwin's book (1859) is entitled: 'Struggle for life most severe between individuals and varieties of the same species; often severe between species of the same genus'. Since the contents of the section in part were empirical observations, the quotation serves to emphasize two points: firstly, that competition between different individuals presupposes not only temporal, but also spatial, coexistence, a stipulation which in turn requires that the organisms in question must be so closely related that they may share the same niche, and secondly, that competition may be either *intraspecific* or *interspecific*, two topics that will also be dealt with in the present section.

Adaptability

As an attempt to explain the observable course of evolution, it has been proposed that 'evolution *is* progressive adaptation, and consists in nothing else' (Fisher, quoted by Grene, 1969, p. 66). Since it is obvious that there is more than just a grain of truth in Fisher's statement, the idea has been generally adopted. Yet, to the critical mind, there are many difficulties involved in the very concept of 'adaptation', which rather preclude the acceptance of adaptation as such as the guiding principle in evolution.

The criticism has been voiced from many different angles. Among these are the epistemological assertion that 'adaptation' introduces teleology in the theory of evolution. This point was formulated by Grene (1969, p. 66) in the following way:

'For one thing, the concept of evolution as identified with progressive adaptation is basically ambiguous. On the one hand, such adaptations are supposed to be mechanically self-generating—through mutation, natural selection, recombination and isolation—and so to entail no teleological reference. Yet on the other hand, adaptation, like its Victorian twin, "utility", is itself a teleological concept: it is adjustment of something to something for some end. To interpret organisms as adaptation machines, as neo-Darwinism does, therefore, is to interpret them as complexes of means for ends, and therefore teleologically. Secondly, if the means–end reference is admitted, one must ask further: means to what end? But the "end", for Darwinism, dare not be some "higher" form of life, the next "level" to which evolution aspires: that would be to reintroduce a forbidden version of "unscientific" teleology. It must then, and is usually said to, be *survival* that adaptation is "for".'

Another shortcoming is the vagueness of the concept. As has been repeatedly argued, the statement that an animal is adapted to its environment does not imply anything more than that it can and does live in that environment. Thus, it is not possible to quantitate adaptation and therefore it is in principle impossible to decide which, among a jellyfish, a squid or a dolphin, is best adapted to marine life, even if the three definitely represent a phylogenetic sequence. To vouch for the mammal because it originated last is clearly a circular argument.

It is evident, however, that an environment is no constant thing, but is subjected to continuous changes. It is also clear that the better an organism is able to cope with such environmental changes, the greater are the chances of survival, for the individual as well as for the offspring. The power to sustain environmental changes must be termed 'adaptability' rather than 'adaptation'. If it is postulated that selection acts to further adaptability then the example mentioned above meets with no difficulties, for even without any thorough analysis it is quite certain that the level of adaptability increases parallel with the phylogenetic advance in the three marine animals.

Sommerhoff (1950) has stated that the Cartesian tradition in biological thought is one reason why the phenomenon of adaptation often has been associated with a notion of purpose. To circumvent that inherited handicap, this author introduced the concept of *directive correlation* which, although enjoying wider generality, includes the phenomenon of biological adaptation, without any teleological implications.

Directive correlation is defined in the following way: 'any event or state of affairs R_{t_1} occurring at a time t_1 is *directively correlated* to a given simultaneous event or state of affairs E_{t_1} in respect of the subsequent occurrence of an event or state of affairs $[FC_{t_2}]$ if the physical [or chemical] system of which these are part is objectively so conditioned that there exists an event or state of affairs CV_{t_0} prior to t_1, and a set of possible alternative values of CV_{t_0}, such that

(a) under the given circumstances any variation of CV_{t_0} within this set implies variations of both R_{t_1}, and E_{t_1};

(b) any such pair of varied values of R_{t_1}, E_{t_1} (as well as the pair of their actual values) is a pair of corresponding members to two correlated sets of possible values R'_{t_1}, R''_{t_1}, R'''_{t_1}, . . . and E'_{t_1}, E''_{t_1}, E'''_{t_1}, . . . which are such that under the circumstances all pairs of corresponding members, but no other parts, cause the subsequent occurrence of $[FC_{t_2}]$' (Sommerhoff, 1950, pp. 54–55).

In a biological situation, the parameter E_{t_1} represents an environmental variable, and R_{t_1} corresponds to the appropriate response on behalf of an organism. CV_{t_0}, the *coenetic variable*, is in this situation also an environmental variable, antecedent to, but causally related to, E_{t_1}, the one to which adaptation occurs. The same relation holds between CV_{t_0} and R_{t_1}, for since R_{t_1} and E_{t_1} are simultaneous, they cannot be causally dependent. FC_{t_2}, the *focal condition*, represents the goal of the adaptation. The correlations between the various parameters are shown in Figure 255.

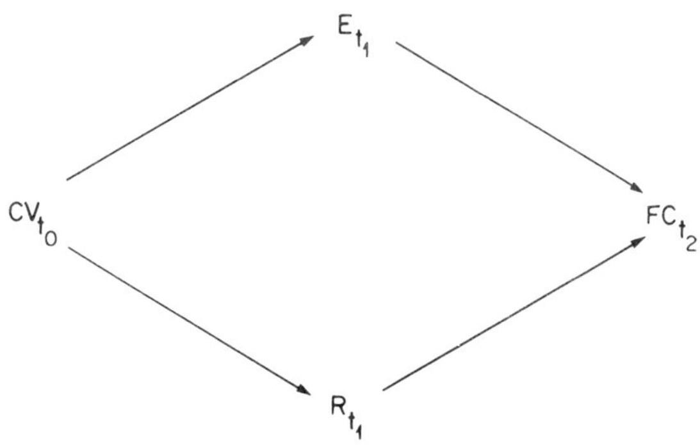

Figure 255. Diagram of directive correlation. CV, coenetic variable; E, environmental circumstance; R, response; FC, focal condition; t_0, t_1 and t_2, successive points of time. (Reproduced from G. Sommerhoff (1950), *Analytical Biology*, by permission of The Clarendon Press, Oxford)

According to Sommerhoff (1950, p. 140): 'it is possible for directive correlations to form entire hierarchies. In these hierarchies the focal condition of each directive correlation on one level of the hierachy figures as either a directively correlated variable or a parameter of specialization in the next higher level. Such hierarchic relationships are particularly characteristic of living organisms . . . In short, the characteristic unity of the higher living organisms derives from the fact that their activities are united by hierarchies of directive correlations. The biological unity of a living organism is essentially the unity of such an hierarchy, at whose apex we have the focal condition of self-preservation'.

In a discussion of Sommerhoff's work, Waddington writes: 'Now the normal use of the word adaptation does not necessarily involve . . . modifications [appropriate to the particular environmental situation]. We might say that an animal with a thick fur is adapted to a cold climate without meaning to imply that the thickness of the fur would alter in relation to the intensity of the frosts to which it was subjected' (1957, p. 144). This quotation shows better than anything the shortcoming of the concept of 'adaptation', that it always involves a goal. From the point of view of adaptability, the animal in question possesses a means, the homoiothermic apparatus, which serves to cope with variations in a coenetic variable, viz., the environmental temperature. The essential difference between this animal and a less furry one is that in the former the permissible range of the coenetic variable has been displaced towards a lower part of the temperature scale.

As justly pointed out by Waddington (1957), directive correlation as defined above corresponds to *adaptability*, the capacity for adaptive response. I do not believe that Sommerhoff's notions, when applied to living organisms, go beyond the implications more or less intuitively invoked when other authors use 'adaptability', but it seems to be an advance that the concept is clearly defined, and in a way that excludes the suspicion that any kind of purpose is involved. For that reason the following discussion will be based on the ideas advanced by Sommerhoff. However, since 'adaptability' is more familiar than 'directive correlation', the former word will be used consistently from now on.

Since adaptability is concerned with responses to environmental changes, the properties of the environment must be incorporated in the definition of 'adaptability'. The following statements represent a tentative proposal as to how this may be achieved:

(1) A *niche* is a spatially continuous and circumscribed segment of the biosphere defined by a number of physical, chemical and biological parameters. The sum total of the defining parameters is called the *extension* of the niche, and the variation possible in each parameter is called its *range*. Summation of all parameters in the extension with proper regard to their range is called the *variability* (V) of the niche.

(2) The *absolute power of adaptability* (APA) of a given kind of living organisms is represented by the maximum niche variability compatible with their subsistence.

(3) The *relative power of adaptability* (RPA) of a given kind of living organisms is the ratio between their absolute power of adaptability and the variability of the niche they inhabit. Thus, $RPA = APA/V \geqq 1$.

(4) The *relative power of adaptability of a population* (RPAP) of living organisms is proportional to APA (or RPA) and to the number of individuals, at least for small values of this number.

The concepts defined here have the merit relative to 'fitness' that, in theory at least, if not necessarily in practice, it should be possible to estimate them quantitatively.

Among the differences obtaining between different kinds of animals, some are concerned with competitive success. At least three factors may contribute to this end, namely *individual survival*, *mating success* and *fecundity* (cf. Ayala, 1970). It may be justified to postulate that increases in the power of adaptability with respect to each of these factors will entail improvements enhancing the success of reproduction. This point is obvious as far as individual survival is concerned, but I believe that reflection will show it to be valid also in the other cases.

It will be noted that the apical focal condition of the hierarchy of adaptability represented by a living organism may well, as stated by Sommerhoff, be self-preservation, i.e. individual survival. In evolutionary contexts the relevant unit is not the individual, but the 'Mendelian population,' the community of potentially interbreeding individuals to which it belongs, and the reproductive success responsible for the progression and survival of the latter comprises other factors than individual survival. Thus, the natural history of Invertebrata encompasses numerous examples where the adult individual is sacrificed in the service of reproduction.

The circumstance that living beings live side by side does not imply that competition obtains; rather, it is necessary further to assume, with Darwin, that the available resources are limited, i.e. that the niche in which they subsist is *sated*. Hence the relation between competition, adaptability and selection may be stated in the following postulate or *rule of selection*:

When the populations P_i and P_j inhabit the same sated niche, and $RPAP_i \geq RPAP_j$, then selection will entail the progression of P_i and the regression of P_j.

Since the populations P_i and P_j may belong to the same or to different species the statement is valid for both intra- and interspecific competition. It follows that two populations with the same RPAP may survive in the same niche.

The essential aspect of phylogenetic evolution being the continuous replacement of less by more competitively successful forms, it is important that competition may lead to extinction of one of the combatants. The situation covered by the premise stated above clearly allows for this possibility; the two populations are supposed to be independent of each other and when the process of regression is protracted in time it must inevitably lead to the extinction of P_j. When organisms are dependent upon each other, as holds for the relation predator and prey, as well as between parasite and host, their interaction cannot lead to the extinction of one form without entailing that of the other also, as appears for instance through the analyses of Volterra (1931) and Gause (1931), which deal with various forms of interspecific competition. In a formal way it may be postulated that populations of this kind cannot compete because they occupy separate niches which only overlap marginally.

Intraspecific competition

Only slight differences in RPA may be presumed to obtain between members of the same species. This circumstance implies that intraspecific is in several respects quite different from interspecific competition. Thus, although 'extinction' in a certain sense may be said to occur, this is not a conspicuous feature, because the forms that succumb are so similar to those that survive that it may be impossible to tell the difference between them, and hence to establish that competition and selection take place. This point is typically demonstrated by the fact that when Darwin (1859) wanted to illustrate his thesis that the severest struggle for existence obtains between members of the same species, the examples he chose all represented competition between different species.

The essential aspect of intraspecific competition, as least as far as bisexually reproducing organisms are concerned, is covered by the neo-Darwinian theory according to which the various genotypes that will materialize under these conditions give rise to phenotypes differing with respect to fitness. Owing to their differential in reproductive success, as implied by 'fitness', the result of intraspecific competition will be a slow and gradual change in the gene pool in successive generations of a given species, positive genes increasing and negative decreasing in number. Under constant external conditions an optimum assemblage of genes may eventually be approached.

The corresponding phenotypical change is often described as 'adaptation'. This phenomenon need not involve an increase in the power of adaptability; thus, it is imaginable that it concerns the adjustment of the range of one or more of the parameters to better conformance with the corresponding variations occurring in the niche. The optimum genomic composition can, according to the theory, be changed only through modifications of the values of the selection coefficients, and this in turn requires alterations in the environmental conditions. The latter event therefore becomes the major source of evolutionary progress.

This process is not easily demonstrated because the effects of individual mutations usually are rather slight. The genes responsible for the synthesis of pigmented differentiation products form an exception to this rule, as evidenced by the so-called 'industrial melanism', observed in several species of European Lepidoptera, in which industrial pollution appears to have conferred a selective advantage upon a dominant melanic mutant, the conspicuity of which appears to be reduced under 'civilized' conditions compared to the light-coloured variant (cf. Sheppard, 1961).

Interspecific competition

Between animals differing significantly with respect to their power of adaptability, without therefore being too distantly related, competition is a matter of life and death, ensuing in the extinction of the inferior type. This situation thus rather corresponds to Darwin's Malthusian vision. Examples of interspecific competition have been recorded on many occasions, both palaeontologically and historically, the fates of various metatherian and eutherian orders in Australia and South America being probably the best-known cases. There is no doubt that adverse modifications of environmental conditions and interspecific competition are the primary factors responsible for all the palaeontologically registered cases of extinction.

The existence of sympatric species, i.e. closely related organisms living in contiguous niches which potentially may be inhabited by both species, may represent a case of interspecific competition not issuing in extinction because the relative powers of adaptability of the populations or the individuals are identical.

In the neo-Darwinian theory, interspecific competition is accounted for as 'niche deterioration'. Hence, invasion of the territory occupied by an animal or plant by a superior form is simply equated with the annihilation of the niche, a solution against which no formal criticism can be raised. It seems, however, that this phenomenon for other reasons is difficult to explain. Thus, if evolutionary progress is the outcome of selection from an inexhaustible pool of random mutations, then it must be anticipated that, in any environment, intraspecific competition would result in the establishment of the, under the given conditions, optimal phenotype or epiphenotype. The creation of a 'superior' form under different conditions is not to be excluded, but it is very hard to grasp how it can be superior when transferred to another environment, for if it is, then the particular combination of genes represented by this type should have been selected also in the indigenous population. I am well aware of the fact that this phenomenon can be explained by means of *ad hoc* hypotheses, but in my opinion the observed outcome of interspecific competition rather supports the notion that mutations, particularly those which lead to substantial evolutionary progress, are extremely infrequent phenomena.

ENVIRONMENT

In the discussion of the environmental effect upon ontogenesis in the introduction to this book it was proposed that for the successful outcome of epigenesis a certain specified milieu is a necessary condition. If variations beyond the stated limits occur, most likely the epigenetic processes will go astray, resulting in ontogenetic failure. It was admitted, however, that under certain exceptional conditions environmental changes possibly may lead to felicitous modifications of the epiphenotype.

As regards the influence of the environment upon phylogenesis it has already been intimated that according to neo-Darwinism this factor plays a decisive role in directing the course of evolution. The discussion of this question rightly belong to the conclusions of the theory and we may therefore confine ourselves to outlining the various ways in which the environment may, and must, have played a role in evolution.

First of all, the inorganic world has made available the physical space required for the enactment of the whole drama. But furthermore, as shown in the well-known book by Henderson (1913) on the 'fitness of the environment', many of the physical and chemical parameters are such that they were uniquely favourable for the origin and survival of living beings.

Secondly, the circumstance that the environment may exist in the gaseous, the fluid and the solid state, and that great variations obtain in a number of physical parameters, light, temperature, etc., has of course been of great consequence for the course of evolution because radically different properties are of importance for subsistence under each of these multifarious conditions. The plants, besides serving as food, have also contributed to the diversification of the environment of animals. It requires little imagination to envisage the kinds of creatures that would be absent if one or more of these various factors were missing.

Thirdly, owing to the fact that the space on the earth is bounded, there exists a limit to the number of organisms which can survive at any moment; this holds both for the planet as a whole and for smaller compartments, comprising local ecosystems and ecological niches. This fact is, as has already been discussed, the basic condition for the phenomenon of competition and hence of selection.

Many examples from contemporaneous natural history demonstrate that occasionally members of 'obsolete' taxa of animals and plants have managed to survive through avoidance of interspecific competition. These events are examples of the more general phenomenon of isolation, the realization of which is the fourth contribution of the environment to the furtherance of evolution. In the following pages an attempt will be made to show that for phylogenetic divergence isolation may be even more important than selection. The basis of isolation is partly geographical features like islands, lakes, etc., and partly the fact that very large variations occur in diverse physical parameters, a circumstance that has allowed for isolation through specialization for life under extreme conditions. With the advance of evolution even the animate environment has afforded possibilities for isolation.

Conclusions

The subject of evolution is, unfortunately, so vast and complex that it has been impossible so far to erect an axiomatic system that allows for the derivation of hard and fast conclusions pertaining to all aspects of this event. The conclusions which above were shown to follow from the Darwinian premises are stated in such a general form that they relate only to the gross aspects of the process and outcome of evolution. The specifications regarding the various premises which have been set forth do, however, carry certain implications which may be corroborated empirically. We shall, in the present context, discuss the six conclusions, isolation, selection, evolution, progress, divergence and direction, with regard to their consequences for the mechanism of evolution and, when possible, evaluate the support the conclusions command through confrontation with observational evidence. Since the phenomena of selection and progress are linked together as cause and effect they will be treated together in one section.

ISOLATION

The importance of isolation for taxonomic divergence has been recognized by many evolutionists, thus by Darwin (1859), although his stand was ambiguous, but also, to mention only a few outstanding examples, by Wright (1931; 1949), Cuénot (1936), Robson and Richards (1936), Dobzhansky (1937), J. S. Huxley (1940; 1943), Timofeeff-Ressovsky (1940), Mayr (1942; 1963), Simpson (1953a), Ludwig (1959) and Rensch (1960). From a molecular-biological point of view the significance has been stressed recently by Ohno, who claims that 'isolation is *conditio sine qua non* of speciation' (1970, p. 44). The primary concern of most, if not all, of the authors mentioned is the suppression of gene flow between two populations which is necessary if they are to evolve into separate species. This standpoint, which is based upon neo-Darwinian population genetics, thus asserts that isolation is a precondition for *speciation*.

In the present context the principal implication of the word 'isolation' is 'exemption from competition'. Clearly, this state of affairs is of little avail to populations with a higher RPAP than any other populations which actually inhabit or potentially may invade their niche: they are under any circumstances certain to progress or survive, as the case may be. For populations that do not fulfil this demand the situation is different: they can avoid extinction only through isolation; their progression requires that neither intra- nor interspecific competition obtains, whereas their survival is ensured in the absence of the latter. These two situations will be considered in the following sections.

Evolutionists distinguish three kinds of isolation, which may be called *geographical*, *ecological* and *reproductive*, respectively. These isolation mechanisms do not stand in one-to-one relationships to the three types of processes that may be affected by isolation.

Speciation

From the Darwinian point of view taxonomic divergence always consists of speciation, i.e. the breaking up of a species into two, of which at least one is a new species. According to the neo-Darwinian theory the critical step in this event is the cessation of gene interchange with the ancestral population and this is accomplished through isolation (cf.

Wright, 1949; Mayr, 1963). The latter author contends that in this process geographical isolation is the most important factor, and this seems plausible since ecological isolation may be difficult to achieve between members of populations that are closely related and since sexual isolation of certain individuals from the rest of the Mendelian population to which they belong surely reduces their chance of leaving offspring, no matter the virtues they may otherwise possess. It must be emphasized, though, that Mayr's generalization may have special applicability for large animals like birds and mammals; for small animals, as for instance insects, ecological isolation may still play an important role. Many cases of speciation—or at least subspeciation—which can be referred to natural geographical barriers, on isolated islands, in isolated lakes, valleys, etc., have been recorded by Mayr (1963).

According to the theory of Wright (1931) dealing with the genetics of small populations, isolation as such may, through genetic drift, serve to establish some divergence from the ancestral stock, at least during the first stages of independence. Even the acquisition of genetic incompatibility between the parental and offspring populations, usually employed as a species criterion, can be explained in this way, although hardly as a regular feature.

Subsequent divergence must, according to the neo-Darwinian theory, depend on differences in the respective environments. This disparity may entail that the phenotypic expression of certain mutant genes becomes of greater advantage, the resulting changes in their selection coefficients ensuing in their eventual fixation.

Many observations on animal distribution may be interpreted in this way, thus possibly that of the two subspecies of the crow, *Corvus corone corone* and *C. corone corvix* (Meise, 1928). The implication of this mechanism is that the animals can, and do, live in the new environment before they become subjected to changes, the latter being just improvements in the adjustment between organism and environment. In other words, isolation occurs *before* mutation, or at least before the selection of the latter.

Progression

From the preceding section it appears that neo-Darwinism requires isolation only for the purpose of taxonomic divergence; all other progress is supposed to occur in large interbreeding populations. As to the number of animals required in an individual case of isolation, opinions are divided. To Darwin (1859), and to biogeographers in general, who have to account for the colonization of isolated localities like oceanic islands, mountain lakes, etc., it has been natural to assume that if 'the environment afforded them a favourable opening, the propagation of the race would require either two individuals of different sex or a gravid female' (Matthew, 1915, p. 38). The 'favourable opening' might in some cases consist of fortuitous transportation to a new tract. This view may be contrasted with the population-genetical one: 'individuals are always members of a population and cannot singly form the basis for any evolutionary changes' (Bock and von Wahlert, 1963, p. 144).

Although neo-Darwinian in his outlook, Mayr (1942; 1963) has advocated the 'founder principle', essentially identical with the biogeographical one quoted above. The theory of Wright (1931) may also be said to endorse this alternative, and the same presumably holds for occasional observations made by breeders of animals and plants. Also from the molecular-biological point of view this contention seems acceptable, as evidenced from the following citation: 'Thus, only one individual in a population of hairy apes must have originally acquired a hairless trait. A species of hairless apes must have arisen from the descendants of this particular mutant as a result of intense inbreeding which established the homozygous state for hairlessness. Thus, the process of speciation usually requires the reproductive "isolation" of a minority from the majority' (Ohno, 1970, p. 55).

Since the mutations, which in the neo-Darwinian theory are responsible for phylogenetic advance, appear with regular frequency, isolation is not necessary for their fixation. According to the view advocated here, the crucial evolutionary innovations are the accomplishment of large or intermediate mutations. Their occurrence being unique events, isolation is an absolute necessity for their establishment (cf. Schindewolf, 1950; Whyte, 1965). This inference may be made intuitively, but follows also from the postulate that the RPAP is proportional to the number of individuals in a given population, and hence must be very low for populations consisting of one or a few individuals which are the bearers of a mutation involving a phylogenetic novelty.

Isolation may thus, through the elimination of 'interspecific' competition, ensure the progression and survival of a population of individuals characterized by the possession of properties differing from those of the population from which they stem. It is possible to distinguish two cases of isolation which differ in some important aspects. These, called *non-random* and *random isolation*, respectively, will be discussed in the following sections.

Non-random isolation

A mutation may confer upon an organism the faculty to colonize a new, hitherto inaccessible, ecological niche. It is seldom, presumably, that small, and even most intermediate, mutations warrant ecological isolation, but when this possibility materializes it allows for a very rapid establishment of new populations or taxa, owing to the high frequency of such mutations. This mechanism possibly may be responsible for the taxonomic subdivision of Darwin's finches, the Geospizinae (cf. Lack, 1947; 1949).

It seems evident that immediate contact with the new niche is a precondition for the usefulness of the new property, and thus also for its employment for the purpose of isolation. A mutation giving rise to webbed feet may be of great advantage to shore birds, but hardly to a prairie bird. Likewise, the limb reductions which have occurred in the quadruped forebears of marine mammals would certainly have been crippling deformities if these animals had not already invaded the coastal ranges of the sea. Under any conditions it is obvious that in this case the isolation must occur *after* mutation, a circumstance that may warrant its being designated as 'preadaptation', although apparently 'ecological isolation' may serve as well. It goes without saying that if 'adaptation' is an unfortunate word in evolutionary contexts, 'preadaptation' is still more so, being teleological to the second power.

In the literature of phylogenetic evolution much discussion has been centred around this concept which implies that, without further changes, 'an organism . . . is able to shift into a new habitat [or] a structure . . . can assume a new function without interference with the original function' (Mayr, 1963, p. 593). It may be presumed that in the latter case also the preadaptation entails invasion of a new niche. In point of fact it appears that preadaptation pure and simple, i.e. without isolation, is of no explanatory value for evolutionary divergence.

Thus, if it is imagined that a population of organisms are genetically preadapted to the extent that they, as envisaged by Mayr, can invade a new niche then this event will not entail divergence unless isolation obtains, preventing the continuous exchange of individuals and genes between the old and new habitat, an occurrence which surely must counteract the fixation of the modifications that might otherwise arise, partly as a change in the environmental conditions. Hence, the characteristic feature of preadaptation appears to be that it directly qualifies for the occurrence of isolation, a circumstance that surely justifies the latter's being designated 'non-random'.

Simpson (1953a) has outlined nine different cases of 'preadaptation' which, it seems, may be subsumed under two headings, one of which refers to the *genetic* 'preadaptation' discussed above. The other mechanism involves that intensification or modification of an existing feature may come to serve a new function. However, if this alteration is not caused by direct selection, it seems necessary to assume that its establishment is causally linked to some other property, which is selected because it is useful under the prevailing conditions. Many different mechanisms presumably may be included in this class of *epigenetic* 'preadaptations', but the best-known instance is found in the phenomenon of allometric growth, which implies that some particular feature, for instance antlers in various Ruminantia, may evolve in association with a general increase in body size.

In the present context may be mentioned the phenomenon of *specialization*, i.e. 'adaptation' to life under exceptional conditions, extreme temperatures, high altitudes, etc., or in a narrowly circumscribed ecological niche, as exemplified for instance by many parasites. In the former case it is possible that preadaptation and non-random isolation have been involved, in the latter instance it is most likely that isolation was initially random, and that the specialization has occurred later. Specialization is often, particularly among parasites, associated with a remarkable degeneration, involving a definite decrease in the absolute power of adaptability. According to neo-Darwinism these changes, to be selected, must entail advantages. However this may be, it is evident that in the new niche the environmental variability is much reduced so there is good reason to believe that no change occurs in the relative power of adaptability. In particular, it may be presumed that the faculties that are lost are such as do not contribute to the power of adaptability in the chosen environment.

That specialization, as suggested by Cope (1896) and by many others after him, is an evolutionary blind alley seems to follow from the fact that it permits the occurrence of losses, which may be irreversible, in the absolute power of adaptability.

Random isolation

Preadaptation, or ecological isolation, seems to be the only existing possibility to account for non-random isolation, and as isolation is a prerequisite for taxonomic divergence it follows that either preadaptation is always instrumental in isolation or else that random isolation may occur. Since, as discussed above, biogeographers find it necessary to invoke the latter mechanism in order to explain at least some aspects of the observable distribution of animals and plants it seems to be justified to submit that random isolation is indeed an evolutionary expedient, and one which may be of general occurrence and great importance.

Random isolation may involve isolation before mutation, as discussed above, but also isolation *of* a mutation. From a classical neo-Darwinian point of view this makes no great difference, but if more than small mutations are allowed for, then the possibility obtains that a mutant, embodying a disadvantageous gene, may progress and survive in isolation, being exempt from both intra- and interspecific competition initially, and from the latter kind later on when the new niche is sated.

Wright's neo-Darwinian theory, valid for small populations (1931), allows for 'random genetic drift'. This phenomenon 'has been misused in a way Wright himself never intended, namely, as a spurious "explanation" of evolutionary changes that seem to be devoid of adaptive significance, and therefore hard to explain by natural selection' (Dobzhansky, 1970, p. 231). As appears from this quotation, the survival of anything but the best in unacceptable to orthodox neo-Darwinism. The stand of Wright is evidenced from the following citation: 'The view that these species differences are largely due to mutations that were not only adaptive from the first but were carried directly to fixation by favourable selection is undoubtedly too simple, as indeed has probably been recognized by all who accept the general hypothesis of transformation. It makes the occurrence of adaptive mutations the limiting factor in evolution to an extent that severely restricts the possible rate of evolutionary change. There is evidence that evolution can proceed with great rapidity, given an adequate ecological opportunity' (Wright, 1949, p. 367). The importance of isolation for the preservation of disadvantageous features has also been stressed by Ludwig (1959).

A phenomenon which in principle can be classed as 'random isolation' may occur in connection with the geographical dispersion of a successful taxon. In this instance the expanse beyond the boundaries of already occupied territory is essentially an empty niche as long as it is not inhabited by an equally successful potential competitor; if and when this situation materializes the dispersion comes to a halt. While it occurs, however, intra- and interspecific competition is absent, thus allowing for the progression of new forms at the periphery of the 'niche'. Whether or not survival will ensue depends on the intrinsic properties of the 'mutant'; if they are not advantageous it can survive only in isolation.

This mechanism of isolation of a mutation, or *divergence through dispersion*, would be ideally suited for the inception of animal types which differ only trivially, or through the possession of some bizarre character, from the parental type. The disparities between certain species and subspecies are known to be so slight that, did not the neo-Darwinian equations disprove this assumption, one might be inclined to assert that the affiliated populations do not differ significantly with respect to ecological specialization. Should this notion nevertheless be correct in some instances, then the observed state of affairs may be explained if the aberrant animals have become randomly isolated during a process of dispersion. Under any circumstances the phenomenon is somewhat corroborated by the fact that the formation of new subspecies usually occurs at the margins of the area occupied by the parental species (cf. Mayr, 1963; Hennig, 1966).

Survival

Although it cannot be excluded that important advances, including transspecific evolution, can be accomplished through intraspecific competition, it is nonetheless certain that this situation does not obtain ubiquitously, as demonstrated by the survival of many relic forms, e.g. *Neopilina* and *Latimeria*, to mention two recent startling discoveries.

Clearly, evolution has long ago bypassed the level represented by these animals, and if they had not managed to escape interspecific competition by seeking refuge in the depths of the sea they would surely have suffered the fate of their congeners, extinction. These examples, as well as the history of Marsupialia in Australia and in the Americas, respectively, may serve to illustrate the importance of random isolation for the survival of less 'competitive successful' animals. Similar instances are known from the vegetal kingdom.

Even in this case non-random isolation may be of importance, namely in those cases where preadaptation has permitted the colonization of a niche which is inaccessible to subsequently evolved forms.

SELECTION AND PROGRESS

The conclusion that selection is instrumental in evolution follows from the premises that among the variations obtaining between living beings some pertain to their competitive success and that, under certain specified conditions, they are actually forced to participate in competition. As a consequence the competively successful organisms will increase numerically, i.e. be selected, at the cost of the less successful one. This process may be equated with evolutionary progress if the acquisition of properties ensuring competitive success is defined as 'progress'.

Since the existence of two kinds of competition, intra- and interspecific, respectively, has been postulated, it follows that there must be two corresponding kinds of selection. We shall deal with these phenomena and with the role of selection after an introductory discussion of certain *rules of selection*.

Selection rules

In the selection postulate stated above (p. 423) it was submitted that, when two populations with different RPAP are present in the same niche, selection will ensure that the one with the higher RPAP undergoes progression, the other regression. From this it follows that evolutionary progress is associated with increase in RPAP. Now this parameter is a function of the number of individuals in the population, and hence, in order to evaluate the change in APA resulting from selection, we may assume as the normal case that the ultimately victorious organism starts with the numerically smallest population. If, in spite of this handicap, it manages to establish itself, then we may infer that its RPA, and since the competing populations are in the same niche, even the APA is higher than that of the competitors. Hence we may state as the *rule of progressive evolution*:

Selection acts to maximize the absolute power of adaptability in living organisms.

As a corollary of this statement we can paraphrase the *rule of exclusion*, first enunciated, albeit in different terms, by Darwin (1859):

Two populations of living organisms, differing significantly with respect to their absolute power of adaptability, cannot coexist in the same ecological niche.

Since only members of related taxa can ever share a niche, it follows that a taxon must become extinguished unless its members are isolated from representatives of a kindred and more advanced form.

If two populations living in the same niche do not differ with respect to their APA, then clearly neither of them can suppress the other one. We may therefore formulate a *rule of coexistence* as follows:

Two populations of living organisms, differing insignificantly with respect to their absolute power of adaptability, may coexist in the same ecological niche.

In theory this rule may be applied to subgroups within a Mendelian population, to geographical subspecies or to sympatric species. In the first case it is valid only if selection has gone on long enough to establish equilibrium between the various genotypes realizable under the prevailing conditions. In the second case it is also possible to explain the coexistence if the hybrids are less 'fit' than their parent or if each subspecies is ecologically specialized to its particular niche. Observations supporting both of these alternatives have been made (cf. Mayr, 1963). In the case of sympatric species it is also possible to presume that they are ecologically isolated from each other. However, in all those cases where this explanation is unlikely, their occurrence together is accounted for by the principle of coexistence.

The rule of progressive evolution cannot claim to cover all aspects of this process since, according to the preceding discussion, various evolutionary modifications occurring in association with isolation evidently imply reduction in the APA. It seems that we must distinguish in this context between non-random and random isolation, respectively. Non-random isolation, which is represented by various cases of preadaptation and specialization, is often associated with losses in APA, but also with a reduction in the variability of the environment, hence the possibility exists that the RPA is increased. If the directive influence of selection is involved in this case it seems necessary to assume that this situation obtains, and we may hence tentatively state the *rule of directed evolution*, comprising both progressive evolution and specialization:

Selection acts to maximize the relative power of adaptability in living organisms.

Evolution occurring as a consequence of random isolation is not subject to any directive moderation. About this process we may submit the following *rule of undirected evolution*:

Nothing can be predicted about the changes in the power of adaptability occurring after random isolation.

Intraspecific selection

The outcome of intraspecific selection is a change in the relative frequency of various genes. This mechanism is particularly efficient in organisms with bisexual reproduction, inefficient when parthenogenesis and asexual reproduction obtain. Three different kinds of selection may be distinguished (Dobzhansky, 1970, p. 96):

(1) '[Normalizing selection] protects the adaptive norm by the elimination of harmful mutants, malformations, and weaknesses of various sorts'.

(2) 'Directional selection enhances the frequencies of some gene alleles or gene combinations, and depresses the frequencies of others. Usually it acts in response to environmental changes'.

(3) 'Balancing selection is a complex of selective processes that maintain, enhance, or regulate genetic variability, most of which is adaptively beneficial'.

It is important to emphasize that geneticists know what they talk about when discussing these kinds of selection since all of them have been demonstrated by empirical observations. Thus, there is no doubt that normalizing and directional selection under normal conditions always will ensure that the gene distribution in a given population is, or is approaching, the optimum under the given external condition and that changes in the latter will entail further selection resulting in phenotypes that are better adjusted to the new environment.

So far nobody questions the importance of intraspecific selection. The crucial point in this neo-Darwinian mechanism is how far this kind of selection can carry the process of evolution, whether, indeed, the infinite number of gene combinations and the, if not infinite, then at least very large, variability in the environmental parameters suffice to account for the variation of living forms, past and present. We shall return to this question in the section about the role of selection and later in this chapter.

Interspecific selection

It is typical of interspecific selection that it is associated neither with sex nor, usually, with 'tooth and claw'. If the latter situation prevails, a predator–prey relation obtains which cannot involve competition. Rather the competition is concerned with space, food and other necessities of life and the outcome is always that 'obsolete' organisms will become extinct, leaving the resources of the niche at the disposal of animals and plants possessing higher APA. Thus interspecific selection is always directional, never normalizing, and hence the only incontestable means of ensuring evolutionary progress.

As appears, extinction is a very important factor in evolution. This event may ensue also through adverse changes in the inanimate environment, and although not directed in this case, it may not be quite random since, presumably, the likelihood of surviving upheavals of this kind will increase with the APA. Also during the subsequent phases of reoccupation of devastated niches this parameter must be of importance for the progression and survival of individual populations.

Role of selection

The living world displays an almost infinite series of marvellous 'adaptations' to its physical environment. However, many organisms exhibit complications of structure and function which, to human perception at least, appear to be disadvantageous or, at most, neutral features. There may be a tendency to evaluate these two classes of characters differently. The successful employment of the organic substratum for the exertion of functions increasing the mastery of organisms over their environment easily evokes our awe for the resourcefulness of Nature, whereas the significance of the apparently less fortunate features may be ignored or explained away.

Yet, as stated above, it is a matter of course that to survive any organism must be adjusted to its physical environment, and the gradual improvement of the means of interaction with the external milieu is certainly a prediction of any theory postulating that progressive evolution is the outcome of selection of advantageous characters. The existence of more or less adverse structures and functions suggests that once survival was ensured, Nature permitted the persistence of many traits of this kind, and this fact is much more difficult to explain. Actually, contemplation of the prevailing state of affairs intimates that 'a major feature in understanding the evolution of adaptive form and function is the realization that this is essentially based on the survival of the least inadequate. A successful species is not one which is adapted to a particular environment but rather one which is least badly adapted to that environment. Evolution is based, not on superlatives, but on adequacies' (Fraser, 1967, p. 107).

However this may be, it is obvious that the very fact that the 'adaptation' is concerned with the environment shows that this factor somehow has been involved in the evolution of the animal organism. Yet, as regards the relative importance ascribed to the two participants, environment and organism, different opinions are advocated by Neo-Darwinism and by the complete theory.

The neo-Darwinian view has been forcefully propounded by Simpson (1944, p. 80), who states that it has been 'urged as disproof of the importance of selection that its role is purely negative, that only genetic mutation creates, selection prunes and chooses among the indefinitely large supply of creations. If this were entirely true, selection would still be a most essential factor in evolution, but it might be proper to rule out adaptation in nature except in the qualified sense of preadaptation. Only mutation supplies the materials of creation . . . , but in the theories of population genetics it is selection that is truly creative, building new organisms with these materials. In most natural populations the number of possible combinations of discrete hereditary units into individual zygotes is enormous, and only a small proportion of them can be realized. To the extent to which selection controls the frequencies of the different units in the population, it determines, within limits, which combinations will be realized and in what proportions (Fisher, 1930). It does not, then, simply accept or reject fixed types of organisms delivered to it by random action of the laws of heredity, as seems to be inherent in the cruder forms of the preadaptation theory of evolution. Selection also determines which among the millions of possible types of organisms will actually arise, and it is therefore a truly creative factor in evolution'.

It is obvious that, according to the statement quoted, the role of the organism is to present to the environment a random mixture of mutations, the effect of which, of course, is somehow determined by the nature of the organic substrate. The much more important function, namely the selection of the combinations of myriads of small, almost imperceptible variations (this adjective refers not only to the human eye, but also to the environment, as seen from the low values of the selection coefficients), is the burden of the inanimate and animate environment, for it is against its exigencies that each of the mutations is tested.

An example illustrating this mechanism has been worked out by Simpson (1953). Thus, on the basis of various assumptions, this author arrived at the estimate that for any character about 300 progressive steps were required for the evolution from *Hyracotherium* to *Equus*. This event was associated with an increase in the ectoloph length of the upper cheek teeth from about 8 to 400 mm, suggesting that in this instance each step would amount to a change of 0·1 mm. However, even in the smallest species the variability was about 3 mm, thus corresponding to 30 evolutionary steps. Since this variation presumably is genetically based, it would *a priori* be anticipated that whenever the environment favoured an increase in length, a rather rapid selective change ought to occur, confining the variation to its previous upper range, and that this state would be preserved, new mutations for increased ectoloph length being incorporated as soon as they arose. According to Simpson this did not happen; rather evolution proceeded in many small steps, as mentioned above. It must be submitted that an interpretation claiming that a variation corresponding to only 1/30th of the existing range can have selective value makes great demands on human imagination as well as on the discriminatory powers of natural selection.

A consequence of the neo-Darwinian view is that in a constant environment the gene frequency must approach an equilibrium, representing the optimum under the given conditions (genetic homeostasis, cf. Lerner, 1954; 1958). Under these circumstances only normalizing selection occurs, no innovations of form, pattern, function or behaviour being possible without preceding changes in the environment. Thus, in a certain sense it is the environment that is the driving force in evolution.

The theory propounded here does not contest the importance of intraspecific selection to the extent it has been demonstrated empirically, and even beyond this range. It does, however, assert that the large and decisive events in phylogenetic evolutions have their origin in random mutations occurring at an extremely low frequency. Owing to this fact they cannot be established by any population-genetical mechanisms, but only in isolation, random or non-random, as the case may be. In the latter case, represented by preadaptation or specialization, there is generally no possibility of returning to the original habitat, but if the randomly isolated organism embodies a gain in APA it may subsequently invade the niche containing the population from which it sprang, and other niches as well, causing the extinction of their inhabitants through interspecific competition.

The main contrast between the two theories is thus that the former proposes to explain the entire course of phylogenetic evolution with one, the latter with three different mechanisms. It may be objected that isolation is actually accepted by neo-Darwinism, but in this case it is invoked only as a logically necessary element for taxonomic divergence. According to the complete theory, then, neither selection or environment are the creative agents in evolution, except for the changes that may be achieved within a Mendelian population. The real Creator is indeed genetic— and epigenetic—mutation, operating with the proficient and indispensable assistance of isolation and interspecific competition. Among the two, the former is responsible for the constructive, but undirected, the latter for the destructive, but progressive, phase of evolution.

EVOLUTION

To paraphrase Croizat (1958), the organic world has arisen through a process of 'evolution of form and pattern, function and behaviour in space and time'. Hence, the evaluation of the course and outcome of this grandiose operation is the concern of comparative biology. The evolution of form and pattern is registered by morphologists and systematists, of function by physiologists and biochemists and of behaviour by ethologists, the evolution in space by biogeographers and palaeontologists and the evolution in time by the latter. The present section will be devoted mainly to a discussion of the evolution of form since this parameter primarily is the one involved in the major steps of evolution. A comprehensive discussion of the evolution in space and time is beyond the scope of the present book, and of my competence too, but since, in the present context, complete omission of these topics would be most unsatisfactory they will be dealt with in two brief sections.

Evolution of form

The present subject is best approached by dealing with the features which distinguish the largest groups of animals, this being, presumably, the only way to avoid becoming lost in trivial details. We shall therefore in this section discuss the properties used for superphyletic distinction and those characterizing the major phyletic archetypes and finally, on the basis of the epigenetic theory advanced above, we shall, in a few selected examples, outline the mechanisms that may be invoked in the ontogenetic, and hence in phylogenetic, changes of form.

Superphyletic distinctions

The animal kingdom has customarily been subdivided into Protozoa and Metazoa, and the latter in turn into Invertebrata and Vertebrate (Chordata), but since the first and the last of these taxa have been accorded phyletic rank, this distinction does not amount to a true superphyletic classification. For this purpose, several criteria have been employed, the most important being symmetry relations, the nature of the coelom, the relation of the blastopore to the mouth and anus and the number of germ layers. A fifth alternative is to use what may be called the 'grade' of the body plan.

All these charactcrs are seen to be related to morphological traits of a rather formal nature. There can be no doubt that the aims of this formanalysis are laudable, but some of the groupings thus established are partly overlapping, suggesting that the criteria adopted are unsatisfactory from a phylogenetic point of view. For this reason it may be justified to discuss the validity of this approach.

Symmetry relations

The distinction involving symmetry relations hinges upon the presence or absence of bilateral symmetry, implying that the Porifera, Cnidaria and Ctenophora, which have either radial or biradial symmetry, are set apart from all other animals. Anticipating the subsequent discussion, it may be pointed out that this has the consequence that the phyla exhibiting the gastrula grade of organization are separated, the Platyhelminthes being counted among the Bilateria. Without maintaining that the classification according to body plan is necessarily superior to the one based on symmetry, it yet seems appropriate to point out that bilateral symmetry may be arrived at in different ways. Thus, from our previous discussion it is possible to distinguish at least two mechanisms, viz., the cases where an apical polarity through a morphogenetic process is converted into a median polarity and those where a median polarity exerts its action through interference with cell division and cell diversification. The first of these mechanisms is found in Turbellaria and Echinodermata, and possibly in other taxa, the second in quite a few of the higher phyla, e.g. Annelida, Mollusca and Chordata, as well as in the Arthropoda, although the latter represents a special case. I would venture to suggest that no morphological feature is a 'good' character for classification, except when the mechanism responsible for its origination is the same in all cases. In other words, bilateral symmetry in Platyhelminthes and in the larvae of Echinodermata does not in itself warrant any conclusions about the kinship of these phyla with the others exhibiting bilateral symmetry.

Nature of coelom

Three different types of coelom are distinguished. If the primary body cavity is filled with mucopolysaccharide, we have what, using an unorthodox word, may be called an *acoelom*. If the cavity is filled with fluid, it is a *pseudocoelom*, unless its walls are covered by a layer of cells, in which case it is a true *coelom*. There may be a gradual transition between the pseudocoelom and the coelom, thus in one of the most important pseudocoelomate taxa, the Nematoda, a nucleated network of cells often cover the body wall and the digestive tract (cf. Hyman, 1951b). If these cells were flattened out to make contact, the class would presumably be placed among the coelomates.

For the purpose of our discussion it may be illuminating to consider first the conventional opinion regarding the origins of the coelom:

'(1) *Schizocoel*. In the teloblastic, derived teloblastic, and lamellar modes of mesoderm formation, the coelom arises as a split inside the mesodermic bands, plates or masses. The split expands until the mesoderm comes in contact with the body wall on one side and the gut wall on the other.

(2) *Enterocoel*. In the enterocoelous method of mesoderm formation, the mesodermal sacs evaginated from the archenteron become cut off and lie in the blastocoel. Their cavity is the coelom, their walls the mesoderm. These sacs expand until they touch the body and gut wall, and the end result is therefore the same as by the schizocoelous method.

(3) *Mesenchymal coelom*. The mesenchyme may rearrange itself so as to enclose a space which is then a true coelom since it is bounded on all sides by mesoderm' (Hyman, 1951a, p. 22).

Morphologically, the coelom is, in the last analysis, a cellular vesicle. As such, it may arise in two different ways, viz., as a spontaneous formation or as a topological deformation of a two-dimensional cell aggregate. In the latter case the presence of sf-cells in a morphogenetic space will be required. The former principle clearly corresponds to methods (1) and (3) outlined by Hyman, the latter to method (2). In the enterocoelous method is included the segmentation of the myotomes in the Chordata which, as I have suggested above, occurs according to a mechanism essentially identical with that taking place in the Annelida, and in both cases it is proposed to involve a spontaneous formation of vesicles. The only case I have observed in the literature which does not conform to either of these two methods is represented by the Chaetognatha, but the published illustrations of the embryology of this phyla are so schematized that it is very difficult to form a well-founded idea about the mechanism of coelom formation.

It has been suggested in an earlier chapter that formation of a coelom by the enterocoelous method (e.g. in Echinodermata) occurs by a mechanism which is identical with that responsible for gut diverticulation in Turbellaria and for the formation of the gastrovascular system in Cnidaria and Ctenophora. As a matter of fact, according to rules governing the morphogenetic mechanisms established here, a requirement for the execution of this event is that the blastocoele is filled with mucopolysaccharide. In other words, the formation of an enterocoelom presupposes that the blastocoele, temporarily at least, is an acoelom.

As far as the coelom is concerned, the only difference between pseudocoelomate and coelomate animals, excepting of course the cases where the additional phenomenon of segmentation is involved, is the presence of a peritoneum in the latter case. The function of the coelom proper is presumably independent of this covering, and the primary significance of the peritoneum may be to reveal the presence of the particular cell-type, presumably sf-cells, responsible for its existence. If, as suggested above, this very same cell-type is involved in the formation of blood vessels, it thus would follow that, as a general rule, blood vessels are a distinguishing feature of coelomate animals. On the whole, I believe this proposition is borne out.

On the basis of the preceding discussion it appears possible to distinguish two groups of animals both of which contain 'coelomate' and 'acoelomate' phyla. The first of these are those derived from a mesenchyme blastula, and it comprises the acoelomate phyla (e.g. Ctenophora and Platyhelminthes) and those with enterocoeloms (e.g. Echinodermata). The other group includes the pseudocoelomate phyla and those with spontaneous coelom formation. Among the latter the Mollusca are included, although the coelom usually is represented only by the pericardial cavity. However, as will be argued later, the basis for the particular morphological organization in this phylum is that most of the body cavity is filled with mucopolysaccharide, which appears to warrant its classification as acoelomate. Even in the other coelomate phyla it holds that the anterior part of the blastocoele contains mucopolysaccharide, this being a precondition for the morphogenetic processes involved in the formation of the head.

The acoelomate condition clearly is more primitive than the pseudocoelomate one, and the enterocoelom is more primitive than the schizocoelom. It is also apparent, however, that the classification on the basis of the nature of the coelom is most ambiguous. As regards the various notions concerning the phylogenetic significance of the coelom, as embodied in the enterocoele, gonocoele, nephrocoele and schizocoele theories, it seems that knowledge about the morphogenetic mechanisms involved tends to impair their validity.

Blastopore–mouth–anus relations

According to Grobben (1908), animals may be classed into two groups, Protostomia and Deuterostomia, according to whether the blastopore becomes mouth or anus, respectively.

From our discussion of morphogenetic processes it may be inferred that in most Deuterostomia the tip of the archenteron, after traversing the blastocoele, fuses with the ectoderm to form the mouth. Since the gastrulation, initiated by sf-cells, always begins at the dorsal–vegetal side of the embryo, it follows that the gastrulation must end at the ventral–animal side. As the nerve cells, generally formed near the animal pole, serve to establish the antero–posterior axis, the opening arising by the fusion must be the mouth.

This argument has to be slightly modified in the case of echinoderm larvae, in which the flagellated ventral bands take the place of the brain. In every case of gastrulation corresponding to this description it is a precondition that the cells are kept together by a hyaline membrane, in Invertebrata and Vertebrata alike. In the absence of this entity gastrulation can occur only by epiauxesis, in which case the blastopore is not defined with the same precision as in the former case. It may fuse completely and disappear, fuse in the middle, etc.

Even though the definition of the Protostomia is somewhat doubtful because of this ambiguity, the classification may still be valid. However, the formation of a hyaline membrane, which seems to be a necessary and sufficient condition for the particular orientation of the digestive tract, can hardly be considered a more important criterion than the presence or absence of a brain (cephalic ganglion). The absence of a brain in the larvae of Echinodermata would seem to indicate that this phylum is further away from the vertebrates than, for instance, Annelida and Mollusca. In these taxa the mechanism of gastrulation is epiauxetic, presumably because their cortical vesicles fail to rupture and thus to ensure the formation of a hyaline membrane, but they have, instead, head ganglia exhibiting similarities with a brain. The present argument does not intend to advocate a specific relationship with the mentioned invertebrate phyla and the vertebrates, but rather to stress that such relatively simple processes as the mechanism of gastrulation may be used for classificatory purposes only with great discrimination.

The consequence of accepting this classification is that the main taxa in Deuterostomia are the Chordata and the Echinodermata, the intermediate link between these being primarily some lowly worms like Chaetognatha and Hemichordata. Most other Invertebrata are classed as Protostomia. This astounding idea has been accepted almost unchallenged by more than two generations of zoologists.

It must be admitted that various cogent reasons may be advanced to explain the peculiar lack in the fossil record of the genetic relations thus indicated. Therefore, if the criterion used for this classification was precise and unambiguous, it would presumably be necessary to accept these relationships. However, as pointed out by Kerkut (1960), there are numerous cases where it is impossible to decide the fate of the blastopore, and there are other instances where direct observations contradict the established classification. It therefore appears justified to suggest that Grobben's criterion is not 'good' for classificatory purposes, and that it consequently may be of very little value for phylogenetic generalizations.

Germ layers

We have already discussed the 'germ layers' above and found that it is impossible to establish an unambiguous definition of this concept. There is therefore no reason to anticipate that a classification on the basis of the number of germ layers can lead to anything but confusion. This statement is well illustrated by the Cnidaria. In the class Hydrozoa the layer of mesogloe is so thin that, with some justification, these animals are classed as *diploblastic*. However, in the Anthozoa the mesogloe assumes substantial dimensions and, being apparently not too different from the mesenchyme–mesoderm in Ctenophora and Platyhelminthes, it has with reason been questioned whether this class should not be considered *triploblastic*. It is true that the mesogloe does not lend itself to the kind of morphogenetic activities observed in the mesoderm of the phyla mentioned, presumably as a consequence of differences in their mechanical properties, but this feature does not enter in the definition of the germ layers.

From the preceding discussion it appears that morphology may easily lead astray. This has probably happened with respect to the distinction between Protostomia and Deuterostomia and between diplobastic and triploblastic phyla. There seems to be more firm ground as regards the Bilateria and the classification based upon the nature of the body cavity, even though certain ambiguities exist. According to my unorthodox application of these criteria, two different groups may be distinguished in each case. These coincide for both types of classification, at least with respect to the phyla discussed here.

Grade of body plan

As far as the general body plan is concerned, I believe that it is possible to distinguish three types, which may be called the *gastrula*, the *tubula* and the *neurula grade*, respectively. The implications of these designations are obvious except for 'tubula', which is defined as representing a gastrula in which the endoderm has established two openings with the exterior. It is seen that the gastrula grade comprises Cnidaria, Ctenophora, Turbellaria and some other taxa among the lower Invertebrata. The Porifera occupy a unique position, but may be included in this group. The tubula grade encompasses most other Invertebrata, whereas the neurula grade contains only Cephalochordata and Vertebrata. Various parasites, which have no digestive tract, form a special group, which possibly might be called the *blastula* grade.

Although this classification probably has some phylogenetic significance, it is not suggested that each group is monophyletic; this is almost certain not to be the case. Nonetheless, the three classes do represent a graded advance in the complexity of the topological transformation of the embryogenetic substrate as it prevails in the form of a cellular vesicle at the end of blastulation. Being less ambiguous than the other classifications discussed above, it will be adopted in the present context.

Phyletic archetypes

The arguments advanced so far are, with few exceptions, time-honoured and well known. I do not believe that a proposition gains persuasiveness by simple reiteration, even if it is correct from the outset, so in order to strengthen the epigenetic approach to phylogenetic evolution it will be necessary to mobilize new evidence. I believe it is possible, on the basis of the ideas on epigenesis advocated here, to outline the mechanisms responsible for the establishment of the most important archetypes. If this is correct, then we shall be able to specify the nature of the mutations required for the appearance of new archetypes. With this knowledge available, it will be much easier to argue about the existence and importance of large mutations in phylogenesis.

The difficulties involved in accounting for morphogenetic processes are of various kinds. For one thing, these events are to a certain extent simply topological changes of cell aggregates, and for some peculiar reason many embryologists, including myself, have great difficulties in thinking in three dimensions. It is for this reason that in some very important cases, namely Amphibia and Cephalochordata as discussed above, the views held about the changes occurring during gastrulation have been topologically impossible. I think that, particularly in the former case, this circumstance has not only hampered our understanding of what really goes on during gastrulation, but also the interpretation of other experimental data.

Another difficulty is that it is imperative to keep track of morphogenetic processes in the whole embryo at once. I do not intend to maintain that morphogenetic processes in the head must necessarily influence those of the tail, but this possibility cannot be excluded. What I mean, however, is that when a certain morphogenetic agent, for instance collagenocytes, becomes available, then it may simultaneously instigate morphogenetic processes everywhere, including the extremes mentioned. From a superficial point of view, these processes may be very different, and it is important at any particular stage to survey all the activities taking place and try to see how much of the totality can be referred to one kind of activity or at least to one kind of morphogenetic agent.

It would seem that the most logical approach towards the solution of morphogenetic problems would be to establish for a given case all the morphogenetic elements present, as well as their mutual spatial relations, trying on this basis to predict the further course of ontogenesis. In case of variance between prediction and outcome, it would, of course, be necessary to change the premises. Evidently, this approach does not require detailed knowledge about the mechanisms involved in the generation of the various substrate elements, as long as we can take their specific contribution to morphogenesis for granted. I have no doubt that systematic application of this method could be successfully pursued far beyond the stages analysed above. Unquestionably, one would sooner or later arrive at a point where none of the already established mechanisms will suffice to account for further events. Very often, in such cases, it will be possible 'inductively' to envisage the nature and contribution of the elements responsible.

As a matter of fact, generalizations and hypotheses of this kind are not unusual at all; lots of them can be found in the literature, dealing with such phenomena as the influence of the nervous system on the outgrowth of limbs, the influence of skeletal growth on other morphogenetic events, etc. This is indeed very stimulating, but unfortunately no attempt has been made to unify these various ideas to account comprehensively for one specific case of ontogenesis.

The following discussion of the effect of mutations on morphogenesis will take a historical view and, beginning with a discussion of the first Metazoon, we shall deal with a number of subsequent highlights in the phylogenesis of animal morphology, within the framework of the subdivision of grades of body plan suggested above.

The first Metazoon

Metazoa are multicellular animals containing more than one kind of cells. On the basis of our preceding discussion of cell diversification it therefore follows that at least one cell transformation is required to account for the genesis of the first Metazoon. Among the three existing cell transformations, it seems safe to exclude the sc-transformation since colligocytes are not found in the phylum Porifera. Among the solocytes, the sf-cells are extremely motile, tending to scatter unless they become attached to colligocyte aggregates, or, as in Porifera, are imbedded in a copious matrix. We may thus also exclude the lf-transformation, since it would be too much to demand a well-developed differentiation pattern in the first Metazoon. Furthermore, it would hardly be possible to construct a viable organism with sl- and sf-cells. This leaves us with the nv-transformation, and the postulate that the first Metazoon consisted of nsl- and vsl-cells, amoebocytes and flagellocytes.

This proposal obviously gains credibility from the fact that the nv-transformation may be observed in present-day Protozoa, e.g. *Naegleria* (cf. Willmer, 1960). These two cell-types are solocytes and do not normally adhere together. In order to account for their coexistence it must therefore be presumed that some of the cells secreted a matrix of mucopolysaccharide exceeding that corresponding to the normal surface coat. From observations on various larvae it appears that this expedient might suffice to keep the cells together.

In order to develop true to metazoan standard, the animal should arise from a fertilized egg. The question of fertilization we may leave for the moment. We have already previously pointed out that the egg typically is a large cell, which can subsequently be divided into a number of somatic cells of normal size. Among the large cells which have been proposed to account for the origin of the Metazoa is the multinucleate ciliate (Hadži, 1963). This suggestion appears to be unacceptable, since no cilia are present in Metazoa except in colligocytes. Rather, it seems to be necessary to look for the ancestor among amoebocytes which transform spontaneously to flagellocytes, like, for instance, *Naegleria*.

The classification of *Naegleria* appears to be somewhat disputed; it may belong to the class Rhizopoda, order Amoebina, but more likely it belongs to the Myxomycetes (cf. Hyman, 1940). In view of the close relationship which in other respects is known to prevail between fungi and animals, this point of ambiguity is of little consequence in the present context. It is known that amoeboid cells, among Myxomycetes as well as Protozoa, tend to form syncytia, and this circumstance may account for the formation of the large cell that could serve as a presumptive egg. This phenomenon is actually observed in various lower phyla, where the egg is formed by one cell through engulfment of its peers. A principal difference between syncytia and eggs appears to be that only one nucleus is preserved in the latter.

With proper stimulation the egg may begin to divide, the blastomeres being kept together with extracellular mucopolysaccharide. For reasons outlined above, some of the mucopolysaccharides between the interior cells might absorb water, thus forming a blastocoele. By this means a normal polarity would come into existence and hence, according to the generalizations established above, the differential required for the nv-transformation. The two cell-types present would be distributed with the flagellocytes towards the exterior and the amoebocytes towards the interior. The establishment of this arrangement must formally, if not necessarily in the actual execution, correspond to a multipolar ingression of the 'endoderm' which, I submit, is the most primitive mechanism of 'gastrulation'. The organism established so far is a *Volvox*-like creature, a *Blastaea* which, owing to the possession of the external flagella, would be able to lead a pelagic existence.

A blastula-like stage is, quite apart from any consideration of recapitulation, a natural, if not a necessary, intermediate stage in the epigenesis of Metazoa. An organism of this kind is perfectly viable when autotrophic, as *Volvox*, but the latter, as well as the *Blastaea*, 'the serious dearth reveals, that both are quite incapable of taking solid meals!' (Garstang, 1951, p. 30). An opening in the body wall therefore was imperative before the first Metazoan could utilize the nutritional provisions represented by the various pelagic microorganisms which, we may presume, teemed in the primeval ocean.

The formation of a gastropore–mouth may not, in itself, be too difficult to account for; processes of this kind are found in certain primitive larvae. A more worrying problem is the location, which may be either random or at a predetermined point. The latter possibility implies one further constraint on the likelihood of the formation of the first Metazoon. However, it is possible that, supposing fertilization was involved, the entrance of the sperm established the apical polarity required for the determination.

With the formation of a mouth, the tension in the cellular layer is no longer uniform, and a deviation from the spherical shape involving a certain elongation may therefore be anticipated. In this way we have arrived at an organism in which division of labour obtains between two cell-types, the first condition for acknowledging an animal as a Metazoon, viz., the external locomotory cells, and the internal digestive cells. It is easily seen that this animal corresponds closely to Haeckel's *Gastraea* (1874), and it is therefore fully justified to apply this name to the first Metazoon, as envisioned here.

As far as reproduction is concerned, meiosis and syngamy is known among Protozoa, if not exactly in the groups suggested to be closest to the Metazoa. There is therefore no reason to believe that this point would be of special difficulty in *Gastraea*. It is seen that the two basic types of gametes, the amoeboid egg (nsl-cell) and the flagellated spermatozoan (vsl-cell), were already present.

The gastrula grade

Gastraea appears to be a perfectly viable organism, which may have persisted for eons. This animal may have been the point of departure for several evolutionary experiments, each of which involved that one or more diversification patterns were added to the modest repertoire present in *Gastraea*. We shall in the present and the following sections discuss some cases where the correlation between the introduction of a new diversification pattern and the origin of a new morphological archetype is very evident. It has already been pointed out that the gastrula grade comprises the phyla Cnidaria and Ctenophora and the class Turbellaria in phylum Platyhelminthes and that, with some justification, Porifera may also be included.

Porifera. It is obvious from our previous discussion of diversification patterns that any advance over and above that represented by *Gastraea* must involve the introduction of one or both of the remaining cell transformations. As appears from the absence of typical epithelia, colligocytes are completely absent in Porifera. This statement is not contradicted by the presence of the so-called 'epidermis' in certain sponges, for the cells involved in this structure are clearly of the flattened type encountered in mesothelia and elsewhere which, for a number of reasons, has been proposed to be sf-cells. Among the solocytes three types are distinct, the archaeocytes (nsl-cells), the so-called 'amoebocytes' of the mesenchyme (nsf-cells) and the flagellated choanocytes which presumably are vsl-cells. Originally the nsf-cells were undoubtedly myxocytes, but later differentiation patterns comprise several kinds of sclerocytes. The total absence of a nervous system, in spite of the possession of sf-cells, may be referred to the lack of colligocytes for, as we have seen, the basic feature in primitive nervous conduction is that sf-cells, suspended between two colligocyte aggregates, conduct messages in the form of electric potentials between the cell layers.

The unique feature of the Porifera, their very low structural organization, which has earned them a systematic position apart from all other Metazoa, may and must be related to the lack of colligocytes, for this cell class is found in the remaining animal phyla. Since apparently no animal exists possessing colligocytes but no filocytes, it follows that we may expect to find all cell orders, at least transiently, in the remaining phyla.

Turbellaria. Among recent evolutionists there seems to be a general tendency to consider the Turbellaria, and notably the Acoela, as being close to the origin of the Metazoa. If we compare a typical member of this order with *Gastraea*, we shall find that the main changes involve the substitution of the superficial flagellocytes with ciliated colligocytes and that the interior digestive cells, imbedded in a mucopolysaccharide matrix, form a syncytial mass occupying most of the space confined by the superficial cells. However, there are some other and very important changes. First of all, there is a nervous system, originating in the former animal pole. This circumstance indicates the presence of an apical polarity, and there is reason to believe that already at this stage the mechanism of cell division, through its influence upon the cleavage pattern, is involved in the determination and the distribution of the cell diversification patterns. Another feature worth noting is the typical flattening of the body which, initially at least, depends upon sf-cells spanning the original blastocoele. The establishment of the bilateral symmetry seems to be intimately associated with this feature, and presumably involves a transformation of an apical to a median polarity. The phenomenon is evidently very similar to that discussed in Echinoidea above.

It is seen that the evolution from *Gastraea* to the very simplified representative of Acoela sketched here involves at least three important steps, viz., the lf- and the sc-transformations plus a 'determinate' cleavage pattern. The advances most often observed in the higher orders of Turbellaria include two new filocyte differentiation patterns, collagenocytes forming a basement membrane, and myocytes, forming a well-organized subcutaneous muscular system, partly a consequence of the former innovation, and a real gut, owing to the transformation of the digestive cells into colligocytes. The two former changes will usually entail an elongation of the body. A typical feature in some of these orders is the elaborate branching of the gut (Figure 153). As we have already discussed, this may be related to the fact that the body cavity is an acoelom, permitting sf-cells to span between the body wall and the gut during the early embryonic stages.

Ctenophora. The biradial symmetry of Ctenophora may, as we have seen, be referred to the cleavage pattern established by the three first cleavage divisions. The following divisions, through which the micromeres are given off, give rise to two different kinds of cells, epithelial cells and comb plates. It was suggested above that this very peculiar cleavage pattern, and hence a number of morphological features of the Ctenophora, are causally related to the inability of the astral microtubules to leave the ectoplasm and penetrate the interior of the egg. Through the gradual spreading of the micromeres over the macromeres, the cleavage pattern determines an apical polarity (the presumptive oral–aboral axis), but the latter apparently does not influence the distribution of cell diversification patterns along this axis.

It was proposed that in Turbellaria sf-cells are initially responsible for the flattening of the body. A similar flattening does occur in some Ctenophora (Beroïda). Whether or not this takes place may be decided by the temporal relation between the dynamic and the chemical activities in the blastocoele. If the swelling begins at an early stage, the typical cydippid form is likely to ensue, but if sf-cells in sufficient number are attached to the

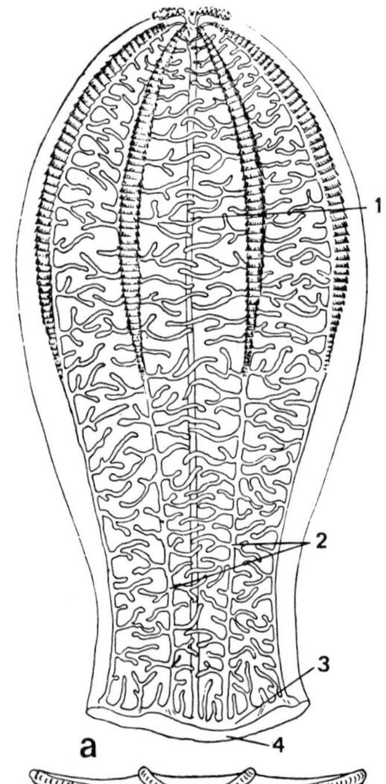

Figure 256. Diverticulation of the gastrovascular system in Beroë. (a), lateral view; (b), aboral view. 1, pharyngeal canals; 2, meridional canals; 3, canal along mouth rim; 4, mouth. (Reproduced from L. H. Hyman (1940), *The Invertebrates*, Vol. 1, by permission of McGraw-Hill, New York)

body wall before the water uptake becomes extensive, then they may prevent the assumption of the otherwise predictable globular shape.

The shaping of the digestive tract in Cydippida has been discussed above. Here it may be pointed out that if flattening occurs, as in Beroïda, or where other special conditions allow for it, as in Platyctenea, a very extensive branching of the gastrovascular system may occur (Figure 256). From what we know, it is not difficult to account for this feature through the activity of sf-cells, working in the mucopolysaccharide-filled body cavity. This and other similarities between the orders mentioned and various Platyhelminthes must depend primarily on the presence of the same substrate elements. Since the latter are very widespread, the resemblances hardly warrant too far-reaching inferences concerning the phylogenetic affinities between the two phyla.

Cnidaria. Oncotically active mucopolysaccharides, and thus myxoblasts, play an evident role in the morphogenesis of the tentacles in both medusae and polyps, and in that of the umbrella in the former. Traces of their activity are seen in the gastrovascular system, which in certain cases may be very similar to that found in some Polycladida (Figure 257).

The outstanding feature of this phylum is nonetheless that the cells overwhelmingly

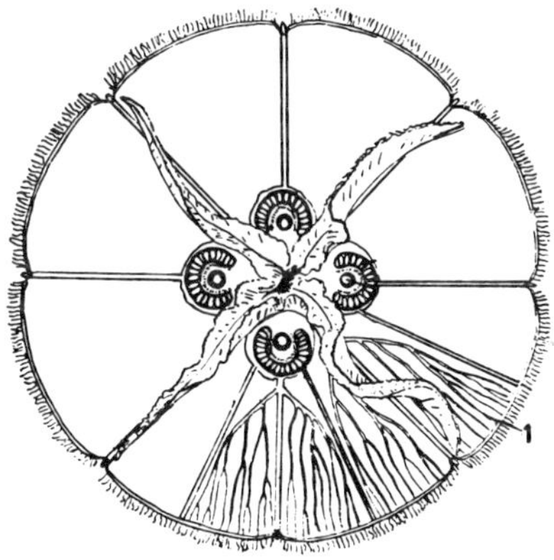

Figure 257. Diverticulation of the gastrovascular canals in *Aurelia*. 1, gastrovascular canals. (Reproduced from L. H. Hyman (1940), *The Invertebrates*, Vol. 1, by permission of McGraw-Hill, New York)

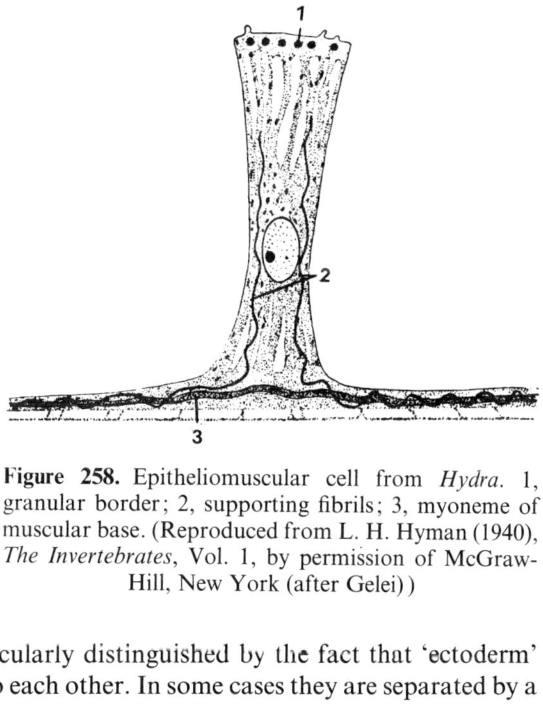

Figure 258. Epitheliomuscular cell from *Hydra*. 1, granular border; 2, supporting fibrils; 3, myoneme of muscular base. (Reproduced from L. H. Hyman (1940), *The Invertebrates*, Vol. 1, by permission of McGraw-Hill, New York (after Gelei))

are colligocytes. The polyps are particularly distinguished by the fact that 'ectoderm' and 'endoderm' are lined up parallel to each other. In some cases they are separated by a substantial layer of mesogloe, elaborated by sf-cells of the myxocyte and the sclerocyte families. This unusual dominance of colligocytes may be reflected by the presence of epitheliomuscular cells (Figure 258), even if the more usual type of muscle cell is also found in this phylum.

The possession of nematocysts may presumably be taken as a phyletic character and this, together with the various other substrate elements required for the formation of even the lowest member of the phylum, suggests that in spite of the primitive body plan the Cnidaria are highly developed animals, far removed from the origin of the Metazoa. On the basis of the principle of recapitulation, much speculation has centred around the significance of the planula larvae. However this may be, the evolution of Cnidaria must have passed through many steps, among them possibly a pelagic stage, histologically much simpler than present-day forms but maybe similar to the ephyra larva.

The tubula grade

The animals possessing the tubula body plan comprise the remaining invertebrate phyla, with some few exceptions represented primarily by parasites. It would be completely impossible to discuss all these groups with respect to the substrate elements which are involved in the elaboration of their bodies, and we shall therefore confine the attention to certain selected taxa, Echinodermata, Nematoda, Annelida, Mollusca and Arthropoda.

Before this discussion is entered upon, it may be pointed out that there is no causal relation between the mechanism of gastrulation and the grade of the body plan. This may be inferred from the fact that among the animals discussed in the preceding section gastrulation occurs by multipolar and unipolar ingression and epiauxesis as well as by invagination. Also, in the present group different mechanisms are encountered; suffice it to mention that in Echinodermata gastrulation occurs by invagination, and in Annelida and Mollusca typically by epiauxesis.

Echinodermata. From the above discussion of the process of gastrulation and the formation of the water–vascular system and the coeloms in the Echinodermata it follows that at this stage of development the embryo is clearly acoelomate, this being a precondition for the particular morphogenetic processes going on. There is reason to believe that this situation prevails for some time, since for instance the formation of the respiratory tree in Holothuroidea must require the activity of sf-cells working in a morphogenetic space. However, in the adult animal the walls of the body cavity are covered by an epithelium composed mostly of flattened cells. From the rules established earlier this situation cannot obtain if the cavity is filled with mesenchyme, and it may therefore be presumed that the production of mucopolysaccharide ceases at some stage.

We have in an earlier chapter outlined the mechanisms involved in the formation of the pluteus larva, and the larval morphogenesis is closely similar in the other classes, except that a calcified skeleton is found only in Ophiuroidea.

Nematoda. With the advent of the sc-transformation ciliated colligocytes became available to the Metazoa. These cells are very useful for locomotion. It is likely that in the earliest animals possessing ciliated cells, the latter were used for swimming (e.g. Acoela), but when the animals reach beyond a certain size, the force developed by the cilia can no longer keep the animal suspended in the water, and the cilia can be used only for creeping on a solid substrate, as is observed in Tricladida and Polycladida.

The locomotory efficiency of cilia is proportional to the ciliated area and thus, presumably, to the square of the linear dimensions of the animal, whereas the mass to be moved about varies with the cube. The imbalance resulting from increase in size may to some extent, as pointed out by Clark (1964), be circumvented if the body is flattened, since in that case both mass and surface area will increase approximately in proportion to the square of the linear dimension. This circumstance may partly explain the particular success of ciliary movement in even relatively large Turbellaria, especially since the flattening of the body also promotes the gaseous exchange.

However, while granting the usefulness of the compression of the body, I believe it is necessary to be explicit about the cause of its existence. The Turbellaria are not flattened 'partly as a locomotory adaptation, and partly due to their lack of a blood vascular system and the slow rate of diffusion of oxygen through the tissues, making it necessary for all organs to be close to the surface of the body' (Clark, 1964, p. 44), but because, during their ontogenesis, conditions prevail which allow for this flattening. That this body form has been preserved is, of course, to some unknown extent due to the ensuing advantages, as listed in the above quotation.

With further increase in size, locomotion by means of cilia becomes inefficient even in the flatworms, and in many of the large Turbellaria movement is accomplished by pedal locomotory waves, established by rhythmical 'contraction' and relaxation of the ventral longitudinal muscles of the body cavity, and the other muscles of the body wall are involved in maintaining the turgor pressure required for elongation of the relaxed longitudinal muscles. In this type of movement a certain, although slight, transmission of fluid pressure from one end of the body to the other is required, and it seems that in spite of its viscosity this function may be exerted by the mesenchyme in the body cavity of Turbellaria. For more efficient movement, involving the subcutaneous muscle layers, the acoelomate condition is a disadvantage. Thus, loss of the mucopolysaccharide in the body cavity clearly will improve the dynamic properties.

Many animals, pseudocoelomate and coelomate alike, have made use of the body cavity as a hydrostatic skeleton for locomotory purposes. In Nematoda, where only longitudinal muscles are present in the body wall, the movements are largely confined to flexions of the body in the dorso–ventral plane, resulting from antagonizing dorsal and ventral muscles (cf. Clark, 1964).

The present discussion of the correlation between the properties of the coelom and the power of motility shows that one mutation, involving the loss of some extracellular element, may have far-reaching consequences. It can be argued that the transition from acoelom to pseudocoelom does not entail the transformation of a turbellarian to a nematoid worm. Without contending the opposite, I venture to submit that a quite striking approximation may occur. Thus, with no mucopolysaccharide in the body cavity, no parenchymal muscle fibres can occur, and consequently the body form will be determined only through the interaction between the pressure of the hydrostatic skeleton, the basement membrane and the body wall musculature. Granting a growth process entailing elongation of the body, the typical spindle-formed body of Nematoda must ensue from this cooperation. It will be realized that, on the premises outlined, no diverticulation of the gut can take place.

Annelida. Many annelid worms utilize the hydrostatic skeleton for locomotion. In this context the segmentation of the body cavity by means of septa is of great advantage, because it prevents the transmission of locally established pressures to other parts of the body. The chaetae present in these animals serve to increase the friction to the substrate.

We have already discussed the mechanism of segmentation in the Annelida. In most invertebrates the presumptive muscle cells migrate freely as sf-cells, and settle at the inner surface of the body wall. In the segmented worms, however, they accumulate at the posterior end of the embryonic or larval body, where they transform into cf-cells according to a temporal gradient which ensures that successive batches of cells form a vesicle. Between the vesicles septa are formed. It is primarily, but not exclusively, the cells in contact with the outer body wall which become myocytes, and it is therefore likely that the contact with the collagenous basement membrane is of importance for their differentiation. The muscle cells in each segment are arranged in longitudinal and circular layers.

The loss of mesenchyme in the primary body cavity, although of some value for the development of a hydrostatic skeleton, may be a handicap insofar as it reduces the possibilities for morphogenetic activities. It is interesting to observe that in all higher animals at least part of the body preserves the acoelomate condition. Most often the region concerned is the one close to the original animal pole which, as we have seen, is distinguished by a copious production of neurocytes and myxoblasts. The latter, through their differential activity, give rise to a swelling of the anterior end of the body, which in turn allows for the aggregation of the neurocytes into ganglia. If present, myocytes may form muscles, sclerocytes cartilage, etc. In the absence of a basement membrane, the myxoblasts may engage in the formation of tentacles. A very clear demonstration of the acoelomate condition in the anterior region of an annelid larva is shown in Figure 259.

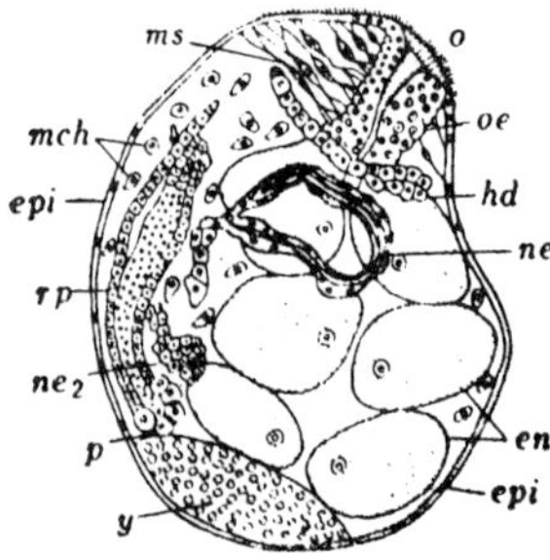

Figure 259. Longitudinal section of *Nephelis* larva. end, endoderm; epi, epithelium; hd, head blastema; mch, mesenchyme cells; ms, muscle cells; ne_1, primary archinephridium; ne_2, secondary archinephridium; o, mouth; oe, oesophagus; p, teloblast of 'ectodermal' band; rp, trunk blastema; y, yolk cell. (Reproduced from K. Okada in *Invertebrate Embryology* (1968), after Bergh, by permission of Nolit, Belgrade and National Library of Medicine, Bethesda)

It was suggested above that the absence of mucopolysaccharide in the blastocoele is a precondition for the formation of the hydrostatic skeleton used for locomotion by many worm-like animals. However, from the rules established it follows that the pseudocoelomate and the coelomate conditions imply that no real muscles can be formed, since all myocytes must settle on the surfaces of colligocyte aggregates. Originally, movements by means of a segmented hydrostatic skeleton undoubtedly constituted an enormous advance, but the extensive changes of the body wall which it requires probably involve many disadvantages. The locomotion by means of parapodia acquired by the Polychaeta may therefore represent a mechanical and energetic gain (cf. Clark, 1964). The parapodia are moved by true muscles, and to allow for their formation it may consequently be presumed that, during some formative period, an acoelomate condition prevails even in the trunk region.

Mollusca. In Mollusca the only coelomic cavity is that enclosed by the pericardium. Particularly characteristic is the fact that myxoblasts are present at the posterior ventral side in the early embryo where they give rise to the swelling that becomes the foot. From a locomotory point of view the foot is not a very successful expedient, movement generally being confined to the pedal waves otherwise typical of acoelomate animals. The morphogenetic space of the body cavity does, however, allow for the formation of muscles (e.g. pedal and adductor muscles); the latter may in Bivalvia be used for locomotion.

The peculiarities of the Mollusca may to some extent be accounted for by the fact that the acoelomate condition, elsewhere confined to the most anterior part of the body, in them characterizes most of it. It is therefore only a natural consequence that, as their name indicates, in Cephalopoda the feet (or arms) are directly attached to the head.

The distinguishing feature in molluscan systematics is the presence or absence of shells, and in the former case their number and form. It has already been suggested that the difference between the Monoplacophora and Polyplacophora and the remaining classes, except Aplacophora, appears to be that in the former the formation of the shell gland is relatively delayed, thus allowing the larva to flatten out on the substrate before the shell is formed. The difference between these two classes may be accounted for by a mechanism of segmentation of the shell gland as outlined above.

In the other classes the shell gland, when present, is subjected to differentiation at a very early larval stage, thereby ensuring its influence on the body form of the adult. We shall not enter upon a discussion of this point here, but it may be noted that apparently the literature does not contain a satisfactory description of the morphogenetic mechanism responsible for the subdivision of the shell gland in Bivalvia.

Arthropoda. The Arthropoda evidently belong to the tubula grade. Yet, generally this body plan is realized only at a very late developmental stage, and not in immediate association with the process of gastrulation as is usual in most other phyla of this group. This phenomenon is clearly associated with the fact that epigenesis in the Arthropoda is so radically different from anything found elsewhere in Nature.

The most characteristic feature of the development of the relatively large eggs is that, as a consequence of the superficial cleavage, all morphogenetic activity takes place at, or from, the surface. This circumstance is the more puzzling when it is realized that, as we have seen above, the yolk-filled interior is a potential morphogenetic space. The greater part of the typical features of the external form, shape in general and segmentation in particular, is clearly closely associated with the diversification of the ectoderm, partly involving mechanisms outlined earlier.

From our survey of locomotory mechanisms it appears that the establishment of powerful means of propulsion, especially on land, has been a very important challenge for the early life on our planet. It was observed that, so far, the best solution involves external appendages, moved by real muscles. With the acquisition of an exoskeleton of suitable mechanical properties the Arthropoda made the first important step towards the generation of the segmented limbs which have been so instrumental in their success.

The neurula grade

This group comprises the subphyla Cephalochordata and Vertebrata, and is characterized by the presence of a chorda, at least during some phase of development. The formation of a chorda involves the participation of several morphogenetic factors, but when the prerequisites are analysed in detail, they turn out to be far less exacting than might be anticipated. Therefore, even if the establishment of the proper conditions, i.e. the simultaneous presence of certain substrate elements in a certain part of the body, may be difficult to achieve, there is no *a priori* reason to presume that, as a phylogenetic innovation, it is a unique event. It may have occurred only once, as is generally conceded, but in my opinion we should not hastily dismiss the possibility that it has occurred independently in the two taxa mentioned above.

The Chordata exhibit a number of other characters in common, elongated body, mesoderm segmentation, gill slits, etc. However, if, as contended here, each developmental stage is a necessary and sufficient condition for the following one, one may suspect that the similarities mentioned are causally related to the presence of a notochord. As a matter of fact, the elongation of the body certainly is thus conditioned, and since the process of segmentation purportedly is based upon a gradient in the transformation of the mesodermal sl-cells, there can be no doubt that elongation of the body, and thus of the particular cell conglomerations, favours the establishment of such a gradient. The formation of gill slits is also a type of segmentation process, accomplished by sf-cells present in the mesenchyme substratum enclosed between the ectodermal and endodermal aggregates. There is good reason to presume that the body elongation may be somehow involved in the establishment of a linear series of slits.

If this view is correct, then it is entirely unwarranted to accord any particular significance to most of the morphological similarities between the two groups of chordates, since the source of the former may be exclusively referred to the elaboration of a notochord. This circumstance must obviously enhance the likelihood of a separate origin of the various groups of Chordata. Admittedly, the appearance in the two taxa, subsequent to notochord formation, of a number of other similar features does, of course, imply certain further prerequisites but, as intimated, these may be relatively simple. The notochord clearly warrants the inclusion of the Cephalochordata and the Vertebrata in the neurula grade but, unless several other weighty arguments can be mobilized in favour of a close relationship between Cephalochordata and Vertebrata, there may be some justification for the supposition that these two taxa have arisen independently from invertebrate ancestors, and consequently should be ranked as separate phyla.

This contention does away with certain difficulties involved in accounting for various 'invertebrate' features in the Cephalochordata. It is usual to explain these characters as marks of 'degeneration', i.e. loss in the power of adaptability, but even if the existence of this phenomenon must be granted, it nonetheless seems dangerous to use this 'anti-evolutionary' concept as a general expedient to account for observations which do not fit into a preconceived pattern.

The Urochordata have not been mentioned in the present discussion. The reason for this is that if a notochord is defined as a linear aggregate of cells, surrounded by a collagenous sheath, then this structure is conspicuously absent in this subphylum of the Chordata. To me it appears that the shape of the larvae in this taxon is one of the caprices of Nature, and that the phylogenetic hypothesis based on this feature is probably completely unfounded.

Mechanisms

In the preceding pages some distinctive features of a number of phyletic archetypes have been outlined. The morphogenetic preconditions for their establishment are mostly relatively simple epigenetic events, involving presence or absence of mucopolysaccharide in the body cavity, non-adhesive or adhesive muscle cells, presence or absence of a transformation gradient in the latter case, formation of a notochord, etc. As far as I can see, all of these features are characterized by working in an all-or-none fashion. Therefore, although the effects of the mutations responsible for the transition from one archetype to another may be trivial on the chemical level, they have had a very profound effect upon the epigenetic outcome of the morphogenetic processes. Mutations of this kind, involving for instance suppression, confinement or retardation of myxoblast activity or of the sc-transformation of myocytes, may thus indeed be characterized as large mutations.

The following steps in ontogenesis can be described as modifications of the phyletic archetype, which characteristically vary from class to class, and from order to order within each phylum. In the present section we shall try to outline a few of the most important mechanisms and principles which have been involved in the further modifications of the archetypes, with reference to each of the three epigenetic mechanisms.

Cell division

The size of the metazoan body is ultimately determined by the number and the size of the individual cells, even if allowance must be made for the volume occupied by extracellular elements. With certain remarkable exceptions, some of which we shall discuss presently, the cell size is relatively constant throughout the animal kingdom.

It appears that the cell volume bears a constant relation to the amount of DNA in the nucleus, as borne out particularly by observations on experimentally produced amphibian polyploids (cf. Fankhauser, 1955). It is possible that most of the variation in cell size observable in the animal kingdom may be related to changes in the DNA content of the genome. This point has not been established, but a correlation between the amount of DNA per cell and phylogenetic evolution has been demonstrated in the vertebrates (Goin and Goin, 1968; Hinegardner, 1968). Very little evidence is available concerning the phylogenetic importance of variations in cell volume, but a very interesting start has been made (cf. Szarski, 1968).

For evolutionary increases in body size only two roads are open, enlargement of cell size or increase in cell number. For reasons already mentioned, the former expedient does not seem to be generally accessible, but it is realized in the Aschelminthes. In some classes of this phylum, cell division (except for the reproductive cells) comes to an end during larval development, and all further growth involves increase in the volume of the cells, which as a consequence may become very large. It seems likely that this mechanism is of questionable value for phylogenetic progress, as evidenced for instance by the fact that it is not encountered in the most developed phyla.

Since differentiated cells are distinguished by the loss of their power of propagation, the other possibility implies that a reservoir of undifferentiated cells must be available as long as growth goes on. There is no doubt that this expedient is exploited in all cases where the adults reach more than microscopic size. It is likely that several different mechanisms are involved; presumably the lymph glands and the thymus are suppliers of undifferentiated cells in Vertebrata. In Mammalia the cessation of growth is, of course, mainly a question of ossification of the epiphyses, preventing further increase in body length. However, the increase in cell number of other organs and tissues also ceases, suggesting that the supply of cells is controlled by a regulatory mechanism. As is well known, the latter is mainly endocrinological, the most important participants presumably being the pituitary growth hormone and the sex hormones.

The fact that the cell number is limited may have an interesting consequence in the rule of 'compensation of the body material', dating back to Goethe and Geoffroy St. Hilaire. It was stated by the latter in 'a "loi de balancement", according to which the strongly growing parts of an organism consume so much body material that the less strongly growing parts remain small or become more or less reduced' (Rensch, 1960, p. 180). Doubts have been raised concerning the validity of this rule when applied as a generalization in comparative anatomy. These may be justified, since a number of compensating mechanisms are possible. However, it appears that various experimental observations lend rather comfortable support to the rule (cf. Rensch, 1960). Its application to the wings of a bat leads to the prediction that the change in one parameter, viz., the rate of longitudinal growth, will ensue in the formation of digits that are not only very long, but also very delicate.

Since cell size is a relatively constant parameter, it follows that small organisms are composed of a very limited number of cells. This has the consequence that the number of cells available for elaboration of a structural entity present in larger animals of the same taxon may not suffice for the purpose, with the result that the structure in question becomes vestigial or suppressed. As an example we may mention the Archiannelida, which have been placed in a separate class apart from the Polychaeta because, in spite of many similarities, a number of morphological details present in the latter are absent in the former. However, except for *Polygordius*, these animals are all minute (0·25–15 mm), and Remane (1932) therefore proposed that differences between the Archiannelida and the Polychaeta were to be referred primarily to the size difference.

As regards the Gastrotricha, which are still smaller (0·06–1·5 mm), Remane (1932) pointed out that a number of features suggest these to be related to the Annelida. The absence of segmentation and of a coelom in the former would thus, according to this approach, be due to the fact that neither space nor cell number allow for the morphogenetic mechanism of segmentation. Without vouching for the classification suggested by Remane, I think that the reasoning is based on very sound principles.

The lower size limit of a given archetype may thus be determined by the space and the material available for morphogenesis. However, the range of permissible body sizes is often determined by physical factors. In Invertebrata the acquisition of blood circulation was the first requirement for growth beyond a certain size, except for those cases where the body is filamentous or greatly flattened. Circulation, together with branchial respiration, has permitted the creation of animals of considerable size among Crustacea and Mollusca. The trachael respiration in Insecta has severely limited body enlargement in this class. The difference in size between the largest marine homoiothermic and poikilothermic vertebrates demonstrates the superiority of lung respiration. In terrestrial animals increase in size is mainly limited for mechanical reasons. On the other hand, the minimum size in homoiothermic animals (Aves and Mammalia) is largely a matter determined by the temperature regulation. Poikilothermic vertebrates (e.g. Pisces) may be much smaller, but they never reach the size of small invertebrates. Clearly, construction of the complex vertebrate body entails a relatively large minimum size.

Cell diversification

The embryogenetic substrates arising through cell diversification are important agents of epigenesis and thus of phylogenesis. In the present section we are going to deal partly with mechanisms serving to control the processes of cell diversification, and partly with cell differentiation patterns which directly control morphogenesis.

Regulation of cell diversification. In the most primitive animals the number of diversification patterns is very limited. The first expedient employed by Nature to secure an increase of cell-differentiation patterns in the embryo was to establish a strictly controlled determinate cleavage pattern. I hasten to emphasize that this mechanism involves changes both in the genome corresponding to the new diversification patterns and in the mechanisms of nucleo-cytoplasmic interaction. However, even if this contrivance has been turned to good use, its exploitation is severely circumscribed, since clearly there is a limit to the number of cytoplasmic states which can be realized through unequal divisions.

Two further developments have improved upon this situation, the first one being the mechanism of induction, through which the assumption of a diversification pattern can be communicated from one cell to another. This means that a certain cell does not need to establish in its own cytoplasm the conditions required for the occurrence of a certain cell diversification, and by this expedient the number of realizable differentiation patterns undoubtedly has been increased. The other mechanism is that of dynamic determination of differentiation patterns, rather than the static one involved in determinate cleavage. This may not necessarily increase the number of differentiation patterns, but it increases the possibilities for variation in their temporal and spatial distribution, the consequences of which are beautifully demonstrated in the development of Amphibia.

The mechanisms discussed in this book cannot account for all the differentiation patterns realized in the higher animals. We can envisage that they arise through recurrent ramification from the original selfdifferentiation patterns, but we do not have any idea about the agents responsible for the branching. It may be surmised that in certain cases induction from neighbouring cells occurs, and in other cases it is almost certain that hormones are involved.

Only few observations have a bearing upon this problem, among which are the studies of Burnett (1968) on *Hydra*. For all we know, the cleavage pattern in Cnidaria is not determinate to any significant extent. There is undoubtedly an apical, and presumably also a normal, polarity. These cannot, by any means, answer for the establishment of the diversification patterns in the adult body, which by far exceed ten in number. Clearly ramifications must occur, implying control mechanisms of various kinds. It is therefore of great importance that complicated interactions between the various differentiation patterns could be demonstrated in *Hydra*, controlling the cellular activities, including cell division. It is interesting to note that neurocytes play an important role in this context. It is possible that some clue to the control of cell differentiation in higher organisms is buried in these observations; as a matter of fact, it does not seem unlikely that the phenomenon observed by Burnett represents a primitive type of neurosecretion. This proposition finds support in the fact that this regulatory mechanism is common in Invertebrata (cf. Gabe, 1966; Durchon, 1967).

Regulative cell-differentiation patterns. It has been stressed repeatedly by Goldschmidt (cf. 1940) that, with the advent of differentiation patterns, the products of which were able to influence humorally the growth processes, Nature came in possession of an extremely powerful tool of evolutionary innovation. The reason for this is quite obvious, for a single mutation, involving either the elaboration of a new hormone or the change in efficiency of an already existing one, may have far-reaching effects upon the animal body. Therefore, with the mobilization of this kind of agent, the tempo of evolution could be speeded up substantially.

Unquestionably, several hormones are involved in the influence of vertebrate morphogenesis, their cooperation being controlled through the pituitary hormones. There is, on the other hand, no doubt that the sex hormones, thyroxin, and the pituitary growth hormone are the most important ones. The various hormones may act as antagonists, but each of them have morphogenetic effects of their own. The action of the sex hormones, in Arthropoda as well as in Vertebrata, is particularly clearly demonstrated in the phenomenon of sexual dimorphism. Although the sex chromosomes are known to contain genes relating to other properties than those concerned with the synthesis of sex hormones, it is not possible to refer the morphological differences to these other genes since, by appropriate application of sex hormones, it is possible to change the epiphenotype so as to be at variance with the genotype. There is thus no doubt that these hormones are directly involved as regulators of morphogenetic processes. The extreme degree of sexual dimorphisms found in certain Arthropoda presumably may be interpreted as the result of interference by sex hormones at a very early ontogenetic stage.

Although body size usually is determined by the cooperation of many genes, several single gene mutations are known which directly influence body size, for instance one causing dwarfism in mice through suppression of the pituitary growth hormone (cf. Grüneberg, 1943). This is a clear-cut example of a mutation with extensive morphological effects, but unfortunately it represents a loss. There is no doubt, however, that mutations may enhance the activity of the growth hormone. From the syndrome of acromegaly we know that in human beings effects of this kind will increase the rate of growth of various bones, notably in limbs, nose and chin. The type of response clearly depends, in the first approximation, upon the reacting tissue. However, certain regularities in the response indicate that a growth centre resides in the anterior end of the body (Klatt, 1948), and there is indeed no reason why it could not be the pituitary. In that case the high concentration of growth hormone around its source might be the cause of the high allometric coefficients for various cephalic structures like jaws, antlers, etc. It is worth pointing out that if the growth hormone affects the growth of the pituitary, a kind of 'autocatalysis' may prevail. An interpretation of this kind directly accounts for J. S. Huxley's law of anterior–posterior development (1932).

One of Thompson's famous Cartesian transformations (1942), that from *Diodon* to *Mola* (*Orthagoriscus*) (Figure 260), is an apparent contradiction of the above law, suggesting the presence of a postero–anterior rather than antero–posterior growth 'gradient'. It does not seem possible to account for this phenomenon for the time being, but it may be mentioned that the abnormal reduction of the spinal cord in *Mola* (cf. Weichert, 1965) presumably is another consequence of the unusual growth mechanism involved in the epigenesis in this species.

The accelerating effect on taxonomic divergence of hormones is particularly well borne out in two animal classes. The first of these, Insecta, is distinguished by a great variety of forms. It is well known, not the least from the work of Goldschmidt (1940), that hormones play a very important role in the establishment of many of the taxonomic characters in this class. The large number of species may, of course, primarily be referred to the small size of the animals, which implies that isolation is easily accomplished. The other class, Mammalia, comprises a limited number of species, most of which are relatively large animals with long-lasting reproductive cycles. The large size must perforce entail that the total number of individuals is low and this, together with the low reproductive rate, has the consequence that the likelihood of new mutations will be substantially reduced. Therefore, the rate of evolutionary progression might be antici-

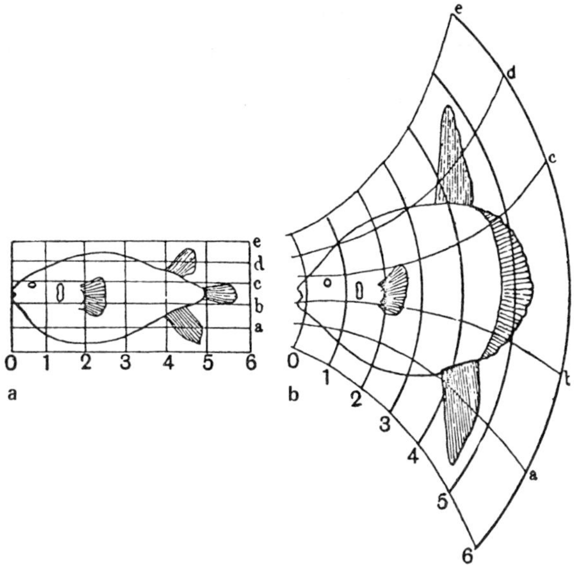

Figure 260. Transformation of the body form of *Diodon* to that of *Mola* through deformation of the Cartesian coordinates. (a), *Diodon*; (b), *Mola*. (From D'Arcy W. Thompson (1942), *On Growth and Form*. © Cambridge University Press)

pated to be very slow in Mammalia. In point of fact, the fossil record demonstrates that, compared with many other animal groups, this class has evolved with remarkable speed. I fail to see how this course of events can be explained except on the assumption that evolution has proceeded in large steps under hormonal control. This holds in particular for the process of foetalization (Bolk, 1926), so important for the evolution of man.

Hormones may also affect taxonomic divergence at the lower levels of the hierarchy. Thus, as has been observed in animals living in archipelagos, the variations seem to follow a similar pattern on many different islands, e.g. 'dwarf or giant races, or races with peculiar colour characters (albinism, melanism), or with peculiar structures (long [and otherwise varied] bills in birds), or other peculiar characters (loss of special plumage in birds)' (Mayr, 1942, p. 236). This description fits also the pattern of divergence in the Geospizinae (Lack, 1947). Apart from albinism, which represents a loss mutation, all the other modifications are such that they can be accounted for by changes in the activity of either growth or sexual hormones.

Morphogenetic processes

A great number of different topics might be included under the present heading, demonstrating the influence of modifications in the course of morphogenetic processes upon the final body form. Since a comprehensive survey is out of the question, I shall confine myself to two subjects, viz., the importance of shell growth for evolution in Mollusca and the phenomenon of recapitulation.

Shell growth in Mollusca. The mathematical description of the various forms of shells has for a long time stimulated the curiosity of scientists (cf. Thompson, 1942). This author strongly emphasized that the proper understanding of the shape of the various shell types can be arrived at only when it is realized that they are the outcome of growth processes. Thompson established a number of parameters which may be used for the definition of the equiangular spiral which approximately describes the shells in Mollusca, and in Brachiopoda as well. The study of this problem has in recent years been resumed anew (Raup, 1961; Raup and Michelson, 1965) and I shall here follow these authors' approach.

According to Raup (1961, p. 606), the basic form of gastropod shell can be described by four parameters: 'shape of the generating curve, the rate of expansion of this curve, amount of overlap between successive whorls, and the position of the whorls relative to the axis' (Figure 261). Obviously, if Thompson's notions are correct, it must be possible

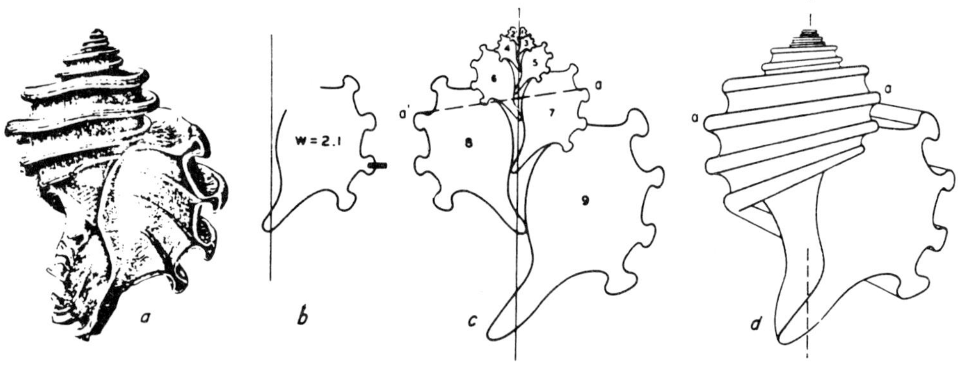

Figure 261. Reconstruction of the basic form of the shell of *Ecphora* from its coiling characteristics. (a), the authentic shell; (b), presentation of the four coiling characteristics, generating curve, rate of expansion (w), amount of overlap (indicated by a tick) and position of the whorl relative to the axis; (c), reconstruction of a cross section of the shell, with indication of the overlap angle; (d), final reconstruction. (Reproduced from D. M. Raup, *Proc. Natl. Acad. Sci.*, **47** (1961), by permission of the National Academy of Sciences, Washington)

to establish the connection between these parameters and factors or events relating to the living animals and their ontogenesis. For one thing, the generating curve, i.e. the cross section of any whorl in the shell, is determined by the form of the animal. If the general shape is preserved as it grows in size, the generating curve will remain unchanged.

The rate of expansion is indicated by a number representing the ratio, in linear dimensions, between any whorl and the one separated from it by 360° (cf. Figure 261). It is easily realized that this parameter is composite, depending partly on the growth of the animal and partly on the rate at which new shell is formed by the mantle. If the rate of the former is slow compared to the latter, the shell will be acute, as in *Epitonium*; if the opposite relation holds, it will be obtuse, as in *Natica* (cf. Figure 262).

The amount of overlap can be expressed in various ways, for instance by the position of the suture line on successive coils (Figure 261b), but it may be more precisely defined as the angle between the axis of coiling and a line passing through the 'centres' of successive whorls (Figure 261c). It is easily seen that if this angle is 90°, we get a discoid shell, as found in *Nautilus*, while the outcome will be a helix if the angle is 0°. As we have noted in a previous chapter, the direction of the coiling in Gastropoda is determined by the polarities established by the cleavage pattern. It may be presumed that the value of the angle is also controlled by this means.

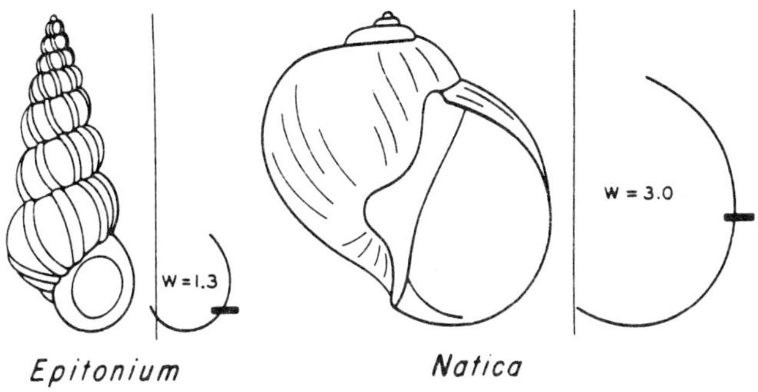

Figure 262. Coiling characteristics of an acute (*Epitonium*) and an obtuse (*Natica*) shell. (Reproduced from D. M. Raup, *Proc. Natl. Acad. Sci.*, **47** (1961), by permission of the National Academy of Sciences, Washington)

Finally, as far as the position of the whorls relative to the axis is concerned, this parameter also appears to be composite. It is partly determined by the constant angle by which the curvature of the equiangular spiral is determined, i.e. the angle between a radius vector and the tangent in the point where it intersects the spiral. If this angle is 90°, the spiral becomes a circle, if it is 0°, a straight line (cf. Figure 263).

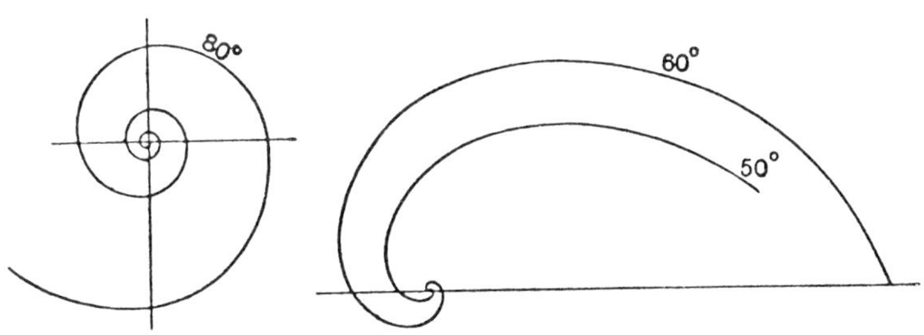

Figure 263. The influence of the value of the constant angle upon the shape of equiangular spirals. (Reproduced from D'Arcy W. Thompson (1942), *On Growth and Form*. © Cambridge University Press)

As pointed out by Thompson (1942), this angle of curvature depends on the location of the zone of active growth at the edge of the shell. If growth occurs uniformly all around its circumference, the shell will be straight, otherwise it will be curved, the angle depending upon the dorso–ventral differential in growth rate. However, the distance from the axis must also be influenced by the rate of expansion, in which the absolute rate of shell growth is involved. Thus, for a given value of this parameter, there is a minimum value of the angle of curvature which still allows for a tight coil, i.e. with the distance to the axis being zero. For values below this one, the coil will be open. When the rate of expansion increases, the minimum angle value increases too. The growth may occur at the dorsal or the ventral edge. The former situation is realized in Cephalopoda, the second in Gastropoda. In the latter case we have seen that this growth process leads to intestinal flexure and torsion, with the result that the shell is transferred to a dorsal position.

Since both the torsion and the formation of a turbinate shell are causally linked to the presence of a dextro–sinistral polarity, it seems to follow that the genesis of the class Gastropoda depends upon the utilization of the potentialities of this particular feature. The overlap angle is most often kept constant, but it may approach 90° in certain species, a circumstance which does not influence the phenomenon of torsion. If, however, shell formation is suppressed completely, no torsion will occur, and if it is greatly reduced, particularly the coiling, detorsion may take place. I may just add that I cannot subscribe to the notion that torsion is an adaptation to larval life as suggested by Garstang (1928). For one thing, torsion is an inherent aspect of the formation of the turbinate shell in Gastropoda, and secondly, I believe it would be just as difficult, if not more so, for the adult to survive, with the edge of shell trailing on the ground just behind the foot.

By the points established here we can account for various forms of shells found among the major molluscan classes. Thus, the shells of Scaphopoda are distinguished by a rather low rate of expansion, and a very small spiral angle. In *Nautilus* and certain Gastropoda the latter parameter has increased substantially, but the angle of overlap is 90°. In most Gastropoda, however, the angle is less than that, with the resulting formation of a turbinate shell. Finally, it may be noted that the shells of Bivalvia and Monoplacophora are characterized by a very high rate of expansion. Many shells are distinguished by ornaments of various kinds like spines, grooves, etc. The formation of these does, of course, involve parameters over and above those discussed here.

The phylum Mollusca comprises a large number of species, the taxonomic distinction being to a large extent based on the geometry of the shells. It appears that much of this variation must have been accomplished by slight variations in the biological parameters involved in the growth of the shells. Once more it is thus seen that, if the epigenetic interpretation of evolution is accepted, it is possible to account for substantial modifications as the outcome of relatively trivial mutations. The advantage of this approach is partly that it affords an understanding of the evolutionary mechanism, but furthermore 'the burden on natural selection is also appreciably lightened' (Huxley, 1932, pp. 221–222).

Recapitulation. The concept of recapitulation in embryogenesis is of ancient date (cf. Holmes, 1948; Oppenheimer, 1955), but its formulation as a scientifically acceptable hypothesis is generally credited to von Baer (1828; 1837). The importance of this phenomenon from an evolutionary point of view is quite obvious, and it is therefore understandable that it was embraced with enthusiasm by Haeckel (1866), who used it to establish his 'biogenetic law'. Much verbiage has been spent upon arguments about the validity of this hypothesis. It appears that while the principle of recapitulation remains uncontested, there is today general consensus on the point that Haeckel's generalization is completely unwarranted, as shown most comprehensively and convincingly by Garstang (1922). We shall in the present section try to outline the significance of the phenomenon of recapitulation, and give some illustrations of the evolutionary consequence of modifications in recapitulatory sequence.

In most cases the morphogenetic prerequisites for elaboration of the phyletic archetype are established at gastrulation or shortly thereafter. Since the archetypes differ fundamentally from each other, it follows that from this stage on recapitulation is no longer possible between, but only within, the phyla. In most Invertebrata development is so fast, or else comprises so few stages, that the characters distinguishing the lower taxa are soon established, and hence recapitulation can no longer occur. As an example may be mentioned the flagellated larva which, with some few exceptions, is an archetypical trait in Echinodermata, but with the formation of a calcified skeleton, affiliation to either Ophiuroidea or Echinoidea is indicated. Likewise, in Mollusca the early appearance and activity of the shell gland is typical for most members of the phylum, but the position and time of differentiation will at a very early stage determine the class to which the developing organism belongs.

It is not feasible to pursue this argument in further detail, but it seems possible to state as a generalization, which may not hold for Arthropoda, that within each particular phylum the morphogenetic processes occur in a succession such that first the phyletic body plan is established and subsequently the modifications of the archetype which distinguish the successively lower taxonomic levels. This principle, which was established by von Baer (1828; 1837), follows directly from the fact that, as pointed out above, the various epigenetic processes are causally connected. To mention an example from the vertebrates, it appears that the location of the limbs is intimately related to the process of segmentation, being situated immediately in front of and behind the segmented part of trunk. Hence, the limbs cannot be formed before the segmentation has occurred. In various lower taxa, modifications of the limbs are characteristic features. These modifications obviously cannot be established before the elaboration of the limbs is well under way. Thus, on the morphogenetic level there is, and must be, in each separate case of ontogenesis, a process of 'recapitulation', which certainly reflects also the course of phylogenetic evolution.

There are some noteworthy exceptions to the principle stated here which, it seems, may be accounted for by the generalization that characters or events that are not causally involved in the overall epigenetic process may be lost or appear at a stage different from that which might otherwise be anticipated. This holds for the larval stage, the loss of which is a rather irregular occurrence in Invertebrata, and for such class characters as scutes, feathers and hairs, usually making their appearance at a much later stage than suggested by von Baer's rule.

The phenomenon of recapitulation is most easily observed in Vertebrata because the ontogenesis in this case comprises a long sequence of events which are causally related. One of the most conspicuous features, and of great importance in all arguments on recapitulation, is the presence of gill slits, or at least gill pouches, in all vertebrate embryos, although gills have lost their function during embryogenesis from the Reptilia and in adult life from the Amphibia. Their persistence is without any exception, and they may therefore be considered a characteristic feature of an embryonic stage in the neurula grade. The preservation of the gill arches might be a coincidence, but it is not, for as pointed out by Sedgwick (1894), the branchial arches are causally involved in the establishment of the pulmonary circulation and of various endocrine glands. Their presence is therefore an epigenetic necessity.

The phenomenon of recapitulation is thus an unquestionable reality. However, since the archetypes are generally established around gastrulation, the path common for two different phyla ends at this stage. Since the events occurring during this early part of embryogenesis are rather trivial, it follows that studies of recapitulation cannot be of importance for the determination of the affiliation between different phyla.

The problem concerning deviations from the principle of recapitulation is the theme of a book written by de Beer (1951). These phenomena have some bearing upon the mechanisms involved in phylogenetic evolution and therefore it is necessary to discuss them briefly. I shall not quote the extensive literature here, but rather refer to the work mentioned.

De Beer has divided these phenomena into eight classes, the first of which is called *caenogenesis*. It corresponds to the situation where different animals are similar during the early and the late stages of ontogenesis, but exhibit marked dissimilarities at an intermediate stage (Figure 264a). Examples illustrating this phenomenon comprise a number of instances where extremely similar adults have widely diverging larva, and also of course the many cases where direct and indirect development occurs within the same taxon. The embryonic membranes in Amniota are also mentioned as structures which cannot be recapitulatory. It is not certain that all these examples are equivalent, but in any case we may conclude that these various caenogenetic acquisitions (or losses) may be of functional significance, contributing to the survival of the embryo, but that they cannot participate in the chain of causally connected epigenetic processes.

Another example of caenogenesis, and of recapitulatory thinking, has been given by Goodrich (1958, p. 612): 'Indeed, the lung seems to represent a remarkably clear case of an ontogeny which is not recapitulative. For the accumulation of a thick layer of mesenchyme round the narrow endodermal rudiment of thick columnar epithelium can surely in no way repeat a primitive stage, but is rather due to the precocious gathering

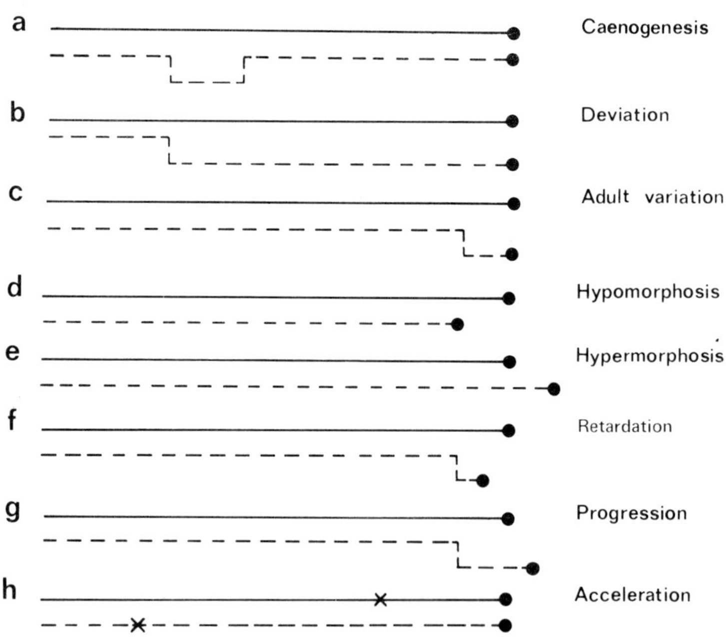

Figure 264. Diagram of various kinds of modification in the phenomenon of recapitulation

together of the building material for future differentiation. On the contrary, the lung, to be an efficient organ of respiration, must from the first have had a thin wall abundantly supplied with a superficial network of blood-vessels, and have become progressively folded and pocketed to form the parenchyma of air-cells in higher forms. Thus, the respiratory surface, which is the last to develop in the embryo, must have been present from the first and throughout the phylogeny, although doubtless less perfect in the lower than in the higher terrestrial vertebrates'.

This reasoning is not based on an understanding of the workings of morphogenetic mechanisms. A primitive lung may well have been a simple air sac of vascular epithelium similar to the air bladder of Pisces, but once it happened that during ontogenesis the presumptive lung epithelium was exposed to a morphogenetic space containing mucopolysaccharide and sf-cells, the latter would immediately exert their activity, leading to a lung with the typical ramifications of the epithelial aggregate. Incidentally, it will be realized that the continued branching directly reflects the growth process of the organ. This description does not preclude that phylogenetic improvements in the structure and function of the lungs may have occurred, for instance by the incorporation of new differentiation patterns. However, from the first time a lung was ever formed, the principal mechanisms must have been the same as those observed today, and the question of recapitulation is therefore irrelevant.

If evolutionary innovation is to take place, it is obviously necessary to change some part of the recapitulatory succession of events. This may happen by the substitution of a new feature for an old one within a given span of ontogenesis. De Beer (1951) distinguishes two cases, depending upon the time at which the innovation is introduced. If this occurs at an early stage, it is called *deviation*, if at a late stage, *adult variation* (Figure 264b and c). A special case, reduction, involving loss of a feature, may be included under deviation or adult variation.

An example of deviation is found in the gill pouches in Vertebrata which occur even in those classes which have lost the gills. This is thus a clear case of reduction, but as we have seen, the presence of the gill pouches ensures that other properties are gained. According to de Beer all the genetic variations studied by geneticists may be classified as adult variations, and I suppose one might add to these many of the properties distinguishing separate species.

Another mechanism used by Nature to create phylogenetic innovation is to vary the span of ontogenesis, i.e. stop the development at a stage which in another group represents an immature stage, or prolong it beyond its former scope (Figure 264d and e). The former case may be called *hypomorphosis*, the latter *hypermorphosis*. In hypomorphosis two separate phenomena are distinguished, paedogenesis if a larval stage becomes sexually mature, and neoteny if only certain larval features are preserved in the adult. The most famous case of neoteny is presumably the phenomenon of foetalization in man, named so because human beings exhibit many features distinguishing a foetal ape. Most other examples of this kind are observed in Arthropoda and Vertebrata, clearly suggesting that neoteny is a matter of mutations interfering with the action of hormones. In certain cases this proposition has been clearly borne out by experiment and observation. If neoteny is hormonally controlled the same might be expected to hold for hypermorphosis, and this proposition is supported by the fact that the most well-known example of this phenomenon is the so-called 'allometric growth'. As we have discussed above, there is good reason to believe that this process is controlled primarily by the growth hormone.

In association with hypomorphosis it may be anticipated that, as a consequence of the shortening of the range of ontogenesis, certain adult characters of the ancestor become reduced or suppressed. This phenomenon, called *retardation*, may be difficult to distinguish from adult variation (Figure 264f). From a purely logical point of view one may infer the existence of a mechanism, *progression*, corresponding to hypermorphosis (Figure 264g). This case does not appear in de Beer's list; if occurring, it will also be very hard to separate from adult variation.

The last phenomenon discussed by de Beer (1951) is called *acceleration*, characterized by the appearance of a special feature at an earlier stage of development (Figure 264h). Since those characters which are truly recapitulatory, i.e. those which form part of a causally connected chain of events, cannot be displaced without upsetting the whole process, there is reason to believe that acceleration rather represents the earlier appearance of some differentiation pattern. This interpretation does not account for all the examples quoted by de Beer in this context, but I am not sure that they are hard and fast examples of acceleration. However, the precocious development of callocities in the young ostrich, one of the examples, fits quite well with my interpretation.

Much discussion has been devoted to the problem of recapitulation, its causes and exceptions. The contribution of de Beer (1951) has been an important step towards clearer understanding of this problem which, although of great importance, does not have quite the significance believed by our predecessors.

Evolution in space

The interrelation between geographical distribution and evolution was clearly stated by Darwin (1859) and has been further stressed by a number of biogeographers and evolutionists, notably Willis (1922), Mayr (1942; 1963), Croizat (1958), Brundin (1966) and Hennig (1966).

It may first be mentioned that the geographic details of the planet afford innumerable opportunities for isolation, random and non-random as well. The various changes to which the earth's crust has been subjected, e.g. separation of land masses, have of course also involved random isolation of communities of living organisms. The progression and survival of many forms, including relics, may be accounted for in terms of geography.

The factors favouring isolation, e.g. natural barriers and low motility, also reduce the likelihood that it occurs in the first place. For this reason it has been proposed by Darwin (1859), Matthew (1915) and many others that natural rafts and meteorological forces may incidentally be responsible for random isolation. However, in many cases the power of dispersal seems to be of remarkably slight importance for geographical isolation. Thus, various forms of the terrestrial bird *Zosterops rendovae* isolated on islands in the Solomon archipelago are separated by distances which could be covered by the birds in a few minutes (Mayr, 1942).

The concept of 'divergence through dispersion', involving the chance of random isolation, is also of importance in biogeographical contexts. If, namely, taxonomic divergence is associated with the colonization of new regions, then it affords a simple explanation of the phenomenon of vicariance (cf. Croizat, 1958; Hennig, 1966), according to which related groups, at various taxonomic levels, replace each other in different geographical areas. It also supports Willis' concept of 'age and area' (1922), which postulates that old taxa must occupy larger territories than young ones. This idea is largely borne out by empirical facts, but it must be admitted that the actual dispersion depends upon other factors besides age, for instance the means of dispersal and chance. Willis' theory is generally not accepted by neo-Darwinists, primarily because he, rightly it seems, inferred that the geographical distribution of animals and plants indicates that evolution has involved large-step mutations (cf. Mayr, 1963), a circumstance which greatly reduces the importance of environment and selection for the progress of evolution.

According to this conception it may be envisaged that when a new form arose, embodying properties which today are used to define a taxon of high rank, then the competitive advantages inherent in these novel characters were such that it could spread through interspecific competition over vast expanses. Indeed, only natural barriers or equally successful forms could arrest its advance. In the course of this event modifications might occur at different geographical locations which are now employed as taxonomic characters for taxa of lower rank. Most often the new subdivisions presumably would remain in a given area, either because their peculiarities represented specializations to a particular habitat, or because their APA did not permit their invasion of neighbouring regions populated by related forms. In some cases, however, the innovations might increase the power of adaptability so much that a new phase of dispersion, partly retrograde this time, might take place. It seems that many biogeographical observations may be accounted for on the basis of such reiterated waves of expansion.

Biogeography is of very great importance for phylogenetic systematics, particularly for the deciding of cases of convergence and parallelism, involving the independent acquisition of the same characters in two or more distinct taxa. If similarities of this kind obtain in geographically widely separated taxa it is common to invoke convergence or parallelism as explanations of this fact. Although unquestionably these phenomena have occurred on many occasions, and even may have involved a considerable number of causally connected features, it is clearly an interpretation that is at variance with Occam's razor and hence it should be used only with great discrimination.

In this context biogeography becomes of great importance. This point is brought home with great clarity by the regrettable fact that most texts on biogeography (e.g. Matthew, 1915; Simpson, 1953b; P. J. Darlington, 1957; Croizat, 1958) reject or ignore Wegener's hypothesis of continental drift. According to evidence accumulated in recent years this notion apparently is well substantiated, and many biogeographical and systematic problems yield to a simple solution, reducing the need for convergence or parallelism, on the assumption that the continents of the southern hemisphere were once in contact (cf. Brundin, 1966).

Evolution in time

The present topic, evolution in time, may be said to cover two aspects, viz., the rate of evolution and the nature of the diverse extinct forms. Both of these points are elucidated by the fossil record, together with information about the geographical distribution of fossil animals and plants.

Among fossils there are few forms that may plausibly serve as transitional links between the higher taxa. This fact partly depends, of course, on the circumstance that organisms without hard tissues do not become fossilized. Over and above this explanation two different opinions have been voiced to account for this state of affairs. The first one, voiced by evolutionists from Darwin (1859) to Simpson (1953a), claims that the fossil record is anything but complete, that many intermediate forms have not been found, and often may not even have been preserved. The other view, advocated by Schindewolf (1950) in particular, asserts that the palaeontological evidence is reasonably complete, and that the links are missing because they never existed. This standpoint obviously requires the occurrence of large-step mutations, and hence has been rejected by neo-Darwinian evolutionists (cf. Simpson, 1944; 1953a).

Following the ideas contended here, taxonomic divergence, and hence the speed of evolution, depends, besides the generation time and the size of the population, upon two factors, mutation and isolation. As regards the former it has been proposed that the mutations responsible for the origination of high-ranking taxa are extremely rare events which, more often than not, act through interfering with the course of ontogenesis. This kind of event is so rare that no valid predictions about the rate of evolution can be made. Stagnation may ensue simply because for millions of years no such mutations occur. On the other hand, because of their way of action, it may be anticipated that, for instance, a new phyletic body plan may open up a number of new opportunities for epigenetic modifications. This may be the main reason for the 'explosive' phase of 'adaptative' radiation during the early stages of existence of new high-ranking taxa which repeatedly has been evidenced by the fossil record.

With respect to the second factor, isolation, it may suffice here to point out that this phenomenon is furthered by specialization and preadaptation, ensuring non-random isolation. Random isolation is particularly likely to occur in small organisms, requiring small niches, and when the power of motility is low. The actual taxonomic divergence in animals and plants corroborates this generalization when proper allowance is made for the effect of the power of adaptability.

No new grade of body plan has arisen since the appearance of the Agnatha in the Ordovican, and these Chordata belong to the most recent phylum which has left a fossil record, suggesting that the origin of new phyla ended many hundred million years ago. Since the creation of a new grade or phylum must imply modification of the earliest embryonic stages, it seems that the likelihood of such an event would become smaller as the epigenesis of existing animals became more and more complex. Furthermore, the chances of isolation would also decrease as different animal forms multiplied with time. There is therefore reason to agree with Cuénot (1940) and Woltereck (1940) in their view that the generation of new phyla is inconceivable. However, if the extinction of taxa is due to competition with new forms with higher power of adaptability, there might be no ground to share the pessimism as regards the fate of the existing animal kingdom, as expressed by Schindewolf (1950) and Müller (1955); unless man interferes with the process we may expect that higher powers of adaptability may be reached in some lines, although within the framework of the existing classificatory system. It should be mentioned that Ohno (1970) has arrived at a rather gloomy view as regards future evolution on the basis of molecular-biological considerations.

DIVERGENCE

Even to the naive observer of Nature it is evident that animals and plants can be grouped according to the similarities and dissimilarities they display, and the theory of evolution implies that a systematic arrangement of living beings on this basis may reveal their phylogenetic kinship. This expedient actually is a necessary requirement for the appraisal of the course of evolution. That it is not a sufficient one is shown by the fact that the work of Linnaeus did not convert him into an evolutionist.

In any case, the erection of classificatory hierarchies serves to outline the degree of taxonomic divergence which has been accomplished through the process of evolution, and it is therefore of great importance to establish the basis on which such work rests. In the following section an attempt will be made to approach this goal through axiomatization. This may appear a very ambitious undertaking but it is at least justified by the fact that biological thought is being subjected to the same laws of reasoning as all other intellectual endeavours. Since this problem is of great general importance an axiomatized presentation may have the advantage that anyone who disagrees with my conclusions may discover the mistakes I have made more easily than otherwise.

The following discussion will be divided into three sections, dealing with some general principles of classification and with phylogenetic and Linnean classification, respectively.

Principles of classification

The present section is concerned with some general rules of classification which in principle may be applied to any group of objects exhibiting varying degrees of similarity, although the axioms, for expediency, have been phrased largely with respect to biological classification.

(P1)* A system of classification is a scientific theory.

This premise implies that a classificatory system must obey the rules of deductive logic, that it must be testable through confrontation with empirical observation and that it must be falsifiable (Popper, 1968).

(P2) Classification involves the distinction and naming of sets (or taxa) of organisms defined by the possession of certain properties (or taxonomic characters) in common.

Simpson (1961, p. 18) states that 'an individual never is and cannot be classified. Classification involves only groups'. As a general rule this assertion is invalid, for a set may certainly contain only one member. This, presumably, is the intended outcome of naming human beings and pet animals. It is true that in the classification of living animals this situation usually does not obtain, but it might happen, for instance, in a taxon on the verge of extinction. As regards fossils, I am convinced that the occurrence of several specimens of a certain kind is no prerequisite for the naming and classification of an extinct animal.

If a taxon T is exhaustively and unambigously defined, then it is not possible to erect another taxon comprising the members of T and only these. One or more additional names may be applied to T, but no new taxon is thereby defined. In general set theory the definition of a set s implies the simultaneous determination of the complementary set $\bar{s}$ or non-s, because the two sets are united by the relations $s+\bar{s}=U$, where U is the logical universe. In classification a restriction is necessary as regards the meaning of U, a point that will be taken up in the sequel.

According to Simpson (1961, p. 71) classification requires that 'the whole organism shall be considered in all its parts and aspects'. As it stands, this statement may appear to represent an ultimate goal which is far from being reached. Indeed, if complete knowledge of all characters was required before classification could be attempted, then

*P, D and C stand for 'premise', 'definition' and 'conclusion', respectively.

we should have no systematics today. It must be emphasized, however, that even the information we possess today in many cases probably is redundant for the purpose of classification; a complete listing of properties in common might in such instances make the taxonomic definitions unwieldy. It is therefore necessary for the sake of expediency to introduce the amendment that although ideally all properties possibly should be used, the number actually employed as taxonomic characters may be reduced to that required for the establishment of correct taxa. The implications of the word 'correct' will appear from the following discussion.

(C1) All properties possessed by an organism are either actual or potential taxonomic characters.

From P1 and P2 it follows that all new information of comparative nature which becomes available may, and must, be used for testing a classificatory system. If agreement obtains between empirical observations and theoretical predictions, the classification is corroborated, and there is no reason to employ the investigated properties as taxonomic characters. However, if the observed data do not fit, the system is falsified and must be changed, and in that case the investigated properties presumably must be employed for the required modifications of the classification. Hence it follows that all properties that are not actually taxonomic characters potentially may be so.

(C2) Classification must be based upon extant organisms.

This conclusion follows from C1 because only contemporaneous organisms can be subjected to the exhaustive scrutiny which, potentially at least, is needed for their classification. Thus, although fossil evidence has been of enormous importance for demonstrating that evolution has in fact occurred, and for outlining the course of this event, the information available from this source does not suffice to establish the definite taxonomic relationship between extinct and still living organisms. It seems warranted, therefore, to classify fossils in an independent system, as suggested by Crowson (1970).

(P3) Taxa can be arranged hierarchically.

Experience has shown that some properties are found in a certain set of organisms, thus defining a taxon, while other properties may be common only to subsets within the set, thereby defining taxa that are included in the larger one, etc. A hierarchy may hence be formed by placing at the apex the taxon comprising all living organisms, and at the base those taxa between which the members have no taxonomic characters in common other than those defining taxa at higher levels in the hierarchy. From this premise it does not follow that living organisms can be arranged in a unique hierarchy. As will be shown presently this is not the case; for the erection of unique systems it is necessary to make further specifications with respect to the properties of the classificatory system.

(C3) Each organism is the bearer of the characters defining all the taxa to which it belongs plus a number of properties that are not taxonomic characters.

This conclusion follows from P2 and P3. The properties which are not taxonomic characters are either redundant ones, distinguishing the defined taxa, potential taxonomic characters that may lead to revisions of the classification, or subtaxonomic properties or characters, i.e. features common to sets of organisms within the lowest-ranking accepted taxa. If sets of one organism obtain we may be dealing with 'individual variations'.

The establishment of a hierarchical system allows for the ranking of the various levels.

The rank of a taxon may be defined in either of the following ways:
(D1.1) A taxon T_{j+1}, of rank j+1, is a set of n_j taxa T_j, of rank j ($j \geqq 1$).
(D1.2) A taxon T_j, of rank j, is a set of n_{j+1} taxa T_{j+1}, of rank j+1 ($j \geqq 1$).

These definitions differ with respect to the starting point from which the taxonomic ranks are numbered. According to D1.1 the numbering begins at the base of the hierarchy, as is wont in Linnean classification, implying that T_1 is a species or a subspecies and, in animal classification, T_n the animal kingdom (cf. Beckner, 1959). It will be shown below, however, that this approach cannot be applied in phylogenetic classification, where inconsistencies are encountered unless the numbering begins at the hierarchical apex, as proposed by D1.2. It is not proposed that this expedient should be adopted in ordinary classification, on the contrary D1.1 will be employed in the discussion of Linnean classification, but it must be emphasized that the implication of 'taxonomic rank' consequently is different in the two systems. To avoid misunderstandings, we shall use the adjectives 'high' and 'low' only when referring to taxonomic levels in the traditional classification, i.e. to ranks distinguished by high and low numbers, respectively. In the phylogenetic classification the epithets 'superior' and 'inferior' will be used to designate the corresponding positions in the hierarchy, a superior taxon hence being of low numerical rank and *vice versa*.

(P4) The complete definition of the taxa T_j included in the taxon $T_{j\pm1}$ comprises specification of the differences between the taxa.

As stated in P2, similarities of various kinds are used for the erection of taxa. However, when a taxon $T_{j\pm1}$ is subdivided into a number of taxa T_j, then it is necessary to establish not only the characters that distinguish the individual taxa T_j, but also the differences between them. This premise bears upon the notion of complementary sets mentioned above and implies that if one of the taxa T_j is defined by the character x, then the remaining ones of this rank are defined by $\bar{x}$. In classification the notion corresponding to U in set theory is thus represented by all the taxa T_j in $T_{j\pm1}$, or by $T_{j\pm1}$.

A consequence of this premise is that if, in a taxon T_j the properties of, say, some morphological entity are used as taxonomic characters, then the peculiarities of the corresponding or homologous structure are taxonomic characters in the other taxa T_j within the taxon $T_{j\pm1}$. Thus, since the shapes of the feet and the beak vary between the diverse avian orders, the exhaustive definition of any order should in principle include characterization of these parts. This expedient is seldom adhered to in practice, presumably because this measure would be redundant.

(C4) The taxonomic level of rank j is defined only if $n_j \geqq 2$.

From P4 it follows that if there are not at least two taxa T_j, then it is not possible to establish their differences, and hence their taxonomic characters. A single taxon T_j is undefinable and quite simply identical with the $T_{j\pm1}$, provided the latter is defined.

(C5) An organism cannot belong to more than one taxon T_j of the rank j.

This conclusion also follows from P4, owing to the demand that the differences between the taxa should be clearly specified.

(C6) The only taxonomic characters in common between the taxa T_j included in $T_{j\pm1}$ define the taxa $T_{j\pm1}, T_{j\pm2} \ldots T_{j\pm n}$ ($j - n \geqq 1$).

The special implication of this conclusion, also deduced from P4, is that if two or more of the taxa have properties in common, then this group of taxa should be a T_j taxon, and the included taxa allotted the rank $T_{j\mp1}$.

Phylogenetic classification

Since the properties possessed by living organisms, and hence the taxonomic characters, have been acquired in the course of evolution, some correspondence must obtain with respect to the phylogenetic kinship between various organisms and the relationships established through their classification. The points mentioned here we may state in the following premise which embodies the essential aspects of Darwin's theory of evolution:
(P5) All organisms, living and extinct, have a common phylogenetic origin. The varying degrees of similarity obtaining between different kinds of organisms are expressions of phylogenetic kinship.

It was mentioned above that it may be possible to establish different classificatory hierarchies, and that further specification is required to choose between these alternatives. Under these circumstances the statement advanced in P5 obviously suggests that the classification is required to reveal the phylogenetic relationships between the various living organisms. A classification of this kind is in the present context called *phylogenetic classification*; a synonym is 'natural system'. The stipulation made upon the classification discussed in the present section may hence be formulated as follows:
(P6) The hierarchy of classification must be phylogenetic.

Since, as implied in P5, all living organisms have arisen through one unique process, it follows:
(C7) There is but one possible hierarchy in phylogenetic classification.

The one hierarchy mentioned in C7 comprises all living beings, but within this it is possible to distinguish several subsystems, for instance five, namely, the kingdoms Monera, Protista, Plantae, Fungi and Animalia (cf. R. H. Whittaker, 1969).

It is obviously implied by P6 that the properties used as taxonomic characters in phylogenetic classification must be more than mere similarities, they must be similarities that are genetically related. Hence we may conclude:
(C8) In phylogenetic classification a property can be used as a taxonomic character for the taxon T_j only if its presence in all members of the taxon can be referred to its acquisition at one occasion in the course of evolution.

The particular importance of C8 is that it directs attention to a circumstance that in some cases may constitute a serious impediment for classificatory work, namely, that evolution has had a capricious course, entailing among other things that the same property has arisen independently at more than one occasion. Other kinds of events may also tend to obscure classification as will be discussed at a later stage. Suffice it to mention here that problems of this kind can be solved only through the approach envisaged in C1, viz., by including further properties as taxonomic characters.

A taxon which obeys the criterion established in C8 is called a *monophyletic taxon*, a concept that may be defined as follows:
(D2) A monophyletic taxon T_j is a set of organisms comprising all the descendants of the first individuals endowed with all the actual and potential taxonomic characters of T_j.

From D2, in combination with P6 and C8, we may deduce:
(C9) The hierarchy of phylogenetic classification comprises only monophyletic taxa.

In order to establish the correspondence between phylogenetic classification and the process of evolution it is necessary to introduce a postulate concerning the mechanism through which new taxa arise. As appears from phylogenetic trees and other illustrations of the kinship between living and extinct organisms, descendent forms often, but not always, are depicted as arising through bifurcations from ancestral ones. Since the word 'form', as used here, for living organisms may be replaced by 'taxon', it is seen that for any taxon it must hold that if its pedigree could be traced back in time, one would sooner or later come to a point of bifurcation from which the other line also led to a taxon, i.e. a set of living organisms. I believe that this view is generally accepted, although it has been propounded with particular emphasis by the cladists (Hennig, 1966; Brundin, 1968). Of more importance in the present context is the other contention submitted by cladism, the seemingly obvious and innocent postulate that two 'sister groups', i.e. taxa which can be traced to the same point of bifurcation, should be allotted the same taxonomic rank. The extent to which this notion has been granted by other systematists is not easy to ascertain, but it will be adopted here as the following premise:

(P7) The taxa T_j have arisen through bifurcation from the taxon T_{j-1} in which they are included.

The ranking according to D1.2 has been introduced at this stage, although the justification for this expedient will be demonstrated only later.

Before we proceed to deduce the consequences of this very important axiom it may be warranted to state my disagreement with another cladistic principle, the rule of deviation, which submits that 'when a species splits, one of the two daughter species tends to deviate more strongly than the other from the common stem species' (Hennig, 1966, p. 207; cf. also Sharov, 1955). As long as this rule is applied only to species it is probably valid, since then it implies only that the changes associated with the splitting were confined primarily to one of the two separating groups. If, however, this reasoning is extrapolated to higher taxonomic levels, then there is no *a priori* reason to accept this rule, since many bifurcations may have been involved in the creation of each of two sister taxa. That the rule nevertheless is correct in many cases is easily demonstrated by practical examples, e.g. the purported sister taxa Crocodylia and Aves.

As regards the terminology applied to the two filiated taxa, I suggest calling them *twin taxa* rather than 'sister taxa', the reason being that the proposed name is noncommittal with respect to the sex of the taxa and, more important, that it stresses that there are only two sister taxa. It must be admitted that 'twin' usually implies a significant degree of similarity, but against this it may be rejoindered that dizygotic twins are no more alike than are brothers and sisters.

I have stated above that the theory advanced here is based on the theory of cladism, but this is to some extent a misrepresentation as regards P7, the premise asserting the principle of dichotomy, for Hennig (1966) has nowhere claimed that sister groups must always occur in numbers of two, even though reading his book one gets the impression that this situation normally prevails.

In my opinion the principle of dichotomy is next to a natural law, being a consequence of the mechanism through which taxonomic diversification occurs. The latter phenomenon involves that, in some members of a given population (taxon), an innovation arises which distinguishes an incipient new taxon, the twin of the original one. This case clearly represents a bifurcation. In order to account for trichotomy it seems necessary to make the very improbable presumption that two independent innovations originate simultaneously in organisms situated at the same location, since separation in either time or space may be associated with the acquisition of new properties. Hence, even if we cannot exclude the occurrence of true trifurcations, tetrafurcations, etc., it seems justified to rule them out as events of any practical significance.

It is conceivable, however, that two bifurcations may happen under such conditions that it is impossible for the taxonomist to distinguish the result from that ensuing from a true trichotomy. Various schemes may be thought out to account for this occurrence, the simplest probably being the following one: if the members of a certain taxon are completely homogenous with respect to all actual and potential taxonomic characters, if at two or more locations innovations arise which subsequently become taxonomic characters of new taxa and if all these taxa survive, then clearly we are facing situations which must be interpreted by the taxonomist as instances of trichotomy, tetrachotomy, etc.

The if's suggest that even this kind of event is rather unlikely but, should a taxonomist encounter a group of three or more separate taxa in which it is impossible, in spite of all efforts, to discover any 'apomorphous' characters uniting any of them, then he is obviously entitled to introduce the taxa as 'sisters' or triplets, quadruplets, etc. This must, of course, be regarded as a provisional expedient, for nothing prevents that future discoveries may furnish the missing information about the kinship between the taxa.

Owing to the clause about survival stated above, it appears most likely that the systematist, as opposed to the phylogeneticist, will expose deviations from the principle of dichotomy. Under any circumstances, relying upon Occam's razor, I believe that it is justified in either case to proceed on the presumption that dichotomy always obtains and accept plurichotomy only when no other alternative is available. For that reason the present analysis has been based entirely on the principle of dichotomy, but it must be emphasized that the potential non-conformity with this principle in no case significantly changes the implications of the various conclusions arrived at below.

We are now ready to extract the information contained in P7. First we shall conclude:

(C10) In phylogenetic classification there are two, and only two, taxa T_j included in the taxon T_{j-1}.

Clearly, this conclusion implies that the hierarchy of phylogenetic classification is a dichotomous dendrogram.

A further conclusion is:

(C11) Every fossil group has a twin taxon which, however, may be common to two or more fossil groups.

This point is brought out by Figure 265. In the formulation of this conclusion it should be noted that the fossil groups are not classed as taxa, and that consequently they cannot be ranked. On this point I am in disagreement with cladism which apparently ranks even the splittings where one of the branches is represented by a fossil group.

We can further infer:

(C12) The absolute rank of a taxon T_j can be objectively determined.

From P7 it also follows:

(C13) The taxon T_{j-1} is older than the taxa T_j which it includes.

It should be noted in this context that we do not, and cannot, conclude that the two taxa T_j are of the same age. On this point, which will be discussed presently, I am in disagreement with the cladists (cf. Hennig, 1966).

Since a property originated at one occasion must have been present before the splitting occurred, we may deduce from C8 and P7:

(C14) No taxonomic characters defining the taxon T_{j-1} can have come into existence later than any of those defining the included taxa T_j.

We also have, however, that:

(C15) No taxonomic characters defining the taxa T_j included in the taxon T_{j-1} can have come into existence earlier than any of those defining T_{j-1}.

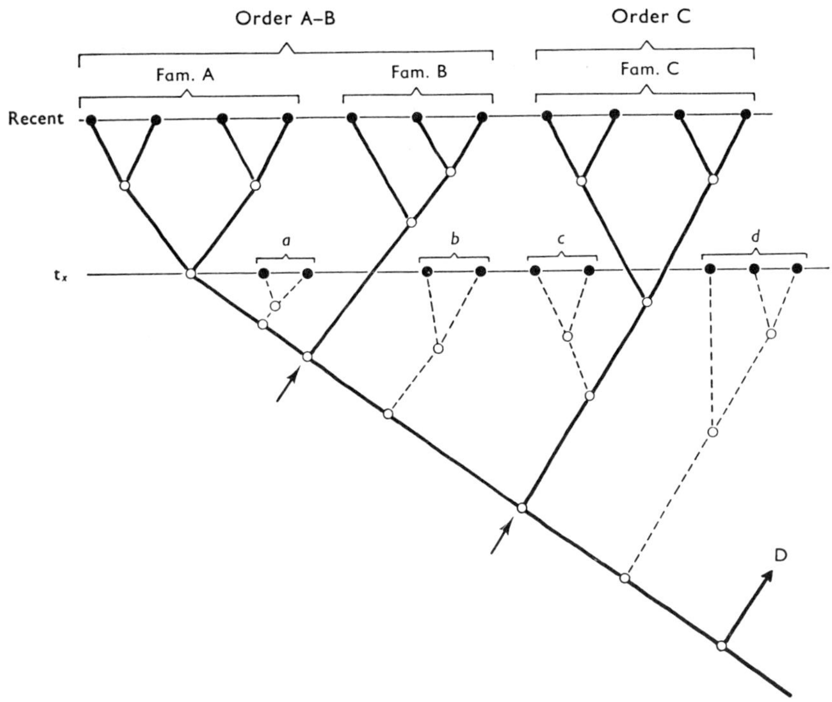

Figure 265. Hypothetical diagram of phylogenetic classification. The 'family' A is seen to be the twin taxon to the extinct groups a, the 'order' A–B to the groups b. It will be realized that if one or more extinct groups were introduced between the arrows, they would all have A–B as their twin taxon. (From L. Brundin, in *Current Problems of Lower Vertebrate Phylogeny* (1968), Almquist & Wicksell)

Since, in C13, the taxon symbolized by T_{j+1} is nearer the apex than are the taxa T_j, it follows:

(C16) Phylogenetic evolution has proceeded from the apex to the base in the hierarchy of phylogenetic classification.

We may further deduce:

(C17) Phylogenetic classification must begin at the apex of the hierarchy.

This conclusion is implied by P6, stating that phylogenetic classification must reveal the genealogical relations between the various taxa, and by C16, stating the direction of the process through which the latter arose.

We have seen above that phylogenetic classification involves the establishment of a continuous series of taxonomic bifurcations. There might be reason to anticipate that the whims of Nature would ensure that the phenomenon of branching has proceeded at different speeds in disparate parts of the hierarchy, and empirical observations rather support this notion. Hence it is warranted to postulate:

(P8) The frequency of bifurcation varies from one region to another in the hierarchy of phylogenetic classification.

It is easily seen that if, in two twin taxa near the base of the hierarchy, no splitting has occurred in one line since the separation from the other taxon, while the latter has been subject to this event two or more times, then the taxa at the base of the hierarchy cannot be of equal rank. Owing to this circumstance clearly there is no sense in using their ordinary categorical names, and we shall therefore introduce the name *terminal taxa* for the taxa situated at the terminal branches of the hierarchy. Noting that the effect illustrated by the practical example may be either amplified or reduced as we pass upwards in the system we shall conclude from P7 and P8:

(C18) The terminal taxa in the hierarchy of phylogenetic classification do not have the same rank.

We may now touch upon the question of the number of taxonomic levels in the phylogenetic classification. From C10 and C18 it is possible to deduce:

(C19) If, in the taxon $T_{j\div 1}$, there are n terminal taxa, then $j > \log n/\log 2$.

It is readily seen that if the bifurcation occurred regularly, then n would be a power of 2, and $j = \log n/\log 2$. The irregularity in the occurrence of splitting warrants the inequality in C19.

It might be objected to C18 that we, by convention, could make the terminal taxa of equal rank, but in that case the superior taxa higher up in the hierarchy would be of different rank when approached from the separate sister taxa. Hence, from C16 and C18 we deduce:

(C20) The numbering of the taxonomic levels in the hierarchy of phylogenetic classification is consistent only when it begins from the apex.

This conclusion thus justifies the decision taken above concerning the numbering of the taxonomic levels in phylogenetic classification. Before we proceed with the argumentation we shall define the concept *phylogenetic line*, or 'line', when no ambiguities obtain, as follows:

(D3) A phylogenetic line is a sequence of taxa in the hierarchy of phylogenetic classification, $T_1, T_2, \ldots T_n$, each of which is included in the preceding one.

It is easily visualized that a phylogenetic line will be step-like, the horizontal parts representing the event of isolation, whereas the vertical parts represent periods of time during which, as we shall presently discuss, taxonomic characters have been acquired.

In C13 we have been able to deduce a relation between taxonomic rank and age. This conclusion may be generalized as follows:

(C21) In a line in the hierarchy of phylogenetic classification T_j is older than T_{j+1}, which in turn is older than T_{j+2}, etc. Nothing can be predicted about the relation between age and rank of taxa in different lines.

We may further conclude from C19:

(C22) It is not possible *a priori*, i.e. without knowledge of n, the number of terminal taxa that must be accommodated within the phylogenetic classification, to establish the minimum number of necessary taxonomic levels.

In connection with this conclusion it may be mentioned that the terminal taxa in the phylogenetic hierarchy by no means must coincide with the lowest taxa, species and subspecies, the ordinary classification. On the contrary, if it is possible, on the basis of the variation obtaining within a taxon of this kind, to establish subgroups which have arisen through bifurcations as envisaged in P7, then it is clearly not only permissible, but even mandatory, to introduce these as terminal taxa in the system.

It is a well-known fact that many taxa, particular those of superior rank, differ from their twin taxon in a number of taxonomic characters, as demonstrated with unusual clarity in the taxa Crocodylia and Aves. Nobody would presumably claim that in general an assemblage of characters is acquired at one step, and for the mentioned case the fossil *Archaeopteryx*, intermediate between birds and reptiles, clearly demonstrates

that several steps must have been involved. Hence we may submit the following premise:
(P9) Some phylogenetic taxa T_j are distinguished by several properties, including taxonomic characters, some of which have been acquired at independent occasions.

The second clause in the premise allows for the possibility that certain characters may be causally related and thus may have originated at the same occasion, a point which has been discussed above on several occasions.

Since two twin taxa have followed the same phylogenetic line from the origin of life until they became separated, it may be warranted, even with allowance for the phenomena of loss and convergence, to draw the following inference:
(C23) No other taxon can have as many characters in common with the taxon T_j as its twin taxon.

This certainly is a reassurance for the one who, following the cladistic dictum, sets out to look for the 'sister' of a given taxon, even if, of course, only some of the characters define the level j. The circumstance that, as stated in P9, phylogenetic taxa often are distinguished by several independent characters emphasizes the importance of the approach of numerical taxonomy, involving that as many characters as possible should be employed (Sokal and Sneath, 1963). The value of the features used by these authors may be questioned in some instances; in general, I believe that most weight should be put on qualitative rather than on quantitative properties. In the future it may be anticipated that physiological and biochemical properties will become of increasing importance for classification, particularly in those cases where the traditional morphological characters have failed to establish satisfactorily the phylogenetic relationship between various taxa.

From the principle of dichotomy stated in P7, together with P9, we may infer:
(C24) If the taxon T_j is distinguished by n actual and potential taxonomic characters, then the formation of the taxon, from the time of separation from its twin taxon till the origin of the taxa T_{j+1}, has involved n steps of bifurcation.

Since these various steps have had no taxonomic consequences we conclude from C11 and C24:
(C25) In the steps of bifurcation involved in the formation of a taxon T_j, all the branches that did not acquire T_j characters became extinct.

The general knowledge about the number of characters which distinguish various taxa may give a fair impression of the enormous degree of extinction that has been associated with phylogenesis. Surely, the road of evolution is strewn with extinguished taxa.

Since we cannot erect taxa for extinct organisms, it is not possible to state formally the consequences of extinctions which have occurred in the past. However, allowing for the possibility that a contemporaneous taxon becomes extinct, we may, on the basis of C25, formulate the taxonomic consequences of this event in the following way:
(C26) The extinction of a taxon T_{j+1} requires that its twin taxon is raised to the level j, its $j+1$ characters becoming j characters, and that the numerical value of the rank of the taxa included in the twin taxon is lowered by one.

The event opposite to extinction is the recording of a new taxon, the effect of which upon the system may be stated in the following terms:
(C27) The discovery of a taxon T_{j+1} requires that, from the taxon T_j which includes the new taxon, a twin taxon to the latter is erected, defined by some of the previous T_j characters, and that the numerical value of the ranks of the taxa included in the twin taxon is raised by one.

The preceding discussion shows that it is possible to delimit the period during which a taxon T_j came into existence, viz., from the time of subdivision of the taxon T_{j-1}, in which it is included, until the time when it itself becomes subdivided into two extant taxa T_{j+1}. Since the events which have taken place in two twin taxa during this interval are subjected to chance, both as regards innovation and extinction, it is obviously impossible to make any predictions about the relative age of two sister taxa T_j. What happened when the taxon T_{j-1}, from which they arose, was bifurcated was not that the taxa T_j, but only the taxonomic level j, became defined.

At this stage it may be warranted to discuss certain concepts established by Hennig (1966). This author 'will call the characters or character conditions from which transformation started . . . in a monophyletic group *plesiomorphous*, and the derived conditions . . . *apomorphous*' and 'the presence of plesiomorphous characters in different species *symplesiomorphy*, the presence of apomorphous characters *synapomorphy*' (Hennig, 1966, p. 89). It is easily visualized that, in a phylogenetic line originating at the top of the hierarchy and ending with a terminal taxon, the sum of the characters accumulated in each and all of the vertical parts of the line constitute the total of all the actual and potential taxonomic characters possessed by the members of the terminal taxon. It is also seen that among all these characters those that were accumulated during the formation of the taxon T_j, as stated in C24, are apomorphous for this taxonomic level. Those properties that were acquired before the separation of T_j from its twin are all plesiomorphous, and those that originated after the separation of the two incipient taxa $T_j{}^+{}_1$ cannot be characterized by either of these two epithets. Hence it follows that all characters possessed by a living organism are apomorphous for one, and only one, taxonomic level j, and plesiomorphous for all taxa of superior rank.

Empirically it has been found that a number of irregularities obtain in living organisms which are difficult to account for in classificatory systems. I do not pretend to be able to give an exhaustive analysis of this phenomenon, but some important points may be covered by the following axioms.

(P10) Identical or similar properties can be acquired independently in different taxa.

From a logical point of view this premise is uninteresting. The only inference that can be drawn is that if this situation obtains in taxa that are not too distantly related it may be embarrassing to the systematist and that other characters must be used to entangle the proper phylogenetic relationships.

The phenomenon described by P10 is called 'convergence' or 'parallelism'. It should be pointed out that a situation which may be interpreted as parallelism can arise if a new taxon arises through hybridization of two independent monophyletic taxa. If this occurs the hybrid clearly may possess taxonomic characters from both parent taxa. Species formation through hybridization is well known from plants; in some cases it has even been possible to create new artificial species through this expedient (cf. Dobzhansky, 1955).

I propose not to introduce hybridization in an axiom, primarily because it would contradict P7, allowing, as it does, for the unification of separate monophyletic lines. Hybridization may be anticipated only between groups which have become separated recently, since genetic compatibility is bound to develop sooner or later between isolated populations. Hence in the hierarchy of phylogenetic classification, hybridization is likely to occur only between inferior taxa, a circumstance entailing that this phenomenon will not seriously upset the classification. However, the superior taxa were once young, and it cannot be excluded *a priori* that, far back in time, hybridization occurred between some of them, even if this possibility seems rather unlikely.

One further 'irregularity' is accounted for by the following premise:

(P11) Taxonomic characters can be lost independently in different taxa.

This postulate has the following consequence:

(C28) If a T_j character x is lost in a taxon T_{j+n} ($n \geqq 1$), then x and $\bar{x}$ become T_{j+n} characters in the lines to which the latter and its twin taxon belong.

The loss of wings in birds and of legs in various reptiles are examples illustrating the situation envisaged by P11.

A supposedly usual phenomenon in Linnean classification is the so-called 'polytypic' or 'polythetic' taxa, i.e. taxa which are classed together because they exhibit a number of similarities, but in which no character is common to the members of all taxa (cf. Beckner, 1959; Mayr, 1969). It may be pointed out that losses of taxonomic characters potentially can account for the existence of polythetic taxa, but the most obvious explanation seems to be that more than two taxa are placed together in one taxon of higher rank. This situation might not arise if they were classified in a phylogenetic hierarchy.

The loss of a character in a subordinate taxon does not change the status of the character as defining a superior taxon. Hence we may conclude:

(C29) A taxonomic character may define taxa of different rank.

The many interesting conclusions that have been derived above, particularly on the basis of P7, demonstrate the significance of cladism for the problem of phylogenetic classification, even though, apparently, the present approach has permitted us to draw several inferences overlooked in the presentations of the cladistic theory (Hennig, 1966; Brundin, 1968).

Indeed, some of the views contended by these authors seem to be in disagreement with the logical consequences of the theory. Thus, as we have shown above, the principle of bifurcation leads to a classification which is so fundamentally different from the classical systematics that, for a number of reasons, it is impossible to work with its categories of taxa. As illustrated in Figure 265, this break has not been made by the cladists; Figure 266 shows how the same classification would look if the consequences of the theory were accepted.

Furthermore, Hennig (1966) distinguishes between the relative and the absolute rank

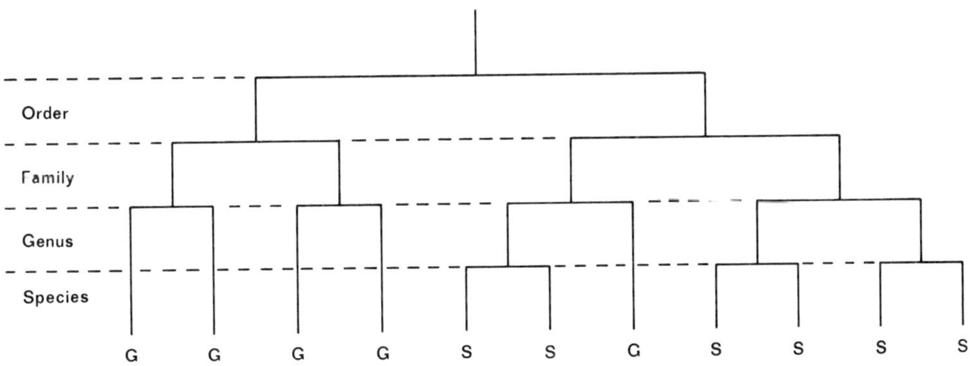

Figure 266. The consequence of preserving Linnean categories in phylogenetical classification. The figure represents the same taxa and the same genetic relationships as in Figure 265. As indicated, some of the terminal taxa are species, others genera

of a taxon. The relative rank is of course determined by its place in the hierarchy, but the establishment of the absolute rank is to Hennig a question of determining the category to which it should be allotted. If we disregard for a moment that, as the above discussion has revealed, this issue cannot be settled in phylogenetic classification, it would still seem to be a rather moot problem if it was only concerned with the categorical name given to a special taxon. However, to Hennig this problem is of great importance because the absolute rank, in his opinion, should be correlated with the age of the taxon, an idea which, incidentally, dates back to Säve-Söderbergh (1934).

The argumentation on this point is not clear, but an idea may be got from the following quotation (Hennig, 1966, pp. 155–156): 'The Acrania [Cephalochordata] and Craniota [Vertebrata] are certainly sister groups. Consequently they are given the same rank in the system, and are (usually) designated subphyla. The Acrania include only about fourteen species, the Craniota tens of thousands distributed over several classes, orders, families, etc. The question now arises whether all the species of the Acrania together correspond to one class of the Craniota, to several orders, to an order with several families, or even to only a family. The fourteen species of the Acrania are generally included in one family, probably because of their slight morphological divergence. Of the two rank levels—family and subphylum—into which the monophyletic groups of these fourteen species are classified, the former is based on morphological divergences among the species themselves, the latter on its relationship to its sister group, the Craniota'.

Without taking a stand on the important question of the phylogenetic kinship between Cephalochordata and Vertebrata it must be admitted that if, indeed, they are twin taxa, then they must be given equal rank. From that stage phylogenetic classification requires only that the animals in each group are classified in a system of bifurcations until the terminal taxa are reached. That the latter will have a much higher rank in the Cephalochordata than in Vertebrata is quite obvious, but the method of classification does not allow for any other approach. Phylogenetic classification allows for the establishment of age relations within, but not between, phylogenetic lines. As pointed out by various authors (e.g. Mayr, 1969; P. J. Darlington, 1970) the flaw in Hennig's reasoning on this point is the assumption that the rate of taxonomic divergence is similar in the different lines, a point which is refuted by various kinds of palaeontological evidence. As it is, the age of taxa can be determined only by independent observations in the fossil record, if at all.

It would be unfair to end this criticism of cladism without stressing once more that Hennig's thesis, that every taxon has a sister taxon of the same rank, the goal of the phylogenetic systematist being to identify the latter, is a principle which unquestionably will be of great importance for phylogenetic classification in the future.

Linnean classification

The name *Linnean classification* has here been given to the traditional classification. The reason for this is not that the current approach is based solely upon the ideas advanced by Linnaeus; as we shall presently see, this is indeed not the case. Yet, in most essentials the classification is still based upon the principles outlined by him, a circumstance which may justify the choice of name. Among the axioms of Linnean classification, although not necessarily stated explicitly by its originator, are those set forth in the first section, P1, P2, P3, P4 and D1.1, and also the conclusions derived from them, namely C1, C2, C3, C4, C5 and C6.

Besides the premises mentioned there are some which are dictated by convention, for expediency or other reasons. The first of these is concerned with the number of taxa which should be included in a taxon of next higher rank, a question that in specific cases may be a matter of deciding how many taxonomic levels should be employed. As we shall presently discuss there are many instances where Nature has left little choice for the systematist. However, when option is possible, certain rules are supposed to be followed. These rules have not been stated specifically, and hence it is necessary to quote an authority on this topic (Mayr, 1969, p.235): 'Since a classification functions like a filing system, its subdivisions (the taxa) should fulfill the demands of greatest possible efficiency, that is, they should ideally be of approximately equal size in order to facilitate information retrieval. The amount of material included under one heading should be neither too voluminous nor too scanty'. We shall later try to formulate a premise which may serve the aim stated by Mayr, but, since the established classification shows the last stipulation to imply that the number of taxa at each level preferably should be greater than two, we may for the time being express this rule in the following premise:

(P12) Whenever possible, the number n_j of taxa T^*_j, included in the taxon T^*_{j+1}, should fulfill the condition $n_j > 2$.

In this premise a special symbol, T^*_j, is introduced for a taxon in the Linnean classification. The reason for this expedient is that, as already apparent, the taxa are different in various respects in the two classificatory systems.

According to the practice introduced by Linnaeus the taxa on the various levels in the hierarchy are given categorical names. Thus, within the animal kingdom the highest category is the phylum, the lowest the species or subspecies, as the case may be. As regards these categories, their use is dictated by certain conventions. Thus, it is required that the categories established by Linnaeus, i.e. six within the animal kingdom, should be applied in classification. Furthermore, the fact that the various levels are designated by names, and not merely by numbers, implies that the total number of categories is fixed by convention. According to Simpson (1961) a complete animal hierarchy comprises twenty-one levels, ranging from the kingdom to the subspecies. These conventions can be stated in the following premise:

(P13) In Linnean classification the minimum and maximum numbers of taxonomic levels are established *a priori*.

Confronting this premise with C22, we are entitled to infer:

(C30) The hierarchies established in phylogenetic and Linnean classification are not identical.

Among the rules of convention dealt with in P12 and P13 only the one stating that all categories should be employed is imperative; the other clearly permit options dictated by taste and preference. This circumstance clearly warrants the following conclusion:

(C31) The rules of Linnean classification do not ensure the erection of a unique hierarchy.

Comparison between C7 and C31 demonstrates a further difference between the two classificatory systems.

The rule that all Linnean categories should be used generally does not pose any problems; most often this number of taxonomic levels is rather too small. However, in certain cases the opposite situation arises, as exemplified by the following quotation: 'the order Tubulidentata (aardvarks) . . . includes only one living species, so that (excluding a few extinct forms) the order Tubulidentata, family Orycteropodidae, and genus *Orycteropus* are all monotypic. In other words, all have identically the same organisms as members, since all include the single set (species) *Orycteropus afer* and no other sets' (Simpson, 1961, pp. 20–21).

According to C4 none of the mentioned taxa, except the order, are definable. The convention in this case thus involves nothing but naming the order three more times. The problem posed by the classification of Tubulidentata bears on 'Gregg's paradox', according to which it is necessary, in cases like the one discussed here, to assume overlapping between the categories in order to avoid the logical inconsistency that a taxon is included in and equal to one of higher rank (Gregg, 1954; 1968). From our point of view the problem is fictitious. The establishment of undefinable taxa may be accepted as a tribute, and even a logical consequence of adherence, to taxonomic convention, but it is an illogical expedient and thus cannot be expected to obey the rules of logic.

It is customarily asserted by systematists that classification begins from the base of the hierarchy, as witnessed by the following quotation: 'In the first stage, often called *alpha taxonomy*, emphasis is on the description of new species and their preliminary arrangement in comprehensive genera. In *beta taxonomy* relationships are worked out more carefully on the species level and on that of the higher categories; emphasis is placed on the development of a sound classification' (Mayr, 1969, p. 15).

This may indeed faithfully represent the actual approach employed in systematics, but it seems a tacit assumption that the affiliation with higher taxonomic categories is no challenge at all to the specialist. If an animal was caught, the relationship of which to the existing phyla and classes could not be established, then naming of the species would seem to be a trivial matter compared to the other systematic consequences of such a discovery.

To return to the aardvarks discussed above, it is not difficult to establish that these animals belong to Vertebrata, Mammalia and a particular order, Tubulidentata, within this class. The decisions regarding these points can, as a matter of fact, be made on the basis of a single specimen, a circumstance suggesting that naming in Linnean systematics also begins at the top of the hierarchy. The convention that the Linnean categories must all be employed, and the species given a name, implies, as we have seen, nothing but inventing synonyms for the name of the order, an activity which hardly justifies the contention that classification passes from the lowest towards the highest ranks.

From the preceding discussion it thus appears that even in traditional systematics classification begins at the apex of the hierarchy. However, this does not follow from any of the axioms involved, hence this requirement must be stated as a premise:

(P14) Linnean classification must begin at the apex of the hierarchy.

With the advent and adoption of Darwin's theory of evolution it has been realized by systematists that the characters used for classification might be expressions of phylogenetic kinship, a standpoint which amounts to the acceptance of P5. The aspiration has even been stated that the classification should reflect the phylogenetic relationships between the living organisms. As we have seen, the straitjacket of conventions in Linnean classification do not permit this goal to be reached, but this fact does not exclude that certain aspects or premises of phylogenetic classification are incorporated. Thus, for one thing, the following stipulation may be made:

(P15) The hierarchy of Linnean classification shall comprise only monophyletic taxa.

This premise is compatible with the conventions outlined above, if a slight change is made in the definition of 'monophyletic taxon', to wit:

(D4) A monophyletic taxon T_j^* is a set of organisms all of which are descendants of the first individuals endowed with all the actual and potential taxonomic characters of T_j^*.

Before the classificatory consequences of P15 and D4 are evaluated we shall replace P12 by an axiom which possibly, when properly applied, may abolish the above-mentioned ambiguity of Linnean classification:

(P16) If two or more phylogenetic taxa T possess actual or potential taxonomic characters in common, they must be united in a Linnean taxon T*.

Reservations are made as regards the generality of the formulation of this premise; instead the implications will be illustrated by means of an example. A tentative phylogenetic classification of the higher Vertebrata is shown in Figure 267(a). The taxonomic characters defining the taxon Reptilia + Aves are seen to comprise only those common to both. However, the taxon Crocodylia, being the primitive or 'plesiomorphous' group in Archosauria, possesses a number of characters in common with the other reptilian orders which, for phylogenetic reasons, can be employed only as taxonomic characters for a taxon T^*_{j+4}. If such characters are employed as proposed by P16, they will serve to define a monophyletic Linnean taxon, Reptilia, of the same rank as Mammalia. Yet, by this expedient Aves becomes suspended in the air, the only way to ensure contact with *terra firma*, a vital necessity even for birds, being to extend the line representing the taxon down, or up in the dendrogram, to the one connecting Reptilia and Aves (Figure

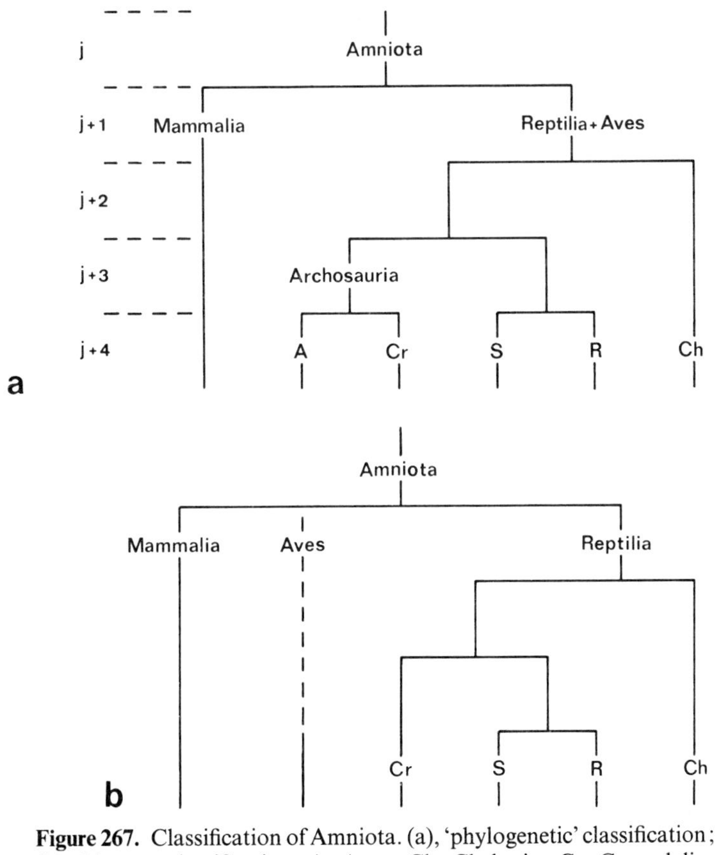

Figure 267. Classification of Amniota. (a), 'phylogenetic' classification; (b), Linnean classification. A, Aves; Ch, Chelonia; Cr, Crocodylia; R, Rhyncocephalia; S, Squamata

267b). In this way Aves thus becomes a monophyletic Linnean taxon of the same rank as Reptilia and Mammalia within the monophyletic taxon Amniota.

The example quoted here shows that in the ordinary classification the relative rank between phylogenetic taxa may be changed, a fact which warrants the following conclusion from C12 and P16, and also from C30 and C31.

(C32) The Linnean categories do not represent absolute ranks.

This conclusion emphasizes that even if the allocation of categorical rank to a given taxon must obey specifiable rules, entailing that the ranking is not arbitrary, the concept of 'taxonomic category' is nonetheless of limited value in phylogenetic contexts. In particular, it is obviously impossible to make absolute inferences about the age of a taxon from its categorical rank. In contrast to what holds for phylogenetic classification this statement is valid even within a monophyletic taxon, as is evidenced by the fact that the latter in Linnean classification may comprise more than two taxa of the same rank. The points discussed here do not exclude, of course, that, in a Linnean 'phylogenetic line', there is a correlation between taxonal age and rank.

As envisaged by C26, the numerical value of the rank of a phylogenetic taxon is changed as a consequence of extinctions. In Linnean classification, where, according to P13, the number of ranks is determined *a priori*, no change of category for a certain taxon follows as a logical consequence of the extinction of a related taxon.This does not preclude that the present classification might have been different if a number of extinct organisms had survived, but this question is completely conjectural since classification according to C2 must be based on living animals. Hence, it may be possible to introduce the following postulate:

(P17) The category of a taxon T_j^* defined by a given set of taxonomic characters has remained unchanged throughout the course of evolution.

According to P9 a phylogenetic taxon may be distinguished by several, independently acquired characters. This clearly must hold for Linnean taxa also, and we may hence deduce from P9 and P17:

(C33) The origination of the taxon T_j^* dates back to the event through which the first organisms came into existence, possessing all the actual and potential taxonomic characters defining T_j^*.

This conclusion demonstrates an important difference between phylogenetic and Linnean classification with respect to extinct organisms. Whereas, in the former case, the ranking of these is excluded, it holds that if, and only if, we are sure that an organism represented by fossil remains possessed of all the characters defining the taxon T_j^*; then it did belong to this taxon. This statement appears contradictory to C2, but may not be so since, in my opinion, it seems quite impossible ever to be sure about *all* characters of extinct organisms.

We may further infer from C4, P14, P15 and C33:

(C34) A taxon T_j^* always arises from another taxon T_j^* within T_{j+1}^*, except the first one which arises from T_{j+1}^*.

Clearly, if the taxon T_{j+1}^* is monophyletic, the included taxa must be genetically related. The only way to account for their origination is that in certain individuals belonging to the original T_{j+1}^* properties arose which today are used as taxonomic characters x, distinguishing a taxon T_j^*. By this expedient two taxa T_j^* were established, defined by x and $\bar{x}$, respectively. Since the categorical level j became established by this event, all further changes involved in the creation of the present organisms belonging to taxa T_j^* within T_{j+1}^* could occur between taxa of the lower rank.

The fossil record has revealed a number of intermediate forms, thus the transition between Pisces and Amphibia is represented by *Ichtyostega*, between Amphibia and Reptilia by *Seymouria*, between Reptilia and Aves by *Archaeopteryx* and between Reptilia and Mammalia by *Ictidosauria*.

In a discussion of these fossils de Beer (1954) points out that they are all mosaics insofar as they possess features typical of both the ancestral and the descendant class, a phenomenon first described by Watson (1919). Some of the properties distinguishing a certain taxon may be causally related, but for others this cannot hold, as indicated by the incomplete list of reptilian and avian characters in Table 13. Hence it follows, as postulated in P7, that the transition between two taxa may occur in many independent steps. According to the suggestions advanced here, the intermediate extinct forms involved in the transition do not belong to any taxon T^*_j, yet according to C34 they must have arisen from a taxon T^*_j—or exceptionally T^*_{j+1}—and have given rise to another taxon T^*_j. Hence:

(C35) Extinct forms, intermediary between two taxa T^*_j, belong to the categorical level j.

Whether or not this conclusion should be used in the classification of extinct animals may be a matter of dispute. However, for the reasons given here, I think that the discussion about *Archaeopteryx* being a bird or a reptile stems from a lack of appreciation of the logics of taxonomy, for this animal surely belongs to a class intermediary between Reptilia and Aves. An appropriate revision must accordingly also be made in Garstang's famous dictum (1922, p. 99): 'The first Bird was hatched from a Reptile's egg', a statement which otherwise, in my opinion, contains more than a grain of truth.

The modifications involved in the transition from an old to a new taxon must occur in a 'straight' line. However, various branchings presumably have occurred in several of the intermediate forms, and it cannot be excluded that the same innovations have arisen in more than one line. Fossil evidence suggests that the phenomenon dealt with here has occurred in the taxa 'uniting' Reptilia and Mammalia, causing some embarrassment in the classification of the latter group (cf. Simpson, 1959). Ideas similar to those submitted here have been discussed by Brink (1967) in a paper on the monophyletic and polyphyletic origins of animal taxa.

The progress of evolution has not been regular to the extent that all diversification at the categorial level j took place before that at the lower levels. Hence, in a certain line phylogenesis may have proceeded with the formation of organisms representing contemporaneous taxa at the levels j–1, j–2, etc., before, among members of one of these taxa, changes occurred which involved substitution of their T^*_j characters with others defining a new taxon T_j. It therefore follows from C15, which must hold also for Linnean taxa, and C34:

(C36) If members of the taxon T^*_j give rise to another taxon T^*_j, then all properties defining taxa of lower rank in the original T^*_j become T^*_j characters.

The origin of the birds is a case in question. Clearly, no change in the ordinal characters of Crocodylia, with preservation of its class characters, would ever lead to the creation of a bird. Yet a four-chambered heart is an ordinal character in Crocodylia and a class character in Aves, and if its presence in the latter case is an inheritance from its reptilian ancestors, then the character has been changed with respect to the level it defines.

No discussion of Linnean classification is complete which does not deal with the species concept. In the present context we shall attempt to approach even this difficult problem through axiomatization, beginning with a premise concerning the origin of taxonomic characters:

(P18) All properties, including actual and potential taxonomic characters, have arisen through mutations. The latter are chance events, the appearance and establishment of which depend upon various intrinsic and extrinsic factors.

The premise coincides with one of the premises of neo-Darwinism, although it does not follow that implications as regards the possible effects of mutations necessarily are the same as those inherent in the theory mentioned.

From P18 we may arrive at the following conclusion:

(C37) The rate of evolution is unpredictable, and varies in different phylogenetic lines.

This statement is accepted by most evolutionists, and it is empirically corroborated by many palaeontological observations (cf. Simpson, 1944). It is of special interest in the present context when confronted with the cladistic proposition that there is, or ought to be, a correlation between the rank and the age of a Linnean taxon, a situation that, according to C21, does not obtain in phylogenetic classification.

In organisms with sexual reproduction the genetic compatibility between different individuals plays an important role in attempts to define the 'species'. The relation between genetic compatibility and mutations is stated in the following premise:

(P19) Some mutations entail genetic incompatibility between different individuals.

From P18 and P19 we have:

(C38) With the passage of time genetic incompatibility must arise between the members of different populations or taxa.

The word 'population' is introduced in C38 to allow for the possibility that genetic incompatibility may arise without therefore giving rise to a new taxon, and the word 'taxa' implies that genetic compatibility may obtain between different taxa. Whether one or both of these alternatives is accepted clearly depends upon the relation between taxon definition and genetic compatibility one chooses to adopt.

If the taxa are defined independently of interfertility, for instance on the basis of morphological characters, then the circumstance that mutations are chance events and that those responsible for genetic compatibility may occur independently of those responsible for changes in other properties (P18 and C38) may allow for the following conclusion:

(C39) If taxa are defined without regard to genetic compatibility, then the taxonomic level at which the latter condition arises in a phylogenetic line cannot be predicted.

As a consequence of C39 it may be anticipated that during the early history of a new taxon its members are interfertile with those of the ancestral one. Since, according to C38, genetic incompatibility must arise with the passage of time, it holds that in general interfertility at most obtains between members of closely related taxa. However, it cannot be excluded that genetic incompatibility obtains between members of the same species, if the latter are defined solely on the basis of morphological characters.

Hence we may deduce from C34 and C39:

(C40) Genetic compatibility and genetic incompatibility can obtain between members of young taxa of all ranks.

This conclusion cannot be refuted by reference to the indisputable fact that the members belonging to different of the existing taxa of high rank are distinguished by genetic incompatibility, since none of these taxa reasonably can be postulated to be in the early stages of their history. Rather it may be asserted that the observations, albeit few, reporting hydridization between members of different species or genera corroborate the conclusion, and the same holds for the occasionally reported infertility between members of morphologically indistinguishable populations.

The preceding discussion has been based on the assumption that taxa were defined without regard to genetic compatibility. Turning to the question about the use of the latter criterion for classificatory purposes it is obvious from C40 that:

(C41) Neither genetic compatibility nor genetic incompatibility can be used as the sole distinction between two taxa.

As far as the high-ranking Linnean taxa are concerned this conclusion clearly is of little interest, since genetical incompatibility always prevails in such instances. Indeed, it is only on the species level that the question of the use of interfertility as a taxonomic character has been brought up (cf. Simpson, 1961; Mayr, 1942; 1969). The fact that reproduction usually occurs between members of the same species has prompted biologists to presume that the species assumes a special position of greater reality than the other taxa, which are more readily accepted to be 'abstract, non-spatiotemporal entities' (Gregg, 1950, p. 426). This notion, as advanced by the advocates of the new systematics on the basis of the theory of neo-Darwinism (cf. J. S. Huxley, 1940), was stated by Simpson (1961, p. 150) in the following way: 'species are in fact evolutionary units'.

Admittedly, the crossings observable between members of different taxa, whether under natural or artificial conditions, are so rare that they may be devoid of phylogenetic importance, except when hybridization leads to the formation of a new species. Nevertheless, the use of genetic compatibility as a species criterion involves certain difficulties; for one thing, it applies only to organisms with bisexual reproduction and, secondly, it would be necessary to make a number of systematic changes involving that individuals belonging to otherwise distinct populations would have to be united in the same species (cf. Burma, 1949; de Laubenfels, 1953; Simpson, 1961).

The situation is different as concerns genetic incompatibility. Many instances have been reported where organisms which to all appearances are identical cannot be crossed or, if they can, give rise to infertile offspring. Clearly in this case it is possible to introduce genetic incompatibility as a species criterion as proposed in the following premise:

(P20) When other criteria fail, genetic incompatibility can be used as the distinction between different species.

The adoption of P20 implies that mating experiments are performed whenever members of the same 'species' are collected from localities so far apart that interbreeding between the populations is unlikely to occur normally. Yet, unless one is under the spell of neo-Darwinian orthodoxy, it is difficult to see any important distinction between two isolated populations of morphologically identical organisms which incidentally have lost genetic compatability, and another pair that has preserved the capacity for interbreeding without actually exploiting this potentiality. The possibility for participation in neo-Darwinian evolution is zero in the one, and infinitely small in the other case.

It appears that the whole problem is simpler when treated from the viewpoint of phylogenetic classification, where many more taxonomic levels are allowed for. In this case genetic incompatibility between two otherwise identical organisms may without further ado be accepted as a criterion of a taxonomic bifurcation, not because it concerns the process of reproduction, but because it is a manifest difference.

Hence, whether or not P20 is accepted as a classificatory convention, it turns out that the definition of 'species' is the real stumbling-block in Linnean classification. Maybe the only definition which does not lead to ambiguities is the following one:

(D5) The species is the lowest category in the original Linnean system.

It was suggested above that the terminal taxa in the phylogenetic hierarchy do not coincide with the Linnean species or subspecies but that, rather, the process of bifurcation may be pursued far below this categorical level. Hence it may be warranted to conclude:

(C42) The species is a set of one to several terminal and subterminal phylogenetic taxa.

As for the criteria applicable for the definition of individual species, we turn to another classic in proposing the following premise:

(P21) 'Hence, in determining whether a form should be ranked as a species . . ., the opinion of naturalists having sound judgement and experience seems the only guide to follow' (Darwin, 1859, p. 47).

Thus, as appears from the preceding discussion, quite apart from the fact that the species in certain cases is undefinable in principle, it probably is the least 'real', i.e. circumscribed, of all taxa, an eventuality which may explain all the controversy about its proper definition. This state of affairs has long been recognized, thus for instance by de Candolle (quoted by Darwin, 1928, pp. 57–58): 'They are mistaken, who repeat that the greater part of our species are clearly limited, and that the doubtful species are in a feeble minority. This seemed to be true, so long as a genus was imperfectly known, and its species were founded upon a few specimens, that is to say, were provisional. Just as we come to know them better, intermediate forms flow in, and doubts as to specific limits augment'. And the reason for this is, in turn, that the species is defined so as to comprise the terminal phylogenetic taxa. The latter represent the ultimate stage reached by phylogenetic evolution and also the only source of further progress. With this reservation it may still be correct to consider the species as 'evolution in the making'.

The discussion of classification above suggests that Linnean classification may give a fair impression of the course taken by taxonomic divergence during phylogenetic evolution, but that a correct representation of the kinship between living organisms can be obtained only through phylogenetic classification. These two systems are, however, incompatible and so long as systematists, with justification it seems, require that their classification must warrant efficient information retrieval there are no chances of integration.

DIRECTION

The question of the direction of evolution has, as a matter of undisputable fact, been settled in the preceding section: phylogenetic evolution has proceeded from the apex to the base of the systematic hierarchies. As pointed out by some authors, e.g. Schindewolf (1950), evolution involving divergence on the species level must result in chaos rather than system in the organic world.

Yet, although the name of his book may leave doubt about his view on this problem, Darwin is not faltering when he states his opinion that the direction has been the opposite: 'And the two new families, or orders, will have descended from two species of the original genus; and these two species are supposed to have descended from one species of a still more ancient and unknown genus' (1859, p. 125).

It is not always easy to get clear-cut answers to this question from members of the neo-Darwinian school, but a couple of examples will show that apparently Darwin's stand is still upheld. Thus: 'the origin of the higher categories is a process which is nothing but an extrapolation of speciation. All the processes and phenomena of macroevolution and of the origin of the higher categories can be traced back to intraspecific variation' (Mayr, 1942, p. 28) and 'the accumulation of the changes of the speciational mode tends towards segregations of higher category' (Simpson, 1944, p. 202).

Taken at face value, these statements are not particularly convincing, for even a very superficial consideration of the nature of intraspecific and interspecific variations shows that they seldom are of the same kind as those distinguishing taxa of higher—or superior—rank. It is evidently impossible to justify the view that the 'individual', subspecific and specific variations known to occur within the classes Aves and Reptilia could ever, through accumulation, be concerned with origination of such features as feathers, wings or beak. Likewise, the outstanding distinction between Chordata and Invertebrata, the formation of a notochord at an early stage of development, surely is an all-or-none phenomenon, arising through one specific epigenetic mechanism. As has been brought out above, the main phyletic character is the unique archetypical body plan, whereas the distinguishing marks of the successive taxa of lower rank consist of a gradual series of less and less comprehensive modifications of this body plan, ending with the specific characters which usually are minor changes in form, size and pattern.

However, irrespective of the views held on the question of direction the phylogenetic trees drawn by various authors usually agree in showing the subdivisions of phyla into classes, classes into orders, etc., in agreement with the course of events recorded by palaeontology. Clearly, then, the disagreement must be concerned with semantics rather than fact. The source of this difference of opinion seems to be two premises basic to neo-Darwinian thought, viz.,

(1) All organisms, past and present, belong to a species.

(2) Reproduction, and hence evolution, at least in bisexual organisms, occurs only between members of the same species.

These premises are clearly influenced by the Linnean tradition, according to which the terminal and subterminal phylogenetic taxa are called species, and given the same categorical rank. As shown above, this convention involves a serious distortion of the ranking in the phylogenetic classification of living animals, suggesting that the absolute value given to the species category is fictitious. Secondly, it was found that we cannot classify fossils in the same system as living beings, entailing that it makes no sense to make assertions about the systematic affiliation of extinct organisms. Yet, it may be argued that, at the time of origin of a major taxon, say a class, taxonomic divergence had progressed so far that a contemporaneous systematist would have been obliged to carry the classification of the organism involved to the species level. Against this argument it must be countered that whatever orders, families, genera and species existed at the time, they have nothing to do with the corresponding taxa which now are included in the class; the latter have not arisen, and could not arise, before the class in question was established with all its actual and potential taxonomic characters. Furthermore, as we have seen, characters defining possible lower taxa became, with the incipient formation of the new class, automatically class characters.

As regards the level at which reproduction occurs it may first be pointed out that it seems to make more sense to say that reproduction occurs between members of the definable order Tubulidentata than that members of the undefinable species *Orycteropus afer* are engaged in this activity. In general it may be stated that since individuals are bearers of the taxonomic characters of all the taxa to which they belong, then they also participate in the sexual act as members of all these taxa. And if, in the past, modifications were concerned with what today are accepted as taxonomic characters for a high-ranking taxon then, indeed, the individuals were representatives of the taxon in question.

The reason why this point is differently interpreted on the basis of Linnean classification lies in the fact that the taxonal rank depends on the degree of taxonomic divergence. A phylogenetic innovation occurring today may well be accounted for by the creation of a new species, but should it turn out to lead to a large number of subordinate taxa it might, in the distant future, be necessary to make the feature a taxonomic character for a taxon of higher rank. If the question is viewed from the standpoint of phylogenetic classification it may be answered by stating that reproduction is possible between members of one, or sometimes a limited number of related, terminal taxa. But once in the past, the present superior taxa were terminal taxa and hence their members might be genetically compatible, at least during the initial phases of separation between the taxa.

The question of the direction of evolution may thus be answered by stating that, irrespective of the manner of classification, the general course of evolution is from the apex to the base of the hierarchies. However, as far as the details are concerned, a difference obtains between the two systems of classification. In phylogenetic classification the direction is always from levels with lower to levels with higher numerical values, thus $n \rightarrow n+1 \rightarrow n+2$, etc. A similar situation obtains within a Linnean monophyletic taxon, but since convention permits that an entity of this kind is taken out of its phylogenetic context and raised to a higher categorical level, it may happen that a class is derived from an order, a genus from a species, etc. So far, therefore, it is possible to endorse the views advanced by Darwin. The application of this principle on contemporaneous evolutionary events is not necessarily warranted. To be sure, if genetic incompability is adopted as a species criterion, then many observations may indeed demonstrate the origination of one species from another one, but it must be recalled that, owing to the specific rules adopted by Linnean classification, only the future divergence of the new 'species' can decide whether or not its descendants will be raised to a higher taxonomic category.

Neo-Darwinism and the complete theory

In a very stimulating analysis of the evolutionary theories of Schindewolf (1950) and Simpson (1953a), Grene (1958, p. 116) stated that these authors 'agree on their major premiss: *traditional morphology and neo-Darwinism are incompatible*'. That two such outstanding evolutionists should reach this impasse is deplorable indeed, for surely we shall never understand the workings of phylogenetic evolution outside the frames of morphology or without the help of population-genetic analysis. The present section represents an attempt to evaluate the extent to which the theory advanced here may serve to bridge this gap.

Before we embark upon this project it may be stressed that nobody questions the importance of the neo-Darwinian theory for intraspecific and interspecific evolution; on this point corroboration obtains to a degree that is rare in biology, and which surely makes neo-Darwinism one of the most successful biological theories. When, however, it is dogmatically stated that this principle may be extrapolated to explain the whole course of evolution, the theory runs into a number of difficulties which to a large extent lie on the morphological level, a circumstance serving to explain that morphologists have been its foremost critics.

For one thing, the premise that selection occurs between interbreeding animals implies that the process is incessant, always striving to reach or maintain an optimal phenotype under given external conditions. Also, since the mutations are always there to be selected, any change in the selection coefficients ensuing from environmental changes will immediately elicit a selectional response. Presumably, if the alterations in the surroundings occur in a sufficiently sophisticated manner any evolutionary result may obtain. This mechanism, the 'progressive adaptation' of Fisher (1930), is contradicted by numerous palaeontological observations for which 'no selectionist interpretation . . . has as yet been given' (Huxley, 1942, p. 508).

Famous examples of this kind are *Gryphaea* and *Megoceros*, the Irish elk. In the former case an excessive growth of the shells occurred in this extinct bivalve mollusc, entailing that the right valve 'could be opened only slightly if at all. Such a state of affairs could only lead to the death of the individual and the extinction of the race' (Swinnerton, quoted from Simpson, 1944, p. 170). In reply to this sweeping conclusion it seems necessary to point out that both the individuals and the race survived long enough to leave fossils and also that extinction has befallen many forms for which such suicidal trends are not demonstrable. The enormous antlers of the Irish elk became, it is surmised, at long last so embarrassing to their possessors that the species became extinct. However, according to the neo-Darwinian interpretation, their growth must have been selected covetously for up to the final slight step of increase which tipped the cart.

The explanation of phenomena of this kind must account for both the initial progression and the ultimate lack of survival of these bizarre animals, as well as for their coming into existence in the first place. Since we have the word of J. S. Huxley that selection, i.e. of the intraspecific type, is of no avail in this context, we may make the assumption that, as proposed by the complete theory, intraspecific selection is not the ubiquitous force asserted by neo-Darwinism. On this premise it is not difficult to account for the progression and temporary survival of such animals by means of random isolation, involving the absence of interspecific competition. Since their peculiarities presumably have involved reduction in the absolute power of adaptability, their extinction may have been caused either by adverse changes in the inanimate environment or the later invasion in their respective niches of animals possessing a higher APA.

Apparently the selection theory may be saved if it is postulated that selection is concerned with an advantageous increase in size, the growth of the antlers being an unavoidable associated phenomenon. However, if this was the case, surely selection should put an end to the growth before the disadvantage of the larger antlers outweighed the advantage of the large body size. As it is, *Megoceros* seems to violate the principle of progressive adaptation. Furthermore, although the existence of pleiotropic genes is acknowledged by geneticists, so is that of modifying genes; it seems that, had not the inventiveness of selection slackened direly in this and similar cases, it ought to have been possible to pick out some mutations retarding the growth of the antlers without affecting the body size.

Convergence or parallel evolution is a very general phenomenon. As examples may be mentioned coloration patterns in the wings of various butterflies and moths, most impressive of course in the cases involving mimicry, and in the furs of diverse mammals.

Mimicry, which, we may presume, has the function ascribed to it, is impossible to explain on the basis of a selection mechanism, if the establishment of a certain colour pattern involves a large number of infinitesimal steps. However, since this event can be accounted for, in large outline at least, through processes involving a limited number of epigenetic factors (cf. Goldschmidt, 1940), the likelihood that a similar pattern may arise spontaneously and independently in different taxa, although undoubtedly very slight, need not be so small that the occurrence of this phenomenon must be ruled out.

On the morphological level convergence may involve growth of the upper canines and the lower jaw (Figure 268) and body shape (Figure 269) in various Vertebrata.

As regards the examples illustrated by these figures it may be stated that if they were universal the selective force of the environment might be invoked as the directive agent. Yet, all carnivores do not have enormous canines, and all aquatic vertebrates do not have a torpedo-shaped body. In the former case the size of the teeth may probably be referred to allometric growth. The shape of the body is intimately associated with certain aspects of the mechanism of segmentation. If this process is continued regularly along the antero–posterior axis, the body will become worm-like, an outcome that usually is associated with suppression of limb formation. Although, with justification, it may be argued that the elongation of the body reduces the usefulness of the limbs it may not be correct to postulate that these two processes have taken place, hand in hand, through a large sequence of minute steps. This explanation is hardly acceptable if it can be shown that the mechanism of uniform segmentation generally has, as a causal effect, the obliteration of the limbs.

Under any circumstances the worm-like shape is the primitive vertebrate body form, as clearly evidenced by the Cyclostomata. The torpedo shape requires a second element in the mechanism of segmentation, involving primarily an antero–posterior gradient in the size of the myotomes and the ribs. But this expedient, which may or may not involve the cooperation of several genes, can with some confidence be considered a unitary mechanism, being one of the two major modes of segmentation in Vertebrata.

Another point which is difficult to explain by neo-Darwinism is the lack of transitional forms in the fossil record. As mentioned above, two different opinions have been voiced to explain this fact, namely, either that the transitional forms never existed, or that the fossil record is anything but complete, and that many intermediate types have not been found, and in many cases may not even have been preserved.

The latter contention undoubtedly contains some truth, thus there certainly must have been many more transitional forms between reptiles and birds than the few ones known to us. This does not, however, allow for the conclusion that transitions between systematics have proceeded in infinitesimal steps. This alternative cannot but give rise to 'hopeless monsters'. As such I would consider a reptile with 117/300 of the forelimbs being feet and the remainder wings, or a gastropod turned 209/300 into a cephalopod. If these animals are the ones the neo-Darwinian evolutionsts are so anxiously hoping to find in the fossil record, then I am sure that Goldschmidt (1940) and Schindewolf (1950) are correct when they contend that this expectation will never be fulfilled. Such animals never did and never could exist, for their formation would be a morphological, their survival an ecological, impossibility. Rather, as I have tried to show, when evolution is considered from an epigenetic point of view, then the problem facing neo-Darwinism is resolved, for most of the systematic differences on all taxonomic levels appear to be all-or-none steps. I therefore believe that, in spite of its incompleteness, the fossil record is in better agreement with the views advocated here than with the neo-Darwinian theory.

Another merit of the complete theory, and a very important one, I believe, is that, as

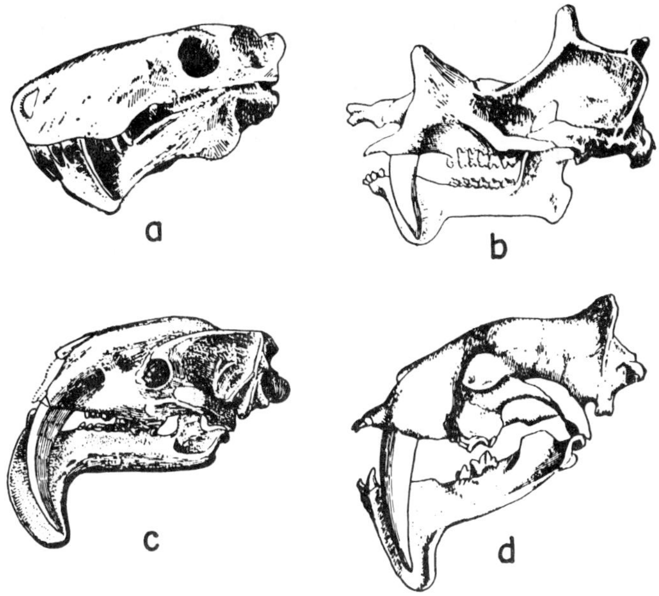

Figure 268. Convergent evolution of excessively long upper canines and corresponding modifications in the lower jaw. (a), *Inostranzewia* (Sauria); (b), *Uintatherium* (Amblypoda); (c), *Thylacosmilus* (Marsupialia); (d), *Eusmilus* (Felidae). (Reproduced from B. Rensch (1960), *Evolution Above the Species Level*, by permission of Ferdinand Enke, Stuttgart)

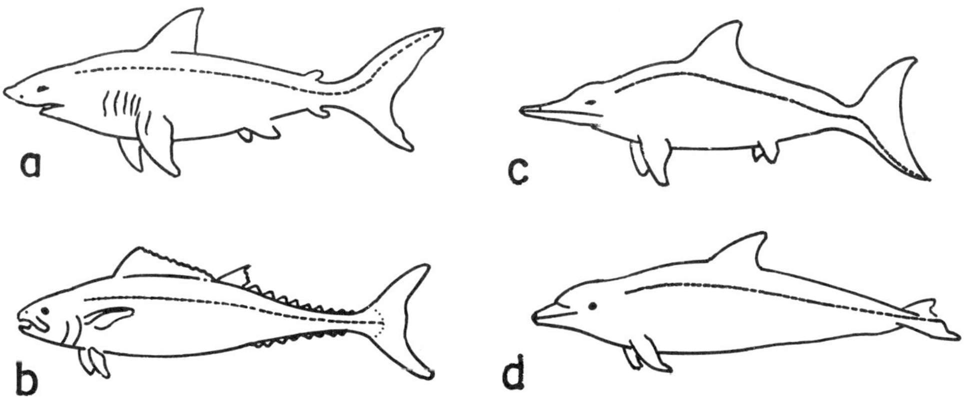

Figure 269. Convergent evolution in the form of the body and the limbs in aquatic animals belonging to different vertebrate classes. (a), Selachii; (b), Pisces; (c), Ichtyosauria; (d), Mammalia. (Reproduced from O. Schindewolf (1950), *Grundfragen der Paläontologie*, after Jacobs, by permission of E. Schweizerbart'sche Verlagsbuchhandlung, Stuttgart)

pointed out by Waddington in the discussion of the paper of Schützenberger (1967), epigenetics constitutes the missing link between the two topologically different spaces occupied by the genotypes and the organisms, respectively.

Dobzhansky (1951) has stated that 'any evolution theory which disregards the established genetic principles is faulty at its source'. I think that nobody would contest this assertion, but the following paraphrase appears to be equally true: Any evolution theory which disregards the established epigenetic principles is faulty at its source. Genetics is a necessary, but not a sufficient, basis for the explanation of phylogenetic evolution, and the same holds for epigenetics. Taken together they seem to be sufficient, and therefore a complete evolutionary theory must encompass both of these disciplines.

The 'complete' theory propounded here is an eclectic theory, which owes its completeness only to the fact that it rests on a much broader base than neo-Darwinism, including a large number of ideas which have been advanced by outstanding biologists in the course of time, both before and after Darwin published his theory. As I, and many others before me, have tried to show, this 'complete' theory accounts effortlessly for most, if not all, of the empirical observations which, to the unbiased mind, constitute falsifications of the synthetic theory.

It is, under these circumstances, very difficult to understand the completely deprecatory attitude exhibited by the neo-Darwinians towards any attempt to extend and modify their theory. The noblest effort a scientist can indulge in is, as Popper (1968) has shown us, after all to attempt to falsify prevalent theories, for success in this pursuit paves the way for the only possible kind of scientific progress, the substitution of old theories for new and better ones.

I shall not try to answer the question concerning the deviating attitude of the neo-Darwinists, but one possible reason may be pointed out. If, namely, the complete theory is correct, then the supreme creative forces of the synthetic theory, intraspecific selection and environmental pressure, assume a much more modest, if not unimportant, role in the drama of phylogenetic evolution. Rather, the real creative agent becomes the mutations responsible for innovations, particularly through interaction with epigenesis, aptly assisted by isolation. Selection answers for the progress of evolution even in the complete theory, but interspecific selection is invoked to account for more than trivial strides. A particular consequence of this view on the effect of selection is that the implication of fitness becomes comparative rather than superlative. 'Fittest' no longer means the 'fittest possible', but only the 'fittest that ever lived in a particular niche'.

As a conclusion it may be pointed out that the occurrence of the large and very rare mutations postulated by the theory advocated here shakes the 'democratic' foundation of neo-Darwinism, according to which almost every individual may contribute to the advance of the tribe by adding its genes to the common pool. As far as I can see, everything suggests that the chore of ensuring that evolution is progressive is the prerogative and the burden of the chosen few.

'I make no claim whatsoever having invented anything new on my own. All I have done, and still am to do, is to put together old ideas'.
(L. Croizat, 1958, Vol. 2, p. 183)

'It is my aim to do today what ought to have been done at least forty years ago, that is, to place before the naturalist a body of thought . . . of essentially deductive nature'.
(L. Croizat, 1958, Vol. 1, p. XIV)

CONCLUSION

After the pretentious statements on 'deductive sciences', 'theoretical biology', etc., that I made in the introduction to this book, its contents may possibly be considered a failure, falling vastly short of reaching the established goal. Yet, admitting this, I still venture to contend that there is no rival theoretical work which, on the basis of so relatively few premises and assumptions, can account for so many observations on ontogenesis and phylogenesis, among which several have proven refractory to theoretical interpretation so far. Therefore, even if there are no doubt many aspects of my work which offhand can be rejected by specialists, I nevertheless assert that even if only the main lines are correct, then it may still be of value. I know from my own experience that some of the hypotheses advocated by me fulfill the foremost requirement to be made concerning intellectual constructions, namely, that they deductively lead to predictions which can be tested by experimental observations. Thus, even if the predictions should turn out to be false, my endeavours might at least contribute to the encouragement of planned biological research, and possibly to the formulation of new and better hypotheses. Indeed, this much a theorist is bound to hope that his work may achieve, for it is regrettable if a theory is wrong, but it is unforgivable if it is sterile. In this context some consolation may actually be found in de Morgan's paradox: 'Wrong hypotheses, rightly worked from, have produced more useful results than unguided observation' (quoted from Medawar, 1967, p. 150).

However, there is another, and more weighty, reason why the views advocated by me deserve consideration, namely, that to a very large extent they are not mine. As I have tried to show, most of the ideas which are part and parcel of the theory of epigenetics have been expounded by outstanding biologists through more than a century, beginning with von Baer, His, Roux and Davenport, and ending with Thompson, Goldschmidt, Woodger, Holtfreter, Schindewolf, Brien, Waddington, Willmer and Gustafson, to mention only a very few of the contributors to biological thought.

At this point I may refer to the statement by Croizat quoted at the head of this section. It is my serious conviction that the theory of epigenesis presented here could have been advanced forty years ago or even earlier. At that time it would not have been possible to substantiate the views with so many experimental observations as today. However, even if many more premises would have had to be stated as unsupported postulates, it would have been feasible to establish a theory very similar to the present one. Since the study of embryology presumably would have benefited from such an effort it is most deplorable that it was not made.

It is tempting to try to find the reason for this lack of enterprise in the field of theoretical biology. This state of affairs has not always prevailed, for in the wake of Darwin's work, but even before, a lot of biological hypotheses were published and subjected to heated controversies (cf. Rádl, 1930). Much of this work was undoubtedly completely sterile from a biological point of view, for instance the vitalism–mechanism debate centred around Driesch, but the Morgan–Weismann debacle turned out to be very fruitful. Nonetheless, weariness and suspicion towards the theorist developed around the beginning of this century. One possible reason for this course of events is the quite miraculous empirical advances which were first made at this time and which, with the constant improvement of the experimental techniques, have continued to pour forth in a never-ending and steadily growing stream. It is justifiable that biologists are proudly and primarily concerned with this enormous treasure of knowledge. However, the reverse of the coin is that the technical sophistication and the enormous quantities of information have forced upon the practical worker a specialization which entails that very few ever manage to gain a comprehensive mastery of any biological discipline, let alone the whole field of biology. Clearly, the situation is ripe for the training of a corps of theoretical biologists who would take upon themselves the task of surveying, digesting and generalizing the information collected by the practical workers, similarly to what has already been realized within the fields of physics and chemistry.

It may be argued that there are already a lot of people working with theoretical biology. This is true, but as appears from a perusal of their work, it is mostly confined to Woodger's third level, as defined in the introduction, implying that their efforts may be of great value and inspiration for theoretical biologists working on the second level, but less for the practical biologist. Unfortunately, it is on the second and vital level that the shortage of contributors is most appalling. It is sometimes suggested that the practical biologists cannot appreciate theoretical work of the kind which can inspire and guide their work. I do not think that this accusation is fair, judging from the particular esteem enjoyed by Mendelism, neo-Darwinism and molecular biology. As we have seen in the preceding pages, the situation is rather the opposite, namely, that biologists stretch the application of existing theories far beyond their range of validity.

The essence of the criticism of the established theories which I have tried to intimate in the present book is not that these theories are wrong. On the contrary, they have made marvellous contributions to our knowledge. However, these theories are not comprehensive enough: there is more to life than Mendelian inheritance and molecular biology. What primarily is missing is the physical aspect. Morphogenesis, the very core of ontogenesis and phylogenesis, is largely a matter of physical forces and events which cannot be analysed in chemical, let alone in genetical, terms. I hope that the underscoring of this fact will be an encouragement to those biologists who, more or less intuitively, have felt that the fashionable trends in biology for many years have neglected the aspects of biology most cherished by them.

'I ask charity for . . . imperfections in view of the labor involved'.
(L. H. Hyman, 1940, p. IX)

EPILOGUE

The present book is not the one I would have liked to write. For that purpose I might have needed at least twenty years of reading, thinking and writing. However, if this approach was possible once upon a time, it certainly is no more, for one thing because the pertinent literature grows at a speed which exceeds the rate at which it can be digested. The present book touches on problems of biophysics, biochemistry, genetics, cytology and morphology in general, embryology, taxonomy, palaeontology and biogeography. No living person can possess an adequate knowledge of all these subjects, and the most sensible course would undoubtedly have been to set up a team of specialists to write the book. This would have been a feasible project if it had been concerned with accepted theories and ideas, but to find people willing to participate in the challenge to prevailing orthodoxy is wellnigh impossible.

Therefore, being captured by the idea that it is possible to make biology a 'well-developed' science by establishing a theory encompassing the undeveloped aspects, I have had to go the way alone. This cannot but influence the quality of the work. For one thing, I have been forced to be very restrictive in my reading, concentrating on reviews and monographs. This was particularly necessary because I was forced to plough through much of the older literature, in which was to be found valuable information which had been buried and forgotten. One particular source of worry is that I have not always paid due credit to my predecessors. As I have said above, I cannot claim originality for many of the ideas advocated here. Whenever possible, I have mentioned the original source, but I know that in many cases this does not hold true. Often the reason may be that I have read about the particular point a long time ago, incorporating it unconsciously in my own frame of references. In other cases I may know that the idea has been advanced by someone else but without being able to find the reference. Finally, some ideas may be original to the extent that I have come upon them, often by deduction, without knowing that someone else had got there before me. Maybe coincidences of this kind support the soundness of my approach.

The writing of the present book was begun in 1966, and the parts concerning ontogenesis were largely finished in 1969. Since then much evidence has been published and come to my knowledge, some of which did, and some of which did not, support the views I held at the time. The latter kind of information, as well as such implying supplementation and complementation of my ideas, has been incorporated in the text, after proper modifications of the theory. Data which in my opinion are confirmatory have largely been left unmentioned.

REFERENCES

O. Abel, 1929, *Paläobiologie und Stammesgeschichte*, Gustav Fischer Verlag, Jena.

P. H. Abelson, 1947, 'Permeability of eggs of Arbacia punctulata to radioactive phosphorus', *Biol. Bull.*, **93**, 203.

H. Abercrombie and J. E. M. Heaysman, 1953, 'Observations on the social behaviour of cells in tissue culture I. Speed of movement of chick heart fibroblasts in relation to their mutual contacts', *Exptl. Cell Res.*, **5**, 111–131.

E. Adams and C. V. Finnegan, 1965, 'An investigation of lactate dehydrogenase activity in early amphibian development', *J. Exptl. Zool.*, **158**, 241–252.

M. R. Adelman, G. G. Borisy, M. L. Shelansky, R. C. Weisenberg and E. W. Taylor, 1968, 'Cytoplasmic filaments and tubules', *Feder. Proc.*, **27**, 1186–1193.

J. W. Adelson, 1971, 'Enterosecretory proteins', *Nature*, **229**, 321–325.

B. A. Afzelius, 1956, 'The ultrastructure of the cortical granules and their products in the sea urchin egg as studied with the electron microscope', *Exptl. Cell Res.*, **10**, 257–285.

I. Agrell, 1964, 'Natural division synchrony and mitotic gradients in metazoan tissues', in *Synchrony in Cell Division and Growth* (ed. E. Zeuthen), Wiley, New York, pp. 39–67.

E. R. Allen and F. A. Pepe, 1965, 'Ultrastructure of developing muscle cells in the chick embryo', *Am. J. Anat.*, **116**, 115–147.

R. D. Allen, 1958, 'The initiation of development', in *A Symposium on the Chemical Basis of Development* (ed. W. D. McElroy and B. Glass), Johns Hopkins Press, Baltimore, pp. 17–72.

R. D. Allen and J. L. Griffin, 1958, 'The time sequence of early events in the fertilization of sea urchin eggs. I. The latent period and the cortical reaction', *Exptl. Cell Res.*, **15**, 163–173.

E. J. Ambrose, 1961, 'The movements of fibrocytes', *Exptl. Cell Res.*, Suppl. **8**, 54–73.

P. Ancel and P. Vintemberger, 1948, 'Recherches sur le déterminisme de la symétrie bilatérale dans l'oeuf des amphibiens', *Bull. Biol.*, Suppl. **31**, 1–182.

G. F. Andrews, 1897, 'Some spinning activities of protoplasm in starfish and Echinus eggs', *J. Morphol.*, **12**, 367–389.

R. Assheton, 1896, 'Notes on the ciliation of the ectoderm of the amphibian embryo', *Quart. J. Micros. Sci.*, **38**, 465–484.

W. T. Astbury, E. Beighton and C. Weibull, 1955, 'The structure of bacterial flagella', *Symp. Soc. Exptl. Biol.*, **9**, 282–305.

K.-B. Augustinsson and T. Gustafson, 1949, 'Cholinesterase in developing sea-urchin eggs', *J. Cell. Comp. Physiol.*, **34**, 311–321.

F. J. Ayala, 1970, 'Competition, coexistence and evolution', in *Essays in Evolution and Genetics in Honor of Theodosius Dobzhansky* (ed. M. K. Hecht and W. C. Steere), North-Holland, Amsterdam, pp. 121–158.

S. Bäckström, 1959, 'Activity of glucose –6–phosphate dehydrogenase in sea urchin embryos of different developmental trends', *Exptl. Cell Res.*, **18**, 347–356.

'I ask charity for . . . imperfections in view of the labor involved'.
(L. H. Hyman, 1940, p. IX)

——, 1963, '6-Phosphogluconate dehydrogenase in sea urchin embryos', *Exptl. Cell Res.*, **32**, 566–569.
S. Bäckström, K. Hultin and T. Hultin, 1960, 'Pathways of glucose metabolism in early sea urchin development', *Exptl. Cell Res.*, **19**, 634–636.
K. E. von Baer, 1828, 1837, *Ueber Entwickelungsgeschichte der Thiere: Beobachtung und Reflexion*, Gebrüder Kornträger, Königsberg.
A. Bajer and J. Molè-Bajer, 1956, 'Cine-micrographic studies on mitosis in endosperm II. Chromosome, cytoplasmic and Brownian movements', *Chromosoma*, **7**, 558–607.
P. C. Baker and T. E. Schroeder, 1967, 'Cytoplasmic filaments and morphogenetic movements in the amphibian neural tube', *Develop. Biol.*, **15**, 432–450.
R. C. Baker, 1927, 'The early development of the ventral part of the neural plate of Amblystoma', *J. Comp. Neurol.*, **44**, 1–27.
F. M. Balfour, 1885, *The Works of Francis Maitland Balfour*, Vol. 2 (ed. M. Foster and A. Sedgwick). Macmillan, London.
B. I. Balinsky, 1959, 'An electron microscopic investigation of the mechanism of adhesion of the cells in a sea urchin blastula and gastrula', *Exptl. Cell Res.*, **16**, 429–433.
——, 1965, *An Introduction to Embryology*, 2nd ed., Saunders, Philadelphia.
E. G. Ball and B. Meyerhof, 1940, 'On the occurrence of iron-porphyrin compounds and succinic dehydrogenase in marine organisms possessing the copper blood pigment hemocyanin', *J. Biol. Chem.*, **134**, 483–493.
F. Baltzer and P. S. Chen, 1960, 'Über das cytologische Verhalten und die Synthese der Nukleinsäuren bei den Seeigelbastarden *Paracentrotus* ♀ x *Arbacia* ♂ und *Paracentrotus* ♀ x *Sphaerechinus*♂', *Rev. Suisse Zool.*, **67**, 183–194.
F. D. Barbieri and H. Salomon, 1962, 'Cetoacidos en huevos de Bufo arenarum en desarrollo. II Analisis cuantitativo', *Arch. Biochim. Quim. y Farm. Tucumán*, **10**, 85–92.
——, 1963, 'Some experimental studies on glycogen and lactic acid in the developing eggs of *Bufo arenarum*', *Acta Embryol. Morphol. Exptl.*, **6**, 304–310.
L. G. Barth, 1941, 'Neural differentiation without organizer', *J. Exptl. Zool.*, **87**, 371–383.
——, 1946, 'Studies on the metabolism of development', *J. Exptl. Zool.*, **103**, 463–486.
H. Barthélemy and R. Bonnet, 1926, 'L'énergie de croissance. IX. Influence de la température sur l'utilization de l'énergie an cours du développement de l'oeuf de Grenouille rousse (*Rana fusca*)', *Bull. Soc. Chim. Biol.*, **8**, 1071–1080.
E. Bataillon, 1929, 'Analyse de la fécondation par la parthénogenèse expérimentale', *Arch. Entw. mech.*, **115**, 707–778.
M. Beckner, 1959, *The Biological Way of Thought*, Columbia University Press, New York.
G. R. de Beer, 1947, 'The differentiation of neural crest cells into visceral cartilages and odontoblasts in *Amblystoma*, and a re-examination of the germ-layer theory', *Proc. Roy. Soc.* [B], **134**, 377–398.
——, 1951, *Embryos and Ancestors*, 2nd ed., Clarendon, Oxford.
——, 1954, 'Archaeopteryx and evolution', *Adv. Sci.*, **11**, 160–170.
R. Bellairs, 1963, 'Differentiation of the yolk sac of the chick studied by electron microscopy', *J. Embryol. Exptl. Morphol.*, **11**, 201–225.
W. E. Berg and N. D. Long, 1964, 'Regional differences of mitochondrial size in the sea urchin embryo', *Exptl. Cell Res.*, **33**, 422–437.
W. E. Berg, D. A. Taylor and W. J. Humphreys, 1962, 'Distribution of mitochondria in echinoderm embryos as determined by electron microscopy', *Develop. Biol.*, **4**, 165–176.
T. Bergfors, K. Hansson-Mild and S. Løvtrup, 1970, 'Modifications in the automatic diver balance technique', *J. Exptl. Biol.*, **53**, 187–193.
K. -E. Berntsson, B. Haglund and S. Løvtrup, 1964, 'Water permeation at different tonicities in the amphibian egg', *J. Exptl. Zool.*, **155**, 317–324.
L. von Bertalanffy, 1952, *Problems of Life. An Evaluation of Modern Biological Thought*, Watts, London.
K. Białaszewicz and M. Mincówna, 1921, 'O przemianie tluszczowej i azotowej we wezesnym rozwoju żaby', *Prace Inst. im M. Menckiego*, **1** (11), 1–23.
W. J. Bock and G. von Wahlert, 1963, 'Two evolutionary theories—A discussion', *Brit. J. Philos. Sci.*, **14**, 140–146.

E. J. Boell and R. Weber, 1955, 'Cytochrome oxidase activity in mitochondria during amphibian development', *Exptl. Cell Res.*, **9**, 559–567.
L. Bolk, 1926, *Das Problem der Menschwerdung*, Gustav Fischer Verlag, Jena.
R. Bonfig, 1925, 'Die Determination der Hauptrichtungen des Embryos von Ascaris megalocephala', *Z. Wiss. Zool.*, **124**, 407–456.
H. Borei, 1948, 'Respiration of oocytes, unfertilized eggs and fertilized eggs from Psammechinus and Asterias', *Biol. Bull.*, **95**, 124–150.
——, 1951, 'Cytochrome c in sea-urchin eggs', *Acta Chem. Scand.*, **4**, 1607–1610.
H. P. Bottelier, H. Holter and K. Linderstrøm-Lang, 1943, 'Studies on enzymatic histochemistry. XXXVI. Determination of peptidase activity, nitrogen content and reduced weight in roots of the barley, *Hordeum vulgare*', *C. R. Lab. Carlsberg, Sér. chim.*, **24**, 289–313.
Th. Boveri, 1888, *Zellen-Studien. 2. Die Befruchtung und Teilung des Eies von Ascaris megalocephala*, Gustav Fischer Verlag, Jena.
——, 1896, 'Zur Physiologie der Kern- und Zelltheilung', *Sitz.-Ber. Physik. Med Ges. Würzburg*, **1896**, 133–151.
——, 1899, *Die Entwickelung von Ascaris megalocephala mit besonderer Rücksicht auf die Kernverhältnisse. Festschr. siebenzigsten Geburtstag C. von Kupffer*, Gustav Fischer Verlag, Jena, pp. 383–430.
——, 1910, *Die Potenzen der Ascaris-Blastomeren bei abgeänderter Furchung. Zugleich ein Beitrag zur Frage qualitativ-ungleicher Chromosomen-Teilung. Festschr. sechszigsten Geburtstag R. Hertwig*, Gustav Fischer Verlag, Jena, pp. 133–214.
A. E. Boycott and C. Diver, 1923, 'On the inheritance of sinistrality in *Limnaea peregra*', *Proc. Roy. Soc.* [B], **95**, 207–213.
A. Brachet, 1921, *Traité d'Embryologie des Vertébrés*, Masson, Paris.
J. Brachet, 1933, 'Recherches sur la synthèse de l'acide thymonucléique pendant le développment de l'oeuf d'Oursin', *Arch. Biol.*, **44**, 519–576.
——, 1939, 'Étude du métabolisme d'oeuf de grenouille (*Rana fusca*) au cours de développement. V. Le métabolisme protéique et hydrocarboné de l'oeuf en relation avec le problème de l'organisateur', *Arch. Biol.*, **50**, 233–267.
J. Brachet and J. Needham, 1935, 'Étude du métabolisme de l'oeuf de grenouille (*Rana fusca*) au cours du développement. IV. Le teneur en glycogène de l'oeuf de la ségmentation a l'éclosion', *Arch. Biol.*, **46**, 821–835.
A. N. Bragg, 1938, 'The organization of the early embryo of Bufo cognatus as revealed especially by the mitotic index', *Z. Zellforsch*, **28**, 154–178.
P. Brien, 1969, 'Polymorphisme intraspécifique et évolution épigénétique', *Bull. Soc. Roy. Sci. Liège*, **38**, 718–734.
R. Briggs and T. J. King, 1959, 'Nucleocytoplasmic interactions in eggs and embryos', in *The Cell* (ed. J. Brachet and A. E. Mirsky), Academic Press, New York, Vol. 1, pp. 537–617.
J. S. Brimacombe and J. M. Webber, 1964, *Mucopolysaccharides*, Elsevier, Amsterdam.
A. S. Brink, 1967, 'On the monophyletic and polyphyletic origins of taxonomic groups', in *Problèmes Actuels de Paléontologie Evolution des Vertébrés* (ed. J.-P. Lehman), Centre National de Recherche Scientifique, Paris, pp. 301–313.
R. J. Britten and E. H. Davidson, 1969, 'Gene regulation for higher cells: A theory', *Science*, **165**, 349–357.
——, 1971, 'Repetitive and non-repetitive DNA sequences and a speculation on the origins of evolutionary novelty', *Quart. Rev. Biol.*, **46**, 111–138.
D. D. Brown, 1964, 'RNA synthesis during amphibian development', *J. Exptl. Zool.*, **157**, 101–114.
D. D. Brown and J. B. Gurdon, 1964, 'Absence of ribosomal RNA synthesis in the anucleolate mutant of *Xenopus laevis*', *Proc. Natl. Acad. Sci.*, **51**, 139–146.
D. D. Brown and E. Littna, 1966, 'Synthesis and accumulation of low molecular weight RNA during embryogenesis of *Xenopus laevis*', *J. Mol. Biol.*, **20**, 95–112.
L. Brundin, 1966, 'Transantarctic relationships and their significance, as evidenced by chironomid midges. With a monograph of the subfamilies Podonominae and Aphroteniinae and the austral Heptagyiae', *Kungl. Svenska Vetenskapsakad. Handl.* [4], **11** (1), 3–472.
——, 1968, 'Application of phylogenetic principles in systematics and evolutionary theory', in *Current Problems of Lower Vertebrate Phylogeny* (ed. T. Ørvig), Almqvist & Wicksell, Stockholm, pp. 473–495.

R. C. Buck and J. M. Tisdale, 1962, 'An electron microscopic study of the development of the cleavage furrow in mammalian cells', *J. Cell Biol.*, **13**, 117–125.

I. K. Buckley and K. R. Porter, 1967, 'Cytoplasmic fibrils in living cultured cells. A light and electron microscope study', *Protoplasma*, **64**, 349–380.

B. H. Burma, 1949, 'The species concept: A semantic review', *Evolution*, **3**, 369–370.

A. L. Burnett, 1968, 'The acquisition, maintenance, and lability of the differentiated state in Hydra', in *The Stability of the Differentiated State* (ed. H. Ursprung), Springer Verlag, Berlin, pp. 109–127.

G. A. Buznikov, I. V. Chudakova and N. D. Zvezdina, 1964, 'The rôle of neurohumours in early embryogenesis. I. Serotonin content of developing embryos of sea urchin and loach', *J. Embryol. Exptl. Morphol.*, **12**, 563–573.

B. Byers and K. R. Porter, 1964, 'Oriented microtubules in elongating cells of the developing lens rudiment', *Proc. Natl. Acad. Sci.*, **52**, 1091–1099.

H. Bytinski-Salz, 1931, 'Untersuchungen über die Induktionsfähigkeit der hinteren Medullarplattenbezirke', *Arch. Entw.mech.*, **123**, 518–564.

——, 1938, 'Chromatophorenstudien II. Struktur und Determination des adepidermalen Melanophorennetzes bei Bombina', *Arch. Exptl. Zellforsch*, **22**, 132–170.

R. D. Cahn, 1968, 'Factors affecting inheritance and expression of differentiation: Some methods of analysis', in *The Stability of the Differentiated State* (ed. H. Ursprung), Springer Verlag, Berlin, pp. 58–84.

D. Carazzi, 1900, 'L'embriologia dell' *Aplysia limacina* L. fino alla formazione delle strisce mesodermiche. Le prima fasi dello sviluppo del *Pneumodermon mediterraneum* van Ben.', *Anat. Anz.*, **17**, 77–102.

G. S. Carter, 1951, *Animal Evolution*, Sidgwick & Jackson, London.

S. B. Carter, 1965, 'Principles of cell motility: The direction of cell movement and cancer invasion', *Nature*, **208**, 1183–1187.

——, 1967, 'Effects of cytochalasins on mammalian cells', *Nature*, **213**, 261–264.

C. T. Caskey, A. Beaudet and M. Nierenberg, 1968, 'RNA codons and protein synthesis. 15. Dissimilar response of mammalian and bacterial transfer RNA fractions to messenger RNA codons', *J. Mol. Biol.*, **37**, 99–118.

P. Cerfontaine, 1906, 'Recherches sur le développement de l'Amphioxus', *Arch. Biol.*, **22**, 229–418.

L. De Cesaris Coromaldi, 1955, 'Studio delle fosfatasi nello sviluppo embrionale e larvale di "Bufo vulgaris"', *Ric. Sci.*, **25**, 2323–2333.

E. L. Chambers, 1939, 'The movement of the egg nucleus in relation to the sperm aster in the Echinoderm egg', *J. Exptl. Biol.*, **16**, 409–424.

R. Chambers and E. L. Chambers, 1961, *Explorations into the Nature of the Living Cell*, Harvard University Press, Cambridge, Mass.

C. H. Child, 1900, 'The early development of Arenicola and Sternaspis', *Arch. Entw.mech.*, **9**, 587–723.

C. M. Clapp, 1899, 'Relation of the axis of the embryo to the first cleavage plane', *Biol. Lectures Wood's Holl 1898*, 139–151.

R. B. Clark, 1964, *Dynamics in Metazoan Evolution. The Origin of the Coelom and Segments*, Oxford University Press, Oxford.

E. R. Clark, W. J. Hitschler, H. T. Kirby-Smith, R. O. Rex and J. H. Smith, 1931, 'General observations on the ingrowth of new blood vessels into standardized chambers in the rabbit's ear, and the subsequent changes in the newly grown vessels over a period of months', *Anat Rec.*, **50**, 129–167.

K. W. Cleland and [N. M. V.] Rotschild, 1952a, 'The metabolism of the sea-urchin egg. Anaerobic breakdown of carbohydrate', *J. Exptl. Biol.*, **29**, 285–294.

——, 1952b, 'The metabolism of the sea-urchin egg. Oxidation of carbohydrate', *J. Exptl. Biol.*, **29**, 416–428.

A. I. Cohen, 1954, 'Studies in glycolysis during the early development of the Rana pipiens embryo', *Physiol. Zoöl.*, **27**, 128–141.

S. Cohen, 1954, 'The metabolism of $C^{14}O_2$ during amphibian development', *J. Biol. Chem.*, **211**, 337–354.

R. K. Cole, 1967, 'Ametapodia, a dominant mutation in the fowl', *J. Heredity*, **58**, 141–146.

J. R. Coleman, 1962, 'Deoxyribonuclease activities in the development of the leopard frog, Rana pipiens', *Develop. Biol.*, **5**, 232–251.

J. R. Collier, 1965, 'Morphogenetic significance of biochemical patterns in mosaic embryos', in *The Biochemistry of Animal Development*, Vol. 1, Academic Press, New York, pp. 203–244.

J. R. Collier and M. McCann-Collier, 1962, 'The deoxyribonucleic acid content of the egg and sperm of *Ilyanassa obsoleta*', *Exptl. Cell Res.*, **27**, 553–559.

E. G. Conklin, 1897, 'The embryology of Crepidula, a contribution to the cell lineage and early development of some marine gasteropods', *J. Morphol.*, **13**, 1–226.

——, 1902, 'Karyokinesis and cytokinesis in the maturation, fertilization and cleavage of Crepidula and other Gasteropoda', *J. Acad. Nat. Sci. Philadelphia* [2], **12**, 3–122.

——, 1903, 'The cause of inverse symmetry', *Anat. Anz.*, **23**, 577–588.

——, 1905, 'The organization and cell-lineage of the ascidian egg', *J. Acad. Nat. Sci. Philadelphia* [2], **13**, 3–119.

——, 1917, 'Effects of centrifugal force on the structure and development of Crepidula', *J. Exptl. Zool.*, **22**, 311–417.

——, 1924, 'Cellular differentiation', in *General Cytology* (ed. E. V. Cowdry), University of Chicago Press, Chicago, pp. 539–607.

——, 1932, 'The embryology of Amphioxus', *J. Morphol.*, **54**, 69–151.

——, 1933, 'The development of isolated and partially separated blastomeres of Amphioxus', *J. Exptl. Zool.*, **64**, 303–375.

G. W. Cooper, 1965, 'Induction of somite chondrogenesis by cartilage and notochord: A correlation between inductive activity and specific stages of cytodifferentiation', *Develop. Biol.*, **12**, 185–212.

E. D. Cope, 1896, *The Primary Factors of Organic Evolution*, Open Court, Chicago.

H. E. Crampton, 1894, 'Reversal of cleavage in a sinistral gastropod', *Ann. N.Y. Acad. Sci.*, **8**, 167–170.

R. K. Crane and A. K. Keltch, 1949, 'Dinitrocresol and phosphate stimulation of the oxygen consumption of a cell-free oxidative system obtained from sea urchin eggs', *J. Gener. Physiol.*, **32**, 503–509.

D. R. Crofts, 1955, 'Muscle morphogenesis in primitive gastropods and its relation to torsion', *Proc. Zool. Soc. London*, **125**, 711–750.

L. Croizat, 1958, *Panbiogeography*, Caracas.

R. A. Crowson, 1970, *Classification and Biology*, Heinemann, London.

L. Cuénot, 1936, *L'espèce*, Doin, Paris.

——, 1940, 'Remarques sur un essai d'arbre généalogique du règne animal', *C.R. Acad. Sci. Paris*, **210**, 23–27.

A. S. G. Curtis, 1961, 'Timing mechanisms in the specific adhesion of cells', *Exptl. Cell Res.*, Suppl. **8**, 107–122.

——, 1966, 'Cell adhesion', *Sci. Progr.*, **54**, 61–86.

G. Cuvier, 1816, *Le Règne Animal Distribué d'aprés son Organisation*, Deterville, Paris.

A. Dalcq, 1928, *Les Bases Physiologiques de la Fécondation et de la Parthénogénèse*, Les Presses Universitaires de France, Paris.

——, 1946, 'Recent experimental contributions to brain morphogenesis in amphibians', *Growth*, **10**, Suppl., 85–119.

——, 1949, 'L'apport de l'embryologie causale au problème de l'évolution', *Portug. Acta Biol.* [A], Vol. R. B. Goldschmidt, 367–400.

A. M. Dalcq, 1954, 'Les ontomutations a l'origine des Mammifères', *Bull. Soc. Zool. France*, **79**, 240–255.

——, 1957, *Introduction to General Embryology*, Oxford University Press, Oxford.

A. Dalcq and A. Dollander, 1948, 'Sur les phénomènes de regulation chez le Triton après séparation des deux premiers blastomères et sur la disposition de la pellicule (coat) dans l'oeuf fécondé et segmenté', *C.R. Soc. Biol.*, **142**, 1307–1312.

J. C. Dan, 1948, 'On the mechanism of astral cleavage', *Physiol. Zoöl.*, **21**, 191–218.

——, 1950, 'Sperm entrance in Echinoderms, observed with the phase contrast microscope', *Biol. Bull.*, **99**, 399–411.

K. Dan, 1943, 'Behaviour of the cell surface during cleavage. VI. On the mechanism of cell division', *J. Fac. Sci. Tokyo Univ.* [4], **6**, 323–368.

——, 1960, 'Cyto-embryology of echinoderms and amphibia', *Internat. Rev. Cytol.*, **9**, 321–367.

K. Dan and J. C. Dan, 1947, 'Behavior of the cell surface during cleavage. VIII. On the cleavage of medusan eggs', *Biol. Bull.*, **93**, 163–188.

K. Dan and K. Okazaki, 1956, 'Cyto-embryological studies of sea urchins. III. Role of the secondary mesenchyme cells in the formation of the primitive gut in sea urchin larvae', *Biol. Bull.*, **110**, 29–42.
J. F. Danielli and H. Davson, 1935, 'A contribution to the theory of permeability of thin films'. *J. Cell. Comp. Physiol.*, **5**, 495–508.
P. J. Darlington, 1957, *Zoogeography: The Geographical Distribution of Animals*, Wiley, New York.
——, 1970, 'A practical criticism of Hennig-Brundin "Phylogenetic systematics" and antarctic biogeography', *Syst. Zool.*, **19**, 1–18.
C. Darwin, 1859, *The Origin of Species*. A facsimile of the first edition with an introduction by Ernst Mayr, 1964, Harvard University Press, Cambridge, Mass.
——, 1928, *The Origin of Species*, 6th ed., Everyman's Library, Dent, London.
C. B. Davenport, 1895, 'Studies in morphogenesis. IV. A preliminary catalogue of the processes concerned in ontogeny', *Bull. Mus. Comp. Zool. Harvard*, **27**, 173–199.
E. H. Davidson, 1968, *Gene Activity in Early Development*, Academic Press, New York.
J. N. Davidson and I. Leslie, 1950, 'A new approach in the biochemistry of growth and development', *Nature*, **165**, 49–53.
B. D. Davis, 1964, 'Theoretical mechanisms of differentiation', *Medicine*, **43**, 639–649.
D. D. Davis, 1949, 'Comparative anatomy and the evolution of the vertebrates', in *Genetics, Paleontology and Evolution* (ed. G. L. Jepsen, E. Mayr and G. G. Simpson), Princeton University Press, Princeton, pp. 64–89.
I. B. Dawid, 1965, 'Deoxyribonucleic acid in amphibian eggs', *J. Mol. Biol.*, **12**, 581–599.
H. Dennis, 1968, 'Role of messenger ribonucleic acid in embryonic development', *Adv. Morphogen.*, **7**, 115–150.
T. A. Detlaff, 1964, 'Cell divisions, duration of interkinetic states and differentiation in early stages of embryonic development', *Adv. Morphogen.*, **3**, 323–362.
H. F. Deutsch and T. Gustafson, 1952, 'The changes in catalase and cytochrome oxidase activity in developing sea urchin eggs', *Ark. Kemi*, **4**, 221–231.
C. Devillers, 1961, 'Mouvements cellulaires dans le développement de l'embryon des amphibiens et des poissons', *Exptl. Cell Res.*, Suppl. **8**, 221–233.
Th. Dobzhansky, 1937, 'Genetic nature of species differences', *Am. Nat.*, **71**, 404–420.
——, 1951, *Genetics and the Origin of Species*, 3rd ed., Columbia University Press, New York.
——, 1970, *Genetics of the Evolutionary Process*, Columbia University Press, New York.
A. Dollander, 1951, 'Observations concernant la structure du cortex de l'oeuf de certains amphibiens urodèles. Mise en évidence du "coat" de Holtfreter', *C.R. Ass. Anat.*, **38**, 430–435.
A. Dollander and G. Debry, 1949, 'Données complementaires sur la régulation consécutive à la ligature frontale de l'oeuf aux stades jeunes chez le triton (*T. helveticus*)', *Ann. Soc. Roy. Zool. Belg.*, **80**, 9–20.
F. Dorris, 1935, 'The development of structure and function in the digestive tract in Amblystoma punctatum', *J. Exptl. Zool.*, **70**, 491–523.
A. H. Drew, 1923, 'Growth and differentiation in tissue cultures', *Brit. J. Exptl. Pathol.*, **4**, 46–52.
H. Driesch, 1892, 'Entwicklungsmechanische Studien. I. Der Werth der beiden ersten Furchungszellen in der Echinodermenentwicklung. Experimentelle Erzeugung von Theil-und Doppelbildungen II. Über die Beziehungen des Lichtes zur ersten Etappe det thierischen Formbildung', *Z. Wiss. Zool.*, **53**, 160–184.
——, 1897, 'Betrachtungen über die Organisation des Eies und ihre Genese', *Arch. Entw.mech.*, **4**, 75–124.
——, 1898, 'Über rein-mütterliche Charaktere an Bastardlarven von Echiniden', *Arch. Entw. mech.*, **7**, 65–102.
O. Duboscq and O. Tuzet, 1937, 'L'ovogénèse, la fécondation, et les premiers stades du développement des éponges calcaire', *Arch. Zool. Exptl. Géner.*, **79**, 157–316.
M. Durand, 1961, 'L'acide déoxyribonucléique (ADN) dans les gamètes et au cours des premiers stades de l'embryogénie des Gryllides (Orth.)', *Bull. Biol.*, **95**, 28–123.
M. Durchon, 1967, *L'endochrinologie des Vers et des Mollusques*, Masson, Paris.
C. de Duve, 1963, 'The lysosome concept', in *Lysosomes* (ed. A. V. S. de Reuck and M. P. Cameron), Churchill, London, pp. 1–35.
C. de Duve, R. Wattiaux and P. Baudhuin, 1962, 'Distribution of enzymes between subcellular fractions in animal tissues', *Adv. Enzymol.*, **24**, 291–358.

R. M. Eakin and J. A. Westfall, 1962, 'Fine structure of the notochord of Amphioxus', *J. Cell Biol.*, **12**, 646–651.
N. B. Eales, 1950, 'Torsion in Gastropoda', *Proc. Malacol. Soc.*, **28**, 53–61.
A. H. Ebeling and A. Fischer, 1922, 'Mixed cultures of pure strains of fibroblasts and epithelial cells', *J. Exptl. Med.*, **36**, 285–289.
M. V. Edds, 1958, 'Origin and structure of the intercellular matrix', in *A Symposium on the Chemical Basis of Development* (ed. W. D. McElroy and B. Glass), Johns Hopkins Press, Baltimore, pp. 157–173.
——, 1964, 'The basement lamella of developing amphibian skin', in *Small Blood Vessel Involvement in Diabetes mellitus* (ed. M. D. Siperstein, A. R. Colwell and K. Meyer), American Institute of Biological Sciences, Washington, D.C., pp. 245–252.
M. Eden, 1967, 'Inadequacies of neo-Darwinian evolution as a scientific theory', in *Mathematical Challenges to the neo-Darwinian Interpretation of Evolution* (ed. P. S. Moorhead and M. M. Kaplan), Wistar Institute Press, Philadelphia, pp. 5–19.
M. Edman, G. Hermerén and S. Løvtrup, 1973, *The Logic of Darwinism*, (in preparation).
H. Eisig, 1898, 'Zur Entwicklungsgeschichte de Capitelliden', *Mitth. Zool. Stat. Neapel*, **13**, 1–292.
G. Ekman, 1925, 'Experimentelle Beiträge zur Herzentwicklung der Amphibien', *Arch. Entw.mech.*, **106**, 320–352.
D. Elson, T. Gustafson and E. Chargaff, 1952, 'The nucleic acids of the sea-urchin during embryonic development', *J. Biol. Chem.*, **209**, 285–294.
Y. Endo, 1952, 'The role of the cortical granules in the formation of the fertilization membrane in eggs from Japanese sea urchins. I', *Exptl. Cell Res.*, **3**, 406–418.
B. Ephrussi and L. Rapkine, 1928, 'Composition chimique de l'oeuf d'oursin *Paracentrotus lividus* Lk. et ses variations au cours du développement', *Ann. Physiol. Physicochim. Biol.*, **3**, 386–398.
L. Ernster and F. Navazio, 1956, 'The cytoplasmic distribution of *isocitric* dehydrogenases', *Exptl. Cell Res.*, **11**, 483–510.
——, 1957, 'Studies on TPN-linked oxidations I. Pathways of *iso*citrate oxidation in rat liver mitochondria', *Biochim. Biophys. Acta*, **26**, 408–415.
C. Evola-Maltese, 1957, 'Histochemical localization of alkaline phosphatase in the sea urchin embryo', *Acta Embryol. Morphol. Exptl.*, **1**, 98–104.
G. Fankhauser, 1955, 'The role of nucleus and cytoplasm', in *Analysis of Development* (ed. B. H. Willier, P. A. Weiss and V. Hamburger), Saunders, Philadelphia, pp. 126–150.
G. Farfaglio, 1963, 'Experiments on the formation of the ciliated plates in ctenophores', *Acta Embryol. Morphol. Exptl.*, **6**, 191–203.
E. Fauré-Fremiet, 1924, 'L'oeuf de Sabellaria alveolata L.', *Arch. Anat. Micros.*, **20**, 210–342.
E. Fauré-Fremiet and J. Dragoiu, 1923, 'Le premier cycle de croissance du têtard de "Rana temporaria" ', *Arch. Internat. Physiol.*, **21**, 403–437.
D. W. Fawcett, 1966, *An Atlas of Fine Structure. The Cell: Its Organelles and Inclusions*, Saunders, Philadelphia.
H. B. Fell and E. Mellanby, 1953, 'Metaplasia produced in cultures of chick ectoderm by high vitamin A', *J. Physiol.*, **119**, 470–488.
R. A. Fisher, 1930, *The Genetical Theory of Natural Selection*, Clarendon, Oxford.
——, 1932, 'The bearing of genetics on theories of evolution', *Sci. Progr.*, **27**, 273–287.
A. Fraser, 1967, 'Comments on mathematical challenges to the neo-Darwinian concept of evolution', in *Mathematical Challenges to the neo-Darwinian Interpretation of Evolution* (ed. P. S. Moorhead and M. M. Kaplan), Wistar Institute Press, Philadelphia, p. 107.
M. Gabe, 1966, *Neurosecretion*, Pergamon, Oxford.
W. Garstang, 1922, 'The theory of recapitulation: A critical re-statement of the biogenetic law', *J. Linn. Soc. London Zool.*, **35**, 81–101.
——, 1928, 'The origin and evolution of larval forms', *Rep. Brit. Ass. Adv. Sci.*, 77–98.
——, 1951, *Larval Forms and other Zoological Verses*, Blackwell, Oxford.
G. F. Gause, 1934, *The Struggle for Existence*, Williams & Wilkins, Baltimore.
C. Gegenbaur, 1870, *Grundzüge der vergleichenden Anatomie*, 2nd ed., Engelmann Verlag, Leipzig.
I. R. Gibbons, 1968, 'The biochemistry of motility', *Ann. Rev. Biochem.*, **37**, 521–546.
D. Gingell, 1970, 'Contractile responses at the surface of an amphibian egg', *J. Embryol. Exptl. Morphol.*, **23**, 583–609.

M. J. Glimcher, 1960, 'Specificity of the molecular structure of organic matrices in mineralization', in *Calcification in Biological Systems* (ed. R. F. Sognnaes), American Association for the Advancement of Science, Washington, D.C., pp. 421–488.
A. Glücksmann, 1951, 'Cell deaths in normal vertebrate ontogeny', *Biol. Rev.*, **26**, 59–86.
O. B. Goin and C. J. Goin, 1968, 'DNA and the evolution of the vertebrates', *Amer. Midland Nat.*, **80**, 289–298.
R. J. Goldacre, 1964, 'On the mechanism and control of amoeboid movement', in *Primitive Motile Systems in Cell Biology* (ed. R. D. Allen and N. Kamiya), Academic Press, New York, pp. 237–255.
R. Goldschmidt, 1940, *The Material Basis of Evolution*, Yale University Press, New Haven.
E. S. Goodrich, 1958, *Studies on the Structure and Development of Vertebrates*, Dover, New York.
R. A. Gortner, 1914, 'Studies on the chemistry of embryonic growth. II. Comparative analysis of eggs and the newly-hatched larvae of the giant salamander, *Cryptobranchus allegheniensis*', *J. Am. Chem. Soc.*, **36**, 1556–1566.
P. Grant, 1958, 'The synthesis of deoxyribonucleic acid during early embryonic development of *Rana pipiens*', *J. Cell Comp. Physiol.*, **52**, 227–248.
——, 1960, 'The influence of folic acid analogs on development and nucleic acid metabolism in *Rana pipiens* embryos', *Develop. Biol.*, **2**, 197–251.
J. Gray, 1924, 'The mechanism of cell-division. I. The forces which control the form and cleavage of the eggs of *Echinus esculentus*', *Proc. Cambridge Philos. Soc. Biol. Sci.*, **1**, 164–188.
——, 1927, 'The mechanism of cell-division. III. The relationship between cell-division and growth in segmenting eggs', *J. Exptl. Biol.*, **4**, 313–321.
D. E. Green and R. F. Goldberger, 1967, *Molecular Insights into the Living Process*, Academic Press, New York.
J. R. Gregg, 1948, 'Carbohydrate metabolism of normal and of hybrid amphibian embryos', *J. Exptl. Zool.*, **109**, 119–134.
——, 1950, 'Taxonomy, language and reality', *Amer. Nat.*, **84**, 419–435.
——, 1954, *The Language of Taxonomy. An Application of Symbolic Logic to the Study of Classificatory Systems*, Columbia University Press, New York.
——, 1957, 'Morphogenesis and metabolism of gastrula—arrested embryos in the hybrid *Rana pipiens* ♀ x *Rana sylvatica* ♂', in *The Beginnings of Embryonic Development* (ed. A. Tyler, R. C. von Borstel and C. B. Metz), American Association for the Advancement of Science, Washington, D.C., pp. 231–261.
——, 1959, 'On deciding whether Protistans are cells', *Philos. Sci.*, **26**, 338–346.
——, 1960, 'Respiratory regulation in amphibian development', *Biol. Bull.*, **119**, 428–439.
——, 1962, 'Anaerobic glycolysis in amphibian development', *Biol. Bull.*, **123**, 555–561.
——, 1968, 'Buck and Hull: A critical rejoinder', *System. Zool.*, **17**, 342–344.
J. R. Gregg and R. Ballentine, 1946, 'Nitrogen metabolism of Rana pipiens during embryonic development', *J. Exptl. Zool.*, **103**, 143–168.
J. R. Gregg and S. Lovtrup, 1950, 'Biochemical gradients in the axolotl gastrula', *C.R. Lab. Carlsberg, Sér chim.*, **27**, 307–324.
——, 1955, 'Synthesis of deoxyribonucleic acid in lethal amphibian hybrids', *Biol. Bull.*, **108**, 29–34.
——, 1960, 'A reinvestigation of DNA synthesis in lethal amphibian hybrids', *Exptl. Cell. Res.*, **19**, 621–623.
M. Grene, 1958, 'Two evolutionary theories', *Brit. J. Philos. Sci.*, **9**, 110–127, 185–193.
——, 1969, 'Bohm's metaphysics and biology. Notes on Maynard Smith's "Status of neo-Darwinism" ', in *Towards a Theoretical Biology 2. Sketches* (ed. C. H. Waddington), Edinburgh University Press, Edinburgh, pp. 61–69, 97–98.
K. Grobben, 1908, 'Die systematische Einteilung des Tierreiches', *Verh. Zool. Bot. Ges. Wien*, **58**, 491–511.
C. Grobstein, 1953, 'Epithelio-mesenchymal specificity in the morphogenesis of mouse submandibular rudiments in vitro', *J. Exptl. Zool.*, **124**, 383–413.
——, 1964, 'Cytodifferentiation and its controls', *Science*, **143**, 643–650.
C. Grobstein and J. Cohen, 1965, 'Collagenase: Effect on the morphogenesis of embryonic salivary epithelium *in vitro*', *Science*, **150**, 626–628.
J. Gross, 1956, 'The behavior of collagen units as a model in morphogenesis', *J. Biophys. Biochem. Cytol.*, **2**, Suppl., 261–274.

K. Grüneberg, 1943, *The Genetics of the Mouse*, Cambridge University Press, Cambridge.
J. B. Gurdon, 1964, 'The transplantation of living cell nuclei', *Adv. Morphogen.*, **5**, 1–43.
T. Gustafson, 1952, 'Studies on the determination of the oral side of the sea urchin egg II. The dorso-ventral structure of the unfertilized egg', *Arkiv Zool.* [2], **3**, 273–282.
——, 1961, 'Studies on the cellular basis of morphogenesis in the sea urchin', *Proc. 1st IUB/IUBS Int. Symp.*, Vol. 2, London, pp. 497–508.
——, 1963, 'Cellular mechanisms in the morphogenesis of the sea urchin embryo. Cell contacts within the ectoderm and between mesenchyme and ectoderm cells', *Exptl. Cell Res.*, **32**, 570–589.
——, 1964, 'The role and activities of pseudopodia during morphogenesis of the sea urchin larva', in *Primitive Motile Systems in Cell Biology* (ed. R. D. Allen and N. Kamiya), Academic Press, New York, pp. 333–349.
——, 1965, 'Morphogenetic significance of biochemical patterns in sea urchin embryos', in *The Biochemistry of Animal Development* (ed. R. Weber), Academic Press, New York, pp. 139–202.
T. U. Gustafson, 1969, 'Cell recognition and cell contacts during sea urchin development', in *Cellular Recognition* (ed. R. T. Smith and R. A. Good), Appleton-Century-Crofts, New York, pp. 47–60.
T. Gustafson and I. Hasselberg, 1950, 'Alkaline phosphatase activity in developing sea urchin eggs', *Exptl. Cell Res.*, **1**, 371–375.
——, 1951, 'Studies on enzymes in the developing sea urchin egg', *Exptl. Cell Res.*, **2**, 642–672.
T. Gustafson and H. Kinnander, 1956, 'Microaquaria for time-lapse cinematographic studies of morphogenesis in swimming larvae, and observations on sea urchin gastrulation', *Exptl. Cell Res.*, **11**, 36–51.
T. Gustafson and P. Lenique, 1952, 'Studies on mitochondria in the developing sea urchin eggs', *Exptl. Cell Res.*, **3**, 251–274.
——, 1955, 'Studies on mitochondria in early cleavage stages of the sea urchin egg', *Exptl. Cell Res.*, **8**, 114–117.
T. Gustafson and L. Wolpert, 1961a, 'Cellular mechanisms in the morphogenesis of the sea urchin larva. The formation of arms', *Exptl. Cell Res.*, **22**, 509–520.
——, 1961b, 'The forces that shape the embryo', *Discovery*, **22**, 470–477.
——, 1963a, 'The cellular basis of morphogenesis in sea urchin development', *Internat. Rev. Cytol.*, **15**, 139–214.
——, 1963b, 'Studies on the cellular basis of morphogenesis in the sea urchin embryo. Formation of the coelom, the mouth and the primary pore-canal', *Exptl. Cell Res.*, **29**, 561–582.
——, 1967, 'Cellular movement and contact in sea urchin morphogenesis', *Biol. Rev.*, **42**, 442–498.
A. C. Guyton, 1971, *Textbook of Medical Physiology*, 4th ed., Saunders, Philadelphia.
J. Hadźi, 1963, *The Evolution of the Metazoa*, Pergamon, New York.
E. Haeckel, 1866, *Generelle Morphologie der Organismen*, G. Reimer Verlag, Berlin.
——, 1874, 'Die Gastraea—Theorie, die phylogenetische Klassification des Thierreichs und die Homologie der Keimblätter', *Jena Z. Naturwiss*, **8**, 1–55.
B. E. Hagström, 1955, 'Studies in the cortical dark field colour change of the sea urchin egg', *Exptl. Cell Res.*, **9**, 407–413.
J. B. S. Haldane, 1932a, *The Causes of Evolution*, Longmans Green, London.
——, 1932b, 'The time of action of genes, and its bearing on some evolutionary problems', *Amer. Nat.*, **66**, 5–24.
A. W. Ham, 1953, *Histology*, 2nd ed., Lippincott, Philadelphia.
W. J. Hamilton, J. D. Boyd and H. W. Mossman, 1972, *Human Embryology Prenatal Development of Form and Function*, 4th ed., Heffer, Cambridge.
N. M. Hancox, 1946, 'On the occurrence in vitro of cells resembling osteoclasts', *J. Physiol.*, **105**, 66–71.
J. Hanson and H. E. Huxley, 1955, 'The structural basis of contraction in striated muscle', *Symp. Soc. Exptl. Biol.*, **9**, 228–264.
J. Hanson and J. Lowy, 1960, 'Structure and function of the contractile apparatus in the muscles of invertebrate animals', in *The Structure and Function of Muscle* (ed. G. H. Bourne), Vol. 1, Academic Press, New York, pp. 265–335.
——, 1965, 'Molecular basis of contractility in muscle', *Brit Med. Bull.*, **21**, 264–271.
K. Harm, 1903, 'Die Entwicklungsgeschichte von Clava squamata', *Z. Wiss. Zool.*, **73**, 115–165.
H. Harris, 1961, 'Chemotaxis', *Exptl. Cell Res.*, Suppl. **8**, 199–208.

R. G. Harrison, 1910, 'The outgrowth of the nerve fiber as a mode of protoplasmic movement', *J. Exptl. Zool.*, **9**, 787–848.

——, 1921, 'Experiments on the development of the gills in the amphibian embryo', *Biol. Bull.*, **41**, 156–170.

——, 1925, 'The development of the balancer in Amblystoma, studied by the method of transplantation and in relation to the connective-tissue problem', *J. Exptl. Zool.*, **41**, 349–427.

E. de Harven, 1968, 'The centriole and the mitotic spindle', in *The Nucleus* (ed. A. J. Dalton and F. Haguenau), Academic Press, New York, pp. 197–227.

S. C. Harvey, 1949, 'The carbohydrate metabolism of Trypanosoma hippicum', *J. Biol. Chem.*, **179**, 435–453.

S. Hatano and F. Oosawa, 1966, 'Isolation and characterization of plasmodium actin', *Biochim. Biophys. Acta*, **127**, 488–498.

B. Hatschek, 1881, 'Studien über die Entwicklung des Amphioxus', *Arb. Zool. Inst. Wien*, **4**, 1–88.

S. D. Hauscka, 1968, 'Clonal aspects of muscle development and the stability of the differentiated state', in *The Stability of the Differentiated State* (ed. H. Ursprung), Springer Verlag, Berlin, pp. 37–57.

E. D. Hay, 1964, 'Secretion of a connective tissue protein by developing epidermis', in *The Epidermis* (ed. W. Montagna and W. C. Lobitz), Academic Press, New York, pp. 97–116.

E. D. Hay and J.-P. Revel, 1963, 'Autoradiographic studies of the origin of the basement lamella in *Ambystoma*', *Develop. Biol.*, **7**, 152–168.

F. R. Hayes, 1938, 'The relation of fat changes to the general chemical embryology of the sea urchin', *Biol. Bull.*, **74**, 267–277.

H. Heath, 1899, 'The development of Ischnochiton', *Zool. Jahrb. Abt. Anat.*, **12**, 567–656.

N. G. Heatley and P. E. Lindahl, 1937, 'Studies on the nature of the amphibian organizer center. 5. Distribution and nature of glycogen in the amphibian embryo', *Proc. Roy. Soc.* [B], **122**, 395–402.

L. V. Heilbrunn, 1920, 'Studies in artificial parthenogenesis. III. Cortical change and the initiation of maturation in the egg of Cumingia', *Biol. Bull.*, **38**, 317–339.

——, 1956, *An Outline of General Physiology*, 3rd ed., Saunders, Philadelphia.

L. J. Henderson, 1913, *The Fitness of the Environment*, Macmillan, New York.

W. Hennig, 1966, *Phylogenetic Systematics*, University of Illinois Press, Urbana.

C. Herbst, 1895, 'Experimentelle Untersuchungen über den Einfluss der veränderten chemischen Zusammensetzung des umgebenden Mediums auf die Entwicklung der Thiere. 2. Thiel. Weiteres über die morphologische Wirkung der Lithiumsalze und ihre theoretische Bedeutung', *Mitt. Zool. Stat. Neapel*, **11**, 136–220.

H. Herrmann and M. L. Tootle, 1964, 'Specific and general aspects of the development of enzymes and metabolic pathways', *Physiol. Rev.*, **44**, 289–371.

O. Hess, 1956, 'Die Entwicklung von Exogastrulakeimen bei dem Süsswasser-Prosobranchier *Bithynia tentaculata* L', *Arch. Entw.mech.*, **148**, 474–488.

R. Hinegardner, 1968, 'Evolution of cellular DNA content in teleost fishes', *Amer. Nat.*, **102**, 517–523.

Y. Hiramoto, 1956, 'Cell division without mitotic apparatus in sea urchin eggs', *Exptl. Cell Res.*, **11**, 630–636.

——, 1958, 'A quantitative description of protoplasmic movement during cleavage in the sea-urchin egg', *J. Exptl. Biol.*, **35**, 407–424.

W. His, 1874, *Unsere Körperform und das physiologische Problem ihrer Entstehung. Briefe an einem befreundeten Naturforscher*, F.C.W. Vogel Verlag, Leipzig.

W. S. Hoar, 1966, *General and Comparative Physiology*, Prentice-Hall, Englewood Cliffs.

E. Hoff-Jørgensen, 1954, 'Deoxynucleic acid in some gametes and embryos', *Colston Papers*, **7**, 79–88.

E. Hoff-Jørgensen and E. Zeuthen, 1952, 'Evidence of cytoplasmic deoxyribosides in the frog's egg', *Nature*, **169**, 245–246.

H. Hoffmann-Berling, 1960, 'Other mechanisms producing movements', in *Comparative Biochemistry* (ed. M. Florkin and H. S. Mason), Vol. 2, Academic Press, New York, pp. 341–370.

S. J. Holmes, 1900, 'The early development of Planorbis', *J. Morphol.*, **16**, 369–458.

——, 1944, 'Recapitulation and its supposed causes', *Quart. Rev. Biol.*, **18**, 319–331.

H. Holmgren, 1940, 'Studien über Verbreitung und Bedeutung der chromotropen Substanz', *Z. Mikros. Anat. Forsch.*, **47**, 489–521.

H. Holter and P. E. Lindahl, 1940, 'Beiträge zur enzymatischen Histochemie. XXXII. Über die Verteilung der Peptidase in *Paracentrotus*-Keimen', *C.R. Lab. Carlsberg, Sér. chim.*, **23**, 249–256.

H. Holter and B. A. Lowy, 1959, 'A study of the properties and localization of acid phosphatase in the amoeba *Chaos chaos* L', *C.R. Lab. Carlsberg, Sér. chim.*, **31**, 105–127.

H. Holter and E. Zeuthen, 1948, 'Metabolism and reduced weight in starving *Chaos chaos*', *C.R. Lab. Carlsberg, Sér. chim.*, **26**, 277–296.

J. Holtfreter, 1933, 'Der Einfluss von Wirtsalter und verschiedenen Organbezirken auf die Differenzierung von angelagertem Gastrulaektoderm', *Arch. Entw.mech.*, **127**, 619–775.

——, 1934, 'Formative Reize in der Embryonalentwicklung der Amphibien, dargestellt an Explantationsversuchen', *Arch. Exptl. Zellforsch.*, **15**, 281–301.

——, 1936, 'Regionale Induktionen in xenoplastisch zusammengesetzten Explantaten', *Arch. Entw. mech.*, **134**, 466–550.

——, 1938a, 'Differenzierungspotenzen isolierter Teile der Urodelengastrula', *Arch. Entw.mech.*, **138**, 522–656.

——, 1938b, 'Differenzierungspotenzen isolierter Teile der Anurengastrula', *Arch. Entw.mech.*, **138**, 657–738.

——, 1939, 'Studien zur Ermittlung der Gestaltungsfaktoren in der Organentwicklung der Amphibien II. Dynamische Vorgänge an einigen mesodermalen Organanlagen', *Arch. Entw. mech.*, **139**, 227–273.

——, 1943a, 'Properties and functions of the surface coat in the amphibian embryo', *J. Exptl. Zool.*, **93**, 251–323.

——, 1943b, 'A study of the mechanics of gastrulation. Part I', *J. Exptl. Zool.*, **94**, 261–318.

——, 1944a, 'A study of the mechanics of gastrulation. Part II', *J. Exptl. Zool.*, **95**, 171–212.

——, 1944b, 'Neural differentiation of ectoderm through exposure to saline solution', *J. Exptl. Zool.*, **95**, 307–341.

——, 1945, 'Neuralization and epidermization of gastrula ectoderm', *J. Exptl. Zool.*, **98**, 161–209.

——, 1946, 'Structure, motility and locomotion in isolated embryonic amphibian cells', *J. Morphol.*, **79**, 27–62.

——, 1947, 'Neural induction in explants which have passed through a sublethal cytolysis', *J. Exptl. Zool.*, **106**, 197–222.

——, 1948, 'Significance of the cell membrane in embryonic processes', *Ann. N.Y. Acad. Sci.*, **49**, 709–760.

——, 1968, 'Mesenchyme and epithelia in inductive and morphogenetic process', in *Epithelial–mesenchymal Interactions* (ed. R. Fleischmajer and R. E. Billingham), Williams & Wilkins, Baltimore, pp. 1–30.

H. Holtzer, 1961, 'Aspects of chondrogenesis and myogenesis', in *Synthesis of Molecular and Cellular Structures* (ed. D. Rudnick), Ronald Press, New York, pp. 35–87.

H. Holzer and J. Abbott, 1968, 'Oscillations of the chondrogenic phenotype *in vitro*', in *The Stability of the Differentiated State* (ed. H. Ursprung), Springer Verlag, Berlin, pp. 1–16.

H. Holtzer and S. Detwiler, 1953, 'An experimental analysis of the development of the spinal column. III. Induction of skeletogenous cells', *J. Exptl. Zool.*, **123**, 335–369.

B. L. Horecker and H. H. Hiatt, 1958, 'Pathways of carbohydrate metabolism in normal and neoplastic cells', *New Engl. J. Med.*, **258**, 177–184, 225–232.

S. Hörstadius, 1935, 'Über die Determination im Verlaufe der Eiachse bei Seeigeln', *Pubbl. Staz. Zool. Napoli*, **14**, 251–479.

——, 1937, 'Experiments on determination in the early development of Cerebratulus lacteus', *Biol. Bull.*, **73**, 317–342.

——, 1939, 'The mechanics of sea urchin development, studied by operative methods', *Biol. Rev.*, **18**, 132–179.

——, 1944, 'Über die Folgen von Chordaextirpation an späten Gastrulae und Neurulae von Ambystoma punctatum', *Acta Zool.*, **25**, 75–87.

——, 1950, *The Neural Crest. Its Properties and Derivatives in the Light of Experimental Research*, Oxford University Press, Oxford.

S. Hörstadius and S. Sellman, 1946, 'Experimentelle Untersuchungen über die Determination des knorpeligen Kopfskelettes bei Urodelen', *Nova Acta Soc. Sci. Upsaliensis* [4], **13** (8), 1–170.

S. Hörstadius and A. Wolsky, 1936, 'Studien über die Determination der Bilateralsymmetrie des jungen Seeigelkeimes', *Arch. Entw.mech.*, **135**, 69–113.

S. C. Hsiao and W. K. Fujii, 1963, 'Early ontogenetic changes in the concentration of alkaline phosphatase in Hawaiian sea urchins', *Exptl. Cell Res.*, **32**, 217–231.

W. C. Hueper and M. A. Russel, 1932, ' "Capillary-like formations" in tissue cultures of leucocytes', *Arch. Exptl. Zellforsch.*, **12**, 407–424.

S. Hughes-Schrader, 1924, 'Reproduction in Acroschismus Wheeleri Pierce', *J. Morphol.*, **39**, 157–205.

T. Hultin, 1953a, 'The amino acid metabolism of sea urchin embryos studied by means of N^{15}-labelled ammonium chloride and alanine', *Ark. Kemi*, **5**, 543–552.

——, 1953b, 'Incorporation of N^{15}-DL-alanine into protein fractions of sea urchin embryos', *Ark. Kemi*, **5**, 559–564.

——, 1953c, 'Incorporation of C^{14}-labelled carbonate and acetate into sea urchin embryos', *Ark. Kemi*, **6**, 195–200.

W. J. Humphreys, 1964, 'Electron microscope studies of the fertilized egg and the two-cell stage of *Mytilus edulis*', *J. Ultrastr. Res.*, **10**, 244–262.

S. Hunt, 1970, *Polysaccharide–Protein Complexes in Invertebrates*, Academic Press, London.

J. O. Hutchens, M. J. Kopac and M. E. Krahl, 1942, 'The cytochrome oxidase content of centrifugally separated fractions of unfertilized Arbacia eggs', *J. Cell. Comp. Physiol.*, **20**, 113–116.

J. O. Hutchens, A. K. Keltch, M. E. Krahl and G. H. A. Clowes, 1942, 'Studies on cell metabolism and cell division VI. Observations on the glycogen content, carbohydrate consumption, lactic acid production, and ammonia production of eggs of Arbacia punctulata', *J. Gener. Physiol.*, **25**, 717–731.

J. S. Huxley, 1924, 'Early embryonic differentiation', *Nature*, **113**, 276–278.

——, 1932, *Problems of Relative Growth*, Dial Press, New York.

——, 1940, 'Towards the New Systematics' in *The New Systematics* (ed. J. Huxley), Clarendon, Oxford, pp. 1–46.

——, 1942, *Evolution: The Modern Synthesis*, Harper, New York.

J. S. Huxley and G. R. de Beer, 1934, *The Elements of Experimental Embryology*, Cambridge University Press, Cambridge.

L. H. Hyman, 1940, *The Invertebrates: Protozoa through Ctenophora*, Vol. 1, McGraw-Hill, New York.

——, 1951a, *The Invertebrates: Platyhelminthes and Rhynchocoela, The Acoelomate Bilateria*, Vol. 2, McGraw-Hill, New York.

——, 1951b, *The Invertebrates: Acanthocephala, Aschelminthes and Entoprocta, The Pseudocoelomate Bilateria*, Vol. 3, McGraw-Hill, New York.

——, 1955, *The Invertebrates: Echinodermata, The Coelomate Bilateria*, Vol. 4, McGraw-Hill, New York.

J. Immers, 1961, 'Comparative study of the localization of incorporated ^{14}C-labelled amino acids and $^{35}SO_4$ in the sea urchin ovary, egg and embryo', *Exptl. Cell Res.*, **24**, 356–378.

——, 1962, *Investigations on Macromolecular Sulfated Polysaccharides in Sea Urchin Development*, Uppsala.

J. Immers and J. Runnström, 1960, 'Release of respiratory control by 2, 4-dinitrophenol in different stages of sea urchin development', *Develop. Biol.*, **2**, 90–104.

F. Inaba, 1936, 'Studies on the artificial parthenogenesis of *Ostrea gigas* Thunberg', *J. Sci. Hirosima Univ.* [B1], **5**, 29–46.

S. Inoué, 1964, 'Organization and function of the mitotic spindle', in *Primitive Motile Systems in Cell Biology* (ed. R. D. Allen and K. Kamiya), Academic Press, New York, pp. 589–598.

H. Ishikawa, R. Bischoff and H. Holtzer, 1969, 'Formation of arrowhead complexes with heavy meromyosin in a variety of cell types', *J. Cell Biol.*, **43**, 312–328.

N. Isono, 1962, 'Carbohydrate metabolism in sea urchin eggs II. Pentose phosphate cycle in developing eggs', *J. Fac. Sci. Univ. Tokyo* [4], **9**, 369–377.

——, 1963, 'Carbohydrate metabolism in sea urchin eggs III. Changes in respiratory quotient during early embryonic development', *Annot. Zool. Jap.*, **36**, 126–132.

N. Isono and J. Ishida, 1962, 'Carbohydrate metabolism in sea urchin eggs. I. Pentose phosphate cycle in unfertilized sea urchin eggs', *J. Fac. Sci. Univ. Tokyo* [4], **9**, 357–367.

F. Jacob and J. Monod, 1961, 'Genetic regulatory mechanisms in the synthesis of proteins', *J. Mol. Biol.*, **3**, 318–356.

——, 1963, 'Genetic repression, allosteric inhibition, and cellular differentiation' in *Cytodifferentiation and Macromolecular Synthesis* (ed. M. Locke), Academic Press, New York, pp. 30–64.
E. Jacobshagen, 1927, *Zur Reform der allgemeinen vergleichenden Formlehre der Tiere*, Gustav Fischer Verlag, Jena.
C.-O. Jacobson, 1962, 'Cell migration in the neural plate and the process of neurulation in the axolotl larva', *Zool. Bidrag, Uppsala*, **35**, 433–449.
B. J. Jandorf and M. E. Krahl, 1942, 'Studies on cell metabolism and cell division. VIII. The diphosphopyridine (cozymase) content of eggs of Arbacia punctulata', *J. Gener. Physiol.*, **25**, 749–754.
R. Jarosch, 1964, 'Screw-mechanical basis of protoplasmic movement', in *Primitive Motile Systems in Cell Biology* (ed. R. D. Allen and N. Kamiya), Academic Press, New York, pp. 599–622.
P. M. Johnston and C. L. Comar, 1957, 'Autoradiographic studies of the utilization of S^{35}-sulfate by the chick embryo', *J. Biophys. Biochem. Cytol.*, **3**, 231–238.
T. H. Jukes, 1966, *Molecules and Evolution*, Columbia University Press, New York.
E. E. Just, 1939, *The Biology of the Cell Surface*, Technical Press, London.
H. Kacser, 1955, 'The cortical changes on fertilization of the sea-urchin egg', *J. Exptl. Biol.*, **32**, 451–467.
J. A. Kälin, 1945, *Ganzheitliche Morphologie und Homologie*, Freiburg.
F. Kallman and C. Grobstein, 1965, 'Source of collagen at epitheliomesenchymal interfaces during inductive interaction', *Develop. Biol.*, **11**, 169–183.
N. Kamiya, 1964, 'The motive force of endoplasmic streaming in the amoeba', in *Primitive Motile Systems in Cell Biology* (ed. R. D. Allen and N. Kamiya), Academic Press, New York, pp. 257–277.
N. O. Kaplan, 1965, 'Evolution of dehydrogenases', in *Evolving Genes and Proteins* (ed. V. Bryson and H. J. Vogel), Academic Press, New York, pp. 243–277.
S. Karasaki, 1959, 'Electron microscopic studies on cytoplasmic structures of ectoderm cells of the *Triturus* embryo during the early phase of differentiation', *Embryologia*, **4**, 247–272.
H. E. Karrer, 1956, 'The ultrastructure of mouse lung. Some remarks regarding the fine structure of the alveolar basement membrane', *J. Biophys. Biochem. Cytol.*, **2**, Suppl., 287–292.
R. L. Katzman and R. W. Jeanloz, 1970, 'Are acid polysaccharides involved in collagen fibril formation or stabilization?', *Biochim. Biophys. Acta*, **220**, 516–521.
D. Keilin and E. F. Hartree, 1938. 'Cytochrome a and cytochrome oxidase', *Nature*, **141**, 870–871.
N. E. Kemp, 1956, 'Electron microscopy of growing oocytes of Rana pipiens', *J. Biophys. Biochem. Cytol.*, **2**, 281–292.
——, 1959, 'Development of the basement lamella of larval anuran skin', *Develop. Biol.*, **1**, 459–476.
G. A. Kerkut, 1960, *The Implications of Evolution*, Pergamon, New York.
R. G. Kessel, 1960, 'The role of cell division in gastrulation of *Fundulus heteroclitus*', *Exptl. Cell Res.*, **20**, 277–282.
R. Kilkson, 1968, 'Evolution of chromosomes and viruses, derivation of structure', *Quart. Rev. Biophys.*, **1**, 265–308.
H. D. King, 1901, 'The maturation and fertilization of the egg of Bufo lentiginosus', *J. Morphol.*, **17**, 293–350.
J. L. King and T. H. Jukes, 1969, 'Non-Darwinian evolution', *Science*, **164**, 788–798.
B. F. Kingsbury, 1924, 'The developmental significance of the notochord (Chorda dorsalis)', *Z. Morphol. Antropol.*, **24**, 59–73.
B. Klatt, 1948, 'Messend-anatomische Untersuchungen und gegensätzlichen Wuchsformtypen', *Arch. Entw.mech.*, **143**, 573–592.
L. Klein and J. D. Currey, 1970, 'Echinoid skeleton: Absence of a collagenous matrix', *Science*, **169**, 1209–1210.
S. Kochav and H. Eyal-Giladi, 1971, 'Bilateral symmetry in chick embryo determination by gravity', *Science*, **171**, 1027–1029.
C. A. Kofoid, 1894, 'On some laws of cleavage in Limax: A preliminary notice', *Proc. Am. Acad. Arts Sci.*, **29**, 180–203.
——, 1895, 'On the early development of Limax', *Bull. Mus. Comp. Zool. Harvard*, **27**, 35–118.
F. Kopsch, 1901, 'Die Entstehung des Dottersackentoblasts und die Furchung bei Belone acus', *Internat. Monatschr. Anat. Physiol.*, **18**, 43–127.
E. Korschelt and K. Heider, 1936, *Vergleichende Entwicklungsgeschichte der Tiere* (ed. E. Korschelt), Gustav Fischer Verlag, Jena.

M. E. Krahl, 1950, 'Metabolic activities and cleavage of eggs of the sea urchin, Arbacia punctulata A review, 1932–1949', *Biol. Bull.*, **98**, 175–217.

——, 1956, 'Oxidative pathways for glucose in eggs of the sea urchin', *Biochim. Biophys. Acta*, **20**, 27–32.

M. E. Krahl, A. K. Keltch, C. E. Neubeck and G. H. A. Clowes, 1941, 'Studies on cell metabolism and cell division. V. Cytochrome oxidase activity in the eggs of Arbacia punctulata', *J. Gener. Physiol.*, **24**, 597–617.

M. E. Krahl, A. K. Keltch, C. P. Walters and G. H. A. Clowes, 1955a, 'Hexokinase activity from eggs of the sea urchin Arbacia punctulata', *J. Gener. Physiol.*, **38**, 31–39.

——, 1955b, 'Glucose-6-phosphate and 6-phosphogluconate dehydrogenases from eggs of the sea urchin Arbacia punctulata', *J. Gener. Physiol.*, **38**, 431–439.

K. Kratochwil, 1969, 'Organ specificity in mesenchymal induction demonstrated in the embryonic development of the mammary gland of the mouse', *Develop. Biol.*, **20**, 46–71

B. J. Krijgsman, 1925, 'Beiträge zum Problem der Geisselbewegung', *Arch Protistenk.*, **52**, 478–488.

E. J. Krugelis, 1950, 'Properties and changes of alkaline phosphatase activity during amphibian development', *C.R. Lab. Carlsberg, Sér. chim.*, **27**, 273–290.

E. J. Krugelis, J. S. Nicholas and M. E. Vosgian, 1952, 'Alkaline phosphatase activity and nucleic acids during embryonic development of Amblystoma punctatum at different temperatures', *J. Exptl. Zool.*, **121**, 484–504.

H. Kuhlenbeck, 1967, *The Central Nervous System of Vertebrates. Vol. 1. Propaedeutics to Comparative Neurology. Vol. 2. Invertebrates and Origin of the Vertebrates*, Karger, Basel.

A. Kühn, 1955, *Vorlesungen über Entwicklungsphysiologie*, Springer Verlag, Berlin.

Y. Kuriki and R. Okazaki, 1959, 'Deoxyribonucleic acid and acid-soluble deoxyribosidic compounds in amphibian embryos', *Embryologia*, **4**, 337–348.

D. Lack, 1947, *Darwin's Finches*, Cambridge University Press, Cambridge.

——, 1949, 'The significance of ecological isolation', in *Genetics, Paleontology, and Evolution* (ed. G. L. Jepson, E. Mayr and G. G. Simpson), Princeton University Press, Princeton, pp. 299–308.

C. A. Lang and P. Grant, 1961, 'Respiratory enzyme changes during frog embryogenesis', *Proc. Nat. Acad. Sci.*, **47**, 1236–1244.

S. Larsson and S. Løvtrup, 1966, 'An automatic diver balance', *J. Exptl. Biol.*, **44**, 47–58.

I. Lasnitzki, 1965, 'The action of hormones on cell and organ cultures', in *Cells and Tissues in Culture* (ed. E. N. Willmer), Vol. 1, Academic Press, London, pp. 591–658.

M. W. de Laubenfels, 1953, 'Trivial names', *System. Zool.*, **2**, 42–45.

F. E. Lehmann, 1928, 'Die Bedeutung der Unterlagerung für die Entwicklung der Medullarplatte von Triton', *Arch. Entw.mech.*, **113**, 123–171.

——, 1929, 'Die Entwicklung des Anlagenmusters im Ektoderm der Tritongastrula', *Arch. Entw. mech.*, **117**, 312–383.

——, 1945, *Einführung in die physiologische Embryologie*, Verlag Birkhäuser, Basel.

——, 1956, 'Plasmatische Eiorganisation und Entwicklungsleistung bei Keim von Tubifex (Spiralia)', *Naturwiss.*, **43**, 289–296.

F. E Lehmann and H. Ris, 1938, 'Weitere Untersuchungen über die Entwicklung der Achsenorgane bei partiell chordalosen Tritonlarven', *Rev. Suisse Zool.*, **45**, 419–423.

A. L. Lehninger, 1965, *The Mitochondrion. Molecular Basis of Structure and Function*, Benjamin, New York.

Å Lennerstrand, 1933, 'Aerobe und anaerobe Glycolyse bei der Entwicklung des Froscheies (*Rana temporaria* L.)', *Z. Vergl. Physiol.*, **20**, 287–290.

T. L. Lentz and J. P. Trinkhaus, 1967, 'A fine structural study of cytodifferentiation during cleavage, blastula, and gastrula stages of *Fundulus heteroclitus*', *J. Cell Biol.*, **32**, 121–128.

I. M. Lerner, 1954, *Genetic Homeostasis*, Oliver & Boyd, Edinburgh.

——, 1958, *The Genetic Basis of Selection*, Wiley, New York.

E. B. Lewis, 1951, 'Pseudoallelism and gene evolution', *Cold Spring Harbor Symp. Quant. Biol.*, **16**, 159–174.

W. H. Lewis, 1951, 'Cell division with special reference to cells in tissue cultures', *Ann. N.Y. Acad. Sci.*, **51**, 1287–1294.

F. R. Lillie, 1895, 'The embryology of the Unionidae. A study in cell-lineage', *J. Morphol.*, **10**, 1–100.

——, 1899, 'Adaptation in cleavage', *Biol. Lectures Wood's Holl 1898*, 43–67.

——, 1901, 'The organization of the egg of Unio, based on a study of its maturation, fertilization, and cleavage', *J. Morphol.*, **17**, 227–292.

——, 1902, 'Differentiation without cleavage in the egg of the annelid Chaetopterus pergamentaceus', *Arch. Entw.mech.*, **14**, 477–499.

——, 1906, 'Observations and experiments concerning the elementary phenomena of embryonic development in Chaetopterus', *J. Exptl. Zool.*, **3**, 153–268.

——, 1929, 'Embryonic segregation and its role in the life history', *Arch. Entw.mech.*, **118**, 499–533.

P. E. Lindahl, 1932a, 'Zur Kenntnis der Ovarialeies bei dem Seeigel', *Arch. Entw.mech.*, **126**, 373–390.

——, 1932b, 'Zur experimentellen Analyse der Determination der Dorsoventrallachse beim Seeigelkeim', *Arch. Entw.mech.*, **127**, 300–322.

——, 1936, 'Zur Kenntnis der physiologischen Grundlagen der Determination im Seeigelkeim', *Acta Zool.*, **17**, 179–365.

——, 1939a, 'Über die biologische Sauerstoffaktivierung nach Versuchen mit Kohlenmonoxid an Seeigelleiern und Keimen', *Z. Vergl. Physiol.*, **27**, 136–168.

——, 1939b, 'Zur Kenntnis der Entwicklungsphysiologie der Seeigeleies', *Z. Vergl. Physiol.*, **27**, 233–250.

——, 1941, 'Physiologische Probleme der Entwicklung und Formbildung des Seeigelkeimes', *Naturwiss.*, **29**, 673–685.

O. Lindberg, 1943, 'Studien über das Problem des Kohlehydratabbaus und der Säurebildung bei der Befruchtung des Seeigeleies', *Ark. Kemi* [A], **16** (15), 1–20.

O. Lindberg and L. Ernster, 1948, 'On carbohydrate metabolism in homogenized sea urchin eggs', *Biochim. Biophys. Acta*, **2**, 471–477.

K. Linderstrøm-Lang and H. Holter, 1940, 'Die enzymatische Histochimie', in *Methoden der Fermantforschung* (ed. E. Bamann and K. Myrbäck), Georg Thieme Verlag, Leipzig, pp. 1132–1162.

J. Loeb, 1964, *The Mechanistic Conception of Life* (ed. Donald Fleming), Belknap Press of Harvard University Press, Cambridge, Mass.

S. Løvtrup, 1953a, 'Energy sources of amphibian embryogenesis', *C.R. Lab. Carlsberg, Sér. chim.*, **28**, 371–399.

——, 1953b, 'Utilization of reserve material during amphibian embryogenesis at different temperatures', *C.R. Lab. Carlsberg, Sér. chim.*, **28**, 400–425.

——, 1953c, 'Changes in the content of peptidases during amphibian embryogenesis at different temperatures', *C.R. Lab. Carlsberg, Sér. chim.*, **28**, 426–443.

——, 1955, 'Chemical differentiation during amphibian embryogenesis', *C.R. Lab. Carlsberg, Sér. chim.*, **29**, 261–314.

——, 1956a, 'The induced synthesis of β-galactosidase in *E.coli* I. Synthesis of enzymes under various experimental conditions', *Biochim. Biophys. Acta*, **19**, 247–255.

——, 1956b, 'The induced synthesis of β-galactosidase in *E.coli* II. Analysis of the accompanying synthetic activity by means of isotopes', *Biochim. Biophys. Acta*, **19**, 433–439.

——, 1958, 'Synthesis of alkaline phosphatase during the embryonic development of *Rana platyrrhinus*', *Acta Embryol. Morphol. Exptl.*, **2**, 54–57.

——, 1959a, 'On the causes of variations in the respiratory curves during amphibian embryogenesis', *J. Exptl. Zool.*, **140**, 231–246.

——, 1959b, 'Utilization of energy sources during amphibian embryogenesis at low temperatures', *J. Exptl. Zool.*, **140**, 383–394.

——, 1959c, 'Synthesis of alkaline phosphatase during embryonic development of *Rana platyrrhina* at low temperatures', *J. Exptl. Zool.*, **140**, 395–404.

——, 1959d, 'Biochemical indices of embryonic differentiation', *4th Internat. Congr. Biochem.*, **6**, 105–120.

——, 1961, 'Morphological development and chemical differentiation during amphibian embryogenesis at low temperatures', *J. Exptl. Zool.*, **147**, 227–232.

——, 1962, 'Permeability changes in fertilized and activated amphibian eggs', *J. Exptl. Zool.*, **151**, 79–84.

——, 1965a, 'Morphogenesis in the amphibian embryo. Gastrulation and neurulation', *Zoologica Gothoburgensia*, **1**, 3–139.

——, 1965b, 'Morphogenesis in the amphibian embryo. Determination of the site of invagination', *Acta Zool.*, **46**, 119–165.

——, 1965c, 'Morphogenesis in the amphibian embryo. Fertilization and blastula formation', *Arch. Entw.mech.*, **156**, 204–248.

——, 1966a, 'Morphogenesis in the amphibian embryo. Cell type distribution, germ layers, and fate maps', *Acta Zool.*, **47**, 209–276.

——, 1966b, 'The chemical basis of sea urchin embryogenesis', *Bull. Swiss Acad. Med. Sci.*, **22**, 201–276.

——, 1966c, 'DNA synthesis during the development of *Rana temporaria*', *Exptl. Cell Res.*, **43**, 65–71.

——, 1967a, 'Amphibian embryogenesis. Induction, cell transformation and cell differentiation', *Acta Zool.*, **48**, 193–267.

——, 1967b, 'Amphibian embryogenesis Morphogenetic mechanisms', *Arkiv Zool.*, **20**, 1–53.

——, 1973, 'The construction of a microrespirometer for the determination of respiratory rates of eggs and small embryos', in *Experiments in Physiology and Biochemistry* (ed. G. A. Kerkut), Vol. 6, Academic Press, London.

S. Løvtrup and R. M. Iverson, 1969, 'Respiratory phases during early sea urchin development, measured with the automatic diver balance', *Exptl. Cell Res.*, **55**, 25–32.

S. Løvtrup and A. Pigon, 1958, 'Inversion of the dorso-ventral axis in amphibian embryos by unilateral restriction of oxygen supply', *J. Embryol. Exptl. Morphol.*, **6**, 486–490.

——, 1969, 'Observations on the gill development in frogs', *Ann. Embryol. Morphogen.*, **2**, 3–13.

S. Løvtrup and B. Werdinius, 1957, 'Metabolic phases during amphibian embryogenesis', *J. Exptl. Zool.*, **135**, 203–220.

W. Ludwig, 1959, 'Die Selektionstheorie', in *Die Evolution der Organismen Ergebnisse und Probleme der Abstammungslehre* (ed. G. Heberer), 2nd ed., Gustav Fischer Verlag, Stuttgart, pp. 662–712.

E. W. MacBride, 1910, 'The formation of the layers in *Amphioxus* and its bearing on the interpretation of the early ontogenetic processes in other vertebrates', *Quart. J. Micros. Sci.*, **54**, 279–345.

R. D. McMaster, 1955, 'Deoxyribose nucleic acid in cleavage and larval stages of the sea urchin', *J. Exptl. Zool.*, **130**, 1–28.

J. P. McMurrich, 1886, 'A contribution to the embryology of the prosobranch gasteropods', *Stud. Biol. Lab. Johns Hopkins Univ.*, **3**, 403–450.

R. Maggio, 1959, 'Cytochrome oxidase activity in the mitochondria of unfertilized and fertilized sea urchin eggs', *Exptl. Cell Res.*, **16**, 272–278.

R. Maggio and A. Ghiretti-Magaldi, 1958, 'The cytochrome system in mitochondria of the unfertilized sea urchin eggs', *Exptl. Cell Res.*, **15**, 95–102.

S. M. Manton, 1928, 'On the embryology of a mysid crustacean, *Hemimysis Ramornae*', *Philos. Trans. Roy. Soc.* [B], **216**, 363–463.

B. N. Manukhin and G. A. Buznikov, 1965, 'The physiological role of chemical transmitters in ontogenesis', in *Essays on Physiological Evolution* (ed. J. W. S. Pringle), Pergamon, Oxford, pp. 190–199.

V. T. Marchesi and E. Steers, 1968, 'Selective solubilization of a protein component of the red cell membrane', *Science*, **159**, 203–204.

E. Margoliash and E. L. Smith, 1965, 'Structural and functional aspects of cytochrome c in relation to evolution', in *Evolving Genes and Proteins* (ed. V. Bryson and H. J. Vogel), Academic Press, New York, pp. 221–242.

C. L. Markert, 1963, 'Epigenetic control of specific protein synthesis in differentiating cells', in *Cytodifferentiation and Macromolecular Synthesis* (ed. M. Locke), Academic Press, New York, pp. 65–84.

D. A. Marsland and J. V. Landau, 1954, 'The mechanisms of cytokinesis: Temperature–pressure studies on the cortical gel system in various marine eggs', *J. Exptl. Zool.*, **125**, 507–539.

W. D. Matthew, 1915, 'Climate and evolution', *Ann. N.Y. Acad. Sci.*, **24**, 171–318. (Reprinted, Special Publications, Vol. 1, N.Y. Acad. Sci. 1939.).

E. Mayr, 1942, *Systematics and the Origin of Species. From the Viewpoint of a Zoologist*, Columbia University Press, New York.

——, 1949, 'The species concept: Semantics versus semantics', *Evolution*, **3**, 371–372.

——, 1963, *Animal Species and Evolution*, Belknap Press of Harvard University Press, Cambridge, Mass.

——, 1969, *Principles of Systematic Zoology*, McGraw-Hill, New York.

D. Mazia, 1961, 'Mitosis and the physiology of cell division', in *The Cell. Biochemistry, Physiology, Morphology* (ed. J. Brachet and A. E. Mirsky), Vol. 2, Academic Press, New York, pp. 77–422.

D. Mazia and A. Ruby, 1968, 'Dissolution of erythrocyte membranes in water and comparison of the membrane protein with other structural proteins', *Proc. Nat. Acad. Sci.*, **61**, 1005–1012.

A. D. Mead, 1897, 'The early development of marine annelids', *J. Morphol.*, **13**, 227–326.

——, 1898, 'The origin and behaviour of the centrosomes in the annelid egg', *J. Morphol.*, **14**, 181–218.

P. Medawar, 1967, *The Art of the Soluble*, Methuen, London.

W. Meise, 1928, 'Die Verbreitung der Aaskrähe (Formenkreis *Corvus corone* L.)', *J. Ortnithol.*, **76**, 1–203.

E. H. Mercer, 1961, *Keratin and Keratinization: An Essay in Molecular Biology*, Pergamon, New York.

——, 1964, 'The external surface of the cell and intercellular adhesion', *Recent Progr. Surface Sci.*, **1**, 360–378.

E. H. Mercer and L. Wolpert, 1962, 'An electron microscope study of the cortex of the sea urchin (*Psammechinus miliaris*)', *Exptl. Cell Res.*, **27**, 1–13.

E. Metschnikoff, 1884, 'Embryologische Mittheilungen über Echinodermen', *Zool. Anz.*, **7**, 43–47, 62–65.

——, 1886, *Embryologische Studien an Medusen. Ein Beitrag zur Genealogie der Primitiv- Organe*, Hölder Verlag, Wien.

A. Meyer, 1928, 'Die Furchung nebst Eibildung, Reifung und Befruchtung des Gigantorhynchus gigas', *Zool. Jahrb. Abt. Anat.*, **50**, 117–218.

E. Meyer, 1901, 'Studien über den Körperbau der Anneliden. V. Das Mesoderm der Ringelwürmer', *Mitt. Zool. Stat. Neapel*, **14**, 247–585.

K. Meyer, 1957, 'The chemistry of the mesodermal ground substances', *Harvey Lectures*, **51**, 88–112.

J. M. Mitchison and M. M. Swann, 1954, 'The mechanical properties of the cell surface I. The cell elastimeter', *J. Exptl. Biol.*, **31**, 443–460.

——, 1955, 'The mechanical properties of the cell surface III. The sea-urchin egg from fertilization to cleavage', *J. Exptl. Biol.*, **32**, 734–750.

L. Monné and S. Hårde, 1950a, 'On the formation of the blastocoele and similar embryonic cavities', *Ark. Zool.* [2], **1**, 463–469.

——, 1950b, 'On the cortical granules of the sea urchin egg', *Ark. Zool.* [2], **1**, 487–498.

L. Monné and D. B. Slautterback, 1950, 'Differential staining of various polysaccharides in sea urchin eggs', *Exptl. Cell Res.*, **1**, 477–491.

J. Monod, 1947, 'The phenomenon of enzymatic adaptation and its bearing on problems of genetics and cellular differentiation', *Growth*, **11**, 223–289.

J. Monod, G. Cohen-Bazire and M. Cohn, 1951, 'Sur la biosynthèse de la β-galactosidase (lactase) chez Eschericia coli. La spécificité de l'induction', *Biochim. Biophys. Acta*, **7**, 585–599.

A. Monroy and A. Tyler, 1967, 'The activation of the egg', in *Fertilization* (ed. C. B. Metz), Vol. 1, Academic Press, New York, pp. 369–412.

S. Mookerjee, 1953, 'An experimental study of the development of the notochordal sheath', *J. Embryol. Exptl. Morphol.*, **1**, 411–416.

S. Mookerjee, E. M. Deuchar and C. H. Waddington, 1953, 'The morphogenesis of the notochord in Amphibia', *J. Embryol. Exptl. Morphol.*, **1**, 399–409.

A. R. Moore, 1945, *The Individual in Simpler Forms*, The University Press, Oregon.

C. R. Moore, W. Hughes and T. F. Gallagher, 1930, 'Rat seminal vesicle cytology as a testis hormone indicator and the prevention of castration changes by testis extract injection', *Amer. J. Anat.*, **45**, 109–135.

J. A. Moore, 1946, 'Studies in the development of frog hybrids. I. Embryonic development in the cross Rana pipiens ♀ x Rana sylvatica ♂', *J. Exptl. Zool.*, **101**, 173–219.

T. H. Morgan, 1927, *Experimental Embryology*, Columbia University Press, New York.

T. H. Morgan and A. Tyler, 1930, 'The point of entrance of the spermatozoön in relation to the orientation of the embryo in eggs with spiral cleavage', *Biol. Bull.*, **58**, 59–73.

——, 1938, 'The relation between entrance point of the spermatozoön and bilaterality of the egg of *Chaetopterus*', *Biol. Bull.*, **74**, 401–402.

M. Morris, 1917, 'A cytological study of artificial parthenogenesis in Cumingia', *J. Exptl. Zool.*, **22**, 1–51.

A. A. Moscona, 1962, 'Analysis of cell recombinations in experimental synthesis of tissues *in vitro*', *J. Cell. Comp. Physiol.*, **60**, Suppl. 1, 65–80.

——, 1965, 'Recombination of dissociated cells and the development of cell aggregates', in *Cells and Tissues in Culture* (ed. E. N. Willmer), Vol. 1, Academic Press, London, pp. 489–529.

M. L. Moss, 1964, 'The phylogeny of mineralized tissues', *Internat. Rev. Gener. Exptl. Zool.*, **1**, 297–331.

M. L. Moss and M. M. Meehan, 1967, 'Sutural connective tissues in the test of an echinoid *Arbacia punctulata*', *Acta Anat.*, **66**, 279–304.

I. Motomura, 1941, 'Materials of the fertilization membrane in the eggs of echinoderms', *Sci. Repts. Tôhoku Imp. Univ.* [4], **16**, 345–364.

——, 1950, 'Studies of cleavage. V. The role of vacuoles in the cleavage plane formation in sea-urchin eggs', *Sci. Repts. Tôhoku Imp. Univ.* [4], **18**, 255–261.

W. B. Muchmore, 1957, 'Differentiation of the trunk mesoderm in Ambystoma maculatum. II. Relation of the size of presumptive somite explants to subsequent differentiation', *J. Exptl. Zool.*, **134**, 293–314.

——, 1958, 'The influence of embryonic neural tissues on differentiation of striated muscle in Ambystoma', *J. Exptl. Zool.*, **139**, 181–188.

A. H. Müller, 1955, *Der Grossablauf der stammes-geschichtlichen Entwicklung*, Gustav Fischer Verlag, Jena.

H. J. Muller, 1927, 'Artificial transmutation of the gene', *Science*, **66**, 84–87.

J. B. Nadol, J. R. Gibbins and K. R. Porter, 1969, 'A reinterpretation of the structure and development of the basement lamella: An ordered array of collagen in fish skin', *Develop. Biol.*, **20**, 304–331.

A. Naef, 1931, 'Die Gestalt als Begriff und Idee', in *Handbuch der vergleichenden Anatomie der Wirbeltiere* (ed. L. Bolk, E. Göppert, E. Kallius and W. Lubosch), Vol. 1, Urban & Schwarzenberg Verlag, Berlin, pp. 77–118.

O. Nakamura, 1942, 'Die Entwicklung der hinteren Körperhälfte bei Urodelen', *Annot. Zool. Jap.*, **21**, 169–235.

M. Nameroff and H. Holtzer, 1967, 'The loss of phenotypic traits by differentiated cells IV. Changes in polysaccharides produced by dividing chondrocytes', *Develop. Biol.*, **16**, 250–281.

J. Needham, 1942, *Biochemistry and Morphogenesis*, Cambridge University Press, Cambridge.

——, 1959, *A History of Embryology*, 2nd ed., Cambridge University Press, Cambridge.

O. E. Nelsen, 1953, *Comparative Embryology of the Vertebrates*, Blakiston, New York.

D. A. T. New, 1959, 'The adhesive properties and expansion of the chick blastoderm', *J. Embryol. Exptl. Morphol.*, **7**, 146–164.

P. D. Nieuwkoop and P. A. Florschütz, 1950, 'Quelques caractères spéciaux de la gastrulation et de la neurulation de l'oeuf de *Xenopus laevis* Daud., et de quelques autres anoures. 1ere partie.—Étude descriptive', *Arch. Biol.*, **61**, 113–150.

K.-G. Nyholm, 1943, 'Zur Entwicklung und Entwicklungsbiologie der Ceriantharien und Actinien', *Zool. Bidrag, Uppsala*, **22**, 87–248.

D. H. O'Day and C. V. Finnegan, 1970, 'Alkaline phosphatase activity and histogenesis in two species of urodele amphibians', *Can. J. Zool.*, **48**, 859–866.

A. G. Ogston, 1966, 'When is pressure osmotic?', *Feder. Proc.*, **25**, 1112–1114.

S. Ohno, 1970, *Evolution by Gene Duplication*, Springer Verlag, Berlin.

K. Okada, 1968, 'Annelida', in *Invertebrate Embryology* (ed. M. Kumé and K. Dan), NOLIT Publishing House, Belgrade, pp. 192–241.

K. Okazaki, 1960, 'Skeleton formation of sea urchin larvae II. Organic matrix of the spicule', *Embryologia*, **5**, 283–320.

——, 1965, 'Skeleton formation of sea urchin larvae V. Continuous observation of the process of matrix formation', *Exptl. Cell Res.*, **40**, 585–596.

K. Okazaki, T. Fukushi and K. Dan, 1962, 'Cyto-embryological studies of sea urchins IV. Correlation between the shape of the ectodermal cells and the arrangement of the primary mesenchyme cells in sea urchin larvae', *Acta Embryol. Morphol. Exptl.*, **5**, 17–31.

F. Oosawa and S. Higashi, 1967, 'Statistical thermodynamics of polymerization and polymorphism of protein', *Progr. Theor. Biol.*, **1**, 79–164.

J. M. Oppenheimer, 1936, 'Processes of localization in developing Fundulus', *J. Exptl. Zool.*, **73**, 405–444.

——, 1940, 'The non-specificity of the germ-layers', *Quart. Rev. Biol.*, **15**, 1–27.

——, 1955, 'Problems, concepts and their history', in *Analysis of Development* (ed. B. H. Willier, P. A. Weiss and V. Hamburger), Saunders, Philadelphia, pp. 1–24.

Å. Örström and O. Lindberg, 1940, 'Über den Kohlenhydratstoffwechsel bei der Befruchtung des Seeigeleies', *Enzymologia*, **8**, 367–384.

G. Ortolani, 1963, 'Sulla origine del mesoderma nei Ctenofori', *Atti. Acad. Naz. Lincei*, **34**, 434–435.

R. Owen, 1847, 'Report on the archetype and homologies of the vertebrate skeleton', *Rep. Brit. Assoc. Adv. Sci.*, 169–340.

A. M. Pappenheimer and C. M. Williams, 1954, 'Cytochrome b_5 and the dihydrocoenzyme I—oxidase system in the Cecropia silkworm', *J. Biol. Chem.*, **209**, 915–929.

C. L. Parmenter, 1940, 'Chromosome numbers in Rana fusca parthenogenetically developed from eggs with known polar body and cleavage histories', *J. Morphol.*, **66**, 241–260.

J. K. Parnas and Z. Krasinska, 1921, 'Über den Stoffwechsel der Amphibienlarven', *Biochem. Z.*, **116**, 108–137.

A. K. Parpart, 1964, 'Echinochrome granule motion in the egg of *Arbacia punctulata*', in *Primitive Motile Systems in Cell Biology* (ed. R. D. Allen and N. Kamiya), Academic Press, New York, pp. 471–483.

J. J. Pasteels, 1958, 'Nouvelles recherches sur la métachromasie *in vivo* et sur l'histochimie de l'oeuf normal et centrifugé de *Paracentrotus lividus*', *Arch. Biol.*, **69**, 591–619.

——, 1964, 'The morphogenetic role of the cortex of the amphibian egg', *Adv. Morphogen.*, **3**, 363–388.

J. J. Pasteels and E. de Harven, 1962, 'Étude au microscope électronique du cortex de l'oeuf de Barnea candida (mollusque bivalve), et son évolution au moment de la fécondation, de la maturation, et de la segmentation', *Arch. Biol.*, **73**, 465–490.

J. J. Pasteels and J. Mulnard, 1957, 'La Métachromasie *in vivo* au bleu de toluidine et son analyse cytochimique dans les oeufs de *Barnea candida*, *Gryphaea angulata* (Lamellibranches) et de *Psammechinus miliaris*', *Arch. Biol.*, **68**, 115–163.

T. L. Pearce and J. Zwaan, 1970, 'A light and electron microscopic study of cell behavior and microtubules in the embryonic chicken lens using Colcemid', *J. Embryol. Exptl. Morphol.*, **23**, 491–507.

D. C. Pease, 1942, 'Echinoderm bilateral determination in chemical concentration gradients II. The effects of azide, pilocarpine, pyocyanine, diamine, cysteine, glutathione, and lithium', *J. Exptl. Zool.*, **89**, 329–345.

P. Pelseneer, 1906, 'Mollusca', in *A Treatise on Zoology* (ed. E. Ray Lankaster), Vol. 5, Adam & Charles Black, London.

A. Penners, 1925, 'Experimentelle Untersuchungen zum Determinationsproblem am Keim von Tubifex rivulorum Lam. II. Die Entwicklung teilweise abgetöteter Keime', *Z. Wiss. Zool.*, **127**, 1–140.

A. Penners and W. Schleip, 1928, 'Die Entwicklung der Schultze'schen Doppelbildungen aus dem Ei von *Rana fusca*. I–IV', *Z. Wiss. Zool.*, **130**, 305–454.

W. A. Perlzweig and E. S. G. Barron, 1928, 'Lactic acid and carbohydrate in sea urchin eggs under aerobic and anaerobic conditions', *J. Biol. Chem.*, **79**, 19–26.

M. M. Perry and C. H. Waddington, 1966, 'Ultrastructure of the blastopore cells in the newt', *J. Embryol. Exptl. Morphol.*, **15**, 317–330.

B. A. Pethica, 1961, 'The physical chemistry of cell adhesion', *Exptl. Cell Res.*, Suppl. **8**, 123–140.

G. B. Pierce, 1966, 'The development of basement membranes of the mouse embryo', *Develop. Biol.*, **13**, 231–249.

A. Pigon, 1962, 'Changes of enzyme activity during starvation and encystment of a ciliate (Urostyla). Cathepsin, acid phosphatase', *Acta Biol. Cracov.*, *Ser. Zool.*, **5**, 175–197.

N. W. Pirie, 1952, 'Concepts out of context: The pied pipers of science', *Brit. J. Philos. Sci.*, **2**, 269–280.

G. W. E. Plaut and S.-C. Sung, 1955, 'Diphosphopyridine nucleotide isocitric dehydrogenase from animal tissues', in *Methods in Enzymology* (ed. S. P. Colowick and N. O. Kaplan), Vol. 1, Academic Press, New York, pp. 710–714.

A. W. Pollister and P. F. Pollister, 1943, 'The relation between centriole and centromere in atypical spermatogenesis of viviparid snails', *Ann. N.Y. Acad. Sci.*, **45**, 1–48.

K. R. Popper, 1968, *The Logic of Scientific Discovery* (rev. ed.), Hutchinson, London.

K. R. Porter, 1956a, 'Changes in cell fine structure accompanying mitosis', *I.U.B.S. Publ.* [B], **21**, 236–250.

——, 1956b, 'Observations on the fine structure of animal epidermis', in *Proc. 3rd Intern. Conf. Electron Microscopy*, London, pp. 539–546.

——, 1966, 'Cytoplasmic microtubules and their functions', in *Principles of Biomolecular Organization* (ed. G. E. W. Wolstenholme and M. O'Connor), Churchill, London, pp. 308–356.

K. R. Porter and G. D. Pappas, 1959, 'Collagen formation by fibroblasts of the chick embryo dermis', *J. Biophys. Biochem. Cytol.*, **5**, 153–166.

W. B. Provine, 1971, *The Origins of Theoretical Population Genetics*, University of Chicago Press, Chicago.

C. Rabl, 1898, 'Über den Bau und die Entwicklung der Linse', *Z. Wiss. Zool.*, **63**, 496–572.

E. Rádl, 1930, *The History of Biological Theories*, Oxford University Press, Oxford.

R. Rappaport, 1965, 'Geometrical relations of the cleavage stimulus in invertebrate eggs', *J. Theor. Biol.*, **9**, 51–66.

——, 1966, 'Experiments concerning the cleavage furrow in invertebrate eggs', *J. Exptl. Zool.*, **161**, 1–8.

D. M. Raup, 1961, 'The geometry of coiling in gastropods', *Proc. Natl. Acad. Sci.*, **47**, 602–609.

D. M. Raup and A. Michelson, 1965, 'Theoretical morphology of the coiled shell', *Science*, **147**, 1294–1295.

Chr. P. Raven, 1936, 'Zur Entwicklung der Ganglienleiste. V. Über die Differenzierung des Rumpfganglienleistenmaterials', *Arch. Entw.mech.*, **134**, 122–146.

——, 1937, 'Experiments on the origin of the sheath cells and sympathetic neuroblasts in Amphibia', *J. Comp. Neurol.*, **67**, 221–240.

——, 1966, *Morphogenesis: The Analysis of Molluscan Development*, 2nd ed., Pergamon, Oxford.

L. I. Rebhun, 1961, 'Endoplasmic reticulum in the asters of invertebrate eggs', *Biol. Bull.*, **121**, 367–368.

——, 1962, 'Electron microscope studies on the vitelline membrane of the surf clam, *Spisula solidissima*', *J. Ultrastruct. Res.*, **6**, 107–122.

——, 1963, 'Induced amoeboid movement in eggs of the surf-clam *Spisula solidissima*', *Exptl. Cell Res.*, **28**, 593–602.

——, 1964, 'Saltatory particle movements in cells', in *Primitive Motile Systems in Cell Biology* (ed. R. D. Allen and N. Kamiya), Academic Press, New York, pp. 503–525.

A. Remane, 1932, 'Archiannelida', in *Tierwelt der Nord- und Ostsee* (ed. G. Grimpe), Lief. **22**, Teil VIa, Akademische Verlagsgesellschaft, Leipzig, pp. 1–36.

F. L. Renaud, A. J. Rove and I. R. Gibbons, 1968, 'Some properties of the protein forming the outer fibers of cilia', *J. Cell Biol.*, **36**, 79–90.

B. Rensch, 1960, *Evolution Above the Species Level*, Columbia University Press, New York.

G. Reverberi, 1961, 'The embryology of ascidians', *Adv. Morphogen.*, **1**, 55–101.

G. Reverberi and G. Ortolani, 1963, 'On the origin of the ciliated plates and of the mesoderm in the ctenophores', *Acta Embryol. Morphol. Exptl.*, **6**, 175–190.

L. Rhumbler, 1898, 'Physikalische Analyse von Lebenserscheinungen der Zelle. I. Bewegung, Nahrungsaufnahme, Defäkation, Vakuolen-Pulsation und Gehäusebau bei lobosen Rhizopoden', *Arch. Entw.mech.*, **7**, 103–350.

A. Richards and R. P. Porter, 1935, 'Analysis of early development of fish embryos by means of the mitotic index II. The mitotic index in pre-neural tube stages of Fundulus heraclitus', *Amer. J. Anat.*, **56**, 365–393.

P. del Rio-Hortega, 1917, 'Contribución al conocimiento de las epiteliofibrillas', *Trab. Lab. Invest. Biol. Univ. Madrid*, **15**, 201–299.

H. Ris, 1955, 'Cell division', in *Analysis of Development* (ed. B. H. Willier, P. A. Weiss and V. Hamburger), Saunders, Philadelphia, pp. 91–125.

W. A. Robbie, 1946, 'The effect of cyanide on the oxygen consumption and cleavage of the sea urchin egg', *J. Cell. Comp. Physiol.*, **28**, 305–324.

A. Robert, 1902, 'Recherches sur le développement des Troques', *Arch. Zool. Exptl. Géner.* [3], **10**, 269–538.

H. S. Roberts, 1961, 'Mechanisms of cytokinesis: A critical review', *Quart. Rev. Biol.*, **36**, 155–177.

J. D. Robertson, 1959, 'The ultrastructure of the cell membranes and their derivatives', *Biochem. Soc. Symp.*, **16**, 3–43.

——, 1967, 'The organization of cellular membranes', in *Molecular Organization and Biological Function* (ed. J. M. Allen), Harper & Row, New York, pp. 65–106.

G. C. Robson and O. W. Richards, 1936, *The Variations of Animals in Nature*, Longmans Green, London.

S. M. Rose, 1955, 'Specific inhibition during differentiation', *Ann. N.Y. Acad. Sci.*, **60**, 1136–1159.

R. M. Rosenbaum, 1958, 'Histochemical observations on the cortical region of the oocytes of *Rana pipiens*', *Quart. J. Micros. Sci.*, **99**, 159–169.

S. Rosin, 1944, 'Über die orthogonale Struktur der Grenzlamelle der Cutis bei Amphibienlarven und ihre Verbreitung bei andern Chordaten', *Rev. Suisse Zool.*, **51** 376–382.

[N. M. V.] Rothschild, 1949, 'The metabolism of fertilized and unfertilized sea urchin eggs. The action of light and carbon monoxide', *J. Exptl. Biol.*, **26**, 100–111.

——, 1956, *Fertilization*, Methuen, London.

——, 1965, *A Classification of Living Animals*, 2nd ed., Longmans Green, London.

W. Roux, 1895, 'The problems, methods, and scope of developmental mechanics', *Biol. Lectures Woods Holl 1894*, 149–190.

R. Rugh, 1948, *Experimental Embryology. A Manual of Techniques and Procedures*, Burgess, Minneapolis.

——, 1951, *The Frog. Its Reproduction and Development*, McGraw-Hill, New York.

J. Runnström, 1926, 'Experimentelle Bestimmung der Dorso-Ventralachse bei dem Seeigelkeim', *Arkiv Zool.* [A], **18** (4), 1–6

——, 1928a, 'Die Veränderungen der Plasmakolloide bei der Entvicklungserregung des Seeigeleies', *Protoplasma*, **4**, 388–514.

——, 1928b, 'Plasmabau und Determination bei dem Ei von Paracentrotus lividus Lk.', *Arch. Entw. mech.*, **113**, 556–581.

——, 1929, 'Über Selbstdifferenzierung und Induktion bei dem Seeigelkeim', *Arch. Entw.mech.*, **117**, 123–145.

——, 1933, 'Zur Kenntnis der Stoffwechselvorgänge bei der Entwicklungserregung des Seeigeleies', *Biochem. Z.*, **258**, 257–279.

——, 1935, 'Influence of iodoacetate on activation and development of the eggs of Arbacia punctulata', *Biol. Bull.*, **69**, 351–355.

——, 1956, 'Some considerations on metabolic changes occurring at fertilization and during early development of the sea urchin egg', *Pubbl. Staz. Zool. Napoli*, **28**, 315–340.

——, 1966, 'The vitelline membrane and cortical particles in sea urchin eggs and their function in maturation and fertilization', *Adv. Morphogen.*, **5**, 221–325.

J. Runnström and L. Monné, 1945, 'On some properties of the surface-layers of immature and mature sea-urchin eggs, especially the changes accompanying nuclear and cytoplasmic maturation', *Arkiv Zool.* [A], **36** (18), 1–26.

J. Runnström, B. E. Hagström and P. Perlmann, 1959, 'Fertilization', in *The Cell* (ed. J. Brachet and A. E. Mirsky), Vol. 1, Academic Press, New York, pp. 327–397.

W. J. Rutter, 1965, 'Enzymatic homology and analogy in phylogeny', in *Evolving Genes and Proteins* (ed. V. Bryson and H. J. Vogel), Academic Press, New York, pp. 279–291.

W. J. Rutter, W. D. Ball, W. S. Bradshaw, W. R. Clark and T. G. Sanders, 1967, 'Levels of regulation in cytodifferentiation', *Exptl. Biol. Med.*, **1**, 110–124.

G. Ruud, 1925, 'Die Entwicklung isolierter Keimfragmente frühester Stadien von Triton taeniatus', *Arch. Entw.mech.*, **105**, 209–293.

H. Sakai, 1962, 'Studies on sulfhydryl groups during cell division of sea urchin egg IV. Contractile properties of the thread model of KCl-soluble protein from the sea urchin egg', *J. Gener. Physiol.*, **45**, 411–425.

F. C. Sauer, 1935, 'Mitosis in the neural tube', *J. Comp. Neurol.*, **62**, 377–405.

J. W. Saunders and J. F. Fallon, 1966, 'Cell death in morphogenesis', in *Major Problems in Developmental Biology* (ed. M. Locke), Academic Press, New York, pp. 289–314.

G. Säve-Söderbergh, 1934, 'Some points of view concerning the evolution of the vertebrates and the classification of this group', *Arkiv Zool.* [A], **26** (17), 1–20.

C. H. Sawyer, 1943, 'Cholinesterase and the behavior problem in Amblystoma. I. The relationship between development of the enzyme and early mobility', *J. Exptl. Zool.*, **92**, 1–29.

L. Saxén, O. Koskimies, A. Lahti, H. Miettinen, J. Rapola and J. Wartiovaara, 1968, 'Differentiation of kidney mesencyme in an experimental model system', *Adv. Morphogen.*, **7**, 251–293.

E. Scarano, 1961, 'An example of cell differentiation at the enzymic level', *Symposium on Germ Cells and Earliest Stages of Development, Pallanza 1960*, Fondazione A. Baselli-Istituto Lombardi, Milan, pp. 402–410.

C. Schierholz, 1889, 'Über Entwicklung der Unioniden', *Denkschr. Kais. Akad. Wiss. Math.-Naturw. Classe*, **55**, 183–214.

O. H. Schindewolf, 1950, *Grundfragen der Paläontologie*, E. Schweizerbart'sche Verlagsbuchhandlung, Stuttgart.

W. Schleip, 1929a, *Die Determination der Primitiventwicklung*, Akademische Verlagsgesellschaft, Leipzig.

——, 1929b, 'Die Herkunft der Polarität des Eies von Ascaris megalocephala', *Arch. Mikros. Anat. Entw.mech.*, **100**, 573–598.

E. Schlottke, 1934, 'Unterschiede in der Entwicklung des phagocytierenden und des resorbierenden Darmepithels', *Biol. Zbl..*, **54**, 51–64.

I. I. Schmalhausen, 1949, *Factors of Evolution. The Theory of Stabilizing Selection*, Blakiston, Philadelphia.

P. F. Scholander, C. L. Claff, S. L. Sveinsson and S. I. Scholander, 1952, 'Respiratory studies of single cells. III. Oxygen consumption during cell division', *Biol. Bull.*, **102**, 185–199.

W. Schönmann, 1938, 'Der diploide Bastard Triton palmatus ♀ x Salamandra salamandra ♂', *Arch. Entw.mech.*, **138**, 345–375.

F. Schrader, 1941, 'The spermatogenesis of the earwig *Anisolabis maritima* Bon. with reference to the mechanism of chromosomal movement', *J. Morphol.*, **68**, 123–147.

——, 1953, *Mitosis. The Movements of Chromosomes in Cell Division*, Columbia University Press, New York.

T. E. Schroeder, 1968, 'Cytokinesis: Filaments in the cleavage furrow', *Exptl. Cell Res.*, **53**, 272–276.

O. Schultze, 1895, 'Die bilaterale Symmetrie des Amphibieneies', *Anat. Anz.*, **10**, 23–29.

M. P. Schützenberger, 1967, 'Algorithms and the neo-Darwinian theory of evolution', in *Mathematical Challenges to the neo-Darwinian Interpretation of Evolution* (ed. P. S. Moorhead and M. M. Kaplan), Wistar Institute Press, Philadelphia, pp. 73–80.

G. Scollo-Lavizzari, 1956, 'Il comportamento delle alfa e beta amilasi durante lo sviluppo embrionale e larvale di "Bufo vulgaris" ', *Ric. Sci.*, **26**, 2146–2152.

A. Sedgwick, 1894, 'On the law of development commonly known as Von Baer's Law, and on the significance of ancestral rudiments in embryonic development', *Quart. J. Micros. Sci.*, **36**, 35–52.

G. G. Selman, 1955, 'Studies on the forces producing neural closure in Amphibia', *Proc. Roy. Phys. Soc. Edinburgh*, **24**, 24–27.

G. G. Selman and M. M. Perry, 1970, 'Ultrastructural changes in the surface layers of the newt's egg in relation to the mechanism of its cleavage', *J. Cell Sci.*, **6**, 207–227.

G. G. Selman and C. H. Waddington, 1955, 'The mechanism of cell division in the cleavage of the newt's egg', *J. Exptl. Biol.*, **32**, 700–733.

R. Semon, 1888, 'Die Entwickelung der Synapta digitata und ihre Bedetung für die Phylogenie der Echinodermen', *Jena Z. Naturwiss.*, **22**, 175–309.

A. G. Sharov, 1965, 'Evolution and taxonomy', *Z. Zool. Syst. Evol.*, **3**, 349–358.

J. R. Shaver, 1956, 'Mitochondrial populations during development of the sea urchin', *Exptl. Cell Res.*, **11**, 548–559.

——, 1957, 'Some observations on cytoplasmic particles in early echinoderm development', in *The Beginnings of Embryonic Development* (ed. A. Tyler, R. C. von Borstel and C. B. Metz), American Association for the Advancement of Science, Washington, D.C., pp. 263–290.

P. M. Sheppard, 1961, 'Some contributions to population genetics resulting from the study of the Lepidoptera', *Adv. Genet.*, **10**, 165–216.

R. Siewing, 1969, *Lehrbuch der vergleichenden Entwicklungsgeschichte*, Verlag Paul Parey, Hamburg.

G. G. Simpson, 1944, *Tempo and Mode in Evolution*, Columbia University Press, New York.

——, 1953a, *The Major Features of Evolution*, Columbia University Press, New York.

——, 1953b, *Evolution and Geography An Essay on Historical Biogeography with Special Reference to Mammals*, Eugene.

——, 1959, 'Mesozoic animals and the polyphyletic origin of mammals', *Evolution*, **13**, 405–414.

——, 1961, *Principles of Animal Taxonomy*, Columbia University Press, New York.

J. Sobotta, 1897, 'Die Reifung und Befruchtung des Eies von Amphioxus lanceolatus', *Arch. Mikros. Anat. Entw.gesch.*, **50**, 15–71.

R. R. Sokal and P. H. A. Sneath, 1963, *Numerical Taxonomy*, W. H. Freeman, San Francisco.

G. Sommerhoff, 1950, *Analytical Biology*, Oxford University Press, Oxford.

T. M. Sonneborn, 1965, 'Degeneracy of the genetic code: Extent, nature and genetic implications', in *Evolving Genes and Proteins* (ed. V. Bryson and H. J. Vogel), Academic Press, New York, pp. 377–397.

H. Spemann, 1938, *Embryonic Development and Induction*, Yale University Press, New Haven.

S. Spiegelman, 1948, 'Differentiation as the controlled production of unique enzymatic patterns', *Symp. Soc. Exptl. Biol.*, **2**, 286–325.

S. Spiegelman and H. B. Steinbach, 1945, 'Substrate–enzyme orientation during embryonic development', *Biol. Bull.*, **88**, 254–268.

B. S. Spooner and N. K. Wessells, 1970, 'Mammalian lung development: Interactions in primordium formation and bronchial morphogenesis', *J. Exptl. Zool.*, **175**, 445–454.

M. S. Steinberg, 1963a, 'Tissue reconstruction by dissociated cells', *Science*, **141**, 401–408.

——, 1963b, '"ECM": Its nature, origin and function in cell aggregation', *Exptl. Cell Res.*, **30**, 257–279.

——, 1964, 'The problem of adhesive selectivity in cellular interactions', in *Cellular Membranes in Development* (ed. M. Locke), Academic Press, New York, pp. 321–366.

R. E. Stephens, 1968, 'Reassociation of microtubule protein', *J. Mol. Biol.*, **33**, 517–519.

——, 1970, 'On the apparent homology of actin and tubulin', *Science*, **168**, 845–847.

R. E. Stephens and R. W. Linck, 1969, 'A comparison of muscle actin and ciliary microtubule protein in the mollusc *Pecten irradians*', *J. Mol. Biol.*, **40**, 497–501.

R. Stephens, F. L. Renaud and I. R. Gibbons, 1967, 'Guanine nucleotide associated with the protein of the outer fibers of flagella and cilia', *Science*, **156**, 1606–1608.

O. zur Strassen, 1896, 'Embryonalentwicklung der Ascaris megalocephala', *Arch. Entw.mech.*, **3**, 27–105, 133–190.

C. F. Strittmatter, 1961, 'Studies on microsomal cytochromes and related substances', in *Haematin Enzymes* (ed. J. E. Falk, R. Lemberg and R. K. Morton), Pergamon, Oxford, pp. 461–478.

A. H. Sturtevant, 1923, 'Inheritance of direction of coiling in Limnaea', *Science*, **58**, 269–270.

M. Sugiyama, 1951, 'Re-fertilization of the fertilized eggs of the sea urchin', *Biol. Bull.*, **100**, 335–344.

F. M. Surface, 1907, 'The early development of a polyclad, Planocera inquilina Wh.', *Proc. Acad. Nat. Sci. Philadelphia*, **59**, 514–559.

M. M. Swann and J. M. Mitchison, 1953, 'Cleavage of sea-urchin eggs in colchicine', *J. Exptl. Biol.*, **30**, 506–514.

——, 1958, 'The mechanism of cleavage in animal cells', *Biol. Rev.*, **33**, 103–135.

H. Szarski, 1968, 'Evolution of cell size in lower vertebrates', in *Current Problems of Lower Vertebrate Phylogeny* (ed. T. Ørvig), Almqvist & Wicksell, Stockholm, pp. 445–453.

D. Szollozi, 1968, 'The contractile ring and changes of the cell surface during cleavage', *J. Cell Biol.*, **39**, 133a.

G. Ten Cate, 1952, 'The formation of enzymes during embryogenesis', *Arch. Néerl. Zool.*, **10**, Suppl. 1, 109–126.

——, 1953, *The Intrinsic Development of Amphibian Embryos*, North-Holland, Amsterdam.

G. Ten Cate, J. S. Kooij and M. H. J. Zuidweg, 1951, 'The effect of temperature on the synthesis of the enzyme cholinesterase in amphibian embryos', *Proc. Kon. Ned. Akad. Wet.* [C], **54**, 157–170.

D'Arcy W. Thompson, 1942, *On Growth and Form*, 2nd ed., Cambridge University Press, Cambridge.

W. Thomson, 1887, 'On the division of space with minimum partition area', *Philos. Mag.*, **24**, 503–514.

H. Tiedemann, 1963, 'Biochemische Untersuchungen über die Induktionsstoffe und die Determination der ersten Organanlagen bei Amphibien', *13 Coll. Ges. Physiol. Chem.*, 177–204.

——, 1966, 'Embryonale Induktion und Differenzierung RNS und Proteinstoftwechsel in Triturus-embryonen', *Bull. Swiss Acad. Med. Sci.*, **22**, 89–96.

L. G. Tilney and J. R. Gibbins, 1969, 'Microtubules in the formation and development of the primary mesenchyme in *Arbacia punctulata* II. An experimental analysis of their role in development and maintenance of cell shape', *J. Cell Biol.*, **41**, 227–250.

N. W. Timofeeff-Ressovsky, 1940, 'Mutations and geographical variation', in *The New Systematics* (ed. J. Huxley), Clarendon, Oxford, pp. 73–136.

G. M. Tomkins, T. D. Gelehrter, D. Granner, D. Martin, H. H. Samuels and E. B. Thompson, 1969, 'Control of specific gene expression in higher organisms', *Science*, **166**, 1474–1480.

J. C. Torrey, 1903, 'The early embryology of Thalassema mellita (Conn)', *Ann. N.Y. Acad. Sci.*, **14**, 165–246.

P. L. Townes and J. Holtfreter, 1955, 'Directed movements and selective adhesion of embryonic amphibian cells', *J. Exptl. Zool.*, **128**, 53–120.

D. F. Travis, 1960, 'Matrix and mineral deposition in skeletal structures of decapod crustacea', in *Calcification in Biological Systems* (ed. R. F. Sognnaes), American Association for the Advancement of Science, Washington, D.C., pp. 57–116.

——, 1970, 'The comparative ultrastructure and organization of five calcified tissues', in *Biological Calcification: Cellular and Molecular Aspects* (ed. H. Schraer), North-Holland, Amsterdam, pp. 203–311.

D. F. Travis, C. J. François, L. C. Bonar and M. J. Glimcher, 1967, 'Comparative studies of the organic matrices of invertebrate mineralized tissues', *J. Ultrastr. Res.*, **18**, 519–550.

A. L. Treadwell, 1901, 'The cytogeny of Podarke obscura Verrill', *J. Morphol.*, **17**, 399–486.

R. L. Trelstad, E. D. Hay and J.-P. Revel, 1967, 'Cell contact during early morphogenesis in the chick embryo', *Develop. Biol.*, **16**, 78–106.

J. P. Trinkaus, 1966, 'Morphogenetic cell movements', in *Major Problems in Developmental Biology* (ed. M. Locke), Academic Press, New York, pp. 125–176.

——, 1969, *Cells into Organs. The Forces that Shape the Embryo*, Prentice-Hall, Englewood Cliffs.

J. P. Trinkaus and T. L. Lentz, 1967, 'Surface specializations of *Fundulus* cells and their relation to cell movements during gastrulation', *J. Cell Biol.*, **32**, 139–153.

G. H. Tucker, 1942, 'The histology of the gonads and development of the egg envelopes of an ascidian (Styela plicata Lesueur)', *J. Morphol.*, **70**, 81–113.

P. M. Tuft, 1962, 'The uptake and distribution of water in the embryo of *Xenopus laevis* (Daudin)', *J. Exptl. Biol.*, **39**, 1–19.

T. C. Tung, S. C. Wu and Y. Y. F. Tung, 1958, 'The development of isolated blastomeres of Amphioxus', *Scientia Sinica*, **7**, 1280–1320.

——, 1960, 'The developmental potencies of the blastomere layers in Amphioxus egg at the 32 cell stage', *Scientia Sinica*, **9**, 119–141.

——, 1962a, 'The presumptive areas of the egg of Amphioxus', *Scientia Sinica*, **11**, 629–644.

——, 1962b, 'Experimental studies on the neural induction in Amphioxus', *Scientia Sinica*, **11**, 805–820.

V. C. Twitty, 1949, 'Developmental analysis of amphibian pigmentation', *Growth*, **13**, Suppl., 133–161.

V. C. Twitty and D. Bodenstein, 1941, 'Experiments on the determination problem', *J. Exptl. Zool.*, **86**, 343–380.

A. Tyler, 1931, 'The production of normal embryos by artificial parthenogenesis in the echiuroid, *Urechis*', *Biol. Bull.*, **60**, 187–211.

——, 1932, 'Production of cleavage by suppression of the polar bodies in artificially activated eggs of Urechis', *Biol. Bull.*, **63**, 218–223.

——, 1941, 'Artificial parthenogenesis', *Biol. Rev..*, **16**, 291–336.

——, 1947, 'An auto-antibody concept of cell structure, growth and differentiation', *Growth*, **10**, Suppl., 7–19.

A. Tyler and A. Monroy, 1959, 'Changes in the rate of transfer of potassium across the membrane upon fertilization of the egg of Arbacia punctulata', *J. Exptl. Zool.*, **142**, 675–690.

A. Tyler, A. Monroy, C. Y. Kao and H. Grundfest, 1956, 'Membrane potential and resistance of the starfish egg before and after fertilization', *Biol. Bull.*, **111**, 153–177.

L. von Ubisch, 1937, 'Die normale Skelettbildung bei Echinocyamus pusillus und Psammechinus miliaris und die Bedeutung dieser Vorgänge für die Analyse der Skelette von Keimblatt-Chimären', *Z. Wiss. Zool.*, **149**, 402–476.

——, 1950, 'Die Entwicklung der Echiniden', *Kon. Ned. Akad. Wet.*, *Verh. Afd. Nat.* [2], **47** (2), 1–50.

E. Urbani, 1957, 'Studi di embryologia e zoologia chimica degli Anfibi', *Istituto Lombardo, Rend. Sci.* [B], **92**, 69–179.

E. Urbani and L. De Cesaris Coromaldi, 1954a, 'Dipeptidasi e acidi nucleici nello sviluppo embrionale e larvale di "Bufo vulgaris"', *Ric. Sci.*, **24**, 1275–1284.

——, 1954b, 'Studio delle proteinasi nello sviluppo embrionale e larvale di "Rana esculenta"', *Ric. Sci.*, **24**, 2364–2374.

——, 1955, 'Studio delle proteinasi nello sviluppo embrionale e larvale di "Bufo vulgaris"', *Atti Accad. Naz. Lincei* [8], **18**, 341–347.

E. Urbani and G. Scollo-Lavizzari, 1955, 'Studio delle amilasi nello sviluppo embrionale e larvale di "Bufo vulgaris"', *Ric. Sci.*, **25**, 1151–1160.

E. Urbani, L. De Cesaris Coromaldi and G. Scollo-Lavizzari, 1957, 'Studio delle dipeptidase e delle lipasi nello sviluppo embrionale di *Bufo viridis* e di *Rana dalmatina*', *Atti Accad. Naz. Lincei* [8], **23**, 82–86.

A. Vandel, 1963, 'Évolution et auto-régulation', *Ann. Biol.*, **2**, 179–197.

E. Vasseur, 1952, *The Chemistry and Physiology of the Jelly Coat of the Sea Urchin Egg*, Stockholm.

C. Vendrely and R. Vendrely, 1949, 'Sur la teneur individuelle en acide désoxyribonucléique des gamètes d'oursins *Arbacia* et *Paracentrotus*', *C.R. Soc. Biol.*, **143**, 1386–1387.

C. A. Villee, 1942, 'The phenomenon of homoeosis', *Amer. Nat.*, **76**, 494–506.

H. J. Vogel, 1965, 'Lysine biosynthesis and evolution', in *Evolving Genes and Proteins* (ed. V. Bryson and H. J. Vogel), Academic Press, New York, pp. 25–40.

W. Vogt, 1909, 'Ueber rückschreitende Veränderungen von Kernen und Zellen jungen Entwicklungsstadien von Triton cristatus', *Ges. Bef. Ges. Naturw. Marburg*, 109–124.

——, 1925, 'Gestaltungsanalyse am Amphibienkeim mit örtlicher Vitalfärbung. Vorwort über Wege und Ziele. I. Teil. Methodik und Wirkungsweise der örtlichen Vitalfärbung mit Agar als Farbträger', *Arch. Entw.mech.*, **106**, 542–610.

——, 1929a, 'Gestaltungsanalyse am Amphibienkeim mit örtlicher Vitalfärbung. II. Teil. Gastrulation und Mesodermbildung bei Urodelen und Anuren', *Arch. Entw.mech.*, **120**, 384–706.

——, 1929b, 'Chorda, Hypochorda und Darmentoderm bei anuren Amphibien', *Anat. Anz.*, **67**, Suppl., 153–163.

V. Volterra, 1931, *Leçons sur la Theorie Mathematique de la Lutte pour la Vie*, Gauthiers-Villars, Paris.

C. H. Waddington, 1932, 'Experiments on the development of chick and duck embryos, cultivated *in vitro*', *Philos. Trans. Roy. Soc.* [B], **221**, 179–230.

——, 1956, *Principles of Embryology*, Allen & Unwin, London.

——, 1957, *The Strategy of the Genes. A Discussion of Some Aspects of Theoretical Biology. With an Appendix by H. Kacser*, Allen & Unwin, London.

——, 1967, 'The principle of archetypes in evolution', in *Mathematical Challenges to the neo-Darwinian Interpretation of Evolution* (ed. P. S. Moorhead and M. M. Kaplan), Wistar Institute Press, Philadelphia, pp. 113–115.

——, 1968, 'The basic ideas of biology', in *Towards a Theoretical Biology 1. Prolegomena* (ed. C. H. Waddington), Edinburgh University Press, Edinburgh, pp. 1–32.

——, 1969, 'Paradigm for an evolutionary process', in *Towards a Theoretical Biology 2. Sketches* (ed. C. H. Waddington), Edinburgh University Press, Edinburgh, pp. 106–128.

J. B. Walker, 1965, 'Negative feedback control of enzyme levels during embryonic development', in *Development and Metabolic Control Mechanisms and Neoplasia*, Williams & Wilkins, Baltimore, pp. 317–334.

R. A. Wallace, 1961, 'Enzymatic patterns in the developing frog embryo', *Develop. Biol.*, **3**, 486–515.

H. Wartenberg and W. Schmidt, 1961, 'Elektronenmikroskopische Untersuchungen der strukturellen Veränderungen im Rindenbereich des Amphibieneies im Ovar und nach der Befruchtung', *Z. Zellforsch.*, **54**, 118–146.

S. Watasé, 1890, 'Studies on Cephalopoda. I. Cleavage of the ovum', *J. Morphol.*, **4**, 247–302.

——, 1895, 'Origin of the centrosome', *Biol. Lectures Wood's Holl 1894*, 273–287.

A. J. Waterman, 1934, 'Observations on reproduction, prematuration and fertilization in Sabellaria vulgaris', *Biol. Bull.*, **67**, 97–114.

D. M. S. Watson, 1919, 'On Seymouria, the most primitive known reptile', *Proc. Zool. Soc. London*, 267–301.

R. Weber, 1967, 'Biochemistry of amphibian metamorphosis', in *The Biochemistry of Animal Development* (ed. R. Weber), Vol. 2, Academic Press, New York, pp. 227–301.

C. K. Weichert, 1965, *Anatomy of the Chordates*, 3rd ed., McGraw-Hill, New York.
L. Weiss, 1967, *The Cell Periphery, Metastasis and Other Contact Phenomena*, North-Holland, Amsterdam.
L. P. Weiss and D. W. Fawcett, 1953, 'Cytochemical observations on chicken monocytes macrophages and giant cells in tissue culture', *J. Histochem. Cytochem.*, **1**, 47–65.
P. Weiss, 1934, '*In vitro* experiments on the factors determining the course of the outgrowing nerve fiber', *J. Exptl. Zool.*, **68**, 393–448.
——, 1939, *Principles of Development*, Holt, New York.
——, 1959, 'Interaction between cells', *Rev. Modern Phys.*, **31**, 449–454.
P. Weiss and W. Ferris, 1956, 'The basement lamella of amphibian skin. Its reconstruction after wounding', *J. Biophys. Biochem. Cytol.*, Suppl., **2**, 275–282.
N. K. Wessells, 1970, 'Mammalian lung development: Interaction in formation and morphogenesis of tracheal buds', *J. Exptl. Zool.*, **175**, 455–466.
N. K. Wessells, B. S. Spooner, J. F. Ash, M. O. Bradley, M. A. Luduena, E. L. Taylor, J. T. Wrenn and K. M. Yamada, 1971, 'Microfilaments in cellular and developmental processes', *Science*, **171**, 135–143.
A. H. Whiteley and F. Baltzer, 1958, 'Development, respiratory rate and content of desoxyribonucleic acid in the hybrid Paracentrotus ♀ x Arbacia ♂', *Pubbl. Staz. Zool. Napoli*, **30**, 402–457.
C. O. Whitman, 1878, 'The embryology of Clepsine', *Quart. J. Micros. Sci.*, **18**, 215–315.
——, 1895, 'Evolution and epigenesis', *Biol. Lectures Wood's Holl 1894*, 205–224.
J. R. Whittaker, 1968, 'The nature and probable cause of modulations in pigment cell cultures', in *The Stability of the Differentiated State* (ed. H. Ursprung), Springer Verlag, Berlin, pp. 25–36.
R. H. Whittaker, 1969, 'New concepts of kingdoms of organisms', *Science*, **163**, 150–160.
J. Whitten, 1968, 'Metamorphic changes in insects', in *Metamorphosis* (ed. W. Etkin and L. J. Gilbert), North-Holland, Amsterdam, pp. 43–105.
L. L. Whyte, 1965, *Internal Factors in Evolution*, Brazillier, New York.
A. Wierzejski, 1905, 'Embryologie von Physa fontinalis L.', *Z. Wiss. Zool.*, **83**, 502–706.
C. E. Wilde, 1955, 'The urodele neuroepithelium. II. The relationship between phenylalanine metabolism and the differentiation of neural crest cells', *J. Morphol.*, **97**, 313–344.
——, 1961, 'The differentiation of vertebrate pigment cells', *Adv. Morphogen.*, **1**, 267–300.
C. E. Wilde and R. B. Crawford, 1968, 'Epithelial–mesenchymal interactions in the lower vertebrates', in *Epithelial–Mesenchymal Interactions* (ed. R. Fleischmajer and R. E. Billingham), Williams & Wilkins, Baltimore, pp. 98–113.
J. Williams, 1965, 'Chemical constitution and metabolic activities of animal eggs', in *The Biochemistry of Animal Development* (ed. R. Weber), Vol. 1, Academic Press, New York, pp. 13–71.
M. B. Williams, 1970, 'Deducing the consequences of evolution: A mathematical model', *J. Theoret. Biol.*, **29**, 343–385.
J. C. Willis, 1922, *Age and Area. A Study in Geographical Distribution and Origin of Species*, Cambridge University Press, Cambridge.
E. N. Willmer, 1960, *Cytology and Evolution*, Academic Press, New York.
J. A. Wills, 1936, 'The respiratory rate of developing amphibia with special reference to sex differentiation', *J. Exptl. Zool.*, **73**, 481–510.
E. B. Wilson, 1889, 'The embryology of the earthworm, *J. Morphol.*, **3**, 387–462.
——, 1892, 'The cell-lineage of Nereis. A contribution to the cytogeny of the annelid body', *J. Morphol.*, **6**, 361–480.
——, 1893, 'Amphioxus, and the mosaic theory of development', *J. Morphol.*, **8**, 579–638.
——, 1899a, 'The structure of protoplasm', *Biol. Lectures Wood's Holl 1898*, 1–20.
——, 1899b, 'Cell-lineage and ancestral reminiscence', *Biol. Lectures Wood's Holl 1898*, 21–42.
——, 1903a, 'Notes on merogony and regeneration in Renilla', *Biol. Bull.*, **4**, 215–226.
——, 1903b, 'Experiments on cleavage and localization in the Nemertine-egg', *Arch. Entw.mech.*, **16**, 411–460.
——, 1904, 'Experimental studies on germinal localization. I. The germ-regions in the egg of Dentalium', *J. Exptl. Zool.*, **1**, 1–72.
——, 1928, *The Cell in Development and Heredity*, 3rd ed., Macmillan, New York.
S. Wischnitzer, 1966, 'The ultrastructure of the cytoplasm of the developing amphibian egg', *Adv. Morphogen.*, **5**, 131–179.

K. E. Wohlfarth-Bottermann, 1964a, 'Cell structures and their significance for amoeboid movement', *Internat. Rev. Cytol.*, **16**, 61–131.

——, 1964b, 'Differentiations of the ground cytoplasm and their significance for the generation of the motive force of amoeboid movement', in *Primitive Motile Systems in Cell Biology* (ed. R. D. Allen and N. Kamiya), Academic Press, New York, pp. 79–109.

L. Wolpert, 1960, 'The mechanics and mechanism of cleavage', *Internat. Rev. Cytol.*, **10**, 163–216.

——, 1965, 'Cytoplasmic streaming and amoeboid movement', *Symp. Soc. Gener. Microbiol.*, **15**, 270–293.

——, 1966, 'The mechanical properties of the membrane of the sea urchin egg during cleavage', *Exptl. Cell Res.*, **41**, 385–396.

L. Wolpert and T. Gustafson, 1961a, 'Studies on the cellular basis of morphogenesis of the sea urchin embryo. Development of the skeletal pattern', *Exptl. Cell Res.*, **25**, 311–325.

——, 1961b, 'Studies on the cellular basis of morphogenesis of the sea urchin embryo. The formation of the blastula', *Exptl. Cell Res.*, **25**, 374–382.

L. Wolpert and E. H. Mercer, 1963, 'An electron microscope study of the development of the blastula of the sea urchin embryo and its radial polarity', *Exptl. Cell Res.*, **30**, 280–300.

R. Woltereck, 1904, 'Beiträge zur praktischen Analyse der Polygordius. Entwicklung nach dem "Nordsee-" und dem "Mittelmeertypus", I. Der für beide Typen gleichverlaufende Entwicklungsabschnitt: von Ei bis zum jüngsten Trochophora-Stadium', *Arch. Entw.mech.*, **18**, 377–403.

——, 1940, *Grundzüge einer allgemeinen Biologie*, 2nd ed., Enke Verlag, Stuttgart.

J. H. Woodger, 1929, *Biological Principles*, Routledge & Kegan Paul, London.

——, 1930, 'The "concept of organism" and the relation between embryology and genetics. Parts I and II', *Quart. Rev. Biol.*, **5**, 1–22, 438–463.

——, 1931, 'The "concept of organism" and the relation between embryology and genetics. Part III', *Quart. Rev. Biol.*, **6**, 178–207.

——, 1939, 'The technique of theory construction', *Internat. Encyclop. Unif. Sci.*, Vol. 2, **5**, 1–81.

——, 1948, 'Observations on the present state of embryology', *Symp. Soc. Exptl. Biol.*, **2**, 351–365.

——, 1952, *Biology and Language. An Introduction to the Methodology of the Biological Sciences including Medicine*, Cambridge University Press, Cambridge.

——, 1962, 'Abstraction in natural science', in *Logic, Methodology and Philosophy of Science* (ed. E. Nagel, P. Suppes and A. Tarski), Stanford University Press, Stanford, pp. 293–302.

S. Wright, 1931, 'Evolution in Mendelian populations', *Genetics*, **16**, 97–159.

——, 1949, 'Adaptation and selection', in *Genetics, Paleontology, and Evolution* (ed. G. L. Jepsen, E. Mayr and G. G. Simpson), Princeton University Press, Princeton, pp. 365–389.

——, 1955, 'Classification of factors of evolution', *Cold Spring Harbor Symp. Quant. Biol.*, **20**, 16–24.

E. Yamada, 1960, 'Collagen fibrils within the renal glomerulus', *J. Biophys. Biochem. Cytol.*, **7**, 407–408.

T. Yamada, 1961, 'A chemical approach to the problem of the organizer', *Adv. Morphogen.*, **1**, 1–53.

M. Yčas, 1954, 'The respiratory and glycolytic enzymes of sea-urchin eggs', *J. Exptl. Biol.*, **31**, 208–217.

C. L. Yntema, 1943, 'An experimental study on the origin of the sensory neurones and sheath cells of the IXth and Xth cranial nerves in Amblystoma punctatum', *J. Exptl. Zool.*, **92**, 93–119.

M. Yoneda, 1964, 'Tension at the surface of sea-urchin egg: A critical examination of Cole's experiment', *J. Exptl. Biol.*, **41**, 893–906.

R. S. Young, 1958, 'Fluoride inhibition of skeleton formation in the sea urchin', *Exptl. Cell Res.*, **15**, 398–404.

K. A. Youngstrom, 1938, 'On the relationship between choline esterase and the development of behaviour in Amphibia', *J. Neurophysiol.*, **1**, 357–363.

R. Zangerl, 1948, 'The methods of comparative anatomy and its contribution to the study of evolution', *Evolution*, **2**, 351–374.

E. Zeuthen, 1948, 'A Cartesian diver balance weighing reduced weights (R.W.) with an accuracy of $\pm 0{\cdot}01\ \gamma$', *C.R. Lab. Carlsberg, Sér. chim.*, **26**, 243–266.

——, 1951, 'Segmentation, nuclear growth and cytoplasmic storage in eggs of echinoderms and Amphibia', *Pubbl. Staz. Zool. Napoli*, **23**, Suppl., 47–69.

H. E. Ziegler, 1902, *Lehrbuch der vergleichenden Entwickelungsgeschichte der niederen Wirbeltiere*, Gustav Fischer Verlag, Jena.

M. A. Zieliński, 1939, 'Carbohydrate metabolism and phosphorus in the fertilized eggs of the sea-urchin (Paracentrotus lividus Lm.)', *Acta Biol. Exptl.*, **13**, 35–48.

A. M. Zimmerman, J. V. Landau and D. Marsland, 1957, 'Cell division: A pressure–temperature analysis of the effects of sulfhydryl reagents on the cortical plasmagel structure and furrowing strength of dividing eggs (Arbacia and Chaetopterus)', *J. Cell. Comp. Physiol.*, **49**, 395–435.

A. I. Zotin, 1962, 'The mechanism of cleavage furrow formation in eggs of ganoid fishes and of amphibians', *Acta Biol. Cracov., Ser. Zool.*, **5**, 215–228.

E. Zuckerkandl and L. Pauling, 1965, 'Evolutionary divergence and convergence in proteins', in *Evolving Genes and Proteins* (ed. V. Bryson and H. J. Vogel), Academic Press, New York, pp. 97–166.

AUTHOR INDEX

SUBJECT INDEX

Notes added in proof

1. In Caudata the neural folds meet at the ventral median point of the blastopore. The latter is thus formally completely surrounded by neural tissue and must be so, for topological reasons, also before gastrulation. This implies that the neural primordium must border the line of gastrulation which, in Figure 114, represents the presumptive blastopore. The strip of epidermis indicated in the figure between the neural and notochordal areas is a mistake. The arguments based on its purported existence should therefore be modified. The fact remains that the spreading of the If-induction is blocked in the dorsal area and can reach the neural plate only after contact is established between the presumptive neural plate and the invaginated notochordal cells.

Another modification should be noted in this figure. The notochordal area should have an extension corresponding rather closely to the grey crescent region. A rather broad strip in the lower part of the area designated 'chorda' actually is endoderm. This one may realize through a closer scrutiny of Vogt's results than was originally made.

2. This fate map represents the postulated *origin* of the mesodermal cells. But already in the late blastula most of the free interior cells are collected in the marginal zone between the ectoderm and the endoderm, a location which is indicated by the vital staining experiments. Essential agreement obtains between Vogt's fate map and the one shown in Figure 115 if, in the former case, the fate map is projected on the free interior cells and, in the latter case, the various lines are pushed down towards the marginal zone.

3. It is a general principle in deductive biology that as few explanatory principles as possible should be introduced. In the present case it seems possible to dispense with the asserted 'activity differential'.

The basis for this postulate is the following: the presence of the pigmented ectoplasmic layer shows that linear TP polymers are preferentially attached to the external surface in the membrane-bound ecto-mesodermal cells. But that must imply that when these cells divide paratangentially, the second rule of cell division predicts an unequal division. The proximal daughter cell will be relatively large and contain relatively more yolk. The distal membrane-bound cell will be smaller and contain more hyaloplasm, which means that it will be more 'exhausted'.

There are, however, two populations among the pigmented cells, the slightly pigmented ones in the grey crescent region, and the more strongly pigmented ones in the remaining part of the animal hemisphere. If, as suggested, pigment granules interfere with the attachment of microtubules, then the less pigmented cells must be smaller and more hyaline than the ones discussed above; in other words, they will be the most 'exhausted' among the ecto-mesodermal cells.

According to the proposed fate map the notochordal cells arise from the membrane-bound cells in the grey crescent area (see Note **1**), the nerve cells from the more pigmented dorsal cells, and the mesoderm from the interior cells. The sequence of 'exhaustion' in these three populations of cells thus corresponds exactly to the sequence of exhaustion purportedly arising through an activity differential. Hence the latter is unnecessary. The segregation of the ecto-mesodermal cells may therefore occur through interaction between cell division and cell diversification, a mechanism which later will be proposed to prevail in various invertebrate embryos.

It must be noted that the present discussion concerns only the cells which undergo the If-transformation; as is well known, the remaining ones become epidermal cells.